# Biosensors Nanotechnology
# 2nd Edition

**Scrivener Publishing**
100 Cummings Center, Suite 541J
Beverly, MA 01915-6106

*Publishers at Scrivener*
Martin Scrivener (martin@scrivenerpublishing.com)
Phillip Carmical (pcarmical@scrivenerpublishing.com)

# Biosensors Nanotechnology
# 2nd Edition


Edited by

**Inamuddin**

*Department of Applied Chemistry, Aligarh Muslim University, India*

and

**Tariq Altalhi**

*Department of Chemistry, College of Science, Taif University, Saudi Arabia*



Scrivener Publishing

WILEY

This edition first published 2023 by John Wiley & Sons, Inc., 111 River Street, Hoboken, NJ 07030, USA and Scrivener Publishing LLC, 100 Cummings Center, Suite 541J, Beverly, MA 01915, USA

For more information about Scrivener publications please visit www.scrivenerpublishing.com.

**1st edition (2014), 2nd edition (2023)**


**Wiley Global Headquarters**
111 River Street, Hoboken, NJ 07030, USA

For details of our global editorial offices, customer services, and more information about Wiley products visit us at www.wiley.com.



***Library of Congress Cataloging-in-Publication Data***

ISBN 978-1-394-16624-4

Cover images: Pixabay.Com
Cover design by Russell Richardson

Set in size of 11pt and Minion Pro by Manila Typesetting Company, Makati, Philippines

Printed in the USA

10 9 8 7 6 5 4 3 2 1

# Contents

# Preface

The biosensor industry began as a small, niche activity in the 1980s and has since developed into a large, global industry. Nanomaterials have substantially improved not only non-pharmaceutical and healthcare uses, but also telecommunications, paper, and textile manufacture. Biological sensing aids in the understanding of living systems and may be used in a variety of sectors, including medicine, drug discovery, process control, environmental monitoring, food safety, military, and personal protection. It brings up new opportunities in bionics, power generation, and computing, all of which will benefit from a greater understanding of the bio-electronic relationship, as advances in communications and computational modeling are forcing us to reconsider how we offer healthcare and perform R&D and manufacturing to the modern world.

As a result of the customization of everything from health to environmental control, new payment structures and commercial models will arise. Wearable, mobile, and integrated sensors are being used in an increasing variety of products, but the majority of these devices still rely on physical sensors to measure elements like temperature and pressure. There is a conspicuous dearth of sensors that are both robust and convenient in the field of body chemistry sensors. This book examines emerging technologies that are accelerating scientific study and laying the foundation for new goods meant to extend and improve the quality of our lives. In this newly evolving discipline, the combination of nanoscale materials with biosensor technology is gaining a lot of traction. Nanostructures have been used to increase the adherence of biosensor materials to electrode surfaces, print nano barcodes on biomaterials, increase the pace of bio-responses, and amplify the electric signal. Finally, nanomaterial-based biosensors may be employed in a wide range of medical diagnostics and environmental monitoring applications due to their better response speed, greater sensitivity, simple design, specificity, and cost-effectiveness.

The book covers the major materials employed in the development of biosensors such as nanoparticles, nanowires, nanotubes, nanoribbons, nanorods, nanosheets, and many more nanostructures.

**Chapter 1** discusses how the innovative techniques used in biosensors evolved from cell engineering and 3D cell immobilization. The various parts of biosensors based on cell detection using the cell's bioreceptors are discussed in detail along with their working mechanism and applications.

**Chapter 2** discusses the bioreceptors for enzymatic interactions as part of an enzyme-based biosensor to detect the analyte from a sample with applications. Materials used for the construction of various types of biosensors and types of bioreceptors, immobilization of enzymes for biosensors, and types of transducers for enzymatic interactions are also described.

**Chapter 3** discusses the history, structure, synthesis, types, physical, and chemical properties including the merits and demerits of dendrimers in detail. This chapter focuses on dendrimers as drug delivery via electrochemical, enzymatic, optic, QCM, and glucose-based biosensors.

**Chapter 4** details the importance and need for various 2D Photonic Crystal biosensors. The chapter discusses, in detail, how disease identification is done by measuring the effective change in the refraction of different analytes, and as well as various Photonic Crystal structures. It also focuses on the scope and future development of biosensors using 2D structures.

**Chapter 5** explains that the applications of bioreceptors are numerous, ranging from benchtop analysis to point-of-care diagnosis and treatment. Hence, this chapter is an overview of common bioreceptors, especially affinity-binding receptors that are crucial in theranostic applications.

**Chapter 6** details the brief history, basic working principles, and the present developments in glucose biosensors. The chapter focuses on fabrication methods and recent trends in the application of nanomaterials and nano/microfabrication for the development of paper analytical devices or wearable glucose biosensors.

**Chapter 7** is focused on the recent progress in the preparation of metal-free quantum-sized sensors. The biosensing and detection applications of carbonaceous quantum dots such as carbon, graphene, and carbon nitride quantum dots are investigated. Optical and electrochemical techniques are discussed with a consideration for the limit of detection values.

**Chapter 8** details the latest research progress on bioreceptors for microbial biosensors. In addition, it summarizes the types and applications of microbial biosensors: the recent trend and the future challenges of microbial biosensor technology.

**Chapter 9** explains the use of several noble plasmonic materials including gold, silver, copper, niobium, and aluminum, and their widespread applications in optical sensing and sensors. Also, the advantages, disadvantages, and prospects of some highly used plasmonic nanomaterials in sensors are discussed in this chapter.

**Chapter 10** explains the differences between various magnetic sensors and their accompanying sensitivity responses and consequences. This chapter discusses how magnetic sensors use magnetic fluids for studying the impact on a range of particles and various sensor architectures for tracking biological interactions with distinct magnetic strength change.

**Chapter 11** deals in detail with the various salivary biomarkers and the biosensors associated with several neurological disorders including Alzheimer's disease, Parkinson's disease, Huntington's disease, ALS, multiple sclerosis, autism spectrum disorders, and neuropsychiatric disorders, and various cancers affecting breasts, lungs, pancreas, and the gastrointestinal system.

**Chapter 12** deals with the different molecules for biological amine detection by fluorescent chemosensors and its specific emphasis on the use of bioimaging applications. In addition, the sustained improvements in fluorescent biosensors are anticipated to result in universal biosensors for essential biological amines, which can be detected in real-time analysis.

**Chapter 13** provides concise information about chromo-fluorogenic biosensing applications of diverse nanomaterials including zero, one, two, and, three-dimensional nanomaterials. Biosensing from complicated bio-matrices along with intracellular imaging is also discussed. Finally, loopholes of present research and future research directions are also outlined to stay current with medical diagnosis.

**Chapter 14** explores present and prospective breakthroughs in nanotechnology-based biosensors for real-time assessment of several analytes and the toxicity mechanisms in living creatures, using primary datasets from 2018 onwards. Innovative biosensing technologies centered on unique sensing components and transduction concepts receive special attention. The chapter also discusses the opportunities and future considerations for the utilization of NMs-based biosensors for enhanced environmental and food-sensing devices.

**Chapter 15** deals with the introduction and roadmap of biosensors, followed by the use of nanotechnology in cancer therapy. Additionally, it explores nanomaterial infused with a biosensor, the fabrication of nano biosensor, and the diagnosis of breast cancer, including point-of-care and wearable analysis.

**Chapter 16** provides a brief overview of antibody-based biosensors, also called immunosensors. The chapter includes a discussion on antigen-antibody interactions and the application of immunosensors in various areas of the health and food industries. Lastly, some new approaches to antibody modifications that offer several advantages over classical antigen-antibody receptors are also discussed.

**Chapter 17** describes the use of biosensors for various paints and pigment analysis. It discusses the biosensor's components, history, and working principle in regard to paint and pigments. For future benefit, the chapter describes the characteristics of paints and pigments, with analytical methods, and various applications of paints and pigments-based biosensors.

**Chapter 18** discusses biological devices that are incorporated into various animal and plant tissues. The applicability in varied physiology along with its distinguished classification based on different principles is covered also. Furthermore, the chapter enumerates the increasing exigencies for these devices in a broad range of areas including various medical ailments.

**Chapter 19** details various methods for pesticide detection with simple, highly selective and sensitive, fast response, cost-effective and portably sized biosensors. The chapter also discusses recent results from the use of nanomaterials in the fabrication of pesticide biosensors in food and environmental applications.

**Chapter 20** discusses the applications of biosensors for monitoring the behavior of agro-products. Two categories of biosensors are highlighted, and their roles in detecting contaminants in dairy processing are discussed. The application of receptor-based biosensors was proposed for monitoring the survival rate of bacteria in milk processing.

**The Editors**
August 2023

# 1

# Bioreceptors for Cells

**Vipul Prajapati[1]* and Salona Roy[2]**

[1]*Department of Pharmaceutics, SSR College of Pharmacy (Permanently Affiliated to Savitribai Phule Pune University), Sayli-Silvassa Road, Sayli, Silvassa, Union Territory of Dadra Nagar Haveli & Daman Diu, India*
[2]*Department of Pharmacology and Toxicology, NIPER Hajipur, Export Promotions Industrial Park (EPIP), Industrial Area Hajipur, Dist. Vaishali, Bihar, India*

***Abstract***
A biosensor is a tool that quantitatively determines the disturbance in the homeostatic equilibrium in a system in an extremely low concentration in the healthcare sector, such as diagnosis, treatment, and mitigation. Biosensors utilize specific biomarkers to aid in an accurate diagnosis based on its sensitivity, reproducibility, biocompatibility, and robustness, which has several advantages over conventional diagnosis, including onsite diagnosis in less time. This article covers various techniques involved in the pretreatment of the cell to modify certain bioreceptors, types of transducers, and their wide arena of application. Cells for biosensors are often labeled with certain enzymes or secondary substances producing an intensifying response of intrinsic signal transduction. The efficacy of a single mini device had attracted the attention of many researchers to aid in the early diagnosis of life-threatening diseases. Although enhancement in the performance of biosensor of cell has been going on, it has provided a gateway to next-generation approaches in the healthcare system.

***Keywords:*** Cell, biosensor, diagnosis, treatment, bioreceptor

## 1.1 Introduction

Living cells are the biorecognition elements since they can detect any unknown stimuli or perception from their environment as a method of adapting, as well as surviving. Cell biosensors can detect analytes with utmost accuracy, high sensitivity, and specificity in a cost-effective, invasive or non-invasive way. They are appropriate bioreceptor elements because they provide versatility in sensing tactics and make production relatively straightforward and inexpensive when compared to pure enzymes, DNA, and antibodies-based detection techniques, such as ELISA, RIA, etc. [1].

Biosensors utilize a mixture of biological, chemical, and physical technologies to measure micro physiological signals in real-time and on-site detection. A biosensor system is made up of a few crucial elements like biological sample receptors, transducers,

**Corresponding author*: vippra2000@yahoo.com

Inamuddin and Tariq Altalhi (eds.) Biosensors Nanotechnology, 2nd Edition, (1–32) © 2023 Scrivener Publishing LLC

backing laminate, and display systems that use electrical, chemical, or photonic components. It is to detect the results and then turn the cascading event of recognition into a measurable signal strength, which can be grouped in conjugated and integrated biosensors [2].

A biosensor's basic principle is to detect the bio-element at the molecular recognition level and convert it into a different sort of signal strength using a transducer [3]. Living cells, bioactive substrates, and transducers make up cell-based biosensors. The impact of biochemical or pharmacological compounds on cells might be measured by changes in cellular polarity or physiological characteristics like cell membrane permeability or ligand expression on biosensors after treatment [4]. Rapid analysis of small amounts of data as compared to conventional techniques, with the added benefit, may be used for the basis of clinical evaluation that integrated design platforms. The ability of sensor cells to detect specific analogs of targets while distinguishing them from structural counterparts that had no functional similarities within them provided the basis for accurate results [5].

Biosensors of cells have sparked attention as a potential alternative to traditional sensing methods due to several benefits, which include affordability and mobility with the major advantage of the absence of equipment and trained staff. Another pleasing characteristic of biosensors of cells is their versatility in terms of design and outputs, which made them capable of being adjusted to the unique feature whenever needed for desired outputs. They provide a diverse platform for analytical applications in various arenas, such as food, biomedical science, environmental, and society healthcare by merging disciplinary technologies and expertise. Environmental monitoring, bioproduction, biomedical applications in diagnostics, and health monitoring are all possible scopes for cell-based biosensors [6].

## 1.2 Classification of the Cell as a Bioreceptor

With the greater selectivity to target molecules, bio-elements, such as functional enzymes, serum antibodies, DNA, and other elements, have traditionally been used as bioreceptors in biosensing. Living cells, on the other hand, provide an intriguing alternative to these molecular bioreceptors due to their broad diversity of biomolecular processes [59]. Cell biosensors have been classified by many researchers based on biorecognition element and signal transduction but in this section, in another way they can be classified based on the structural system and extensive work done in the lab to convert a cell into a probe, i.e., cell pretreated in the lab or which utilizes the invasive method of detection and cell without pretreatment or utilizes non-invasive method as shown in Figure 1.1, which is sensitive enough to record and display on-the screen with the maximum level of accuracy.

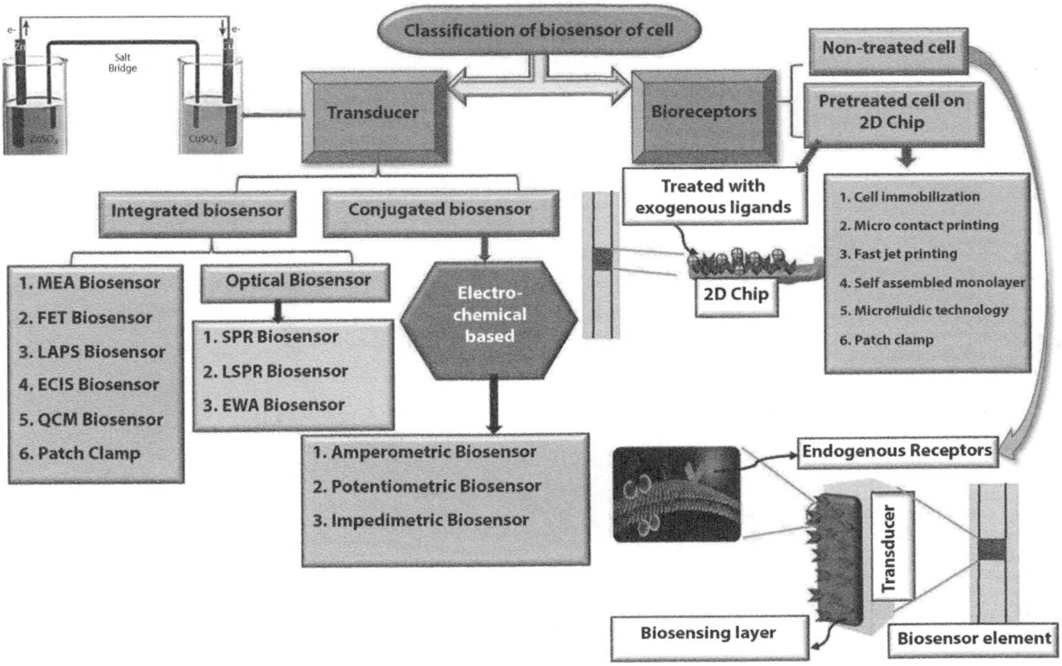


**Figure 1.1** Schematic classification of the biosensor of the cell.

### 1.2.1 On the Basis of Cell Origin

#### *1.2.1.1 Mammalian Cell*

The key transducers for signal creation in mammalian cell-based sensing systems are mammalian cells. When analytes excite living cells, the transducer converts changes in physiological characteristics or biological responses into a measurable and processable signal, fulfilling the detection and analysis goal [60].

Biosensors based on mammalian cells have been touted as potential instruments for pharmacology and toxicology, drug discovery, bioassay of drug substance, pathogen and toxin screening, environmental monitoring,, and biosafety research. The binding of cellular receptors to external substances triggers the cell–analyte interaction. The future of next level of generation biosensing techniques that use natural bio cellular receptors like GPCRs or the nicotinic in cholinergic receptor to measure the ligand-based secondary cellular response have been considered as shown in Figure 1.2 [61].

Hematopoietic and nervous system stem cells have been identified and grown *in vitro*. They develop into cell types, including neurons and myocardiocytes, which spontaneously contract in culture when induced to differentiate. Understanding neural network function from cells structured *in vivo* has been highly beneficial using brain tissue slices. Using isolated populations of cells as sensors ignores the analyte's involvement in the *in vivo* cell metabolism developing coculture systems to expand the operational sensitivity of excitable cells to include metabolites would improve the current capabilities [62].

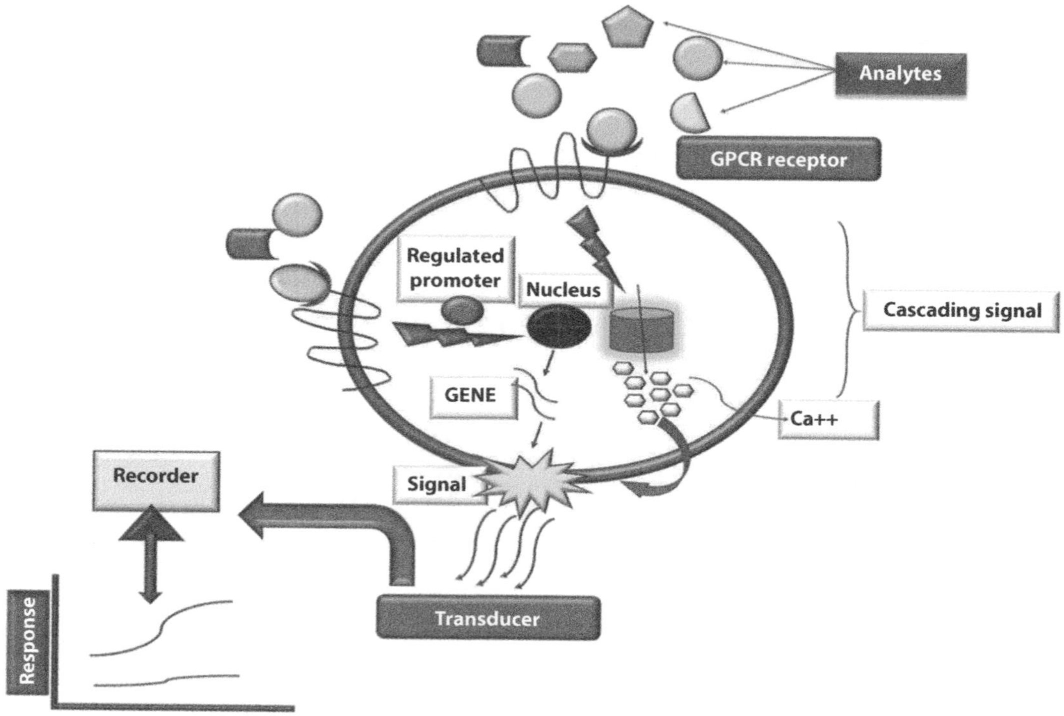


**Figure 1.2** Pictorial view of detection mechanism of cell biosensors through GPCR receptor.

### *1.2.1.2 Microbial Cell*

In the creation of biosensors, microbes are a good substitute for enzymes. They may be manufactured in vast quantities biotechnologically, and microbial cells contain multiple enzymes with the requisite cofactors or coenzymes, allowing them to detect a wide range of substrates. Furthermore, enzymes are very stable in their natural environment and do not require isolation [63].

In a variety of biological systems, biosensors of microbial detection are based on illumination from bacteria that are being used as a sensitive, fast, and non-invasive test. Bioluminescent bacteria may be found in a variety of habitats, from the sea to the land. Bioluminescent-like organism biosensors have also been created for the detection of organic, pesticide, and lead or mercury-like heavy metal pollution utilizing genetically engineered microorganisms (GEM). Cellular organelles can be thought of as multi-oriented biocatalysts that fall between complete cells and enzymes in terms of complexity [64]. In the manufacturing of biosensors, enzymes are the most extensively employed biological sensing element. Although pure enzymes have a high selectivity for their substrate, their use in biosensors may be restricted by the time and expense of enzyme purification, the necessity for numerous enzymes to make the measurable product, or the requirement for a coenzyme [65].

Viable cells and nonviable cells are the two types of cells found in microbes. The assessment of biological oxygen demand (BOD) or the consumption of other growth is two of the most common uses of living cells. Pollutants and hazardous substrates can also be detected by viable cells. Both virulent and dead microbial cells are utilized in the immobilization of microbial-based biosensors, although their immobilization needs varied. In case of utilization of an external light source, bacterial-based fluorescence occurs in an entire cell, and the fluorescence emission or radiating intensity is precisely proportional to the amount of analyte detection at very lower concentration [66].

Immobilized yeast was used to detect formaldehyde and assess the toxicity of cholanic acids, with changes in metabolism indicative of the analyte measured by $O_2$ electrode readings or extracellular acidification rates. Organophosphate hydrolase is an enzyme that creates protons during the breakdown of organophosphate insecticides or nerve agents including sarin, soman, and VX. Analyte detection has been proposed using pH changes in the effluent from these immobilized cells [62].

### 1.2.2 On the Basis of Cell Treatment

#### *1.2.2.1 Cell Pretreated in the Lab or Invasive Detection as Bioreceptor*

The cell biosensors can be further classified on the basis of treatment of the cell or modified cell for construction of the biosensor. The methods for cultivating biological live cells on two-dimensional (2D) chip technologies are known as cell culture on-chip. Due to the advent of the scale of the grown environment within the micro-sized chip is tailored to the length of the cells, microchip approaches can give several benefits for cell culture systems.

##### 1.2.2.1.1 Cell Immobilization Techniques

Cell culture of different cell types-based detection involving biosensing techniques use the stimulated response to the external stimuli mediated by a transducer through the modified signal. Immobilized enzymes can be utilized again after the completion of one process and they have higher stability in terms of catalytic activity than mobile enzymes. Although they have a lower catalytic rate when compared to mobile enzymes and require extensive treatment steps, immobilized enzymes are commonly used in medical and industrial plants because of benefits such as fast and efficient control by reusable enzymes i.e., ease of separation of the enzymes from the final product, maintaining purity of the product and high stability of the immobilized enzyme. There are mainly five methods of enzyme immobilization that can be classified as: adsorption, encapsulation, covalent bonding, entrapment as well as cross-linking, among all five, adsorption and covalent bonding method is mainly preferred. Adsorption, encapsulation, and entrapment are termed physical methods of immobilization whereas cross-linked covalent bonding is termed chemical methods or ways of immobilization [31]. Adsorption is one of the simplest techniques of immobilizations, it relies on the only weak force of attraction between enzyme and carrier such as London forces, electrostatic or ionic contacts, and hydrophobic or lipophilic interactions between them. As an extra reagent is not required, the adsorption approach is simple and affordable, moreover it is less damaging to enzyme function than other methods. On the other hand, because of their weak interaction, the enzyme immobilized by this approach are quickly deposited by changes in experimental setups like temperature, pH, or ionization potential.

Furthermore, contamination and signal interference may result from the non-targeted adsorption of various substrates onto a surface. Second, one of the most extensively utilized ways is covalent bonding, which creates stable complex compounds between enzymes and supports or carriers. The covalently immobilized enzymes have stronger binding than the adsorption and this method can provide more stable enzyme immobilization. Despite its advantages, the development of covalent bonds reduces the activity of the immobilized enzymes, and this approach necessitates a considerable amount of bioreagent to be used with it. Third, entrapment is not directly coupled to substrates or products; rather, it is entrapped or encased in polymers, which creates room for substrates and end-products to freely spread. Polymerization can take place in a combination of enzymes and monomeric units to entrap enzymes. Entrapment, like covalent bonding, is a physical interaction that offers the enzymes excellent stability and reduces leaching. Fourth, cross-linking provides strength to the bonded enzymes and ensures the leakage of enzymes during the utilization. It gives one of the very strong and robust connections between enzymes, the immobilization method via cross-linking enhances efficiency and stability. The use of cross-linking reagents like GTA (Glutaraldehyde), on the other hand, can result in a loss of activity due to severe modifications of the functional and non-functional enzymes caused by covalent bonding linkage. Fifth, encapsulation, which is more similar to that of the entrapment process, the only difference, lies in the arrangement of enzymes within the system. The encapsulation method involves the collection of enzymes in the semipermeable membrane and entrapment involves the arrangement of enzymes in the matrix forming structures. These biosensing devices are capable of reliably detecting water toxicity, assuring human safety and aquatic life welfare [32–34].

#### 1.2.2.1.2 Microcontact Printing

The modification of surface technique that uses custom inks to modify surface chemical and biological signals as well as manufacture particular topographical characteristics is widely known as Micro-contact printing (MCP). In 1993, it had been proposed that the combination of MCP as a matter for regulating the concentration and area distribution of proteins will be adsorb onto to the designed self-assembled monolayers (SAMs). SAM is a part of the soft lithography method family, which is one of the most widely utilized surface modification techniques in biomedical applications. MCP has considerable benefits over other surface patterning methods in terms of cheap cost, high dependability, and adaptability. Two of MCP's main duties in adjusting the surface are the creation of organized geometric or characteristic patterns and 2D surface designing. The similarity with nanoimprinted lithography lies in creating intricate patterns as in the case of MCP. MCP is a flexible surface modification technology that may manipulate surface chemical and biological signals as well as produce specific topographical features using specialized inks.

The primary distinction is that in nanoimprinted lithography, the stamp is made out of surface structured materials that have been created using processes like photolithography and another process known as the femtosecond laser ablation technique. The primary merit of MCP in comparison with the other surface modification methods lies in its good two-dimensional surface patterning capability, which allows varied functional groups or polysaccharides, proteins, and biological signals to be cast or transferred onto the surface of the substrate in precisely specified patterns. MCP has considerable benefits over other surface patterning methods in terms of cheap cost, high dependability and adaptability. Even though mechanical action regulated by a preset program ensures sample reproducibility and speed, it also allows for large-scale tailored manufacturing. This approach had been studied and implemented in a variety of sectors, and it has shown great promise in facilitating multidisciplinary study in science, engineering and medicine, with new applications being developed all the time. MCP's operating settings may be gently tweaked and improved to generate excellent surface patterns on a range of substrates, despite of its basic premise [35].

#### 1.2.2.1.3 Fast Ink-Jet Printing

Ink-jet printing is commonly used to accumulate a range of patterns on flexible and tangible substrates for conveniency, environment friendly, high process control, and low-cost electronic devices, among the various nanostructure deposition procedures on electrode surfaces. This method combines spatial resolution, printing speed, repeatability, inexpensive initial investment, and reduced waste into an attractive package. Inkjet printing also eliminates the need for dyes, masks, coloured chemical etchants using etching, and other patterning issues by printing the desired device pattern directly to the substrate. On a number of substrates, inkjet printing has been used to manufacture a range of electrical devices, sometimes in concert with other deposition or patterning processes [36].

Inkjet printing is an old practice that has lately been resurrected for the manufacturing of low-cost and simple electrical sensing and biosensing devices. This technology may be used to create flexible microsystems on ecologically friendly substrates like polymers and paper. Inkjet printing stands out among them because it blends an old-fashioned printing technology with cutting-edge nanoparticle-based inks to provide the printed device important qualities including conductivity, hydrophobicity or hydrophilicity and resistance or insulating properties. The technology was even utilized to use other bio-element such as enzyme, antibody and genetic material-based inks to functionalize the printed devices. This approach has been investigated in the biosensor sector for the past ten years, resulting in a vivid large number of papers as well as intriguing prospects and applications. Inkjet printing has several benefits over other printing processes, including the ability to print on both solid and flexible substrates, the lack of extra components other than inks, and the ability to reduce the time from concept to prototype to a few minutes and therefore it has been produced for electrochemical and optical biosensing by some companies [37].

#### 1.2.2.1.4 Self-Assembled Monolayer

The two-dimensional molecular level structures that form or cast spontaneously on the surface of different kinds of substrates, hence known as Self assembled monolayer (SAMs). Self-assembly is the most versatile concept in nature as such around all biomolecules of larger size to smaller size, such as proteins, peptides, amino acids combine and self-assemble to produce functionally relevant structured and ordered structures. Proteins or short chain polypeptides have been used by nature to build a range of materials, including shells, pearls, and keratin. It may be used to create whole new molecular structures. SAMs are gaining popularity due to their applications in electrical devices such as biosensors, thin film transistors, micropatterning, etc. Molecules by molecules are assembled in self-assembly, thereby "bottom up" approach language is used. The notion of SAMs is gaining prominence in the field of modification of surface using biomaterials or biological compounds. SAM has a number of advantages over other physical surface modification techniques including UV irradiation and electron beam. Since the SAM surface modification enables for covalent bonding of molecules, one of the core advantages is their long shelf life or functional period. SAMs employing the aforementioned physical approaches, on the other hand it has shown low stability. In some cases, results in a loss of surface chemical activity as a result of irradiation. Furthermore, conventional approaches such as physical or chemical adsorption, London forces, and cross-linking of biomolecules had suffered from some stability issues, but the SAM approach has become widely used for biomolecule immobilization. There are two types of SAM which are small molecule SAM and polymer SAM and also there are two types of SAM production processes, substrate coupled and substrate decoupled, based on interactions between the substrate and the molecule of SAM. For instance, in the first stage, SAM headgroups chemisorb on certain areas of the surface to form an ordered monolayer which have been done experimentally in lab involves saturated thiols on a gold surface. In this case, the substrate's crystalline structure is crucial, and single crystals such as gold are often used to create the ordered and linear assembly. In a substrate-decoupled process like the synthesis of alkyl siloxane on a substrate of silicon which is in hydrated form, there is indirect contact with the surface molecules, and the building of monolayer process is entirely controlled by intermolecular force of interactions [38–40].

#### 1.2.2.1.5 Microfluidic Technology

Microfluidics are techniques for regulating small-scale fluids in devices and systems, and many microfluidic based devices have been actively and widely used in the replication and evaluation of specific and targeted biological processes in tiny devices with a limited quantity of material. Using microfluidics as an instrument to replace the existing conventional research equipment at a very minimal cost, cell counting, sorting and trapping have been reduced to a good extent. In various research, organizations have developed their specific microfluidic cell culture technique that had imitated like the exact organs of human beings in order to study targeted biological processes or evaluate the efficacy or toxic potential of drugs. Microfluidic based impedance virus biosensors have been intensively investigated because they rely on a simpler and faster method of identifying particular viruses than existing methods. One of the main advantages of microfluidic devices lies in the intrinsic capacity of the device to install many analytical modalities or elements; as a result, MFT biosensors have been created for applications ranging from utilization in an analytical instrument, in research to healthcare and associated industry. Some researchers developed a technique that can be used to test influenza of avian or bird virus with high degree of sensitivity using a portable impedance-based biosensor with twenty-five pairs of microelectrodes were gathered and a microfluidic technology, as well as magnetic nano particles coated with antibodies [24].

### *1.2.2.2 Cell Without Pre-Treatment or Label-Free or Non-Invasive Detection*

Live cells carefully perceive and respond to outside information. For quite some time, cell transmission was thought to be linear, with an ambient stimulus triggering a series of steps that culminated in a well-defined detection response. However, there is compelling evidence that are more complex when cellular reactions take place in response to external stimuli. The spatial and temporal targeting of proteins to suitable locations, which dictates the timing and strength of cell signals and responses, influences the specificity and effectiveness of various protein interactions. Cell signaling mediated by a cellular target like the G protein-coupled receptor (GPCR) is generally well-ordered and regulated, including a succession or series of spatial (space) and temporal(time) events, many of which result in alteration in localized mass density or redistribution of cell-based contents. When these changes or redistributions within the cell occur with the detecting volume, this process was followed by the optical sensors. This is because non-specific optical responses of cells recorded with a label-free biosensor are often overlooked. Optical biosensors can detect receptor signaling in real time and provide a kinetic optical signal when the ligand is administered. Since a receptor may interact with several signaling molecules and proteins scaffolding such as arrestins while simultaneously linking to multiple G protein subtype. Important cellular decisions like cytoskeletal remodeling, cell cycle checkpoints, and death need precise temporal or time specific control and relative spatial moreover space dimension distribution of active signal-transducers [28]. Due to its non-invasiveness at the micron scale, another use might be to track the evolution of a disorder or disease and find sensitive chemicals that would ordinarily be metabolized or inactivated in various biological samples utilized for clinical testing, such as serum, urine, or faeces [29]. A typical electrochemical sensor device for non-invasive epidermal consists of an adhesive single top layer membrane, a flexible or stretchy substrate positioned over the adhesive single membrane, and an anodic electrode assembly containing an iontophoretic electrode deployed over the flexible or stretchable substrate. Above the flexible or stretchy substrate, an iontophoretic electrode and a cathodic electrode assembly were kept close to the anodic electrode assembly. The device has an electrode interface assembly with several electrically conducting connections [30]. In most label-free biosensors, a transducer is employed to change a stimulus-induced biological interactive reaction into a readable and countable signal. Depending on the principle of the transducers, the label-free biosensors for whole cell sensing technique are classified as electric biosensors [29].

## 1.3 Types of Nanomaterials Used in Cell Biosensor

The size and dimensions of nanomaterials are used to classify them. Nanomaterials come in four different dimensions: zero, one, two, and three. The existence of three dimensions of materials in nanoscale of nano particles (NPs) of inert elements such as gold, palladium, platinum, silver, or quantum dots are nanomaterials having zero dimension. With a diameter of 1electron 50 nm, NPs can be spherical in shape. Zero-dimensional nanomaterials have been discovered in the form of cube and polygon forms. One dimension of 1D nanomaterials is in the region of 1electron 100 nm, whereas the other two dimensions might reach macroscale. One-dimensional (1D) nanomaterials include nanowires, nanofibers, nanorods, and nanotubes. 1D nanostructures may be made from inert metals, metal oxides, quantum dots, and other materials. The two dimensions are nanoscale whereas one dimension is macroscale in the 2D class of nanomaterials. Two-dimensional (2D) nanomaterials include nano-sheets, and nano-walls, nano-thin films, thin-film multilayers. 2D nanomaterials can have a surface area of several square micrometres while maintaining a thickness in the nanoscale range. There are no nanoscale dimensions in three-dimensional (3D) nanomaterials but all dimensions are macroscale in nature. Loaded materials are collection of 3D nanomaterials which is made up of individual blocks that can be as small as a nanometre or larger than that as well [58].

## 1.4 Classification of Biosensors Based on Transducers

### 1.4.1 Conjugated Biosensor

#### *1.4.1.1 Electrochemical Biosensors*

The biosensor which is a common sensing device that works by converting biological processes into electrical signals on the principle of electrochemical conversion. An electrode serves as stable base for biomolecule immobilization and electron flow which is considered to be a critical component. Various biosensing approaches for inexpensive and small analytical instruments for surface analysis have recently been developed [8]. Depending on type of electrical information is to be measured to get appropriate information, the electrical biosensors are subdivided into three main categories such as potentiometric, amperometric, and impedance sensors. Recent biosensor development has concentrated on critical aspects such as quick detection, detection limit, operation practicality, and low cost. Some electrochemical based pathogen detections have been performed by some researchers like Salmonella was identified which utilises a carbon electrode in screen printed fashion conjugated with immunomagnetic beads and an amperometric biosensor. A nanofiber based on light potentiometric sensor which can detect E. coli at very lower concentrations range under one hour had been developed. Some have reported an AuNP-based signal-off impedimetric immune-biosensor for detecting *E. coli* [9].

##### 1.4.1.1.1 Amperometric Biosensors

Amperometric biosensing technologies have long been employed in medical settings. The sensing transducer in such sensors is made up of three electrodes termed as working, reference, and counter electrodes. Bioreceptors are immobilized on the working electrode whereas all electrodes are immersed in a specific range buffer solution. One electrode serves as working while other serves as the reference as well as counter electrodes in a two-electrode sensor. In most cases, the sensor is connected directly to a signal system that consist of a battery of specific voltage and a circuit working system. The circuit system controls the voltage of a battery to ensure that the sensor receives sufficient power.

Few researchers have worked on the amperometric biosensors where the sensing transducer's two electrodes are linked to the battery's cathode and anode. Lactate oxidation is catalyzed on the working electrode by immobilized lactate oxidase, which produces hydrogen peroxide ($H_2O_2$). $H_2O_2$ is oxidized further, resulting in a current signal that may be detected using a current metre. Rather of a sophisticated and bulky instrument like a potentiostat, an amperometric biosensing elements directly connected to a button cell battery which provides better overall performance with sufficient reproducibility [10, 11].

#### 1.4.1.1.2 Potentiometric Biosensors

A potentiometric biosensor is a chemically oriented sensor that is used to determine the concentration or amount of certain and specific analytes. The major function is when no voltage is increased, it measures the potential difference of an electrode and when the stimuli hit the transducer, the change in difference in potential is measured. Environmental, medicinal, and industrial sectors have all employed potentiometric biosensors. Potentiometric redox based performance is a low-cost, non-active method for determining the aggregate reduction and oxidation status of biological as well as environmental solutions. The functional ability of the biosensor to perform minute recording in quite small volume with comparable high surface area, nano porous electrodes are of utmost importance. They increase the electrical conductivity and diminish unsuitable bioeffects on the electrochemical signal, hence accuracy of detection is maintained [12]. It is a common electroanalytical method for determining the concentration of some tiny ions in the solution. The current flow does not occur in this technique when compared with other electrochemical sensing techniques like voltammetry sensors, differential pulse voltammetry (DPV) sensors, chrono–amperometry sensors, etc. As a result, the contact remains unaffected and there occur no changes in the chemical structure of the sample. Traditionally, there are membrane-based ion specific electrodes which are likely to do billions of potentiometric measurements per year. Redox potentiometry has been designed with metallic electrodes that demonstrates to be a valuable method for determining the concentration of minute compounds and evaluating the redox characteristics of complicated environmental or biological samples. Although electrode areas can impact measurement in complicated solutions in an indirect way. The surface of electrode is easily structured where a response is often not proportional to electrode area. This passivation can slow down electron flow between the electrode and the oxidative and reductive species, lowering sensitivity, especially at low concentrations. Redox potential tests may provide vital details about a sample's overall redox status, whether it is a simple solution or a more complicated sample like blood. The simplicity and flexibility of this approach to in-field measurements and minute quantities, however, are the merits of potentiometric analysis. It is a non-mediator, non-label, non-membrane technique when the total redox state needs to be determined rapidly and effectively with low cost or expense utilizing simple gear, redox potentiometry becomes the choice [13].

#### 1.4.1.1.3 Impedimetric Biosensors

The current response to an AC voltage, impedance-based electrochemical biosensors work by assessing the alteration in both resistive and conductive characteristics at the phase boundaries of electrodes can be monitored. As a result, these biosensors can be considered a step forward from conductometric biosensors. These biosensors have the main advantage of being very sensitive, low-power devices sensing capabilities. For its low cost, simplicity of downsizing, and availability of basic equipment, impedance biosensors can be used in emergent healthcare applications. In electrochemical impedance biosensors, electrode manufacturing is a critical parameter. For instance, glass and silicon substrates with gold coatings are commonly used as microelectrode substrates. The materials used in electrode manufacturing have undergone significant transition. The ratio of high surface to the volume and surface energy of gold nanoparticles enables for effective interaction in the immobilization techniques. They have utilized several techniques including electrochemical deposition and covalent linkage with self-assessed monolayers that are used to immobilize gold nanoparticles on electrode surfaces.

Electrochemical impedances are made with a dual mode galvanostatic device that consist of either a spectrum analyser or a frequency response analyser as a detecting agent. Potentiostat is often employed wherein the set-up of instrument feeds the input of varying voltage to the electrodes and monitors the associated response of current simultaneously. Bioreceptor molecules such as enzymes, aptamers, lectins, and peptides can be recognized by impedance biosensors. Impedance biosensors, on the other hand, are not confined to these categories but a variety of different biorecognition molecules and detection techniques are also available. Enzymes catalyze the half-cell redox reaction in the enzymatic based impedance biosensors, allowing charge transfer takes place between electrode and analyte. These devices have had a substantial influence on a variety of areas, including pharmaceuticals, food, agriculture, and healthcare, as well as the environmental analysis and monitoring [14].

### *1.4.1.2 Integrated Biosensor*

#### 1.4.1.2.1 Optical Biosensor

Optical biosensors can be easily downsized and compressed and have chip integration potential all together in a single chip. Other features such as microfluidics, can be integrated onto a single platform to create a lab on a chip system to be made available for various onsite detection. A conventional solitary fibre mode has a fairly large and circular mode field, whereas integrated devices are generally micron and nanoscale structures with a restricted mode field. According to the concept of evanescent based detection technique involved, the propagation of optical wave through the waveguide film and the cladding layer is penetrated by the evanescent wave, which is covered by dense part or the bulk of solution and senses the change in the refractive index in the sample solution. The difference between the guiding films and cladding films produces an enhanced evanescent field view in a waveguide with a small dimension and comparatively a large refractive index. Apart from silicon on insulator, which has a well-established processing technology, wave propagation loss is lower in silicon nitride and a stronger field in the near infrared area, indicating that it has a greater potential for application in waveguide construction systems. The mechanism of evanescent waves aids the construction of optical waveguides as biosensors, with the most often employed designs comprising slot, rib, and photonic waveguide [15].

#### 1.4.1.2.2 Surface Plasmon Resonance

Surface plasmon resonance (SPR) is the extensively used and well-known technique for detecting the shift in the refractive index led by chemical interaction via a surface plasmon wave at the metal surface. The operating concept is based on photon interaction, which happens when photons or discrete packets of energy from an electromagnetic wave strike with the electrons on a metal surface. It is a phenomenon that occurs when a surface plasmon wave is formed when the plasmon propagates in a parallel fashion to the surface of a metal. The plasmon wave travels along the metal dielectric barrier, where it is exponentially attenuated. At a metal dielectric contact, plasmon occurs only when the optical wave vector overlaps with that of the plasmon optically. As the plasmon mode's wave vector is often larger than that of the incoming wave, it cannot be directly stimulated by light in most cases. To increase the light wavenumber, other optical phenomena such as diffraction phenomena or attenuated total reflection (ATR) can be exploited in a coupling device [53]. Prism couplers, grating couplers, and fibre couplers are the most often utilized coupling structures in this kind of resonance [15, 27]. The most recent and well-known application of SPR is applied in the detection of Alzheimer's disease biomarker in the brain. Increased elongation of a peptide known as Aβ42 in brain cells, is considered the commencement of Alzheimer's disease. As a result, Aβ42 is a crucial biomarker for predicting the prognosis of people with Alzheimer's disease. Modification of specific SPR chips with Silver (Ag) in SPR-based systems is an effective way to boost response intensity [57].

#### 1.4.1.2.3 Localized Surface Plasmon Resonance

In reaction to incoming light, electrons in noble metal nanoparticles collectively oscillate, which is known as localized surface plasmon resonance (LSPR). This phenomenon can be utilized as a sensor since it is particularly sensitive to a little fluctuation in the refractive index near nanoparticles of metal. LSPR is the most promising biosensors because they may increase biomolecular detection sensitivity by avoiding the time-consuming step that reduces molecule bonding disruptions. Another type of LSPR is FOLSPR (fibre-based optic localized surface plasmon resonance), which uses a fibre-optic platform to provide miniaturization, distant sensing, and lossless signal transport [55]. Focused ion beam (FIB) nanopatterning methods were utilized to construct the FOLSPR sensor to increase its performance. To develop an integrated FOLSPR sensor, metal colloids are employed to immobilise on the surface of an optical fibre by metal nanoparticles. This can be a phenomenon utilized as a sensor since it is particularly sensitive to any fluctuation in the refractive index near metal nanoparticles. This analytical platform enables simultaneous screening of a large quantities of proteins or polypeptide polymeric materials from a protein library. In comparison to traditional bioanalytical techniques, the microarray format requires a small amount of chemicals, cells, and reagents [17].

#### 1.4.1.2.4 Evanescent Wave Absorbance

When propagation of light takes place through a fundamental high-index optical fibre based on the principle of total internal reflection (TIR), an electromagnetic field may be evanescent wave is created at the waveguide or sample surface. This field gradually decays as one moves away from the interface. An evanescent wave can therefore generate fluorescence in close proximity to the detecting surface, such as fluorescently tagged proteins bonded to the optical sensor surface. Evanescent field-based optical fibre or waveguide biosensors detect any fluctuation in refractive index directly affecting the optical conditions for forming an evanescent field produced by binding events of an analyte using selective biorecognition components. The wave absorption pattern of evanescent approach has been promising, and widely used in optical fibre biosensors due to its advantage of having higher sensitivity, greater resolution, and low limit of detection. An evanescent wave is formed when light decays linearly into a small area of the coating in optical fibre, and this field is utilized to manufacture cladding modified intrinsic optical fibre sensors. It is used in optical fibre sensors as a modified coating to respond to chemical and biological activities on the surface. In certain field such as water and food sources, food industries, and homeland security activities, pathogen detection employing evanescent field fibre optic biosensors is becoming increasingly frequent which has enabled the vast application at the same time. These biosensors might be used to identify many infections at the same time [56].

#### 1.4.1.2.5 Microelectrode's Array

Cell microarrays are also a lab on a chip device that enables for high throughput screening of cell material surface interactions. Various similar kinds of cells are bound to a flat surface or a particle to conduct multiple tests in a high throughput way in cell microarrays technique, which have been produced or manufactured using various microfabrication processes. Cell microarrays are then paired with electrical or optical sensor technologies and incorporated into system of microfluidic to track changes in cellular environment more efficiently as a result of external factors. The different types of cells can attach and grow on flat 2D chip-like carrier systems, known as positional arrays, or on suspended particles, known as suspension arrays, to perform multiple system and high-throughput tests, particularly cell-based assays [16]. Microscale patches on glass or silicon type substrates are printed by biomolecules or bioreceptors with the remainder of the substrate passivated to prevent or avoid non-specific cell and protein adhesion on the surface. Every printed spot in the collection of printed spots is considered to be an independent experimental duplicate due to its spatial isolation. This analytical platform allows for the screening of a large number of selected proteins or polypeptide polymeric materials from a library at the same time. The microarray format need the modest number of chemicals, cells, and reagents in compared to typical bioanalytical techniques or procedures [17]. Microarrays work on the natural concept that complementary strands of DNA will bind to each other. The pieces of DNA from molecules are obtained with the help of restriction endonucleases which are then labelled with fluorescent markers. The prepared probe is then reacted to a targeted complementary sequence of DNA with the help of endonuclease enzymes and the remaining DNA fragments are washed away. Passing a laser on the target DNA fragments allows them to be recognized by their unique fluorescence emission pattern. A computer or equipment is used to record both the fluorescence emission pattern and the DNA identification. This approach of recognizing multiple DNA fragments at the same time in a short period of time utilizing DNA chips is extraordinarily rapid, as well as sensitive and specific. MEA recordings have become more popular as an in vitro approach for detecting and characterizing the capacity of chemicals and medicines to produce neurotoxicity in general. With the advent of facilitatory MEA systems, rapid and easy approaches to assess compound impacts on health of the cell are necessary for successful chemical screening using MEAs [18].

#### 1.4.1.2.6 Positional Array

Multiple cellular micro spots on flat substrates are the basic construction method which is used to create positional arrays. Micropatterning technologies of surface such as photolithography and soft-lithography are commonly utilized to manufacture microarrays of cellular component on diverse substrates which has contributed majorly to breakthroughs in microelectromechanical systems (MEMS).

Photopatterning is one of the procedures for creating a cell microarray system, with the micropatterns that act as the templates. Only in areas of a photopatterning process exposed to UV light, covalent type of bonding prevails between substrates and photoreactive groups. The requisite cellular micropatterns are generated after removing any non-reacted species using a solvent and seeding the cells. Photopatterned PEG (Polyethylene glycol) hydrogels, as well as alternative hydrogels based on hyaluronic acid and gelatine, were employed to construct cellular micropatterns. Soft lithography, which uses soft elastomeric materials instead of photolithography to transfer patterns, was discovered to be a good fit for biological applications. Soft-lithographic techniques such as microcontact printing and polydimethylsiloxane-based microfluidic channels are becoming increasingly popular in cell patterning (PDMS). The generation of cellular micropatterns by moving cells to a specific location using external pressures is possible, in addition to photolithography and soft lithography techniques that employ cell adhesive and non-adhesive microdomains to passively guide cell attachment. Although a variety of regulatory forces may be used to trap and locate cells, dielectrophoretic (DEP) has sparked a lot of attention as a method for creating cellular micropatterns [16].

##### 1.4.1.2.7 Suspension Array

Even though the popularity and utilization of positional arrays for high yielding experiments, suspension arrays play a major role as an alternative microarray structure because they are expected to provide greater multiplexity flexibility than positional arrays. Patterning distinct cells on stand-alone substrates in the case of positional microarrays is tough and needs a sophisticated procedure. Multiplex assay systems may be easily created with suspension arrays by simply combining the separately manufactured microparticles containing various cells. Self-encoded microparticles are used in the recognition of the cells in this case which include the cells as just an array component. Microparticles contain the encoding elements for discriminatory optical detection: Quantum dots with fluorescent dyes [16].

#### *1.4.1.3 Field Effect Transistor*

A field effect transistor (FET) is a device found in solid state in which the electroconductivity of the semiconductor placed between the source and drain terminals is regulated by a third electrode gate and an insulator. Despite the fact that nanostructures are produced and put into their structure to reframe their surface and boost their sensitivity, they require very little postprocessing. The source and drain electrodes are linked to the technique of immobilization of certain natural receptors such as nucleic acids, aptamers, enzymes, antibodies, cells or artificial biomaterials. When biosensors are subjected to a certain biological analyte, they produce complexes, such as immunogenic, enzyme substrate, DNA structures, and so on, which are then converted into detectable signals by the transducer system. Attaching the charged molecules to the surface of the dielectric of gate generates a voltage, resulting in cut-off voltage fluctuations. As a result, the FET biosensors approach relies on adsorbed species influencing conductance. The two basic types of FETs are n-type device and p-type device, which are distinguished by their operating principles, with electrons and holes serving as the primary charge carriers, respectively. The n-type FET detector will respond with a spike in conductance due to electron aggregation. The conductivity will be reduced if the target is a negatively charged molecule and vice versa if the target molecule is positively charge. The p-type FET system uses cell-based FETs, which is opposite of the current trend.

The gate voltage is applied differently, the design is different, and the material utilized in the gate and channel regions is different [19].

#### 1.4.1.4 Light Addressable Potentiometric

A light-addressable potentiometric sensor (LAPS) is based on semiconductor type chemical sensor that uses lighting to produce a point of measurement on the surface of sensor. It is an effect based on field sensor from the electrolyte-insulator-semiconductor (EIS) family, with the insulating layer sensing surface in contact with the solution under examination. With substantial applications in chemistry, biology and medicine, light addressability might be used to investigate the pH distribution or the free concentration of hydronium ion in a specific chemical species. Chemical sensors of the LAPS series include the transistor of ion-sensitive field-effect and an EIS capacitive sensor. The advantages of employing a LAPS sensor plate, one may change the pixel layout for zooming in and out on the fly. The number of pixels is limitless, and the measuring area with any form should be as large as needed. Second, no wires linked pixels, and wires are shielded from the solution by passivation, making it long-lasting. Third, surface area of sensing region is flat which makes it straightforward to connect a LAPS to microfluidics. Finally, unlike photolithography, a LAPS sensor plate does not need any fabrication, making it affordable and ideal for disposable applications. When the light is turned on, photon absorption creates photocarriers, or pairs of electrons and holes, in the semiconductor layer. The electric field within the depletion layer separates electrons and holes arriving through diffusion, resulting in a transitory current despite the loss of some photocarriers owing to recombination. When the light is turned off, the surplus electrons and holes recombine and form a short current in the opposite direction. The current, which is dependent on the capacitance of the depletion layer, may be used to determine the quantitative analysis of an analyte concentration [20]. Patio-temporal pictures of chemical or biological analytes samples, electrical potential differences and impedance may all be created using LAPS. By recording photocurrents created locally in the semiconductor, a focussed, intensity-controlled light source may be utilized in light addressability to detect various chemical or biological analytes such as pH, ions, cells, enzyme activities, etc. In bioimaging applications, such as monitoring the quantitative concentrations in microfluidic channels (MFC) or analyzing metabolic and signaling processes in live cell, LAP sensors are of tremendous interest in this area of work. The electrolyte insulator semiconductor mechanism is used in LAPS, where the surface of sensor on the insulating layer is in direct contact with the solution to be investigated. Chemical pictures of the concentration distribution pattern on top of the sensor substrate may be created using LAPS. The LAPS reaction is triggered by either a change in the surface potential of an insulator or a change in the impedance of electrical run [21].

#### 1.4.1.5 Patch Clamp Chip

Patch-clamp experiments can be used to assess the biochemical action of a protein while controlling the gaseous and biochemical environment of the cells. a novel method for combining lab-on-a-chip with patch-clamp methods the lab-on-a-chip technology offers a potential way to do patch clamp measurements in a well-controlled environment for the cells being studied. The electrical impulses across the plasma membrane of single neurons were recorded using patch-clamp. The newly designed closed microfluidic chamber allowed researchers to examine individual cells in a controlled environment with little oxygen penetration into the microchannels. Patch-clamp is a highly effective technique which is based on the principle of electrophysiology for examining certain ion channel specific behaviour. Ion channels are the fastest channels and are found in practically all cells; nevertheless, neurons, muscle fibres, cardiomyocytes, and oocytes are the most often studied cells employing patch-clamp methods because they overexpress single ion channels. For measuring the single ion channel conductance, a portion or part of the cell membrane that contain the ion channel of interest is excised, and a microelectrode creates a highly resistant bond with the cellular membrane. Alternatively, when the microelectrode is encased in the cell membrane, eruption in the small patch allows the electrode electrical access to the whole cell. After adding voltage to create a voltage gated clamp, the membrane current is monitored. To monitor changes in membrane voltage, a current clamp is used to detect membrane potential. To change the voltage or current within cell membranes, compounds that block or open channels can be utilized [22, 23].

#### *1.4.1.6 Electric Cell-Substrate Impedance Sensor*

By monitoring the current response against the modest AC voltage at the solid or liquid interface, impedance-based biosensors detect the changes in both resistive and conductive phenomena of electrodes. As a result, these biosensors represent a step forward from conductometric biosensors. These biosensors have also demonstrated to be very sensitive, low-power devices capable of real-time sensing. Due to the evident based on affinity interaction and binding with biorecognition molecules, such as the lock and key paradigm, they are also known as affinity-based biosensors. Labelled and label-free affinity-based impedance biosensors are available. In labelled biosensors, the analyte attaches to the surface of electrode ligands, followed by the targeted attachment of secondary ligands or double ligand coupled with labels to immobilised analytes. Impedance biosensors measure electrical impedance by applying sinusoidal voltage to certain frequencies using alternating current flow. Electrical impedance measurements allow impedance sensors to assess minute changes in chemicals on the working electrodes. This characteristic has been widely investigated in impedance biosensors for such a selected biomaterial can assess to the minute amounts of specimens present in the sample. The composite impedance is determined by evaluating the wave characteristics such as sine wave response signal of the amplitude and phase change as it passes through the target material using a sinusoidal alternating current signal that flows through the electrode. Impedance biosensors can be employed at outside the lab as an application due to their low cost, ease of shrinking, and availability of basic equipment [14, 24]. Impedance spectroscopy, which employs technology featuring microfluidic flow of cytometer, necessitates a smaller sample size than traditional or conventional methods while retaining its accuracy in sensitivity. It also simplifies the process of initial stage, which might have an impact on cell properties during sorting. To correctly differentiate between types of lymphocytes such as T-lymphocytes, monocytes, and neutrophils in blood, few of researchers have used microfluidic impedance cytometry. Electrical impedance can be used to sort cells with certain characteristics for further investigation. The researchers also created a useful cytometer that can monitor a wide range of frequency and recognized yeast cells consisting of two types of features based on their dielectric characteristics at four distinct frequencies. Impedimetric-based biosensing is a type of electrochemical biosensor that has shown to be a promising technology for detecting food related or borne or food-infected pathogens by changing the impedance of the electrodes by trapping bacterial cells on them. Cell size, membrane and cytoplasm electrical characteristics, and an electrical impedance recorded at different frequencies all give different forms of information about cells. At lower frequencies, about a few hundred kilohertz, impedance exposes cell size, but it may also be used to interpret membrane reactance [25].

#### *1.4.1.7 Quartz Crystal Microbalance*

Quartz crystal microbalances (QCMs) are operating instruments based on a crystal made up of quartz material that vibrates or harmonises under the influence of an alternating current-voltage resonator and whose resonating frequency is exactly proportionate to its entire mass (QCMs). Small molecules are detected using acoustic sensors, which may also be utilized to monitor specific interactions between complete cells. The sonic sensor can detect changes in the viscous nature of the receptor-ligand complex in addition to mass variations. Compared to optical sensors, this technique shows low sensitivity to any changes in the fluid surrounding. The utilization of a live endothelial cell coupled to the QCM to detect nocodazole activity is one of the more exciting recent instances. Quality screening of the biological cell lines and their susceptibility to anticancer class such as taxane plant derivative medications has been demonstrated using QCM coated epithelial tumour cells. Acoustic type of biosensors has also been described for immune-sensing, as proven by the sensitive detection of the harmful substance cocaine in recent publications [26]. The classification of various types of cell biosensor based on transducer is shown in Table 1.1 with their principle and application.

**Table 1.1** Classification of transducer and its application.

| Biosensor | Transducer | Principle | Applications | References |
|---|---|---|---|---|
| Electrochemical | Potentiometric | Determine the redox homeostasis of biological and environmental solutions overall. | Sugars, urea, antibiotics, neurotransmitters, insecticides, ammonia, carbon dioxide, and a variety of ionic species can all be determined. | [12, 13, 20, 21] |

*(Continued)*

**Table 1.1** Classification of transducer and its application. (*Continued*)

| Biosensor | Transducer | Principle | Applications | References |
|---|---|---|---|---|
| | Amperometric | When a potential is introduced between two electrodes and the analyte, a current is produced. | Analysis of ethanol, glucose, and lactate in wine. | [7, 67] |
| | Impedimetric | Examine changes in resistive and reactive properties at electrode phase boundaries. | Detecting Vascular Endothelial Growth Factor. | [68] |
| Optical | Luminescent | Produce a signal proportional to a measured substance's concentration (analyte). | A wide range of analytes, including viruses, poisons, medicines, antibodies, tumour biomarkers, and tumour cells, may be detected selectively. | [15, 54, 56] |
| Microelectrode array | Positional array, Suspension array | Microarrays work on the concept that complimentary sequences bind to one other and then respond with DNA chip probes. | Detection of neurochemicals in interstitial fluid in both animals and humans. | [18] |

(*Continued*)

**Table 1.1** Classification of transducer and its application. (*Continued*)

| Biosensor | Transducer | Principle | Applications | References |
|---|---|---|---|---|
| Light addressable potentiometer | Semiconductor | Photocarriers, or pairs of electrons and holes, are created in the semiconductor layer by photon absorption when the light is switched on. Despite the loss of some photocarriers due to recombination, the electric field inside the depletion layer separates electrons and holes coming by diffusion, resulting in a transient current. | Analysis of metals, cyanide, ammonia. | [20, 21] |
| Electric cell substrate impedance | Electrodes | They measure electrical impedance by applying sinusoidal voltage to certain frequencies using alternating current flow. | Analyzing the activities and morphologies of cells. | [21] |
| Patch clamp clip | Microfluidic chamber | To monitor changes in membrane voltage, a current clamp is used to detect membrane potential. | Ionic currents in isolated live cells, tissue slices, or patches of cell membrane are studied. | [23] |

(*Continued*)

**Table 1.1** Classification of transducer and its application. (*Continued*)

| Biosensor | Transducer | Principle | Applications | References |
|---|---|---|---|---|
| Field effect transistor | Semiconductor | The two main kinds of FETs are n-type and p-type devices, with electrons and holes as the predominant charge carriers, respectively. An n-type FET sensor will respond with an increase in conductance owing to electron aggregation if the target molecule is positively charged. | Application in cardiovascular diseases, cancers, diabetes, HIV, and DNA sequence. | [69] |
| Quartz crystal microbalance | Quartz crystal | Changes in the viscosity of the receptor-ligand complex may be detected using a quartz crystal that vibrates under the influence of an AC voltage resonator and has a resonance frequency proportionate to its total mass. | Metals, vapours, chemical analytes, environmental contaminants, biomolecules, illness biomarkers, organisms, and infections may all be detected in the vacuum. | [70] |

(*Continued*)

**Table 1.1** Classification of transducer and its application. (*Continued*)

| Biosensor | Transducer | Principle | Applications | References |
|---|---|---|---|---|
| Surface plasmon resonance | Coupling structures | It uses a surface plasmon wave to detect the shift in refractive index caused by chemical interaction at a metal surface. | Visualization of virus-like particles (human immunodeficiency virus (HIV)-based virus-like particles) attaching to round-shaped viruses (inactivated influenza A virus). | [71] |

# 1.5 Application of Biosensors of Cells

## 1.5.1 Quantitative Assessment-Based Application

### *1.5.1.1 Biomedical Application*

Electrochemical redox or pair of reductive-oxidative phenomena induced a difference in the current flow in the electrodes employed in biosensors in the amperometry technique. The electrochemical contact or participation between the target analyte in the sample and the biological component causes changes in the electrical conductivity of a sample solution or any other solution medium, such as nanotechnology nanowires, nanomembranes and so on, in the conductometric method. The electrochemical interaction between the biological recognition element and the analyte causes a voltage between the electrodes of the biosensors in the potentiometry approach. The production and identification of peroxide, as well as the utilization of natural oxygen as a co-substrate are necessary for first-generation glucose biosensors. In case of the second-generation glucose biosensors, it uses a synthetic reductive substance or electron acceptor material instead of oxygen molecule. In third-generation reagent-free glucose biosensors, the use of mediators or any other reagent is no longer required. The electrochemical biosensors such as biocatalytic one includes glucose sensors, lactate sensors, etc [41]. Wearable cell-based biosensors have recently advanced to the proof-of-concept level, owing to the growing interest in point-of-care diagnostics and health monitoring. The advent of advancements in technology such as microfluidics, biosensors may now be employed in a high-throughput way, which is critical for identifying novel medications or drug resistance. The detection of a pathogen can also be connected to downstream processes, such as the manufacture of a therapy, using cell-based biosensors [42, 43]. One of the most successful ways for assessing biomolecule affinity binding and screening druggable chemicals appears to be SPR biosensing. SPR-type sensors are rapidly being used in biomedical research to analyse a variety of biological substances, including DNAs, RNAs, proteins, carbohydrates, lipids, and cells.

### *1.5.1.2 Microbial Application*

Microbial fuel cells (MFCs) are a type of green biotechnology that has mostly been employed in waste water treatment and energy or power generation. Despite its low power output, which limits or restricts its usage for directly controlling most electrical equipment, MFC applications have extended to include chemical production, biological treatment of polluted soil, and water deionization. Chemical production, bioremediation of polluted soils, water desalination, and biosensors have all benefited from advancements in MFC's chemical, electrochemical, and microbiological components. Due to their simplicity and long-term survival, MFC-based biosensors have a lot of attention in recent decades, with applications ranging from water impurity monitoring, such as biochemical oxygen demand (BOD), to air quality detection [44, 46].

### *1.5.1.3 Environmental Monitoring*

Even though they are based on the toxicants' biotoxicity effects directly, microbial fuel cell-based biosensors can be a smart alternative. Toxic contaminants can prevent electrogene from working properly, causing MFC current to be interrupted. The more dangerous the material is to bacteria, the lower the current. As a result of the relationship between dangerous substances and current reduction amplitude, different toxicity sensors may be developed. Toxicity sensors are typically used to determine if an effluent's concentration of dangerous substances exceeds the permissible maximum concentration. As a result, the MFC sensor's toxicity testing concentration is directed at the pollutants' detection limit rather than the linear range as in BOD. Depending on the target pollutants, heavy metals biosensors, antibiotics biosensors, organic pollutant biosensors and acidic toxicity biosensors are the four primary categories of MFC-based toxicity biosensors [44, 45].

#### 1.5.1.3.1 Heavy Metals Biosensor

Heavy metals have a lengthy half-life and are difficult for microorganisms to eliminate or decrease. They would gather in human bodies as they moved up the food chain. Heavy metal ions may compete for the electrons at the anode in some MFCs designed for specific target chemicals, resulting in less electrons being transported to the cathode in the cathodic chamber. For instance, chromium ion Cr6+ is a terminal electron acceptor that may be reduced in anaerobic conditions by Cr6+-reducing anaerobes [85]. When an MFC is constructed employing Cr6+-reducing anaerobes, the cell voltage is predicted to drop as the Cr6+ concentration rises [44].

#### 1.5.1.3.2 Antibiotics Biosensor

A cathode-shared MFC sensor array was employed in another study to detect acidic toxicity. As the cathode performance fluctuation was minimized, the detection credibility of this sensor array functioning in continuous mode could be assured.

#### 1.5.1.3.3 Organic Pollutant Biosensor

Organic toxins, such as organic-based nitrogen compounds, organic based phosphate compounds, polycyclic aromatic hydrocarbons (PHA) and heteronuclear compounds like polychlorinated biphenyls (PCBs) are commonly found in water and can cause eutrophication. Researchers developed a single-chamber MFC to detect formaldehyde in water. To maintain an ideal anodic potential and prevent air bubbles formation at microscale from entering the MFC biosensor, Ag/AgCl reference electrode and a microscale air bubble trap were utilized in this micro-sized device. While the anode potential was maintained constant at certain voltage against the reference electrode, the current decreased according to the formaldehyde content in the medium, which varied from 0.001 percent to 0.10 percent [44].

#### 1.5.1.3.4 Acidic Toxicity Biosensors

As some types of harmful compounds in wastewater, such as mine drainage, generate a fast change in pH, acidic toxicity should be checked on a regular basis. Low pH inhibits microbial activity as well as aquatic animal and plant development, reducing the ability of the water bodies to self-purify and lower the water quality. Due to part of a research, a cathode MFC containing a single-chamber air was built and operated in a continuous batch mode. Hydrochloric acid was used to change the pH of the effluent. The output voltage fell fast when the hydrochloric acid was turned off while the pH was held at three or four. Another work used to detect acidic toxicity which was based on a cathode-shared MFC sensor array. The detection credibility of this sensor array working in continuous mode could be ensured since the cathode performance fluctuation was avoided. After the MFC array attained a steady state, acidified anolyte was used to provide an acidic toxicity shock. The voltage dropped from higher to lower value very instantly as the pH dropped from six to four. Although the peak or threshold pH value varies depending on the biofilm constitution, this phenomenon offers for a viable means of accessing the pH in water based on the interruption of MFC cell voltage [44].

## 1.5.2 Qualitative Assessment-Based Application

### *1.5.2.1 Diagnostic Application*

The quantity of lactate in the blood correlates with the degree of circulatory failure in Diabetes Mellitus, and a lactate excess of 4 mM predicts mortality. Lactate blood concentration is especially important for athletes because high levels of blood lactate cause muscular weakness and cramps by lowering blood pH. Cholesterol is another biologically important analyte, since excessive levels have been associated with circulatory system problems. The blood urea level is also an essential indication because of the disorders associated with it, such as majorly including gout, Lesch–Nyhan syndrome, and Fanconi syndrome. Changes in the values of numerous neurotransmitters are integrated with a number of brain diseases. Glutamate has been related to schizophrenia, Parkinson's disease, stroke, and epilepsy, among other mental diseases. Creatinine is produced by muscle and subsequently released into the circulation. It is a crucial analyte for diagnosing kidney and muscle problems. Detecting ethanol in human breath, saliva or blood serum is a crucial tool in clinical forensics [6, 43].

##### 1.5.2.1.1 Cancer Diagnosis

Cell-adhesion receptors on the surface of multicellular organisms allow them to develop strong interactions with surrounding cells and the extracellular matrix (ECM), which are necessary for the creation of different types of cell linings and its layers, tissues and organs. Some of the most important protein families identified in cell-adhesion receptors made up of numerous proteins include integrins, cadherins, immunoglobulin superfamily cell-adhesion molecules (IgCAMs), selectins, and syndecans. Several dangerous microorganisms infiltrate mammalian cells via interacting with ECM proteins and adhesion receptors. For instance, L. monocytogenes, an intracellular foodborne pathogen, encodes a set of adhesion and invasion proteins, including members of the internalin (Inl) family, that help it invade a range of cell types by interacting with cell surface receptors. Researchers have used modified L. monocytogenes to deliver luciferase-gene-loaded nanoparticles, often known as microbots, into cancer cells. This system based on microbot potential might be used to not only locate cancer cells and tissues, but also to deliver anticancer drugs to specific cells [61].

#### *1.5.2.2 Food Industry Application*

The food industry has always had a need for analysis of food composition, freshness, and microbial contamination. Vitamins, essential amino acids, colourants, emulsifiers, flavouring agents, preservatives, antibacterial agents, and allergies, among other things, might all be measured using biosensors. Biosensors for trimethylamine, hypoxanthine, and potassium levels have all been produced. Food raised toxins, allergens, viruses, heavy metals, and other contaminants may now be detected more effectively and quickly, thanks to the use of microfluidics in food safety studies. Due to its miniaturization, mobility, and small sample and reagent quantities, microfluidics is an excellent option for food sustainability research. The combination of electrochemical-based microfluidic technology and cell culture technology is a revolutionary food analysis technology. It can detect dietary induced changes due to the presence of various types of allergens in cell structure and metabolism, allowing for faster food safety detection [47].

## 1.6 Analytical Method for Biosensors of Cells

The cell-based biosensors are justified or evaluated using a set of benchmarks of the analytical findings or investigations. A preliminary experiment was conducted to determine the sensing sensitivity of the surface and selectivity for the target analyte. Based on the output of the signal with various quantitative amount of analyte molecules present, a regression plot can be created, which is subsequently utilized as a standard for future sensing applications. The plot of regression is now used to assess these biosensors' analytical capabilities.

### 1.6.1 Sensitivity

The slope of the regression line or straight line obtained, which indicates a variability in the output signal due to variations in analyte concentration, represents the sensitivity of a biosensor. A biosensor with a steeper slope may detect even little changes in amount or quantity of analyte in a solution. The following is the equation of sensitivity:

$$\text{Sensitivity} = \frac{YQ - YP}{XQ - XP}$$

In the calibration figure, XP and XQ are two analyte concentrations, while YP and YQ are the corresponding or respective signals.

### 1.6.2 Limit of Detection

The least quantity or amount of the analyte that can give a detectable signal in a biosensing device is referred to as the limit of detection (LOD). The following equation is commonly used to compute LOD:

$$\text{LOD} = \frac{3SD\ BLANK}{SLOPE}$$

where LOD denotes the detection limit and SD denotes the standard deviation. The slope indicates the sensitivity of the probe for the specific analyte, and blank represents the standard deviation of blank.

### 1.6.3 Limit of Quantification

A biosensor's limit of quantification (LOQ) is the least amount of target analyte it can detect. The following equation may be used to express LOQ mathematically:

$$\text{LOQ} = \frac{3SD\ BLANK}{SLOPE}$$

It should be preference that the LOQ might be the same as or greater than the LOD but it cannot be lower than that value.

### 1.6.4 Linear Dynamic Range

The linear dynamic range (LDR) is the analyte concentration range in the diagram for regression analysis, where the changing signal is directly proportional to the variation in analyte concentrations. Biosensors having a higher LDR are intended to detect analyte concentrations that are clinically significant.

### 1.6.5 Selectivity

Selectivity refers to a biosensor's ability to identify the targeted analyte from other biomolecules and provide a signal that is specific to the target analyte. In the biosensor literature, the terms specificity and selectivity are frequently interchanged. Biosensors recognise just one target analyte, which is desired even in a complex solution. Aptamers, antibodies, and biotin-avidin-like conjugation pairs can all be used to achieve specificity. Selectivity refers to the ability to recognise a set of closely similar chemicals.

### 1.6.6 Response Time

The time it takes a biosensor to create a signal in response to analyte amount or quantity or to collect a number of analyte molecules on its surface which can be detected easily. Some of the methods or ways used to shorten the reaction time of a biosensor include the microfluidic based technology, analyte collection by beads scattered in solution, and nano or micromotors.

## 1.7 Recovery Time

The recovery time is the time it takes for a biosensor to be ready for the next sensing cycle after the original one. A good biosensor has a wide linear dynamic range, a low value of LOD, high sensitivity, high selectivity, quick response, quick recovery, and a long shelf life. Biosensors are designed to detect analytes in the therapeutically relevant range with excellent sensitivity.

### 1.7.1 Real Sample Analysis

In a wide spectrum of biological fluids and samples, biosensors can detect therapeutically important analytes. When the biosensors' analytical capabilities have been established, clinically examined samples are analyzed by biosensors. The analyte concentration is determined using the calibration plot that was previously created. When these samples are unavailable, the spike of the recovery and classic addition methods are used to reproduce the genuine sample. Coexisting or structurally similar molecules make up the imitating solution.

$$\%\text{Recovery} = \frac{\textit{Observed} \times 100}{\textit{Expected}}$$

where observed is the value of analyte concentration observed after the spiking of analyte; and expected is the exact concentration of the analyte spiked into the sample.

Coexisting molecules have little influence on signal production if the recovery response is near to the spiking value. It is because of false positive findings that the recovery reaction is significantly bigger than the spike value. The sensing probe interacting with a non-specific analyte causes a false positive result, whereas the sensing probe interacting with a non-specific analyte causes an abnormally low recovery response. With the addition of a known amount or quantity of analyte from the sample to the sensing solution, the standard addition technique is utilized to create a calibration plot. By connecting the dots in the standard plot to zero concentrations after treating the unknown sample with the sensor, the analyte concentration in it may be estimated [48–52].

## 1.8 Conclusion

The advancement in biosensor of cell have reached the half way since the innovation of this prestigious technology for on point utilization of device which have replaced many conventional diagnostic or detection techniques due to its tremendous response against minute concentration in no time. These biosensors have evolved from various innovative techniques involving cell engineering, 3D cell immobilization, so on and so forth for rapid and efficient biosensing through bioreceptor and analyte interaction. This has enabled the high throughput screening of drugs and monitoring distinguishable cell parameters which has opened the way to study unknown interaction of analyte with cell and cellular responses which will impact more on diverging sectors, such as environmental monitoring, diagnostic, food and eatable quality control, drug discovery process.

## References

1. Islam, M., Fangxin, D., Baohua, L., Guobao, X., Chapter 7: Cell-based biosensors, in: *World Scientific Series: From Biomaterials towards Medical Devices, Biochemical Sensors – Fundamental and Development*, vol. 4, pp. 299–357, World Scientific Publishing Co. Pte. Ltd., Singapore, 2021.
2. Yi-Chen, E.L. and I-Chi, L., The current trends of biosensors in tissue engineering. *Biosensors*, 10(8), 88, 1–22, 2020.
3. Chandran, K., Raju, R., Kalpana, B., Chapter 1: Introduction to biosensors, in: *Biosens. Bioelectron.*, 2, 3, 16, 2015.
4. Maria, E.I., Mark, M., Timothy, K.L., Cell-based biosensors for immunology, inflammation, and allergy. *J. Allergy Clin. Immunol.*, 144, 645–646, 2019.
5. Gary, C.H.M., Brian, R., Fabian, H., Premashis, M., Xinxing, Y., Eric, G., Chris, B., Ashlee, M.P., Brian, T., Zan, C., Yuxiao, W., Eileen, J.K., Philip, A.C., Karen, G.F., Amy, P., Ralph, J., Jie, X., Peter, D., Jin, Z., Genetically encoded biosensors for visualizing live-cell biochemical activity at super-resolution. *Nat. Methods*, 14, 427–434, 2017.
6. Maggie, H., Till, T.B., Baojun, W., Synthetic biology enables programmable cell-based biosensors. *Chem. Phys.*, 21, 133–134, 2019.
7. Koji, S., Tomohiko, Y., Inyoung, L., Takuya, H., Wakako, T., BioCapacitor: A novel principle for biosensors. *Biosens. Bioelectron.*, 76, 20–28, 2016.
8. Il-Hoon, C., Dong, H.K., Sangsoo, P., Electrochemical biosensors: Perspective on functional nanomaterials for on-site analysis. *Biomater. Res.*, 24, 1–12, 2020.
9. Jiayu, L., Ibrahem, J., Amjed, A., Zhenyu, S., Lu, Z., Majed, E.D., Shuping, Z., Mahmoud, A., An integrated impedance biosensor platform for detection of pathogens in poultry products. *Sci. Rep.*, 8, 16109, 1–10, 2018.
10. Xiaojin, L., Xuesong, Y., Yalei, Z., Xingwen, Z., Guangming, X., Yue, C., Amperometric biosensing system directly powered by button cell battery for lactate. *PLoS One*, 14, 1–14, 2019.
11. Emre, C., Alaaddin, C., Huseyin, T., Hussein, S., Huseyin, B.Y., Electrochemical glucose biosensors: Whole cell microbial and enzymatic determination based on 10-(4H-Dithieno[3,2-b:2′,3′-d]pyrrol-4-yl)decan-1-amine interfaced glassy carbon electrodes. *Anal. Lett.*, 52, 1138–1152, 2019.
12. Jeganathan, S., Lakshmanan, R., Mariappan, U.M., A theoretical model of pH-based potentiometric biosensor based on immobilized enzyme membrane. *Am. J. Anal. Chem.*, 7, 363–377, 2016.
13. Freeman, C.J., Ullah, B., Islam, M.S., Collinson, M.M., Potentiometric biosensing of ascorbic acid, uric acid, and cysteine in microliter volumes using miniaturized nanoporous gold electrodes. *Biosensors*, 11, 10, 1–14, 2021.

14. Avishek, C., Dewaki, N.T., Ananya, B., Chapter 5, Impedance-based biosensors, in: *Bioelectronics and Medical Devices*, Kunal, P., Kratz, H.B., Anwesha, K., Sandip, B., Indranil, B., Usha, K. (Eds.), pp. 97–122, Elsevier Ltd., United States, 2019.
15. Chen, C. and Junsheng, W., Optical biosensors: An exhaustive and comprehensive review. *Analyst.*, 8, 15, 2020.
16. Hye, J.H., Woong, S.K., Won-Gun, K., Cell microarray technologies for high-throughput cell-based biosensors. *Sensors*, 2, 3, 5, 2017.
17. Adel, D., Bahman, D., Frances, J.H., Michaelia, P.C., Claudine, S.B., Nicolas, H.V., Porous silicon based cell microarrays: Optimizing human endothelial cell-material surface interactions and bioactive release. *Biomacromolecules*, 17, 11, 3724–3731, 2016.
18. Kathleen, W., Jenna, D.S., Pablo, V., William, R.M., Timothy, J.S., A multiplexed assay for determination of neurotoxicant effects on spontaneous network activity and viability from microelectrode arrays. *Neurotoxicology*, 49, 79–85, 2015.
19. Deniz, S., Mohammad, H., Ebrahim, G.-Z., Biosensing based on field-effect transistors (FET): Recent progress and challenges. *Trends Anal. Chem.*, 133, 1–16, 116067, 2020.
20. Tatsuo, Y., Ko-ichiro, M., Carl, F.W., Arshak, P., Torsten, W., Michael, J.S., Light-addressable potentiometric sensors for quantitative spatial imaging of chemical species. *Annu. Rev. Anal. Chem.*, 3, 5, 2017.
21. Ying, T., Norlaily, A., Joe, B., De-Wen, Z., Steffi, K., Light-addressable potentiometric sensors using ZnO nanorods as the sensor substrate for bioanalytical applications. *Anal. Chem.*, 90, 8708–8715, 2018.
22. Min-Jeong, S., Qui, A.L., Joon-Chul, K., Kyoung-Hee, K., Sun-Hee, W., Measurement of cellular ATP efflux by establishment of biosensor cells. *Yakhak Hoeji, Journal of the Pharmaceutical Society of Korea,* 63, 30–36, 2019.
23. Asako, N., Yuji, I., Nobuyoshi, M., In vivo whole-cell patch-clamp methods: Recent technical progress and future perspectives. *Sensors*, 21, 1448, 1–21, 2021.
24. Soojung, K., Hyerin, S., Heesan, A., Taeyeon, K., Jihyun, J., Soo, K.C., Dong-Myeong, S., Jong-ryul, C., Yoon-Hwae, H., Kyujung, K., A review of advanced impedance biosensors with microfluidic chips for single-cell analysis. *Biosensors*, 11, 412, 1–14, 2021.
25. Jianye, S., Fatima, F., Debashish, B., Mehdi, J., Electrical impedance as an indicator of microalgal cell health. *Sci. Rep.*, 10, 1251, 1–2, 2020.
26. Hui, J.L., Tridib, S., Beng, T.T., Wen, S.T., Chien, W.O., Quartz crystal microbalance- based biosensors as rapid diagnostic for infectious diseases. *Biosens. Bioelectron.*, 168, 112513, 1–15, 2020.
27. Anand, M.S., Uros, C., Ibrahim, A., A comprehensive review on plasmonic-based biosensors used in viral diagnostics. *Commun. Biol.*, 2, 8, 2021.
28. Guido, C., Jan, P., Martin, P., Sarka, J., Vladimir, R., Petr, S., Roberto, R., Non-invasive electromechanical cell-based biosensors for improved investigation of 3D cardiac models. *Biosens. Bioelectron.*, 2, 4, 2018.
29. Maria, E.I., Mark, M., Timothy, K.L., Cell-based biosensors for immunology, inflammation, and allergy. *J. Allergy Clin. Immunol.*, 144, 645–647, 2019.
30. Joseph, W., Amay, J.B., Patrick, M., Non-invasive and wearable chemical sensors and biosensors, Published active US Patent, US-10722160-B2, Current Assignee 'University of California', https://pubchem.ncbi.nlm.nih.gov/patent/US-10722160-B2, Accessed on December 15, 2022.
31. Lu, T. and Kristin, S., Cell culture-based biosensing techniques for detecting toxicity in water. *Curr. Opin. Biotechnol.*, 45, 59–68, 2017.
32. Hoang, H.N., Sun, H.L., Ui, J.L., Cesar, D.F., Moonil, K., Immobilized enzymes in biosensor applications. *Mater.*, 12, 121, 1–34, 2019.
33. Hoang, H.N. and Moonil, K., An overview of techniques in enzyme immobilization. *Appl. Sci. Converg. Technol.*, 26, 157–163, 2017.

34. Ping, L., Xiaoqing, Q., Xinmin, L., Lu, F., Xinyu, L., Daxiang, C., Yurong, Y., An enzyme-free electrochemical biosensor based on localized DNA cascade displacement reaction and versatile DNA nanosheet for ultrasensitive detection of exosomal MicroRNA. *ACS Appl. Mater. Inter.*, 12, 40, 45648–45656, 2020.
35. Shi, Q., Jiawen, J., Wei, S., Jia, P., Jian, H., Yang, L., Jiao, J.L., Guocheng, W., Recent advances in surface manipulation using micro-contact printing for biomedical applications. *Smart Mater. Med.*, 2, 65–73, 2021.
36. Kiesar, S.B., Rafiq, A., Jin-Young, Y., Yoon-Bong, H., Nozzle-jet printed flexible field-effect transistor biosensor for high performance glucose detection. *J. Colloid Interface Sci.*, 506, 188–196, 2017.
37. Giulio, R., Marco, R., Matteo, S., Alessandro, D.T., Alessandro, P., Silver nanoparticles ink-jet-printed flexible biosensor for rapid label-free antibiotic detection in milk. *Sens. Actuators B Chem.*, 280, 280–289, 2019.
38. Shabarni, G., Ratna Nurmalasari, Y., Yeni, W.H., Voltammetric DNA biosensor using gold electrode modified by self assembled monolayer of thiol for detection of mycobacterium tuberculosis. *Proc. Technol.*, 27, 74–80, 2017.
39. Mandeep, S., Navpreet, K., Elisabetta, C., Role of self-assembled monolayers in electronic devices. *J. Mater. Chem. C*, 8, 3938–3955, 2020.
40. Muamer, D., Esma, D., Mehmet, S., Design of amperometric urea biosensor based on self-assembled monolayer of cystamine/PAMAM-grafted MWCNT/Urease. *Sens. Actuators B Chem.*, 254, 93–101, 2018.
41. Mohankumar, P., Ajayan, J., Mohanraj, T., Yasodharan, R., Recent developments in biosensors for healthcare and biomedical applications: A review. *Meas.*, 167, 108293, 1–28, 2021.
42. Liming, B., Cristina, G.E., Weiqi, L., Ping, Y., Junjie, F., Lanqun, M., Biological applications of organic electrochemical transistors: Electrochemical biosensors and electrophysiology recording. *Front. Chem.*, 7, 313, 1–16, 2019.
43. Hicks, M., Bachmann, T.T., Wang, B., Synthetic biology enables programmable cell-based biosensors. *ChemPhysChem.*, 21, 132–144, 2020.
44. Yang, C., Bin, L., Xinhua, T., Microbial fuel cell-based biosensors for toxicity determination: A review. *Biosensors*, 1, 6–9, 2019.
45. Qingyuan, G., Tom, L., Suyan, S., Lu, Y., Yong, L., The application of whole cell-based biosensors for use in environmental analysis and in medical diagnostics. *Sensors*, 2, 7, 2017.
46. Chenxi, Q., Haotian, Z., Jin, H., Biosensors design in yeast and applications in metabolic engineering. *FEMS Yeast Res.*, 19, 1–12, 2019.
47. Suresh, N., Vasanth, R., Xuan, W., Rohit, C., Biosensors for sustainable food engineering: Challenges and perspectives. *Biosensors*, 8, 23, 1–34, 2018.
48. Buddhadev, P., Pramod, R.V., Nagaraj, P.S., Pranjal, C., Biosensor nanoengineering: Design, operation, and implementation for biomolecular analysis. *Sens. Int.*, 1, 100040, 1–19, 2020.
49. Soleymani, L. and Li, F., Mechanistic challenges and advantages of biosensor miniaturization into the nanoscale. *ACS Sens.*, 2, 4, 458–467, 2017.
50. Varnakavi, N. and Nohyun, L., A review on biosensors and recent development of nanostructured materials-enabled biosensors. *Sensors*, 21, 4, 1109, 2021.
51. Kumar, A., Purohit, B., Mahato, K., Roy, S., Srivastava, A., Chandra, P., Design and development of ultrafast sinapic acid sensor based on electrochemically nanotuned gold nanoparticles and solvothermally reduced graphene oxide. *Electroanal.*, 32, 59, 2020.
52. Purohit, B., Kumar, A., Mahato, K., Chandra, P., Novel sensing assembly comprising engineered gold dendrites and MWCNT-AuNPs nanohybrid for acetaminophen detection in human urine. *Electroanal.*, 32, 562, 2020.

53. Hyeong, M.K., Minhee, U., Dae, H.J., Ho-Young, L., Jae-Hyoung, P., Seung-Ki, L., Localized surface plasmon resonance biosensor using nanopatterned gold particles on the surface of an optical fiber. *Sens. Actuators B Chem.*, 280, 183–191, 2019.
54. Guo, Z., Lokendra, S., Yu, W., Ragini, S., Bingyuan, Z., Fengzhen, L., Brajesh, K.K., Santosh, K., Tapered optical fiber-based LSPR biosensor for ascorbic acid detection. *Photonic Sens.*, 11, 418–434, 2020.
55. Mohamed, E.B., Abdellatif, A., Abdellah, M., Design of silver nanoparticles with graphene coatings layer used for LSPR biosensor applications. *Vacuum*, 180, 109497, 1–11, 2020.
56. Botewad, S.N., Pahurkar, V.G., Muley, G.G., Gaikwad, D.K., Gajanan, A., Bodkhe, Shirsat, M.D., Pawar, P.P., PANI-ZnO cladding-modified optical fiber biosensor for urea sensing based on evanescent wave absorption. *Front. Mater.*, 7, 184, 1–9, 2020.
57. Aysa, R., Reza, R., Farzaneh, F., Surface plasmon resonance biosensors for detection of Alzheimer's biomarkers; An effective step in early and accurate diagnosis. *Biosens. Bioelectron.*, 167, 112511, 1–12, 2020.
58. Chapter 1, *Nanomaterials in biosensors: Fundamentals and applications, Nanomaterials for Biosensors*, pp. 1–74, Elsevier Ltd., United Kingdom, 2018. doi:10.1016/B978-0-323-44923-6.00001-7. Epub 2017 Oct 21.
59. Niharika, G., Venkatesan, R., Dorian, L., Ramasamy, P., Bansi, D.M., Cell-based biosensors: Recent trends, challenges and future perspectives. *Biosens. Bioelectron.*, 1, 3, 4, 2019.
60. Xin, L., Yongli, Y., Yinzhi, Z., Xiulan, S., Current research progress of mammalian cell-based biosensors on the detection of foodborne pathogens and toxins. *Crit. Rev. Food Sci. Nutr.*, 61, 3819–3835, 2021.
61. Pratik, B. and Arun, K.B., Mammalian cell-based biosensors for pathogens and toxins. *Trends Biotechnol.*, 27, 183, 2009.
62. Pancrazio, J.J., Whelan, J.P., Borkholder, D.A., Stenger, D.A., Development and application of cell-based biosensors. *Ann. Biomed. Eng.*, 27, 700, 1999.
63. Jana, S. and Jan, T., Application of nanomaterials in microbial-cell biosensor construction. *Chem. Pap.*, 69, 42–53, 2015.
64. D'Souza, S.F., Microbial biosensors. *Biosens. Bioelectron.*, 338, 341, 342, 2001.
65. Yu, L., Wilfred, C., Ashok, M., Microbial biosensors. *Anal. Chim. Acta*, 568, 200–210, 2006.
66. Xia, X. and Yibin, Y., Microbial biosensors for environmental monitoring and food analysis. *Food Rev. Int.*, 27, 300–329, 2011.
67. Tatiana, B.G., Alexey, P.S., Sergei, V.D., Application of amperometric biosensors for analysis of ethanol, glucose, and lactate in wine. *J. Agric. Food Chem.*, 57, 6528–6535, 2009.
68. Kim, M., Iezzi, R.J., Shim, B.S., Martin, D.C., Impedimetric biosensors for detecting vascular endothelial growth factor (VEGF) based on poly(3,4-ethylene dioxythiophene) (PEDOT)/gold nanoparticle (Au NP) composites. *Front. Chem.*, 7, 2, 2019.
69. Syu, Y.C., Hsu, W.E., Lin, C.T., Review—Field-effect transistor biosensing: Devices and clinical applications. *ECS J. Solid State Sci. Technol.*, 7, Q3196–Q3207, 2018.
70. Somayeh., H. and Gholam, H.H., Application of nanoparticles in quartz crystal microbalance biosensors. *J. Sens. Technol.*, 4, 81–100, 2014.
71. Victoria, S., Vladimir, T., Mikhail, M., Joachim, H., Julia, S., Pascal, L., Dominic, S., Frank, W., Peter, M., Heinrich, M., Klaus, U., Roland, H., Alexander, Z., Application of surface plasmon resonance imaging technique for the detection of single spherical biological submicrometer particles. *Anal. Biochem.*, 486, 62–69, 2015.

# 2

# Bioreceptors for Enzymatic Interactions

**Vipul Prajapati[1]* and Shraddha Shinde[2]**

*[1]Department of Pharmaceutics, SSR College of Pharmacy (Permanently Affiliated to Savitribai Phule Pune University), Sayli, Union Territory of Dadra Nagar Haveli & Daman Diu, India*
*[2]Department of Pharmacology, Sinhgad College of Pharmacy (Affiliated to Savitribai Phule Pune University), Vadgaon Budruk, Pune (Maharashtra), India*

### *Abstract*
Upcoming disease rate and high-end health crisis has made it essential to build up modern techniques which would have real time fast response monitoring, such as immune assays or PCR-based diagnosis for life threatening events like SARS COV2, cancer, TB. Biosensors are the devices that combine the physiological response with electronic system to generate the signal for analyte. This chapter focuses on enzyme-based biosensing methodology and tackling setbacks of conventional biosensors by enzymatic interaction. Also provides with application of this technology on commercial and large-scale expanding stage. Preferentially, bioreceptors that can be used for enzymatic interactions. It explains about various technologies that includes enzyme as bioreceptor and works to interpret signals. Applications of bioreceptors enzymatic interactions in the diversified area of health sciences, scientific research, daily life prognosis, etc. Implementation of enzyme-based nanotechnological modification for superior biosensing elements. Elaborates on commercially used enzyme-based biosensing devices implemented globally. Various modified techniques, nanomaterials, biotechnological inventions inclusion in recognition by enzymes.

***Keywords:*** Bioreceptors, enzyme biosensor, signal transduction, biorecognition molecule, nanoparticle

## 2.1 Introduction

Bioreceptors are biological elements of a biosensor that recognize the analyte of interest and binds with it to produce a specific signal by means of heat [1], light [2], liberation of electrons [3–6, 35]. Biosensing process detects signals of biological interaction between biomolecules and analyte, the devices that works on this principle is called biosensor [3–9]. Biosensor can be defined as an analytical device, which with the help of signal from immobilized material–analyte interaction converts it to electrical, physical or chemical indications, which can be analyzed and measured [10]. The measured amount of output in the electrical device is directly proportional to the concentration of analyte present in the sample [10]. Biosensors are evolved because of its capability of rapid analysis, detection

**Corresponding author*: vippra2000@yahoo.com

Inamuddin and Tariq Altalhi (eds.) Biosensors Nanotechnology, 2nd Edition, (33–60) © 2023 Scrivener Publishing LLC

of different analyte in medium in comparison to the traditional methods which were time consuming, extensive and expensive [14]. The objective of biosensor technology is to initiate a faster, accurate method of response about specific analyte and establish simple, cost-effective analytical system without complexity and high valuation issues [14]. They have been brought in market for determining disease at initial stages, identification of viruses, toxins, increased blood levels. Biosensor comprises of analyte, the material to be recognized and which is key element in biosensing complicated organic molecules of human system (saliva, sweat, blood, urine). Bioreceptor, the most essential aspect of the biosensor which detects the analyte from sample, signal generation in biosensor is a result of interaction with protein, enzymes, antibodies, nucleic acid, cells, hormones, DNA/RNA [10, 14, 39]. Transducer, transforms energy from biomolecule–analyte interaction signal to a calculative electrical or optical signal [10, 14]. Electronic device, shows measured signal in the display. Biosensor can be classified based on transduction signal, biorecognition factor used or interaction between analyte and biological material in biosensor which can be further classified in catalytic and affinity types. Emphasizing on the bioreceptor, different interactions are produced such as antigen-antibody, enzyme substrate, cellular, nucleic acid, biomimetic interactions [10]. Enzyme-based biosensor is type of biosensor device where enzyme is preferred as recognizing material and is immobilized on the surface of the transducer matrix to build up enzyme activity. Also, it is considered widely due to its more specificity and selectivity [14, 22]. Reduction oxidation enzyme in the enzyme biosensor is applied as it catalyzes by electron transfer. Horseradish peroxidase and glucose oxidase are extensively used enzymes for enzymatic interactions. Laccase enzyme has ability to oxidase wide variety of substrates. Immobilization is an important step for stabilization of enzyme and improving its reusability [14]. To increase the characteristics of enzymes, artificial enzymes are developed called as nanoparticle or nanozyme. Eventually, they are small sized biosensors denoted as Nano biosensor. Recent advancements in the microelectronic technology have led to foundation of nanotechnology which is way cheaper, accurate, portable, selective, biocompatible, non-toxic, sensitive, rapid and multiple sample analyser. It has become easier to create, manipulate biosensor by minimalist use of materials [5].

Biosensors are implemented in the development of biological-electrical detection of food allergens, contamination's, antioxidant power in food industry [7]. Also, it is used in environmental analysis of pollution of air and water purity. Various applications are seen in the field of agriculture, medicine, bioprocess, national defence [3].

## 2.2 History of Biosensors

Biosensing came into picture in 1906 during demonstration by Scientist Cremer who expressed the amount of $H^+$ ion in mixture equalizes to electrical potential [10]. During 1909, Soren Peder Lauritz Sorenson brought the concept of pH. For pH determination W S Hughes developed electrode in 1922 [10]. Arrest of invertase onto aluminium or charcoal was performed by Griffin and Nelson [5, 10]. First biosensor was a glucose molecule which was oxidized, reduced by using glucose oxidase enzyme. Principle it followed was measurement of electric current obtained from the rate of concentration of oxygen. To conclude decrease in electrical output is proportional to glucose concentration [10].

1. First-generation biosensor: Potentiometric urea electrode was designed.
2. Second-generation biosensor: Simplified, quality analysis done by auxiliary enzymes and co-reactants that are co-immobilized with analyte converting enzyme.
3. Third-generation biosensor: Innovation of SPR (Surface Plasmon Resonance) biosensor, consisted of biomolecules as biosensing material [5].
4. Fourth-generation biosensor: Development of nanotechnology and biotechnology by MEMS/NEMS/BioNEMS (Micro, Nano, BioNano, Electro-Mechanical System) proved to have versatile and muti-applicable [5]. Current biosensors used are BioNEMS, nanoparticles, nanowire, nanotube, quantum dots [10]. Figure 2.1 shows the revolution in the biosensors [10].

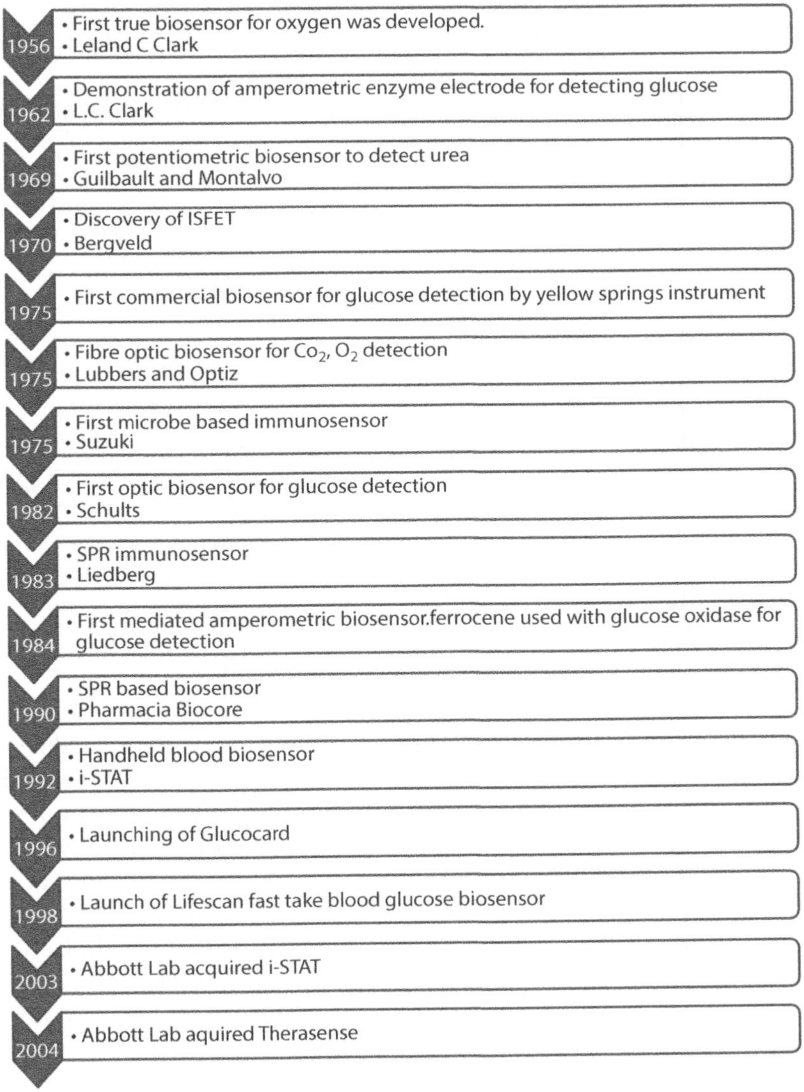


**Figure 2.1** Progression of inventions in biosensors from first innovation to recent modifications.

## 2.3 Biosensors

### 2.3.1 Characteristics of Biosensors

a. Selectivity and sensitivity: selectivity is the tendency of biomolecule to detect the specific analyte from a complex sample whereas, sensitivity is the property of the biosensor to detect response of a unit concentration of analyte [3, 10].
b. Stability: It is the safety and durability parameter, deals with the continuous long-term usage, stable along the whole process and maintaining shelf life [3, 10, 35].
c. Accuracy and precision: Accuracy is the comparison between the acquired value to the actual value. It is represented by % recovery. Precision is standard deviation in response of biosensor when checked at different points of same sample.
d. Reproducibility: The ability of sensor to bring about constant response for multiple setups of sample experiments. It considers robustness of analytical system.
e. Linearity: It is accuracy, where measure of straight line shows proportionality to intensity of ions in substrate [5].
f. Response time: The time taken by the biosensor to sense fraction of change in the biological response, can also be called as time taken by device to gain 95% response.
g. Detection limits and Quantification limit (LOD & LOQ): It is a quantitative term where sensitivity plays a major role in identifying the amount of analyte detected by biosensor. Biosensing can detect minimal analyte concentration, it determines its presence in sample. So, lesser concentration of LOQ would be analyzed and evaluated precisely by accurate method [2, 3].

An ideal biosensor should be accurate, specific, quick responsive, stable, reliable, with low service charge and capital costs, portable, safe and low power consuming.

### 2.3.2 Design and Principle of Biosensors

Biosensor is an integrated system of biomolecule (antibody, enzyme), transducer and electronic device which detects and measures response and transforms it to electrical signal [2, 7]. It can detect the minute change in the analyte and according give indication so it is been used in identification of chemical, biological matters of the environment, low concentration of toxins, pathogens, chemicals. Biosensor measuring devices come in wide variety of sizes shapes, materials [14, 24].

Principle components of biosensor encompass analyte, bioreceptor, transducer, electrical device and display. Analyte is the substance whose elements (glucose, urea, lactose) are to be determined by binding to specific bioreceptor [2, 3, 6]. Bioreceptor (enzyme, DNA, aptamer, cells) is a factor by which it identifies the target analyte and transmit signal (biorecognition) by means of heat, light, pH, mass [10, 19]. Transducer, device that converts the signal from analyte–biomolecule interaction to a measurable entity. Measurable

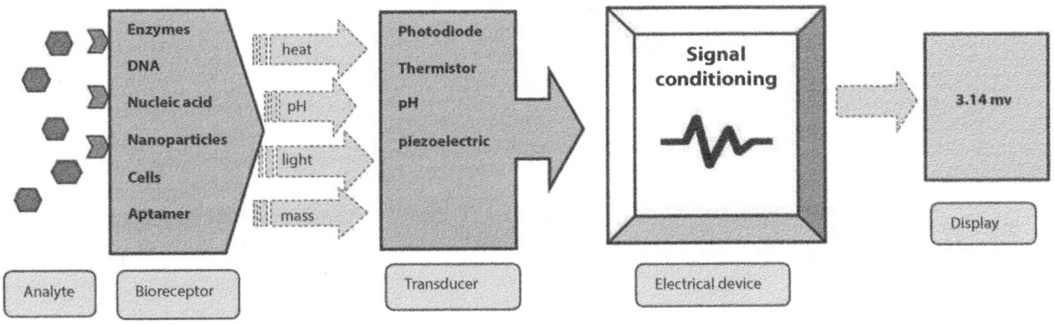


**Figure 2.2** Schematic representation of working of biosensor which indicates the binding of analyte to the respective receptor site and production of an indication for tranducer through various factors for interpreting signal in electrical device and finally reaches the display as signal converts in into calculative data.

quantity of concentration leads to respective signaling [33]. This conversion process is called signalization. Transducer brings about optical or electrical signal that depends on the number of analytes present in the sample [24]. Electronic is a part for displaying signal by processing of electric signal. They are amplified and converted into digital display form. Display unit is interpretation from the electrical that shows readable response monitoring in computer device [6]. Figure 2.2 shows a flow chart for working of biosensor [5, 35].

### 2.3.3 Materials Used for Constructing Biosensors

Analyte, bioreceptor are made of the endogenous substances and biological, chemical moieties respectively whereas the transducer, electronic device and display are constructed by specific, sensitive signal transformer with electrodes, computerized integrated system and framework of signal interpretation [20]. Interface is the linking between bioreceptor and transducer, made up of variety of material according to the characteristic of bioreceptor, biocompatibility, stability, selectivity, etc [35]. Material used in them are nanoparticle, metallic nanomaterial, Carbon-based nanomaterial, polymers like chitosan, agarose, PEG (for coating and immobilization), copolymers, conducting polymer, Metal organic Frameworks (MoF) to detect DNA, RNA, enzyme [31, 48].

## 2.4 Classification of Biosensors

Bioreceptor is the major element in biosensing technology development. It can be classified on the basis of mechanism of transductions, interaction of molecules, biorecognition process [10, 29, 33]. According to bioreceptor-based biosensor, classification consist of enzymatic biosensor (mostly used type of biosensor), aptamer-based biosensor (nucleic acid-based biosensor device), microbial or whole cell, immunosensor (consist of increased specificity and selectivity) [12].

Based on the transducing mechanism it is categorized as electrochemical biosensor with subtypes (amperometric, conductometric, impedance, potentiometric.), electrical biosensing, optical biosensing, thermal biosensor, gravimetric biosensor [18, 24].

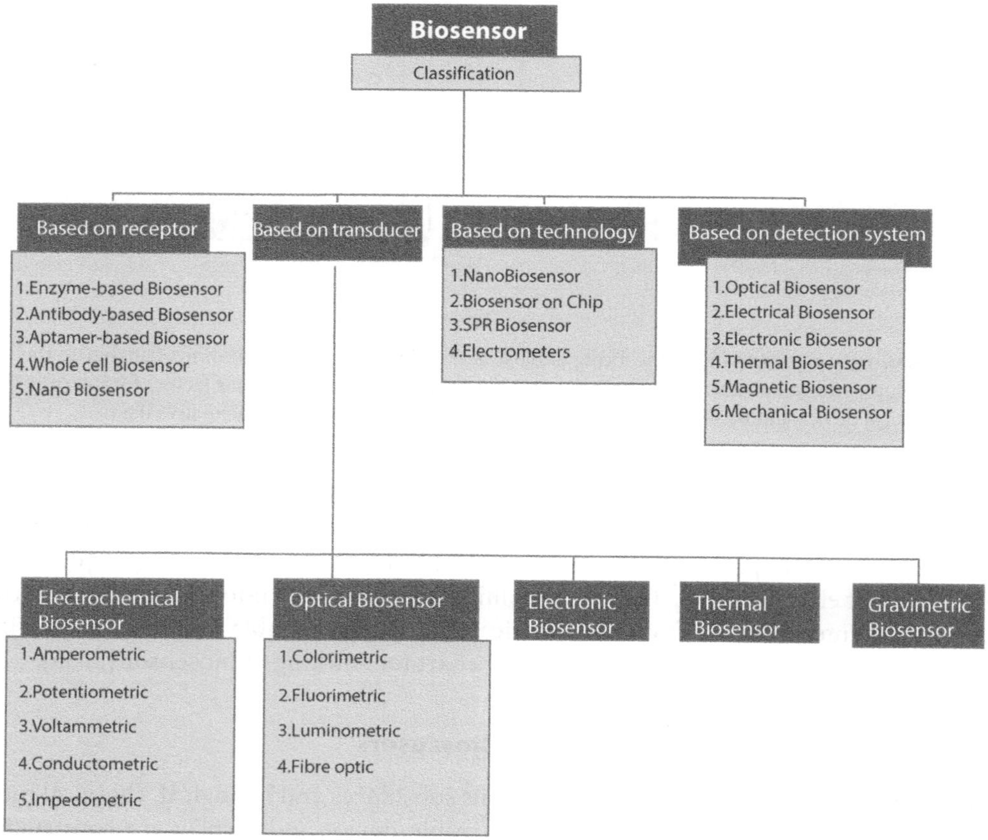


**Figure 2.3** Flowchart of classification of biosensors based on receptor, transducer, technology and detection type with their subtypes.

Also, it is classified on basis of recognition organ into electronic, thermal, optical, magnetic, mechanical biosensor and equipment dependent are nanosensor, SPR technology, biosensor placed above chips and electrometers [38]. Lastly, by biorecognition process, it is classified into two categories catalytic biosensor in which the interaction between analyte and biomolecule produces a new chemical reaction product (comprises of biosensor like enzyme, tissue, whole cell, microorganism.) and non-catalytic biosensor also called as affinity biosensor, has irreversible bounded analyte to receptor that forms no new chemical reaction product during interaction (consist of cells, antibodies, receptors, nucleic acids are focused molecules for sensing.) [35]. Figure 2.3 denotes the detailed classes of biosensor created by focusing on its several parameters [35].

## 2.5 Types of Bioreceptors

### 2.5.1 Microbial or Whole Cell

Cell is used as bioreceptor as it is self-replicating, reproducing, easy to handle, interactive with different analytes, highly sensitive and selective, detection capability, stability [10].

Microbes, fungi, bacteria, virus are used because they produce accurate response. It is used for food analysis, detection of plant constituents, drug screening, heavy metals, pesticides, contaminants, environment monitoring [14].

### 2.5.2 Aptamer

It comprises of a DNA or RNA molecule single stranded with binding ability to the target compound. DNA biosensor works by hybridization of nucleic acid further adjoining the DNA-RNA base pair, DNA-RNA, DNA-DNA, RNA-RNA strand with H linkage over the A-G (adenine-guanine) or C-T/U (cytosine-thiamine/uracil) pairing [35]. Suitability of aptamer is due to structural and functional aspects of nucleic acid character, which makes it stable for diverse range of temperature and storage condition. Modification can be done in chemical property to enhance detection of target moiety. Thermal folding can be adapted and chemical synthesis makes it stale at pH 2-12 [33]. It can be classified into labeled or label-free aptasensor according to transduction. Real time biosystem tracking is guided by fluorescent nanoparticles comprising labeled optical aptasensor. System plasmon resonance is a label free optical sensor. Gold nanoparticles are used conferring to biostability, lower toxicity than QDs. Detection of cancer cells bacteria, proteins, by aptamer-quantum dots conjugates. Techniques mostly used for detection are electrochemical, optical, piezoelectric [12].

### 2.5.3 Antibody

Biomolecules acting on the principle of affinity and interaction between the antigen–antibody. Binding to analyte of interest (pathogens, viruses) generates immunological response. Structure of antibody has immunoglobulins (Ig) in 'Y' shape consisting of two light and two heavy chains of polypeptides with disulfide bond. Ig is classified as IgG, IgA, IgM, IgE, IgD. Immunosensors are the antibody embedded biosensor system to produce antigen–antibody reaction. They are categorized as labeled where detectable label is introduced and non-labeled which detects the physical change from antigen–antibody complex. Transduction is through optical and potentiometric methods. Immunosensor functions for determining ovarian cancer, acute leukaemia's, aflatoxins, immunophenotyping [10].

### 2.5.4 Nanoparticle

Newer developed concept for binding and identifying the analyte with advanced nanotechnology and nanoscience [5]. Nanomaterial works as bioreceptor and transducer. It exhibits catalytic activity in cerium oxide, favorable as bioreceptor. Transducing capacity shoots due to inorganic material like carbon nanotubes and graphene [36]. Gold Nanoparticle is efficiently used because of characteristic compatibility features [12, 36].

### 2.5.5 Enzymes

They are specified proteins involved in bio-catalyzation of biochemical reaction and increase their rate of reaction. Enzyme-based biosensors work by binding to target molecule and catalyzing reaction [35]. Enzyme adopted detection is mechanized by two basic process 1. Enzymatic conversion of analyte by catalyzation. 2. Inhibition of enzyme activity by analyte biorecognition [42]. The activity can be determined by structural component, chemical composition and catalyst specific chemicals. Analyte detection is followed by three mechanisms: (i) Enzyme metabolizes the analyte, catalytic transformation of analyte by enzyme is measured which determines the amount of concentration of enzyme. (ii) Activation or Inhibition of enzyme by the analyte, concentration of analyte decreases the enzymatic product formed [3]. (iii) Interaction of analyte–enzyme alters enzyme characteristic that is tracked. Biological macromolecule used in enzyme-based biosensor has catalytic activity, increased specificity, speeds up a biochemical reaction [5]. Enzyme specific selectivity binds macromolecule to it specific analyte and not to the other molecule. Biosensors are developed by looking at selectivity, specificity stability, adaptability. Nevertheless, configuration of enzyme is very complex so is pricey [10]. Enzyme bioreceptors are advantageous doe to more accuracy, responds faster than that of cell-based biosensor. Till 2016, enzyme biosensor showed poor stability, critical operational condition, high production cost, pH and temperature variation that were its drawbacks.

In enzyme interaction, analyte recognition useful in (a) Identification of enzymatic conversion of analyte into product, (b) Analyte-based determination of enzyme interaction, (c) Properties of enzyme recorded. Enzyme-based biosensors results in precise and continuous monitoring method for analyte owing to its increased characteristics [14]. Particularly, detection of low concentration analyte due to advanced specificity. Catalytic action of the enzymes can the affected by pH, temperature, substrate concentration and presence of inhibitor [38]. Function of enzyme is to produce electroactive species or reactants consumption by oxidation-reduction reaction and leading to measuring of analyte [14]. Transducers mainly used are electrochemical (amperometric, potentiometric, or optical (absorbance, fluorescence, luminescence) type [34].

Evolved researches have shown close proximity with enzyme-based biosensor coupled with amperometer in instant glucose level revelations within central nervous system [24]. Urea biosensor using nanoparticle by potentiometric method [6]. Nanoparticle integrated enzyme biosensor regards to increase use of enzyme biosensor as recognition factors [5]. Development of enzymatic biosensors involves selection of enzyme, selecting transducer, selection of material for matrix, technique of immobilization, optimization, packaging, commercialization [8]. Enzyme biosensor products that have been used largely on a commercial scale are glucose biosensor and microfabricated electrophoresis chips [41]. Enzymes are mostly used as a tool in the manufacturing of ELISA applied kits [42].

Critical parameter in enzyme-based biosensor device restricts enzyme upon matrix platform present on probe surface of transducer [23]. Immobilization factor is directly related to accessibility to active site, stability, reusability of enzyme. Enzyme are to be assembled on the surface of transduction system to enhance reproducibility and stability [4]. Selection of material is key concern for biosensor construction due to enzyme activity, stability characteristics Therefore support material used should be inert, resistant, stable [16]. Different reaction in the catalyzes base on their mechanism consist enzymes that be classified as; oxidoreductases (oxidase, peroxidase, dehydrogenase, oxygenase) catalyzes oxidation and reduction reaction by substrate through hydrogen or electrons, are applied in biosensor technologies [14].

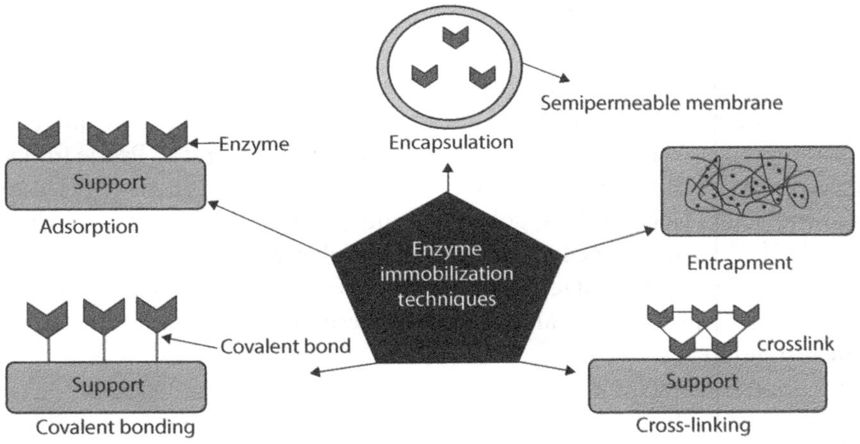


**Figure 2.4** Types of enzyme immobilization techniques used in adaptation of receptor to support.

### *2.5.5.1 Immobilization of Enzymes for Biosensors*

A process by which the enzymes are attached on the solid material or matrix converting substrate adhered to the transducers to obtain certain output [13]. Designed in such a way that catalyst imprisoned on matrix. Activity initiation in biosensor is dependent on enzyme attachment with model [16]. High amount of enzyme is loaded in order to provide sufficient biocatalyst adherence to surface. Also, biocatalyst needs suitable environment for their enzymatic actions [10]. Immobilization should satisfy stability conditions. Immobilized enzymes are reusable, continuously maintains their activity [45, 49]. Life span of biosensor is only of 2-4 weeks due to loss of activity by interacting enzyme with electrode surface [22]. Figure 2.4 shows a diagrammatic representation of enzyme immobilization techniques.

#### 2.5.5.1.1 Strategies to Improve Enzyme Usage in Biosensor

I. Biomodification—Site-Directed Mutagenesis (Enzyme amino acid substitution, enzyme amino acid removal, non-natural amino acid incorporation. enzymatic addition of genetic.
II. Chemical modification—Site-specific chemical modification, non-specific chemical modification on enzyme surface, chemical crosslinking, use of polymers [13, 45, 48].

### *2.5.5.2 Immobilization Techniques*

a. Adsorption

It is a simple, inexpensive method where a molecule is chemically or physically adsorbed on the support matrix. Different chemical or physical molecules are used to immobilize enzyme on their surface. Used in biosensor due to low cost, no reagent requirement, no modification. Not widely preferred technique as weak enzyme-support bond, low efficacy. Material used to build are silica, alumina, charcoal, cellulose. Physical adsorption is used in glucose-based biosensor [11].

b. Entrapment

Straining of enzyme with deep layers of the matrix in a way that it allows a path of substrate with pertaining biomolecules. Increased stability of biosensor, cheap, mild condition requirement, fast method and diffuses to enzyme active site are advantages of entrapment [9]. While matrix interference with the substrates, Prone for leakage of enzyme results in difficultly to build at industrial stage. Gel matrix s used to entrap desired molecule. Gels used are polyacrylamide, conducting materials, nylon, starch, gels, etc [16]. Does not directly bind to matrix. Polymerization reaction is used sometimes but sets inactivity. Galactose oxidase enzyme biosensor is entrapped in polyvinyl matrix and transduced by amperometric method [24].

c. Encapsulation

Placement of the enzyme on semipermeable membrane or gel. Membrane is specific and only allows substrates and enzyme in cavity [22]. Enzymes are stable for long period in capsulated form. Confirmation of enzyme is maintained and minimizes the leakage but less concentrated enzyme required and mass transfer is less to active site [23]. Membrane like cellulose-based, dialysis membrane, ultra-filtration membrane is preferred [38].

d. Covalent binding

Complex of biomolecule and matrix results in bond formation between functional group of enzyme and matrix. Simple and reusable technique. There is no leak of enzyme and different supports are available [23]. Chemical modification of enzyme lead to loss of activity. Detection of phenols is done by amperometric biosensor tyrosinase boron-doped electrons [14].

e. Cross-linking

Along with matrix two or more functional from biomolecule binds to two different materials with different condition. Chemical bonding is seen between biomaterial and support. Mostly used enzyme is glutaraldehyde. Enzyme has activity toward its own carrier and activity is not change by carriers used [35]. Whereas, cross-linking reagent is needed for immobilization and there is possibility of structural modification by cross-linking. Glutaraldehyde presence in electrically prepared film of polyaniline is necessary to form cross bonds of peroxide reductase and oxidiser of cholesterol [16].

## 2.6 Transducers for Enzymatic Interactions

Table 2.1 determines biosensing device based on transducer and their commercially available types.

### 2.6.1 Electrochemical

It is technology derived from analyte and transducer electrical and chemical properties. Applicability of such sensors are because of high selectivity, sensitivity and detection capability. Reaction happens at transducer surface between biomolecule and analyte and initiates a response in voltage, current, capacitance forms [6, 44].

**Table 2.1** Biosensing device based on transducer and their commercially available types.

| Sr. no. | Biosensing device | Commercially available examples |
|---|---|---|
| 1 | Optical transducer [42] | • Classical monodimensional fluorescence spectroscopy<br>• Time Resolved FIA<br>• Flow cytometer microsphere-based immunoassay<br>• Surface Plasmon Resonance (SPR) spectroscopy<br>• Localized SPR<br>• Diffractive Optical Technique<br>• Optical Fibre<br>• Chemiluminescence Immunoassay<br>i. Intra laboratory chemiluminescent biosensor<br>ii. Commercial chemiluminescent biosensor<br>1. The Munich *Chip Reader system* of the third generation (MCR-3)<br>2. Evidence Investigator® system<br>• Luminescent Bacteria |
| 2. | Thermal Transducer [42] | • HEATSENS®<br>• Thermometric Elisa (TELISA) |
| 3. | Mass Based Transducer | • Piezoelectric<br>• Quartz crystal, microbalance, SAW<br>• Magnetoelastic |
| 4. | Magnetic Based | • NMR (Nucleic Magnetic Resonance) Based |
| 5. | Nanotechnologies [42] | • Micro-electro mechanical systems (MEMS)<br>• BIO-MEMS<br>• Microfluidics<br>• Microdevice LOC<br>• Lab-in-a-tube system (LIT) (LOBAS Liat®)<br>• Miniaturized *Total* Chemical *Analysis Systems* (μ TAS) |

There are differentiated into subtypes:

**(1) Conductometric**: Determines variation in conductance from two pair of electrodes because of electrochemical reaction. Monitoring of metabolic process in biological system is performed by conductometric biosensor [12]. It is mostly used in proceeding a biochemical reaction which uses specifically Pt or Au electrodes and produces or consumes ions which are equal or less than measure of electrical conductivity of a solution. It is effective because of a simple mechanism, cut down costs, biocompatibility, no need for reference electrodes and miniaturization possibility. Some setbacks of this process are insignificant specificity and less selectivity [41].

**(2) Potentiometric**: Transducer measures the amount of electric charge obtained from the analyte-bioreceptor reaction at working electrode and compares it to the reference electrode. Selective ion electrodes and ion sensitive transistors are employed for transmitting signal from reaction [24]. Redox or ion-selective electrodes or FET are used in transduction. Detection by enzymatic reaction of ion is initiated and it is equal or less than measure of potential between two electrodes [49]. It has some drawbacks in signal transduction as it lacks specificity, sensitivity and is slower than amperometric sensor and effect of ionic strength at high concentration [42].
**(3) Amperometric**: Principle of biosensor relies on two working electrodes that are made off Au, C, Pt. Current is measured based on the reaction between bioreceptor and analyte which can be oxidation reduction at working electrode [34]. Current produced is directly proportional to conc. of analyte in sample. It is sensitive, fast, precise, linear response when compared with potentiometric and is applicable in mass production. Studies have proven its low selectivity and interferences, which is disadvantage of biosensor [24].
**(4) Impedimetric**: Impedance generated at the electrode surface is measured when small signal is produced. Further transmits the signal and output measured in form of voltage, current concentration [6]. It measures resistance deviation and capacity of interface within probe support and analyte mixture that requires no marking, high cost or tedious processes. Demerits for this mechanism is the lower reproducibility [42].
**(5) Voltametric**: Current is measured at applied potential in controlled variation for analyte identification. It is highly sensitive method with multi analyte detection [3].

### 2.6.2 Optical Biosensor

Biomaterial (enzymes, antibodies, aptamers, cell) are integrated into optical transducer system. Produces a signal that is equal to conc. of target molecule. Gives a label free and parallel detection in biorecognition through optical biosensor [3]. Transduction arise the alteration of absorbance, reflectance, amplitude, frequency, light response. Principle of optical biosensor is characterized into label (fluorescence, calorimetric) and Label free (signal generated from the molecules and transducer) [31, 44]. Subclasses of optical biosensors are (1) Fluorescence: works on the labeling type of mechanism and used in medical diagnosis, food quality check, environmental factors. Material used in transducer is fluorescent dyes, fluorescent proteins, quantum dots [5]. Usually, fluorescence quenching, fluorescence resonance energy transfer and fluorescent enhancement are implied for signal generation. (2) Chemiluminescence: During the biochemical reaction light energy is released and measured for detection DNA [12].

### 2.6.3 Gravimetric Biosensor

Process is related to mass of analyte; it has property of signalizing small variation in mass of analyte. Mass of antibodies, proteins can be measured. Gravimetric biosensors are developed with piezoelectric quartz crystal [10].

### 2.6.4 Thermal Biosensor

They cause the endothermic and exothermic reaction of biological element when the analyte is determined and response produce by biochemical reaction release heat is used in estimation of urea, glucose, penicillin [6].

### 2.6.5 Electronic Biosensor

Principle is based on field effect transistors. Works by direct translation of interaction between FET surface and analyte. Field Effect Transistor functions by maintaining flow of electric current within it [5].

### 2.6.6 Acoustic Biosensor

Sound vibration-based biosensor. Piezoelectric crystal or surface acoustic device measures the vibrational frequencies of molecule. These frequencies are differentiated into negative or positive charge. Change in frequency can be initiated by mass change. Inhibitor binding onto surface can alter frequency. Ex. Determination of cocaine gas by attaching antibodies on crystal surface [6, 44].

## 2.7 Enzymes and Enzymatic Interactions in Biosensor

Table 2.2 shows the analytes and various biosensor used for their detection.

**Table 2.2** Analytes and various biosensor used fo their detection.

| Sr. no. | Analyte | Biosensor used |
|---|---|---|
| 1. | SARS-COV-2 {rapid antigen, FELUDA, qRT-PCR, printed circuit board.} [43] | 1. Magnetosensor<br>2. Electrochemical sensor<br>3. Optical biosensor<br>4. Field effect transistor<br>5. Breath sensor<br>6. Piezoelectric immunosensor<br>7. Laser scanning microscopy<br>8. SPR biosensor<br>9. Paper based biosensor<br>10. SPR based colorimetric biosensor<br>11. Electrochemical biosensor |
| 2. | Glucose and BSA [45] | Colorimetric bioassay |
| 3. | Inorganic phosphate (urine) [46] | Acid phosphatase immobilized on agarose-jellose by fibre based optical biosensor |

(*Continued*)

**Table 2.2** Analytes and various biosensor used fo their detection. (*Continued*)

| Sr. no. | Analyte | Biosensor used |
|---|---|---|
| 4. | Glucose, lactate, uric acid [48] | Paper based screen printing electrodes |
| 5. | Penicillin [49] | Bacteria immobilized on electrode of pH meter. |
| 6. | Glutamate (ECF) [51] | 1. Micro dialysis perfusion followed by *ex situ* analysis<br>2. Implanted amperometric biosensor<br>3. Electrochemical oxidation of enzymatically produced $H_2O_2$<br>4. Stereospecific deamination of glutamate oxidase catalyzes |
| 7. | Dopamine [52]<br>(Real *in vivo*) | Immobilization of tyrosine onto surface of implantable microelectrode |
| 8. | Human serum albumin [52] | CIE 1976 L*a*b* color coordinate system |
| 9. | Acetylcholinesterase, paraoxon and Aflatoxin [52] | Bioactive paper 'Dipstick'<br>Ellman's colorimetric assay |
| 10. | Glucose, cholesterol, lactate, alcohol [52] | Microfluidic paper based conventional electrochemical readers |
| 11. | Protease, tumour, endogenous caspase 1 | NIR Reporter Probes |
| 12. | TB diagnosis<br>Sputum<br>Blood<br>Urine | Whole MTB cells, DNA, Volatiles from MTB cells, secreted and cell internal antigens<br>Host antigens against TB, other Tb Biomarkers<br>Whole MTB cells, DNA, secreted and cell internal antigens, Volatiles from MTB cells |
| 13. | Oxygen, lithium and organophosphates [52] | Fibre optic fluorescence sensors |
| 14. | Protease enzyme activity, Alzheimer's [52] | Micro and nanoparticle platforms |
| 15. | Protease activity, caspase 1 and caspase 3, ovarian cancer, hepatitis C virus [52] | Fluorescent proteins in FRET Reporters |
| 16. | Cancerous breast cells [52] | Quantum dots |
| 17. | Proteases, stroke, cancer, Alzheimer [52] | Optical enzyme sensor |
| 18. | Protein, prostate specific antigen [52] | Microfluidic Chip based Nanoelectrode Array |

### 2.7.1 Horseradish Peroxidase

It is the principal enzyme extracted from the root of horseradish. As it is easily obtained from natural source and simple process for purification and isolation, used in various enzymatic interactions. Enzyme works by oxidation reduction so called as oxidoreductases [3]. High operability, selectivity, high activity, inhibition of substance on high concentration, relatability to treat several conditions makes it extensively applied enzyme. Also, functions in catalyzing organic, inorganic compounds by oxidation, reacting with $H_2O_2$ and thus used for hydrogen peroxidase determination [27].

$H_2O_2$ detection is facilitated by the electrodes or mediators in the biosensors. Mediators carries out transfer of electron between the electrode and enzyme as bioreceptor [9]. $H_2O_2$ in the analyte undergoes reduction by the HRP. In opposition, the mediator is oxidized due to enzymatic interactions and later reduction of oxidized intermediate occurs lastly onto the electrode [8]. Whereas, if mediator is absent HRP is oxidized after the initial reduction at electrode. $H_2O_2$ performs as an intermediate for enzymatic reaction and applied in health diagnosis and also in environment maintenance [14].

Enzymatic reaction example;

Hydrogen peroxide: $H_2O_2$ + e – donor → $2H_2O$ + oxidized donor

(Electron donors—phenols, amines etc.) [8]

#### *2.7.1.1 Applications of Enzymatic Interaction of Hydrogen Peroxide Bioreceptors/ Biosensors*

- Biosensor for detection of 1,4-dehydrobenzene (DHB) was developed on basis of copper sulphide and on which HRP was immobilized on tin oxide glass slide.
- Biosensor can also be used to measure $H_2O_2$ in human plasma. Also, applications are seen in bioserum identification, cancer detection by cancer derived exosomes [14].
- DHB detection by photoelectrochemical system from fluorine doped tin oxide glass plate.
- Screening of cancer derived exomes by colorimetric method from encapsulated DNA of magnesium phosphate crystal.
- Quantification of $H_2O_2$ by chemiluminescent from polymethyl siloxane polystyrene tube [12].
- $H_2O_2$ detection by electrochemical transducer from carbon nanotubule, $\gamma$ aminobutyric acid implemented for environment, health and food [8].
- 17β-estradiol determination with transduction by electro-chemical method on electrode made with Pt. shield with poly (4,5-bis (5-bromothiophen-2-yl) benzothiadiazole).

#### *2.7.1.2 Glucose Oxidase*

Blood glucose monitoring has become essential for healthcare and disease estimation. Glucose oxidase is the enzyme that is immobilized on the transducer for constructing glucose detection biosensor device [4]. Glucose oxidase contains glycoprotein that have orthophosphate proteins so it has characteristic feature like dispersibility, resistance, stability [36, 45]. So, glucose oxidase proves successful in measuring glucose in blood and saliva. Continuous glucose monitoring is a requirement for diabetes mellitus. Glucose oxidase is used for glucose detection because of its stability, reliability, specificity [3]. The working of this enzyme is by oxidizing glucose molecules and further oxidizing electrochemically hydrogen peroxide.

Electrochemical biosensor is mainly used for glucose estimation by glucose oxidase [14]. Glucose oxidase enzyme converts the glucose molecule into gluconic-d lactone. Conversion is carried out through reduction to $C_{27}H_{33}P_2N_9O_{15}$. Further hydrogen peroxide is synthesized because of flavin adenine dinucleotide, ultimately $H_2O_2$ is oxidized to form $O_2$. At last current is produced after potential is applied to biosensor and current is directly proportional to glucose conc. in analyte [10, 45].

Adapting nanomaterial in constructing biosensor can improve characteristic and generate ideal enzyme-based biosensor [13]. Enzymatic reaction is;

Glucose oxidase: Glucose + $O_2$ → gluconic acid + $H_2O_2$

#### 2.7.1.3 *Applications of Enzymatic Interaction of Glucose Oxidase Bioreceptors/Biosensors*

- ZnO nanorods with chitosan for glucose determination, here glucose oxidase enzyme is used for immobilization by adsorption technique on the surface of ZnO nanorods [4].
- Carbon nanotubes also used for glucose detection [31].
- Enzyme electrode based on gold nanoparticle, poly-norepinephrine and glucose oxidase for glucose biosensor by green method [37].
- Glucose in real sample is measured from electrode of carbon blended with PVA nanofibers and polyaziridine by electro-chemical method through glucose oxidase enzyme [18].
- Glucose levels in food is estimated by carbon electrode with pt.-nanoparticle [1].

### 2.7.2 Laccase

It is type of green catalyst, a benzenediol: oxidoreductase or multi copper oxidase. Laccase is a natural enzyme obtained from the plants, bacteria, insects, fungi. Applicable in biotechnology over an extreme range of substrates as it has catalyzing power and oxidizing property of different compound [33]. It has superior catalytic property. Laccase is third type of biosensor. Principle for reaction is transfer of electrons between electrode and enzyme, where laccase is immobilized on the surface of electrodes. Later the oxygen atom is oxidized and eventually reduction by substrates as electron donor produces oxidized form of enzyme [8]. Finally, reduction current is obtained that is similar to the product concentration [10, 47].

#### 2.7.2.1 *Applications of Enzymatic Interaction of Laccase Bioreceptors/Biosensors*

- Quantification of epinephrine is done by laccase hybrid micro flowers which are produced with copper phosphate.$3H_2O$ by photo sensing method in medical diagnostics [21].
- Detection of dopamine through optical method in diagnosis of Alzheimer and Parkinson's disease by involving carbon dots biofunctionalized with amino silicane derivative [14].
- Multiwalled carbon nanotubule-based glass carbon electrode also used to detect dopamine by electrochemical method.
- Spironolactone is detected by carbon electrode layered with carbon nanotube with laccase immobilization by electrochemical transducer [13].
- Phenolics and catechol in water is detected by electrochemical method for environmental applications using laccase immobilized enzymes [14].

### 2.7.3 Other Enzymes

Other enzymes are other than above enzyme used in detection of different compounds [14]. Tyrosinase, polyphenol, alkaline phosphatase, urease are other enzymes used in biosensors [1].

Tyrosinase is extracted from bacteria, fungi, mammals, plants. Enzyme works by catalysis and oxidation [8]. Phenols are majorly detected by voltamperometric transducer. Studies has discovered detection of bisphenols by tyrosinase. They are applied in field of health, food, environment [1, 47].

$$\text{Tyrosinase: phenol} + \frac{1}{2}O_2 \text{ quinolones} + H_2O$$

Alkaline phosphatase is enzyme that catalyzes dephosphorylation of proteins, biomolecules and nucleic acids. Increase in alkaline phosphatase causes diabetes and tumor [10]. Also used in detection of organophosphate pesticide [7].

Urease is applied to detect urea in urine sample. Here the level of urea is directly proportional to protein uptake [14].

Protein phosphates types like PP1A and PP2A are implemented in genetic engineering and also in chemical reactivity in cell physiology. The electrochemical examination of MC-LR in cell of cyanobacterial sample are done by PP2A inhibition-based biosensor. Stability and activity of the enzyme is augmented by catalytic site that is genetically modification by PP2A [41].

For detection of carbamates and organophosphates in insecticides and natural neurotoxic can be followed by acetylcholinesterase and their mutants that are obtained from the electric eel applied in bioassay and biosensing devices [41].

### 2.7.4 Applications of Enzymatic Interaction of Other Enzymes Bioreceptors/ Biosensors

- Lipase is used in detection of methyl parathion by electrochemical method for environmental analysis. Also used for triglyceride determination by optical method in health field [28].
- Lipase: triglyceride + $3H_2O \rightarrow$ glycerol + 3 fatty acids
- Urease is enzyme used in determination of urea in sample by electrochemical transducer in health field.
- Tyrosinase enzyme is immobilized for caffeic acid for food factors by electrochemical method. Bisphenol is detected by electrochemical method in food analysis. Benzoic acid is determined by electrochemical transduction for environments [14, 47].
- Lactate dehydrogenase identifies pyruvate for healthcare needs.
- Lactate dehydrogenase: Lactate + NAD + $\rightarrow$ pyruvate + NADH
- Lactose oxidase: Lactose + $O_2 \rightarrow$ pyruvate + $H_2O_2$
- Alkaline phosphatase detects pesticides by electrochemical technique for environmental purpose [24].
- Detection of anatoxin-a in water sample can be performed by genetically modified enzyme.
- Aflatoxin B1 is identified by recombinant enzyme. Myco-cystein analogues and nodularin are determined by recombinant PP2A in baculovirus.
- For AFB1 analysis genetically modified Acetylcholinesterase inhibited by mycotoxin in SPR analysis.

### 2.7.5 Bienzymes

Detection of multiple analytes in sample can be performed by biosensors with two or more enzymes. determination do not take time consuming, complicated processing and expensive test whereas as size is reduced which makes it easy to work on [16]. Optimization is a must requirement to see that all enzymes are selective, stable, reactive on its site of action. Co-immobilization, process for more than two enzymes placed in single space with multi enzyme immobilization. Detection of multiple analytes by multiple enzymes on electrode is possible [16].

#### *2.7.5.1 Applications of Enzymatic Interaction of Bienzymes Bioreceptors/Biosensors*

- Glucose oxidase and horseradish peroxidase enzyme is used for glucose detection [16].
- Lactate oxidase and HRP enzyme for lactate determination [14].
- Laccase and tyrosinase works for exposing phenols, gallic acid, caffeic acid, catechin [48, 50].
- Alcohol oxidase and HRP enzyme identifies methyl salicylate in plants [47].
- D. amino acid oxidase and HRP recognises total content of D. amino acid in sample [8].

- Urea and penicillin together reveal presence of urea and penicillin in sample [6].
- Acetylcholinesterase with choline esterase determines presence of dichlorvos and demeton.
- Lactate oxidase and glucose oxidase produces signal for glucose and lactate [50].

### 2.7.6 Ribozymes/DNAzymes

In 1982, ribozyme term was invented with feature of natural catalytic property of RNA. While dioxyribozyme was introduced in 1994, Breaker and Joyce detailed about artificial DNA catalyst. In biosensor engineering, the detection of metabolites, toxins, antibiotic is proceeded by catalytic RNA structure hairpin ribozyme. Colorimetric biosensor was developed by DNAzyme coupled with gold nanoparticle. The well-designed DNAzyme is preferred in detection of ochratoxin A functioning by catalyzing luminol in existence of hemin mimicking peroxidase to get required chemiluminescence-based signal [41]. Incorporation of DNAzyme in electrochemical and SERS biosensor have proven successful. Colorimetric aptasensor is made by coupling OTA specific aptamer and HRP-mimicking DNAzyme sequence. Determination of sensitive ochratoxin A is done with aptamer linked DNAzyme by resonance energy transfer. A problem of degradation in body fluids prevailed for DNAzyme but was later tackled by invention of modified nucleotides [41].

### 2.7.7 Nanozymes

Nanoparticles have physicochemical property and imitates the character of enzymes [8]. It shows similar catalytic properties and features as that of bio enzyme [39]. They are categorized according to activity into peroxidase like activity, oxidase like activity and laccase like activity [4]. For conducting chemical analysis, alternatives to natural enzyme catalyst have developed that has mimicking behavior. Mostly seen in emerging nanomaterial-based enzyme. Bioassays, biotechnology and biomedical field are explored by the materials by acting as biomaterials. Suitability of nanozyme has aroused due to its higher stability, production ease and reasonability of catalytic activity. Replacement of horseradish peroxidase enzyme is done by nanozymes to evaluate amounts of hydrogen peroxide, glucose, cholesterol, d-alanine, antioxidants and few toxin analytes [51]. Oxidase like nanoparticle were used in catechol and dopamine estimation by colorimetric method. They do not have exact characteristic as that of real enzyme because it is critical to remake the platform for determination of analyte and thus withhold its approach in amperometric biosensors. Also, nanoparticle has a higher reactivity rate toward interfering molecules so maintaining control over it is challenging [41].

#### *2.7.7.1 Applications of Enzymatic Interaction of Nanozymes Bioreceptors/ Biosensors*

- Detection of $H_2O_2$ is done by peroxidase like nanozyme that is yttrium vanadate microstructure [19].

- $H_2O_2$ and dopamine signals are generated by sodium dodecyl benzene sulfonate nanozymes.
- Heparin sodium and platinum nanoparticle are oxidase like nanozyme that interprets and signalizes for pharmaceutical and clinical diagnosis [21, 40].
- Cerium dioxide nanoparticle (laccase like activity) recognizes organophosphorus pesticides [39].
- Cu-tannic acid nanohybrid shows laccase like activity to detect epinephrine and catechol [38].
- Au-nanoparticle with thiol modified oligonucleotides detects SARs-CoV2 by colorimetric method [40].

## 2.8 Applications of Enzyme Biosensor

Biosensor can be applied in the field of food safety evaluation, standards of food freshness, nutrients and additives are maintained. Also, food processing and drink analysis are tested. Environment and agricultural factors can be inspected by biosensor, soil and water monitoring are facilitated by enzyme-based biosensors [26]. In industry, it is extensively applied for bioprocessing, safety monitoring, etc. In the healthcare field, it is applied for real time estimation of glucose levels, prosthetic devices, disease detection. For research, toxins defence action and drug discovery biosensors are helpful. Nanozymes for the cancer cells and bacterial infections [9]. Au-nanoparticle proved successful or COVID-19 diagnosis. The diagnosis of COVID-19 also depends on enzyme labeled antibodies the particularly recognises the analyte and generates signal to record the amplification. According to the reports obtained from study in 2020 alkaline phosphatase detection by electrochemical method was progressed for SARS-COV-2 protein with help of secondary antibody catalyzing substrate from screen printed carbon electrode [43].

For security reasons, biosensor applied in bioterrorism and in forensic sciences for cyber security, criminal investigations [1, 37]. Preparation of artificial enzyme for cancer detection, neuroprotection, stem cell growth, pollutants removal. Au-nanoparticle enzyme for multianalyte detection of glucose, fructose, vitamin C, cortisol, paracetamol. Enzyme facilitated amplification for DNA/RNA response by electrochemical device [9, 30, 37]. *In-Vivo* electrochemical sensor for blood and tissue specificity. Heart rate tracking, cardiovascular disease monitoring by screen-printed electrodes wearable device [15]. Bio-luminance-based biosensor were developed for detection of $SARS\text{-}CoV_2$ diagnosis [17, 32]. Table 2.3 represents the application of enzyme-based biosensor in different enzymatic interactions with it use in various fields of science.

**Table 2.3** Application of enzyme biosensors in various enzymatic interactions for different purpose.

| Sr. no. | Applications | Reactant | Enzyme | Product | Purpose |
|---|---|---|---|---|---|
| 1. | Enzyme biosensor for food analysis | | | | |
| | 1. Acids | | | | |
| | a. Lactate [9, 19] | L – lactate + $O_2$ [50] | L – lactate oxidase | Pyruvate + $H_2O_2$ [50] | Juice, wine, milk, sauces, fruits |
| | | L – lactate + $NAD^+$ | L – lactate Dehydrogenase | Pyuvrate + NADH + $H^+$ | |
| | | NADH + H + $[Med]_{oxi}$ | Diaphorase | $[Med]_{red.}$ + $NAD^+$ | |
| | b. Malate [26] | L – lactate + $NAD^+$ | Malate dehydrogenase | Oxaloacetate + $NADH^+$ + $H_2$ | Juice, fruits, vegetables |
| | | L – Malate + $NADP^+$ | Malic enzyme | Pyruvate + $Co_2$ + NADPH + $H^+$ | |
| | c. L-Ascorbic | 2L – ascorbate + $O_2$ | L – ascorbate oxidase | 2L – dehydroascorbate + $2H_2O$ | Antioxidant |
| | d. Acetic acid | Acetate + ATP | Acetic kinase | Acetyl – P +ADP | Soya sauce, vinegar, alcoholic drinks |
| | | Phosphoenol pyruvate + ADP | Acetic kinase | Pyruvate + ATP | |
| | | Pyruvate + Phosphate + $O_2$ | Pyruvate kinase | Acetyl phosphate | |
| | e. Citric acid | Citrate | Citrate lycase | Oxaloacetate decarboxylase | Lemons, citrous fruits |
| | | Oxaloacetate | Oxaloacetate decarboxylase | Pyruvate + $CO_2$ | |
| | | Pyruvate + phosphate + $O_2$ | Pyruvate oxidase | Acetyl phosphate + $CO_2$ + $H_2O$ | |

(*Continued*)

**Table 2.3** Application of enzyme biosensors in various enzymatic interactions for different purpose. (*Continued*)

| Sr. no. | Applications | Reactant | Enzyme | Product | Purpose |
|---|---|---|---|---|---|
| | 2. Sugars | | | | |
| | a. Glucose [8, 45] | D – glucose + $O_2$ | Glucose Oxidase | D – gluconic acid | Wine, dairy products, beer, sugar |
| | | $H_2O_2$ + $[Med]_{red.}$ | Peroxidase | $[Med]_{oxi.}$ + $H_2O$ | |
| | b. Fructose | D – Fructose + $[Med]_{oxi.}$ | D – Fructose dehydrogenase | 5 keto D – Fructose + $[Med]_{red.}$ | Fruits, honey |
| | c. Sucrose | D – Sucrose + $H_2O$ | Invertase | D – Fructose + α, D glucose | Sugarcane, fruits, vegetable |
| | d. Lactose [7] | D – Lactose + $H_2O$ | Lactase | D – Galactose + D – glucose | Milk, dairy products |
| | | D – Galactose + $O_2$ | D – Galactose oxidase | D – Galactohexoldialdose | |
| | 3. Amino acids | D – Amino acid + $O_2$ + $H_2O_2$ | D – Amino acid oxidase | 2 oxoacids + $NH_4$ + $H_2O_2$ | Eggs, mushroom, fish |
| | 4. Alcohol [16] | | | | |
| | a. Ethanol | Ethanol + NAD | Alcohol dehydrogenase | Acetaldehyde + NADH + $H^+$ | Wine, alcoholic preparations, breads |
| | | Ethanol + $O_2$ | Alcohol oxidase | Acetaldehyde + $H_2O_2$ | |
| | b. Glycerol | Glycerol + NAD | Glycerol dehydrogenase | Dihydroxy-acetone | Beverages, humectants, preservative |
| | | Glycerol + $[Med]_{oxi.}$ | Poo-glycerol dehydrogenase | Dihydroxyacetone + $[Med]_{red.}$ | |
| | | Glycerol + ATP | Glycerol kinase | L-glycerolphosphate + ADP | |
| | | L-Glycerol Phosphate + $O_2$ | L-glycerol phosphateoxidase | Dihydroxy acetone phosphate + $H_2O_2$ | |

(*Continued*)

**Table 2.3** Application of enzyme biosensors in various enzymatic interactions for different purpose. (*Continued*)

| Sr. no. | Applications | Reactant | Enzyme | Product | Purpose |
|---|---|---|---|---|---|
| 2. | Enzyme biosensors for contaminants determination [7, 16, 19] | | | | |
| | 1. Acetylcholine | Acetylcholine + $H_2O$ | Acetylcholinesterase | Choline + acetic acid | Enzyme inhibition based pesticides |
| | | Choline + $O_2$ | Cholineoxidase | Betaine + hydrogenperoxide | |
| | 2. Tyrosinase | Monophenol + $O_2$ | Cresolate | Catechol | Inhibition by carbamate pesticides |
| | 3. Alkaline phosphate | Catechol + $O_2$ | Catecholase | O-Quinone | Inhibited by different compounds |
| | | Phosphate monoester + water | Alkaline phosphatase | Alcohol + phosphate | |
| | 4. Organo-phosphorus | Organophosphorus + $H_2O$ | Organo phosphorus hydrolase | Organophosphorus acid + R-OH | Hydrolyse organophosphate pesticides |

## 2.9 Conclusion and Future Expectations

This chapter emphasizes on biorecognition of different analyte from sample by transducer (electrical, chemical, optical and advanced technologies.) or bioreceptor (enzyme, antibody, aptamer, nanoparticle, cell) aspect. It gives detail study of ideal biosensors characteristics and efficient type of biosensor in signal transduction. Centralizing enzyme-based biosensor, strategies to include bioreceptors in enzyme interactions to ease specificity, reliability, biocompatibility of sensor. Mechanism of enzyme biosensor and different bio enzymes involved in the detection of biological compounds. Various enzyme immobilization techniques to build biosensor, inclusive material to fabricate interface for enzyme adherence, challenges and approaches to design perfect biosensor [25]. Increasing detection through enzyme interaction in field of engineering, safety, drug delivery, medicine, toxicology, pharmaceuticals, biomedical, disease progression [51]. Enhancements through combining biotechnology with nanotechnology to produce sophisticated results. Miniatured form nanozyme i.e., nanoparticle support for enzyme biorecognition in cancer, immunoassays, cytotoxicity.

DNA, aptamer, nanoparticle-based enzyme called DNAzymes, aptazymes, nanozymes are evolved strategies in biosensing and futural approaches like advances in microfluidics wearable device for relevant monitoring, 3D printed biosensor, use of 2D nanoparticle in bioprocessing, nanoscience, silicon technology, photoelectrochemical methods, multiarray, nanozyme chips, bioelectronics, digital signalizing through software's and mobile apps interface for health, environment, safety analysis [30].

## References


1. Abid, H., Mohd, J., Ravi, P.S., Rajiv, S., Shanay, R., Biosensors applications in medical field: A brief review. *SI*, 2, 100–120, 2021.
2. Ali, A.E., An introduction to sensors and biosensors, in: *Electochemical Biosensor*, pp. 1–10, Elsevier, United States, 2019.
3. Steve, K., Anastasios, E., Stephanos, K., Georgia, N., Nikolelis, D., Enzyme-based sensors, in: *Advances in Food Diagnostics*, Edited by Fidel, T., Leo, M.L.N., 2nd Ed., pp. 231–250, Wiley-Blackwell, United Kingdom, 2018. 2017.
4. Bagyalakshmi, S., Sivakami, A., Balamurugam, K.S., A Zno nanorods based enzymatic glucose biosensor by immobilization of glucose oxidase on a chitosan film. *Obes. Med.*, 18, 100–229, 2020.
5. Dede, S. and Altay, F., Biosensors from the first generation to nano-biosensors. *IAREJ*, 2, 2, 200–207, 2018.
6. Fiel, W., Borges, P., Lins, V., Faria, R., Recent advances on the electrochemical transduction techniques for biosensing of pharmaceuticals in aquatic environment. *Int. J. Biosens. Bioeletron.*, 5, 4, 119–123, 2019.
7. Firoozeh, P., Fatemeh, M., Farnoush, F., Emerging biosensors in detection of natural products. *Synth. Syst. Biotechnol.*, 5, 4, 293–303, 2019.
8. Hossein, A., Sevinc, K., Ali, E., Bengi, U., Latest trends for biogenic amines detection in foods: Enzymatic biosensors and nanozymes applications. *Trends Food Sci. Technol.*, 112, 9, 75–87, 2021.
9. Jos Antonio, C.P., Juan Eduardo, S.H., Syed, M.H., Bilal, M., Roberto, P.S., Hafiz, M.N.I., Bioinspired biomaterials and enzyme-based biosensors for point-of-care applications with reference to cancer and bio-imaging. *Biocatal. Agric. Biotechnol.*, 17, 168–176, 2019.

10. Kaur, J., Choudhary, S., Chaudhari, R., Joshi, A., Jayant, R.D., Enzyme-based biosensors, in: *Bioelectronics and Medical Devices*, Edited by Pal, K., Kraatz, H.-B., Khasnobis, A., Bag, S., Banerjee, Kuruganti, U., pp. 211–240, Elsevier Ltd., United States, 2019.
11. Kucherenko, I.S., Topolnikova, Y.V., Soldatkin, O.O., Advances in the biosensors for lactate and pyruvate detection for medical applications: A review. *TrAC – Trends Anal. Chem.*, 110, 160–172, 2019.
12. Xiaoping, H., Yufang, Z., Ehsan, K., Nano biosensors: Properies, applications, electrochemical techniques. *J. Mater. Res. Technol.*, 12, 1649–1672, 2021.
13. Lynette, A.R., Magdalena, R.A., Jose, R.R., Carlos, C.Z., Juan, E.S.H., Damia, B., Hafiz, M.N.J., Robert, P.S., Exploring current tendencies in techniques and materials for immobilization of laccases – a review. *Int. J. Biol. Macromol.*, 181, 683–696, 2021.
14. Lynette, R., Magdalena, R.A., Jose, R.R., Juan, E.S.H., Elda, M.M.M., Hafiz, M.N.I., Robeto, P.S., Enzyme (single and multiple) and nanozyme biosensors: Recent developments and their novel applications in the water-food-health nexus. *Biosensors*, 11, 410, 2021.
15. Mahshid, P., Christian, E., Sandro, C., Microfluidics by additive manufacturing for wearable biosensors: A review. *Biosensors*, 20, 4236, 1–28, 2020.
16. Mandake, M.B., Urvashi, D., Laxman, P., Sakshi, B., Application of enzyme immobilization in the food industry. *Int. J. Adv. Res. Chem. Sci.*, 7, 2, 6–10, 2020.
17. Manish, S., Neha, S., Mishra, P.K., Bansi, D.M., Prospects of nanomaterials-enabled biosensors for COVID-19 detection. *Sci. Total Environ.*, 754, 142363, 2021.
18. Sánchez-Paniagua Lopez, M. and Lopez-Ruiz, B., Electrochemical biosensor based on ionic liquid polymeric microparticles. An analytical platform for catechol. *Microchem. J.*, 138, 173–179, 2018.
19. Meshram, B.D., Agrawal, A.K., Shaikh, A., Survartan, R., Sande, K.K., Biosensor and its application in food and dairy industry: A review. *Int. J. Curr. Microbiol. Appl. Sci.*, 7, 2, 3305–3324, 2018.
20. Min, S., Xiaogang, L., Zhijia, P., Shibin, X., Lifeng, J., Xiaodong, Z., Haoyue, L., Materials and methods of biosensor interfaces with stability. *Front. Mater.*, 7, 1–11, 2021.
21. Nataliya, S., Oleh, S., Olha, D., Tetiana, P., Galina, G., Marina, N., Mykhailo, G., Synthesis, catalytic properties and application in biosensorics of nanozymes and electronanocatalysts: A review. *Biosensors*, 20, 16, 4509, 2020.
22. Nguyen, H.H., Lee, S.H., Lee, U.J., Fermin, C.D., Kim, M., Immobilized enzymes in biosensor applications. *Mater. (Basel)*, 12, 121, 1–34, 2019.
23. Nyugen, H.H. and Moonil, K., An overview of techniques in enzyme immobilization. *Appl. Sci. Convergence Technol.*, 26, 6, 157–163, 2017.
24. Paolo, B. and Gorton, L., Enzyme based amperometric biosensors. *Curr. Opin. Electrochem.*, 10, 157–173, 2018.
25. Paolo, B. and Evgeny, K., Biosensors—Recent advances and future challenges. *Biosensors*, 20, 6645, 1–5, 2020.
26. Richard, B., Can, D., Estefania, C.R., Maria, T.F.A., Arben, M.A.M., Gerald, A.U., Firat, G., Disposable sensors in diagnostics, food, and environmental monitoring. *Adv. Mater.*, 31, 30, 1806739, 2019.
27. Ruijin, Z., Jun, W., Qingshui, W., Dianping, T., Horseradish peroxidase-encapsulated DNA nanoflowers: An innovative signal-generation tag for colorimetric biosensor. *Talanta*, 2, 21600, 2021.
28. Rupak, P., Avinash, S., Deepak, K., Soham, S., Soham, M., Fatih, S., Avvaru, P.K., Amalgamation of biosensors and nanotechnology in disease diagnosis: Mini-review. *SI*, 2, 10089, 2021.
29. Sanjay, K.M. and Koyeli, G., Diagnostic biosensors in medicine – a review. *Biocatal. Agric. Biotechnol.*, 17, 271–283, 2019.

30. Sharafeldin, M., Kadimisetty, K., Ketki, S.B., Tianqi, C., James, F.R., 3D-printed immunosensor arrays for cancer diagnostics. *Sensors*, 20, 16, 4514, 2020.
31. Sireesha, M., Babu, V.J., Kiran, S.K., Ramkrishna, S., A review on carbon nanotubes in biosensor devices and their applications. *Nanocomposites*, 4, 2, 36–57, 2018.
32. Taha, A., Raghunath, S., Emily, E.F.B., Zaid, T., Reza, R., Rozanne, A., Stephen, B., Jean, S.D., Carolina, S.I., John, C.B., SARS-CoV-2 S1 NanoBiT: A nanoluciferase complementation-based biosensor to rapidly probe SARS-CoV-2 receptor recognition. *Biosens. Bioelectron.*, 180, 113–122, 2021.
33. Thangavel, L. and Subash, C.B.G., An introduction to biosensors and biomolecules, in: *Nanobiosensors for Biomolecular Targeting*, pp. 1–21, 2019.
34. Thomas, C., Qingshen, J., Sohini, K.N., Biosensors based on mechanical and electrical detection techniques. *Sensors*, 20, 19, 5605, 2020.
35. Varnakavi, N. and Nohyun, L., A review on biosensors and recent development of nanostructured materials-enabled biosensors. *Sensors*, 21, 1109, 1–35, 2021.
36. Vesna, V., Sladjana, D., Milos, O., Martin, F., Enzymatic glucose biosensor based on manganese dioxide nanoparticles decorated on graphene nanoribbons. *J. Electroanal. Chem.*, 823, 610–616, 2018.
37. Winktoria, L., Katarzyna, G., Katarzyna, S., Enzyme immobilization on gold nanoparticles for electrochemical glucose biosensors. *Nanomaterials*, 11, 1156, 2021.
38. Yang, L., Nan, X., Shi, W., Xin, L., Zi, H., Yanan, S., Dongtao, G., A glucose biosensor based on the immobilization of glucose oxidase and Au nanocomposites with polynorepinephrine. *RSC Adv.*, 29, 16439–16446, 2019.
39. Yasmin, A., Danting, Y., Samuel, S.F.K., Guozhen, L., Advances in biosensors for the detection of ochratoxin A: Bio-receptors, nanomaterials, and their applications. *Biosens. Bioelectron.*, 141, 111418, 2019.
40. Chong, Y., Qiang, L., Cuicui, G., Advances in oxidase-mimicking nanozymes: Classification, activity regulation and biomedical applications. *Nanotoday*, 37, 101076, 2021.
41. Ingrid, B., Scherrine, A.T., Akhtar, H., M, J.–L., New biorecognition molecules in biosensors for the detection of toxins. *Biosens. Bioelectron.*, 87, 285–298, 2017.
42. Valerie, G., Advances in biosensor development for the screening of antibiotic residues in food products of animal origin – a comprehensive review. *Biosens. Bioelectron.*, 90, 363–377, 2017.
43. Singh, B. and Datta, B., A comprehensive review on current COVID-19 detection methods: From lab care to point of care diagnosis. *Sens. Int.*, 2, 100119, 2021.
44. Donghui, Y., Bertrand, B., V., J.-C., K., J.-M., Biosensors in drug discovery and drug analysis. *Anal. Lett.*, Taylor & Francis, 38, 11, 1687–1701, 2005.
45. Pavla, M. and Miroslav, P., Biosensors for blood glucose and diabetes diagnosis: Evolution, construction, and current status. *Anal. Lett.*, Taylor & Francis, 48, 16, 2509–2532, 2015.
46. Lino, C., Barrias, S., Chaves., R., Adega, F., Martins, L., Fernandes, J.R., Biosensors as diagnostic tools in clinical applications. *Biochem. Biophys. Acta – Rev. Cancer*, 1877, 3, 1–49, 2022.
47. Paulo, A.R.P.A., Tiago, A.S., Fabio, R.C., Lias, R., Eduardo, Z., Daniela, B., Marcio, F.B., Luiz, H.M., Craig, E.B., Osvaldo, N.O., Bruno, C.J., Orlando, F.F., Polyphenol oxidase-based electrochemical biosensors: A review. *Anal. Chem. Acta*, 1139, 198–221, 2020.
48. Dinorath, O.A.B. and Michael, G.M., Electroactive material-based biosensors for detection and drug delivery. *Adv. Drug Deliv. Rev.*, 170, 396–424, 2021.
49. Valentyna, A., Olexandr, S., Lyudmyla, M., Iryna, K., Viktor, K., Sergei, D., Enzyme biosensor based on pH-sensitive field-effect transistors for assessment of total indole alkaloids content in tissue culture of Rauwolfia serpentina. *Electrochem. Sci. Adv.*, 2, e2100152, 1–8, 2021.

50. Ivan, S.K., Oleksandr, O.S., Yaroslava, V.T., Sergei, V.D., Alexei, P.S., Novel multiplexed biosensor system for the determination of lactate and pyruvate in blood serum. *Electroanalysis*, 31, 1608–1614, 2019.
51. Shunsheng, C., Juanrong, C., Xin, J., Weiwei, W., Zhiyuan, Z., Enzyme - based biosensors: Synthesis and applications, in: *Biosensor Nanomaterials*, pp. 95–115, 2011.
52. Cristina, R.I., Georgeta, C., Silvana, A., Review: Recent developments in enzyme-based biosensors for biomedical analysis. *Anal. Lett.*, 45, 2-3, 168–186, 2012.

3

# Dendrimer-Based Nanomaterials for Biosensors

**Chetna Modi[1], Vipul Prajapati[2]*, Nikita Udhwani[3], Khyati Parekh[3] and Hiteshi Chadha[4]**

*[1]Department of Pharmaceutics, Anand Pharmacy College, Anand, Gujarat, India*
*[2]Department of Pharmaceutics, SSR College of Pharmacy (Permanently Affiliated to Savitribai Phule Pune University), Sayli-Silvassa Road, Sayli, Silvassa, Union Territory of Dadra Nagar Haveli & Daman Diu, India*
*[3]Department of Pharmaceutics, Anand Pharmacy College, Anand, India*
*[4]Department of Pharmaceutical Quality Assurance, Anand Pharmacy College, Anand, India*

### *Abstract*
Biosensors are drawing a major attention toward its development due to its application in clinical diagnosis, targeted drug delivery, optical drug delivery and also in healthcare sectors. Dendrimers are highly branched molecular structure having 3D branching. These branching make great interest for attachment of varied molecules to improve their characteristics such as solubility, bioavailability and many more. In this chapter, synthesis of dendrimers, its types, and application of dendrimers as biosensors are discussed. Dendrimers application was focused in the varieties of drug delivery fields, such as gene delivery, controlled drug delivery, antimicrobial and anticancer therapy. Dendrimers-based biosensors are majorly synthesized by layer-by-layer assembling and designed as glucose sensing device, electrochemical detectors, fluorescence detectors and QCM detectors.

***Keywords*:** Biosensor, biorecognition, dendrimer, sensors, electrode surface

## Abbreviations

| | |
|---|---|
| PAMAM | (polyamidoamine) dendrimers |
| QCM | Quartz Crystal Microbalance |

## 3.1 Introduction

Dendrimers are symmetric radially, nanoscale molecules as well as homogenous, and monodisperse assembly. Their structure is of like the branches of any tree (Figure 3.1). Sometimes, they are also known as 'cascade molecules'. Their branching units are constructed symmetrically around a linear polymer core. They are architectural idea rather than a chemical [1]. Various technological challenges such as biocompatibility, nano-concepts,

**Corresponding author*: vippra2000@yahoo.com

Inamuddin and Tariq Altalhi (eds.) Biosensors Nanotechnology, 2nd Edition, (61–84) © 2023 Scrivener Publishing LLC

high drug loading efficiency, wireless transmission, and therapeutic systems' integration are the focus of contemporary medical research. These are being evaluated through the inclusion of nanotechnology. The term dendrimer has been derived from "dendron" (Greek word) that means "tree."

Their manufacturing is multi-step process which involves up to 10 generations (5–50 nm) of branching groups. Their each generation represents a layer. Distinct features of dendrimer like structural homogeneity, globular shape, high functional group density, hydrophilic nature, and adaptability in synthesis at nanometric size, can all be used to build very sensitive sensors [2]. The guest encapsulation within the dendrimers is mostly via electrostatic contacts, complicating reactions, and other weak forces such as Vander Waals, Hydrogen-bonding etc. Dendrimer of Metallic nanoparticles (NPs) has gained a lot of interest these days because of their probable use in catalysis and sensors.

Au NPs adsorbed PAMAM dendrimers via electrochemical connection and the surface of gold or platinum electrode was researched well. Many research was done on PAMAM dendrimer-stabilized silver (Ag) NPs (an electrochemical sensor used in detecting nitrite) and CNTs, Pt NPs, (quick detection of small amounts of analytes on disposable printed chips) [3].

### 3.1.1 Structure of Dendrimers

As shown in Figure 3.1, they are having the core, the dendron and the functional groups at the terminal end as the three domains in their structure. Dendrimeric crevice is the space between the branches of a dendrimer molecule. The characteristics of any dendrimers are

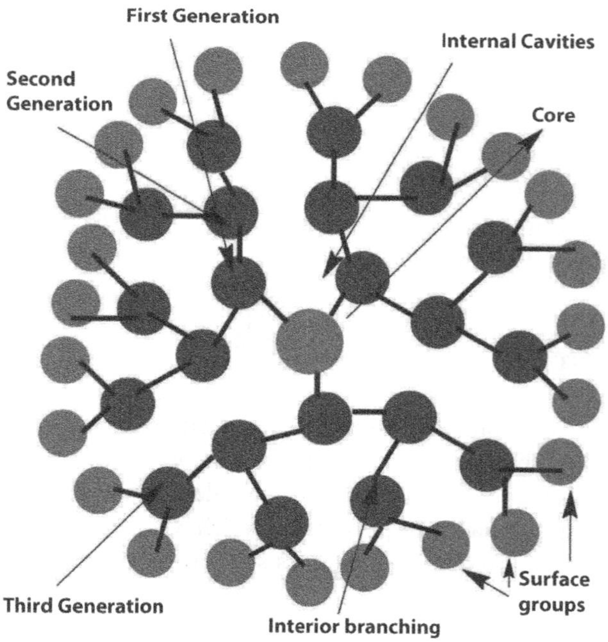


**Figure 3.1** Outlook of dendrimers structure [1].

conquered by a central core, the peripheral functional groups and the dendrimeric crevices [5, 6].

Functional groups are at the outer layer of the dendrimers, particularly for a drug conjugation and to the target specific groups. The entrapment of any drug's molecules is in the inner layer, which help in the enhancement of pharmacological efficacy and it diminishes the drug toxicity. It also helps in controlling the drug release rate. Generally, the inner layers of the dendrimers are made up on the repetition of branches via covalent attachment to the nucleus. These attachments are denoted as 'generation (G*n*)'. Here, *n* can range from 0 to 12. It also represents a digit of the terminal performing groups on the dendrimer's outer shell [1, 7].

#### *3.1.1.1 History*

Dendrimers was discovered in 1978 as the hyperbranched compounds. It was also proved in the 1980s. The second set of the synthesised macromolecules is known as arborols, which means "trees" in Latin [1].

### 3.1.2 Surface Modification Using Dendrimers

Surface modification of the dendrimers as sensor substrates was achieved using a silica (silicon wafers or silica fibres) or gold (glass slides). Such modification techniques include the non-covalent interactions, the Langmuir–Blodgett (LB) method, spin casting and covalent attachment [8, 9]. Table 3.1 shows its detail procedures with examples.

**Table 3.1** Techniques in surface modification of dendrimers.

| Surface modification technique | Mode of bonding | Example | References |
|---|---|---|---|
| Non-covalent immobilization of dendrimers | Hydrogen bond,<br>vander Waal's forces<br>Electrostatic-interaction | • Positively and negatively charged dendrimers silanol sites bind to silica substrates due to ionic interactions as well as hydrogen bonding to some extent.<br>• On silicon wafers, a multilayer construction of PAMAM dendrimer and K2PtCl4 | [50, 51] |

(*Continued*)

**Table 3.1** Techniques in surface modification of dendrimers. (*Continued*)

| **Surface modification technique** | **Mode of bonding** | **Example** | **References** |
|---|---|---|---|
| Covalent immobilization | **Covalent bond:** Functional groups additionally by the means of conventional modification techniques on the surface | ***Silica*:** Glutaraldehyde, a cross linker was taking part for an amino-silanized surface to bind with an amine ended dendrimer. By treating 3-cyanopropyltrichlorosilane coated silica substrates with sulfuric acid, the cyano functional groups were hydrolysed to carboxylic acid terminals, and then transformed to anhydride groups using trifluoroacetic anhydride in $SCCO_2$. LbL approach was used in multilayer films of a PAMAM dendrimer having carboxyl group at terminal end and nitro group of diazoresin on a silica surface. UV-irradiation was used to breakdown the diazonium groups into phenyl cations. Then by covalent ester bonding with the carboxyl group interaction would be formed.<br>***Gold*:** SAMs comprising thiol group containing hetero bi-functional coupling agents at one end were used to graft dendrimers onto gold substrates.<br>*Mercaptoundecanoic acid (MUA) and analogues*<br>Using EDC-NHS driven coupling chemistry, gold surfaces coated with MUA have been fixed by PAMAM dendrimers (numerous generations). | [39, 52–54] |

### 3.1.3 Synthesis of Dendrimers

Few methods of dendrimer production, reported in the different literatures, is explained herewith.

#### *3.1.3.1 Divergent Growth Method*

The nucleus initiates divergent development, and addition of the number of branches is dependent on the nucleus' reactivity. For example, four branches are added with two reactive ends of the nucleus. When new branches (a new generation) were generated, the exposed ends were activated which allows the further branching.

Addition and activation are continued until the required number of generations are attained. To avoid faulty branches, it is critical that the reaction be in order and finished completely. Advantage of divergent approach is to modify the surface of the dendrimer with the necessary functional groups at the completion of the synthesis, a quick synthesis process, and allowing the manufacture of massive dendrimers. Lengthy purification is the disadvantage of divergent method because both the intermediate reagents and the end products have comparable molecular weights, polarity and charges. Even monomers are required in excess amount for their synthesis [7, 10, 11].

#### 3.1.3.2 *Convergent Growth Method*

Convergent approach is completely opposite of the divergent method. Here, synthesis is starting from the branching units of dendrons and synthesised to the inward from the outside direction.

The common thing of both these methods is that the branches are formed in two phases. New branching units are added only on functional group activation. At the final stage, dendrons are generated and then joined to the nucleus to form final structure of the dendrimers.

This method is advantageous because of simpler purification. The intermediate as well as the end products have considerable differences. Lower generation dendrimers are having more homogeneous dispersion and fewer faulty branches in the synthesis. Dendrons that are properly developed are chosen to be joined to the nucleus. Disadvantage of this approach is the poor yield and difficulty in obtaining higher generations [7, 11, 12].

#### 3.1.3.3 *Click Chemistry*

Another technique for the quick and accurate production of dendrimers is the click chemistry approach. The great chemical performance of the reaction is one of the method's most notable qualities. This approach also has simple reaction conditions, commonly available chemicals, and environmentally friendly solvents [10].

#### 3.1.3.4 *Lego Chemistry*

Scientists have proposed the number of approaches to reduce the size of dendrimer during production process. One of the approaches is Lego chemistry. This approach was tested several times which resulted in the rise in the steps of 48 to 250 number terminal groups [10].

### 3.1.4 Types of Dendrimers

According to the structural properties, there are five types of dendrimers such as;

1. Polyester dendrimer
2. Polyacetal dendrimers
3. PAMAM dendrimers
4. Poly-L-lysine dendrimers
5. Propyleneimine dendrimers

Polyester dendrimers are divided into three types, such as (a) polyester dendrimers made up of monomers of "2,2-bis (hydroxylmethyl) propanoic acid (bis-HMPA)" (b) polyester dendrimers made up of monomers sequence alteration and (c) the miscellaneous polyester dendrimers. Acidicity, cost-effectiveness, non-toxicity, biodegradability, and non-immunogenicity are the important properties of the polyester dendrimers [7, 13].

The transactivation approach of protection/deprotection sequences was used to make polyacetal dendrimers containing dendrons of 2, 4, 8, 10-tetraoxaspiro. Polyacetal dendrimers with PEO chains were synthesised in 2011. These dendrimers are water soluble due to seventh generation with peripheral hydroxyls [14].

Because of its biocompatibility, hydrophilicity, and lack of immunogenicity, PAMAM (polyamidoamine) dendrimers have been widely applicable in drug delivery. The divergent approach can be used for the synthesis of these nanocarriers [15]. They contain an ethylene-diamine nucleus. But, diaminohexane, diaminobutane, and diaminododecane are used as the hydrophobic interiors in its synthesis. Medicines, enzymes and antibiotics are loaded on the branches of the dendrimers having amine groups at terminal end. G0.5, G1.5, and G2.5 contain negatively charged -COOH group, whereas G1, G2, G3, and so on contain positively charged -NH2 groups at terminal end [15]. Branches at the surface of PAMAM may be terminated by OH, $NH_2$, CHO, $COOCH_3$, $CH_3$, COONa, etc. functional groups. In most cases, the -$NH_2$ group is taking part to transport the genetic material to the body cells and tissues.

Dendrimers containing Poly-L-Lysine (PLL) are connected through peptide bonds. Both the nucleus and branches are made up of the amino acid lysine. They have two major amines usually changed to increase the therapeutic value of the dendrimers. Biodegradability, biocompatibility, flexibility and hydrophilicity are the important features of PLL dendrimers. They are mostly utilised to transport DNA [16, 17].

Dendrimers made up of PPI (propylene imine) were investigated firstly in 1978. Monomers such as Propylene imine, 1,4-diaminobutane (DAB) or diaminoethane (DAE) were attached as branching units to the centred nuclei. The inclusion of alkyl groups in the branches gives the inner-side 'tris-propylene tertiary amines' of a more hydrophobic interior than the PAMAM dendrimers. Diagnostic purpose is fulfilled by these types of dendrimers [18].

### 3.1.5 Structure, Physical, and Chemical Properties

Synthetic structures of nanomaterials such as polymers, bucky balls or carbon nanotubes are having restricted structural properties as compared to the dendrimers [1]. Polyionic dendrimers cannot maintain their structure over time. They have altered shape, size and flexibility on increment of number of generations. Their physicochemical or biological properties can be altered by functionalization of end-groups of dendrimers. Dendrimers are the artificial macromolecules, which are highly defined containing the large number of functional groups. With increasing dendrimer production, dendritic molecules have a habit of growth in diameter and converts in larger globular form.

Dendrimer molecules have a core containing either a central atom or containing a group of atoms together. Dendrons, or branches of other atoms, sprout from this central structure through a number of chemical processes. The exact structure of dendrimers is still debated, particularly whether they are fully stretched at the surface with maximal density or whether interior is densely packed at the end-groups.

Figure 3.2 explains the structure of repeat units of a dendrimer. Monodisperse and globular macromolecules of the dendrimers may be synthesized with a level of command, having ample number of peripheral groups. They are a novel type of polymeric material. Their chemistry is rapidly expanding branches and most appealing.

Dendrigrafts are nothing but one type of dendritic polymer and are similar to monodisperse dendrimers. Dendrimers have a unique structure that allows for distinct host-guest chemistry (Figure 3.3). They are well suited to multivalent interactions. Biological features of dendrimers are self-assembly, polyvalency, chemical stability, electrostatic interactions with minimal cytotoxicity and greater solubility. Dendrimers are a suitable choice in the medical field because of their numerous features, and this overview covers their many applications [1].

The unique characteristics of the dendrimers is due to the presence of functional groups available on the dendrimers molecular surface. The dendrite replicates and separate internal and exterior functions of active sites in biomaterial, whose molecules will lead to the separation of active moieties. When a dendrimer's end-group is a hydroxyl, or carboxyl, or amino group or any other hydrophilic group, it can be water soluble. It is also possible

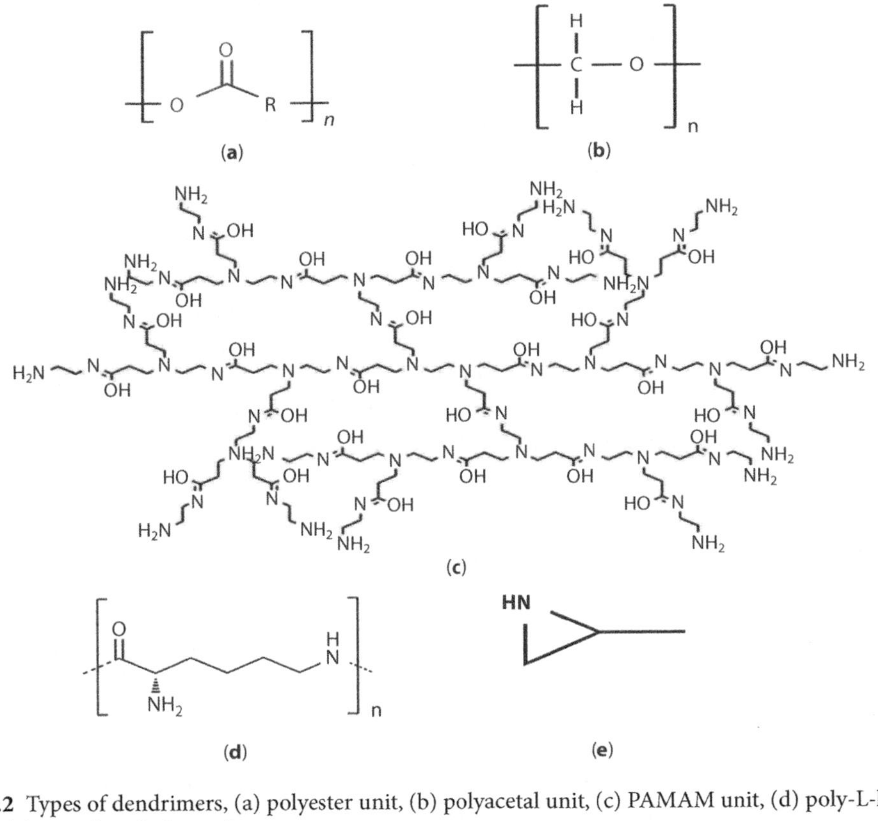


**Figure 3.2** Types of dendrimers, (a) polyester unit, (b) polyacetal unit, (c) PAMAM unit, (d) poly-L-lysine unit, and (e) propylene imine unit.

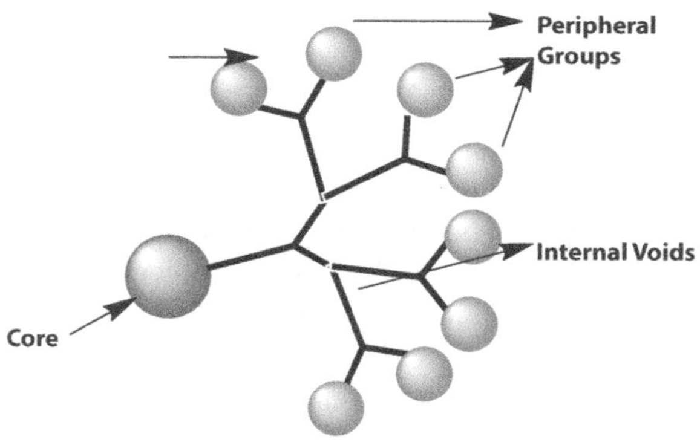


**Figure 3.3** Parts of dendrimer [1].

to build a dendrimer, which is soluble in water with a hydrophobic medication within its boundary. To maintain redox molecule active for a long time, a redox-active-nanoparticle-dendrimer has been also created.

Volume of the positively charged dendrimer is quite huge. Because of this feature, dendrimers can be utilised in a variety of drug delivery systems to carry medicine directly to the infected region of the patient's body. When azo-benzene is covered by dendrimer, for example, it is photo-isomerized by very weak infrared photons. Several dendrimers are produced that absorb light and catch fire, and the energy released is used to excite the molecules within the dendrimers [19].

### 3.1.6 Physicochemical Characterization of Dendrimers

The physicochemical properties of dendrimers have been studied using a variety of analytical methods. Spectroscopic method, dynamic light scattering (DLS), rheology property microscopy, calorimetry, chromatography and electrophoretic characterisation are in a widest use. IR (infrared spectroscopy), ultraviolet (UV)-Visible, mass spectroscopy, Nuclear magnetic resonance (NMR), fluorescence, transmission electron microscopy (TEM), small angle type of X-ray scattering, small angle typed neutron scattering, size exclusion chromatography, laser light scattering, high performance liquid chromatography (HPLC), dielectric spectroscopy, atomic force microscopy (AFM), DSC (Differential scanning calorimetry) and temperature modulated calorimetry are also listed in literature for characterization of dendrimers [8, 20].

### 3.1.7 Merits and Demerits of Dendrimers

One of the benefits provided by dendrimers is that they have a lower toxicity. Their specificity helps in protection of healthy cells and tissues from the toxicity. They have greater bioavailability and better half-life due to altered renal clearance. Encapsulation of drugs in dendrimers reduce the premature enzymatic degradation of drugs. Dendrimers also maintain the levels of plasma drug concentration which help to achieve a controlled drug release in systemic circulation. Furthermore, dendrimers are also playing an important role as the multifunctional excipients and can increase medication solubility [7, 21].

Minimal degradation of dendrimers in the physiological environment is one of the downsides; this can lead to their more retention time in systemic circulation, producing serious adverse effects [7, 16].

## 3.2 Biosensors

Sensor technology is critical to the development of an extensive range of improved-form-factor medical devices having low cost. The biosensors have great promise in industrial processes since they are simple, scalable, and effective [22]. Biosensor is an analytical instrument detecting the changes in the biological processes followed by converting them in an electrical signal. The term biological element or substance, such as enzymes, tissues, microbes, cells, acids, and many others of such types can be used in a biological process. Biosensors are those medical devices which communicates information with the use of two primary component namely "biological" sensors and the second is an "electrical component" which will analyses and delivers signal. Biomolecules may be either made up of proteins, such as antibodies or enzymes, lectins or plant proteins, nucleic acids, or complex materials like bacteria and cell organelles. The constituents from the biological route can interact with the chemical component/drug under investigation, producing a biological reaction that may be converted into an electrical signal [23].

Disease detection, prevention, rehabilitation, patient health surveillance, and human health management have all been effective with biosensors. Bacteria, pathogens, and viral microbes may all be detected with biosensors. Physical activity and step trackers enable people to develop and strengthen healthier habits. These tools open opportunities for the individuals working in clinical science as well. These sensors detect substances without requiring the human body to collect blood [22].

Because of the ageing population and the noticeable increase in diabetes patients as well as chronic illnesses, the biosensor market is attracting a lot of attention. In 2016, the market was worth $15.6 billion, while in 2018, it was worth $12.8 billion [24]. According to recent estimates, the market will be worth "$26 billion" by the end of 2022. Growing demand of tiny devices for monitoring and diagnosis, necessitates material advancements [23].

Biosensors are useful in genotyping individuals in forensic applications, for gene expression investigations and nucleotide mutation detection. Their application is also wide in food safety analysis and environmental analysis in the form of analytical benefits [4]. Evaluation of the bio-sensing devices is done by the surface chemistry employed for immobilisation, micro or nanoscale manufacturing processes and transduction mechanisms.

### 3.2.1 Advantage of Dendrimer as Biosensor

Silanes, Polyaniline (PANI), PLL and alkanethiols are 2-Dimentional configurations of dendrimers. They all have limited probes accessibility, restricted loading efficiency and poor functionality due to irregular probe orientation. This results in responses variation on application of biosensor.

3-D dendrimer, gel-based units are a superior alternative. But they have problem with the entrapped-probe molecules because leaching occurs, which makes its application limited to create sensor systems. Supramolecular dendritic topologies have been showing tremendous capacity in design and development of the sensor systems.

Highly sensitive biosensors from the dendrimers are based on their properties such as uniform structure, nanoscale size, spherical shape, hydrophilicity, versatile design, varied composition with varied and dense functional groups, homogeneity, the presence of dendritic crevices etc [4].

These molecules have enormous intrinsic properties, such as uniform structural similarity, integrity, tunable composition and multidentate homogenous endpoints for sequential bioconjugation processes. These all introduce dendrimers as distinctive macromolecule and stable for an extensive range of applications [4, 25].

## 3.3 Dendrimers in Drug Delivery System

Dendrimers were first used as a therapeutic delivery in the late 1990s. As formulation and nano-construct, in two ways, dendrimers are used for drug delivery.

Medications by non-covalent bonding with a dendrimer are utilising in the formulation strategy. For nano-construct approach, covalent attachment of drugs on dendrimers are needed [6].

### 3.3.1 Dendrimers in Anti-Amyloid Activity

Dendrimer–protein conjugates offer promising biological uses. Proteins and dendrimers undergo biophysical changes as a result of complexation. They have the capacity to impact zeta-potential, protein secondary structure, the charge distributions on the surface, interactions between proteins, and so on. The modifications will open the new paths for the use of these traits for nanotheranostics and biomedicine.

Several therapeutic uses of dendrimers have risen due to interaction between dendrimer and protein. Consequently, the development of stable complexes keeps disordered proteins from aggregating, which is especially crucial in neurodegenerative disorders. Author explained the mechanism for interaction between protein and dendrimer, as well as driving force and type of binding, to understand the genesis of these traits and assess the efficacy of action in his article and also summarises dendrimer anti-amyloid activity and explores the influence of dendrimer architectures and extrinsic variables on anti-amyloid characteristics [26].

### 3.3.2 PAMAM Dendrimers in Improvisation of Drug Characteristics

PAMAM dendrimers are used to improve oral bioavailability, solubility, and stability of the drugs. Drug entrapment in dendrimers and drug release from them may be regulated by altering generations and modification in the surfaces of a dendrimer. They have also been reported to improve transdermal permeability and in targeted drug delivery [6].

Table 3.2 shows a summary of the numerous categories and subcategories used to classify the role of dendrimer with examples.

**Table 3.2** Drug properties improvised by dendrimers.

| Drug properties | Examples |
|---|---|
| Solubility | Indomethacin is a slightly acidic chemical encapsulated in several dendrimers. Research demonstrated that the G4-$NH_2$ shows highest solubility whereas G4-COOH dendrimer shows lowest one [21]. |
| Stability | Resveratrol molecule is a hydrophobic in nature and having poor stability. Its aqueous solubility and storage stability were improved by its encapsulation in dendrimers [55]. |
| Dissolution | For resveratrol, dendrimer-drug complexes demonstrated highly rapid and quick dissolution profile with its comparison to pure hydrophobic drug [55]. |
| Drug release | Indomethacin released slowly with G4-NH2 dendrimer but quickly with G4-COOH dendrimer due to chemical change [56].<br>Cisplatin (CP) release from dendrimers (DD) was faster with a larger CP/DD molar ratio formulation. Reduction in CP/DD molar ratio results in lower drug release on physical modification [57]. |
| Bioavailability | Following oral administration, dendrimer–simvastatin (SMV) complex formulations demonstrated higher oral bioavailability in comparison with pure drug suspension formulation [58]. |
| Multiple drug delivery | Ramipril (RAPL) and hydrochlorothiazide (HCTZ) solubility was affected by dendrimer concentrations and dendrimer solution pH. RAPL and HCTZ combination Dendrimers exhibited quicker solubility in pH 1.2 (simulated gastric fluid) & pH 7.0 (USP Dissolution media) [59]. |

### 3.3.3 Dendrimers in Drug Administration

Bonding propranolol along with lauroyl-3G PAMAM and such 3G dendrimer resulted in the prodrug production. They examined the transport across endocytosis-mediated transepithelial layers to assess the impact. Endocytosis is reduced by the conjugation of propranolol with lauroyl-G3 dendrimers [27].

The researchers demonstrated that conjugating ketoprofen with PAMAM dendrimers corrected a local and systemic issue of the gastrointestinal tract produced by ketoprofen using an acid-induced writhing paradigm in mice [28].

PAMAM dendrimer carriers enhanced the solubility of the anticancer medication 'Camptothecin'. The conjugation of PAMAM dendrimers with the medication SN-38 through glycine or -alanine demonstrated stable in gastric settings and reduced premature drug release in acidic and intestinal environments. Furthermore, owing of its focused action, this combination has less cytotoxicity and higher dosage accumulation at locations [29].

### 3.3.4 Dendrimers in Targeted Drug Delivery

Conjugation of folic acid and PAMAM dendrimer (G5) for targeting of methotrexate revealed receptor-mediated drug delivery with delay in drug release and excellent selectivity for folate receptors of KB cells [4, 8, 30].

Conjugation of an apoptotic sensor (PhiPhiLux G1 D2) to folic acid linked PAMAM dendrimers resulted in five times increment in the fluorescence. This indicates good efficiency in cell-killing and effective drug delivery [31].

Investigation of the amine terminated PAMAM dendrimers, such as $^{14}$C-mannitol dendrimers proved its permeability and the lactate dehydrogenase enzyme leakage which uses paracellular and endocytosis pathways to pass biological membranes [32].

### 3.3.5 Dendrimers in Gene Delivery

Gene therapy can be defined as "group of approaches that allow the RNA or DNA delivery to a target cell in order to modify the production of particular changed proteins, hence correcting cellular defects expressed in the phenotypic and reversing any biological condition that arises". For anticancer treatment, covalent attachment of a DNA molecule and PAMAM dendrimers is reported. The only disadvantage of this treatment is the difficulty in limiting medication release from the dendritic core [27].

Non-Viral and Viral approaches have been used commonly in enabling successful genetic material transfer to target cells. Pandita et al. created gene delivery vectors based on dendrimers, "proton-sponge" action and their cationic nature [33].

Because of their targeting properties, G5 and G6 PAMAM dendrimers were conjugated to RGD domain in variable quantity of peptides to form arginine-glycine-aspartic (RGD) nanoclusters [8, 34].

Different siRNA molecules are used in RNAi therapies to disrupt signalling pathways involved in cell growth and anti-apoptosis. A peptide hormone namely "luteinizing hormone releasing hormone (LHRH)" was effectively conjugated on the quaternate dendrimer G4.0QPAMAM-OH surface in order to transport anti-$BCl_2$ siRNA to the cancer cells of ovary by means of targeting "LHRH" receptor [15].

### 3.3.6 Dendrimers in Anticancer Therapy

Researchers demonstrated cisplatin's lower cytotoxicity and increased accumulation in solid tumours by using degradable connections between cisplatin and dendrimers. The researchers also demonstrated that administering cisplatin-loaded carboxylate-terminated PAMAM dendrimers to an animal with a malignant tumour via intravenous, oral, parenteral, and topical routes resulted in efficient tumour growth suppression [27].

### 3.3.7 Dendrimers in Antimicrobial Therapy

The literature discussed how the salts of silver were encapsulated within PAMAM dendrimers having regulated rate of release, also showed antibacterial action against the *Pseudomonas aeruginosa*, *Staphylococcus aureus*, and *Escherichia-coli* bacteria [35].

### 3.3.8 Dendrimers in Vaccine Delivery

Dendrimers serve as an adjuvant as well as a carrier system for vaccination antigens. The accessible surface groups of a dendrimer facilitate coupling to various functional moieties (e.g., peptides, antibodies, proteins, etc.), which is further aided by the dendrimer's customised charge of surface and the size. This flexibility enables the development of molecularly precise vaccines with highly particular and predictable features, and when combined with the dendrimers immune stimulating (adjuvating) property, making dendrimers an appealing substrate for biomedical applications, including vaccines [36].

Dendrimers' excellent properties make them effective immune-stimulating agents for vaccine administration against infectious illnesses such as HIV1, TB, small pox, and many respiratory and intestinal infectious disorders. Dendrimers are effective because of their multivalent molecular structure, which allows them to bind with antigens.

### 3.3.9 Dendrimers as Diagnostic Tool

In the initial 1990s, Lauterbur *et al.* demonstrated first diagnostic imaging applications, *in-vivo* using MRI contrast agents based on dendrimers. Reagents have been made using the polyamidoamine version of the starburst typed dendrimers that have had free amines which are conjugated to the chelator: "2-(4-isothiocyanatobenzyl)-6-methyl-diethylenetriaminepentaacetic acid". These dendrimer-based compounds improve traditional MR pictures as well as 3-Dimentional time of flight "MR angiograms" [10].

Dendrimers diagnose depending on their size. Dendrimers with a diameter of around 5 nm are utilised as an MRI contrast agent for lymphatic system diagnostics. Dendrimers that are multivalent and highly branched are employed in tissue engineering as a biological application [6].

A good MRI contrast agent should be biocompatible, less poisonous, and more reflexive. Lower molecular weight MRI contrast agents have the ability to rapidly diffuse through blood vessels and have excellent excretion from bodily fluids. In animal models, macromolecules such as Gd (III) conjugates to biomedical polymers demonstrated to be remarkable contrast agents for imagination of cancer and blood pool. However, they have slower excretion rates, which raises the possibility of toxicity during metabolism [27].

### 3.3.10 Dendrimers in Tissue Engineering

Tissue regeneration is feasible when natural healing processes are supplemented by scaffold material. 2-Dimensional surface-like scaffolds are required for cell development because they allow nutrients to enter while restricting or exporting waste from the cells. Natural scaffold materials include collagen, proteins, polysaccharides, and glycoproteins, whereas synthetic scaffold materials may include "poly (glycolic acid) (PGA), poly (caprolactone) (PCL), poly (ethylene glycol) (PEG) and poly (lactic acid) (PLA)" [37].

## 3.4 Dendrimers as Sensors

Multi-layered construction of a dendrimer is required for the biosensors. In this context, dendrimers can play important roles in biosensor construction by the following steps;

a. Formation of surface monolayers as supports for immobilisation of protein.
b. Deposition of Layer by Layer (LbL) multilayer films made up of proteins and dendrimers.
c. Encapsulation of metal particles by covalent bonding and electron transfer mediators.

A biosensing interface is constructed by Au surface deposited via multifaceted enzyme film. Periodate-oxidized glucose oxidase (GOx) and G4 PAMAM dendrimers were

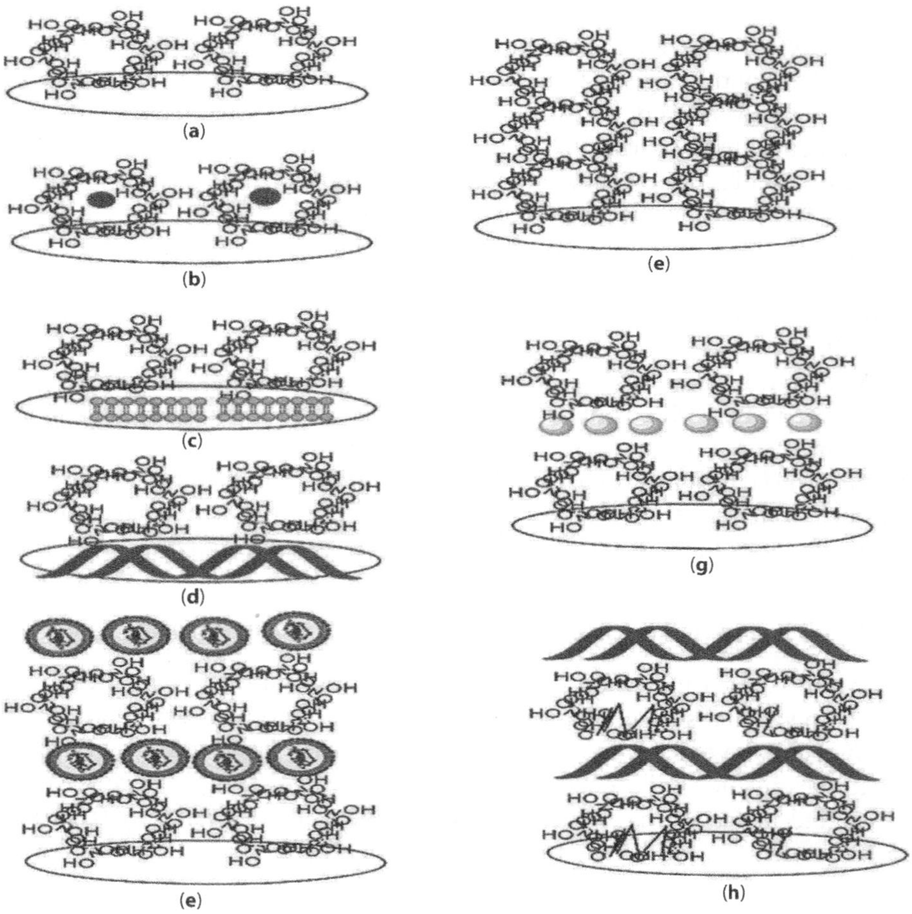


**Figure 3.4** Dendrimers assemblies on the surface of electrode: (a) "monolayer", (b) "monolayer metal nanoparticle-deposited monolayer", (c) "a nanomaterial covered dendrimers" (d) "a on polymer-coated surface dendrimer layer" (e) "dendrimer-dendrimer" LbL assembly, (f) "dendrimer-protein LbL", (g) "dendrimer-nanoparticles LbL", (h) "dendrimer-polymer" bilayers LbL.

alternatively arranged layer-by-layer to create the film. From Gold electrodes, the cyclic voltammograms produced and tailored with GOx–multi-layered dendrimers demonstrated that bio-electrocatalytic response reflects due to the large amount of immobilized active enzymes available on gold electrode surface with respect to the deposited bilayers.

During the multilayer-forming processes the reporting of "per GOx/dendrimer bilayer", active enzyme was assessed using the voltametric signals, indicating that the multilayer is built in an organised way spatially. In addition, the ellipsometry tests revealed that linear increasing in by increasing the thickness of the film, validating the creation of the hypothesised multi-layered structure. The sensitivity of the E5D5 electrode was 14.7 AâmM-1 glucoseâcm-2, which was consistent on calibration over 20 days. The quite simple suggested technique could be relevant to the design thickness and sensitivity controllable bio-sensing interfaces which are made up of single and multi-enzymes [38, 39].

Electrode surface was made over dendrimer based 3-D layer arrangements by many assembly processes such aspiring of polymers and nanomaterials with molecular arrangement of its monolayers and hybrid layers. Layer-by-layer assembling of these molecules were always combined with additional dendrimers, polymers, nanomaterials and proteins [39]. Figure 3.4 illustrates some of these conceivable layouts.

### 3.4.1 Sensor Performance and Dendrimer Characteristics

Dendrimers' fascinating features, including as globular shape, tunable size, hydrophilicity, high surface functionality, as well as chemical and mechanical resilience, address them as an attractive matrix for biomolecule immobilisation. These features improve the biosensors' target capture ability, stability, sensitivity, specificity, and recycling ability. This section describes the usual structural characteristics of dendrimers as well as their impact on sensor performance [4, 40]. Table 3.3 depicts the characteristics of dendrimers as biosensors.

**Table 3.3** Outlook shows the characteristics of dendrimers.

| | |
|---|---|
| The size and shape of dendrimers | • Thickness of the interface<br>• Shape and non-specific binding |
| Composition and high functional density | • High ligand immobilization<br>• Stability and regeneration of biosensor matrix |
| Homogeneous end group distribution | • Homogeneous end group distribution |
| Hydrophilicity and pendent group type | • Type of pendent group and non-specific binding<br>• Ease of bio conjugation<br>• Multilayering |
| Miscellaneous | • Dendrimeric crevices<br>• Flexibility of dendritic chains |

### 3.4.2 Dendrimer-Based Electrochemical Sensors

For electrochemical detection, a conductive surface is ideal. A dendrimer is not recognized as a conductor, but their conductivity can be improved by linking of colloids like functional groups to their structures. "Conductive dendrimers", which have been widely employed in the electrochemical-sensing may include "adequate ferrocenyl ethered dendrimers (Fc-D)" and a "AuNPs nano composites dendrimer" [34, 41]. In general, electrode chemical biosensors are integrated analytical devices for biological recognition. They are converting a biological recognition process into a usable output signal due to direct contact with or integrated with an electrochemical transduction component [42]. As per biorecognition principle, these biosensors may be divided into two broad categories;

(1) *Catalytic biosensors* that is a biorecognition element having capability of both recognising and catalysing the target analyte transformation.

(2) *Affinity biosensors* that is a biorecognition element having capability to identify the target analyte via the affinity-based process but does not result in chemical change of the analyte.

Charge-transfer interactions, hydrogen-bonding interactions, biological interactions, and many more techniques as LbL assembly has got a lot of consideration in construction of "electrochemical-biosensor" [3]. LbL is an extremely helpful method for assembling a varied range of organic and inorganic components such as dendrimers, proteins, DNA, viruses, nanoparticles, and others [39]. The DNA microarray method detects sequence-selective DNA hybridization using electrochemical, optical, and piezo-electric transduction techniques [15]. Dendrimers are homogeneous and multibranched polymers with 3-D structures that are commonly used in electrochemical immunosensor technology as "soft" nanomaterials [24]. Agriculture, Clinical chemistry, environmental monitoring, forensic analysis, food safety, research laboratories, industrial processes, biotechnology, biodefense and homeland security are among areas where it is widely used.

Electrochemical biosensor technology has had a significant influence on biomedical fields, for its application in clinical area and diagnosis at home for selected disorders, as a real time and remote health monitoring [22]. The biorecognition and the electrochemical detection will occur at the sample solution interface and the surface of the electrode where biological recognition element is placed. As a result, the selection of an appropriate electrode surface is critical to the functioning of electrochemical biosensors.

Dendrimers are utilised to make electrodes with desirable surface properties, such as high and low molecular weight chemicals monolayers coating, natural and synthetic polymers functionalization, and micronized and nanosized materials modification.

#### 3.4.2.1 *Manufacturing of Electrochemical Biosensor*

Electrodeposition of first-generation PPI was done at scanning rate of 50 mV/s with 10 cycles of −100 to +1100 mV on the "glassy carbon electrode" using cyclic voltammetry. To produce the biosensor (GCE/G1PPI/ssDNA), a single-stranded, probe DNA (21 mer) (2 M) was electrostatically attracted to the PPI layer by dipping in a 500 L solution. Deeping process was continued at 25°C for 3 hours. Hybridizations were performed using target DNA solutions and blank buffer solutions at 38°C containing complementary DNA, non-complementary DNA, and three bases divergence DNA strands. Every stage of synthesis of biosensor using $(Fe(CN)_6)^{3-/4-}$ as a "redox probe" was investigated using an "Electrochemical Impedance spectroscopy" (impedance, phase angle plots and nyquist). The biosensor had been likewise biased against non-complementary, mismatched target DNA [43].

In biosensors for quick and sensitive detection of cholesterol, reduced Graphene–Polyamidoamine-Ferrocene (rGO-PAMAM-Fc) macromolecule were mixed with gold nanoparticles (Au-NPs). To create the ChOx (cholesterol oxidase) and ChEt(cholesterol esterase) containing Au NPs of rGO-PAMAM-Fc biosensor, both electrode surfaces were coated with ChOx and ChEt. These biosensors exhibit exceptional stability, selectivity, and repeatability [41].

Dendrimers are used as the electrodes modifying agents that act as the electrochemical immunosensors for a variety of reasons, including dendrimers having their perimeter, which can help them adhere to the electrode surface. It can also be used as connecting sites for antibody immobilisation via covalent bonding or non-covalent contact. These hyperbranched nanosized macromolecules are having a permeable spherical shape, resulting in minimal barrier effects for electroactive material transport at the electrode surface.

The presence of chemical group of high density at the surface, shape and composition of dendrimers and their relative conformational rigidity allow them to be used for the controlled design and multi-layered arrangement on various surfaces as respected building blocks.

By properly encapsulating or covalently binding tiny electron transfer mediators and metal nanoparticles, dendrimers may be tailored to increase electrocatalytic and charge transfer activities at the electrode surface. Due to their hydrophilic surfaces (PAMAM and PPI), certain dendrimers, display great compatibility with antibodies like proteins, allowing maintenance of the physiologically active conformation of bioreceptors upon immobilization.

Most importantly, dendrimers may be combined with a wide range of polymers and nanomaterials to create unique hybrid materials for electro-analytical applications.

### 3.4.3 Enzymatic Biosensors

PAMAM dendrimers were coupled with Cytochrome P450s (heme-thiolate enzyme proteins), which are responsible for the inactivation of nearly all xenobiotics. Layering approach and modification of modified gold nanoparticles were used to create a dendrimer-based CYP3A4 biosensor. Electrostatic attraction between oppositely charged moieties was used to layer it. In the construction of biosensors, gold nanoparticles increased the conductivity of PAMAM dendrimers. This biosensor detects caffeine and is always maintained at $4^0$C until it is needed to preserve its efficiency [44, 45].

### 3.4.4 Optic Biosensors

The immobilisation of bioreceptors by a dendrimeric biointerface increases the overall performance of fiber-optic biosensors based on evanescent wave absorbance. Fourth generation poly (amidoamine) dendrimer immobilisation on amine functionalized sensor substrates was compared to substrates silanized conventionally. This amine group of the silanized surface was demonstrated in dendrimer matrix with 2.4 times magnification using X-ray photoelectron spectroscopy and Fluorescein isothiocyanate-based amine assay.

It was 1.3 fold surface area of dendrimer matrix over silane matrix explained by AFM-based topographic analysis [25, 46].

### 3.4.5 QCM Biosensors

The QCM (Quartz Crystal Microbalance) technology detects mass at the quartz crystal's surface by monitoring variations in crystal oscillation frequency. The antigens-antibodies interaction immobilised on a crystal surface were responsible for intensification in crystal growth and reduction in frequency of oscillation.

The literature described a QCM immunosensor for IgG detection in humans.

Anti-IgG were adsorbed on "G4-PAMAM-dendrimers" or Protein A-coated dendrimers with layers of 1-hexadecanethiol. Their work provided SAMs constructed of HDT conjugated G4-PAMAM dendrimer, resulting in the conclusion that it is feasible to build immunosensors with a low detection limit for human IgG detection [44, 46].

### 3.4.6 Dendrimer-Based Glucose Biosensors

A glucose biosensor was created using "chitosan-dendrimers (CHD) and bioactive-polyglycerol (PGLD)" [45]. To create dendrimers with glucose sensing characteristics, bio-conjugation of both was carried with the enzyme (GOx). Because of their strong capacity to induce electron-transfer processes involving "polyaniline nanotubes (PANINTs) and Gox" have been utilised in the form of electron mediators. PANINTs were loaded with PGLD-GOx and CHD-GOx during electrochemical polymerization of aniline template. Conjugated "CHD-GOx/PANINT" and "PGLD-GOx/PANINT" biosensors at lower potential (+100 mV) demonstrate a significant and consistent amperometric response to glucose. These biosensors performed well in the human blood of glucose concentration range. With respect to glucose sensitivity, both biosensors had a good linearity range (0.02 - 10 mM). But "PGLD-GOx/PANINT" was seen greater sensitive in the range of 10.41 versus 7.04 nA.mM$^{-1}$ [47].

### 3.4.7 Xenon Biosensor

It was amplification of the 129Xe NMR signal using Dendrimer-Cage new supramolecular structures. Numerous cages were enclosed carefully by selecting certain dendrimers, enabling water-solubility to the "functionalized" 129Xe and preventing diastereomer interference without extra synthetic processes or cage covalent attachment to dendrimers [48].

### 3.4.8 Penicillin Biosensors

The Literature revealed that alternate adsorption of carboxy group conjugated carbon nanotubes (single-walled) and PAMAM dendrimer onto field effect transistor (FET) gate surface by LbL films were employed to build penicillin biosensors [39].

**Table 3.4** Nanomaterials in biosensors.

| Type of nanomaterial | Analyte used | Transducer attached | Linearity range | Limit of detection | References |
|---|---|---|---|---|---|
| Au Nanoparticles | Uranyl<br>$Pb^{+2}$<br>*E. coli* | Electrochemical<br>Fluorescent | 2.4-480 µg/L<br>50 nm-4 µm<br>10-106 CFU/ml | 0.3 µg/L<br>16.7 nm<br>15 CFU/ml | [60–62] |
| Ag Nanoparticles | Glucose<br>$H_2O_2$<br>$Fe^{+2}$ | Colorimetry | 50 –750 nm<br>1500–3000 nm<br>0.1–90µm | 32 nm<br>290 nm<br>540 nm | [63] |
| Pt Nanoparticles | Epinephrine | Voltammetry | 0.99–21.3×$10^{-5}$ mole/Litre | 2.93×$10^{-4}$ mole/Litre | [64] |
| Pt-$Fe_3O_4$@C nanocomposites | Sarcocine | Amperometry | 500 nm–60 µm | 430 nm | [65] |
| Ni/Cu MOF | Glucose | Field-effect Transistor (FET) | 1µM–20mM | 0.51 µM | [66] |
| $Co_3O_4$-Au | miRNA-141 | Photo-electrical chemical | 1pM–50nM | 0.2 pM | [67] |
| MnO2 NFs | Salmonella | Impedimetric | 3.0–3.0×$10^6$ | 19 CFU/mL | [68] |
| NFs-ZnO | Beta-amyloid | Optic | 2×$10^{-3}$– 2×$10^{-2}$ mL | 2.76 ×$10^{-3}$ mg | [69] |
| Transistor from CNT thin-film | D.N.A | Thin-film transistor | 16×$10^{-5}$– 5 µmole/Ltr | 880 ng/Ltr | [70] |

### 3.4.9 Nanomaterials in Biosensors

Biosensors based on dendrimers improve stability, analytical sensitivity and reproducibility without reducing specific interactions. From literature, it was found that a "PAMAM-dendrimer" integrated tapered-optical fibre typed sensor can be used to detect dengue. The detection limit and the resolution of sensors were 1 pM and 19.53 nM, respectively in nanomaterial concentration ranges between 0.1 pM till 1 M. Table 3.4 shows a list of nanomaterials that are used in biosensors development [49].

## 3.5 Conclusion

Dendrimers exist as hyperbranched, multivalent macromolecules. Their structure properties draw remarkable advantages as sensor development. Mainly "Layer-by-Layer" (LbL) is widely applied biosensors manufacturing. Drug characteristics such as stability, solubility, dissolution, bioavailability is improved by encapsulating them into dendrimers core. Due to these properties, dendrimers are widely used in gene delivery, vaccine delivery, anticancer and antibacterial drug delivery and so on. Catalytic and affinity both types of sensors can be made from dendrimers that based on biorecognition principle. Because of their effectiveness in biorecognition and self-assembling nature, they are used to form electrochemical sensors, QCM sensors, xenon sensors, optic sensors, enzymatic sensors, glucoses biosensors and penicillin sensors.

## References

1. Abbasi, E., Fekri, A.S., Akbarzadeh, A., Milani, M., Nasrabadi, H.T., Woo Joo, S. *et al.*, Dendrimers: Synthesis, applications, and properties. *Nanoscale Res. Lett.*, 9, 1, 247–257, 2014.
2. Narayan, R.J. and Verma, N., Nanomaterials as implantable sensors, in: *Mater. Chem. Sens.*, pp. 123–139, 2017.
3. Govindhan, M., Adhikari, B.R., Chen, A., Nanomaterials-based electrochemical detection of chemical contaminants. *RSC Adv.*, 4, 63741–63760, 2014.
4. Satija, J., Sai, V.V.R., Mukherji, S., Dendrimers in biosensors: Concept and applications. *J. Mater. Chem.*, 21, 38, 14367–14386, 2011.
5. Safari, J. and Zarnegar, Z., Advanced drug delivery systems: Nanotechnology of health design A review. *J. Saudi Chem. Soc.*, 18, 2, 85–99, Elsevier 2014.
6. Chauhan, A.S., Dendrimers for drug delivery. *Molecules*, 23, 4, 2018.
7. González, C.D., Fernández, R.N., Solís, V.G., Santamaría, M.M., Chavarría, R.M., Matarrita, B.D., Rojas, S.M., Madrigal, R.G., Dendrimers and their applications. *J. Drug Deliv. Ther.*, 12, 1-S, 151–158, 2022.
8. Madaan, K., Kumar, S., Poonia, N., Lather, V., Pandita, D., Dendrimers in drug delivery and targeting: Drug-dendrimer interactions and toxicity issues. *J. Pharm. Bioallied Sci.*, 6, 139, 2014.
9. Karatas, O., Keyikoglu, R., Atalay, G.N., Vatanpour, V., Khataee, A., A review on dendrimers in preparation and modification of membranes: Progress, applications, and challenges. *Mater. Today Chem.*, 23, 100683, 2022.
10. Najafi, F., Salami-Kalajahi, M., Roghani-Mamaqani, H., A review on synthesis and applications of dendrimers. *J. Iran. Chem. Soc.*, 18, 503–517, 2020.
11. Prajapati, S.K., Maurya, S.D., Das, M.K., Tilak, V.K., Verma, K.K., Dhakar, R.C., Dendrimers in drug delivery, diagnosis and therapy: Basics and potential application. *J. Drug Deliv. Ther.*, 6, 67–92, 2016.
12. Caminade, A.M., Fruchon, S., Turrin, C.O., Poupot, M., Ouali, A., Maraval, A., Garzoni, M., Maly, M., Furer, V., Kovalenko, V., Majoral, J.P., Pavan, G.M., Poupot, R., The key role of the scaffold on the efficiency of dendrimer nanodrugs. *Nat. Commun.*, 6, 7722, 1–11, 2015.
13. Jain, K., Jain, N.K., Kesharwani, P., Chapter 7: Types of dendrimers, in: *Dendrimer-Based Nanotherapeutics*, P. Kesharwani, (Ed.), pp. 95–123, Academic Press, Cambridge, United States, 2021.
14. Wang, Y., Huang, D., Wang, X., Yang, F., Shen, H., Wu, D., Fabrication of zwitterionic and pH-responsive polyacetal dendrimers for anticancer drug delivery. *Biomater. Sci.*, 7, 3238–3248, 2019.
15. Abedi-Gaballu, F., Dehghan, G., Ghaffari, M., Yekta, R., Abbaspour-Ravasjani, S., Baradaran, B., Dolatabadi, J., Hamblin, M., PAMAM dendrimers as efficient drug and gene delivery nanosystems for cancer therapy. *Appl. Mater.*, 12, 177, 2018.
16. Gorain, B., Choudhury, H., Pandey, M., Mohammed, A.M.C.I., Singh, B., Gupta, U., Kesharwani, P., Chapter 10: Dendrimers as effective carriers for the treatment of brain tumor, in: *Nanotechnology-Based Targeted Drug Delivery Systems for Brain Tumors*, P. Kesharwani, and G. Umesh, (Eds.), pp. 267–305, Academic Press, Cambridge, United States, 2018.
17. Gorzkiewicz, M., Kopeć, O., Janaszewska, A., Konopka, M., Pędziwiatr-Werbicka, E., Tarasenko, I.I., Bezrodnyi, V.V., Neelov, I.M., Klajnert-Maculewicz, B., Poly(lysine) dendrimers

form complexes with siRNA and provide its efficient uptake by myeloid cells: Model studies for therapeutic nucleic acid delivery. *Int. J. Mol. Sci.*, 21, 3138, 2020.
18. Idris, A.O., Mamba, B., Feleni, U., Poly (propylene imine) dendrimer: A potential nanomaterial for electrochemical application. *Mater. Chem. Phys.*, 244, 122641, 2020.
19. Samad, A., Alam, M.I., Saxena, K., Dendrimers: A class of polymer in the nanotechnology for drug delivery. *Curr. Pharm. Des.*, 15, 2958–2969, 2009.
20. Caminade, A.M., Laurent, R., Majoral, J.P., Characterization of dendrimers. *Adv. Drug Deliv. Rev.*, 57, 2130–2146, 2005.
21. Svenson, S. and Chauhan, A.S., Dendrimers for enhanced drug solubilization. *Nanomedicine (Lond)*, 3, 679–702, 2008.
22. Haleem, A., Javaid, M., Singh, R.P., Suman, R., Rab, S., Biosensors applications in medical field: A brief review. *Sens. Int.*, 2, 100100, 2021.
23. Jeff, T., *Biosensors benefiting from newly created protein polymer films*, p. 20, BoydTechnologies, Boyd Biomedical, Inc., Massachusetts, The Northeastern United States, 2018.
24. Sánchez, A., Villalonga, A., Martínez-García, G., Parrado, C., Villalonga, R., Dendrimers as soft nanomaterials for electrochemical immunosensors. *Nanomater. (Basel, Switzerland)*, 9, 12, 2019.
25. Satija, J., Karunakaran, B., Mukherji, S., A dendrimer matrix for performance enhancement of evanescent wave absorption-based fiber-optic biosensors. *RSC Adv.*, 4, 15841–15848, 2014.
26. Sorokina, S.A. and Shifrina, Z.B., Dendrimers as antiamyloid agents. *Pharmaceutics*, 14, 760, 2022.
27. Noriega-Luna, B., Godínez, L.A., Rodríguez, F.J., Rodríguez, A., Zaldívar-Lelo de, L.G., Sosa-Ferreyra, C.F., Mercado-Curiel, R.F., Manríquez, J., Bustos, E., Applications of dendrimers in drug delivery agents, diagnosis, therapy, and detection. *J. Nanomater.*, 2014, 39, 2014.
28. Yiyun, C., Tongwen, X., Rongqiang, F., Polyamidoamine dendrimers used as solubility enhancers of ketoprofen. *Eur. J. Med. Chem.*, 40, 1390–1393, 2005.
29. Sadekar, S., Thiagarajan, G., Bartlett, K., Hubbard, D., Ray, A., McGill, L.D., Ghandehari, H., Poly (amido amine) dendrimers as absorption enhancers for oral delivery of camptothecin. *Int. J. Pharm.*, 456, 175, 2013.
30. Zhang, M., Zhu, J., Zheng, Y., Guo, R., Wang, S., Mignani, S., Caminade, A.M., Majoral, J.P., Shi, X., Doxorubicin-conjugated PAMAM dendrimers for pH-responsive drug release and folic acid-targeted cancer therapy. *Pharm.*, 10, 162, 2018.
31. Myc, A., Majoros, I.J., Thomas, T.P., Baker, J.R., Dendrimer-based targeted delivery of an apoptotic sensor in cancer cells. *Biomacromolecules*, 8, 13–18, 2007.
32. El-Sayed, M., Ginski, M., Rhodes, C.A., Ghandehari, H., Influence of surface chemistry of poly (amidoamine) dendrimers on Caco-2 cell monolayers. *J. Bioact. Compat. Polym.*, 18, 7–22, 2016.
33. Santos, J.L., Oliveira, H., Pandita, D., Rodrigues, J., Pêgo, A.P., Granja, P.L., Tomás, H., Functionalization of poly(amidoamine) dendrimers with hydrophobic chains for improved gene delivery in mesenchymal stem cells. *J. Control. Release*, 144, 55–64, 2010.
34. Kong, L., Alves, C.S., Hou, W., Qiu, J., Möhwald, H., Tomás, H., Shi, X., RGD peptide-modified dendrimer-entrapped gold nanoparticles enable highly efficient and specific gene delivery to stem cells. *ACS Appl. Mater. Interfaces*, 7, 4833–4843, 2015.
35. Winnicka, K., Wroblewska, M., Wieczorek, P., Sacha, P.T., Tryniszewska, E.A., The effect of PAMAM dendrimers on the antibacterial activity of antibiotics with different water solubility. *Molecules*, 18, 8607, 2013.
36. Chowdhury, S., Toth, I., Stephenson, R.J., Dendrimers in vaccine delivery: Recent progress and advances. *Biomaterials*, 280, 121303, 2022.

37. Gorain, B., Tekade, M., Kesharwani, P., Iyer, A.K., Kalia, K., Tekade, R.K., The use of nanoscaffolds and dendrimers in tissue engineering. *Drug Discovery Today*, 22, 652–664, 2017.
38. Yoon, H.C. and Kim, H.S., Multilayered assembly of dendrimers with enzymes on gold: Thickness-controlled biosensing interface. *Anal. Chem.*, 118, 922–926, 1996.
39. Sato, K. and Anzai, J.I., Dendrimers in layer-by-layer assemblies: Synthesis and applications. *Molecules*, 18, 7, 8440–8460, 2013.
40. B.D. Spangler and C.W. Spangler, Biosensors utilizing dendrimer-immobilized ligands and there use thereof. Google Patents WO2005016115A2, 2006, https://patents.google.com/patent/US7138121B2/en.
41. Zhu, J., Ye, Z., Fan, X., Wang, H., Wang, Z., Chen, B., A highly sensitive biosensor based on au NPs/rGo-PaMaM-Fc nanomaterials for detection of cholesterol. *Int. J. Nanomed.*, 14, 835–849, 2019.
42. Thakare, S., Shaikh, A., Bodas, D., Gajbhiye, V., Application of dendrimer-based nanosensors in immunodiagnosis. *Colloids Surf. B Biointerfaces*, 209, 112174, 2022.
43. Arotiba, O.A., Baker, P.G., Mamba, B.B., Iwuoha, E.I., The application of electrodeposited poly(propylene imine) dendrimer as an immobilisation layer in a simple electrochemical DNA biosensor. *Int. J. Electrochem. Sci.*, 6, 673–683, 2011.
44. Müller, M., Agarwal, N., Kim, J., A cytochrome P450 3A4 biosensor based on generation 4.0 PAMAM dendrimers for the detection of caffeine. *Biosensors (Basel)*, 6, 3, 44, 2016.
45. Siriviriyanun, A., Imae, T., Nagatani, N., Electrochemical biosensors for biocontaminant detection consisting of carbon nanotubes, platinum nanoparticles, dendrimers, and enzymes. *Anal. Biochem.*, 443, 169–171, 2013.
46. Erdem, A., Eksin, E., Kesici, E., Yarali, E., Dendrimers integrated biosensors for healthcare applications, in: *Nanotechnol. Biosens.*, pp. 307–317, 2018.
47. Santos, A.N., Soares, D.A.W., De Queiroz, A.A.A., Low potentialstable glucose detection at dendrimers modified polyaniline nanotubes. *Mater. Res.*, 13, 5–10, 2010.
48. Mynar, J.L., Lowery, T.J., Wemmer, D.E., Pines, A., Fré, J.M.J., Xenon biosensor amplification via dendrimer-cage supramolecular constructs. *J. Am. Chem. Soc.*, 128, 6334–6335, 2006.
49. Naresh, V. and Lee, N., A review on biosensors and recent development of nanostructured materials-enabled biosensors. *Sensors*, 21, 4, 1109, 2021.
50. Baker, L.A., Zamborini, F.P., Sun, L., Crooks, R.M., Dendrimer-mediated adhesion between vapor-deposited Au and glass or Si wafers. *Anal. Chem.*, 71, 4403–4406, 1999.
51. Stevelmans, S., Van Hest, J.C.M., Jansen, J.F.G.A., Van Boxtel, D.A.F.J., De Brabander-van Den Berg, E.M.M., Meijer, E.W., Synthesis, characterization, and guest-host properties of inverted unimolecular dendritic micelles. *J. Am. Chem. Soc.*, 118, 31, 7398–7399, 1996.
52. Frey, B.L., Jordan, C.E., Corn, R.M., Komguth, S., Control of the specific adsorption of proteins onto gold surfaces with poly(L-lysine) monolayers. *Anal. Chem.*, 67, 4452–4457, 2002.
53. Puniredd, S.R. and Srinivasan, M.P., Covalent molecular assembly of multilayer dendrimer ultrathin films in supercritical medium. *J. Colloid Interface Sci.*, 306, 118–127, 2007.
54. Satija, J., Shukla, G.M., Mukherji, S., Potential of dendrimeric architecture in surface plasmon resonance biosensor. *Int. Conf. Syst. Med. Biol. ICSMB 2010 – Proc.*, pp. 86–89, 2010.
55. C.A. Singh, N.E. Andrew, G.A. Henry, Compositions comprising a dendrimer-resveratrol complex and methods for making and using the same. Google Patents, US20160206572A1, 2018, https://patents.google.com/patent/US20160206572A1/en.
56. Chauhan, A., Svenson, S., Reyna, L., Tomalia, D., Solubility enhancement propensity of PAMAM nano-constructs. *Mater. Matters Nanomater.*, 2, 24–26, 2007.
57. Kulhari, H., Deep, P., Singh, M.K., Chauhan, A.S., Optimization of carboxylate-terminated poly(amidoamine) dendrimer-mediated cisplatin formulation. *Drug Dev. Ind. Pharm.*, 41, 2, 232–238, 2015.

58. Kulhari, H., Kulhari, D.P., Prajapati, S.K., Chauhan, A.S., Pharmacokinetic and pharmacodynamic studies of poly(amidoamine) dendrimer-based simvastatin oral formulations for the treatment of hypercholesterolemia. *Mol. Pharm.*, 10, 7, 2528–2533, 2013.
59. Singh, M.K., Deep, P., Kulhari, H., Jain, S.K., Sistla, R., Chauhan, A.S., Poly (amidoamine) dendrimer-mediated hybrid formulation for combination therapy of ramipril and hydrochlorothiazide. *Eur. J. Pharm. Sci.*, 96, 84–92, 2017.
60. Shi, S., Wu, H., Zhang, L., Wang, S., Xiong, P., Qin, Z., Chu, M., Liao, J., Gold nanoparticles based electrochemical sensor for sensitive detection of uranyl in natural water. *J. Electroanal. Chem.*, 880, 114884, 2021.
61. Niu, X., Zhong, Y., Chen, R., Wang, F., Liu, Y., Luo, D., A 'turn-on' fluorescence sensor for Pb2+ detection based on graphene quantum dots and gold nanoparticles. *Sens. Actuators B Chem.*, 255, 1577–1581, 2018.
62. Vu, Q.K., Tran, Q.H., Vu, N.P., Anh, Y.L., Dang, T.T.L., Tonezzer, M., Ngyyen, T.H.H., A label-free electrochemical biosensor based on screen-printed electrodes modified with gold nanoparticles for quick detection of bacterial pathogens. *Mater. Today Commun.*, 26, 101726, 2021.
63. Basiri, S., Mehdinia, A., Jabbari, A., A sensitive triple colorimetric sensor based on plasmonic response quenching of green synthesized silver nanoparticles for determination of Fe2+, hydrogen peroxide, and glucose. *Colloids Surf. A Physicochem. Eng. Asp.*, 545, 138–146, 2018.
64. Brondani, D., Scheeren, C.W., Dupont, J., Vieira, I.C., Biosensor based on platinum nanoparticles dispersed in ionic liquid and laccase for determination of adrenaline. *Sens. Actuators B Chem.*, 140, 1, 252–259, 2009.
65. Yang, Q., Li, N., Li, Q., Chen, S., Wang, H.L., Yang, H., Amperometric sarcosine biosensor based on hollow magnetic Pt-$Fe_3O_4$ nanospheres. *Anal. Chim. Acta*, 1078, 161–167, 2019.
66. Wang, B., Luo, Y., Gao, L., Liu, B., Duan, G., High-performance field-effect transistor glucose biosensors based on bimetallic Ni/Cu metal-organic frameworks. *Biosens. Bioelectron.*, 171, 112736, 2021.
67. Zhao, J., Fu, C., Huang, C., Zhang, S., Wang, F., Zhang, Y., Zhang, L., Ge, S., Yu, J., Co3O4-Au polyhedron mimic peroxidase- and cascade enzyme-assisted cycling process-based photoelectrochemical biosensor for monitoring of miRNA-141. *Chem. Eng. J.*, 406, 126892, 2021.
68. Xue, L., Guo, R., Huang, F., Qi, W., Liu, Y., Cai, G., Lin, J., An impedance biosensor based on magnetic nanobead net and $MnO_2$ nanoflowers for rapid and sensitive detection of foodborne bacteria. *Biosens. Bioelectron.*, 173, 112800, 2021.
69. Akhtar, N., Metkar, S.K., Girigoswami, A., Girigoswami, K., ZnO nanoflower based sensitive nano-biosensor for amyloid detection. *Mater. Sci. Eng. C. Mater. Biol. Appl.*, 78, 960–968, 2017.
70. Li, W., Yubo, G., Jiaona, Z., Xiaofang, W., Feng, Y., Zigang, L., Min, Z., Universal DNA detection realized by peptide based carbon nanotube biosensors. *Nanoscale Adv.*, 2, 2, 717–723, 2020.

# 4

# Biosensors in 2D Photonic Crystals

**Gowdhami D. and V. R. Balaji***

*School of Electronics Engineering, Vellore Institute of Technology, Vandalur, Kelambakkam Road, Chennai, Tamil Nadu, India*

***Abstract***
Photonic crystal (PC)-based biosensors are said to be an emerging field in the fields of photonics and biosensing. The PC is an artificially made periodic array of dielectric material that influences the propagating modes of the structure and is said to be a potential platform for sensing applications. The variation of the refractive index of the analyte is the important property that has been used to detect the different biomaterials. Despite the fact that these materials are artificial crystal structures, they have some distinguishing characteristics such as immunity to electromagnetic interference, micro-/nano-sized structures, small/specific sensing regions, and are highly selective and sensitive. The samples required to detect the diseases are very low and respond quickly. Multianalyte detection systems are also said to be developed by many research teams, and integrating the same with the Photonic Integrated system is the major goal. On the whole, the Finite Difference Time Domain method and Plane Wave Expansion methods have been used to study the computation and dispersion characteristics of the photonic crystal.

***Keywords***: Photonic crystal, refractive index, photonic bandgap, analyte, finite difference time domain, planewave expansion method

## 4.1 Introduction

In today's fast moving world, science and technology have taken us a step forward and made things easier. Often, we depend on the most innovative gadgets, such as mobile phones, personal computers, household appliances, microwave ovens, LEDs, etc. to communicate with the physical habitat. Sensors play a major role in our day-to-day lives through these various applications and have been commercialized drastically. What are sensors all about? A sensor is a type of analytical tool capable of detecting and reacting to the changes encountered in the physical environment, for example, light, heat, pressure, temperature, etc., and these changes are obtained as measurable signals which can be further processed and analyzed [1].

**Corresponding author*: photonics.material@gmail.com; ORCID-0000-0002-9239-7724

Inamuddin and Tariq Altalhi (eds.) Biosensors Nanotechnology, 2nd Edition, (85–102) © 2023 Scrivener Publishing LLC

## 4.2 Biosensors

A biosensor is an integrated device that incorporates a biological component, such as cells, antibodies, organelles, etc. with a physiochemical component to generate measurable electrical signals. A biosensor is defined as a receptor-transducer self-accommodate device that utilizes the bio-recognition elements to offer selective and semi-quantitative information for analysis [2]. The first biosensing mechanism was detected by Clark and Lyons in 1962 [3], when the biosensing era started. An ideal biosensor comprises of a sensing and a transducing element, where the sensor plays a major role in detecting the biological elements and a transducer, which converts one form of energy to another.

### 4.2.1 Types of Biosensors

Based on the transduction process they are classified as

1. Electro-chemical bio-sensors [4]
2. Thermal bio-sensors [5]
3. Optical bio-sensors [6–8]
4. Acoustic bio-sensors [9, 10]

#### *4.2.1.1 Optical Biosensors*

Optical biosensing is becoming an important topic of great interest among researchers and others. Due to the spread of various micro and nano organisms that have endangered many lives, it has become an important concern in the identification and detection of such species. It has evolved into an interdisciplinary field with widespread application in a variety of platforms such as microelectronics, organic chemical sensing, environmental monitoring, optical MEMS (Microelectromechanical systems) [11, 12], and so on.

Optical biosensors are also analytical devices that utilizes the analyte to identify the phase variation in the RI (refractive index). The immobilized analyte during the detection phase increases the localization of the electric field within a small region, and it reflects to a small change in the RI of the analyte [13]. A tag-free and real-time analyte's detection is the typical advantage over traditional methods.

##### 4.2.1.1.1 Why Optical Biosensors

- Label free and noninvasive.
- The nature of biological materials is preserved.
- It is more sensitive and less prone to errors.
- Faster response.
- Immune to electromagnetic interference.
- Low detection limit.
- Senses the analyte in a very small area.
- Able to integrate with PC circuitry.

**Figure 4.1** Block diagram of optical biosensor.

### *4.2.1.2 Operation of PC (Photonic Crystal)-Based Optical Biosensors*

The PC biosensor has the following schematic as shown in the block diagram for optical sensing (Figure 4.1). It has an optical source, a PC-based sensor, an optical detector, and a display unit [14].

**Optical sources:** A wide range of optical sources available for optical sensing, sources from a broad range of spectrum to a narrow spectrum are available such as incandescent lamp, gaseous laser ($CO_2$), liquid laser (Dye laser), semiconductor LED, Laser diode (Nd:YAG, Er:YAG), respectively. An optical source should have a good stability and also be able to emit discrete range of frequencies depending upon the requirement of the sensing application.
**PC sensors:** Various PC based optical components where used to bring in the integration of the optical integrated circuits out of which some given below. Pc based optical filters, switches, logic gates, power splitters, channel drop filters, multiplexer, demultiplexer, etc. In all these components, light has been collected from the source and channelled to the specific sensing area where the core sensing happens and the resulting light is collected in the detector section. These sensors are very much selective and sensitive to a very specific wavelength.
**Photodetectors:** The photodetectors must be extremely sensitive in order to detect the optical signal received from the sensor. The other factors that should be considered are directivity, response time, spectral response, etc. PIN and APD are the two detectors used for sensing purposes.
**Signal Processing Unit:** Now the emanated light received in the photo detector is in the form of current or voltage signals. These signals are initially amplified using a preamplifier, sampled, and given to an analog to digital convertor circuit.
**Display:** The output signal is processed by the signal processing unit and displayed to the used in readable form.

### *4.2.1.3 PC-Based Sensors*

Photonic crystal (PC) structures are periodically changing refractive index units that can be used as an alternative for the detection of various diseases. Photonic crystal-based biosensors are highly selective and sensitive. A PC can be created in an ultra-compact size range of micro or nanoscale range and have higher resistance to electromagnetic interference, which makes it a deserving candidate for the disease detection system. Due to their structural peculiarities, i.e., strict periodicity, there is a strong light's localization on the analyte, which in turn converts a very small amounts of analyte into detectable signals with fast responsiveness. The effective refractive index sensing mechanism has been adopted by the biosensor for analyzing the spectral properties of the sensor.

There are different common diseases such as diabetes, anaemia, leukaemia, and pathogen specific diseases such as malaria, dengue, etc., which can be detected by blood analysis, which utilizes traditional laboratory techniques. However, they are said to be complicated, tedious, and time consuming as well. The 2D photonic crystal paves the way for the accurate, fast, and easy detection of such diseases. We utilize the 2D PC structures as they possess very high light confinement, and the propagation modes can be controlled easily by varying the structural parameters [15]. However, 1D PCs have an incomplete bandgap [16] and 3D PCs have a very small lattice structure, and due to their complex nature, it is difficult to fabricate them [17].

PC structures are broadly classified as pillars in air structure and air holes in slab structure. Most commonly, pillars in air structure are used for low optical losses as such [18]. As the PC has the high and low periodicity of dielectric material, certain wavelength bands are forbidden in the structure know as Photonic Band Gap (PBG) [19]. In these disallowed bands, a single mode or a collection of modes are said to be transmitted by creating defects and disturbing the periodicity of the structure. They can be either line, point, or surface defects [20]. Through these defects, the entire structure can be tuned to sense a particular analyte. A significant benefit of such PC-based biosensors is that they transduce a nano scale ranged analyte into measurable signals, which leads to a variety of applications in the industrial and biomedical fields [21].

### *4.2.1.4 Numerical Analysis*

The analysis of the PC can be done using many methods, such as the Transfer Matrix method (TMM) [22], Finite Difference Time Domain Method (FDTD) [23], Plane Wave Expansion Methods (PWE) [24] and the Finite Element Method (FEM) [25]. The FDTD method is used for analysis because it makes use of Yee's mesh [26], and thus, its response is much higher and more accurate than other computation methods. The dispersion behavior of the PC is obtained by the PWE method.

In PWE method, band diagram represents the band gap and total modes propagating through the structure. In the band diagram electric field of the light propagating in the PC structure, the Maxwell equation is governed [20].

$$\nabla \times \left( \frac{1}{\varepsilon(r)} \nabla \times H(r) \right) = \left( \frac{\omega}{c} \right)^2 H(r) \tag{4.1}$$

Equation 4.1 is called the master equation where, $\varepsilon(r)$ is the dielectric constant, c is the velocity of light, and $\omega$ is the resonating frequency. The above equation clearly states that the change in the dielectric constant changes the frequency at resonance which is considered to be the main essence of the effective change in the RI. As the analyte's index of changes, there is a phase change and the corresponding resonating modes inside the PC structure also change, which detects the analyte in the PC based biosensor.

#### *4.2.1.5 Parameters for Sensing*

The mathematical analysis of the biosensor can be measured with the following parameters: sensitivity, quality factor, transmission efficiency, and figure of merit [27].

The quality factor, also called the Q-factor, is the important parameter that emphasizes the total amount of light confined within the structures.

$$Q = \frac{\lambda_0}{\Delta\lambda} \tag{4.2}$$

where $\lambda_0$ is the resonating wavelength, $\Delta\lambda$ is the full width half maximum (FWHM). It is a unit less parameter.

The sensitivity represents how much the resonating wavelength shifts with respect to the refractive index change in the analyte.

$$S = \frac{\Delta\lambda}{\Delta n} \text{ (nm/RIU)} \tag{4.3}$$

where $\Delta\lambda$ and $\Delta n$ is the resonating wavelength and refractive index shift, respectively.

The spectral shift of the resonating wavelength given as resolution.

$$R = \lambda_r - \lambda_s \tag{4.4}$$

where $\lambda_r$ and $\lambda_s$ is the resonating and shifted wavelength.

The limit of detection is given as the ratio between resolution and sensitivity, and it should be as small as possible.

$$D = \frac{R}{S} \text{ (RIU}^{-1}\text{)} \tag{4.5}$$

Figure of merit (FOM) of the sensor is given as

$$FOM = \frac{(S*Q)}{\lambda_r} \tag{4.6}$$

### 4.2.2 Photonic Crystal-Based Bio-Optical Sensor

Gharsallah *et al.* [28] modeled a bio-sensor for detecting the concentration of urea in urine. In this design, they have investigated two structures, both with two photonic crystal waveguides used to transmit the input and output light, and at the centre, sensor1 has a square micro cavity and sensor2 has an elliptical hole. Both have a lattice structure that is hexagonal. The sensor1 pillars are placed in an air background with a lattice parameter of 558 nm, a rod radius (R) of 111.6 nm, and a central square cavity radius twice the radius of R. Figure 4.2a represents sensor 1. Now the second sensor has air holes in a silicon slab configuration. The hole radius is about 111.6 nm with a lattice parameter of 372 nm. As represented in Figure 4.2b, the elliptical hole at the centre has two radii, R1 = 0.558 μm and R2 = 0.148 μm, respectively. Various urea concentrations in urine were analyzed and for the small modification in the refractive index ranges about 0.001, the output intensity shifts for both the sensors. The highest quality factor of about 570 is achieved in sensor 1 and a good sensitivity of about 550 is achieved in sensor 2. Both have a transmission efficiency of 99% and a detection response time of 2 and 0.4 ps is achieved in sensor 1 and 2, respectively.

Asuvaran & Elatharasan [29], proposed an optical sensor for classifying and detect the different abnormal brain tissues during minimally invasive robot assisted surgery. The silicon pillars with hexagonal array configuration in air bed is used. It is basically designed with a two-hexagonal ring resonator with a foot print of 16.2 × 15.4 μm. A lattice parameter of 540 nm and the rod radius are set at 55 nm. Below, Figure 4.3 depicts the schematic of the biosensor. The range of wavelength over which it operates is 1648 to 1794 nm and the sensor shows ON and OFF resonance at 1544 and 1548 nm, respectively, without any analyte. The air medium is replaced with the refractive index of the different tissues, such as normal, infected, and tumour tissues, the resonating wavelength and the normalized output power spectrum change. As a result, it has an average quality factor of 573 and a sensitivity of 4615 nm/RIU with a detection limit of 0.0013 RIU. This sensor remains out of the box as it senses different brain tissue used in the robot assisted minimal inversion surgery, but its other performance parameters can still be improved.

Benmerkhi, A., *et al.* [30] investigated a biosensor for DNA detection. In this design, an innovative method has been adopted by placing 4 slots in the L2 resonant cavity in-between two waveguide couplers. The sensor is designed with an air hole structure with triangular lattice of air holes with a radius of 0.4a. The width and length of the slot are optimized for 160 and 350 nm. The number of slots in the cavity and the total number of functionalized pores in the cavity is optimized to 4 and the analyte (DNA) is infiltrated into the slots and the functionalized holes and the sensor parameters are measured. This design exhibits an outstanding quality factor of $3.7468 \times 10^6$ and a sensitivity of 460 nm/RIU. But it is used to detect only a single analyte.

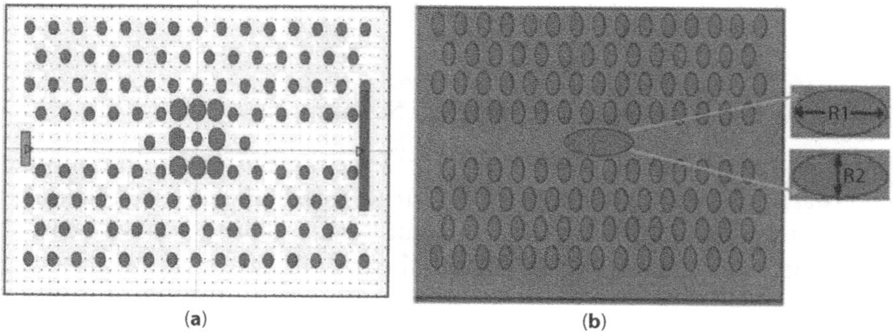


**Figure 4.2** (a) Schematic diagram of sensor 1. (b) Schematic diagram of sensor.

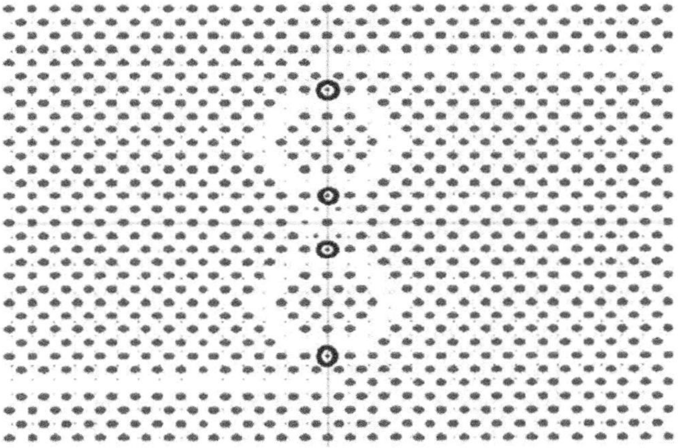

**Figure 4.3** Design of two hexagonal biosensor.

Miyan, H *et al.* [31] proposed a 2D photonic crystal biosensor for detecting different cancer cells, such as cervical, breast, and basal cells. In this sensing platform, a central rod is removed to create a nano cavity with two waveguides. The bandgap for TE mode is calculated and the parameter a/λ is taken as 0.264 from the bandgap. A dielectric substrate with pillars in air background with hexagonal lattice is utilized. The pillar radius is 82 and 410 nm is the lattice constant. The PC-based biosensor has the central blue circle as the sensing region with a radius of 106 nm, and the overall area is reported as 48 $\mu m^2$. PWE and FDTD are used to calculate the optical field distribution and the transmission spectrum of the sensor. The sensing principle is the effective change in refractive index. When a Gaussian wave with a centre frequency of 1550 nm is given at the input side, the sensing region is replaced by the cancerous cells and for the various rod radius and lattice constant the simulation is done. For the malignant breast cell, 15085 nm/RIU and 0.01 is the sensitivity and detection limit obtained. The FOM is acquired as 159.54 $RIU^{-1}$, for the rod radius and lattice constant as mentioned above. The overall quality factor of 81.58 for a = 610 nm and r = 106 nm for the basal cell.

Jindal, Sobti, *et al.* [32] designed a sensor that detects five types of cancer cells namely MDA-MB-231, HeLa, etc. It is coupled with a nanocavity and two waveguide. The rods are placed in air with a triangular lattice. Its lattice constant is about 800 nm and the radii of the silicon rods are 240 nm (0.3 a). The PWE (Plane Wave Expansion) method is used to determine the PBG (Photonic Band Gap) for TE mode, and the frequency range of 1492 to 1908 nm is explored. Now a nanocavity is created at the centre of the waveguide and an optical guided mode of 1661 nm is obtained in the absence of the analyte. The nanocavity is optimized at 128 nm (0.16 a) as it results in good sensitivity and quality factor. The optimization is further enhanced by repositioning the nearby rod(s) and varying the rod radius. Thus, the sensitivity of 388.57 nm/RIU and the highest quality factor of 4856.75 are obtained in the above design structure.

Robinson *et al.*, [33] investigated a bio-sensor with a central ring resonator and two inverted L waveguides. In this, the urine's glucose concentration ranges from zero to fifteen gm/dl, the blood's glucose concentration, urea, bilirubin, and albumin concentration are detected in urine. The sensor is designed with a size of 11.4 × 11.4 μm. This structure adopts a square lattice, and rods are placed in air. The radius and lattice constant of the rods are taken as 100 and 540 nm, respectively. The PBG is acquired as 1241 to 1830 nm for TM mode without any defects. Once the defects are formed in the L and inverted L waveguide with a central ring resonator, the propagation of guided modes takes place the structure. The inner radius of the ring is taken as 50 nm and the outer ring radius of the PC is considered as 100 nm. However, the rods placed in the resonator at the upper and lower end are 86 nm. The resonator shows an ON and OFF resonance at 1545 and 1300 nm, respectively, with a quality factor of 256 without an analyte. If the air bed is supplied with the analyte, say urine, its glucose concentration is varied (0.625 to 5 gm/dl) and the output power spectra is obtained as 55% resonating at 1585 nm with a quality factor of 217 for 5 gm/dL. Similarly, if the air is again replaced with the blood with different glucose concentration, ranging from 0.182 to 0.342 gm/dL, the quality factor of 269 with an output power is about 75% at 1585 nm. The sensor is designed to operate at 1.540 to 1.550 μm. Figure 4.4 represents the biosensor.

Huang *et al.* [34] designed a biochemical sensor with a ring defect coupled resonator (RDCR). This sensor has a triangular lattice configuration with silicon slab etched with air holes. A cavity with two waveguides and ring resonator is formed for designing the structure. The hole radius, lattice constant, and the width of the slab is considered as 118, 348, and 191 nm, respectively. The quasi waveguide is coupled with a polarized light of transverse electric mode at a wavelength of 1552.34 nm. By adjusting the rod's radius at the resonator's central hole and by altering the breadth of the waveguide, the highest quality factor is obtained as $6.93 \times 10^6$. The below Figure 4.5a depicts the biochemical sensor, with the central red rods($r_c$) and width of the waveguide with blue holes is optimized to 0.40 a and 0.985 W1 and the optimized quality factor of $1.86 \times 10^7$. To enhance the sensitivity the number of inner holes in the resonator is increased to 7 and the number of functionalized holes with the ring holes is optimized to 25 as shown in Figure 4.5b. The change in index of the analyte (deuterium water, glycerol solutions) leads to shift in spectral wavelength, and the quality factor of 35,517 is achieved which is said to be the highest, a sensitivity of 330

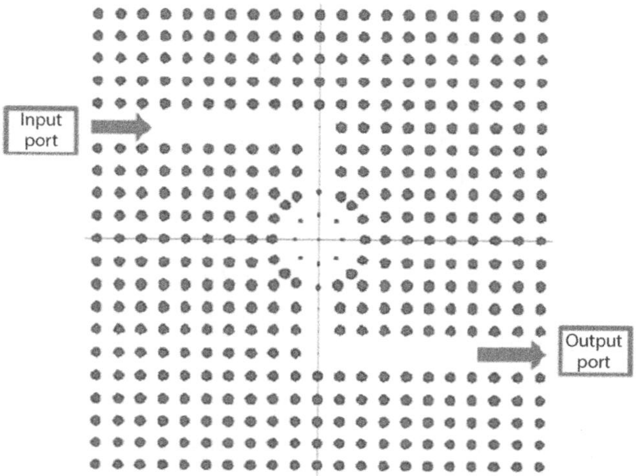


**Figure 4.4** L and inverted L waveguide biosensor.

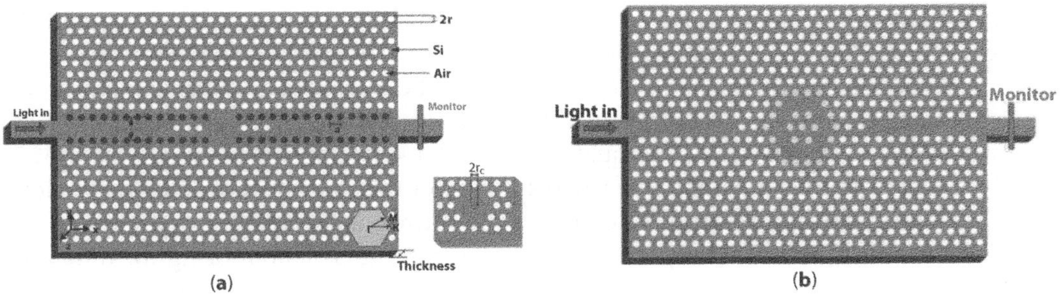


**Figure 4.5** (a) 3D structure of sensing element. (b) Sensor with inner ring holes.

nm/RIU and highest FOM of ~ 8000. The Detection limit is obtained as $1.24 \times 10^{-5}$. The complete computation is carried out in 3D FDTD method.

Arafa *et al.* [35] presented an optical biosensor with an infiltrated photonic crystal cavity. An air hole structure of silicon slab with triangular lattice configuration is used. 196.1 nm is the rod's radii and 530 nm is the lattice parameter of the structure. 2D PWE and 2D FDTD are used to calculate the dispersion characteristics and output spectrum of the designed structure. The PBG was obtained between 1323.7 and 1908.1 nm. The overall sensor has an infiltrated ring of holes etched in the dielectric medium. 126.5 and 247.5 nm is the inner and outer radius, respectively, and the inner air medium is 121 nm. Without any analyte, the sensor acts as a filter and resonates at 1373.7 nm with a reasonable quality factor of $7.06 \times 10^3$. To achieve higher performance parameters, such as sensitivity and quality factor, the inner radius of the in filtered ring is fixed to a radius of 0.255 a. A higher quality factor of $1.112 \times 10^5$ and sensitivity of 462.61 nm is attained by placing the analyte only in four functionalized holes and ring-shaped holes. $3.03 \times 10^{-6}$ is the detection limit observed in this sensor. Figure 4.6 represents the infiltrated biosensor.

Harhouz A. & Hocini A. [36] designed a sensor to determine different liquid components like deionized water, acetone, distilled water, methanol, and isopropyl alcohol. It is constructed with two waveguides and a microcavity on a dielectric Si slab with a hexagonal

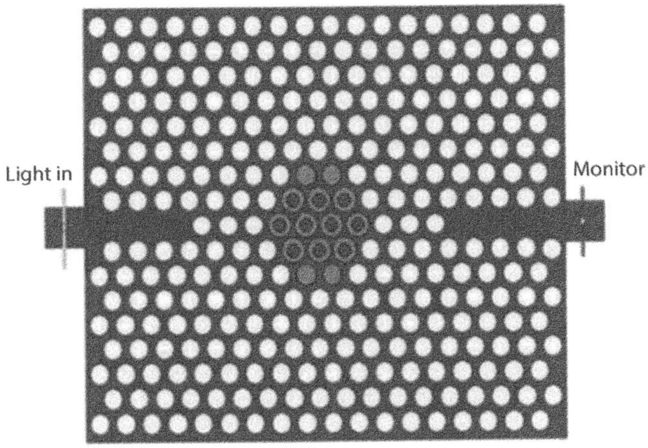


**Figure 4.6** Infiltered photonic crystal biosensor.

lattice. The lattice parameter is 470 nm, and the air hole(s) radius is considered as 190 nm. Before the defect is placed, the PBG is obtained as 1.135 and 1.860 μm for Transverse Magnetic Polarization. The sensitivity of the sensor can be enhanced by optimizing the hole radius (r') along the waveguide as 0.2 μm. Thus, the transmission spectrum ranges between 73% and 84.4% and has a very good quality factor of $1.47 \times 10^4$ with a sensitivity of 425 nm/RIU. The resonant wavelength shifts to 0.435 nm for the shift in RI of 0.001 (detection limit). The FDTD method was used to complete the entire computation.

Suganya and Robinson [14] reported a biosensor for glucose monitoring for different concentration levels ranging from 30 gm/dl to 330 gm/dl. It has a ring resonator in a rhombic shape. It is basically made of circular rods with hexagonal lattice in an air bed. The PBG is extracted before defects, and its wavelength ranges from 1125 to 1725 nm for TE mode. Now biperiodicity has been introduced in a rhombic shaped resonator appears like a ring with input and output ports serving as two waveguides at both the ends. The midway elliptical-shaped rod has a radius of 0.15 μm. The radius of the rods is 0.1 m. The smaller rods inside the resonator have a radius of 50 nm. The lattice constant is about 547 nm. This sensor utilizes FDTD for computation analysis, and the oversize remains at 10 × 12 μm. The sensor resonates at 1545 nm with 100% transmission efficiency with zero glucose level. As the sensor is replaced with a different concentration of glucose, the wavelength shifts to 2.5 nm for every 30 gm/dl of glucose. The overall quality factor and sensitivity are observed to be 1000 nm/RIU and 178.5, respectively. Figure 4.7 represents the biosensor.

Arunkumar *et al.* [37] designed a bio-sensor to identify various components in the blood like RBC, WBC, hemoglobin, etc. This sensor resonates from 1570 to 1610 nm. The sensor consist of square lattice and rods in air configuration with 540 nm as lattice constant and the rods radii as 100 nm. The PBG for TE modes are calculated, and the band gap ranging from 1200 to 1800 nm is used as it lies in the third optical band. The defects are formed in the form of an ellipse with two waveguides as depicted in the Figure 4.8. The dispersion and the transmission characteristics are obtained in the PWE and FDTD method. Before placing the analyte, the sensor produces the output transmission efficiency of 100% with 257.5 as the Q factor at 1590 nm. As the elliptical ring resonator is placed with different blood

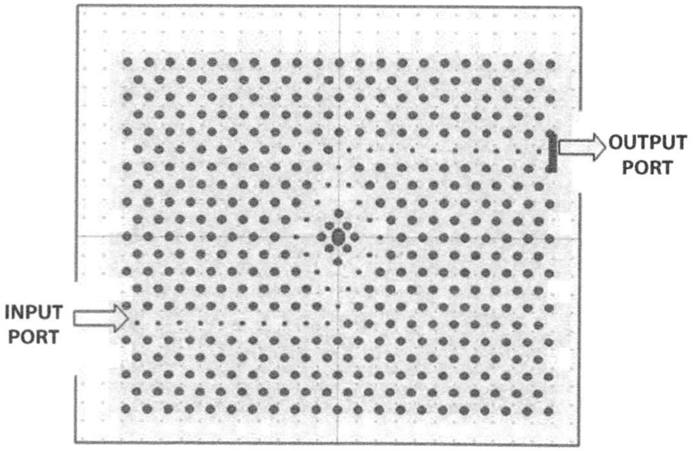


**Figure 4.7** Rhombic resonator for glucose concentration.

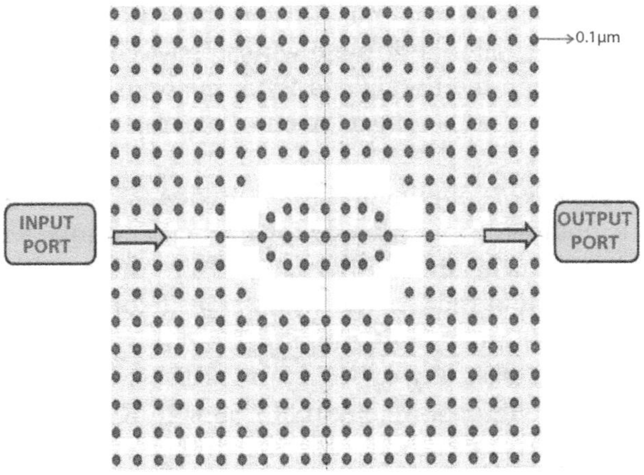


**Figure 4.8** Elliptical resonator as biosensor.

elements the RI changes, and there is a shift in the spectral wavelength. Thus, the detection limit is obtained as 0.002 ($RIU^{-1}$).

Rashidnia *et al.* [38] designed a unique elliptical PC based ring resonator with silicon nitrite and gold as the materials used on a rectangular lattice with an air background to notice malarial diseases. In red blood cells, the cycle of plasmodium falciparum and its different stages are detected by the biosensor. This sensor utilizes a Gaussian continuous wave as an input signal with a wavelength of 0.514 to 1.55 μm. The rod radius and the lattice spacing are given as 200 and 450 nm, respectively. Its overall size is about $9.82 \times 6$ $\mu m^2$. Figure 4.9 represents the biosensor with an elliptical ring resonator. It consists of elliptical rods named A and C, and circular rods named C, and a central elliptical rod named D. The elliptical rod has two radii, given as R and r for each elliptical rod. The yellow rods are gold, and the orange ones are silicon nitride rods. Phosphate-buffer saline has a refractive index

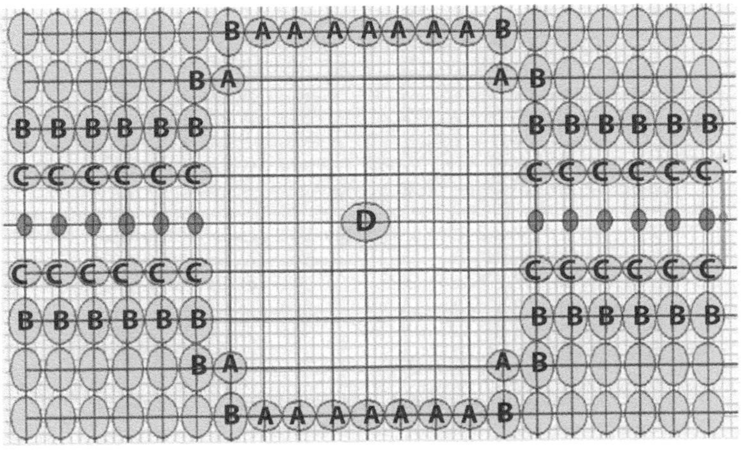

**Figure 4.9** Elliptical biosensor with gold and silicon nitride rods.

of 1.336 and is used as the reference material. This normal RBC and the infected RBC have to be diluted in this reference material through which phase change can be measured. The PBG is calculated both through the FDTD and PWE methods, and it is observed that the same structure with silicon has no band gap, hence replacing the silicon with the metallic PC.

Thus, after defects the PBG is obtained as 460.4 to 584.05 nm with an input of 0.514 μm. Again, the PBG is obtained for TE mode for normal and infected stage of the cells and still 0.514 μm lies within the above PBG range. Here the boundary condition used is Anisotropic PML. The radii of the (Au) gold and (Si) silicon rods are optimized as follows, $R_{SiN4}$=100 nm, $R_{A'C}$=220 nm, $r_A$=140 nm, $R_B$=220 nm, $r_C$=120 nm, $R_D$=300 nm, and $r_D$=180 nm. The maximum sensitivity of 357.21 nm/RIU is obtained. The same design was again simulated with the input wavelength of 1.55 with $R_{SiN4}$=0.18 μm and $r_C$=0.1 μm for different stages of infected and normal cells and the sensitivity of 893.7 was reported in this paper.

Bendib & Bendib [39] proposed a ring resonator-based biosensor with GaAs rods placed in the air. A square lattice is considered with the lattice parameter and rod radius of 590 nm and 100 nm, respectively. The PWE is followed to obtain the PBG of 2099 to 2756 nm. The FDTD method is used for simulating the biosensor. A Gaussian pulse centred at 2.09 μm is applied with PML as the boundary condition. For the RI of regular RBC cells and infected RBC cells in different stages, the sensor resonates and the transmission spectrum shifts for each RI. The transmission spectrum varies from 46.1% to 94.4%. Figure 4.10 represents the biosensor with GaAs rods for the detection of malaria.

Olyaee & Najafgholinezhad [40] introduced a multi-channel microfluidic biosensor with a hexagonal lattice and holes in a slab configuration. The lattice parameter and hole radius are 350 and 110 nm, respectively. For TE mode, the PBG is about 0.25455 to 0.3017. This multichannel waveguide has a central waveguide and three cavities. In channel 1, a cavity is formed in channel 2 and 3 along with the cavity, a hole is shifted along the lattice by 0.2 and 0.3 μm, respectively. 1.2781, 1.2866, and 1.2961 μm are the resonating wavelengths along the three channels and 5000, 6433, 2592.2 is the quality

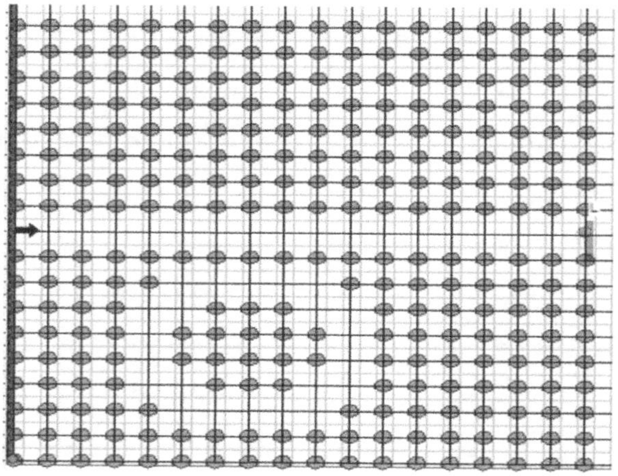

**Figure 4.10** Ring resonator biosensor with GaAs rods.

**Table 4.1** Waveguide and cavity coupled waveguide-based structure.

| Structure | Quality factor (QF) | Sensitivity (S) nm/RIU | Detection limit (DL) | Analyte | Reference |
|---|---|---|---|---|---|
| Waveguide | - | - | - | Glucose | Sharma, P., & Sharan, P. (2014) [41] |
| Waveguide | 23,575 | 638 | - | Glucose | Sharma, P., & Sharan, P. (2015) [42] |
| Waveguide | - | - | - | Cervical, breast and basal cancer cells | Sharma, P., *et al.* (2015, January) [43] |
| L and inverted L Waveguide | 110.53 | 523.81 | - | Cancerous Cervical cell | Sundhar, A., *et al.* (2019, March) [44] |
| Waveguide | - | - | - | Lymphocyte Cancer cells | Sharma, P., & Sharan, P. (2014, September) [45] |
| Waveguide | 118563 | $10^{-6}$ | - | Cancer | Nischita, R., *et al.* (2015, January) [46] |
| Multichannel Waveguide | - | - | - | Glucose concentration | Ameta, S *et al.* (2017, July) [47] |
| Waveguide and a microcavity | - | - | - | Blood Components | Ameta, S. *et al.* (2017, May) [48] |
| Waveguide | - | 2.02 | - | Escherichia coli | Painam, B., *et al.* (2016) [49] |
| Waveguide and a microcavity | 272 | - | - | Glucose | Chhipa, M. K *et al.* (2017, July) [50] |

factor obtained in those three channels, respectively. As the H1 hole is the sensing hole and filled with the R.I from 1.33 to 1.45, there is a phase change in the effective RI and the resonating wavelength changes in channel1 alone. Now the channel 3's sensing hole is filled with the various R.I and its corresponding wavelength changes are noted down. The sensitivity is observed as 1.185 nm/fg for a DNA molecule having an R.I of 1.45 in channel 3 and a detection limit of 0.084 fg for the resolution assumed as 0.1 nm in the measuring tools. The methods used in simulation are PWE for bandgap and FDTD for structural analysis of the sensor.

Comparison of the different types of sensors with line defect and point defect that has been reported in Table 4.1. The performance parameters are also compared, with respect to different analytes.

## 4.3 The Overall Inference

In this chapter, we have discussed the current advancements in 2D photonic crystal sensors, principle of operation, performance parameters, detection of various analytes, etc. The effective change in RI is the primary objective of the PC based sensors which brings in the shift in the resonating wavelength. Further, we have also discussed about the various PC sensors, with respect to cavity, ring resonator, and photonic crystal waveguide. Different sensors to detect different pathogens, blood components and biochemical components are discussed more. Sensors such as analyte infiltrated photonic crystal cavity [35], and analytes placed in the slots [30], multichannel biosensors [40] are some of the unique design structure with a very good impact on its performance. The ring resonators also have a greater impact on the sensing the analytes as the light matter interaction is very much higher in ring resonators [14, 31, 33]. Further, some more sensors utilize different materials such as silicon nitride and gold for designing the sensors [38]. Thus, ultra-compact, multi analyte detection system can still be designed as it can be integrated into the photonic integrated circuit.

## 4.4 Conclusion

The photonic crystal-based biosensor has an important role in real time monitoring, disease detection, life science, etc. Different types of biosensors were designed for different diseases and biochemical compounds with various parameters such as Q-factor, sensitivity, detection of limit, etc. has been reported. As these works are trying to meet the demand in the sensing domains and many have achieved to improve the performance parameters reasonably. But still many requirements are there to be fulfilled to utilize and deliver this technology to a layman. Detection of multiple analytes and monitoring simultaneously at the same time, development of point of care (POC) kits with integrated lab-on-chip are some of the basic demands that need to be developed. Thus, there is wide scope for remodeling and enhancing the designs of the photonic crystal biosensor for biomedical application. In future multianalyte detection will have its pace and with minimum fabrication complexity and improved performance.

## References


1. Naresh, V. and Lee, N., A review on biosensors and recent development of nanostructured materials-enabled biosensors. *Sensors*, *21*, 1109, 2021.
2. Thevenot, D.R., Toth, K., Durst, R.A., Wilson, G.S., Electrochemical biosensors: Recommended definitions and classification. *Pure Appl. Chem.*, *71*, 2333–2348, 1999.
3. Clark, L.C., Jr and Lyons, C., Electrode systems for continuous monitoring in cardiovascular surgery. *Ann. N. Y. Acad. Sci.*, *102*, 29–45, 1962.
4. Grieshaber, D., MacKenzie, R., Vörös, J., Reimhult, E., Electrochemical biosensors-sensor principles and architectures. *Sensors*, *8*, 1400–1458, 2008.
5. Ramanathan, K. and Danielsson, B., Principles and applications of thermal biosensors. *Biosens. Bioelectron.*, *16*, 417–423, 2001.
6. Olyaee, S., Seifouri, M., Mohsenirad, H., Label-free detection of glycated haemoglobin in human blood using silicon-based photonic crystal nanocavity biosensor. *J. Mod. Opt.*, *63*, 1274–1279, 2016.
7. Aly, A.H., Mohamed, D., Mohaseb, M.A., Abd El-Gawaad, N.S., Trabelsi, Y., Biophotonic sensor for the detection of creatinine concentration in blood serum based on 1D photonic crystal. *RSC Adv.*, *10*, 31765–31772, 2020.
8. Painam, B., Kaler, R.S., Kumar, M., On-chip oval-shaped nanocavity photonic crystal waveguide biosensor for detection of foodborne pathogens. *Plasmonics*, *13*, 445–449, 2018.
9. Länge, K., Rapp, B.E., Rapp, M., Surface acoustic wave biosensors: A review. *Anal. Bioanal. Chem. Res.*, *391*, 1509–1519, 2008.
10. Fogel, R., Limson, J., Seshia, A.A., Acoustic biosensors. *Essays Biochem.*, *60*, 101–110, 2016.
11. Soltanian, E., Jafari, K., Abedi, K., A novel differential optical MEMS accelerometer based on intensity modulation, using an optical power splitter. *IEEE Sens.*, *19*, 12024–12030, 2019.
12. Marvi, F. and Jafari, K., A measurement platform for label-free detection of biomolecules based on a novel optical BioMEMS sensor. *IEEE Trans. Instrum. Meas.*, *70*, 1–7, 2021.
13. Liu, Y. and Salemink, H.W.M., Photonic crystal-based all-optical on-chip sensor. *Opt. Express*, *20*, 19912–19920, 2012.
14. Suganya, T. and Robinson, S., 2D photonic crystal based biosensor using rhombic ring resonator for glucose monitoring. *ICTACT Microelectron.*, *3*, 349–353, 2017.
15. Robinson, S. and Nakkeeran, R., PCRR based bandpass filter for C and L+ U bands of ITU-T G. 694.2 CWDM systems. *Opt. Photonics*, *1*, 142, 2011.
16. Suthar, B. and Bhargava, A., Biosensor application of one-dimensional photonic crystal for malaria diagnosis. *Plasmonics*, *16*, 59–63, 2021.
17. Mohanty, S.K., Das, S., Swain, K.P., Bhanja, U., Palai, G., A proposal for testing kit of corona viruses using 3D photonic structure. *Microsyst. Technol.*, *27*, 2823–2827, 2021.
18. Kok, A.A., van der Tol, J.J., Baets, R., Smit, M.K., Reduction of propagation loss in pillar-based photonic crystal waveguides. *J. Light. Technol.*, *27*, 3904–3911, 2009.
19. Joannopoulos, J.D., Villeneuve, P.R., Fan, S., Photonic crystals: Putting a new twist on light. *Nature*, *386*, 143–149, 1997.
20. Joannopoulos, J.D., Johnson, S.G., Winn, J.N., Meade, R.D., *Molding the flow of light*, pp. 66–92, Princeton Univ. Press, Princeton, NJ [ua], 2008.
21. Nair, R.V. and Vijaya, R., Photonic crystal sensors: An overview. *Prog. Quantum. Electron.*, *34*, 89–134, 2010.
22. Pendry, J.B. and MacKinnon, A., Calculation of photon dispersion relations. *Phys. Rev. Lett.*, *69*, 2772, 1992.

23. Taflove, A., Hagness, S.C., Piket-May, M., *Computational electromagnetics: The finite-difference time-domain method*, pp. 629–670, p. 3, The Electrical Engineering Handbook, Boston, London, 2005.
24. Pendry, J.B., Calculating photonic band structure. *J. Phys. Condens. Matter.*, *8*, 1085, 1996.
25. Pelosi, G., Coccioli, R., Selleri, S., *Quick finite elements for electromagnetic waves*, Artech House, US, 2009.
26. Yee, K., Numerical solution of initial boundary value problems involving Maxwell's equations in isotropic media. *IEEE Trans. Antennas Propag.*, *14*, 302–307, 1966.
27. Rajasekar, R., Thavasi Raja, G., Savarimuthu, R., Photonic crystal-based sensors for biosensing applications, in: *Advances in Photonic Crystals and Devices*, N. Kumar and B. Suthar (Eds.), pp. 161–209, CRC Press, India, 2019.
28. Gharsallah, Z., Najjar, M., Suthar, B., Janyani, V., High sensitivity and ultra-compact optical biosensor for detection of UREA concentration. *Opt. Quantum Electron.*, *50*, 1–10, 2018.
29. Asuvaran, A. and Elatharasan, G., Design of two-dimensional photonic crystal-based biosensor for abnormal tissue analysis. *Silicon*, 14, 7203–7210, 2022.
30. Benmerkhi, A., Bouchemat, M., Bouchemat, T., Design of two-dimensional photonic crystal biosensor using DNA detection. *Phosphorus Sulfur Silicon Relat. Elem.*, *195*, 960–964, 2020.
31. Miyan, H., Agrahari, R., Gowre, S.K., Mahto, M., Jain, P.K., Computational study of a compact and high sensitive photonic crystal for cancer cells detection. *IEEE Sens. J.*, *22*, 3298–3305, 2022.
32. Jindal, S., Sobti, S., Kumar, M., Sharma, S., Pal., M.K., Nanocavitycoupled photonic crystal waveguide as highly sensitive platform for cancer detection. *IEEE Sens. J.*, *16*, 3705–3710, 2016.
33. Robinson, S. and Dhanlaksmi, N., Photonic crystal based biosensor for the detection of glucose concentration in urine. *Photonic Sensors*, *7*, 11–19, 2017.
34. Huang, L., Tian, H., Yang, D., Zhou, J., Liu, Q., Zhang, P., Ji, Y., Optimization of figure of merit in label-free biochemical sensors by designing a ring defect coupled resonator. *Opt. Commun.*, *332*, 42–49, 2014.
35. Arafa, S., Bouchemat, M., Bouchemat, T., Benmerkhi, A., Hocini, A., Infiltrated photonic crystal cavity as a highly sensitive platform for glucose concentration detection. *Opt. Commun.*, *384*, 93–100, 2017.
36. Harhouz, A. and Hocini, A., Design of high-sensitive biosensor based on cavity-waveguides coupling in 2D photonic crystal. *J. Electromagn. Waves Appl.*, *29*, 659–667, 2015.
37. Arunkumar, R., Suaganya, T., Robinson, S., Design and analysis of 2D photonic crystal based biosensor to detect different blood components. *Photonic Sensors*, *9*, 69–77, 2019.
38. Rashidnia, A., Pakarzadeh, H., Hatami, M., Ayyanar, N., Photonic crystal-based biosensor for detection of human red blood cells parasitized by plasmodium falciparum. *Opt. Quantum Electron.*, *54*, 1–19, 2022.
39. Bendib, S. and Bendib, C., Photonic crystals for malaria detection. *J. Biosens. Bioelectron.*, *9*, 1000257, 2018.
40. Olyaee, S. and Najafgholinezhad, S., A novel multi-channel photonic crystal waveguide biosensor. *Iranian Conference on Electrical Engineering (ICEE)*, IEEE, pp. 1–4, 2013.
41. Sharma, P. and Sharan, P., Design of photonic crystal-based biosensor for detection of glucose concentration in urine. *IEEE Sens. J.*, *15*, 1035–1042, 2014.
42. Sharma, P. and Sharan, P., An analysis and design of photonic crystal-based biochip for detection of glycosuria. *IEEE Sens. J.*, *15*, 5569–5575, 2015.
43. Sharma, P., Sharan, P., Deshmukh, P., A photonic crystal sensor for analysis and detection of cancer cells. *International Conference on Pervasive Computing (ICPC)*, IEEE, pp. 1–5, 2015.
44. Sundhar, A., Valli, R., Robinson, S., Abinayaa, A., SivaBharathy, C., Two dimensional photonic crystal based bio sensor for cancer cell detection. *IEEE International Conference on System, Computation, Automation and Networking (ICSCAN)*, pp. 1–3, 2019.

45. Sharma, P. and Sharan, P., An optical sensor for propagation analysis of Lymphocyte cell for cancer cell detection. *IEEE Global Humanitarian Technology Conference-South Asia Satellite (GHTC-SAS)*, pp. 93–98, 2014.
46. Nischita, R., Gudagunti, F.D., Sharan, P., Srinivas, T., 2-D photonic crystal based bio-chip for DNA analysis of breast cancer. *International Conference on Pervasive Computing (ICPC)*, IEEE, pp. 1–4, 2015.
47. Ameta, S., Sharma, A., Inaniya, P.K., Designing a multichannel nanocavity coupled photonic crystal biosensor for detection of glucose concentration in blood. *International Conference on Computing, Communication and Networking Technologies (ICCCNT)*, IEEE, pp. 1–4, 2017.
48. Ameta, S., Sharma, A., Inaniya, P.K., Nanocavity coupled waveguide photonic crystal biosensor for detection of different blood components. *International Conference on Computing, Communication and Automation (ICCCA)*, IEEE, pp. 1554–1557, 2017.
49. Painam, B., Kaler, R.S., Kumar, M., Active layer identification of photonic crystal waveguide biosensor chip for the detection of Escherichia coli. *Opt. Eng.*, *55*, 077105, 2016.
50. Chhipa, M.K., Robinson, S., Radhouene, M., Najjar, M., Srimannarayana, K., 2D photonic crystal micro cavity ring resonator based sensor for biomedical applications. *Conference on Lasers and Electro-Optics Pacific Rim (CLEO-PR)*, IEEE, pp. 1–2, 2017.

5

# Bioreceptors for Affinity Binding in Theranostic Development

**Tracy Ann Bruce-Tagoe[1], Jaison Jeevanandam[2] and Michael K. Danquah[3]***

*[1]Department of Chemical Engineering, Kwame Nkrumah University of Science and Technology, Kumasi, Ghana*
*[2]CQM - Centro de Química da Madeira, MMRG, Universidade da Madeira, Campus da Penteada, Funchal, Portugal*
*[3]Department of Chemical Engineering, University of Tennessee, Chattanooga TN, USA*

***Abstract***
Scientists have labored to identify solutions for the effective detection of pathogens, diagnosis, and treatment of diseases, such as cancer and neurodegenerative disorders, as well as quality assurance for several years. Recently, the advancement in theranostics has contributed immensely to the development in medical science and research field. This is evident in the ability to perform diagnosis and treatment simultaneously or successively, the ease with which viruses and bacteria are detected, and the speed with which point of care diagnosis are made. Nonetheless, any of these could not have been possible without the emergence of small biomolecules with excellent specificity, selectivity, and sensitivity toward their targets. These molecules are termed as bioreceptors, having a high binding affinity toward their targets. The applications of bioreceptors are numerous, ranging from bench top analysis to point-of-care diagnosis and treatment. Hence, this chapter is an overview of common bioreceptors, especially affinity-binding receptors, that are crucial in theranostic applications.

***Keywords***: Bioreceptors, affinity, theranostic, antibodies, aptamers

## 5.1 Introduction

Nanosized theranostics are agents on molecular level, that are beneficial for both treatment (therapy) and detection of diseases (diagnosis). The nanomaterials are structurally useful as a novel system for the delivery of drugs, that are incorporated with the agents, that can help in targeted bioimaging applications and are recently termed as nano-theranostics [1]. However, theranostics cannot be discussed without considering bioreceptors as a recognition element is required for any diagnosis, targeted drug delivery or image-guided therapy. Bioreceptors, also known as biological recognition elements, are typically species, that make use of biochemical processes for recognition of analytes (substance of interest). The

**Corresponding author*: michael-danquah@utc.edu

Inamuddin and Tariq Altalhi (eds.) Biosensors Nanotechnology, 2nd Edition, (103–116) © 2023 Scrivener Publishing LLC

bioreceptors contain biologically derived materials or biomimetic components that has potential to bind with or recognize a bio-analyte (enzyme substrate, complementary DNA or antigen) [2]. In the development of theranostics, the bioreceptors to be selected must possess a high binding affinity toward its targets, where the binding affinity is the strength of the binding interaction between a receptor and its target. Thus, a bioreceptor with an excellent binding affinity will selectively and specifically bind with its target [3]. This quality of binding affinity is necessary for the selection or design of a bioreceptor in biosensors and are crucial in drug discovery. Theranostics have proven to be the future of medical science and the ability to produce theranostics with high binding affinity will spearhead the progress of this future. Therefore, this chapter provides an overview of common bioreceptors, especially affinity binding receptors, that are crucial in theranostic applications.

## 5.2 Affinity-Binding Receptors

There is a wide range of affinity binding receptors, that are used for a variety of purposes, from the design of sensors to the diagnosis and treatment of diseases. Affinity binding receptors are molecules with the specific ability to bind with specific target molecules, such as antibodies, enzymes, receptors, biomimetic materials, and nucleic acids [4]. A significant example of a binding phenomenon can be identified in the interaction between an antigen and immune system-produced-antibody.

### 5.2.1 Antibodies

Antibodies are proteins in the globulin, that are formed in the fluids of tissue and serum through; exclusive reaction with the antigen. They are one of the major plasma proteins, commonly referred to as "the first line of defense" against infection. Their significant task is to protect host from microbial pathogens [5] through;

- Prevention of microbial attachment with the host's mucosal surfaces.
- Reduction in the microbial virulence through neutralization of viruses and toxins.
- Facilitation of microbial opsonization through phagocytosis.
- Activation of microbial complement for initiating complement-mediated activities.

Antibodies are Y-shaped molecules with two heavy chains (H chains) and two light chains (L chains), that are held together by disulfide bonds. Each tip of its Y-shaped structure contains a paratope that can specifically bind with an epitope on a corresponding antigen. Each antibody recognizes only a specific antigen due to "antibody specificity," which is an essential feature of antibodies [6]. Furthermore, antibodies are beneficial in identifying very small concentrations of target analyte with their unique antigen-binding properties [7], and this has made antibodies an indispensable tool in the last quarter century.

In recent times, a single antibody type, especially a monoclonal antibody (mAb), is preferred for diagnostic and therapeutic applications [4]. Monoclonal antibodies are types of proteins produced in the laboratory, that can bind to certain specific targets, such as antigens on cancer cell surface. Each mAb is made to bind with only one antigen. In the case of cancer treatment, mAbs can be used, either alone or to carry drugs, toxins, or radioactive substances, directly to the cancer cells [8]. Antibodies are widely applied in the diagnosis and treatment of diseases, drug discovery and delivery, biomedical research, pathogen detection, and biosensor technology.

#### 5.2.1.1 *Advantages of Antibodies*

- They possess high affinity toward their targets.
- They can be created in the laboratory once a host is available (monoclonal antibodies).
- Same quality of the antibody is maintained among the different production batches.

#### 5.2.1.2 *Limitations of Antibodies*

- They are made strictly in animals and cannot be created without the living host (mouse and rat).
- They are larger than other nanomaterials, hence cannot penetrate minute cells.
- They take months to create.
- They are expensive to produce.
- They are difficult to purify and derivatize.
- They work in specific conditions only.

### 5.2.2 Aptamers

Aptamers are short stranded (15 to 60 bases) oligonucleotides with enhanced specificity and affinity toward their molecular target of interest [9]. They are identified from a large random pool of oligonucleotides (RNA and DNA) using the Systematic Evolution of Ligands by Exponential Enrichment (SELEX) method. This well-developed method involves repetitive selection and amplification cycles, where the resultant high affinity oligonucleotides will be enriched and retained for an intended target. This results in the pool enriched to be sequenced and for characterizing the aptamers [10, 11]. Aptamers attain 3-dimensional configurations, which enables them to specifically recognize their target molecules with enhanced affinity, just like antibodies, but with other significant advantages over antibodies [12, 13]. Moreover, aptamers can be used to target a wide range of molecules, such as ions of metal, molecules of small size (10-1000 nm), microbes, peptides, cells, proteins, and tissues. In recent years, aptamers have been under extensive research in various applications, such as detection of cells, food safety, environmental monitoring, diagnosis in clinics, molecular imaging, delivery of drugs, and discovery of biomarkers [12, 14]. Despite the progress made in the production of aptamers, not many aptamers have been approved for use in theranostics [15].

#### 5.2.2.1 *Advantages*

- They are produced chemically produced with few variations in batch-to-batch conversion.

- They are easily modified with unique molecular probes and functional groups.
- In therapeutic applications, it can elicit few immunogenicity.
- They take weeks to produce, which is relatively shorter than it takes to produce antibodies.
- They possess high stability and can be produced against a variety of targets.

#### *5.2.2.2 Limitations*

RNA aptamers possess the ability to degrade rapidly in bio-fluids due to biomolecular interactions. Numerous aptamers have been identified to possess rapid hemo (blood) degradation, which is far too short for most medical applications. The aptamer removal from the bloodstream via renal filtration complicates their therapeutic application. Furthermore, target molecule-based aptamer synthesis is a costly and labor-intensive process, as it must be highly accurate and pure. Similarly, aptamers that recognize specific targets can also bind to molecules with a similar structure in spite of their high specificity [16].

### 5.2.3 DNAzymes

DNAzymes or DNA enzymes also known as deoxyribozymes refer to single-stranded DNA molecules with catalytic capabilities [17]. DNAzymes are not known to exist in nature, however, it can be generated by *in vitro* selection via isolation of rare DNA or RNA sequences from large single-stranded DNA or RNA population with a function of interest [18]. The first DNAzyme was reported in 1994, which was a deoxyribozyme cleaved by RNA, that was lead-dependent. RNA cleavage, ligation of RNA, phosphorylation, modification of peptide sidechains, and hydrolysis of DNA are among the most pursued types of DNA catalyst reactions. Among them, DNAzymes cleaved by RNA are extensively used as antiviral, antibacterial, anti-cancer and anti-inflammatory agents as well as biosensors [19–22]. Various biopharmaceuticals with anti-sense oligonucleotides, small interfering RNAs (siRNAs), ribozymes and DNAzymes have been identified to be beneficial for gene therapy, with messenger RNA (mRNA) as main targets. The key steps involved in this process are recognition of mRNA through hybridization and cleavage [23]. While the DNAzyme field started with cleavage of RNA as its primary goal, its applications in the analytical field have been extensively investigated. These analytical applications can be further classified into metal *in vitro* sensor, cell sensor (bacteria and cancer), sensor to detect nucleic acid and other analyte sensor [23].

DNAzymes are a group of catalysts that possess desirable advantages, such as minimal expense and comfort to fabricate, sturdiness, adaptability and applicability to a wide process range, either standalone or with nanomaterials combination [24, 25]. Similar to other theranostic based on nucleic acid, DNAzymes also encounter significant challenges related to its delivery in *in vivo* condition. Stability and penetration of DNAzymes in the intracellular region are the two factors, that must be of concern for their successful *in vivo* applications. Moreover, DNAzymes are nucleic acids with digestion susceptibility through endonucleases and exonucleases. Additionally, the backbone of phosphate with negative surface charge is a primary cellular uptake barrier due to repulsion from the surface of the cells, electrostatically [26]. It can be noted that the unmodified DNAzyme have relatively short lifetime, however they are used in specific *in vivo* applications, compared to ribozyme [27].

## 5.3 Affinity-Binding Bioreceptors in Theranostic Applications

Researchers have recently established exceptional developments in the medical field however, those upgrades have also led to numerous challenges. The significant attempt for individualized medicine called personalized theranostics was introduced in recent times, where affinity binding receptors serve as a major part [28], as listed in Table 5.1.

### 5.3.1 Cancer

Cancer is a disease with malignancy and is one of the foremost reason for mortality among humans. Initial cancer stages can be cured through resection by surgery with complete prognosis as well as positive response. Nevertheless, sophisticated, and metastatic phases involve supplementary therapeutic intrusion, namely radio/chemotherapy. Notwithstanding substantial advancement in traditional medication for cancer, and its efficacy in exterminating the promptly growing cells of cancer, patients remain sufferers for regrowth of tumor, irrespective of outcomes due to optimistic treatment [29, 30].

Patients with cancer under treatment, through standalone antibodies or in blend with chemo/radiotherapy, or coupled to drugs, extends their survival or a chance at good life [31]. Monoclonal antibodies (mAbs) have been beneficial to the recognition and identification of cancer cell-specific proteins. Hence, distinct mAbs have to be made to target different types of cancer cells. mAb inhibits cancer cells in distinct approaches ways, such as blocking signals that divides cancer cells, act as carrier for drugs toward cancer cells and

**Table 5.1** Summary of theranostic applications of bioreceptors.

| Bioreceptors | Applications |
|---|---|
| Antibodies | • Used as biorecognition elements in biosensors (immunosensors)<br>• Used as therapeutic agents for the treatment of cancer, diabetes, cardiovascular and neurodegenerative diseases<br>• Pregnancy kits<br>• Immunoassays<br>• Tissue typing |
| Aptamers | • Fabrication of sensors (aptasensors)<br>• Production of new drugs (macugen)<br>• Drug delivery system<br>• Bioimaging<br>• Production of reagents |
| DNAzymes | • Catalysis; mRNA degradation (RNA cleavage)<br>• Applied in sensors<br>• Healthcare |

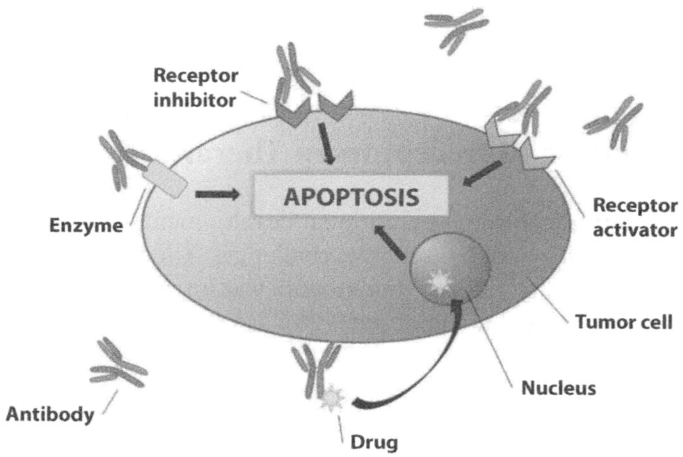


**Figure 5.1** Drug delivery mechanism of antibodies [36], reproduced with permission from © Elsevier (2015).

immunity supporting agent for targeted cancer cell inhibition [32–35]. Figure 5.1 illustrates the ability of antibodies, that are used for drug delivery in the treatment of cancer.

Conjugates of antibody and drug (ADCs) is a novel group of mAb containing target and cytotoxic drugs, as well as linkers. Initially, ADC was intended to elevate the chemotherapy efficacy and lower its toxic reactions. Since the antibody is targeted, the molecules with cytotoxicity can be specifically transferred into the tumor cells to enhance their potential anticancer effect [37]. As of 2016, twenty-five monoclonal antibodies targeted to a total of 16 different antigens had been approved for the treatment of cancer. Currently, new mAbs have been approved and more are under extensive clinical trials [38].

Aptamers are utilized in the treatment of cancer, either for immediate reticence of target molecule activities via targets binding approach, or by targeted delivery of anticancer agents into tumor cells. This leads to a decrement in the normal cell toxicity, decreases the required dose for improving efficacy of the treatment. Furthermore, aptamers can halt or encourage receptors in lymphocytes for the reduction of suppression or boost in the immune response against cancer [39]. Presently, aptamer varieties with ability to target tumor cells have been examined via SELEX and are undergoing laboratory investigations and clinical trials. Examples include Macugen (a United States Food and Drug Administration (USFDA) approved aptamer for age-related macular degeneration) [39], antigen in the membrane with specificity toward anti-prostate (PSMA) aptamer [40], aptamer for anti-nucleolin [41], aptamer with ability to adhere to anti-epithelial cell [42], anti-protein tyrosine kinase 7 (PTK7) aptamer [43], aptamer specific to anti-mucin1 [44]. Figure 5.2 illustrates mechanism of an aptamer-gold nanoparticle interaction with the cancer cells.

DNAzymes have demonstrated a crucial ability to downregulate genes associated with cancer. The DNAzymes with anticancer ability utilizes cancer characteristics, including signals for growth to be self-sufficient, inhibitory growth signals insensitivity, invasion in tissues, continual angiogenesis, metastasis, apoptosis circumvention, and boundless replicative ability to overpower growth of tumor cells [46]. Strategy to intervene gene through a DNAzyme that cleaves RNA has been considered as a favorable substitute for the treatment of cancer due to incredible benefits of DNAzymes. Latest progress on DNAzyme aims at

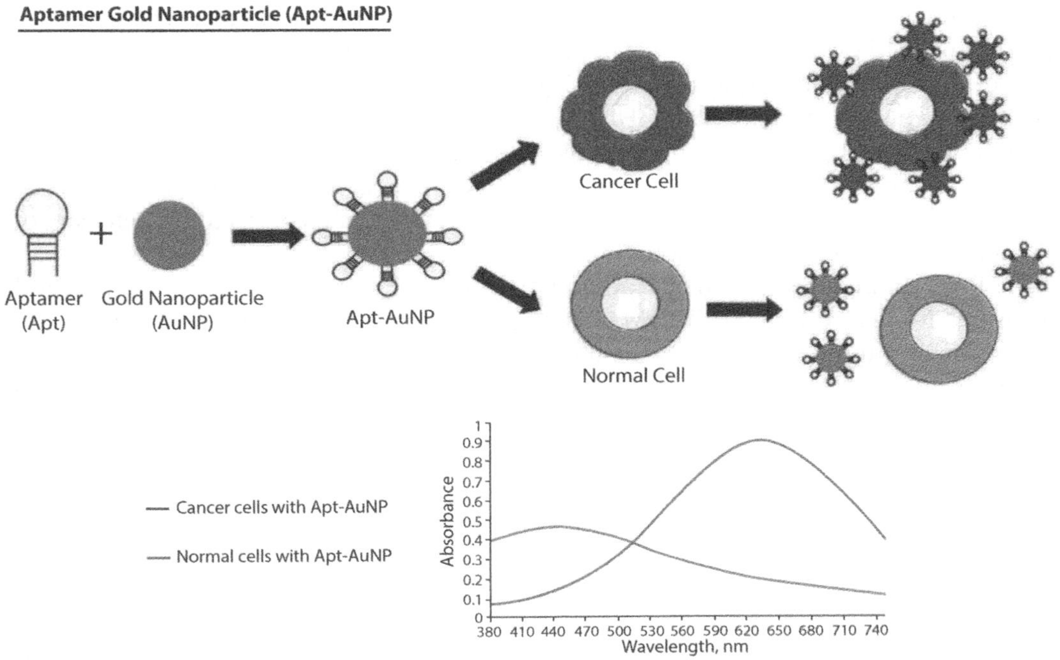


**Figure 5.2** Aptamer-Gold (Ap-Au) nanoparticle used to detect cancer cells [45], Reproduced with permission from © MDPI (2018) [open access].

the development of intelligent nanoplatforms for the effective DNAzyme delivery for the treatment of cancer via gene therapy [47].

### 5.3.2 Diabetes

Diabetes mellitus (DM) is a complicated persistent metabolic disruption of metabolism toward carbohydrate, which is demonstrated through hyperglycemia, during postprandial and/or fasting condition. This disorder has been identified to be due to modified production and secretion of insulin, its resistance in action, or both. Continuing diabetes incidence can lead to impediments including diseases related to cardiovascular pathways, damage to the nerves (neuropathy) and blood vessel damage (angiopathy). Other obstacles, such as renal system damage (nephropathy) and exclusion in the lower limb due to circulation reduction [48]. Type 1 or type 2 are the two exclusive types of diabetes. Type 1 DM (T1DM) will lead to damages through an autoimmune approach to the islet cell secreted-insulin and leads to complete deficiency of insulin [49]. Type 2 DM (TD2M) is a common diabetes type with more than 90–95% cases among diabetes patients. Moreover, insulin resistance is the significant causative factor of T2DM which leads to insensitivity in cells toward glucose existence.

Current type 1 DM treatment methods are concentrated on augmenting insulin secretion for the regulation of glucose however, it does not lead to the destruction of beta cell pathology. A novel affinity binding receptor-based drug named GeNeuro is currently in phase 2 trials, as well as a monoclonal antibody termed as temelimab (GNbAC1), which has a potential for the functional preservation of beta cell without immunity impairment [50].

Autoimmune disorder pathology associated HERV-W, was also related to T1DM and multiple sclerosis (MS) [51]. Recently, T lymphocyte targeting anti-CD3 mAbs shows significant outcomes via clinical trials at T1DM diagnosis. Interesting findings were also expected when B lymphocyte targeting anti-CD20 mAbs were tested. Interleukin-2 blocking mAbs are utilized in pharmaceutical applications, however novel examinations are anticipated with B or T lymphocyte-targeting mAbs. Thus, prevention as well as treatment of diabetes will be efficacious using mAbs in the future, and therefore could constrain the intensity of subcutaneous treatment using insulin [52].

The RNA aptamer utilization has lowered diabetes symptoms in models of animals and are utilized for the treatment of T2DM in humans, due to relationship between murine and the human receptor of insulin. The primary impediment in T2DM is chronic kidney disease (CKD), which appears due to chemokine CCL2 (MCP-1) activity. Aptamer of NOX-E36 was established by NOXXON pharma, which focuses on binding with CCL2, diminishing the kidney ailment progression and is in clinical phase II trial [53–56].

### 5.3.3 Neurodegenerative and Cardiovascular Diseases

Disorders related to neurodegeneration comprise of several circumstances that affect the liberal cell ailment and links in the system of nerves, which is crucial for harmonization, mobility, intensity, cognition, and sensation. These conditions include diseases related to Parkinson's, Alzheimer's, Huntington's, neuron's motor, Ataxia, atrophy in the multiple system and palsy in the gradual supranuclear components [57].

Ailments related to cardiovascular system affects the vessels of the blood and heart function, such as; irregular rhythms of heart (arrhythmias), disease in Aorta and Marfan syndrome, ailment in the artery of coronary position, diseases in the congenital heart, thrombosis in deep-rooted vein, pulmonary embolism, heart attack, cardiomyopathy, incidental vascular disease, vascular disease and stroke [58, 59].

Currently, mAbs have been proven to be safe in disorders related to immune system and oncology. The monoclonal antibodies permitted for cardiological signals therapy are: (1) abciximab (ReoPro®), that are utilized as a treatment modality of antiplatelet during interferences of percutaneous coronary in patients; and (2) Fab digoxin, a monovalent Ig ovine antibody-binding portion was utilized in the digoxin toxicity therapy. Similarly, Simulect® is a mouse-human chimeric mAb to the antigen of CD25, that was appropriate for acute rejection of organ in patients with prophylaxis and transplantation of kidney. It is also beneficial in the acute operational phase as certain initial data to its safety and effectiveness after transplantation of heart [60]. Complete mAbs of humans that were instructed against proprotein convertase subtilisin have proven to be successful in low-density cholesterol lipoprotein (LDL-C) decrement in clinical phase II trials among familial hypercholesterolemia (FH) patients. Additionally, inhibitors of PCSK9 have obtained authorization for FH therapy and disease related to clinical atherosclerosis [61].

Aptamers are also established for brain imaging applications, visualization of neurotransmitter, theranostic of tumors in brain and diseases associated with brain. In various methods, aptamers are radiotracer-labeled for Positron emission tomography (PET), such as fluorine-18 [62, 63]. Precise botulinum neurotoxins recognition, tetrodotoxin, saxitoxin, and paraquat via aptasensors have been developed as well [63]. Activity detection of activities and signaling pathways in neurotransmitter can be utilized for the exploration of functional brain complexities. Likewise, transfer of information between cells by neurotransmitters perform a crucial function in disorders related to neurology. Moreover, dopamine or serotonin specific aptamers have been beneficial for the development of aptasensors for the detection of neurotransmitters, either *in vivo* or *in vitro* [63, 64]. Numerous aptamers can be hypothetically interpreted into theranostic for recognition and supervision of patients affected with cardiovascular disorders. Examples of aptamers investigated for treatment of cardiovascular diseases, include ARC1779 [65]. ARC1779 has been assessed in purpura of thrombocytopenic thrombosis and type 2B von Willebrand disease [66, 67], ARC1172, NU172, ARC183, REG1 (RB006/RB007) [68], and numerous other aptamers are undergoing laboratory investigations and clinical trials. In addition, DNAzymes ameliorates injury due to ischemic myocardial reperfusion and restenosis in instants, which was identified in several *in vivo* studies, that has revealed their benefits in clinical trials [69]. Research also shows that DNAzyme is an effective RNA silencing molecule for potential treatment of multiple polyglutamine (polyQ) diseases [70].

### 5.3.4 Pathogen Detection

Diseases caused by microbes, such as bacteria and virus among humans are critical threat to their health and well-being. Generally, the recognition of these microbes entail various gold standard molecular techniques, such as the culture and reverse transcription-polymerase chain reaction (RT-PCR) [71]. The traditional approach of pathogen recognition typically need separation, biochemical and culturing tests [72]. Hence, the requirement to formulate novel technologies that are highly effective, economical, rapid and simpler to use.

Biosensors for antibody detection have modernized diagnostics for the analyte recognition, such as markers of disease, environmental and food contaminants and illegitimate drugs [73]. The mAbs named 2G1/PPD is a IgG1 that has shown specificity for *Xylella fastidiosa* for the recognition of various strains under distinct subspecies, that can infect different hosts and sequence-Tagged sites. Antiserum 17159-O/IVIA, a polyclonal antibody (PAb) with excellent detection capability has also been produced against *X. fastidiosa* [74]. Various distinct mAbs and recombinant antibodies have been produced for use in biosensors over the years for a variety of viruses, including human immunodeficiency virus (HIV) [75], *Entamoeba histolytica* [76], and *Listeria monocytogenes* [77].

In the last decade, aptamers have been identified to be beneficial in biosensor design for infectious disease diagnosis, due to their exclusive on-site diagnostic ability. Recently, the recognition of infectious ailments based on aptamers' ability has accomplished remarkable advancement and is turning out to be a favorable strategy for detection [78]. Furthermore, aptamers for the detection of infectious bacteria are classified into two groups; (A) pointing prearranged antigens in the surface of the bacterial cell or factors related to bacterial virulence; (B) targeting cells with established or nonestablished target molecules. Thus, aptamers synthesized using bacteria are primarily intended to target *Mycobacterium tuberculosis*, *Listeria monocytogenes*, *Salmonella*, *Escherichia coli* and *Staphylococcus aureus* [78]. Aptamers produced for pathogen detection over the years include; MAA I and MPT64 [79], H63SL2-M6 (HspX, tuberculosis) [80], ssDNA (nucleocapsid protein, SARS-CoV) [81], HPV16 L1 protein targeting ssRNA aptamer in human papillomavirus [82].

DNAzymes have potential applications in the detection of molecules and can be used in the fabrication of sensors for the detection and diagnosis of pathogens. DNAzymes are classified into element for molecular recognition and reporter, which has shown compatibility with technology related to isothermal amplification for enhanced detection of disease-causing bacteria [83]. Recently, studies have shown that RNA-cleaving DNAzymes (RCDs) are stimulated through certain pathogens, such as *Clostridium difficile* and *E. coli.* Furthermore, these RCDs are combined to disparate amplification strategies of DNA to permit optical biosensor development for bacterial detection [84–86]. A group of scientists established a completely disposable and combined device made of paper via integration of protein extraction method sensor, based on DNAzyme and DNA amplified in isothermal condition with portable readout facility with paper. The limit of detection limit was 103 $CFU{\cdot}mL^{-1}$ in the device, where the data can be obtained in 35 min through a smart phone [86]. This shows the prospect of DNAzymes in pathogen detection.

## 5.4 Conclusion

Bioreceptors, such as monoclonal antibodies, aptamers, and DNAzymes, are fast replacing traditional diagnostic and therapeutic technologies, due to their enormous advantages, including ease of manufacturing, low cost, high sensitivity and specificity and the possibility of point-of-care diagnosis. Over the last two decades, scientists and engineers have worked closely to invent novel biomaterials to help solve medical mysteries with huge successes and a few failures. However, extensive research will continue to be made to fix the loopholes in these emerging technologies. Future research will most likely be targeted at making biomaterials with higher stability for *in vivo* therapeutic applications, and those with the ability to generate quicker diagnostic results for point of care diagnosis.

## References

1. Hapuarachchige, S. and Artemov, D., Theranostic pretargeting drug delivery and imaging platforms in cancer precision medicine. *Front. Oncol.*, 10, 1131, 2020.
2. Karunakaran, C., Bhargava, K., Benjamin, R., *Biosensors and bioelectronics*, Elsevier, USA, 2015.
3. Perols, A., Site-specific labeling of affinity molecules for *in vitro* and *in vivo* studies, Doctoral dissertation, KTH Royal Institute of Technology, Stockholm, 2014.
4. Subhash, C.P., *Textbook of microbiology and immunology*, Elsevier, India, 2012.
5. MBL Life Science, Types of antibodies, Medical and Biological Laboratories Co., Ltd, Japan, 2017.
6. Hnasko Robert, M., The biochemical properties of antibodies and their fragments, in: *ELISA*, pp. 1–14, Humana Press, New York, 2015.
7. NCI Dictionaries, National Cancer Institute, National Cancer Institute, 2010, USA, [Online]. Available: https://www.cancer.gov. [Accessed 10 March 2022].
8. Famulok, M. and Mayer, G., Aptamer modules as sensors and detectors. *Acc. Chem. Res.*, 44, 12, 1349–1358, 2011.
9. Sefa, K., Shangguan, D., Xiong, X., O'donoghue, M.B., Tan, W., Development of DNA aptamers using cell-SELEX. *Nat. Protoc.*, 5, 6, 1169–1185, 2010.
10. Ohuchi, S., Cell-SELEX technology, in: *BioResearch Open Access*, pp. 265–272, 2012, www.liebertpub.com.

11. Savory, N., Abe, K., Saito, T., Ikebukuro, K., History of aptamer development, in: *Aptamers Tools for Nanotherapy*, pp. 16–32, Pan Stanford Publishing, Singapore, 2016.
12. Mescalchin, A. and Restle, T., Oligomeric nucleic acids as antivirals. *Molecules*, 16, 2, 1271–1296, 2011.
13. Shigdar, S., Schrand, B., Giangrande, P.H., de Franciscis, V., Aptamers: Cutting edge of cancer therapies. *Mol. Ther.*, 29, 8, 2396–2411, 2021.
14. Lakhin, A.V., Tarantul, V.Z., Gening, L., Aptamer: Problems, solutions and prospects. *Actanature*, V, 4, 34–43, 2013.
15. Xu, J. and Wang, L., *Nano-inspired biosensors for protein assay with clinical applications*, Elsevier, The Netherlands, 2019.
16. Achenbach, J., Chiuman, W., Cruz, R., Li, Y., DNAzymes: From creation *in vitro* to application *in vivo*. *Curr. Pharm. Biotechnol.*, 5, 4, 321–336, 2004.
17. Ponce-Salvatierra, A., Boccaletto, P., Bujnick, J.M., DNAmoreDB, a database of DNAzymes. *Nucleic Acids Res.*, 49, D1, 76–81, 2021.
18. Silverman, S.K., Catalytic DNA: Scope, applications, and biochemistry of deoxyribozymes. *Trends Biochem. Sci.*, 41, 7, 595–609, 2016.
19. Hwang, K., Hosseinzadeh, P., Lu, Y., Biochemical and biophysical understanding of metal ion selectivity of DNAzymes. *Inorg. Chim. Acta*, 452, 12–24, 2016.
20. Fokina, A., Stetsenko, D., François, J., DNA enzymes as potential therapeutics: Towards clinical application of 10-23 DNAzymes. *Expert Opin. Biol. Ther.*, 15, 5, 689–711, 2015.
21. Zhou, W., Ding, J., Liu, J., Theranostic dnazymes. *Theranostics*, 7, 4, 1010, 2017.
22. Silverman, S.K., Catalytic DNA (deoxyribozymes) for synthetic applications-current abilities and future prospects. *Chem. Commun.*, 30, 3467–3485, 2008.
23. Morrison, D., Rothenbroker, M., Li, Y., DNAzymes: Selected for applications. *Small Methods*, 2, 3, 1700319, 2018.
24. Patil, S.D., Rhodes, D.G., Burgess, D.J., DNA-based therapeutics and DNA delivery systems: A comprehensive review. *AAPS J.*, 7, 1, 61–77, 2005.
25. Dass, C.R., Saravolac, E.G., Li, Y., Sun, L.Q., Cellular uptake, distribution, and stability of 10-23 deoxyribozymes. *Antisense Nucleic Acid Drug Dev.*, 12, 5, 289–299, USA, 2002.
26. Tiwari, A., Patra, H.K., Choi, J.W., *Advanced theranostic materials*, Wiley-Scrivener, 2015.
27. Shamaileh, H.A., Xiang, D., Wang, T., Wang, Y., Duan, W., Shigdar, S., Stem-cell-specific aptamers for targeted cancer therapy, in: *Aptamers Tools for Nanotherapy and Molecular Imaging*, pp. 113–138, Pan Stanford Publishing, Singapore, 2016.
28. Modjtahedi, H. and Ali, S., Therapeutic application of monoclonal antibodies in cancer: Advances and challenges. *Br. Med. Bull.*, 104, 1, 41–59, 2012.
29. DeVita, V., Lawrence, T., Weinberg, R., *Devita DePinho, R. Hellman, and Rosenberg's cancer: Principles & practice of oncology*, pp. 2068–2070, Lippincott Williams and Wikins Press, USA, 2015.
30. Seebacher, N., Stacy, A., Porter, G., Merlot, A., Clinical development of targeted and immune based anti-cancer therapies. *J. Exp. Clin. Cancer Res.*, 38, 1, 1–39, 2019.
31. Yang, L., Yu, H., Dong, S., Zhong, Y., Hu, S., Recognizing and managing on toxicities in cancer immunotherapy. *Tumor Biol.*, 39, 3, 1010428317694542, 2017.
32. Chiavenna, S.M., Jaworski, J.P., Vendrell, A., State of the art in anti-cancer mAbs. *J. Biomed. Sci.*, 24, 1, 1–2, 2017.
33. Neves, H. and Kwok, H.F., Recent advances in the field of anti-cancer immunotherapy. *BBA Clin.*, 3, 280–288, 2015.
34. B. PEG, Biopharma PEG, Biopharma PEG, 30 October 2019, [Online]. Available: http://www.biochempeg.com. [Accessed 25 March 2022].
35. Baldo, B.A., Monoclonal antibodies approved for cancer therapy, in: *Safety of Biologics Therapy*, pp. 57–140, Springer, Cham, 2016.

36. Fu, Z. and Xiang, J., Aptamers, the nucleic acid antibodies, in cancer therapy. *Int. J. Mol. Sci.*, 21, 8, 2793, 2020.
37. Lupold, S.E., Hicke, B.J., Lin, Y., Coffey, D.S., Identification and characterization of nuclease-stabilized RNA molecules that bind human prostate cancer cells via the prostate-specific membrane antigen. *Cancer Res.*, 62, 14, 4029–4033, 2002.
38. Shieh, Y.A., Yang, S.J., Wei, M.F., Shieh, M.J., Aptamer-based tumor-targeted drug delivery for photodynamic therapy. *ACS Nano*, 4, 3, 1433–1442, 2010.
39. Xiang, D., Shigdar, S., Qiao, G., Wang, T., Kouzani, A.Z., Zhou, S.F., Kong, L., Li, Y., Pu, C., Duan, W., Nucleic acid aptamer-guided cancer therapeutics and diagnostics: The next generation of cancer medicine. *Theranostics*, 5, 1, 23–42, 2015.
40. Zhu, G., Zheng, J., Song, E., Donovan, M., Zhang, K., Liu, C., Tan, W., Self-assembled, aptamer-tethered DNA nanotrains for targeted transport of molecular drugs in cancer theranostics. *Proc. Natl. Acad. Sci.*, 110, 20, 7998–8003, 2013.
41. Ferreira, C.S.M., Matthews, C.S., Missailidis, S., DNA aptamers that bind to MUC1 tumour marker: Design and characterization of MUC1-binding single-stranded DNA aptamers. *Tumor Biol.*, 27, 6, 289–301, 2006.
42. Ruiz Ciancio, D., Vargas, M.R., Thiel, W.H., Bruno, M.A., Giangrande, P.H., Mestre, M.B., Aptamers as diagnostic tools in cancer. *Pharmaceuticals*, 11, 3, 86, 2018.
43. Thomas, I., Gaminda, K., Jayasinghe, C., Abeysinghe, D., Senthilnithy, R., DNAzymes, novel therapeutic agents in cancer therapy: A review of concepts to applications. *J. Nucleic Acids*, 2021, 9365081, 2021.
44. Huo, W., Li, X., Wang, B., Zhang, H., Zhang, J., Yang, X., Jin, Y., Recent advances of DNAzyme-based nanotherapeutic platform in cancer gene therapy. *Biophys. Rep.*, 6, 6, 256–265, 2020.
45. American Diabetes Association, Standards of medical care in diabetes. *Diabetes Care*, 33, 14–80, 2014.
46. Cooke, D. and Plotnick, L., Type 1 diabetes mellitus in pediatrics. *Pediatr. Rev.*, 29, 11, 374–385, 2008.
47. Curtin, F., Champion, B., Davoren, P., Duke, S., Ekinci, E.I., Gilfillan, C., Morbey, C., Nathow, T., O'Moore-Sullivan, T., O'Neal, D., Roberts, A., A safety and pharmacodynamics study of temelimab, an antipathogenic human endogenous retrovirus type W envelope monoclonal antibody, in patients with type 1 diabetes. *Diabetes Obes. Metab.*, 22, 7, 1111–1121, 2020.
48. Curtin, F., Bernard, C., Levet, S., Perron, H., Porchet, H., Médina, J., Malpass, S., Lloyd, D., Simpson, R., A new therapeutic approach for type 1 diabetes: Rationale for GNbAC1, an anti-HERV-W-Env monoclonal antibody. *Diabetes Obes. Metab.*, 20, 9, 2075–2084, 2018.
49. Philips, J.C., Keymeulen, B., Mathieu, C., Scheen, A.J., Monoclonal antibodies in diabetes: Almost approaching the dream. *Rev. Méd. Liège*, 64, 5–6, 327–333, 2009.
50. Oberthür, D., Achenbach, J., Gabdulkhakov, A., Buchner, K., Maasch, C., Falke, S., Rehders, D., Klussmann, S., Betzel, C., Crystal structure of a mirror-image L-RNA aptamer (Spiegelmer) in complex with the natural L-protein target CCL2. *Nat. Commun.*, 6, 1, 1–11, 2015.
51. Ninichuk, V., Clauss, S., Kulkarni, O., Schmid, H., Segerer, S., Radomska, E., Eulberg, D., Buchner, K., Selve, N., Klussmann, S., Anders, H.J., Late onset of Ccl2 blockade with the Spiegelmer mNOX-E36–3'PEG prevents glomerulosclerosis and improves glomerular filtration rate in db/db mice. *Am. J. Pathol.*, 172, 3, 628–637, 2008.
52. Maasch, C., Buchner, K., Eulberg, D., Vonhoff, S., Klussmann, S., Physicochemical stability of NOX-E36, a 40mer L-RNA (Spiegelmer) for therapeutic applications. *Nucleic Acids Symp. Ser.*, 52, 1, 61–62, 2008.
53. Peter O'Donnell Jr. Brain Institute, *Neurodegenerative disorders*, UT Southwestern Medical Center, Dallas, 2021.

54. Felson, S., WebMD, 2 November 2021. [Online]. Available: http://www.webmd.com. [Accessed 22 March 2022].
55. Segovia, J., Rodríguez-Lambert, J.L., Crespo-Leiro, M.G., Almenar, L., Roig, E., Gómez-Sánchez, M.A., Lage, E., Manito, N., Alonso-Pulpón, L., A randomized multicenter comparison of basiliximab and muromonab (OKT3) in heart transplantation: SIMCOR study. *Transplantation*, 81, 11, 1542–1548, 2006.
56. Gencer, B., Laaksonen, R., Buhayer, A., Mach, F., Use and role of monoclonal antibodies and other biologics in preventive cardiology. *Swiss Med. Wkly.*, 14, w14179, 2015.
57. Lange, C.W., VanBrocklin, H.F., Taylor, S.E., Photoconjugation of 3-azido-5-nitrobenzyl-[18F] fluoride to an oligonucleotide aptamer. *J. Label. Compd. Radiopharm.*, 45, 3, 257–268, 2002.
58. Ozturk, M., Nilsen-Hamilton, M., Ilgu, M., Aptamer applications in neuroscience. *Pharmaceuticals*, 14, 12, 1260, 2021.
59. Si, B. and Song, E., Recent advances in the detection of neurotransmitters. *Chemosensors*, 6, 1, 1, 2018.
60. Arzamendi, D., Dandachli, F., Théorêt, J.F., Ducrocq, G., Chan, M., Mourad, W., Gilbert, J.C., Schaub, R.G., Tanguay, J.F., Merhi, Y., An anti-von Willebrand factor aptamer reduces platelet adhesion among patients receiving aspirin and clopidogrel in an *ex vivo* shear-induced arterial thrombosis. *Clin. Appl. Thromb./Hemost.*, 17, 6, 70–78, 2011.
61. Jilma-Stohlawetz, P., Gorczyca, M.E., Jilma, B., Siller-Matula, J., Gilbert, J.C., Knöbl, P., Inhibition of von Willebrand factor by ARC1779 in patients with acute thrombotic thrombocytopenic purpura. *Thromb. Haemost.*, 105, 3, 545–552, 2011.
62. Mayr, F.B., Knöbl, P., Jilma, B., Siller-Matula, J.M., Wagner, P.G., Schaub, R.G., Gilbert, J.C., Jilma-Stohlawetz, P., The aptamer ARC1779 blocks von Willebrand factor–dependent platelet function in patients with thrombotic thrombocytopenic purpura *ex vivo*. *Transfusion*, 50, 5, 1079–1087, 2010.
63. Wang, P., Yang, Y., Hong, H., Zhang, Y., Cai, W., Fang, D., Aptamers as therapeutics in cardiovascular diseases. *Curr. Med. Chem.*, 18, 27, 4169–4174, 2011.
64. Benson, V.L., Khachigian, L.M., Lowe, H.C., DNAzymes and cardiovascular disease. *Br. J. Pharmacol.*, 154, 4, 741–748, 2008.
65. Zhang, N., Bewick, B., Schultz, J., Tiwari, A., Krencik, R., Zhang, A., Adachi, K., Xia, G., Yun, K., Sarkar, P., Ashizawa, T., DNAzyme cleavage of CAG repeat RNA in polyglutamine diseases. *Neurotherapeutics*, 18, 3, 1710–1728, 2021.
66. Guliy, O.I., Zaitsev, B.D., Larionova, O.S., Borodina, I.A., Virus detection methods and biosensor technologies. *Biophysics*, 64, 6, 890–897, 2019.
67. Farooq, U., Yang, Q., Ullah, M.W., Wang, S., Bacterial biosensing: Recent advances in phage-based bioassays and biosensors. *Biosens. Bioelectron.*, 118, 204–216, 2018.
68. Sharma, S., Byrne, H., O'Kennedy, R.J., Antibodies and antibody-derived analytical biosensors. *Essays Biochem.*, 60, 1, 9–18, 2016.
69. Gorris, M.T., Sanz, A., Peñalver, J., López, M.M., Colomer, M., Marco-Noales, E., Detection and diagnosis of Xylella fastidiosa by specific monoclonal antibodies. *Agronomy*, 11, 1, 48, 2020.
70. Encarnação, J.M., Rosa, L., Rodrigues, R., Pedro, L., da Silva, F.A., Gonçalves, J., Ferreira, G.N., Piezoelectric biosensors for biorecognition analysis: Application to the kinetic study of HIV-1 Vif protein binding to recombinant antibodies. *J. Biotechnol.*, 132, 2, 142–148, 2007.
71. Grewal, Y.S., Shiddiky, M.J., Gray, S.A., Weigel, K.M., Cangelosi, G.A., Trau, M., Label-free electrochemical detection of an *Entamoeba histolytica* antigen using cell-free yeast-scFv probes. *Chem. Commun.*, 49, 15, 1551–1553, 2013.
72. Benhar, I., Eshkenazi, I., Neufeld, T., Opatowsky, J., Shaky, S., Rishpon, J., Recombinant single chain antibodies in bioelectrochemical sensors. *Talanta*, 55, 5, 899–907, 2001.

73. Li, H.Y., Jia, W.N., Li, X.Y., Zhang, L., Liu, C., Wu, J., Advances in detection of infectious agents by aptamer-based technologies. *Emerging Microbes Infect.*, 9, 1, 1671–1681, 2020.
74. Chen, Y., Liu, X., Guo, S., Cao, J., Zhou, J., Zuo, J., Bai, L., A sandwich-type electrochemical aptasensor for Mycobacterium tuberculosis MPT64 antigen detection using C60NPs decorated N-CNTs/GO nanocomposite coupled with conductive PEI-functionalized metal-organic framework. *Biomaterials*, 2, 16, 119–253, 2019.
75. Lavania, S., Das, R., Dhiman, A., Myneedu, V.P., Verma, A., Singh, N., Sharma, T.K., Tyagi, J.S., Aptamer-based TB antigen tests for the rapid diagnosis of pulmonary tuberculosis: Potential utility in screening for tuberculosis. *ACS Infect. Dis.*, 4, 12, 1718–1726, 2018.
76. Zhou, P., Yang, X.L., Wang, X.G., Hu, B., Zhang, L., Zhang, W., Si, H.R., Zhu, Y., Li, B., Huang, C.L., Chen, H.D., A pneumonia outbreak associated with a new coronavirus of probable bat origin. *Nature*, 579, 7798, 270–273, 2020.
77. Zhu, L., Han, J., Wang, Z., Yin, L., Zhang, W., Peng, Y., Nie, Z., Competitive adsorption on gold nanoparticles for human papillomavirus 16 L1 protein detection by LDI-MS. *Analyst*, 144, 22, 6641–6646, 2019.
78. Ma, X., Ding, W., Wang, C., Wu, H., Tian, X., Lyu, M., Wang, S., DNAzyme biosensors for the detection of pathogenic bacteria. *Sens. Actuators B: Chem.*, 331, 129422, 2021.
79. Liu, M., Zhang, Q., Brennan, J.D., Li, Y., Graphene-DNAzyme-based fluorescent biosensor for *Escherichia coli* detection. *MRS Commun.*, 8, 3, 687–694, 2018.
80. Liu, M., Yin, Q., Brennan, J.D., Li, Y., Selection and characterization of DNA aptamers for detection of glutamate dehydrogenase from *Clostridium difficile*. *Biochimie*, 145, 151–157, 2018.
81. Sun, Y., Chang, Y., Zhang, Q., Liu, M., An origami paper-based device printed with DNAzyme-containing DNA superstructures for *Escherichia coli* detection. *Micromachines*, 10, 8, 531, 2019.
82. Zhu, L., Han, J., Wang, Z., Yin, L., Zhang, W., Peng, Y., Nie, Z., Competitive adsorption on gold nanoparticles for human papillomavirus 16 L1 protein detection by LDI-MS. *Analyst*, 144, 22, 6641–6646, 2019.
83. Xie, S., Ji, Z., Suo, T., Li, B., Zhang, X., Advancing sensing technology with CRISPR: From the detection of nucleic acids to a broad range of analytes–A review. *Anal. Chim. Acta*, 1185, 338848, 2021.
84. Chang, D., Zakaria, S., Esmaeili Samani, S., Chang, Y., Filipe, C. D., Soleymani, L., ... Li, Y., Functional nucleic acids for pathogenic bacteria detection. *Acc. Chem. Res.*, 54, 18, 3540–3549, 2021.
85. Song, L., Zhuge, Y., Zuo, X., Li, M., Wang, F., DNA walkers for biosensing development. *Adv. Sci.*, 9, 18, 2200327, 2022.
86. Manochehry, S., Liu, M., Chang, D., Li, Y., Optical biosensors utilizing graphene and functional DNA molecules. *J. Mater. Res.*, 32, 15, 2973–2983, 2017.

# 6

# Biosensors for Glucose Monitoring

**Hoang Vinh Tran**

*School of Chemical Engineering, Hanoi University of Science and Technology, Hanoi, Vietnam*

***Abstract***

Diabetes is a challenging worldwide health problem in the 21st century. In 2021, a total of 537 million people have diabetes, and this number is expected to increase to 643 million in 2030 and 783 million in 2045. The disease is a major cause of cardiovascular or chronic kidney diseases, which can lead to blindness in adults. Unfortunately, it is an intractable disease. Currently, the standard method for screening diabetes cases is measuring glucose concentration in the blood. This method is currently the standard for diabetes management. However, almost half of diabetes cases remain undiagnosed. Therefore, low-cost and effective glucose biosensor-based devices is urgently needed for diabetes screening. Since the invention of first-generation enzyme-based glucose biosensors (GBs) in 1962, continuous efforts have been made to develop enzyme-based GBs mainly based on colorimetric or electrochemical transducers. Colorimetric methods are often used in clinics, and electrochemical GBs are used for personal testing purposes. This chapter briefs the history and basic working principles of GBs and current developments. Recent trends in GB development involving nanomaterials, paper analytical devices, and wearable GBs were discussed.

***Keywords*:** Glucose detection, glucose meter, glucose biosensor, nanomaterial, electrochemistry, colorimetric, paper analytical device, sweat-based sensors

## Abbreviations

| | |
|---|---|
| IUPAC | International Union of Pure and Applied Chemistry |
| WHO | World Health Organisation |
| IDF | International Diabetes Federation |
| GBs | Glucose biosensors |
| ABTS | 2,2′-bis-3-azino-6-Sulfonic acid-ethylbenzothiozoline |
| TMB | 3, 3′, 5, 5′-Tetramethylbenzidine |
| OPD | o-Phenylenediamine |

*Email*: hoang.tranvinh@hust.edu.vn

Inamuddin and Tariq Altalhi (eds.) Biosensors Nanotechnology, 2nd Edition, (117–144) © 2023 Scrivener Publishing LLC

| | |
|---|---|
| 4-AAP | 4-Aminoatipyrene |
| PANi | Polyaniline |
| CS | Chitosan |
| PPy | Polypyrrole |
| FAD | Flavin adenine dinucleotide |
| DNA | Deoxyribonucleic acids |
| GA | Glutaraldehyde |
| ITO | Indium–tin oxide |
| IDE | Interdigitated electrode |
| SPCE | Screen-printed carbon electrode |
| SAM | Self-assembled monolayer |
| GCE | Glassy carbon electrode |
| AuE | Gold electrode |
| FN-CNs | FeOOH/N-doped carbon nanosheets |
| PtNPs | Platinum nanoparticles |
| PdNPs | Palladium nanoparticles |
| CuNPs | Copper nanoparticles |
| AuNPs | Gold nanoparticles |
| Ag/AuNPs | Silver/gold bimetallic nanoparticles |
| AgNPs | Silver nanoparticles |
| GQDs | Graphene quantum dots |
| AgNPs/GQDs | Silver nanoparticles/graphene quantum dots nanocomposite |
| GO | Graphene oxide |
| LoD | Limit of detection |
| HRP | Enzyme horseradish peroxidase |
| GOx | Enzyme glucose oxidase |
| CV | Cyclic voltammetry |
| SWV | Square wave voltammetry |
| SPR | Surface plasmon resonance |
| dPOCT | Disposable platform for point-of-care |
| PADs | Paper analytical devices |

## 6.1 Introduction

Glucose ($C_6H_{12}O_6$, $M_w$ = 180 g $mol^{-1}$), a simple sugar, is the most important source of energy in all organisms. Glucose concentration in the blood or blood glucose level is naturally regulated as a part of metabolic homeostasis, and fluctuations in blood glucose levels are considered biomarkers for several diseases. In a healthy person, blood glucose level is normally in a concentration range of 3 to 8 mM. In a diabetic person, it is about 9 to 40 mM. Diabetes is featured by high glucose concentration in the blood for a long period. Diabetes mellitus is the leading cause of morbidity and mortality. Currently, it is a major health problem in modern societies. Since 1950, the International Diabetes Federation (IDF) was established, with a mission to promote diabetes management, prevention and treatment worldwide. To date, the IDF consists of over 230 national diabetes associations from more than 160 countries and territories. Unfortunately, diabetes prevalence continues to increase. The IDF revealed that 537 million people had diabetes in 2021 and 643 million people are expected to have the disease in 2030; and 783 million, in 2045. Therefore, blood glucose meters for the diabetes screening, also known as glucose biosensors (GBs), have been in high demand [1–4]. The first blood glucose meter was designed in 1965, which was constructed according to color changes in test strips. After blood samples are gently washed off, the strips are inserted into readers. However, this tool is a chemical sensor rather than a biosensor. In 1962, Clark reported the first enzyme-based GB, which was constructed on a platinum (Pt) electrode. Since then, generations of GBs have been developed, which are summarized in Table 6.1 [5]. However, although a variety of GBs are already commercialized, no significant changes in their structures have been observed over the last decades.

**Table 6.1** Generated glucose biosensors. Reprinted from Ref. [5] (copyright 2010 with permission from MDPI).

| Year | Event |
|---|---|
| 1962 | First description of a biosensor by Clark and Lyons |
| 1967 | First practical enzyme electrode by Updike and Hicks |
| 1973 | Glucose enzyme electrode based on detection of hydrogen peroxide |
| 1975 | Relaunch of first commercial biosensor, *i.e.*, YSI analyser |
| 1976 | First bedside artificial pancreas (Miles) |
| 1982 | First needle-type enzyme electrode for subcutaneous implantation by Shichiri |
| 1984 | First ferrocene mediated amperometric glucose biosensor by Cass |
| 1987 | Launch of the MediSense ExacTech blood glucose biosensor |
| 1999 | Launch of a commercial *in vivo* glucose sensor (MiniMed) |
| 2000 | Introduction of a weareable noninvasive glucose monitor (GlocoWatch) |

### 6.1.1 Definitions and Generalities

The International Union of Pure and Applied Chemistry (IUPAC) defined biosensors as analytical devices incorporating biological elements as capture probes combined with physical transducers [6]. Therefore, a biosensor is a self-contained integrated device that can be used in the quantitative or semi-quantitative analysis of targets [7, 8]. Biosensors consist of three major components: a biorecognition element (also called biocapture probe or bioreceptor), a surface for biocapture probe immobilization (also called a carrier surface) and a physical transducer for conversion of the biochemical reaction between a biocapture probe and analytical target (in the recognition reaction) into a measurable signal. In general, a biocapture probe can be a single-strain deoxyribonucleic acid (DNA), enzyme, antibody or aptamer for the specific recognition of a self-analytical target, whereas the immobilization surface can be a polymer film [9–14], nanomaterial layer [15], sol–gel film [16] or self-assembled monolayer [17]. Nearly all glucose biosensors have been constructed by using glucose oxidase enzyme (GOx) as a specific capture probe for glucose and are named enzymatic glucose biosensors. Figure 6.1 illustrates the typical structure of an enzymatic glucose biosensor.

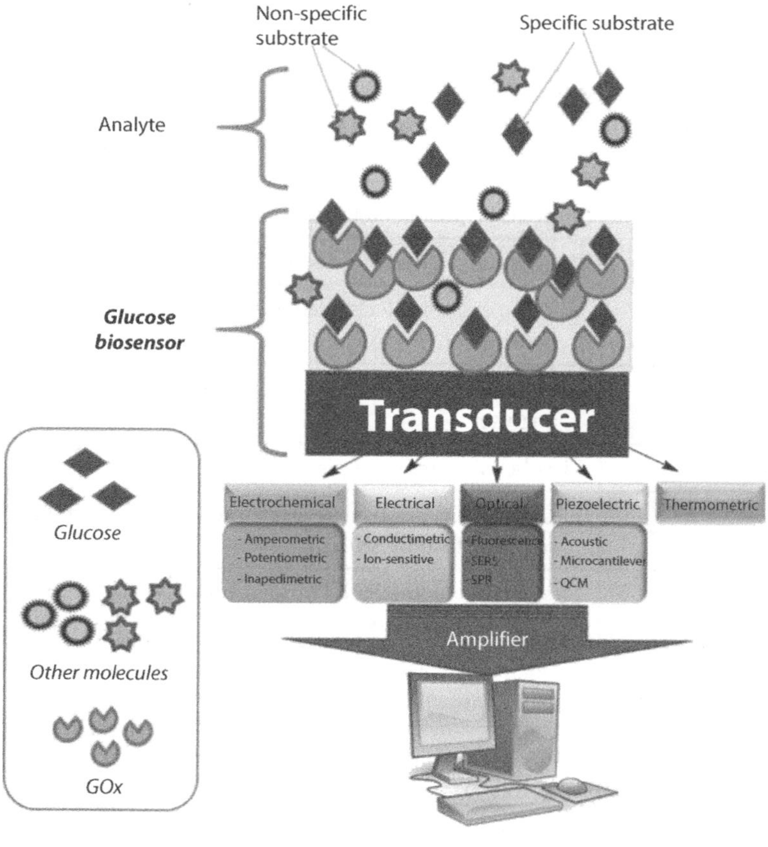


**Figure 6.1** Schematic of the structure of an enzymatic glucose biosensor. This figure is re-drawn.

### 6.1.2 Enzymatic Glucose Biosensor

Enzyme sensors have a specificity for target molecules (their substrates). Various enzymes are used [18–24], such as oxidoreductase for lactate, malate, ascorbate, alcohol, cholesterol, glycerol, and fructose; transferase enzymes for the analysis of acetic acid and determination of xenobiotics, such as captan and atrazine; hydrolase for sucrose; lyase for citric acid; and ligase for DNA point mutation detection [18, 23, 25, 26]. In addition, oxidase enzymes, such as horseradish peroxidase (HRP), is used in determining medicines or drug molecules, such as clozapine [27], phenolthiazines [28], and rifampicin [29]. These applications are summarized in a mini-review [30]. This approach can also be applied for detection of pesticides [31]. Most glucose sensors are fabricated through enzymatic approaches, which rely on using GOx as a specific probe and peroxidase enzyme as a signal probe. This method is commonly used in monitoring glucose in hospitals. The principle of the method is the use of GOx to oxidize glucose in a sample into gluconic acid. This reaction produces hydrogen peroxide ($H_2O_2$), which is in turn reduced by a peroxidase enzyme (HRP is usually used). This subsequent reaction requires a co-substrate, which is the color indicator (Figure 6.2). 2,2′-bis-3-azino-6-Sulfonic acid-ethylbenzothiozoline (ABTS), 3,3′,5,5′-tetramethylbenzidine (TMB), o-phenylenediamine, or 4-aminoatipyrene (4-AAP)/phenol are the most common color indicators.

Color intensity is related to the amount of glucose in a sample and thus used to generate a calibration curve for determining glucose concentration (Table 6.2). Similarly, glucose can be measured electrochemically with GOx and HRP enzymes, but a redox indicator is used instead of a colorimetric one (Table 6.3).

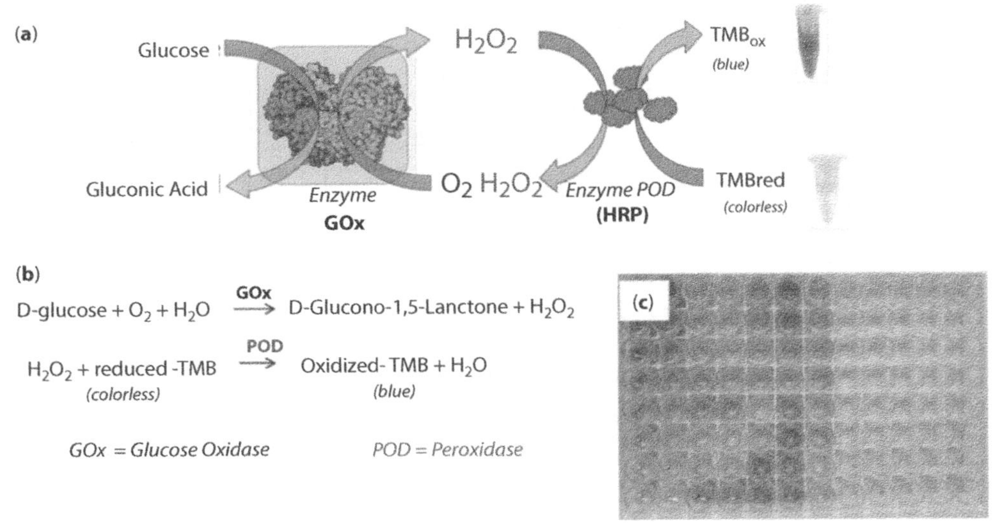


**Figure 6.2** (a) Illustration of the working principle of a colorimetric glucose sensor using two enzymes (GOx and HRP); (b) biochemical reactions in the detection process and (c) color intensity proportional to glucose concentration in sample; the system blue case oxidized TMB. This figure is re-drawn.

**Table 6.2** Enzyme-based colorimetric glucose biosensors.

| Capture molecules | Signal elements | Designing approach | Detection methods | Detection range | LoD | Refs. |
|---|---|---|---|---|---|---|
| GOx | AgNPs solution | Direct | UV-Vis measurement or naked-eyes observation | 0.5–8 mM | 30 μM | [32] |
| GOx | AgNPs@rGO and TMB | Indirect | UV-Vis measurement | 125 μM to 1 mM | 40 μM | [33] |
| GOx | FeOOH/N-doped carbon nanosheets (FN-CNs) and TMB | Indirect | UV-Vis measurement | 8 μM to 0.8 mM | 0.2 μM | [49] |
| GOx | $Co_3O_4$-$CeO_2$ and TMB | Indirect | UV-Vis measurement | 5 μM to 1.5mM | 0.21 μM | [83] |
| GOx | Cu-Pd/rGO and TMB | Indirect | Naked-eyes detection | 0.2–50 μM | 0.29 μM | [82] |
| GOx | Graphene oxide–$Fe_3O_4$ magnetic nanocomposites and TMB | Indirect | UV-Vis measurement | 2–200 μM | 0.74 μM | [48] |
| GOx | $Fe_3O_4$ nanoparticles and ABTS | Indirect | UV-Vis measurement | 50 μM to 1 mM | - | [38] |
| GOx | $MoS_2$@$MgFe_2O_4$ composite and TMB | Indirect | UV-Vis measurement | 5.0–200μM | 2.0 μM | [45] |
| GOx | GQDs/CuO nanocomposites and TMB | In direct | UV-Vis measurement | 2–100 μM | 0.59 μM | [46] |
| GOx | MFe2O4 (M=Mg, Ni, Cu) magnetic nanoparticles (MNPs) and ABTS | Indirect | UV-Vis measurement | 0.94–25.0 μM | 0.45 μM | [47] |
| GOx | Pt nanoclusters (Pt NCs) and TMB | Indirect | UV-Vis measurement | 0–200 μM | 0.28 μM | [56] |
| GOx | Poly(diallyldimethylammonium)-stabilized Au@Ag heterogeneous nanorods (NRs) and ABTS | Indirect | UV-Vis measurement | 0.05–20 mM | 39 μM | [85] |
| GOx | Manganese selenide nanoparticles (MnSe NPs) with graphene-like structures and TMB | Indirect | UV-Vis measurement | 8.0–50 μM | 1.6 μM | [54] |
| GOx | Vanadium disulfide ($VS_2$) nanosheets and TMB | Indirect | UV-Vis measurement | 5–250 μM | 1.5 μM | [52] |
| GOx | Greigite magnetic nanoparticles ($Fe_3S_4$-MNPs) and TMB | Indirect | UV-Vis measurement | 2–100 μM | 0.16 μM | [53] |
| GOx | 2D ultrathin nanosheets of Co–Al layered double Hydroxides and TMB | Indirect | UV-Vis measurement | 0.05–0.5 mM | 0.05 mM | [50] |

**Table 6.3** Enzyme-based electrochemical glucose biosensors.

| Capture molecules | Signal elements | Designing approach | Electrode | Detection methods | Detection range | LoD | Refs. |
|---|---|---|---|---|---|---|---|
| GOx | AgNPs | Direct detection | GCE | CV | 0.1–10 mM | 0.01 mM | [86] |
| GOx | Palladium-helical carbon nanofiber (Pd-HCNF) hybrid nanostructures | Direct detection | GCE | CV | 0.06–6.0 mM | 0.03 mM | [61] |
| GOx | Hydroxyapatite (HAp) nanorods/ reduced graphene oxides | Direct detection | GCE | CV | 0.1–11.5 mM | 0.03 mM | [87] |
| GOx | Os(bpy)PVI redox polymer/ CNTs | Direct detection | Graphite rods | CV | 0–11.5 mM | - | [88] |
| GOx | Au electrode | Direct detection | Gold electrode | - | 0.025–25 mM | 0.055 mM | [89] |
| GOx | Plant Produced Mn Peroxidase (PPMP) | Direct detection | Gold electrode | LSV | 20.0 μM to 15.0 mM | 2.9 μM | [90] |
| GOx | Poly(glycidyl methacrylate-co-vinylferrocene) (poly(GMA-co-VFc) | Direct detection | Pencil graphite electrode | Amperometry | 1–16 mM | 2.7 μM | [91] |
| GOx | Polypyrrole/gold nanoparticles (PPy/AuNPs) | Direct detection | Graphite rod | | 0.09–19.9 mM | 0.2 mM | [92] |
| GOx | AuNPs | Direct detection | Graphite rod (GR) | Amperometry | 0.1–10 mM | 0.08 mM | [93] |
| GOx | Graphene/AuNPs/chitosan nanocomposites | Direct detection | Gold electrode | Amperometry | 2–10 mM | 0.18 mM | [94] |
| GOx | Prussian blue (PB) | Direct detection | Gold electrode | Amperometry and EIS | 0–30 mM and 30–130 mM | - | [95] |
| GOx | Nanostructured copper oxide (Nano-CuO) | Direct detection | FTO | EIS | 0.2–15 mM | 27 μM | [96] |

## 6.2 Development of Enzyme-Based Glucose Biosensors

### 6.2.1 First Generation of Enzyme-Based Electrochemical Glucose Biosensors

The first enzymatic electrochemical glucose biosensor (GB) device was invented by Clark and Lyons and was constructed according to the immobilization of GOx on a platinum (Pt) electrode surface (Figure 6.3a). A GOx enzyme consists of two identical protein subunits and one flavin adenine dinucleotide (FAD) as a coenzyme molecule, which plays a role as an active site. Given that FAD is combined with GOx, Gox is named GOx–FAD. The GOx–FAD works as a glucose oxidation catalyst because of its reversible chemical redox property. FAD is an electron acceptor centre in GOx–FAD, which is reduced to GOx-$FADH_2$ through a redox reaction. In a first-generation enzyme-based electrochemical GB using Gox–FAD, which is used in the oxidation of glucose into glucono-D-lactone (Eq. 6.1), FAD is an electron acceptor centre and thus reduced to $FADH_2$ (GOx–$FADH_2$ form) through the following redox reaction:

$$\text{GOx-FAD} + \text{glucose} \rightarrow \text{GOx-FADH}_2 + \text{Glucono-d-lactone} \tag{6.1}$$

Subsequent reaction with $O_2$ produces $H_2O_2$ and regenerates the FAD cofactor as described in Eq. 6.2.

$$\text{GOx-FADH}_2 + O_2 \rightarrow \text{GOx-FAD} + H_2O_2 \tag{6.2}$$

The glucono-d-lactone product then further undergoes a reaction in a water medium to produce carboxylic acid (gluconic acid; Figure 6.3b). Eq. 6.1 and Eq. 6.2 are normally summarized in total equation (Eq. 6.3), which is usually used for describing the catalytic activity of GOx in the chemical oxidation of glucose.

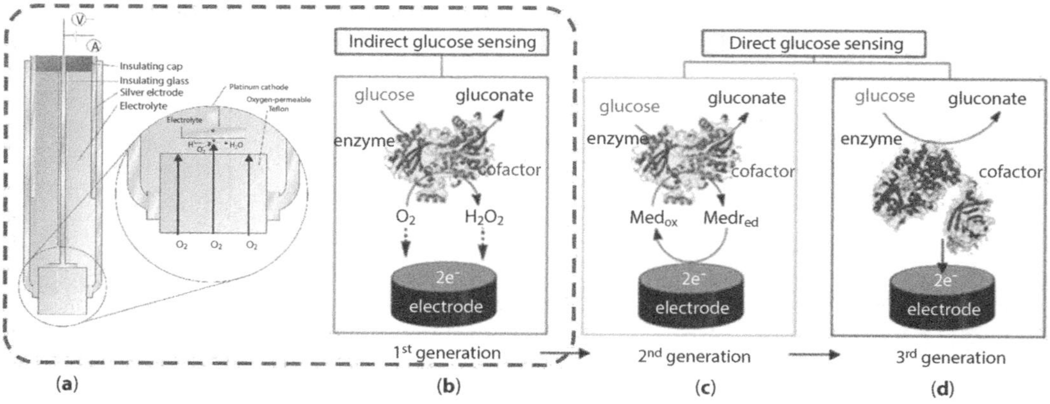


**Figure 6.3** (a) First-generation enzymatic electrochemical glucose biosensor. Reprinted from Ref. [97] (copyright 2021 from Wikipedia); (b–d) Working principle of (b) first-generation GBs, (c) second-generation GBs and (d) third-generation of GBs. This figure is taken from Ref. [98] (copyright 2021 from MDPI).

$$\text{Glucose} + O_2 + H_2O \xrightarrow{GOx} \text{Gluconic acid} + H_2O_2 \tag{6.3}$$

These reactions occur on a Pt working electrode as the anode, where the number of transferred electrons can be correlated to the amount of $H_2O_2$ produced through a process described in Eq. 6.4, which hence reflects glucose concentration.

$$H_2O_2 \xrightarrow{Pt} 2H^+ + O_2 + 2e^- \tag{6.4}$$

The original reactions are based on GOx–glucose, an enzyme–substrate couple, which is responsible for the high specificity of first-generation GBs for glucose sensing. The electrical signal of these sensors are measured according to decrease in $O_2$ concentration (Eq. 6.3) or increase in $H_2O_2$ concentration (which is generated following Eq. 6.2). These concentrations are proportional to glucose concentration in a sample. Therefore, the first-generation GBs can be classified as indirect detection modes and called oxygen electrode sensors. In 1962, Updike and Hicks proposed the use of polyacrylamide gel coatings on Pt electrodes, which is used as a carrier surface for GOx immobilization and stabilization, and fabricated an electrochemical enzymatic glucose assay. Based on the Clark's approach, the first commercial glucose sensors were fabricated in 1975, which were built on the basis of electrochemical transducers through an amperometric technique for $H_2O_2$ detection (Figure 6.3b). The main problem of the first-generation GB-based commercial devices arises from the amperometric technique for determining $H_2O_2$ concentration. This technique requires a high applied potential to enhance selectivity. In addition, first-generation glucose biosensors are expensive because they require platinum electrodes that reduce $H_2O_2$ during signal generation. Given these disadvantages, the use of first-generation GBs devices is strictly confined in the clinical settings. Therefore, second-generation GBs are needed.

### 6.2.2 Second-Generation Enzyme-Based GBs

Since the 1980s, to solve limitations of first-generation GB devices, redox mediators (also called electron mediators) have been used to carry electrons from the FAD center of GOx to a working electrode's surface. These mediators are called second-generation GBs. A variety of electron mediators, including ferrocene ($Fe(C_5H_5)_2$), ferricyanide ($[Fe(CN)_6]^{3-}$), quinine($C_{20}H_{24}N_2O_2$), methylene blue ($C_{16}H_{18}ClN_3S$), tetrathiafulvalene ($(H_2C_2S_2C)_2$), tetracyanoquinodimethane ($(NC)_2CC_6H_4C(CN)_2$), thionine ($C_{12}H_{10}N_3S^+$), and methyl viologen ($[C_5H_4NCH_3)_2]^{2+}$), have been synthesized and used in developing GBs. Moreover, various fabrication techniques have been used to promote electron transfer between GOx–FAD and working electrode surface. For instance, GOx is wired onto an electron-conducting redox and hydrogels or a redox conducting polymer or a coupled of GOx with a group of the electron-relaying or using nanomaterials as electrical connectors (Figure 6.3c).

### 6.2.3 Third-Generation Enzyme-Based GBs

Third-generation enzyme electrochemical GBs emerged in the 1990s, which were called reagentless biosensors because of they were developed on the basis of the concept that electrons are directly transferred from GOx-based redox centers to working electrodes without redox mediators. Conductor-based materials containing charge-transferring complexes were used as the electrode transducers instead of highly toxic redox mediators (as illustrated in Figure 6.3d). Therefore, the third-generation GBs can be used as the implantable GBs or the needle-type GBs devices for continuously monitoring glucose. Without electron mediators were led to these third-generation enzymatic electrochemical GBs having high selectivity.

Currently, most personal glucose meters involve the "finger pricking sampling" approach (also called needle type). They are enzyme-based GBs involving the approach of sampling blood from a finger through pricking. The blood sample is analyzed with *in vitro* methods using GBs, which are test strips constituting three-electrode system. The structure of this three-electrode system is shown in Figure 6.4a and its working principle is illustrated in Figure 6.4b. The electrodes are disposable enzymatic electrode strips. Readout signal processors are miniaturized to a portable equipment and are glucometers (Figure 6.4c and d). Third-generation GBs are mostly based on electrochemical transducers and use GOx as a bioreceptor. Third-generation enzymatic electrochemical GBs are mostly preferred for blood glucose monitoring given their high sensitivity, disposability, and portability.

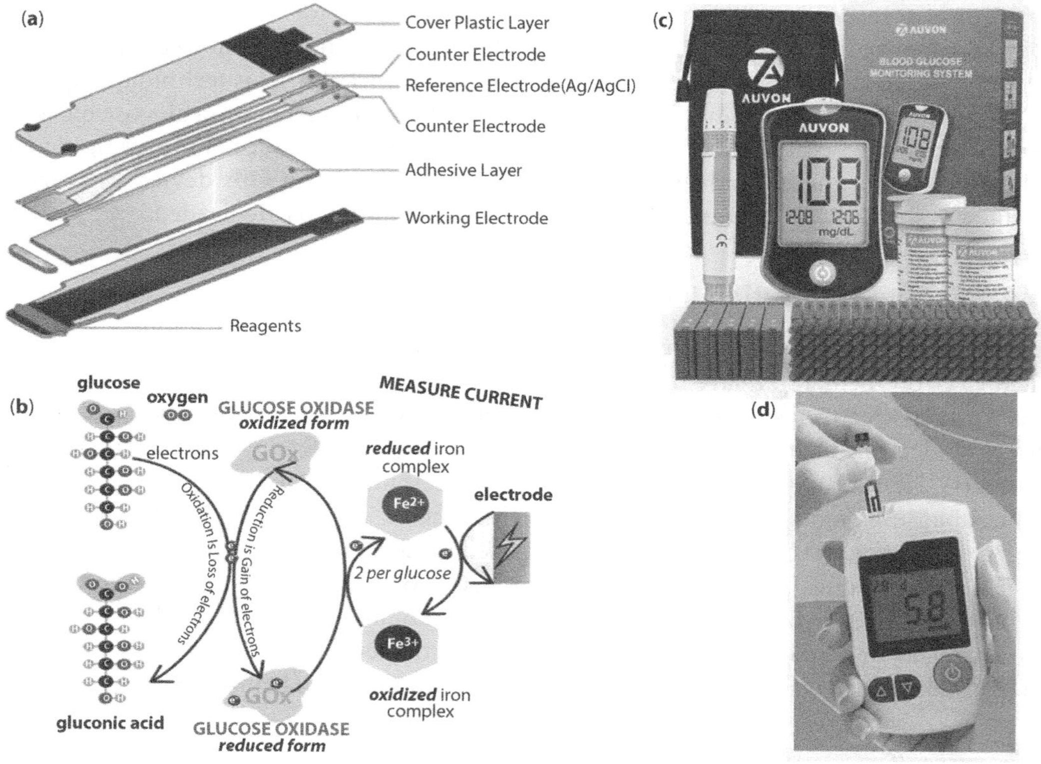


Figure 6.4 (a) A typical structure of an enzymatic electrochemical glucose biosensor, (b) working principle of an enzymatic electrochemical glucose biosensor, (c) commercialized enzymatic electrochemical glucose biosensor system and (d) display of glucose concentration in the sample by a commercial an enzymatic electrochemical glucose biosensor. Figure 1.4.B is taken from Ref. [99] (copyright 2019 from https://thebumblingbiochemist.com; Figure 1.4.A, 1.4.C and 1.4.D are re-drawn.

Recently developed enzymatic electrochemical GBs are test strips with a sensitivity of μM to mM and high reproducibility and are inexpensive, portable and easy to use. Figure 6.4a presents a typical structure of an enzymatic electrochemical GB, in which GOx and HRP enzymes were immobilized onto a screen-printed carbon electrode as a working electrode. As test strips, these components are packed into commercial devices (Figure 6.1a).

## 6.3 Fabrication of Enzymatic Glucose Biosensors

### 6.3.1 Direct and Indirect Detection Modes

GOx-based enzymatic glucose biosensors have sufficient specificity and selectivity for glucose determination in blood, food, urine, and real samples. They are widely used for glucose monitoring in clinics, food control and pharmaceutical quality control. The microfabrication or nanofabrication of advanced nanomaterials are considered to improve test sensitivity and in the development of wearable and on-site monitoring glucose meters. Recent advances in nanotechnology can support many opportunities for developing efficient GB-based glucose monitoring systems. The unique chemical, physical and biophysical properties of nanomaterials, particularly their small sizes, high surface-to-volume ratios and high biocompatibility have been used to amplify accurate signals for glucose detection at low levels and are highly specific for the production of high-precision GBs. In general, enzyme-based GBs can be designed through direct or indirect detection methods. In direct detection methods, electrons can be directly transferred from active GOx–FAD centres to physical transducers. This process leads to changes in the specific physical properties of transducers. Therefore, this approach requires special transducers, which have specific physical properties for directly monitoring response signals.

Using spherical AgNPs as an optical transducer combined with GOx as a bioreceptor, Tran [32] proposed a label-free enzymatic colorimetric GB. Its working principle is illustrated in Figure 6.5. In this strategy, GOx oxidizes glucose into gluconic acid and $H_2O_2$ products according to Eq. 6.3. The generated $H_2O_2$ is recognized by AgNPs via a redox reaction.

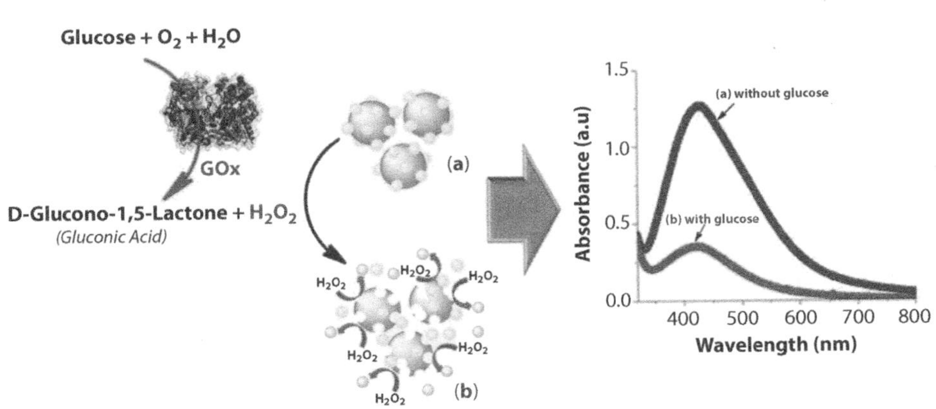


**Figure 6.5** Working principle of a label-free enzyme-based colorimetric biosensor combining GOx with AgNPs/GQDs. This figure is taken from Ref. [32] (copyright 2018 from Elsevier).

The concentration of $H_2O_2$ proportions glucose concentration in samples with principle of response signal was constructed basing on a redox chemical reaction between the Ag with $H_2O_2$ as descried in Eq. 6.5:

$$Ag^0 + H_2O_2 + 2H^+ \rightarrow Ag^+ + 2H_2O \quad (6.5)$$

This reaction is realized on the $\varepsilon^0_{Ag+/Ag} = 0.8$ V (vs. SHE) < $\varepsilon^0_{H_2O_2/H_2O} = 1.77$ V (vs. SHE). Therefore, Ag in Ag nanoparticles (NPs) is etched and converted from $Ag^0$ into free $Ag^+$, decreasing the concentration of AgNPs, resulting in the fading of the color of a AgNP solution and facilitating visualization for the direct determination of glucose level. UV-Vis measurement is used in monitoring decrease in AgNP surface plasmon resonance (SPR) at a wavelength of 425 nm. These strategies facilitate glucose sensing in a concentration range of 0.5÷8 mM and at an LoD of 30 μM.

The immobilization of GOx onto physical transducers leads to direct electron transfer and has been efficiently established. However, the process is not easy to implement. Moreover, the use of fabrication transducers for monitoring sensitive response to direct output signals remains challenging. Thus, redox mediators are used to transfer electrons from FAD (on the GOx enzyme) to transducers. This approach is named indirect detection mode. Recently, most enzymatic colorimetric GBs have been developed using the indirect detection mode, in which added redox mediators are used as signal probes. The developed colorimetric enzymatic glucose biosensors fabricated using the indirect detection mode show excellent sensitivity and high specificity for glucose monitoring.

AgNPs decorated on reduced graphene oxide nanocomposites (AgNPs@rGO) with peroxidase-like catalytic activity have been used to replace HRP in the development of indirect enzymatic colorimetric GBs. The working principle is illustrated in Figure 6.6 [33]. Firstly, the peroxidase-like activity of AgNPs@rGO was investigated during the catalytically oxidation of TMB by $H_2O_2$ in the production of the oxidized form of TMB, which has typical blue color and $\lambda_{max}$ of 655 nm. Then, a colorimetric GB was created by combining AgNPs@rGO

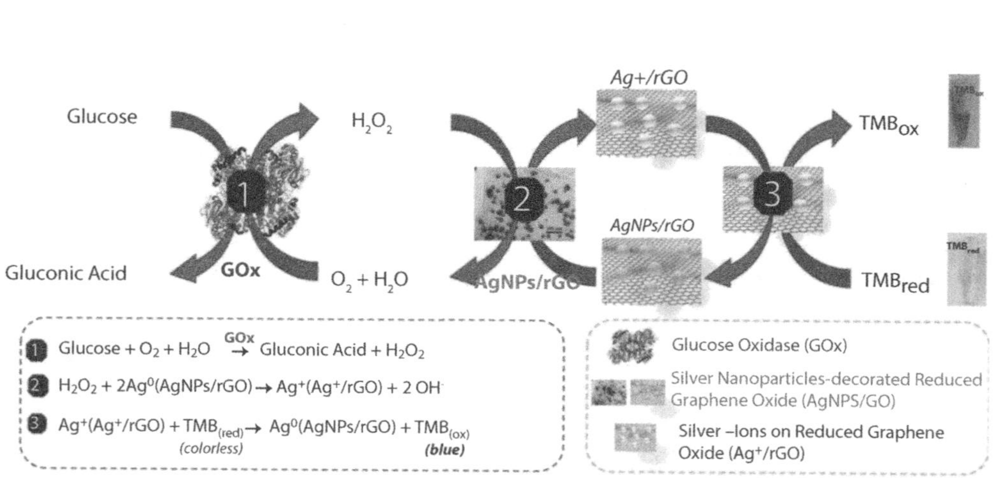


**Figure 6.6** Working principle of an enzyme-based colorimetric GB combining GOx with synthesized AgNPs/rGO. This figure is taken from Ref. [33] (copyright 2020 from Elsevier).

with GOx for the measurement of glucose in human sera. This indirect system facilitates glucose detection at concentrations from 125 μM to 1 mM and has an LoD of 40 μM.

### 6.3.2 Immobilization of Enzyme for the Development of Glucose Biosensors

Enzymes constitute a special class of proteins and serve as the catalysts of chemical reactions that convert substrates into other products. In enzymatic GBs, GOx enzyme is the most often used and immobilized onto a transducer on an immobilization surface (as mentioned in Section 6.1.1). These GBs are used for the selective oxidisation of glucose into gluconic acid and $H_2O_2$ through Eq. 6.3. GOx immobilization allows improving GOx reactivity and stability by protecting GOx from structural denaturation causes by the external environment; thus, GOx activity can be maintained from various working conditions allowing for stability preserving storage. In addition, immobilized GOx can be easily separated from the working solution for washing step and reuse. Normally, the GOx immobilization techniques can be a physical adsorption, covalent bonding, entrapment or cross-linking. The applying method depends on properties of the using transducer and their immobilization surface's materials, in which they can be created from an organic or inorganic material or organic–inorganic nanocomposite.

#### *6.3.2.1 Adsorption Technique*

The easiest method for immobilizing GOx is plain adsorption (Figure 6.7a). In this method, materials in immobilisation surfaces interact with Gox proteins, and this interaction is strong and prevents the release of GOx on immobilisation surfaces under working and washing conditions. For instance, carbon-based electrodes have a large π–π system for GOx through van der Waals interaction force. GOx easily self-assembles on the surfaces of noble metal NPs, such as AuNPs, AgNPs, PtNPs and gold electrode (AuE) surfaces through sulfur–metal binding. Although considered the simplest technique for GOx immobilization, the adsorbed GOx enzymes are sensitive with the changes of working conditions (such as ionic strength, pH, working temperature) leading to desorb GOx from the working surfaces.

#### *6.3.2.2 Covalent Immobilization of GOx*

GOx can attach to a transducer's surface through covalent bonding (Figure 6.7b). In this approach, immobilization surfaces should have functional groups that form covalent bonds with GOx proteins. For example, GOx can be immobilized onto AuE through covalent binding. A layer of carboxylic functional group as immobilization surface is firstly created on AuE through self-assembly using a 3-sulfanylpropanoic acid ($HSCH_2CH_2$-COOH). Herein, $HSCH_2CH_2$-COOH binds with AuE through chemical bonds between thiol groups and Au. Carboxylic groups are then activated to an ester form, which can bind with amine groups on GOx and immobilize GOx to a AuE surface. Chitosan (CS), polypyrrole (PPy), and polyaniline with abundant amine groups can use to modify transducers as immobilization surfaces. Herein, a covalent bond can be created from the amine groups of these polymers and those on GOx cross-linked by glutaraldehyde. Using covalent bonds is the most stable method for GOx immobilization in glucose biosensor fabrication.

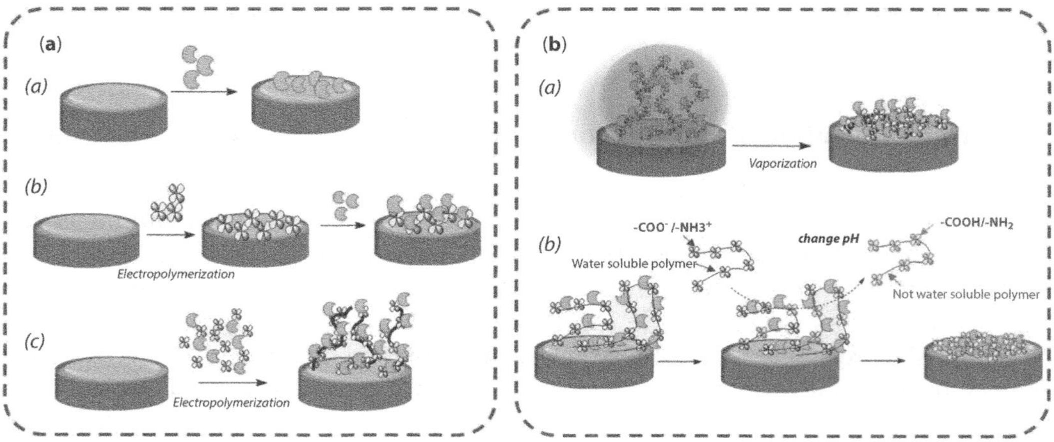


**Figure 6.7** Illustration of various strategies for GOx immobilization onto electrode: (a) (a) an adsorption technique, (b) covalent binding working electrode, (c) electropolymerization and (b) (a) drop coating and (b) co-electrodeposition. This figure is re-drawn.

### 6.3.2.3 *Entrapment of GOx Into a Polymer Matrix*

GOx can immobilize onto a transducer's surface by creating a layer of polymer as an immobilization surface on top of the transducer. GOx is captured inside a polymer-matrix-like fisher net. This technique requires that the polymer does not have any interfere or chemical reactions with GOx. When the polymer deforms GOx or the polymer blocks the diffusion of glucose to the GOx enzyme, it blocks the diffusion of enzymatic reaction's products away from GOx or it prevents the movement of mediators between GOx and the transducer. The GBs cannot function at this point. The immobilization of GOx onto a transducer with an entrapment approach can be carried out through various techniques, such as drop casting, co-precipitation, coating and electro-polymerization. Drop casting is the simplest, in which a polymer and GOx are firstly mixed and then the mixture is drop-cast onto the transducer's surface, as shown in Figure 6.7b (a). The co-precipitation method uses a water-soluble polymer to mix a polymer with GOx. The mixture is then cast onto the transducer's surface after the adjustment of the pH of a medium. The precipitate of the polymer is then obtained (Figure 6.7b [b]). CS can be used to immobilize GOx in the co-precipitation method. Herein, CS was only soluble in an acidic solution subsequently mixed with GOx, and the alkaline medium was required for the co-precipitation of GOx in the CS matrix. In several cases, GOx can be linked with a transducer by using a polymer layer created through electropolymerization. The monomer solution is mixed with GOx enzyme, and then a potential is applied to an electrode for electro-polymerization under continuous stirring. The growth of polymer chains creates a polymer matrix, which automatically entraps GOx during electro-polymerization.

### 6.3.3 Application of Nanomaterials for the Development of a Transducer for Glucose Biosensors

#### *6.3.3.1 Using Nanoparticles as an Artificial Peroxidase for the Fabrication of Indirect Glucose Biosensors*

In GBs, GOx is required to ensure specificity to glucose, and peroxidase plays a role in signal generation. HRP is the mostly used peroxidase in GBs. It is a natural enzyme in organisms and can catalyse $H_2O_2$ reduction through chemical reactions. It is widely used in GB fabrication because of its high substrate specificity and catalytic efficiency. However, as a natural enzyme, HRP has a high cost because of its complicated preparation and purification processes. Specially, HRP has low stability for storage. In addition, the activity of HRP is extremely sensitive to environmental factors, including pH (alkalinity/acidity), working temperature and ionic strength. These limitations hamper the wides application of HRP [34–37]. Therefore, compounds with peroxidase-like catalytic activities are needed to replace HRP in GB development. In particular, nanomaterials with peroxidase-like catalytic activity, also called artificial peroxidases or nanozymes, have attracted considerable interest because of their unique properties, including high catalytic activity, low cost and excellent stability compared with natural peroxidases. Since 2008, $Fe_3O_4$ nanoparticles with peroxidase-mimicking catalytic activity have been used to replace HRP in the fabrication of glucose biosensors [38], which can be classified as follows (Table 6.2 and Table 6.3):

- ✓ Transition metallic oxides, such as $Co_3O_4$ NPs [39], NiO [40], $MnO_2$ [41–43], $MoO_3$ [44], $MoS_2$-decorated $MgFe_2O_4$ NPs [45], CuO [46], $Fe_3O_4$ NPs [38], $MeFe_2O_4$ NPs (Me = Mg, Ni, Cu) [47], graphene/$Fe_3O_4$ [48], FN–CNs [49], Co–Al-layered double hydroxides [50], and $CeO_2$ nanoparticles [51]
- ✓ Transition metallic salts: $VS_2$ nanosheets [52], $Fe_3S_4$ [53], MnSe NPs [54], $Co_3(PO_4)_2$, $Fe_3H_9(PO_4)_6$, and $FeVO_4$
- ✓ Metallic nanoparticles: AgNPs [32, 35], Au@Ag [55], Pt nanoclusters [56], and Au nanocluster
- ✓ Carbon-based nanomaterials: $C_{60}$-carboxyfullerenes [57] and GQDs [58]

FeOOH/N-doped carbon nanosheets (FN–CNs) with peroxidase-like catalytic activity have been used as artificial peroxidases for the fabrication of colorimetric glucose biosensors (Figure 6.8) [49].

These enzymatic GBs facilitate indirect detection with FN–CNs, and TMB serves as a transducer, which is added into the system for signal generation. These GBs are used in glucose bioassays at glucose concentrations ranging from 8 μM to 0.8 mM and have an LoD of 0.2 μM. This developed GB can detect glucose in real human urine samples.

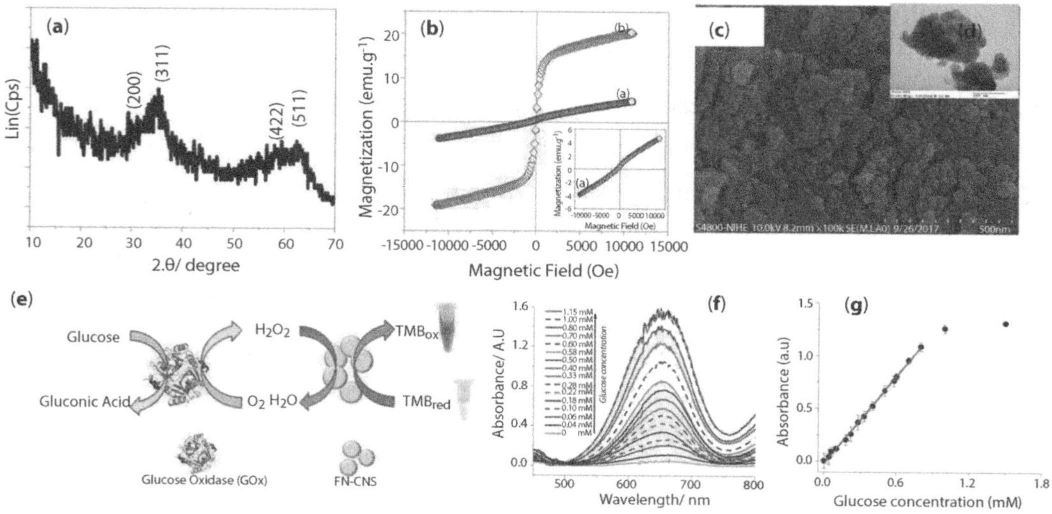

**Figure 6.8** (a) XRD pattern, (b) VSM plots, (c) FE-SEM and (d) TEM of FN-CNs, (e) Working principle of the colorimetric glucose assay following indirect detection mode based on using FN-CN with a peroxidase-like activity to replace of enzyme HRP, (f) Responded UV-vis signal and (g) The calibration curve for glucose monitoring. This figure is taken from Ref. [49] (copyright 2018 from Springer).

### 6.3.3.2 *Using Nanoparticles as Bifunctional Tools for Developing Label-Free Glucose Biosensors*

In direct detection (also called label-free detection) for GB fabrication, no redox mediators, color indicators or electron mediators can be used. Therefore, nanomaterials with unique electrical and optical properties can be used as bifunctional probes for both replacements: (i) a peroxidase and (ii) peroxidase's chromogenic reagents (such as TMB, ABTS and 4-AAP/phenol) for signal readout. Gao [59] coupled AuNPs and AgNPs to assemble a plasmonic sensing platform for glucose detection. The working principle is shown in Figure 6.9a. Herein, AuNPs with particle sizes of approximately 4 nm serve as GOx enzymes that oxidize glucose to produce $H_2O_2$, which then dissolves AgNPs and thereby changes the color of the solution. AgNPs act as artificial HRP enzyme and substrate for signal generation. This colorimetric sensor has a dual readout signal mode: naked-eye observation (color changes from yellow to red; Figure 6.9c) or spectrophotometry (Figure 6.9b) for glucose detection in a range of 5–70 μM. The LoD is 3 μM.

Figure 6.9d illustrates the use of AgNPs/CS nanocomposites as electrochemical transducers for detecting as-generated $H_2O_2$ concentration, which is proportional to glucose concentration in a sample [60]. Herein, AgNPs have been used instead of HRP to reducing $H_2O_2$ and generating electron for electrochemical transducer recognize. Thus, the GCE was firstly modified by AgNPs/CS (AgNPs/CS/GCE), and it was used as working electrode for $H_2O_2$ monitoring. By preparing a detection solution containing of a glucose and GOx, whereas GOx catalytically oxidized glucose to gluconic acid and $H_2O_2$. And $H_2O_2$ then be recognized by AgNPs/CS/GCE using CV technique. Under optimized conditions, this development allows for detection of glucose concentration from 0.1 to 10 mM and an LoD of 0.01 mM. Jia [61] has proposed a fibrous composite made of PdNPs immobilized on helical carbon nanofibers for electrochemical glucose detection, in which, the GOx

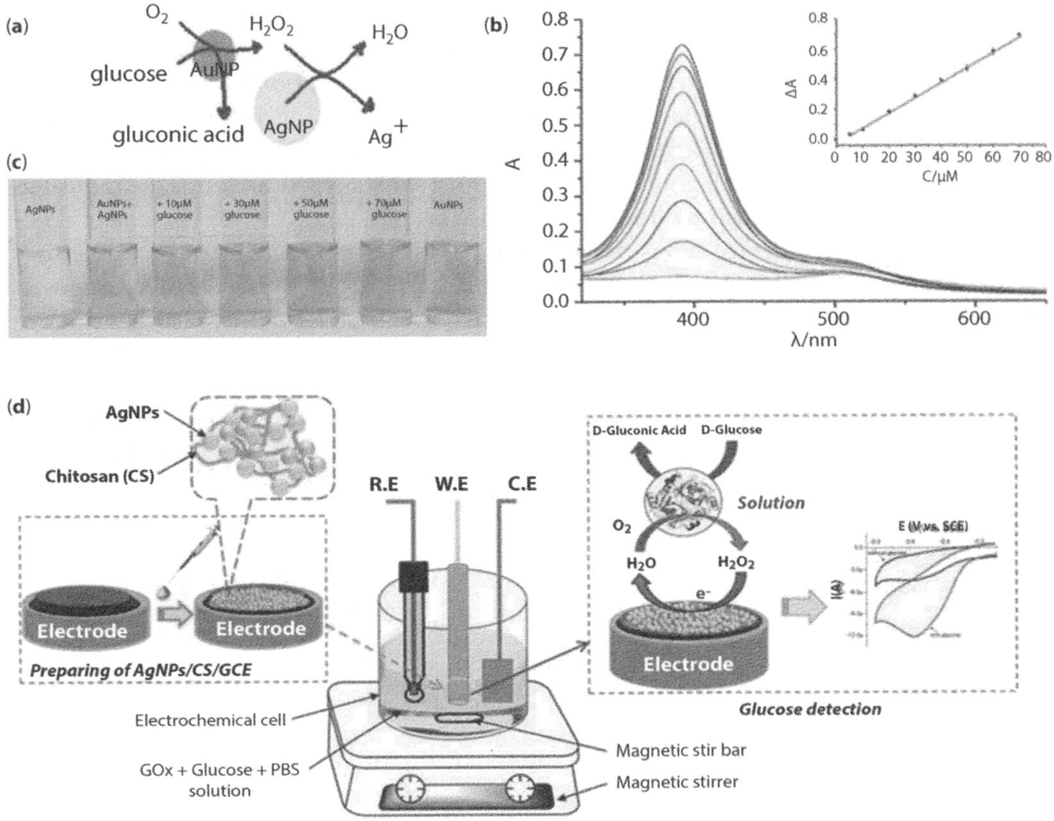


**Figure 6.9** (a) Working principle of an enzymatic colorimetric glucose biosensor based on using AuNPs coupled with AgNPs as colorimetric transducers, (b) responded UV-Vis spectra and calibration curve, (c) color change as a function of glucose concentration. This figure is taken from Ref. [59], copyright 2017 from Elsevier. (d) An enzyme-based electrochemical GB using AgNPs/CS for modifying of GCE. This figure is taken from Ref. [60] (copyright 2019 from IOP).

was immobilized on top of the modified electrode using a Nafion membrane. As the same strategy, a variety of label-free enzymatic electrochemical GBs have been developed using transition metallic oxides ($Fe_3O_4$ [62]; CuO [63], NiO [64, 65]), or metallic nanoparticles (AgNPs [66], PtNPs [67, 68], or PdNPs [61]).

## 6.4 Recent Trends for Development of Glucose Biosensors

To shorten response time, paper analytical devices (PADs), which are simple biosensors, have emerged as disposable platforms for point-of-care testing (dPOCT), which have the following benefits: low cost, minimal sample or reagent consumption, wide applications (including environmental monitoring, food and water analysis, forensic studies, and clinical diagnostics) [69–81]. Recently, for production of simple and cheap PAD-based GBs for dPOCT, most developed PADs are used in combination with colorimetric methods. Colorimetric PAD-based GBs can be integrated with smart phone cameras, digital cameras, scanners or portable microscopes, facilitating image acquisition and processing as the analytical readout systems of analytical instruments. These devices are widely available, portable and globally affordable. In addition, colorimetric PAD-based GBs have good performance, and PADs are sufficiently accurate and reliable for clinical assays, having promising applications.

Figure 6.10a presents a simple glucose PAD based on GOx and HRP enzymes. The working principle is illustrated in Figure 6.10b. It is similar to standard methods, as indicated in Figure 6.2a, except that it provides a simplified view of a colorimetric enzymatic reaction (Figure 6.10c). The device can generate a regression curve (Figure 6.10d) for quantitative analysis and has high reproducibility. Darabdhara [82] used a Cu-Pd/rGO composite instead of HRP to fabricate an enzymatic colorimetric glucose PAD. The working principle is shown in Figure 6.10e. Cu-Pd/rGO has been coated on the head of a paper strip (also named sensing zone) after drop casting with TMB as a chromogenic substrate. The sample zones of the strips are dipped in testing samples. Glucose in a testing solution moves up to the sensing zones of paper strips (circular heads where Cu-Pd/rGO and TMB are already

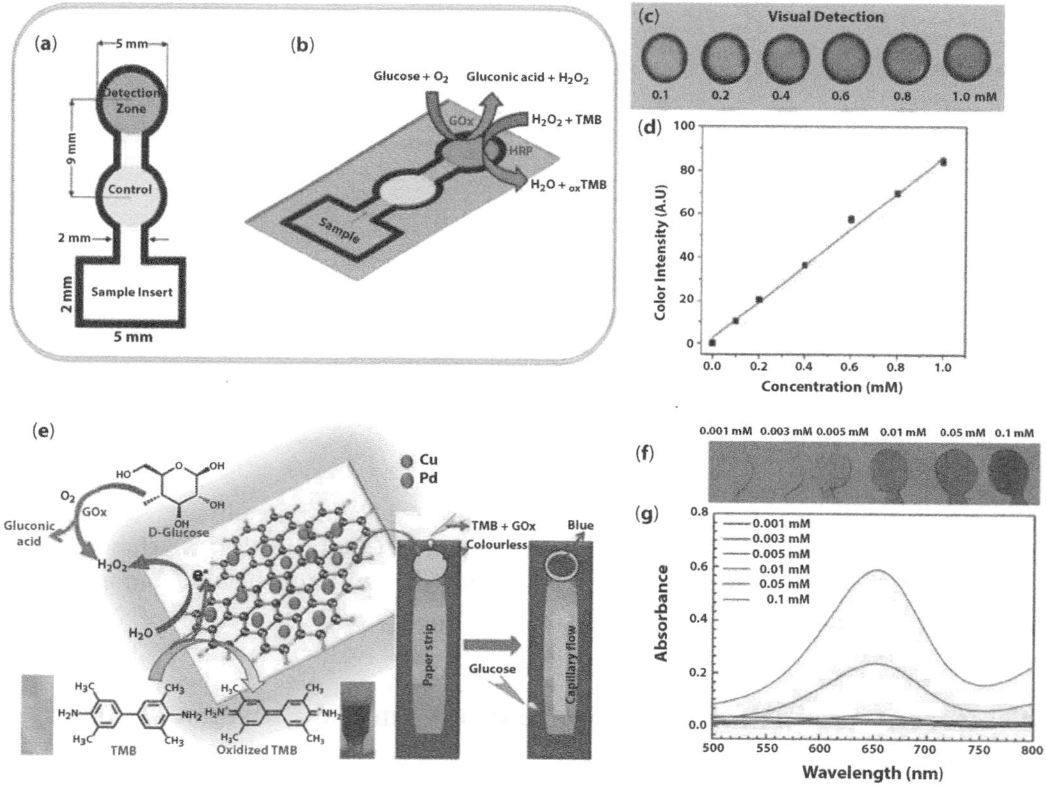


**Figure 6.10** (a) A structure layout of colorimetric-based PAD for glucose sensing, (b) working principle of the colorimetric glucose PAD involved in the presence of GOx, HRP enzymes and chromogenic reagent (TMB) for glucose detection, (c) color scale for visual tear glucose detection and (d) calibration curve for glucose assay. This figure is taken from Ref. [100] (copyright 2017 from MDPI); (e) Working principle of a glucose PAD using GOx combined with Cu-Pd/rGO as artificial peroxidase, (f) color growth on the colorimetric glucose PAD with various glucose concentrations and (g) corresponding UV-visible spectra from (f). This figure is taken from Ref. [82] (copyright 2019 from Springer).

immobilized) through capillary forces. An immediate color change to blue occurs, and its intensity is proportional to glucose concentration, thus allowing naked-eye detection (Figure 6.10f). Detected results on these glucose PADs have been verified by performing spectrophotometry on paper strips (Figure 6.10g), and the selectivity of PADs for glucose detection has been verified by various interference molecules. These developed PADs can recognize glucose concentration from 0.2 to 50 μM and has an LoD of 0.29 μM.

$Co_3O_4$-$CeO_2$ nanocomposites have been utilized as nanozymes along with GOx in the development of enzyme-based colorimetric PADs for glucose sensing from 5 μM to 1.5 mM and have an LoD of 0.21 μM. Color changes in PADs are recorded and analyzed with a smartphone application (Figure 6.11) [83].

GB-based PADs are comfortable to use and have many benefits. They are cost effective, disposable and easy to use and have simple measuring system requirements. However, these PADs requires the "finger pricking" approach and thus has been used only for patients with diabetes, who have undergo blood collection almost several times per day through "finger pricking," which causes pain and discomfort. Therefore, a novel method without involving "finger pricking" in glucose sensing is highly needed.

Various portable, wearable and implantable GB-based sensors for glucose sensing have been commercialized. They exhibit excellent performance and are bendable, robust and biocompatible. Recently, highly flexible GB-based sensors (bendable, portable, wearable, or implantable) with excellent performance (robust, multi-screening capable and highly biocompatible) for glucose sensing have been extensively used. They have attracted considerable interest because of their application potential for glucose monitoring. Moreover, glucose-sensing devices can be combined and packed in a point-of-care medical diagnostic devices, such as multi-screening and wearable devices for glucose sensing and monitoring of other biomarkers. Wearable sweat-based glucose devices are the most useful devices for patients with diabetes. They can be used for the continuous measurement of glucose level in human sweat and are extremely useful media for glucose concentration detection, thereby facilitating with "painless" and "stress free" approach without finger pricking. Using AuNPs/PtNPs alloy NPs deposited on rGO (AuNPs/PtNPs-rGO) combined with

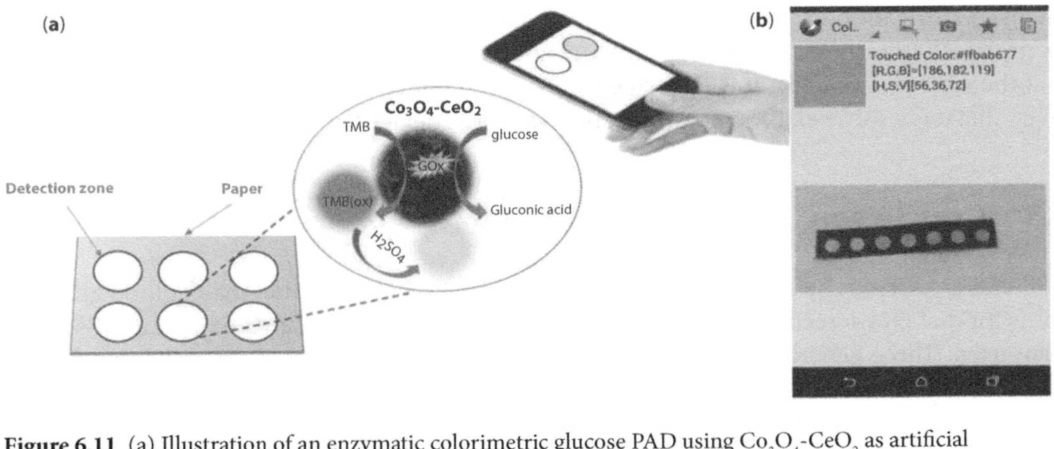


**Figure 6.11** (a) Illustration of an enzymatic colorimetric glucose PAD using $Co_3O_4$-$CeO_2$ as artificial peroxidase, (b) Responded color on glucose PADs, which can be scanned with a smartphone camera for the readout of results. This figure is taken from Ref. [83] (copyright 2019 from Elsevier).

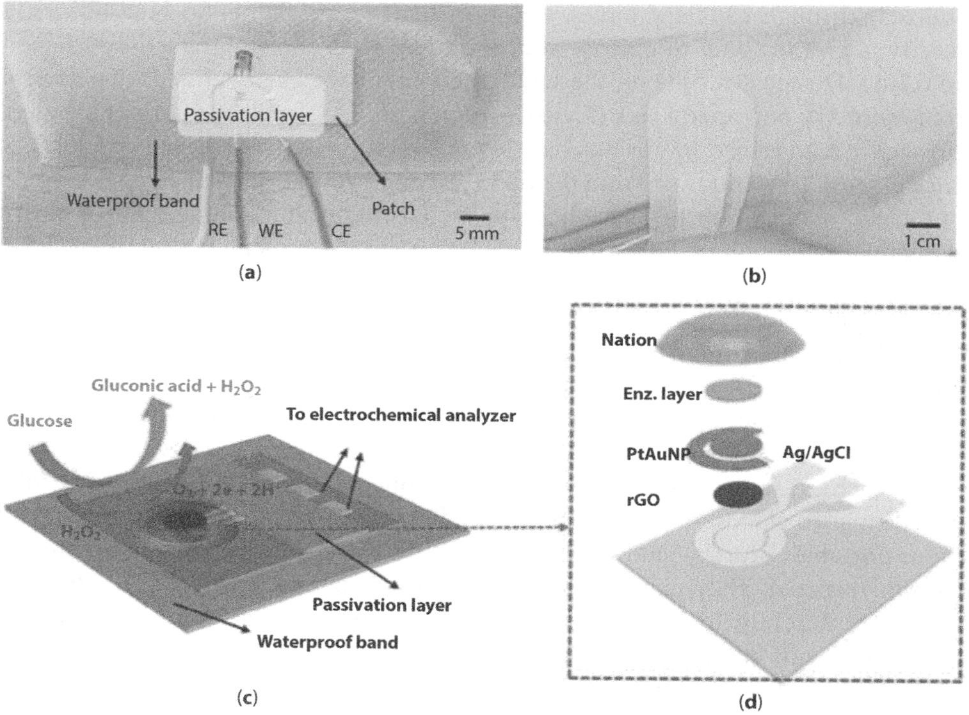


**Figure 6.12** (a) Photograph and schematics of the structure of a sweat-based sensor for continuous glucose monitoring (CGM), (b) photograph of the device attached to the skin, (c) the working mechanism of sweat-based sensors for continuous glucose monitoring and (d) exploded view for the sensor. This figure is taken from Ref. [84] (copyright 2018 from Elsevier).

GOx, a sweat-based enzymatic GB for continuous glucose monitoring was reported [84]. As shown in Figure 6.12a and b, AuNPs/PtNPs-rGO and GOx are embedded in the AuE of a three-electrode system using Nafion polymer as a binder. The working principle of the developed wearable sweat-based enzymatic GB is illustrated in Figure 6.12c, and the involved fabrication layers are presented in Figure 6.12d. Although this GB is a prototype, its practical assessment results for measuring glucose level in human sweat have implied that the system is reliable. It is expected to be used in glucose monitoring in human sweat, given that it is a soft, flexible, and disposable electronic device with robust sensing capability.

## 6.5 Conclusion

Diabetes is a serious health problem that has caused a global epidemic. Using blood glucose levels for diabetes screening is now considered a standard method, and enzymatic GBs have been used. Three generations of enzymatic GBs have been developed since the first fabrication of GBs by Clark and Lyons. Various microtechnologies/nanotechnologies for fabricating platforms, useful technologies for immobilization GOx bioreceptors and advanced materials as nanozymes have facilitated the development of miniaturized, personalized, and portable measuring systems. These systems can be considered revolutionary in terms of sensitivity, selectivity, response, cost and disposability, and are efficient in controlling, on-site monitoring, and managing diabetes. Currently, enzymatic GBs work following the invasive methods for blood sample collection via the "finger pricking"-based techniques, which cause discomfort, pain, and stress. Therefore, developing novel methods noninvasive approaches for glucose monitoring are needed, especially those that can monitor glucose level in body fluids, such as tears, sweat, and saliva instead of blood samples. Recently, scientists and engineers have made considerable efforts to develop novel sensing technologies, such as miniaturized, portable, wearable, and implantable wireless measuring equipment that can continuously sense glucose concentration, aiming to improve sensitivity and shorten response time in small sample volumes. Soft, flexible, and disposable electronic devices with robust sensing capabilities are urgently needed. Sweat-based enzymatic glucose biosensors will be widely available in the healthcare market in the near future.

## Acknowledgment

This work was supported by the Vietnam Ministry of Education and Training (under project number CT 2022.04.BKA.04).

## References

1. Singh, S., Mitra, K., Singh, R., Kumari, A., Gupta, S.K.S., Misra, N., Maiti, P., Ray, B., Colorimetric detection of hydrogen peroxide and glucose using brominated graphene. *Anal. Methods*, 9, 6675–6681, 2017.
2. Jabariyan, S., Zanjanchi, M.A., Arvand, M., Sohrabnezhad, S., Colorimetric detection of glucose using lanthanum-incorporated MCM-41. *Spectrochim. Acta, Part A*, 203, 294–300, 2018.
3. Liu, Q., Ma, K., Wen, D., Wang, Q., Sun, H., Liu, Q., Kong, J., Electrochemically mediated ATRP (eATRP) amplification for ultrasensitive detection of glucose. *J. Electroanal. Chem.*, 823, 20–25, 2018.
4. Meng, A., Sheng, L., Zhao, K., Li, Z., A controllable honeycomb-like amorphous cobalt sulfide architecture directly grown on the reduced graphene oxide–poly(3,4-ethylenedioxythiophene) composite through electrodeposition for non-enzyme glucose sensing. *J. Mater. Chem. B*, 5, 8934–8943, 2017.
5. Yoo, E.-H. and Lee, S.-Y., Glucose biosensors: An overview of use in clinical practice. *Sensors (Basel)*, 10, 5, 4558–4576, 2010.
6. Turner, A.P.F., Biosensors: Sense and sensibility. *Chem. Soc. Rev.*, 42, 184–3196, 2013.
7. Thevenot, D.R., Toth, K., Durst, R.A., Wilson, G.S., Electrochemical biosensors: Recommended definitions and classification. *Pure Appl. Chem.*, 71, 2333–2348, 1999.
8. Mcnaught, A.D. and Wilkinson, A., *(The "Gold Book")-IUPAC- Compendium of chemical terminology*, 2nd ed., Blackwell Scientific Publications, Oxford, 1997.
9. Gerard, M., Chaubey, A., Malhotra, B.D., Application of conducting polymers to biosensors. *Biosens. Bioelectron.*, 17, 345–359, 2002.
10. Reisberg, S., Piro, B., Noël, V., Pham, M.C., DNA Electrochemical sensor based on conducting polymer: Dependence of the "signal-on" detection on the probe sequence localization. *Anal. Chem.*, 77, 10, 3351–3356, 2005.

11. Piro, B., Reisberg, S., Anquetin, G., Duc, H.T., Pham, M.C., Quinone-based polymers for label-free and reagentless electrochemical immunosensors: Application to proteins, antibodies and pesticides detection. *Biosensors*, 3, 1, 58–76, 2013.
12. Piro, B., Zhang, Q.D., Reisberg, S., Noel, V., Dang, L.A., Duc, H.T., Pham, M.C., Direct and rapid electrochemical immunosensing system based on a conducting polymer. *Talanta*, 82, 2, 608–612, 2010.
13. Tran, H.V., Piro, B., Reisberg, S., Tran, L.D., Duc, H.T., Pham, M.C., Label-free and reagentless electrochemical detection of microRNAs using a conducting polymer nanostructured by carbon nanotubes: Application to prostate cancer biomarker miR-141. *Biosens. Bioelectron.*, 49, 0, 164–169, 2013.
14. Tran, H.V., Yougnia, R., Reisberg, S., Piro, B., Serradji, N., Nguyen, T.D., Tran, L.D., Dong, C.Z., Pham, M.C., A label-free electrochemical immunosensor for direct, signal-on and sensitive pesticide detection. *Biosens. Bioelectron.*, 31, 1, 62–68, 2012.
15. Luo, X., Morrin, A., Killard, A.J., Smyth, M.R., Application of nanoparticles in electrochemical sensors and biosensors. *Electroanalysis*, 18, 4, 319–326, 2006.
16. Ansari, A.A., Kaushik, A., Solanki, P.R., Malhotra, B.D., Sol–gel derived nanoporous cerium oxide film for application to cholesterol biosensor. *Electrochem. Commun.*, 10, 9, 1246–1249, 2008.
17. Arya, S.K., Solanki, P.R., Datta, M., Malhotra, B.D., Recent advances in self-assembled monolayers based biomolecular electronic devices. *Biosens. Bioelectron.*, 24, 2810–2817, 2009.
18. Hayat, A., Barthelmebs, L., Marty, J.L., Enzyme-linked immunosensor based on super paramagnetic nanobeads for easy and rapid detection of okadaic acid. *Anal. Chim. Acta*, 690, 2, 248–52, 2011.
19. Malhotra, R., Patel, V., Chikkaveeraiah, B.V., Munge, B.S., Cheong, S.C., Zain, R.B., Abraham, M.T., Dey, D.K., Gutkind, J.S., Rusling, J.F., Ultrasensitive detection of cancer biomarkers in the clinic by use of a nanostructured microfluidic array. *Anal. Chem.*, 84, 14, 6249–55, 2012.
20. Hervas, M., Lopez, M.A., Escarpa, A., Simplified calibration and analysis on screen-printed disposable platforms for electrochemical magnetic bead-based immunosensing of zearalenone in baby food samples. *Biosens. Bioelectron.*, 25, 7, 1755–60, 2010.
21. Haccoun, J., Piro, B., Tran, L.D., Dang, L.A., Pham, M.C., Reagentless amperometric detection of l-lactate on an enzyme-modified conducting copolymer poly(5-hydroxy-1,4-naphthoquinone-co-5-hydroxy-3-thioacetic acid-1,4-naphthoquinone). *Biosens. Bioelectron.*, 19, 10, 1325–1329, 2004.
22. Gamero, M., Pariente, F., Lorenzo, E., Alonso, C., Nanostructured rough gold electrodes for the development of lactate oxidase-based biosensors. *Biosens. Bioelectron.*, 25, 9, 2038–44, 2010.
23. Upadhyay, S., Rao, G.R., Sharma, M.K., Bhattacharya, B.K., Rao, V.K., Vijayaraghavan, R., Immobilization of acetylcholineesterase-choline oxidase on a gold-platinum bimetallic nanoparticles modified glassy carbon electrode for the sensitive detection of organophosphate pesticides, carbamates and nerve agents. *Biosens. Bioelectron.*, 25, 4, 832–8, 2009.
24. Putzbach, W. and Ronkainen, N.J., Immobilization techniques in the fabrication of nanomaterial-based electrochemical biosensors: A review. *Sensors (Basel)*, 13, 4, 4811–40, 2013.
25. Monošík, R., Streďanský, M., Šturdík, E., Biosensors — Classification, characterization and new trends. *Acta Chim. Slov.*, 5, 1, 109–120, 2012.
26. Bhimji, A., Zaragoza, A.A., Live, L.S., Kelley, S.O., Electrochemical enzyme-linked immunosorbent assay featuring proximal reagent generation: Detection of human immunodeficiency virus antibodies in clinical samples. *Anal. Chem.*, 85, 14, 6813–6819, 2013.
27. Yu, D., Blankert, B., Kauffmann, J.M., Development of amperometric horseradish peroxidase based biosensors for clozapine and for the screening of thiol compounds. *Biosens. Bioelectron.*, 22, 11, 2707–2711, 2007.

28. Petit, C., Murakami, K., Erdem, A., Kilinc, E., Borondo, G.O., Liegeois, J.F., Kauffmann, J.M., Horseradish peroxidase immobilized electrode for phenothiazine analysis. *Electroanalysis*, 10, 18, 1241–1248, 1998.
29. Lomillo, M.A.A., Kauffmann, J.M., Martinez, M.J.A., HRP-based biosensor for monitoring rifampicin. *Biosens. Bioelectron.*, 18, 9, 1165–1171, 2003.
30. Yu, D.H., Blankert, B., Vire, J.C., Kauffmann, J.M., Biosensors in drug discovery and drug analysis. *Anal. Lett.*, 38, 11, 1687–1701, 2005.
31. Cabanillas, A.G., Diaz, T.G., Salinas, F., Ortiz, J.M., Kauffmann, J.M., Differential pulse voltammetric determination of fenobucarb at the glassy carbon electrode, after its alkaline hydrolysis to a phenolic product. *Electroanalysis*, 9, 12, 952–955, 1997.
32. Nguyen, N.D., Nguyen, T.V., Chu, A.D., Tran, H.V., Tran, L.T., Huynh, C.D., A label-free colorimetric sensor based on silver nanoparticles directed to hydrogen peroxide and glucose. *Arabian J. Chem.*, 11, 7, 1134–1143, 2018.
33. Tran, H.V., Nguyen, N.D., Tran, C.T.Q., Tran, L.T., Le, T.D., Tran, H.T.T., Piro, B., Huynh, C.D., Nguyen, T.N., Nguyen, N.T.T., Dang, H.T.M., Nguyen, H.L., Tran, L.D., Phan, N.T., Silver nanoparticles-decorated reduced graphene oxide: A novel peroxidase-like activity nanomaterial for development of a colorimetric glucose biosensor. *Arabian J. Chem.*, 13, 7, 6084–6091, 2020.
34. Ding, C., Yan, Y., Xiang, D., Zhang, C., Xian, Y., Magnetic Fe3S4 nanoparticles with peroxidase-like activity, and their use in a photometric enzymatic glucose assay. *Microchim. Acta*, 183, 2, 625–631, 2016.
35. Tran, H.V., Huynh, C.D., Tran, H.V., Piro, B., Cyclic voltammetry, square wave voltammetry, electrochemical impedance spectroscopy and colorimetric method for hydrogen peroxide detection based on chitosan/silver nanocomposite. *Arabian J. Chem.*, 11, 4, 453–459, 2018.
36. Zhang, W., Ma, D., Du, J., Prussian blue nanoparticles as peroxidase mimetic for sensitive colorimetric detection of hydrogen peroxide and glucose. *Talanta*, 10, 362–367, 2014.
37. Xing, Z., Tian, J., Asiri, A.M., Qusti, A.H., Al-Youbi, A.O., Sun, X., Two-dimensional hybrid mesoporous Fe2O3-graphene nanostructures: A highly active and reusable peroxidase mimetic toward rapid, highly sensitive optical detection of glucose. *Biosens. Bioelectron.*, 52, 452–457, 2014.
38. Wei, H. and Wang, E., Fe3O4 Magnetic nanoparticles as peroxidase mimetics and their applications in H2O2 and glucose detection. *Anal. Chem.*, 80, 6, 2250–2254, 2008.
39. Liu, P., Wang, Y., Han, L., Cai, Y., Ren, H., Ma, T., Li, X., Petrenko, V.A., Liu, A., Colorimetric assay of bacterial pathogens based on Co3O4 magnetic nanozymes conjugated with specific fusion phage proteins and magnetophoretic chromatography. *ACS Appl. Mater. Interfaces*, 12, 8, 9090–9097, 2020.
40. Yan, L., Ren, H., Guo, Y., Wang, G., Liu, C., Wang, Y., Liu, X., Zeng, L., Liu, A., Rock salt type NiO assembled on ordered mesoporous carbon as peroxidase mimetic for colorimetric assay of gallic acid. *Talanta*, 301, 406–412, 2019.
41. Han, L., Shi, J., Liu, A., Novel biotemplated MnO2 1D nanozyme with controllable peroxidase-like activity and unique catalytic mechanism and its application for glucose sensing. *Sens. Actuators, B*, 252, 919–926, 2017.
42. Han, L., Liu, P., Zhang, H., Li, F., Liu, A., Phage capsid protein-directed MnO2 nanosheets with peroxidase-like activity for spectrometric biosensing and evaluation of antioxidant behaviour. *Chem. Commun.*, 53, 5216–5219, 2017.
43. Liu, P., Han, L., Wang, F., Li, X., Petrenko, V.A., Liu, A., Sensitive colorimetric immunoassay of Vibrio parahaemolyticus based on specific nonapeptide probe screening from a phage display library conjugated with MnO2 nanosheets with peroxidase-like activity. *Nanoscale*, 10, 2825–2833, 2018.

44. Ren, H., Yan, L., Liu, M., Wang, Y., Liu, X., Liu, C., Liu, K., Zeng, L., Liu, A., Green tide biomass templated synthesis of molybdenum oxide nanorods supported on carbon as efficient nanozyme for sensitive glucose colorimetric assay. *Sens. Actuators, B*, 296, 126517, 2019.
45. Zhang, Y., Zhou, Z., Wen, F., Tan, J., Peng, T., Luo, B., Wang, H., Yin, S., A flower-like MoS2-decorated MgFe2O4 nanocomposite: Mimicking peroxidase and colorimetric detection of H2O2 and glucose. *Sens. Actuators, B*, 275, 155–162, 2018.
46. Zhang, L., Hai, X., Xia, C., Chen, X.-W., Wang, J.-H., Growth of CuO nanoneedles on graphene quantum dots as peroxidase mimics for sensitive colorimetric detection of hydrogen peroxide and glucose. *Sens. Actuators, B*, 248, 374–384, 2017.
47. Su, L., Qin, W., Zhang, H., Rahman, Z.U., Ren, C., Ma, S., Chen, X., The peroxidase/catalase-like activities of MFe2O4 (M=Mg, Ni, Cu) MNPs and their application in colorimetric biosensing of glucose. *Biosens. Bioelectron.*, 63, 384–391, 2015.
48. Dong, Y., Zhang, H., Rahman, Z., Su, L., Chen, X., Hu, J., Chen, X., Graphene oxide-Fe3O4 magnetic nanocomposites with peroxidase-like activity for colorimetric detection of glucose. *Nanoscale*, 4, 13, 3969–3976, 2012.
49. Tran, H.V., Nguyen, T.V., Nguyen, N.D., Piro, B., Huynh, C.D., A nanocomposite prepared from FeOOH and N-doped carbon nanosheets as a peroxidase mimic, and its application to enzymatic sensing of glucose in human urine. *Microchim. Acta*, 185, 270, 2016.
50. Chen, L., Sun, B., Wang, X., Qiao, F., Ai, S., Paper 2D ultrathin nanosheets of Co-Al layered double hydroxides prepared in L-asparagine solution: Enhanced peroxidase-like activity and colorimetric detection of glucose. *J. Mater. Chem. B*, 1, 2268–2274, 2013.
51. Zhao, H., Dong, Y., Jiang, P., Wang, G., Zhang, J., Highly dispersed CeO2 on TiO2 nanotube: A synergistic nanocomposite with superior peroxidase-like activity. *ACS Appl. Mater. Interfaces*, 7, 12, 6451–6461, 2015.
52. Huang, L., Zhu, W., Zhang, W., Wang, K.C., Wang, R., Yang, Q., Hu, N., Suo, Y., Wang, J., Layered vanadium(IV) disulfide nanosheets as a peroxidase-like nanozyme for colorimetric detection of glucose. *Microchim. Acta*, 185, 7, 2018.
53. Ding, C., Yan, Y., Xiang, D., Zhang, C., Xian, Y., Magnetic Fe3S4 nanoparticles with peroxidase-like activity, and their use in a photometric enzymatic glucose assay. *Microchim. Acta*, 183, 2, 625–631, 2016.
54. Qiao, F., Chen, L., Li, X., Li, L., Ai, S., Peroxidase-like activity of manganese selenide nanoparticles and its analytical application for visual detection of hydrogen peroxide and glucose. *Sens. Actuators, B*, 193, 255–262, 2014.
55. Han, L., Li, C., Zhang, T., Lang, Q., Liu, A., Au@Ag heterogeneous nanorods as nanozyme interfaces with peroxidase-like activity and their application for one-pot analysis of glucose at nearly neutral pH. *ACS Appl. Mater. Interfaces*, 7, 26, 14463–14470, 2015.
56. Jin, L., Meng, Z., Zhang, Y., Cai, S., Zhang, Z., Li, C., Shang, L., Shen, Y., Ultrasmall Pt nanoclusters as robust peroxidase mimics for colorimetric detection of glucose in human serum. *ACS Appl. Mater. Interfaces*, 9, 11, 10027–10033, 2017.
57. Li, R., Zhen, M., Guan, M., Chen, D., Zhang, G., Ge, J., Gong, P., Wang, C., Shu, C., A novel glucose colorimetric sensor based on intrinsic peroxidase-like activity of C60-carboxyfullerenes. *Biosens. Bioelectron.*, 47, 502–507, 2013.
58. Lin, L., Song, X., Chen, Y., Rong, M., Zhao, T., Wang, Y., Jiang, Y., Chen, X., Intrinsic peroxidase-like catalytic activity of nitrogen-doped graphene quantum dots and their application in the colorimetric detection of H2O2 and glucose. *Anal. Chim. Acta*, 869, 89–95, 2015.
59. Gao, Y., Wu, Y., Di, J., Colorimetric detection of glucose based on gold nanoparticles coupled with silver nanoparticles. *Spectrochim. Acta, Part A*, 173, 207–212, 2017.

60. Tran, L.T., Tran, H.V., Le, T.D., Bach, G.L., Tran, L.D., Studying Ni(II) adsorption of magnetite/graphene oxide/chitosan nanocomposite. *Adv. Polym. Technol.*, 2019, Article ID 8124351, 9, 2019.
61. Jia, X., Hu, G., Nitze, F., Barzegar, H.R., Sharifi, T., Tai, C.-W., Wågberg, T., Synthesis of palladium/helical carbon nanofiber hybrid nanostructures and their application for hydrogen peroxide and glucose detection. *ACS Appl. Mater. Interfaces*, 5, 22, 12017–12022, 2013.
62. Nguyen, H.L., Nguyen, T.T.M., Nguyen, B.H., Nguyen, T.N., Mai, T.T.T., Huynh, C.D., Pham, N.T., Nguyen, P.X., Nguyen, A.V., Tran, L.D., Covalent immobilization of cholesterol oxidase and poly(styrene-co-acrylic acid) magnetic microspheres on polyaniline films for amperometric cholesterol biosensing. *Anal. Methods*, 5, 1392–1398, 2013.
63. Song, J., Xu, L., Zhou, C., Xing, R., Dai, Q., Liu, D., Song, H., Synthesis of graphene oxide based CuO nanoparticles composite electrode for highly enhanced nonenzymatic glucose detection. *ACS Appl. Mater. Interfaces*, 5, 24, 12928–12934, 2013.
64. Marimuthu, T., Mohamad, S., Alias, Y., Needle-like polypyrrole-NiO composite for non-enzymatic detection of glucose. *Synth. Met.*, 207, 35–41, 2015.
65. Liu, S., Yu, B., Zhang, T., A novel non-enzymatic glucose sensor based on NiO hollow spheres. *Electrochim. Acta*, 102, 104–107, 2013.
66. Tran, H.V., Trinh, A.X., Huynh, C.D., Le, H.Q., Facile hydrothermal synthesis of Silver/Chitosan nanocomposite and application in the electrochemical detection of hydrogen peroxide. *Sens. Lett.*, 13, 1–7, 2015.
67. Zhu, X., Li, C., Zhu, X., Xu, M., Nonenzymatic glucose sensor based on Pt-Au-SWCNTs nanocomposites. *Int. J. Electrochem. Sci.*, 7, 8522–8532, 2012.
68. Su, C., Zhang, C., Lu, G., Ma, C., Nonenzymatic electrochemical glucose sensor based on pt nanoparticles/mesoporous carbon matrix. *Electroanalysis*, 22, 16, 1901–1905, 2010.
69. Yetisen, A.K., Akrama, M.S., Lowea, C.R., Paper-based microfluidic point-of-care diagnostic devices. *Lab. Chip*, 13, 2210–2251, 2013.
70. Martinez, A.W., Phillips, S.T., Whitesides, G.M., Carrilho, E., Diagnostics for the developing world: Microfluidic paper-based analytical devices. *Anal. Chem.*, 82, 3–10, 2010.
71. Martinez, A.W., Phillips, S.T., Butte, M.J., Whitesides, G.M., Patterned paper as a platform for inexpensive, low-volume, portable bioassays. *Angew. Chem. Int. Ed.*, 46, 8, 1318–1320, 2007.
72. Aksorn, J. and Teepoo, S., Development of the simultaneous colorimetric enzymatic detection of sucrose, fructose and glucose using a microfluidic paper-based analytical device. *Talanta*, 207, 120302, 2020.
73. Sriram, G., Bhat, M.P., Patil, P., Uthappa, U.T., Jung, H.-Y., Altalhi, T., Kumeria, T., Aminabhavi, T.M., Pai, R.K., Madhuprasad, Kurkuri, M.D., Paper-based microfluidic analytical devices for colorimetric detection of toxic ions: A review. *TrAC, Trends Anal. Chem.*, 93, 212–227, 2017.
74. Huang, X., Shi, W., Li, J., Bao, N., Yu, C., Gu, H., Determination of salivary uric acid by using poly(3,4-ethylenedioxythipohene) and graphene oxide in a disposable paper-based analytical device. *Anal. Chim. Acta*, 1103, 75–83, 2020.
75. Liu, M.M., Li, S.H., Huang, D.D., Xu, Z.W., Wu, Y.W., Lei, Y., Liu, A.L., MoOx quantum dots with peroxidase-like activity on microfluidic paper-based analytical device for rapid colorimetric detection of H2O2 released from PC12 cells. *Sens. Actuators, B*, 305, 127512, 2020.
76. Nguyen, M.P., Kelly, S.P., Wydallis, J.B., Henry, C.S., Read-by-eye quantification of aluminum (III) in distance-based microfluidic paper-based analytical devices. *Anal. Chim. Acta*, 1100, 156–162, 2020.
77. Noviana, E., Carrão, D.B., Pratiwi, R., Henry, C.S., Emerging applications of paper-based analytical devices for drug analysis: A review. *Anal. Chim. Acta*, 1116, 70–90, 2020.
78. Sia, S.K. and Kricka, L.J., Microfluidics and point-of-care testing. *Lab. Chip*, 8, 1982–1983, 2008.

79. Shishov, A., Trufanov, I., Nechaeva, D., Bulatov, A., A reversed-phase air-assisted dispersive liquid-liquid microextraction coupled with colorimetric paper-based analytical device for the determination of glycerol, calcium and magnesium in biodiesel samples. *Microchem. J.*, 150, 104134, 2019.
80. Yang, X. and Wang, E., A nanoparticle autocatalytic sensor for Ag+ and Cu2+ ions in aqueous solution with high sensitivity and selectivity and its application in test paper. *Anal. Chem.*, 83, 12, 5005–5011, 2011.
81. Zhai, H.M., Zhou, T., Fang, F., Wu, Z.Y., Colorimetric speciation of Cr on paper-based analytical devices based on field amplified stacking. *Talanta*, 210, 120635, 2020.
82. Darabdhara, G., Boruah, P.K., Das, M.R., Colorimetric determination of glucose in solution and via the use of a paper strip by exploiting the peroxidase and oxidase mimicking activity of bimetallic Cu-Pd nanoparticles deposited on reduced graphene oxide, graphitic carbon nitride, or MoS2 nanosheets. *Microchim. Acta*, 186, Article number: 13, 2019.
83. Alizadeh, N., Salimi, A., Hallaj, R., Mimicking peroxidase-like activity of Co3O4-CeO2 nanosheets integrated paper-based analytical devices for detection of glucose with smartphone. *Sens. Actuators, B*, 288, 44–52, 2019.
84. Xuan, X., Yoon, H.S., Park, J.Y., A wearable electrochemical glucose sensor based on simple and low-cost fabrication supported micro-patterned reduced graphene oxide nanocomposite electrode on flexible substrate. *Biosens. Bioelectron.*, 109, 75–82, 2018.
85. Su, L., Qin, W., Zhang, H., Rahman, Z.U., Ren, C., Ma, S., Chen, X., The peroxidase/catalase-like activities of MFe2O4 (M=Mg, Ni, Cu) MNPs and their application in colorimetric biosensing of glucose. *Biosens. Bioelectron.*, 63, 384–391, 2015.
86. Tran, H.V., Le, T.A., Giang, B.L., Piro, B., Tran, L.D., Silver nanoparticles on graphene quantum dots as nanozyme for efficient H2O2 reduction in a glucose biosensor. *Mater. Res. Express*, 6, 11, 115403, 2019.
87. Bharath, G., Madhu, R., Chen, S.–M., Veeramani, V., Balamurugan, A., Mangalaraj, D., Viswanathan, C., Ponpandian, N., Enzymatic electrochemical glucose biosensors by mesoporous 1D hydroxyapatite-on-2D reduced graphene oxide. *J. Mater. Chem. B*, 3, 1360–1370, 2015.
88. Jayakumar, K., Bennett, R., Leech, D., Electrochemical glucose biosensor based on an osmium redox polymer and glucose oxidase grafted to carbon nanotubes: A design-of-experiments optimisation of current density and stability. *Electrochim. Acta*, 371, 137845, 2021.
89. Müsse, A., La Malfa, F., Brunetti, V., Rizzi, F., De Vittorio, M., Flexible enzymatic glucose electrochemical sensor based on polystyrene-gold electrodes. *Micromachines*, 12, 7, 805, 2021.
90. Izadyar, A., Nivan, M., Rodriguez, K.A., Seok, I., Hood, E.E., A bienzymatic amperometric glucose biosensor based on using a novel recombinant Mn peroxidase from corn and glucose oxidase with a Nafion membrane. *J. Electroanal. Chem.*, 895, 115387, 2021.
91. Dervisevic, M., Çevik, E., Şenel, M., Development of glucose biosensor based on reconstitution of glucose oxidase onto polymeric redox mediator coated pencil graphite electrodes. *Enzyme Microb. Technol.*, 68, 69–76, 2015.
92. German, N., Ramanavicius, A., Ramanaviciene, A., Amperometric glucose biosensor based on electrochemically deposited gold nanoparticles covered by polypyrrole. *Electroanalysis*, 29, 5, 1267–1277, 2017.
93. German, N., Ramanavicius, A., Ramanaviciene, A., Electrochemical deposition of gold nanoparticles on graphite rod for glucose biosensing. *Sens. Actuators, B*, 203, 25–34, 2014.
94. Shan, C., Yang, H., Han, D., Zhang, Q., Ivaska, A., Niu, L., Graphene/AuNPs/chitosan nanocomposites film for glucose biosensing. *Biosens. Bioelectron.*, 25, 15, 1070–1074, 2010.

95. Wang, H., Ohnuki, H., Endo, H., Izumi, M., Impedimetric and amperometric bifunctional glucose biosensor based on hybrid organic–inorganic thin films. *Bioelectrochemistry*, 101, 1–7, 2015.
96. Asrami, P.N., Mozaffari, S.A., Tehrani, M.S., Azar, P.A., A novel impedimetric glucose biosensor based on immobilized glucose oxidase on a CuO-Chitosan nanobiocomposite modified FTO electrode. *Int. J. Biol. Macromol.*, 118, 649–660, 2018.
97. Https://En.Wikipedia.Org/Wiki/Clark_Electrode#Cite_Note-4,
98. Jang, C., Lee, H.-J., Yook, J.-G., Radio-frequency biosensors for real-time and continuous glucose detection. *Sensors (Basel)*, 21, 5, 1843, 2021.
99. The bumbling biochemist: Diabetes, insulin, and glucometers, 2019, https://thebumblingbiochemist.com/365-days-of-science/diabetes-insulin-and-glucometers/.
100. Gabriel, E.F.M., Garcia, P.T., Lopes, F.M., Coltro, W.K.T., Paper-based colorimetric biosensor for tear glucose measurements. *Micromachines*, 8, 4, 104, 2017.

7

# Metal-Free Quantum Dots-Based Nanomaterials for Biosensors

**Esra Bilgin Simsek**[1,2]

[1]*Department of Chemical Engineering, Faculty of Engineering, Yalova University, Yalova, Turkey*
[2]*Department of Chemical Engineering, Faculty of Engineering, Gebze Technical University, Gebze, Kocaeli, Turkey*

***Abstract***
Quantum dots (QDs) have gained particular interest owing to their optical, electrochemical, and structural features. Especially, photochemical and florescence features of QDs provide them a favorable platform for sensing areas. These zero-dimensional materials can considerably improve the analytical abilities of sensors, such as limit of detection and sensitivity. However, traditional metal-based QDs have drawbacks such as harmfulness and non-biodegradability, so it is essential to explore metal-free QDs. Carbonaceous QDs, namely carbon QDs (CQDs), graphene QDs (GQDs), and graphitic carbon nitride (CNQDs) QDs have been newly discovered quantum-sized materials biosensing and detection applications owing to their excellent photostability, harmless, biocompatibility, and unique quantum features. In this report, recent progress on the preparation of metal-free quantum sized sensors such as CQDs, GQDs and CNQDs was investigated. The optical and electrochemical techniques employed to determine the bioanalytes were reported by comparing linear concentration range and detection limits. The aim is to guide future investigations into metal-free and safer biosensor design via developing CQD-, GQD-, and CNQD-based biosensors.

***Keywords*:** Biosensor, metal-free, quantum dot, CQDs, g-C3N4, GQDs

## 7.1 Introduction

Quantum dots (QDs) have been discovered as next-generation nanomaterials due to their unique physical, chemical, optical, and electronic transport features. QDs have been widely utilized in several areas such as optical devices, sensor, diagnosis, and lasers, etc., especially they have been recognized as ideal bioprobes and biosensors in the area of biosensing and analyte detection [1, 2]. QDs are classified in the class of fluorescent nanomaterials and they are identified by single absorption emission or wide range of emission wavelength [3]. They show extended fluorescence lifetime and high quantum efficiency with broad excitation spectra, thus, the target biomolecules can be easily detected by QD-based fluorescent

*Email*: ebilgin.simsek@yalova.edu.tr; esrabilgin622@gmail.com

Inamuddin and Tariq Altalhi (eds.) Biosensors Nanotechnology, 2nd Edition, (145–168) © 2023 Scrivener Publishing LLC

probes, even at low concentrations. Moreover, robust anti-photobleaching, high photochemical stability, and fluorescence resistance features of quantum dots enable them to be widely used in DNA and protein detection, cell imaging and screening of tumor cells in living organisms [2]. QDs are consisted of inorganic nuclei with organic molecules such as carbon, silicon, cadmium telluride (CdTe), cadmium sulphide (CdS), or indium arsenide in the nanoparticle range of 1 to 10 nm [4]. Among the class of QDs, CdS, CdTe are generally considered quantum materials on account of band potential and great fluorescence features. Despite these advantages, heavy metal-based QDs have important obstacles such as high toxicity and environmental risks. Hence, finding benign alternatives have become a critical issue. Carbon-based QDs, such as carbon (CQDs), graphene (GQDs), graphitic carbon nitride (g-$C_3N_4$) quantum dots have gained importance as potential non-metallic nanomaterials in biosensing applications. Carbon-based quantum dots possess good biocompatibility, high photostability, low-toxicity and their tunable physicochemical features allow them to be utilized in fluorescent screening, light-emitting devices, sensors, bioimaging and drug delivery [5, 6]. In addition, carbon-based quantum dots have abundant surface functional carboxyl and amide groups, which provide high solubility in aqueous media and improve their conjugation with biological molecules [7].

In literature, many reviews have been documented about monitoring of biomolecules by using different quantum dots [4, 7–11]. However, to the best of our knowledge, no systematic review reported about metal-free quantum dots-based biosensors to detect biological targets. Therefore, this review investigated the applications of carbon-based metal-free QDs namely carbon (CQDs), graphene (GQDs), g-$C_3N_4$ quantum dots (CNQDs). The optical and electrochemical techniques employed to determine the bio-analytes were reported by comparing linear concentration range and detection limits.

## 7.2 Metal-Free Quantum Dots as Biosensors

### 7.2.1 Carbon Quantum Dots as Biosensors

Carbon quantum dots (CQDs) with a size range between 2 and 10 nm have excellent features including great photostability, biocompatibility, anti-photobleaching, low-cost, low toxicity, and outstanding water solubility [12]. Carbon QDs have upconversion photoluminescence features that they have absorption in extended visible region [13]. Especially, non-blinking photoluminescence ability of CQDs provides monitoring of single-molecules while the photo-stability facilitates long-term imaging [14]. Based on these excellent features, CQDs have been considered as ideal candidates for fluorometric indicator in biosensing or chemo-sensing applications. In recent years, metal-fee CQDs have approved their potential as fluorescence, electrochemical, photoelectrochemical probes in monitoring biochemical substances (Table 7.1).

Chai *et al.* [15] investigated CQDs functionalized with dopamine for monitoring tyrosinase activity with a detection limit of 7.0 U/L. They underlined that the conjugated CQDs displayed great biocompatibility and could sensitively detect tyrosinase in melanoma cells, providing favorable utilization in disease examination and medical diagnostics.

On the other hand, a fiber glucose biosensor including glucose oxidase/cellulose acetate-based fluorescent CQDs (CQDs-GOD/CA) was developed by Yu *et al.* [16] through dip-coating technique and the as-synthesized hybrid material with high enzymatic activity

**Table 7.1** CQDs based metal-free materials as biosensors.

| Quantum dots | Analyte | Method | LOD | Ref. |
|---|---|---|---|---|
| NCQD/$TiO_2$ | Chlorpyrifos | Photoelectrochemical | 0.07 ng/mL | [13] |
| Dopamine/CQDs | Tyrosinase | Fluorescence | 7.0 U/L | [15] |
| CQDs-GOD/CA | Glucose | Fluorescence | 25.79 nM | [16] |
| CQDs/AuNPs-GOx | Glucose | Chronoamperometric | 626.06 mA/mM/ $cm^2$ | [17] |
| β-CD-CQDs | α-glucosidase | Fluorescence | 0.6 U/L | [18] |
| β-CD-CQDs | Cholesterol | Fluorescence | $0.7 \pm 0.1$ μM | [19] |
| B-CQDs | Amoxicillin | Fluorescence | 0.835 μmol/L | [20] |
| ssDNA/CQDs | Acrylamide | Fluorescence | $2.41 \times 10^{-8}$ M | [21] |
| CDs-pDNA-/$Fe_3O_4$@PDA | microRNA-167 | Fluorescence | 76 pM | [22] |
| CQDs/AuNPs | Paraoxon | Fluorescence | 0.05 μg/L | [23] |
| CQDs/AuNPs | Maitotoxin | Fluorescence | 0.3 pmol/L | [24] |
| CQDs | Mitochondrial $H_2O_2$ | Fluorescence | - | [25] |
| CQDs | Vitamin B12 | Fluorescence | 0.04 μM | [26] |
| CQDs/o-phenylenediamine | Nitric oxide | Fluorescence | 9.12 nM | [27] |
| | | | | |
| Thionine/CQDs | Cocaine | Electrochemical | 0.26 pM | [28] |
| ITO/CQDs/AuNPs | Salmonella DNA | Electrochemical | 10 fM | [29] |
| N-CQDs-DNA | miRNA-21 | Electrochemiluminescence | 10 aM | [30] |
| CD-PMMA | TNF-α | Electrochemical | 0.05 pg/mL | [31] |
| rGO-MWCNT/CS/CQD | Lysozyme | Electrochemical | 3.7 fmol/L (DPV)<br>1.9 fmol/L (EIS) | [32] |
| CQDs | Biothiols | Colorimetric | 0.4 μM (CySH)<br>0.6 μM (GSH)<br>0.9 μM (HcySH) | [35] |
| GO@$Fe_3O_4$@β-CD@A-Apt | Adenosine | Chemiluminescence | $2.1 \times 10^{-13}$ mol/L | [36] |

was applied as the fluorescence indicator. The biosensor showed good response for detection of glucose evet at low concentrations. The fluorescence intensity linearly changed according to the glucose concentration ranging from 10 μmol/L to 200 μmol/L with a low detection limit (6.43 μM). Moreover, when the linearity range was 10–100 nmol/L, the limit of detection limit was determined as 25.79 nM. On the other hand, chronoamperometric sensing of glucose by using glucose oxidase (GOx) immobilized on CQDs-Au NPs was reported by Buk and Pemble [17] and the hybrid biosensor showed high response toward glucose with a detection limit as 626.06 $mA/mM/cm^2$. In another work, β-Cyclodextrin was covalently attached to CQDs for quantitative determination of α-glucosidase which has an important part in monitoring glycemic levels and maintenances the level of glucose [18]. The functionalization of CQDs with β-Cyclodextrin enabled a unique transduction mechanism through supramolecular assembly in the presence of host-guest recognition, leading sensitive biosensing for α-glucosidase activity. Similarly, fluorometric detection of cholesterol was successfully obtained by using β-cyclodextrin incorporated CQDs (β-CD-CQD) with a detection limit of 0.7±0.1 μM [19].

Zhang *et al.* investigated the fluorometric detection of antibiotic amoxicillin by using boron doped CQDs [20]. In the synthesis (Figure 7.1), citric acid and boric acid were applied as carbon and boron sources, respectively. The molar ratio of C/B was determined as 5:1 in the B-CQDs structure. After hydropyrolysis process in Teflon lined autoclave, the products were filtrated and yellowish B-CQD suspension with high solubility was obtained. The boron doping favored blue fluorescence, decreased the particle diameter with as sized of 2.3 nm and increased the quantum yield to 30.8%. The detection mechanism was based on the interaction of amoxicillin with hydrogen bonds of graphitic functional bonds (-OH or -COOH) which played as probes to sensitively detect antibiotic molecules in a concentration area of 1.43–429.12 μmol/L with a detection limit of 0.825 μmol/L.

A sensitive fluorescence biosensor based on CQDs labelled with DNA was reported by Wei *et al.* [21] for the monitoring of acrylamide which is a potential carcinogen produced in food industry. For this, CQDs were fabricated through hydrothermal process and subsequently they were quenched on single-stranded DNA (ssDNA). The fluorescence intensity at 445 nm was significantly affected after the attachment of ssDNA. After addition of acrylamide, ssDNA was selectively conjugated with acrylamide via hydrogen bonds and the

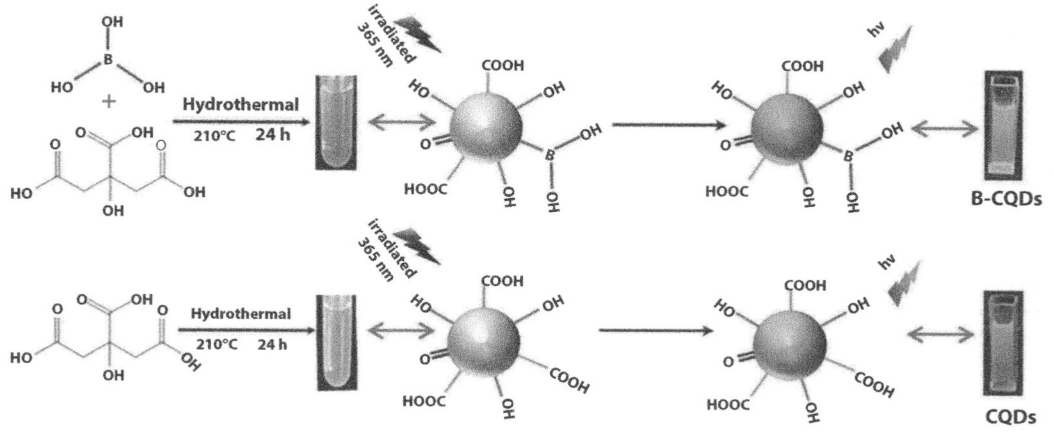


**Figure 7.1** Preparation of CQDs and B-CQDs [20]. Reprinted by permission from Elsevier Publisher.

decrease in fluorescence intensity was found smaller than that without acrylamide. At optimum conditions, the material displayed a linear concentration response between $5\times10^{-3}$ and $1\times10^{-7}$ M with a detection limit of $2.41\times10^{-8}$ M. In addition, the sensor exhibited a well recovery degree (91.36–98.11%) in bread crust revealing its usage in industrial utilizations. A unique approach was reported by Cao *et al.* [22]. They built fluorescent resonant energy transfer mechanism between carbon quantum dots and DNA (pDNA) probe for the measurement of microRNA-167 in A. thaliana seedlings. In the study, CQDs labelled with pDNA was applied as fluorescent donor while the polydopamine (PDA)-coated $Fe_3O_4$ nanoparticles ($Fe_3O_4$@PDA) were utilized as acceptor (Figure 7.2). It was hypothesized that the CQDs-pDNA were absorbed on $Fe_3O_4$@PDA surface via robust $\pi$–$\pi$ stacking relationship between pDNA and PDA. The fluorescence signal of CDs at 445 nm was quenched due to the improved adsorption ability of $Fe_3O_4$@PDA by calcium ions and the biosensor displayed a linear concentration range of 0.5–100 nM toward microRNA-167 molecule with a LOD of 76 pM. Wu *et al.* [23] also built a fluorescent resonant energy transfer strategy between CQDs and gold nanoparticles (Au NPs) and it was stated that the fluorescence emission was greatly hindered by Au NPs. The as-built energy transfer mechanism was applied as determination of paraoxon which is classified in the group of organophosphorus pesticides. According to the results, the linear range of paraoxon concentrations was determined as 0.05–50 μg/L and the limit of detection was obtained as 0.05 μg/L. The CQDs-Au NPs sensor displayed satisfactory recoveries in the range of 91.36–98.11% in bread crust samples. Similarly, Gholami and co-workers [24] reported high sensitivity and selectivity of CQDs coupled with Au NPs for the determination of maitotoxin which is a marine toxin with critical $LD_{50}$, causing severe threat to neuronal and cardiovascular systems. Under certain conditions, the linear concentration range was reported as 1 to 600 pmol/L and the limit of detection was obtained as 0.3 pmol/L. In another work [25], bioimaging and monitoring of mitochondrial $H_2O_2$ was investigated by using CQD-based FRET probe in which

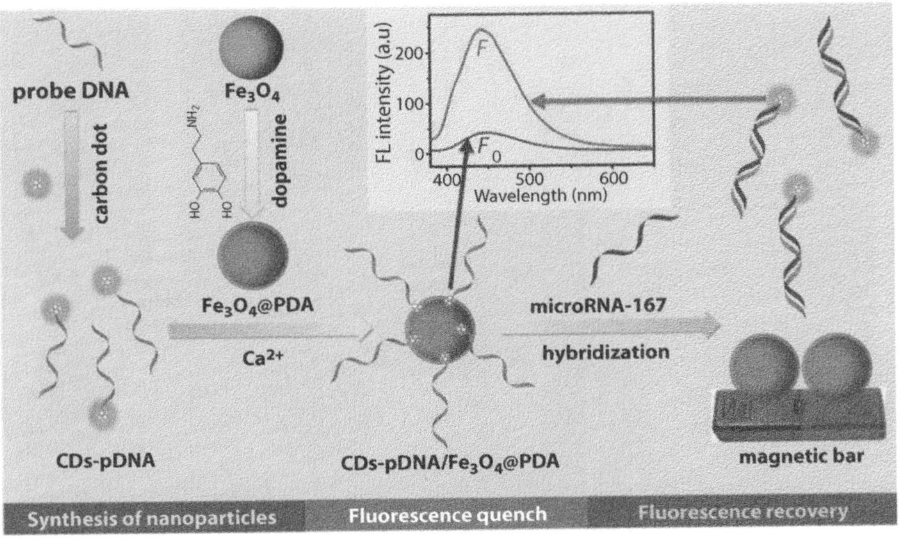


**Figure 7.2** Fabrication of CDs-pDNA-/$Fe_3O_4$@PDA for sensing microRNA-167 [22]. Reprinted by permission from Springer Publisher.

CQDs acted as donor. The nanoprobe with high cell permeability specifically screened the target mitochondria and could effectively determine hydrogen peroxide in L929 cells.

Fluorescent CQDs were also applied for sensitive detection of vitamins. Recently, determination of cobalamin (vitamin B12) was investigated by using CQDs synthesized from curry berries via ultrasonication method [26]. It was underlined that the fluorescence quenching of carbon quantum dots was mainly due to their strong interactions with the $Co^{2+}$ ions which are main element of Vitamin B12. Therefore, the photoluminescence intensity of CQDs significantly diminished with higher vitamin B12 concentration. The oxygen-containing functional groups on the surface resulted in formation of complexation with $Co^{2+}$, promoting the transfer of electrons from CQDs to $Co^{2+}$ (Figure 7.3). Thereby, the fluorescence emission of the CQDs was greatly reduced. Under optimum conditions, the linearity of cobalamin was found in the range of 0–0.40 μM and LOD was reported as 0.04 μM.

CQDs have been also utilized as fluorescent probes in the sensing of biological nitric oxide (NO) which plays a unique role in organisms such as immune processes, nerve conduction and physiological functions [27]. Wu and co-workers [27] modified CQDs with o-phenylenediamine groups through microwave treatment and underlined the fact that the o-phenylenediamine bonds reacted with NO and created an electron-free triazole structure which decreased the fluorescence intensity of the carbon quantum dots. The limit of detection was observed 9.12 nM at laser power of 280 mW.

Very recently, Azizi and co-workers [28] utilized CQDs as electrochemical sensor for sensitive detection of cocaine which is a well-known addictive stimulant drug causing serious health issues. At first, gold nanoparticles were attached on ITO electrode via physical vapor deposition process, and subsequently the electrode surface was modified with thiol-functionalized aptamer sequence. Next, the thionine modified CQDs were covalently attached to the other aptamer chain. During monitoring of cocaine, the aptamer conformational changes occurred and decreased the distance between thionine and electrode which

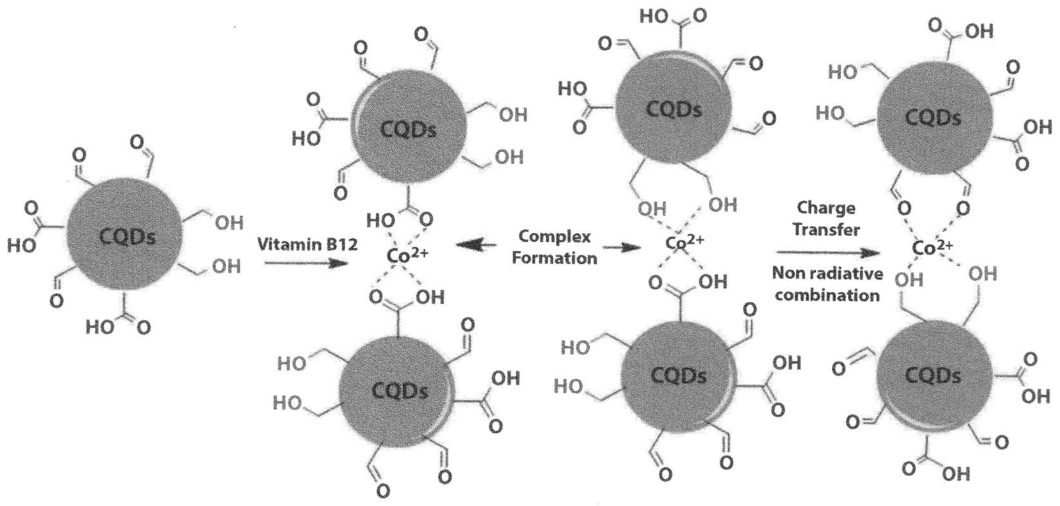


**Figure 7.3** Interaction between CQDs and $Co^{2+}$ ions of Vitamin B12 [26]. Reprinted by permission from Elsevier Publisher.

enhanced charge migration efficiency. Under certain conditions, the biosensor showed a linear response in the concentration range of 10–70 pM and the LOD was reported as 0.26 pM. Moreover, the aptamer-based biosensor displayed good sensitivity of cocaine in human serum and urine samples demonstrating its feasible utilization. In another study conducted by Azizi *et al.* [29] CQDs modified with neutral red were applied as a redox probe on the ITO/AuNPs electrode and the biosensing performance of the hybrid material was explored for the detection of salmonella DNA sequence at low concentrations. According to the voltametric measurements, the linear range was found as 20–240 fM with a detection limit of 10 fM. Liu *et al.* [30] investigated electro-chemiluminescent monitoring of miRNA-21 by using DNA functionalized nitrogen doped CQDs and the ECL increased with increasing miRNA-21 concentration. The constructed biosensor displayed high sensitivity and selectivity toward determination of miRNA-21 in the range of 10 aM –$10^4$ fM with a LOD of 10 aM, demonstrating its usage in clinical diagnosis. Sri *et al.*, [31] synthesized carbon dots supported on poly methyl methacrylate (CD-PMMA) which showed great electrocatalytic conductivity, biocompatibility and high surface area. The as-prepared biosensor was utilized as transducing material to electrochemical sensing of TNF-a molecule in concentration area ranging from 0.05 pg/mL to 160 pg/mL with a LOD of 0.05 pg/mL. Authors demonstrated that the CD-PMMA immunosensor displayed high correlation with industrial Enzyme-Linked Immunosorbent Assay method, for rapid monitoring cancer serum species.

Rezaei and co-workers [32] constructed an aptamer-based material for the electrochemical sensing of lysozyme protein which is very effective molecule in the immune system. In this context, a nanocomposite including graphene oxide, carbon nanotubes, chitosan, and CQDs (rGO-MWCNT/CS/CQD) was prepared and enhanced immobilization of aptamer was obtained owing to the presence of abundant amino and carboxyl functional sites as well as oxygen-vacancies of graphene structure. The sample was characterized by different electrochemical methods namely electrochemical impedance spectroscopy and differential pulse voltammetry. The voltametric results revealed that when the aptamer caught the target lysozyme, a decrease in the differential peak intensity was obtained while the impedance signal was strengthened. Based on the measurements, two very low limit of detections were determined as 3.7 and 1.9 fmol/L in the detection range of 20 fmol/L–10 nmol/L, and 10 fmol/L–100 nmol/L according to the voltammetry and impedance techniques, respectively. Moreover, recovery tests with the proposed aptasensor were applied in different environment such as Urine, Serum, Egg white, and Wine.

Photoelectrochemical analysis (PEC) is based on the photoelectric conversion process in which photoactive material electronically excited and subsequently charge transfer occurs under the light irradiation [33]. However, PEC activity of CQDs is known as relatively weak owing to their low electron transfer features, so modification of CQDs can overcome this obstacle. It was postulated that coupling CQDs with molybdenum carbide nanotubes could be beneficial for sensitive detection of dual microRNAs [34]. In the sensing mechanism, the energy transfer between amino-functionalized carbon quantum dots and gold nanoparticles remarkably affected the PEC response. Moreover, photoelectrochemical detection performance of nitrogen doped CQDs (NCQDs) was investigated by Cheng and co-workers [13]. Chlorpyrifos as an organophosphorus pesticide was chosen as target molecule. Indium tin oxide (ITO) was used as substrate and $TiO_2$ NPs were used as photoactive material while the NCQDs and acetylcholinesterase (AChE) were applied as photosensitizer and biorecognition molecule, respectively. After addition of nitrogen doped CQDs, the photocurrent intensity of the composite electrode (NCQD/$TiO_2$/ITO) significantly enhanced in contrast to $TiO_2$/ITO sample. The presence of acetylcholinesterase on the electrode acted as catalysis for hydrolysis of acetylcholine chloride for thiocholine formation. The biocatalytic activity was hindered by chlorpyrifos and so the formation of thiocholine was inhibited, resulting in a decrease in the photocurrent response (Figure 7.4). The linear concentration range for the detection of chlorpyrifos was reported between 0.001 mg/mL and 1.5 mg/mL with LOD of 0.07 ng/mL.

Colorimetric detection with nano-sized materials has been also applied by researchers for the detection of biomolecules. Yuan and co-workers [35] examined colorimetric detection of biothiols namely cysteine, glutathione and homocysteine which are highly reductant amino acids and they play crucial roles in keeping physiological processes. At first, CQDs were synthesized via hydrothermal technique by using chicken blood which displayed excellent peroxidase-like ability to catalyse the oxidation to produce blue oxidized 3, 3′, 5, 5′-tetramethylbenzidine (TMB). After reduction of TMB with biothiols, resulting in weaker blueness and lower absorbance. Considering this phenomenon, sensitive determination of biothiols by carbon quantum dots was obtained and the detection limits were reported as 0.4, 0.6, and 0.9 μM for cysteine, glutathione and homocysteine, respectively. Authors also demonstrated that the detection of cysteine in human serum samples was achieved almost 109.3% recovery, revealing the potential usage of the as-synthesized CQDs. In a different work [36], chemiluminescence measurement of adenosine was reported by using aptamers functionalized graphene- magnetic β-cyclodextrin polymers and ssDNA modified CQDs. The biosensor was effectively determined adenosine in urine samples which the recoveries were approximately achieved as 100%.

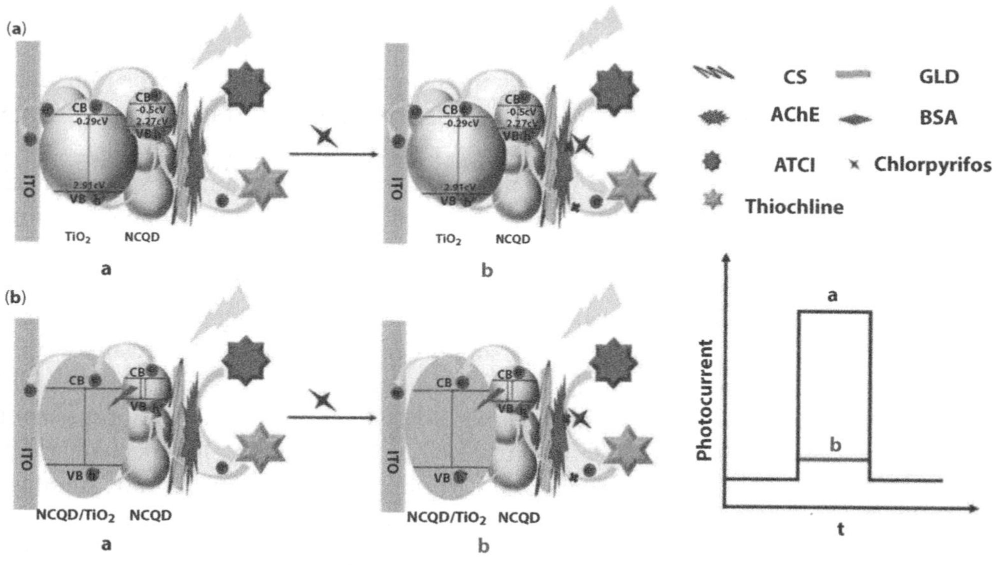


**Figure 7.4** Photoelectrochemical detection of chlorpyrifos on NCQD/$TiO_2$/ITO electrode based on the mechanism of (a) matched energy levels or (b) production of electrons on both materials [13]. Reprinted bypermission from Elsevier Publisher.

### 7.2.2 Graphene Quantum Dots as Biosensors

Graphene quantum dots (GQDs) are a newcomer of graphene family, and they are graphene nanosheets of < 20 nm in lateral dimension including monolayers with $sp^2$-hybridization and interconnected with chemical active sites on the edges. When compared to its two dimensional counterpart, GQDs presents unique properties, such as narrowing bandgap due to quantum confinement, good dispersibility and more abundant active sites [37]. GQDs are biocompatible and environmentally safe materials which are beneficial in industrial utilizations. GQDs show high crystalline structure and they are more crystalline than CQDs [38]. They include π–π stacking bonds in their planar structure in which they can be more or less conducive to stacking according to different synthesis methods [39]. Their excellent quantum confinement effects and photoluminescence features have attracted researchers in many fields. GQDs have photon down altering feature and they can absorb and emit photons in shorter and longer wavelengths, respectively [40]. In the UV spectrum, GQDs have a main absorption peak (~260–320 nm) which is arisen from π–π* bonds in the conjugated network and a shoulder peak (~270–390 nm) is ascribed from n–π* bonds of C=O groups [39]. The photoluminescence color of GQDs can be variable such as blue, green, yellow, or red in context of preparation method, size, chemical bonded groups [38]. GQDs also exhibit upconversion photoluminescence feature, a phenomenon where the emission energy is greater than its excitation [38]. On the other hand, the band gap engineering which is the modification of highest occupied and lowest unoccupied molecular orbital transition energy levels is one of the most essential properties of GQDs [5, 41]. In addition, GQDs serve as ideal platform for immobilization of various molecules due to their large specific surface area. GQDs have also plentiful surface functional groups (carboxyl, hydroxyl etc.) which enable them well dispersed in aqueous media [42]. Especially, the formation of carboxylic acid moieties on the surface endows them to be functionalized via covalent and noncovalent bonds with several biologic or inorganic materials [43]. Considering these merits, GQDs are perfect candidates for detection of biomolecules (Table 7.2).

An early study was reported by He *et al.* [44] who prepared a fluorescent biosensor system consisted of hemin functionalized GQDs through π–π bonds in order to identify glucose in human serum. They stated that the introduction of hemin significantly enhanced the fluorescence quenching of GQDs via modification of surface states of GQDs. The linear range of glucose was found as 9–300 μM with limit of detection as 0.1 μM at optimum conditions. Similarly, Huang *et al.* coupled G-quadruplex/hemin DNAzyme with amino-functionalized GQDs to detect human telomere DNA and they reported that the sensor had a broad range of 0.2 pM ~ 50 pM and the detection limit was found as 25 fM [45]. Chowdhury and Doong [46] synthesized dopamine modified GQDs for monitoring Fe(III) ions at low concentrations. The prepared GQDs displayed high quantum yield (10.2%) and the specific complexation bonds between iron and catechol moiety of dopamine provided high sensitivity for the identification of Fe(III) ions (LOD = 7.6 nM).

A multifunctional GQD@$MnO_2$ sensor based on two-photon excitation mechanism was designed by Meng *et al.* [47] for bio-identification of glutathione (GSH) which is a ubiquitous tripeptide. The as-prepared nano-sized probes displayed high sensitivity for imaging of GSH and the influence of GSH on photodynamic therapy efficiency was investigated *in vitro* tests. It was found that the intracellular GSH decreased the $MnO_2$ nanosheets,

**Table 7.2** GQD-based metal-free materials as biosensors.

| Quantum dots | Analyte | Method | LOD | Ref. |
|---|---|---|---|---|
| GQDs/Hemin | Glucose | Photoluminescence | 0.1 μM | [44] |
| Amin/GQDs/ G-quadruplex/ hemin DNAzyme | Human telomere DNA | Ratiometric fluorescence | 25 fM | [45] |
| GQDs/Dopamine | Fe(III) | | 7.6 nM | [46] |
| Au/GQDs | Glucose | Ratiometric fluorescence | 0.18 μM | [48] |
| rGOQDs | Glucose | Photoluminescence | 3.869 deg/mM | [49] |
| BGQD/NCNTs | Dissolved Oxygen | Electrochemical | 0.011 mA/cm$^2$ ppm | [50] |
| GQDs | Progesterone | Electrochemical | | [51] |
| Au@B-GQDs | Guanine<br>Adenine | Electrochemical | 1.71 μM<br>1.84 μM | [52] |
| GQDs/MWCNTs | Interleukin-6 protein | | 0.0030 pg/ml | [53] |
| | | | | |
| GQDs/SiNWs | Microcystin-LR | Photoelectrochemical | 0.055 μg L$^{-1}$ | [54] |
| QDs-CoPc (ππ)-aptamer | Prostate antigen (PSA) | Electrochemical | 0.66 pM | [60] |
| Au-GQDs | Prostate antigen (PSA) | Electrochemical | 0.14 ng/mL | [61] |
| Au/Ag-rGODs | Prostate antigen (PSA) | Electrochemiluminescence | 0.29 pg/mL | [62] |
| HM-GQDs-AuNPs | Carcinoembryonic antigen (CEA) | Electrochemiluminescence | 0.01 ng/mL | [63] |
| (N, S-GQDs@ Au-PANI | Carcinoembryonic antigen (CEA) | Impedimetric Detection | 0.01 ng/mL | [64] |
| GOQDs | Carcinoembryonic antigen (CEA) | Electrochemiluminescence | 1.0 pg/mL | [65] |
| GOQDs | Carbohydrate antigen 125 (CA125) | Electrochemiluminescence | 0.01-0.05 U/mL | [65] |
| GOQDs | A-fetoprotein (AFP) | Electrochemiluminescence | 1.0 pg/mL | [65] |

*(Continued)*

**Table 7.2** GQD-based metal-free materials as biosensors. (*Continued*)

| Quantum dots | Analyte | Method | LOD | Ref. |
|---|---|---|---|---|
| | | | | |
| Graphdiyne QDs | microRNAs-21 | Ratiometric fluorescence | 0.5 pM | [66] |
| GQDs | HeLa, A-375 cancer cell | Confocal imaging | - | [67] |
| N-GQDs | HeLa cells | Confocal imaging | - | [68] |
| GQDs | MDA-MB-321, T47D cancer cell | Confocal imaging | - | [69] |
| GQDs | A-549 cancer cell | Confocal imaging | - | [70] |
| Conjugated GQDs | MDA-MB-321 | Confocal imaging | - | [71] |
| GO-PEI-GQDs | MDA-MB-321 | Photothermal response | - | [72] |
| GQD | Theophylline | DPV | 0.2 μM | [92] |
| Au/GQDs/pl-cysteine | p53 | Electrochemical | 0.065 fM | [93] |

resulting lower GSH level, thereby the effectiveness of GQDs on photodynamic therapy greatly enhanced.

A similar ratiometric fluorescence approach using GQDs was designed by Hong and co-workers [48] for the determination of glucose. They coupled GQDs with fluorescent gold nanoclusters (AuNCs) and the fluorescence response between 420 and 575 nm was applied to detect glucose and the detection limit was found as 0.18 μM in human serum. Authors demonstrated that the ratiometric fluorescence biosensing provided sensitive and convenient detection for other biomedical applications. Hwang *et al.* [49] also proposed a glucose-sensing detector based on reduced-GQDs (rGOQDs) which was prepared in perfluorotributylamine solution in order to prevent formation of unfavorable functional groups (epoxy, carbonyl, and carboxyl). The butyl fluoride groups bonded with amine had positive impact on the synthesis of high purity of rGOQDs. It was hypothesized that the fluorine atoms provided higher binding and desorption energies for carbon than hydrogen atoms and thereby would decrease the surface carbonyl or carboxylic groups. Authors found that the photoluminescence spectra of rGOQDs with glucose altered the color from blue to green with high sensitivity (3.869 deg/mM) in a concentration area of 0 to 2 mM, ensuring that the as-synthesized material could be utilized as colorimetric biosensor.

Doping heteroatoms has been also regarded as a promising approach for improving electron density and photoluminescence features of GQDs. Researchers have investigated nitrogen doped GQDs by using several techniques for bio-sensing applications. For example, Wang *et al.* [50], investigated electrochemical response of boron doped GQDs and nitrogen doped carbon nanotubes (BGQD/NCNTs) sensor. The BGQDs were incorporated with NCNTs through $\pi$–$\pi$ stacking and the occurrence of oxygen-rich active sites as well as B–N bonding groups provided effective sensing of dissolved oxygen. The electrocatalyst showed long-term stability and the monitoring dissolved oxygen in seawater was sensitively performed with a concentration of 0.011 mA/cm$^2$ ppm. Similarly, electrochemical detection of progesterone was investigated by using GQDs modified glassy carbon electrodes [51]. Authors applied cyclic voltammetry technique for immune-sensing of hormone and the activated electrodes were incubated with 25 mg/mL progesterone antibody solution for 1 h. The results indicated that the progesterone was bonded with related antibody attached on GQDs and the current response was changed with the increased hormone concentration. It was stated that the high surface area of GQDs enabled sensitive detection of progesterone even at low concentrations. In another work [52], gold-loaded B-doped GQDs (Au@B-GQDs) were utilized for the electrochemical monitoring of guanine and adenine biomolecules which are purine bases. The Au@B-GQDs biosensors were fabricated through drop-casting on glass carbon-GC electrodes. The electrochemical measurements showed that the composite sensor displayed a good sensitivity in a wide concentration range (Guanine = 0.5–20 μM, Adenine = 0.1–20 μM), with detection limits of 1.71 μM for guanine and 1.84 μM for adenine.

A molecular imprinting biosensor including GQDs/multi-walled CNTs (GQDs/MWCNTs) was reported by Ozcan and co-workers [53] for the detection of Interleukin-6 (IL-6) protein and it was emphasized that the nanoporous cavities over the composite surface had an effective biosensing platform and the linear concentration range was found as 0.01–2.0 pg/ml and the detection limit was achieved as 0.0030 pg/ml in plasma samples. In another work, GQDs coated on silicon nanowires (GQDs/SiNWs) were tested as a photoelectrochemical immunosensor to detect microcystin-LR [54]. According to the results, the composite GQDs/SiNWs electrode demonstrated higher photoelectrochemical response than that of single nanowires.

Graphene quantum dots have proven the candidacy of biosensors applied in diagnosis of cancer diseases, especially they display high sensitivity and immediate response as cancer biomarkers [41, 55]. GQDs can also be effectively applied for suppression of some cancer types such as colon, breast [56–58]. The detection of biomarkers with GQDs has been usually applied via chemiluminescence or electro-chemiluminescence reagents. The cancer biomarkers are categorized into three main groups (i) diagnostic for highly specific detection of cancer (ii) prognostic for prevention disease recurrence, and (iii) predictive for observation of tumor response toward a specific biomolecule [59]. In general, biomarkers namely carcinoembryonic antigen, prostate-specific antigen, cancer antigens (e.g. CA-125, CA-145), cadherin-17, and interleukin 6/8 have been generally applied with quantum-sized biosensors. It was reported that the electrochemical identification of prostate-specific antigen molecule by using Cobalt phthalocyanine significantly enhanced by covalently bonding with an aptamer and GQDs and the detection limit was determined as 0.66 pM in a good linear range of 1.2–2.0 pM [60]. Notwithstanding the fact that sensitivity of the electrode in the bovine serum albumin, glucose and L-cysteine was very high demonstrating that it could be applied as rapid sensing device for prostate cancer. In this context, researchers have explored doping of different elements of GQDs for modification of electron transfer features in bio-sensing applications. In the study recorded by Srivastava and co-workers [61], a label-free prostate-specific antigen aptasensor and immunosensor was constructed with gold nanorods incorporated with GQDs on printed electrodes (GQDs-AuNRs). The composite electrode displayed boosted electron transfer and the sensors had a limit of detection of 0.14 ng/mL under optimal conditions. Similar approach is made by another group [62] who fabricated electrochemiluminescent immunosensor for PSA determination. GQDs were first functionalized with amine groups and subsequently carboxyl groups. Then, the antibody was adsorbed on the electrode surface through Au/Ag toward proteins which decreased the ECL signal. At optimum conditions, the linear concentration range of PSA was 1 pg/mL ~ 10ng/mL with a limit of detection of 0.29 pg/mL. On the other hand, Dong *et al.* [63] prepared hydrazide-modified GQDs doped with gold nanoparticles via redox reactions. The hybrid sensor exhibited excellent electrochemiluminescence features and the model target analyte was chosen as carcinoembryonic antigen which is applied as a tumor indicator in biomedical studies. The electron migration resistance was increased after immunoreaction and so the luminescence intensity reduced with the increasing antigen concentration. Authors revealed that the linear response was in the range of 0.02 to 80 ng/mL with a LOD of 0.01 ng/mL. The detection of carcinoembryonic antigen was also reported by other researchers [64, 65].

MicroRNAs-21 (miRNA-21) is classified in the group of oncogenic microRNA molecules applied as a target molecule for diagnosis of cancers. For this, a sensitive ratiometric biosensor was developed by using graphdiyne quantum dots (GDQDs) as donor while DNA modified GQDs were applied as acceptor [66]. Forster-resonance-energy transfer process-based fluorescence assay provided effective fluorescence quenching of GDQDs and the fluorescent intensity of GDQD was increased due to the interactions between miRNA-21 molecule and graphdiyne quantum dots. The composite sensor exhibited sensitivity for miRNA-21 molecule in a concentration range of 5–200 nM with a LOD of 0.5 pM. Authors underlined the fact that the as-prepared sensor was able to measure miRNA-21 molecule in the human serum and MCF-7 cell lines and the LOD value was reported as 4 cells/10 μL.

Additionally, GQDs have been applied as confocal imaging of cancer cells such as HeLa [67, 68], A-375 [67], T47D [69], A-549 [70], and MDA-MB-231 [71, 72] cell lines. Rajender *et al.* [67] studied bio-recognition of A-375 and HeLa cancer cells by using GQDs with high quantum yield as 32%. The bio-imaging of cancer cells with GQDs indicated bright blue PL emission and the number of live A-375 cells for GQDs was obtained significantly greater than that of HeLa cells. Liu *et al.* [68], investigated the imaging of Hela cells by applying nitrogen-doped GQDs and demonstrated that the materials displayed high photon absorption range (48 000 GM) revealing that they could be effectively used as fluorescent probes for imaging of tissue cells. Peng *et al.* [69] investigated the application of blue and green fluorescent GQDs for imaging of human breast cancer molecules namely MDAMB-321 and T47D cell lines. Nasrollahi *et al.* [71] evaluated functionalized GQDs for the delivery of Cisplatin which is the most powerful platinum-based drug. The as-prepared GQDs were incorporated with a single chain antibody of scFvB10 which was targeted to epidermal growth factor receptor through amide covalent interactions. The confocal imaging of MDA-MB-231 cancer cells over conjugated GQDs revealed effective application of GQDs in bio-imaging. Similarly, cellular imaging and photothermal response to MDA-MB-231 cancer molecules were examined by Kumawat *et al.* [72] and they functionalized graphene oxide nanosheets with GQDs by using a linkage namely polyethyleneimine (PEI). It was emphasized that the PEI was positively charges leading to negatively charged groups to be high stable. The treatment of cancer cells showed higher photoresponse even at lower densities.

### 7.2.3 g-$C_3N_4$-Based Quantum Dots (gCNQDs) as Biosensors

Graphitic carbon nitride (g-$C_3N_4$) is a polymer-based non-metal substance including carbon and nitrogen elements and it has been widely studied in many fields thanks to its strong thermal and chemical stability, biocompatibility, nontoxicity, and fast response/recovery [73]. Especially, the graphitic form is regarded as the most stable allotrope of $C_3N_4$ due

**Table 7.3** g-$C_3N_4$ QD-based metal-free materials as biosensors.

| Quantum dots | Analyte | Method | LOD | Ref. |
|---|---|---|---|---|
| (N-GQDs)/g-$C_3N_4$ QDs/$TiO_2$ | pcDNA3-HB | Photoelectrochemical | 0.005 fmol/L | [42] |
| $C_3N_4$ QDs/ $C_3N_4$ nanosheet | PDGF-BB | Electrochemiluminescence | 0.013 nM | [83] |
| g-CNQDs/Au NPs | Hairpin DNA (Hai-DNA) | Electrochemiluminescence | 0.01 fM | [84] |
| Ph-g-CNQDs | Fe(II)<br>Fe(III) | Photoluminescence | 58.44 μM<br>25.18 μM | [85] |
| MP-$MoO_3$ QDs/ gCNQDs | Influenza A virus | Magnetoplasmonic-Fluoroimmunosensing | 0.25 pg/mL (deionized water) 0.9 pg/mL (human serum) | [87] |
| gCNQDs-4-AT | Ascorbic acid | Photoluminescence | 0.15 nM | [88] |
| g-CNQDs | Tetracycline | Fluorescence | 0.19 μM | [89] |
| g-CNQDs/PBA | Glucose | Photoluminescence | 16 nm | [90] |
| g-CNQDs/3APBA | Glucose | Photoluminescence | 42 nM | [91] |
| $Ru(bpy)_3^{2+}$/ g-CNQDs | Catechol | Electrochemiluminescence | 2.5 nM | [94] |
| CuS/CQDs/ g-C3N4NS | Diazinon | Electrochemiluminescence | $2.2 \times 10^{-16}$ M | [95] |
| g-CNQDs | Fluoride | Colorimetric | 4.06 μM | [96] |
| g-CNQDs-apt/ CoOOH | Ochratoxin A | Photoluminescence | 0.5 nM | [97] |
| CNQDs | Ascorbic acid | Photoluminescence | 150 nM | [98] |
| g-CNQDs | l-cysteine | Electrochemical | 101.3 μM | [99] |

to the π conjugation between sp$^2$-bonded carbon and nitrogen atoms [74]. The structural chemistry of g-$C_3N_4$ can be modulated by surface modification at atomic level. The presence of –NH and –$NH_2$ bonds enables more surface active sites for attachment of different functional groups [75]. However, the utilization of bulk g-$C_3N_4$ is restricted due to its relatively low specific surface area and weak solubility. Therefore, researchers have aimed to develop nanostructured shapes of g-$C_3N_4$, such as nanosheets [76–78], nanoflakes [79, 80] or quantum dots. g-$C_3N_4$ QDs (gCNQDs) with diameters around 5–20 nm are zero-dimensional nanosized fragments of g-$C_3N_4$ and have relatively wider band gap and faster electron transfer [42]. The quantum confinement effect also provide gCNQDs to have unique photoluminescence features [81, 82]. Based on their characteristic features, gCNQDs have been explored in biosensing applications (Table 7.3).

Xu *et al.* [83] offered that the utilization of g-$C_3N_4$ QDs was more eco-friendly than metals and they impregnated $C_3N_4$ QDs on carbon nitride nanosheets for platelet-derived growth factor-BB determination in an electrochemiluminescence sensor. The nanocomposite improved the ECL response as well as improved intensity of the ECL response when compared to the traditional enzyme-based assay method. The identification of PDGF-BB molecule with a detection limit of 0.013 nM was achieved by incorporating with Fc-labeled aptamer. Besides, an ECL sensing system including g-CNQDs and AuNPs was established for monitoring hairpin DNA (Hai-DNA) [84]. The detection system was established on electrochemiluminescence and resonance energy transfer mechanism between the materials in which g-CNQDs played role as an energy emitter and the Au NPs acted as quenchers during DNA detection. The linearity range of ECL signal was observed between 0.02 fM and 0.1 pM with a low limit of detection (0.01 fM). Similarly, Chen and co-workers [76] examined the electrogenerated chemiluminescence behavior of g-$C_3N_4$ nanosheets coupled with AuNPs which showed sensitive and selective response to carcinoembryonic antigen (CEA).

Pang and co-workers [42] synthesized $TiO_2$ nanopillar N-doped GQDs/g-$C_3N_4$ QDs heterojunction as photoelectrochemical (PEC) sensor for capturing probes for pcDNA3-HBV molecule. At first, $TiO_2$ nanoparticles were deposited on ITO electrode. Then, N-GQDs and gCNQDs were embedded on $TiO_2$ NPs/ITO electrode through activation of EDC/NHS, respectively. After deposition of different amounts of pcDNA3-HBV primers on the electrode surface, the system was treated at 4 °C for 5 h (Figure 7.5). The incorporation of N-GQDs and gCNQDs enhanced the separation of the photoinduced charges and enhanced photocurrent conversion performance. Authors found that the increase in the pcDNA3-HBV concentration, the photocurrent response decreased after the formation of the double helix, and the as-prepared sensor displayed good sensitivity toward pcDNA3-HBV and the limit of detection was found as 0.005 fmol/L.

Vashisht *et al.* [85] investigated the optical detection of ferrous and ferric ions by applying phosphate modified $C_3N_4$ quantum dots (Ph-g-CNQDs) which showed high quantum yield (60.5%) and they reported that the interaction between iron and Ph-g-CNQDs resulted in quenching of blue fluorescence. The sensitive monitoring of ferrous and ferric ions was attributed to the specific bonds after functionalization with phosphate. It was also underlined that the Ph-g-CNQDs material had higher affinity toward Fe(III) due to greater charge density. In another work [86], it was demonstrated that the fluorescent CNQDs prepared by microwave-assisted solvothermal technique had a high quantum yield (27.1%)

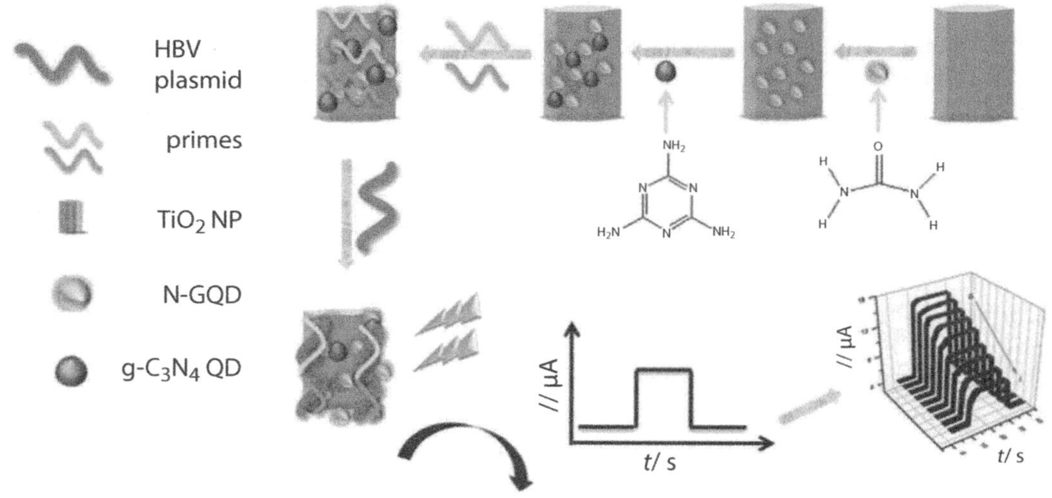


**Figure 7.5** Fabrication of (N-GQDs)/g-$C_3N_4$ biosensor [42]. Reprinted by permission from Elsevier Publisher.

and they could be applied as label-free sensing of mercury ions and the detection limit was found as 0.14 μM.

Achadu and co-workers [87] constructed a fluoro-immunoassay system including molybdenum trioxide quantum dots and g-$C_3N_4$ QDs in order to detect influenza virus. In the system, MP-$MoO_3$ QDs and gCNQDs acted plasmonic/magnetic agent and monitoring probe, respectively. The formation of immunocomplex between antibody-conjugated MP-$MoO_3$ QDs and gCNQDs enabled a gradual improvement fluorescence signal and the detection limits for influenza virus A/New Caledonia were obtained as 0.25 and 0.9 pg/mL in deionized water and human serum samples, respectively. In addition, the liner range of influenza virus A/Yokohama was observed between 45 PFU/mL and 5,000 PFU/mL (LOD = 45 PFU/mL). In another work, Achadu and Nyokong [88] investigated the fluorescent sensing of ascorbic acid which serves crucial roles in human systems and is utilized in disease prevention for cancer therapy. They impregranated 4-amino-2,2,6,6-tetramethyl-(piperidin-1-yl)oxyl) on the g-$C_3N_4$ QDs (gCNQDs-4-AT) through one-pot technique and successfully detected the ascorbic acid in nanomolar ranges.

Bai *et al.* [89] reported the detection of tetracycline in tap water solution and milk powder media by using g-$C_3N_4$ QDs prepared by solvothermal method. In the presence of tetracycline antibiotic, the as-obtained gCNQDs showed "strong quenching" behavior, which was ascribed from the static quenching process due inner filter effect mechanism which is based on the radiation energy transfer originated from the development of ground-state complex. According to this theory, the absorption spectra changes while the fluorescence lifetime keeps stabile for the material with the quencher. The results showed that the absorption peaks of tetracycline shifted, and the relative intensity changed after integrating with the gCNQDs biosensor.

Selective glucose determination in a wide linear range was investigated by using a photoluminescence sensor including phenylboronic acid-functionalized carbon nitride QDs (g-CNQDs/PBA) [90]. The g-CNQDs/PBA sensor displayed high quantum yield of 67%

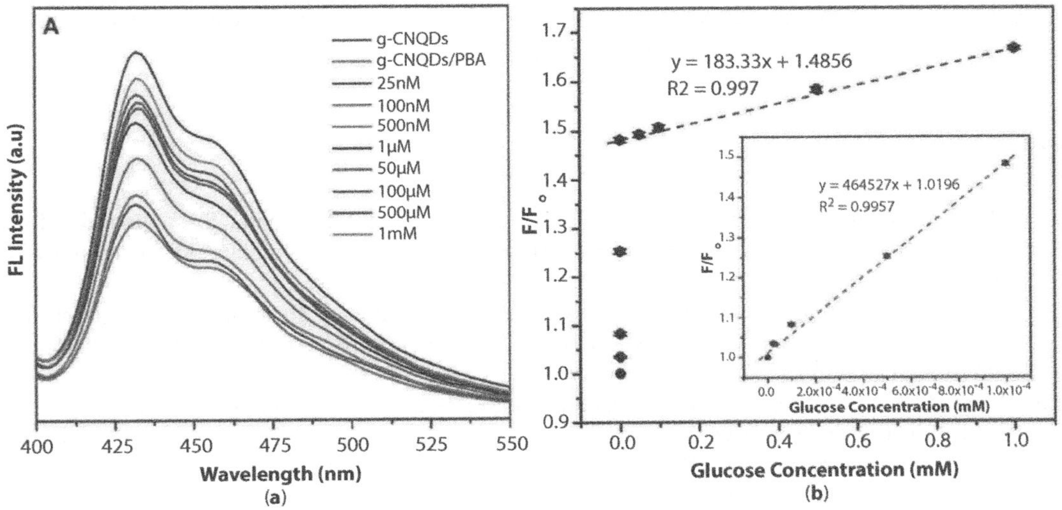


**Figure 7.6** (a) PL spectra of g-CNQDs and g-CNQDs /PBA and (b) change in PL intensity versus glucose concentration [90]. Reprinted by permission from Elsevier Publisher.

that was greater than that of bare g-CNQDs. The PL response progressively increased with higher glucose amount (Figure 7.6). The plots of Stern-Volmer exhibited that the fluorescence signal was mainly related with the concentration. The sensor displayed two linear concentration range as 25 nM–1 μM and 1 μM–1 mM with a low detection limit (16 nM), which was attributed to the specific reactions between phenylboronic acid and glucose. Additionally, the practical utilization of the g-CNQDs/PBA sensor was tested in monitoring glucose amount in horse and rabbit serums and the difference in the glucose concentration was compared with commercial glucometer. According to the results, very small differences were obtained, demonstrating the effective usage of the as-synthesized material. In their another work [91], authors functionalized g-CNQDs with aminoboronic acid (g-CNQDs/3APBA) and demonstrated that the introduction of covalent interactions between g-CNQDs and boronic acid provided effective usage of "on-off-on" based sensor. It was underlined that the g-CNQDs/3APBA sensor has the highest quantum yield (78.5%) comparing with other $C_3N_4$ QD-based fluorescence sensors.

## 7.3 Conclusions

In this review, recent advances of carbon-based metal-free quantum dots (CQDs, GQDs and CNQDs) were reported including the progress of biosensor systems for detection of biological molecules. The electrochemical, photoelectrochemical, photoluminescence and colorimetric detection using these QDs were highlighted. The photoluminescence as well as excellent electronic features of CQDs, GQDs and CNQDs set them apart from other zero-dimensional structures. Their quantum confinement also possess them unique optical properties. Moreover, the π–π adsorption chemistry offers them to be easily functionalized with other substances to improve their sensitivity. Considering high capability of carbon-based QDs in biosensing technologies, challenges still occur for the utilization of these zero-dimensional materials in real systems. The literature survey can serve as a basis for designing metal-free sensors in practical applications.

## References


1. Ahuja, V., Bhatt, A.K., Varjani, S., Choi, K.Y., Kim, S.H., Yang, Y.H., Bhatia, S.K., Quantum dot synthesis from waste biomass and its applications in energy and bioremediation. *Chemosphere*, 293, 133564, 2022.
2. Algar, W.R., Tavares, A.J., Krull, U.J., Beyond labels: A review of the application of quantum dots as integrated components of assays, bioprobes, and biosensors utilizing optical transduction. *Anal. Chim. Acta*, 673, 1, 1–25, 2010.
3. Wang, X., Liu, Z., Gao, P., Li, Y., Qu, X., Surface functionalized quantum dots as biosensor for highly selective and sensitive detection of ppb level of propafenone. *Spectrochim. Acta - Part A Mol. Biomol. Spectrosc.*, 227, 117709, 2020.
4. Ding, R., Chen, Y., Wang, Q., Wu, Z., Zhang, X., Li, B., Lin, L., Recent advances in quantum-dots-based biosensors for antibiotic detection. *J. Pharm. Anal.*, 12, 355, 2021.
5. Molaei, M.J., A review on nanostructured carbon quantum dots and their applications in biotechnology, sensors, and chemiluminescence. *Talanta*, 196, 456–478, 2019.
6. Khojastehnezhad, A., Taghavi, F., Yaghoobi, E., Ramezani, M., Alibolandi, M., Abnous, K., Taghdisi, S.M., Recent achievements and advances in optical and electrochemical aptasensing detection of ATP based on quantum dots. *Talanta*, 235, 122753, 2021.
7. Mohammadi, R., Naderi-Manesh, H., Farzin, L., Vaezi, Z., Ayarri, N., Samandari, L., Shamsipur, M., Fluorescence sensing and imaging with carbon-based quantum dots for early diagnosis of cancer: A review. *J. Pharm. Biomed. Anal.*, 212, 114628, 2022.
8. Gaviria-Arroyave, M.I., Cano, J.B., Peuela, G.A., Nanomaterial-based fluorescent biosensors-for monitoring environmental pollutants: A critical review. *Talanta Open*, 2, 100006, 2020.
9. Farzin, M.A. and Abdoos, H., A critical review on quantum dots: From synthesis toward applications in electrochemical biosensors for determination of disease-related biomolecules. *Talanta*, 224, 121828, 2021.
10. Taghavi, F., Moeinpour, F., Khojastehnezhad, A., Abnous, K., Taghdisi, S.M., Recent applications of quantum dots in optical and electrochemical aptasensing detection of Lysozyme. *Anal. Biochem.*, 630, 114334, 2021.
11. Liu, X. and Luo, Y., Surface modifications technology of quantum dots based biosensors and their medical applications. *Fenxi Huaxue/ Chin. J. Anal. Chem.*, 42, 7, 1061–1069, 2014.
12. Qian, Z.S., Chai, L.J., Huang, Y.Y., Tang, C., Jia Shen, J., Chen, J.R., Feng, H., A real-time fluorescent assay for the detection of alkaline phosphatase activity based on carbon quantum dots. *Biosens. Bioelectron.*, 68, 675–680, 2015.
13. Cheng, W., Zheng, Z., Yang, J., Chen, M., Yao, Q., Chen, Y., Gao, W., The visible light-driven and self-powered photoelectrochemical biosensor for organophosphate pesticides detection based on nitrogen doped carbon quantum dots for the signal amplification. *Electrochim. Acta*, 296, 627–636, 2019.
14. Namdari, P., Negahdari, B., Eatemadi, A., Synthesis, properties and biomedical applications of carbon-based quantum dots: An updated review. *Biomed. Pharmacother.*, 87, 88, 209–222, 2017.
15. Chai, L., Zhou, J., Feng, H., Tang, C., Huang, Y., Qian, Z., Functionalized carbon quantum dots with dopamine for tyrosinase activity monitoring and inhibitor screening: *In vitro* and intracellular investigation. *ACS Appl. Mater. Interfaces*, 7, 42, 23564–23574, 2015.

16. Yu, S., Ding, L., Lin, H., Wu, W., Huang, J., A novel optical fiber glucose biosensor based on carbon quantum dots-glucose oxidase/cellulose acetate complex sensitive film. *Biosens. Bioelectron.*, 146, 111760, 2019.
17. Buk, V. and Pemble, M.E., A highly sensitive glucose biosensor based on a micro disk array electrode design modified with carbon quantum dots and gold nanoparticles. *Electrochim. Acta*, 298, 97–105, 2019.
18. Tang, C., Qian, Z., Qian, Y., Huang, Y., Zhao, M., Ao, H., Feng, H., A fluorometric and real-time assay for A-glucosidase activity through supramolecular self-assembly and its application for inhibitor screening. *Sens. Actuators, B Chem.*, 245, 282–289, 2017.
19. Sun, Q., Fang, S., Fang, Y., Qian, Z., Feng, H., Fluorometric detection of cholesterol based on β-cyclodextrin functionalized carbon quantum dots via competitive host-guest recognition. *Talanta*, 167, 513–519, 2017.
20. Zhang, X., Ren, Y., Ji, Z., Fan, J., Sensitive detection of amoxicillin in aqueous solution with novel fluorescent probes containing boron-doped carbon quantum dots. *J. Mol. Liq.*, 311, 113278, 2020.
21. Wei, Q., Zhang, P., Liu, T., Pu, H., Sun, D.W., A fluorescence biosensor based on single-stranded DNA and carbon quantum dots for acrylamide detection. *Food Chem.*, 356, 129668, 2021.
22. Cao, X., Zhang, K., Yan, W., Xia, Z., He, S., Xu, X., Ye, Y., Wei, Z., Liu, S., Calcium ion assisted fluorescence determination of microRNA-167 using carbon dots-labeled probe DNA andpoly-dopamine-coated $Fe_3O_4$ nanoparticles. *Microchim. Acta*, 187, 4, 212, 2020.
23. Wu, X., Song, Y., Yan, X., Zhu, C., Ma, Y., Du, D., Lin, Y., Carbon quantum dots as fluorescence resonance energy transfer sensors for organophosphate pesticides determination. *Biosens. Bioelectron.*, 94, 292–297, 2017.
24. Gholami, M., Salmasi, M.A., Sohouli, E., Torabi, B., Sohrabi, M.R., Rahimi-Nasrabadi, M., A new nano biosensor for maitotoxin with high sensitivity and selectivity based fluorescence resonance energy transfer between carbon quantum dots and gold nanoparticles. *J. Photochem. Photobiol. A Chem.*, 398, 112523, 2020.
25. Du, F., Min, Y., Zeng, F., Yu, C., Wu, S., A targeted and FRET-based ratiometric fluorescent nanoprobe for imaging mitochondrial hydrogen peroxide in living cells. *Small*, 10, 5, 964–972, 2014.
26. Preethi, M., Viswanathan, C., Ponpandian, N., An environment-friendly route to explore the carbon quantum dots derived from curry berries (Murrayakoenigii L) as a fluorescent biosensor for detecting vitamin B12. *Mater. Lett.*, 303, 130521, 2021.
27. Wu, W., Huang, J., Ding, L., Lin, H., Yu, S., Yuan, F., Liang, B., A real-time and highly sensitive fiber optic biosensor based on the carbon quantum dots for nitric oxide detection. *J. Photochem. Photobiol. A Chem.*, 405, 112963, 2021.
28. Azizi, S., Gholivand, M.B., Amiri, M., Manouchehri, I., Moradian, R., Carbon dots-thionine modified aptamer-based biosensor for highly sensitive cocaine detection. *J. Electroanal. Chem.*, 907, 116062, 2022.
29. Azizi, S., Gholivand, M.B., Amiri, M., Manouchehri, I., DNA biosensor based on surface modification of ITO by physical vapor deposition of gold and carbon quantum dots modified with neutral red as an electrochemical redox probe. *Microchem. J.*, 159, 105523, 2020.
30. Liu, Q., Ma, C., Liu, X.P., Wei, Y.P., Mao, C.J., Zhu, J.J., A novel electrochemiluminescence biosensor for the detection of microRNAs based on a DNA functionalized nitrogen doped carbon quantum dots as signal enhancers. *Biosens. Bioelectron.*, 92, 273–279, 2017.
31. Sri, S., Lakshmi, G.B.V.S., Gulati, P., Chauhan, D., Thakkar, A., Solanki, P.R., Simple and facile carbon dots based electrochemical biosensor for TNF-α targeting in cancer patient's sample. *Anal. Chim. Acta*, 1182, 338909, 2021.

32. Rezaei, B., Jamei, H.R., Ensafi, A.A., An ultrasensitive and selective electrochemical aptasensor based on rGO-MWCNTs/Chitosan/carbon quantum dot for the detection of lysozyme. *Biosens. Bioelectron.*, 115, 37–44, 2018.
33. Li, F., Zhou, Y., Yin, H., Ai, S., Recent advances on signal amplification strategies in photoelectrochemical sensing of microRNAs. *Biosens. Bioelectron.*, 166, 112476, 2020.
34. Wang, M., Yin, H., Zhou, Y., Meng, X., Waterhouse, G.I.N., Ai, S., A novel photoelectrochemical biosensor for the sensitive detection of dual microRNAs using molybdenum carbide nanotubes as nanocarriers and energy transfer between CQDs and AuNPs. *Chem. Eng. J.*, 365, 351–357, 2019.
35. Yuan, C., Qin, X., Xu, Y., Li, X., Chen, Y., Shi, R., Wang, Y., Carbon quantum dots originated from chicken blood as peroxidase mimics for colorimetric detection of biothiols. *J. Photochem. Photobiol. A Chem.*, 396, 112529, 2020.
36. Sun, Y., Ding, C., Lin, Y., Sun, W., Liu, H., Zhu, X., Dai, Y., Luo, C., Highly selective and sensitive chemiluminescence biosensor for adenosine detection based on carbon quantum dots catalyzing luminescence released from aptamers functionalized graphene@magnetic β-cyclodextrin polymers. *Talanta*, 186, 238–247, 2018.
37. Yan, Y., Gong, J., Chen, J., Zeng, Z., Huang, W., Pu, K., Liu, J., Chen, P., Recent advances on graphene quantum dots: From chemistry and physics to applications. *Adv. Mater.*, 31, 21, 1–22, 2019.
38. Zhu, S., Song, Y., Zhao, X., Shao, J., Zhang, J., Yang, B., The photoluminescence mechanism in carbon dots (graphene quantum dots, carbon nanodots, and polymer dots): Current state and future perspective. *Nano Res.*, 8, 2, 355–381, 2015.
39. Walther, B.K., Dinu, C.Z., Guldi, D.M., Sergeyev, V.G., Creager, S.E., Cooke, J.P., Guiseppi-Elie, A., Nanobiosensing with graphene and carbon quantum dots: Recent advances. *Mater. Today*, 39, 23–46, 2020.
40. Tetsuka, H., Nagoya, A., Fukusumi, T., Matsui, T., Molecularly designed, nitrogen-functionalized graphene quantum dots for optoelectronic devices. *Adv. Mater.*, 28, 23, 4632–4638, 2016.
41. Kansara, V., Shukla, R., Flora, S.J.S., Bahadur, P., Tiwari, S., Graphene quantum dots: Synthesis, optical properties and navigational applications against cancer. *Mater. Today Commun.*, 31, 103359, 2022.
42. Pang, X., Bian, H., Wang, W., Liu, C., Khan, M.S., Wang, Q., Qi, J., Wei, Q., Du, B., A bio-chemical application of N-GQDs and g-$C_3N_4$ QDs sensitized $TiO_2$ nanopillars for the quantitative detection of pcDNA3-HBV. *Biosens. Bioelectron.*, 91, 456–464, 2017.
43. Shen, J., Zhu, Y., Yang, X., Li, C., Graphene quantum dots: Emergent nanolights for bioimaging, sensors, catalysis and photovoltaic devices. *Chem. Commun.*, 48, 31, 3686–3699, 2012.
44. He, Y., Wang, X., Sun, J., Jiao, S., Chen, H., Gao, F., Wang, L., Fluorescent blood glucose monitor by hemin-functionalized graphene quantum dots based sensing system. *Anal. Chim. Acta*, 810, 71–78, 2014.
45. Huang, S., Wang, L., Huang, C., Su, W., Xiao, Q., Label-free and ratiometric fluorescent nanosensor based on amino-functionalized graphene quantum dots coupling catalytic G-quadruplex/hemin DNAzyme for ultrasensitive recognition of human telomere DNA. *Sens. Actuators, B Chem.*, 245, 648–655, 2017.
46. Dutta Chowdhury, A. and Doong, R.A., Highly sensitive and selective detection of nanomolar ferric ions using dopamine functionalized graphene quantum dots. *ACS Appl. Mater. Interfaces*, 8, 32, 21002–21010, 2016.
47. Meng, H.M., Zhao, D., Li, N., Chang, J., A graphene quantum dot-based multifunctional two-photon nanoprobe for the detection and imaging of intracellular glutathione and enhanced photodynamic therapy. *Analyst*, 143, 20, 4967–4973, 2018.

48. Hong, G.L., Deng, H.H., Zhao, H.L., Zou, Z.Y., Huang, K.Y., Peng, H.P., Liu, Y.H., Chen, W., Gold nanoclusters/graphene quantum dots complex-based dual-emitting ratiometric fluorescence probe for the determination of glucose. *J. Pharm. Biomed. Anal.*, 189, 113480, 2020.
49. Hwang, J., Le, A.D.D., Trinh, C.T., Le, Q.T., Lee, K.G., Kim, J., Green synthesis of reduced-graphene oxide quantum dots and application for colorimetric biosensor. *Sens. Actuators, A Phys.*, 318, 112495, 2021.
50. Wang, Y.X., Rinawati, M., Huang, W.H., Cheng, Y.S., Lin, P.H., Chen, K.J., Chang, L.Y., Ho, K.C., Su, W.N., Yeh, M.H., Surface-engineered N-doped carbon nanotubes with B-doped graphene quantum dots: Strategies to develop highly-efficient noble metal-free electrocatalyst for online-monitoring dissolved oxygen biosensor. *Carbon N. Y.*, 186, 406–415, 2022.
51. Disha, Kumari, P., Nayak, M.K., Kumar, P., A bio-sensing platform based on graphene quantum dots for label free electrochemical detection of progesterone. *Mater. Today Proc.*, 48, 583–586, 2019.
52. Kaimal, R., Mansukhlal, P.N., Aljafari, B., Anandan, S., Ashokkumar, M., Ultrasound-aided synthesis of gold-loaded boron-doped graphene quantum dots interface towards simultaneous electrochemical determination of guanine and adenine biomolecules. *Ultrason. Sonochem.*, 83, 105921, 2022.
53. Özcan, N., Karaman, C., Atar, N., Karaman, O., Yola, M.L., A novel molecularly imprinting biosensor including graphene quantum dots/multi-walled carbon nanotubes composite for interleukin-6 detection and electrochemical biosensor validation. *ECS J. Solid State Sci. Technol.*, 9, 12, 121010, 2020.
54. Tian, J., Zhao, H., Quan, X., Zhang, Y., Yu, H., Chen, S., Fabrication of graphene quantum dots/silicon nanowires nanohybrids for photoelectrochemical detection of microcystin-LR. *Sens. Actuators, B Chem.*, 196, 532–538, 2014.
55. Arshad, F., Nabi, F., Iqbal, S., Hasan, R., Applications of graphene-based electrochemical and optical biosensors in early detection of cancer biomarkers. *Colloids Surf. B Biointerfaces*, 212, 112356, 2022.
56. Lee, G.Y., Lo, P.Y., Cho, E.C., Zheng, J.H., Li, M., Huang, J.H., Lee, K.C., Integration of PEG and PEI with graphene quantum dots to fabricate pH-responsive nanostars for colon cancer suppression *in vitro* and *in vivo*. *FlatChem*, 31, 134, 100320, 2022.
57. Nene, L.C. and Nyokong, T., Photo-sonodynamic combination activity of cationic morpholino-phthalocyanines conjugated to nitrogen and nitrogen-sulfur doped graphene quantum dots against MCF-7 breast cancer cell line *in vitro*. *Photodiagnosis Photodyn. Ther.*, 36, 102573, 2021.
58. Geng, B., Hu, J., Li, P., Pan, D., Shen, L., DNA binding graphene quantum dots inhibit dual topoisomerases for cancer chemotherapy. *Carbon N. Y.*, 187, 365–374, 2022.
59. Tabish, T.A., Hayat, H., Abbas, A., Narayan, R.J., Graphene quantum dot–based electrochemical biosensing for early cancer detection. *Curr. Opin. Electrochem.*, 30, 100786, 2021.
60. Nxele, S.R. and Nyokong, T., The electrochemical detection of prostate specific antigen on glassy carbon electrode modified with combinations of graphene quantum dots, cobalt phthalocyanine and an aptamer. *J. Inorg. Biochem.*, 221, 111462, 2021.
61. Srivastava, M., Nirala, N.R., Srivastava, S.K., Prakash, R., A comparative study of aptasensor vs immunosensor for label-free PSA cancer detection on GQDs-AuNRs modified screen-printed electrodes. *Sci. Rep.*, 8, 1, 1–11, 2018.
62. Wu, D., Liu, Y., Wang, Y., Hu, L., Ma, H., Wang, G., Wei, Q., Label-free electrochemiluminescent immunosensor for detection of prostate specific antigen based on aminated graphene quantum dots and carboxyl graphene quantum dots. *Sci. Rep.*, 6, 1–7, 2016.
63. Dong, Y., Wu, H., Shang, P., Zeng, X., Chi, Y., Immobilizing water-soluble graphene quantum dots with gold nanoparticles for a low potential electrochemiluminescence immunosensor. *Nanoscale*, 7, 39, 16366–16371, 2015.

64. Ganganboina, A.B. and Doong, R.A., Graphene quantum dots decorated gold-polyaniline nanowire for impedimetric detection of carcinoembryonic antigen. *Sci. Rep.*, 9, 1, 1–11, 2019.
65. Wang, C., Zhang, Y., Tang, W., Wang, C., Han, Y., Qiang, L., Gao, J., Liu, H., Han, L., Ultrasensitive, high-throughput and multiple cancer biomarkers simultaneous detection in serum based on graphene oxide quantum dots integrated microfluidic biosensing platform. *Anal. Chim. Acta*, 1178, 338791, 2021.
66. Bahari, D., Babamiri, B., Salimi, A., Rashidi, A., Graphdiyne/graphene quantum dots for development of FRET ratiometric fluorescent assay toward sensitive detection of miRNA in human serum and bioimaging of living cancer cells. *J. Lumin.*, 239, 118371, 2021.
67. Rajender, G., Goswami, U., Giri, P.K., Solvent dependent synthesis of edge-controlled graphene quantum dots with high photoluminescence quantum yield and their application in confocal imaging of cancer cells. *J. Colloid Interface Sci.*, 541, 387–398, 2019.
68. Liu, Q., Guo, B., Rao, Z., Zhang, B., Gong, J.R., Strong two-photon-induced fluorescence from-photostable, biocompatible nitrogen-doped graphene quantum dots for cellular and deep tissue imaging. *Nano Lett.*, 13, 6, 2436, 2013.
69. Peng, J., Gao, W., Gupta, B.K., Liu, Z., Romero-Aburto, R., Ge, L., Song, L., Alemany, L.B., Zhan, X., Gao, G., Vithayathil, S.A., Kaipparettu, B.A., Marti, A.A., Hayashi, T., Zhu, J.J., Ajayan, P.M., Graphene quantum dots derived from carbon fibers. *Nano Lett.*, 12, 2, 844–849, 2012.
70. Sun, Y., Wang, S., Li, C., Luo, P., Tao, L., Wei, Y., Shi, G., Large scale preparation of graphene quantum dots from graphite with tunable fluorescence properties. *Phys. Chem. Chem. Phys.*, 15, 24, 9907–9913, 2013.
71. Nasrollahi, F., Koh, Y.R., Chen, P., Varshosaz, J., Khodadadi, A.A., Lim, S., Targeting graphene quantum dots to epidermal growth factor receptor for delivery of cisplatin and cellular imaging. *Mater. Sci. Eng. C*, 94, 247–257, 2019.
72. Kumawat, M.K., Thakur, M., Bahadur, R., Kaku, T., Prabhuraj, R.S., Ninawe, A., Srivastava, R., Preparation of graphene oxide-graphene quantum dots hybrid and its application in cancer theranostics. *Mater. Sci. Eng. C*, 103, 109774, 2019.
73. Qu, B., Sun, J., Li, P., Jing, L., Current advances on g-$C_3N_4$-based fluorescence detection for environmental contaminants. *J. Hazard. Mater.*, 425, 127990, 2022.
74. Ghashghaee, M., Azizi, Z., Ghambarian, M., Conductivity tuning of charged triazine and heptazine graphitic carbon nitride (g-$C_3N_4$) quantum dots via nonmetal (B, O, S, P) doping: DFT calculations. *J. Phys. Chem. Solids*, 141, 109422, 2020.
75. Ghanbari, M., Salavati-Niasari, M., Mohandes, F., Injectable hydrogels based on oxidized alginate-gelatin reinforced by carbon nitride quantum dots for tissue engineering. *Int. J. Pharm.*, 602, 120660, 2021.
76. Chen, L., Zeng, X., Si, P., Chen, Y., Chi, Y., Kim, D.H., Chen, G., Gold nanoparticle-graphite-like $C_3N_4$ nanosheet nanohybrids used for electrochemiluminescent immunosensor. *Anal. Chem.*, 86, 9, 4188–4195, 2014.
77. Wang, X., Xiong, H., Chen, T., Xu, Y., Bai, G., Zhang, J., Tian, Y., Xu, S., A phosphorus-doped g-$C_3N_4$ nanosheets as an efficient and sensitive fluorescent probe for $Fe^{3+}$ detection. *Opt. Mater. (Amst)*, 119, 111393, 2021.
78. Zhang, C., Ni, P., Wang, B., Liu, W., Jiang, Y., Chen, C., Sun, J., Lu, Y., Enhanced oxidase-like activity of g-$C_3N_4$ nanosheets supported Pd nanosheets for ratiometric fluorescence detection of acetylcholinesterase activity and its inhibitor. *Chin. Chem. Lett.*, 33, 2, 757–761, 2022.
79. Chen, L., Huang, D., Ren, S., Dong, T., Chi, Y., Chen, G., Preparation of graphite-like carbon nitride nanoflake film with strong fluorescent and electrochemiluminescent activity. *Nanoscale*, 5, 1, 225–230, 2013.
80. Zou, J., Mao, D., Arramel, Li, N., Jiang, J., Reliable and selective lead-ion sensor of sulfur-doped graphitic carbon nitride nanoflakes. *Appl. Surf. Sci.*, 506, 144672, 2020.

81. Perez, M., Vallejo, M.A., Gómez, C., Montez, E., Elias, J., Torres-Castro, A., Vega-Carrillo, H.R., Sosa, M., Dosimetric analysis of graphitic carbon nitride quantum dots exposed to a gamma radiation for a low-dose applications. *Appl. Radiat. Isot.*, 184, 110200, 2022.
82. Dong, Y., Wang, Q., Wu, H., Chen, Y., Lu, C.-H., Chi, Y., Yang, H.-H., Graphitic carbon nitride materials: Sensing, imaging and therapy. *Small*, 12, 39, 5376–5393, 2016.
83. Xu, H., Liang, S., Zhu, X., Wu, X., Dong, Y., Wu, H., Zhang, W., Chi, Y., Enhanced electrogenerated chemiluminescence behavior of $C_3N_4$ QDs@$C_3N_4$ nanosheet and its signal-on aptasensing for platelet derived growth factor. *Biosens. Bioelectron.*, 92, 695–701, 2017.
84. Liu, Z., Zhang, X., Ge, X., Hu, L., Hu, Y., Electrochemiluminescence sensing platform for ultrasensitive DNA analysis based on resonance energy transfer between graphitic carbon nitride quantum dots and gold nanoparticles. *Sens. Actuators, B Chem.*, 297, 126790, 2019.
85. Vashisht, D., Sharma, E., Kaur, M., Vashisht, A., Mehta, S.K., Singh, K., Solvothermal assisted phosphate functionalized graphitic carbon nitride quantum dots for optical sensing of Fe ions and its thermodynamic aspects. *Spectrochim. Acta - Part A Mol. Biomol. Spectrosc.*, 228, 117773, 2020.
86. Cao, X., Ma, J., Lin, Y., Yao, B., Li, F., Weng, W., Lin, X., A facile microwave-assisted fabrication of fluorescent carbon nitride quantum dots and their application in the detection of mercury ions. *Spectrochim. Acta - Part A Mol. Biomol. Spectrosc.*, 151, 875–880, 2015.
87. Achadu, O.J., Takemura, K., Khoris, I.M., Park, E.Y., Plasmonic/magnetic molybdenum trioxide and graphitic carbon nitride quantum dots-based fluoroimmunosensing system for influenza virus. *Sens. Actuators, B Chem.*, 321, 128494, 2020.
88. Achadu, O.J. and Nyokong, T., *In situ* one-pot synthesis of graphitic carbon nitride quantum dots and its 2,2,6,6-tetramethyl(piperidin-1-yl)oxyl derivatives as fluorescent nanosensors for ascorbic acid. *Anal. Chim. Acta*, 991, 113–126, 2017.
89. Bai, R., Sun, H., Jin, P., Li, J., Peng, A., He, J., Facile synthesis of carbon nitride quantum dots as a highly selective and sensitive fluorescent sensor for the tetracycline detection. *RSC Adv.*, 11, 40, 24892–24899, 2021.
90. Ngo, Y.L.T., Choi, W.M., Chung, J.S., Hur, S.H., Highly biocompatible phenylboronic acid-functionalized graphitic carbon nitride quantum dots for the selective glucose sensor. *Sens. Actuators, B Chem.*, 282, 36–44, 2019.
91. Thi Ngo, Y.L., Chung, J.S., Hur, S.H., Aminoboronic acid-functionalized graphitic carbon nitride quantum dots for the photoluminescence multi-chemical sensing probe. *Dyes Pigm.*, 168, 180–188, 2019.
92. Ganjali, M.R., Dourandish, Z., Beitollahi, H., Tajik, S., Hajiaghababaei, L., Larijani, B., Highly sensitive determination of theophylline based on graphene quantum dots modified electrode. *Int. J. Electrochem. Sci.*, 13, 3, 2448–2461, 2018.
93. Hasanzadeh, M., Baghban, H.N., Shadjou, N., Mokhtarzadeh, A., Ultrasensitive electrochemical immunosensing of tumor suppressor protein p53 in unprocessed human plasma and cell lysates using a novel nanocomposite based on poly-cysteine/graphene quantum dots/gold nanoparticle. *Int. J. Biol. Macromol.*, 107, PartA, 1348–1363, 2018.
94. Liu, Z., Wu, H., Ge, X., Zhan, H., Hu, L., A sensitive method to monitor catechol by using graphitic carbon nitride quantum dots as coreactants in Ru(bpy)32+-based electrochemiluminescent system. *J. Electroanal. Chem.*, 860, 113910, 2020.
95. Ali Kamyabi, M. and Moharramnezhad, M., A novel cathodic electrochemiluminescent sensor based on CuS/carbon quantum dots/g-$C_3N_4$ nanosheets and boron nitride quantum dots for the sensitive detection of organophosphate pesticide. *Microchem. J.*, 179, 107421, 2022.
96. Devi, M., Das, P., Boruah, P.K., Deka, M.J., Duarah, R., Gogoi, A., Neog, D., Dutta, H.S., Das, M.R., Fluorescent graphitic carbon nitride and graphene oxide quantum dots as efficient

nanozymes: Colorimetric detection of fluoride ion in water by graphitic carbon nitride quantum dots. *J. Environ. Chem. Eng.*, 9, 1, 104803, 2021.
97. Bi, X., Luo, L., Li, L., Liu, X., Chen, B., You, T., A FRET-based aptasensor for ochratoxin A detection using graphitic carbon nitride quantum dots and CoOOH nanosheets as donor-acceptor pair. *Talanta*, 218, 121159, 2020.
98. Xie, H., Fu, Y., Zhang, Q., Yan, K., Yang, R., Mao, K., Chu, P.K., Liu, L., Wu, X., Selective and high-sensitive label-free detection of ascorbic acid by carbon nitride quantum dots with intense fluorescence from lone pair states. *Talanta*, 196, 530–536, 2019.
99. Nxele, S.R., Oluwole, D.O., Nyokong, T., Electrocatalytic activity of a push pull Co(II) phthalocyanine in the presence of graphitic carbon nitride quantum dots. *Electrochim. Acta*, 326, 134978, 2019.

# 8

# Bioreceptors for Microbial Biosensors


**S. Nalini[1]*, S. Sathiyamurthi[2], P. Ramya[3], R. Sivagamasundari[3], K. Mythili[3] and M. Revathi[4]**

*[1]Department of Microbiology, Bharat Ratna Puratchi Thalaivar Dr.MGR Govt Arts and Science College, Palacode, Tamil Nadu, India*
*[2]Department of Soil Science and Agricultural Chemistry, Faculty of Agriculture, Annamalai University, Annamalai Nagar, Tamil Nadu, India*
*[3]Department of Microbiology, Shree Raghavendra Arts and Science College, Keezhamoongiladi, Chidambaram, Tamil Nadu, India*
*[4]Department of Biochemistry, Shree Raghavendra Arts and Science College, Keezhamoongiladi, Chidambaram, Tamil Nadu, India*



## *Abstract*

A biosensor is a biological recognition element that combines a transducer to generate a signal and a signal processing system to identify specific ions or compounds. Biosensors can detect and recognize specific components within the tissues or the cells. A microbial biosensor is an analytical device that uses microorganisms and a physical transducer to detect a quantifiable signal proportional to specific ions or compounds level. Recently, microbial biosensors have been designed for various purposes, including environmental, food, and biomedical applications. Due to their capacity to examine DNA, microbial biosensors are an excellent and suitable alternative to time saving, sophisticated and costly traditional forensic identification methods. The chapter begins with an overview of several sensing technologies widely employed in microbial biosensing. This chapter summarizes recent advances in the fabrication and application of optical, mechanical, electrochemical, amperometric, potentiometric, and microbial fuel cells. In addition, it discusses the types and applications of microbial biosensors, the recent trend, and the future challenges of microbial biosensor technology.



***Keywords:*** Biosensor, microbial biosensor, whole cell biosensor, microbial fuel cells, microorganisms


## 8.1 Introduction

A biosensor (Bs) is a type of sensor that can detect and recognize the specific component within the tissues or the cells [1]. A biosensor device's functional strategy is highly influenced by the type of the biorecognition molecule (Bioreceptor) used to detect a target analyte (AE). The kind of bioreceptor (BRs) utilized in a Bs can determine biosensing properties like response time, sensitivity (ST) and specificity (SP). Biological molecules such as enzymes (ES), antibodies (ABs),


**Corresponding author*: snalini.msc@gmail.com



Inamuddin and Tariq Altalhi (eds.) Biosensors Nanotechnology, 2nd Edition, (169–184) © 2023 Scrivener Publishing LLC

and nucleic acids are generally used as biosensors for various analytes because of their high sensitivity to target ions or compounds [2]. On the other hand, with their enormous array of biological pathways, living cells (LCs) appeal to these molecular BRs. LCs have been used as biological recognition element (BRE) in Bs since 1970s, owing to their existence in the convergence of biology and electronics. A BRE can detect the AE of interest. They are a fascinating bioreceptor (BRs) choice because they provide you with more options when it comes to the sensing (SN) method, are less expensive than ES and ABs, and production is simple and cost-effective [3].

Microbial biosensors (MBs) are microbe-based analytical instruments that sense the target compounds or substrate and transform the signal (SG) into a measurable quantity in terms of physiological, biochemical, and electrical (EL) approaches. MBs use various conventional and optical detection and recognition techniques (e.g., fluorescence or bioluminescence) [4]. Bs research has acquired significant attention in the last five decades due to its applicability in various fields, including medicine, pollution assessment and monitoring, food safety and processing, and geological exploration etc. [5]. This chapter discusses the latest research progress on bioreceptors for microbial biosensors. In addition, it summarizes the types and applications of microbial biosensors: the recent trend and its future challenges of microbial biosensor technology.

## 8.2 Progression of Biosensor Technology

In recent times, biosensor technology has been advancing. The history of Bs dates back as early as the 1960s. In 1965, Leland C. Clark, Jr. designed the first real "biosensor" to detect oxygen. He is renowned as the "Father of Biosensor" and the oxygen electrode. His invention retains the name "Clark electrode". In 1980, the first affinity Bs were developed which used radiolabelled receptors immobilized on a transducer (TD) surface. Bs based on enzyme-linked immunoassay have also been designed employing labelled antibody/label antigen (AG), in combination with an appropriate TD [6–8].

## 8.3 Biosensors Types

Bs are categorized based on BRE and physiochemical TDs. Biomolecules (BM) have been used as the BRE in Bs technology, including ES, AG/ABs receptors, organelles, DNA, biological tissue and microorganism. Based on the TDs kind, the Bs are classified into various groups, such as optical (OP), mechanical (MN) and electrochemical (EC) [2, 3, 7, 9]. The schematic representation of the biosensor is presented in Fig. 8.1.

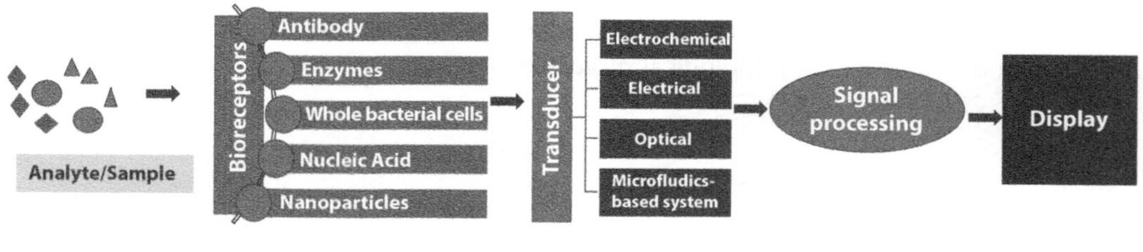


**Fig. 8.1** Schematic representation of biosensor.

## 8.4 Why is a Biosensor Required?

The sensitivity and ST evaluation of diverse compounds of analytical and commercial importance is one of the most important requirements of the modern era. Bs can detect the presence or absence and concentration of certain chemical compounds in a sample. Furthermore, diabetes and obesity are increasing globally, necessitating the development of simple gadgets to monitor the glucose level in diabetic patients. Therefore, the pharmaceutical research sector has needed innovative quick test Bs to speed up drug discovery. The most important application of Bs in military and defence projects is to fight terrorism. In addition, Bs can check food safety and identify environmental pollutants. However, Bs must meet certain criteria, including the significance of output SG to the desired measurement, ST and resolution, accurateness and reproductivity, temperature insensitivity, investment cost, EL and other ET interference, and users' acceptability [10].

## 8.5 Optical Microbial Biosensors

Optical biosensor (OBRs) detects significant optical characteristic variation on sensor surface (SS) due to the binding of the analyte, which is subsequently transmitted to detector (DR). OBRs are commonly classified as fluorescence based (FB) or label free, when analyte recognition occurs. The simplest OBRs measures a variation in florescence, absorbance or luminescence of the Bs surface. Traditional sandwich immunoassays, in which BRE consists of immobilized antibodies that allow for particular AE detection, have evolved into these technologies. The collect AE on the sensor surface is subsequently bound by a secondary reagent, such as a fluorescent (FT) tagged antibody. It produces an OP signal; the strength is proportionate to precise AE interaction. Optical fibres (OFr) were used to identify whole bacterial cells (WBc) to shift the tests from a lab-based 96-well microtiter plate setup to a compact, more transportable BRs device [11, 12].

Fiber optical biosensor (FOBRs) generally contains a light source that travels through OFr, carrying immobilized BRs to a photon DR. A shift in signal at the DR is caused by the binding of AE and the subsequent addition of an appropriate labeling reagent. The FB based biosensor can provide exceptional sensitivity; for example, the FT dye and micelle method was used to identify *Escherichia coli* cells. However, the main disadvantage of using FOBRs is the need to label the samples with FT reagents, which increases the time and cost of the technique [13].

The fluorescent-labelled ABs for imaging the cell give the scientist a better chance to understand the numerous molecular systems inside cell. The FBs genetically encodable Bs is technology that has increased attraction for real time investigation in the systems. PKA (protein kinase A) cAMP sensor molecular switch could be a protein fragment (PF) that changes conformation after enzymatically modified by the signaling enzymes of interest. It could be a protein or PF that hangs its configuration after attaching to a second messenger molecule [14]. Surface plasmon resonance (SPR) was used to determine the range of AE without labeling, while the widely available device was developed by Bacore in 1990 [15]. By using a suitable design, makes it feasible to combine methods and enhance SN competence. In addition, since sample act immediately on many BRE immobilized on a single chip, microfluidics can assist boost a single chip's throughput, allowing for improved sampling stream distribution [16]. *Escherichia coli, Salmonella enterica, Salmonella typhimurium,* and *Listeria monocytogenes* can all be found using SPR sensors in the commercial food industry [17].

Lee *et al.* [18] employed OFr SPR and PCR chip to identify *Salmonella* DNA amplicons. The SPR OFr sensor was fabricated through multi-mode OFr with bimetallic layer of SRP, while the PCR chip was formed of microchannel polymethyl methacrylate substratum. Although incorporated device does not require fluorescent labeling related characteristics such as background signal or size of the sample, when the genetic material sample is nearer to the SS surface, it detects the index of refraction change around the SS induced by the inoculated DNA amplifier, the device's unlabelled feature can ensure its reproducibility.

Morlay *et al.* [19] devised a technique that uses an immune-chip SPR imaging process to identify live bacteria during the bacterial growth phase. In a single day, the sensor could satisfactorily identify the minute amount of *Listeria monocytogenes* with reference to ST and SP. Recently Pebdeni *et al.* [20] reported the OPBRs to identify *E. coli* bacteria in food and water.

## 8.6 Mechanical Microbial Biosensor

Mechanical Bs provide several benefits for usage at the point of care, including higher ST and short production time without additional reagents and sample preparation. Mechanical Bs are classified as either quartz crystal microbalance (QCM) or cantilever technology (CT).

The resonance frequency change caused by the high mass of the SS due to the binding of AE is identified by the quartz crystal microbalance sensor, which is label-free piezoelectric Bs. QCM sensor was developed to identify WBc for *Salmonella enterica serova Typhimurium, E. coli, Bacillus anthracis* and *Campylobacter jejuni.* Hao *et al.* [21] studied a monoclonal antibody functionalized QCMR Bs to detect the vegetative cells or spores in *B. anthracis.* In some conditions, the advancement of sandwich types assessments of using nanomaterials for SG amplify has enabled the detection of several microbial species (10 CFU/mL) [22].

CT is new label-free technique for developing point of care sensors that combine excellent ST, quick reaction times and MT [23]. A Cantilever sensor usually consists of a BRs functionalized microcantilever oscillating at a specific resonance frequency. Due to imposed mechanical bend due to increased mass on the SS, the cantilever (CR) resonant frequency changes.

Using antibodies as BRs, newly developed piezoelectric-excited mm size CR were capable to identify *Escherichia coli* in buffer [24] and *Listeria monocytogenes* cells in milk [25]. Cantilever based system have several drawbacks, including the need to function in the air rather than in physiological media, and are a few instances of cantilever-based sensor being tested in relevant matrices.

## 8.7 Electrochemical Biosensor

Bioelectochemical system (BESs) are gaining attraction as a leading-edge platform with extensive applications in several areas. Electrochemical (EC) Bs encompass potentiometric, amperometric and impedimetric sensing techniques. Amperometric glucose Bs in diabetic monitoring have helped EC Bs have become the most developed commercially. Cost-effective, point of care diagnostics and MT capacity are among their major benefit.

Potentiometric biosensing (PBs) employs ion selective electrodes to detect a solution's potential, determined by ions interactions. The variation in voltage that happens when AE detection occurs at the working electrode is measured using this approach. Despite the fact that PBs is frequently utilized in the Bs area, there are few PBs for the identification of WBc. Potentiometry (PM), in comparison to other approaches like impedance, cannot produce specific and sensitive signals for large AE like microorganisms. However, some new PM applications can be yield reasonable DL. *S. aureus* is a well-known skin commensal and was identified using label free potentiometric detection by Zelada-Guillen *et al.* [26].

Amperometry biosensor measure the current produced by the oxidation of species formed in response to AE-BRs interaction. Amperometric Bs have several benefits, including their relative simpleness and ease of MT. They also have a high level of ST. Low SP depends on the applied voltage. If it is too high, it might cause other redox active species to interact with the K resulting in inaccurate result [27].

Microbial fuel cells (MFCs) are used for electricity generation [28] and microbial electrolysis cells for the production of $H_2$ [29]. MFCs are a type of bio-electrochemical systems that utilizes anaerobically producing biofilms on the surface of an electrode (anode) to oxidize organic molecules (such as waste water) in electricity production [30]. Microbial fuel cells (MFCs) are demonstrated to have outputs that correlate with BOD, making water quality detection a promising application. A hydraulically linked MFCs array has also been devised to boost the top detection limit (DL) of BOD sensors. A linear response up to 720mg $L^{-1}$ $BOD_5$ was obtained from the device [31]. On the other hand, Costa *et al.* [32] developed a microfluidic chemiluminescence system for automated BOD monitoring that uses luminol and oxygen generated by *Saccharomyces cerevisiae.*

Gao *et al.* [33] used double-mediator based whole cell electrochemical Bs to assess the biotoxicity of waste water. The *Saccharomyces cerevisiae* cells were immobilized on chitosan hydrogel polymer with boron-doped nanocrystalline particles on a glassy carbon electrode to assess the biotoxicity of heavy metal ions (Cupric ion, Cadmium ion, Nickle ion and lead ion) and three phenolic compounds in waste water. Yamashita *et al.* [30] designed BOD Bs for *in situ* monitoring of $BOD_s$. Suspended solids in waste water are covered and attached to the anode, resulting in the anaerobic conditions due to BOD loss and identified the *Geobacter* spp. In aerating and non-aerating phase, the current generation of the one type of BOD biosensor significantly rose until BOD concentrations reached 100 mg $L^{-1}$, then plateaued at 250 mg $L^{-1}$. The authors found that detection range of BOD in livestock waste water ranged from 14 to 100 mg $L^{-1}$ for high sensitivity detection. The findings suggested the BOD biosensor will be beneficial in various applications, including automatic aeration intensity control.

Webster *et al.* [34] demonstrated a BESs for the detecting arsenite based on the genetically encoded *Shewanella oneidensis* cells. When exposed to arsenite, The Bs showed a linear range of up to 100μ M and DL of 40μM. Khor *et al.* [35] designed ferricyanide mediated microbial BOD Bs to detect the BOD in a water system. The *C. violaceum* R1 cells were immobilized on the platinum ultra-microelectrode. The biosensor displayed a linear response ranging from 20 to 225 mg $L^{-1}$ of BODs for standard glucose glutamic acid solution. The sensor effectively measured the samples with a high concentration of quickly assimilated compounds. Another study by Hsien and Chung, [36] reported a mediator less MFCs Bs for BOD measurement. A linear response ranging from 8 to 240 $mgL^{-1}$ was obtained by the Bs using anaerobic sludge for anodic biofilm formation. However, the BOD values acquired by these Bs from the industrial effluent didn't match those generated by the usual BOD approach. This research reveals a significant drawback of MFCs Bs: non-correlation with the standard biological oxygen demand technique. Various factors like pH, temperature, the presence of heavy metals and bacterial competition, etc. It might alter the $MFC_S$ output and hence interfere with Bs. As a result, real waste water samples must frequently be diluted before being evaluated using MFCs Bs. An automated, independent and floating Bs for monitoring off-grid regions was designed by Pasternak *et al.* [37]. The Bs detected the urine in the freshwater and activated visual and sound cues (85dB). This Bs was operating autonomously for 5 months. A singled chamber MFCs for measuring chemical oxygen demand (COD) was developed by Di Lorenzon *et al.* [38]. The linear detection ranged from 3-164 mg $L^{-1}$ with a sensitivity of 0.05μ $Mm^{-1}$ $cm^{-2}$, with a remarkable time of 2.8min.

Reshelilov *et al.* [39] developed a *Nitrobacter* strain-based biosensor model for nitrite analyzer under the condition of nitroaromatic compound (NC) biodegradation. The sensor possessed strong selectivity with a nitrite with 10μM detectable levels. Their study suggested that MBs model appears to be promising as a prototype designed for use in a coupled analytical system for detecting NC during a biological treatment.

Aside from biological oxygen demand (BOD) biosensors, amperometric MBs have also been used to test a variety of other compounds. Due to its prominence in the fermentation sector and clinical toxicity [4, 40], MBs for ethanol has received the second most research focus after BOD. Several MBs based on methylotrophic yeast and potentiometric transducers (PT) have been described for the detection of aliphatic alcohols. Korpan and coworkers [41] reported a mutant strain of *P. pinus* and *Hansenula polymorpha* with a PT.

Voronava and coworkers [42] developed a biosensor for ethanol detection on *Pichia angusta* VKM Y-2518. A biosensor based on *Pichia angusta* VKM Y-2518 cells cultured in one percent methanol revealed the most promising for ethanol detection. This sensor was insensitive to carbohydrates and organic acids, with methanol acting as an interfering agent during ethanol detection and 0.012mM was the lower limit of ethanol detection. *Pichia angusta* VKM Y-2518 was immobilized on a Clark oxygen electrode as a transducer to develop an ethanol biosensor. Another study by Reshetilov reported that *Pichia methoanolica* had been immobilized on an oxygen electrode to create an ethanol biosensor (EBs) [43]. Although these Bs have high sensitivity and stability, they tend to have low sensitivity. As a result, there is a significant in developing selective EBs. An enhanced sensitivity was acquired by substituting $O_2$ with FeCN as the electron acceptor for ethanol determination. The *Gluconobacter oxydans* was immobilized on a carbon electrode by cellulose acetate membrane which also confined the accessibility of $C_6H_{12}O_6$ to cells. The biosensor detects ethanol rapidly and accurately, with a DL of 0.85 mM Tkac *et al.* [44].

The BOD is an index measuring the quantity of water pollution by the organic compound. Wang and co-workers developed PdNPs/rGN-COOH modified Ultramicroelectrode array BOD microsensor, comprising of microbial film inserted between cellulose membranes [45]. The microsensor had certain drawbacks, the microorganisms are immobilized between two layers of cellulose acetate membrane in the sandwiched microbial film. Because of slow mass transfer, this sandwiched microbial film structure results in delayed response time. For the lysis of the microbial films, the microsensor (BOD) has a short lifetime. It is hard to control the electrolyte hidden beneath the microbial film structure. Wang *et al.* [46] developed a modified UMAE BOD microsensor to overcome these limitations. The *Bacillus subtilis* cells were electrostatically labelled and immobilized on a UMAE using magnetic $Fe_3O_4$ nanoparticles. Their findings suggested that microsensor was more regenerative and reliable than dissolved oxygen based and mediated BOD sensor with better ST.

Sulphide detection Bs can be divided into two categories. Inhibitive Bs were first developed to detect sulphide based on enzyme activity [47, 48]. Second, microbes serve as biological recognition elements (BRE) in microbial Bs that directly oxidize sulphides. A Bs for sulphide detection was developed using *Thiobacillus thioparus* [49]. Janfada *et al.* [50] employed four sulphide oxidizing microorganisms (*Thiobacillus thioparus*, *Acidithiobacillus thiooxidans* PTCC1717, *Acidithiobacillus ferrrooxidans* PTCC1646, and *Acidithiobacillus ferrooxidans* PTCC1647 to construct a hydrogen sulphide biosensing system. The microbial biosensing device displayed a fast reaction time (<200s) and can be reused and maintained for more than a month if bacterial cells are immobilized. The cells were immobilized on poly vinyl alcohol. *A. thioxidans* and *A. ferrooxidna*s PTCC1646 are suitable bio recognition elements in an $H_2S$ bio-sensing system.

Another study by Liu *et al.* [51] developed microbial Bs using recombinant *Escherichia coli* BL21 strain overexpressing sulfide: quinone oxidoreductase (SQR) for sulphide detection. The *E. coli* cells were immobilized on nanoporous gold to detect sulphide. The E. *coli* NPG/glassy carbon bioelectrode showed a linear response ranging from 50μM to 5mM, high sensitivity of 18.35 μA $mM^{-1}$ $cm^{-1}$ and detection of 2.55 μM. The E. *coli* NPG/glassy carbon electrode was effectively utilized to detect the sulphide in waste water.

Yadav *et al.* [52] reported the electrochemical (EC) Bs for the quantifying streptomycin in the food system. The authors studied the EC Bs developed with several BRE and nanomaterial to detect the streptomycin in food stuffs. The electrochemical Bs can be used to detect antibodies. The ability to interact selectively with specific antigens (AG), antibodies (AB) has been widely used as BRE for detection purposes in Bs. Monoclonal antibodies (MABs) and polyclonal antibodies (PCABs) are extensively used in Bs as bioreceptors. The antibody RE depends mainly on the affinity of the antibody present in the Bs. Currently, antibodies are synthesized using recombination process to increase antibody affinity. Recombinant ABs production is more affordable and time consuming than MABs and PCABs production. Abs fail as a BRE in BS because they react with other ABs.

For the identification of streptomycin residues, Liu *et al.* [53] established a simple and sensitive Electrochemical immunoassay. Bionanolabeles (BNLs) was used as a signal amplifier and 3D redox active organosilica was used as a sensing surface in these Bs. Anti-streptomycin MABs were immobilized on the sensing surface, together with a glassy carbon electrode. Nanogold particles were assembled on mesoporous silica nanostructure to prepare BNLs. Streptomycin was detected in food through competition between the target streptomycin AE and the streptomycin bovine serum albumin conjugated with the existence of $H_2O_2$ as an enzyme substrate. The reduced current of the electrochemical immunosensor reduces as the level of streptomycin in the sample rises. DL of Bs is 0.5pg/mL with a linear range of 0.05-50 pg $mL^{-1}$. This immunosensor is commonly used to quantify streptomycin in samples of honey, milk and kidney.

## 8.8 Impedimetric Microbial Biosensor

Impedimetric Microbial biosensor (IMBs) are an excellent instrument for detecting WBc since they are highly ST, label-free, unhindered by another analyst (AN). ABs are the popular BRs in this method; however, proteins, whole bacteria and viruses are also capable of identifying an extensive analysis [54]. IMBs have the benefit of being able to measure a limitless number of BM without requiring the AN to be an ES substrate or the generation of EC reactions, which would be required for amperometric sensing. Despite significant advancements in impedance technology and growing research in this field, the techniques have failed to attain profitability [55]. Despite its advantage, impedance Bs, have certain drawbacks, including non-specific binding issues, high DL and inconsistent reproducibility. Nevertheless, in recent decades, several studies reporting the impedimetric identification of WBc [55, 56]. Though this technique has been used to identify a variety of microorganism, including sulphate reducing bacteria, *S. typhimurium*, *C.jejuni*, and *S. aureus*, the majority of these research has been performed on *E. coli*. Furthermore, with continuing improvements and advancements in equipment miniaturization (MT), the Electrochemical spectroscopy has a growing interest in Bs applications.

## 8.9 Application of Bs in Various Fields

Biosensor have future use in applications in clinical, environmental monitoring, food industries, agriculture and water treatment (Fig. 8.2). Table 8.1 summarizes various MBs for detecting WBc and environmental (ET) applications. Although MBs have a broad spectrum of applications, clinical diagnosis is essential. To put it another way, clinical diagnostics is known as a big market for Bs. The first glucometer discovered in the market was an enzyme electrode-based glucose analyzer. The scientists were able to design and manufacture a MBs to detect pathogen using rat basophilic leukaemia mast cell. MBs are used to detect the

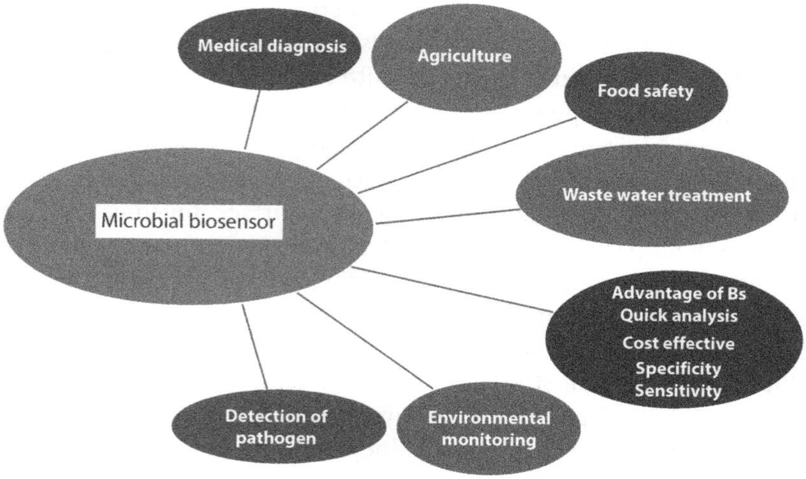


**Fig. 8.2** Application of microbial biosensor in various field.

**Table 8.1** A summarize of various microbial biosensor for detection of whole cell and environmental applications.

| Target AE | Microorganisms | Type of biosensor/ configuration | LD | References |
|---|---|---|---|---|
| Detection of whole cell bacteria | | | | |
| Vegetative cells and spores | *Bacillus anthracis* | Protein A/functionalized SAM on gold | $1\times10^3$CFU | Hao *et al.* [21] |
| Bacteria in buffer | *E.coli* O157;H7 | Antibody-functionalized cantiliever | 1 cells/mL | Campbell *et al.* [24] |
| Bacteria in milk | *Listeria monocytogenes* | Antibody with post capture antibody binding for signal amplification | $1\times10^2$ cells/mL | Sharma *et al.* [25] |
| Environmental applications | | | | |
| BOD | *Geobacter spp* | Bioelectrochemical system | 14-100 mg/L | Yamashita *et al.* [30] |
| BOD | Geobacter spp and Porphyromonadaceae | Single chamber MFCs separated with cation exchange membrane | 720mg/L | Spurr *et al.* [31] |
| BOD | *Saccharomyces cerevisiae* | Sequential injection analysis | 9.53 and 31.75 mg/L | Costa *et al.* [32] |
| Biotoxicity of waste water (Cupric ion, Cadmium ion, Nickle ion, Lead ion and three phenolic compound | *Saccharomyces cerevisiae* | Boron-doped nanocrystalline particles on a glassy carbon electrode | - | Gao *et al.* [33] |
| Biological oxygen demand | *C. iviolaceum.* R1 cells | Electrochemical (chronoamperometry) | 20 to 225mg $O_2$/L | Khor *et al.* [35] |

(*Continued*)

**Table 8.1** A summarize of various microbial biosensor for detection of whole cell and environmental applications. (*Continued*)

| Target AE | Microorganisms | Type of biosensor/ configuration | LD | References |
|---|---|---|---|---|
| BOD | *T. carboxydiphila, P. aeruginosa, Ochrobacterium intermedium, S. frigidimarnia, C. freundii* and *Clostridium* | Two chambered unit (anode and cathide comparments) and separated by Nafion cationic exchange membrane | 8-240 mg/L | Hsien & Chung, [36] |
| BOD | Precultured electroactive bacteria adapted | Single chamber MFCs electrically connected in parallel | - | Pasternak *et al.* [37] |
| COD | Mixed bacterial culture | Air-cathode single chambered with Layer-by layer prototype 3D printer | 3-164 ppm | Di Lorenzo *et al.* [38] |

*AE- Analyte, *LD-limit of detection.

signal in response to environmental factors and most of these sensors are based on bacteria. Bionanolabeles (BNLs) were used as a signal amplifier and 3D redox active organosilica was used as a sensing surface in this Bs. Streptomycin was detected in food through competition between the target streptomycin AE and the streptomycin bovine serum albumin conjugate in the presence of $H_2O_2$ as an enzyme substrate. The *Saccharomyces cerevisiae* cells were immobilized on chitosan hydrogel polymer with BNLs particles to assess the biotoxicity of heavy metal ions. The microfluidic chemiluminescence system for automated BOD monitoring was developed using *S. cerevisiae.* The Bs also have been used to detect the pathogens. Electrochemical Bs have been discovered as viable analytical tool for identifying avian influenza virus in compilated matrices as a result of invention along this line. Bs were capable of determining hazardous heavy metals such as cupric ions and degradable organic and endocrine disruptive chemicals.

## 8.10 Recent Trends, Future Challenges, and Constrains of Biosensor Technology

Morden methods for Bs discovery include integrative approaches including several technologies such as electrochemical, electromechanical and fluorescent-based optical biosensor and genetically engineered microbes. Some of these Bs have many potential applications in clinical diagnosis and medicine. The need and demand for using Bs for quick analysis with economic biofabrication necessitates recognition of cellular to whole animal activity with high accuracy to targeted molecules. The Bs are designed to function in multiplex environments and target and quantify minute ions or molecules of interest with advanced TD, which is necessary for 2D and 3D detection. Table 8.2 summarizes few microbial biosensors described in the literatures.

The next phase of research should focus on developing longer-lasting regenerative Bs. As a result, medical professionals and patients can benefit from new diagnostic Bs for therapeutics, making disease and treatment more holistically understood due to the fact that, Bs based florescence resonance energy transfer proved to be a good diagnostic tool for determining the efficacy of imatinib treatment of chronic myeloid leukaemia [57]. Electrochemical Bs have been discovered as a viable analytical device for identifying pathogen avian influenza virus in compilated matrices due to invention along this line [58]. A recent study found that affinity-based Bs could be used in sports and medicine and doping control studies [59].

Furthermore, scientific progress in the developing MBs using a synthetic biology method will have a significant impact on environmental monitoring and energy demand [60]. The need for MFCs to develop a water treatment technology as well as power source for environmental sensor.

Many authors report that enzymes are one of the main BRE employed in the designing of Bs technology. The purified enzymes have several advantages, for example high SP for those inhibitors. Nonetheless, the time and cost of ES purification, the necessity for various enzymes to yield a quantifiable result, and the demand for cofactors/coenzymes limit their application in Bs synthesis. Microbes are one of the most suitable and optimal solution for these issues [61]. The cost of the sensor system for microbial based Bs is significantly less than that of spectroscopic analysis (Gas and liquid chromatography) [62]. Because microbes can grow in huge quantity using a simple and cost-effective culture procedure, analytical costs can be drastically lowered. The microbial Bs can withstand a wide range of

**Table 8.2** A summarize of various microbial biosensor described in the literatures.

| Target AE | Microbes used | Type of transducer used | LD | References |
|---|---|---|---|---|
| Nitroaromatic compound | *Nitrobacter* strain | Oxygen | 0.1-10mM | Reshelilov *et al.* [39] |
| Ethanol detection | *Gluconobacter oxydans* | glassy carbon electrode | 0.85 mM | Tkac *et al.* [40] |
| Ethanol detection | *Pichia angusta* VKM Y-2518 | Oxygen | 0.012mM | Voronava *et al.* [42] |
| BOD | Bacillus subtilis | Electrochemical (chronoamperometry) | 2-15mg/L | Wang *et al.* [46] |
| sulphide detection | *Escherichia coli* BL21 strain | Nanoporous gold/glassy carbon electrode | 2.55 μM | Liu *et al.* [49] |
| sulphide detection | A. *thioxidans* and *A. ferrooxidnas* PTCC1646 | Dissolved oxygen sensor | 1.09-16.3μM | Janfada *et al.* [50] |

*AE- Analyte, *LD-limit of detection.

eco-friendly conditions. Additionally, microbes are able to react quickly and accurately to changes in conditions because of their small size and rapid growth rate [7]. Using only a few enzymes in the Bs structure increases the risk of altering the molecular structure of the enzymes during immobilization. The main reason for employing complete cells as a BE in the Bs device is the only obtained by LCs.

Despite the fact that MBs offer a number of benefits and are well-known as convenient tools for identifying AE, they nevertheless face number of obstacles that have hindered their adoption and application. According to studies, the use of whole cell Bs for real tine therapeutic monitoring is complicated by creating of a gene(reporter) and an extended response time for cell development.

Synthetic biology is a technology that can handle a variety of difficulties in its applications, including the necessity for multiplex identification and processing of many biomarkers, particularly in medical diagnosis. Because many ligands of interest lack receptors in nature, Bs cannot identify them. As a result, new technologies, particularly ABs for generating synthetic receptors that recognize a wide range of ligands, are critical.

## 8.11 Conclusion

The impact of MBs on medical diagnosis has been categorized based on their qualities such as ST, speed and early identification. Recently, technologies such as microfluidics as well as the finding the biomarker, have been found to augment and enhance the performance of MBs, Given the high spread and occurrence of diseases, as well as the increase death form them, the employment of highly reliable, accurate and fast diagnostic method is essential. For a large section of the world's population, many of the currently available procedures are inaccessible because of the high price and the necessity for skilled expert. MBs have been applied in several application for example environmental, food and diagnostics due to their ST, cost effective and quick response. Lab-on-chip technology has piqued the interest of many researchers, and it's now widely employed in the fabrication of microfluidic devices that allow for the identification of AE at extremely lower concentrations by actively delivering targets to the surface of MBs. The MBs are an excellent and adequate alternative to the time saving, sophisticated and costly traditional forensic identification methods due to their capacity to examine DNA. The importance of handheld, easy to use Bs for WBc identification is evident form market requirements and research trends. Minimization, optimization and clinical studies must all be completed before a product can be released to the market. Finally, the widespread deployment of biosensors will not only mark a turning point in the Bs business but will also have far-reaching implications in the environmental, food, clinical and medical diagnostics.

## References

1. Gui, Q., Lawson, T., Shan, S., Yan, L., Liu, Y., The application of whole cell-based biosensors for use in environmental analysis and in medical diagnostics. *Sensors*, 17, 1623, 2017.
2. Pham, H.T., Biosensors based on lithotrophic microbial fuel cells in relation to heterotrophic counterparts: Research progress, challenges, and opportunities. *AIMS Microbiol.*, 4, 567, 2018.

3. Kumar, J. and D'Souza, S.F., Biosensors for environmental and clinical monitoring. *BARC Newslett.*, 324, 34–38, 2012.
4. D'Souza, S.F., Microbial biosensors. *Biosens. Bioelectron.*, 16, 337, 1997.
5. Dai, C. and Choi, S., Technology and applications of microbial biosensor. *Open J. Appl. Biosens.*, 2, 83, 2013.
6. Clark, L.C. and Lyons, C., Electrode systems for continuous monitoring in cardivascular surgery. *Ann. N. Y. Acad. Sci.*, 102, 29, 1962.
7. McGrath, T.F., Elliott, C.T., Fodey, T.L., Biosensors for the analysis of microbiological and chemical contaminants in food. *Anal. Bioanal. Chem.*, 403, 75, 2012.
8. Rodovalho, V.R., Alves, L.M., Castro, A.C.H., Madurro, J.M., Brito-Madurro, A.G., Santos, A.R., Biosensors applied to diagnosis of infectious diseases—An update. *Biosens. Bioelectron.*, 1, 1, 2015.
9. Malhotra, S., Verma, A., Tyagi, N., Kumar, V., Biosensors: Principle, types and applications. *Int. J. Adv. Res. Innov. Ideas Educ.*, 3, 3639, 2017.
10. Arya, S.K., Chaubey, A., Malhotra, B.D., Fundamentals and applications of biosensors. *Proc. Indian Natl. Sci. Acad.*, 72, 249, 2006.
11. Ligler, F.S., Sapsford, M.A., Barone, S., Myatt, C.J., The array biosensor: Portable, automated systems. *Anal. Sci.*, 23, 5, 2007.
12. Geng, T., Uknalis, J., Tu, S.I., Bhunia, A.K., Fiber-optic biosensor employing alexa-fluor conjugated antibody for detection of *Escherichia coli* O157:H7 from ground beef in four hours. *Sensors*, 6, 796, 2006.
13. Mouffouk, F., Rosa da Costa, A.M., Martins, J., Zourob, M., Abu-Salah, K.M., Alrokayan, S.A., Development of a highly sensitive bacteria detection assay using fluorescent pH-responsive polymeric micelles. *Biosens. Bioelectron.*, 26, 3517–3523, 2011.
14. Oldach, L. and Zhang, J., Genetically encoded fluorescent biosensors for live-cell visualization of protein phosphorylation. *Chem. Biol.*, 21, 186, 2014.
15. Owen, V., Real-time optical immunosensors—A commercial reality. *Biosens. Bioelectron.*, 12, 1–2, 1997.
16. Sonato, A., Agostini, M., Ruffato, G., Gazzola, E., Liuni, D., Greco, G., Travagliati, M., Cecchini, M., Romanato, F., A surface acoustic wave (SAW)-enhanced grating-coupling phase-interrogation surface plasmon resonance (SPR) microfluidic biosensor. *Lab. Chip*, 16, 7, 1224–1233, 2016.
17. Stephen Inbaraj, B. and Chen, B.H., Nanomaterial-based sensors for detection of foodborne bacterial pathogens and toxins as well as pork adulteration in meat products. *J. Food Drug Anal.*, 24, 15, 2016.
18. Lee, K.J., Lee, W.S., Hwang, A., Moon, J., Kang, T., Park, K., Jeong, J., Simple and rapid detection of bacteria using a nuclease-responsive DNA probe. *Analyst*, 143, 332, 2017.
19. Morlay, A., Duquenoy, A., Piat, F., Calemczuk, R., Mercey, T., Livache, T., Roupioz, Y., Label-free immuno-sensors for the fast detection of Listeria in food. *Measurement*, 98, 305–10, 2017.
20. Pebdeni, A.B., Roshani, A., Mirsadoughi, E., Behzadifar, S., Hosseini, M., Recent advances in optical biosensors for specific detection of *E. coli* bacteria in food and water. *Food Control*, 135, 108822, 2022.
21. Hao, R., Wang, D., Zhang Xe Zuo, G., Wei, H., Yang, R., Zhang, Z., Cheng, Z., Guo, Y., Cui, Z., Zhou, Y., Rapid detection of Bacillus anthracis using monoclonal antibody functionalized QCM sensor. *Biosens. Bioelectron.*, 24, 1330–1335, 2009.
22. Salam, F., Uludag, Y., Tothill, I.E., Real-time and sensitive detection of *Salmonella Typhimurium* using an automated quartz crystal microbalance (QCM) instrument with nanoparticles amplification. *Talanta*, 115, 761–767, 2013.
23. Buchapudi, K.R., Huang, X., Yang, X., Ji, H.F., Thundat, T., Microcantilever biosensors for chemicals and bio organisms. *Analyst*, 136, 556, 2011.

24. Campbell, G.A. and Mutharasan, R., A method of measuring *Escherichia coli* O157:H7 at 1 cell/mL in 1 liter sample using antibody functionalized piezoelectric-excited millimeter-sized cantilever sensor. *Environ. Sci. Technol.*, *41*, 1668, 2007.
25. Sharma, H. and Mutharasan, R., Rapid and sensitive immunodetection of Listeria monocytogenes in milk using a novel piezoelectric cantilever sensor. *Biosens. Bioelectron.*, 45, 158, 2013.
26. Zelada-Guillen, G.A., Sebastian-Avila, J.L., Blondeau, P., Riu, J., Rius, F.X., Label-free detection of *Staphylococcus aureus* in skin using realtime potentiometric biosensors based on carbon nanotubes and aptamers. *Biosens. Bioelectron.*, 31, 226, 2012.
27. Higson, S.P., *Biosensors for medical applications*, Woodhead Publishing, Cambridge, United Kingdom, 2012.
28. Liu, H., Ramnarayanan, R., Logan, B.E., Production of electricity during wastewater treatment using a single chamber microbial fuel cell. *Environ. Sci. Technol.*, 38, 2281, 2004.
29. Call, D. and Logan, B.E., Hydrogen production in a single chamber microbial electrolysis cell lacking a membrane. *Environ. Sci. Technol.*, 42, 3401, 2008.
30. Yamashita, T., Ookawa, N., Ishida, M., Kanamori, H., Sasaki, H., Katayose, Y., Yokoyama, H., A novel open-type biosensor for the *in-situ* monitoring of biochemical oxygen demand in an aerobic environment. *Sci. Rep.*, 6, 38552, 2016.
31. Spurr, M.W.A., Yu, E.H., Scott, K., Head, I.M., Extending the dynamic range of biochemical oxygen demand sensing with multi-stage microbial fuel cells. *Environ. Sci.: Water Res. Technol.*, 4, 12, 2029–2040,2018.
32. Costa, S.P.F., Cunha, E., Azevedo, A.M.O., Pereira, S.A.P., Neves, A.F.D.C., Vilaranda, A.G., Araujo, A.R.T.S., Passos, M.L.C., Pinto, P.C.A.G., Saraiva, M.L.M.F.D.S., Microfluidic chemiluminescence system with yeast *Saccharomyces cerevisiae* for rapid biochemical oxygen demand measurement. *ACS Sustainable Chem. Eng.*, 6, 5, 2018.
33. Gao, G., Fang, D., Yu, Y., Wu, L., Wang, Y., Zhi, J., A double-mediator based whole cell electrochemical biosensor for acute biotoxicity assessment of wastewater. *Talanta*, 15, 167–216, 2017.
34. Webster, D.P., TerAvest, M.A., Doud, D.F.R., Chakravorty, A., Holmes, E.C., Radens, C.M., Sureka, S., Gralnick, J.A., Angenent, L.T., An arsenic-specific biosensor with genetically engineered *Shewanellaoneidensis* in a bioelectrochemical system. *Biosens. Bioelectron.*, 62, 320, 2014.
35. Khor, B.H., Ismaild, A.K., Ahamad, R., Shafinaz, S., A redox mediated UME biosensor using immobilized *Chromobacterium violaceum* strain R1 for rapid biochemical oxygen demand measurement. *Electrochim. Acta*, 176, 777, 2015.
36. Oldach L., Zhang J.. Genetically encoded fluorescent biosensors for live-cell visualization of protein phosphorylation. *Chem Biol.*, 20, 21, 2, 186–97, 2014.
37. Pasternak, G., Greenman, J., Ieropoulos, L., Self-powered, autonomous biological oxygen demand biosensor for online water quality monitoring. *Sens. Actuators B: Chem.*, 244, 815, 2017.
38. Di Lorenzo, B.M., Thomson, A.R., Schneider, K., Cameron, P., Ieropoulos, I., A small-scale air-cathode microbial fuel cell for on-line monitoring of water quality. *Biosens. Bioelectron.*, 62, 182–188, 2014.
39. Reshetilov, A.N., Iliasov, P.V., Knackmuss, H.J., Boronin, A.M., The nitrite oxidizing activity of *nitrobacter* strains as a base of microbial biosensor for nitrite detection. *Anal. Lett.*, 33, 29, 2000.
40. Trosok, S.P., Driscoll, B.T., Luong, J.H.T., Mediated microbial biosensor using a novel yeast strain for wastewater BOD measurement. *Appl. Microbiol. Biotechnol.*, 56, 550, 2001.
41. Korpan, Y.I., Gonchar, M.V., Soldatkin, A.P., Starodub, N.F., Sandrovskijj, A.K., Sibirnyjj, A.A., Elskaja, A.V., Conductometric biosensor for ethanol detection based on whole yeast cells. *Ukr. Biokhim. Zh.*, 64, 96, 1992.

42. Voronova, E.A., Iliasov, P.V., Reshetilov, A.N., Development, investigation of parameters and estimation of possibility of adaptation of *pichia angusta* based microbial sensor for ethanol detection. *Anal. Lett.*, 41, 377, 2008.
43. Reshetilov, A.N., Trotsenko, J.A., Morozova, N.O., Iliasov, P.V., Ashin, V.V., Characteristics of *Gluconobacter oxydans* B-1280 and *Pichia methanolica* MN4 cell based biosensors for detection of ethanol. *Process Biochem.*, 36, 1015, 2001.
44. Tkac, I., Vostiar, L., Gorton, P., Gemeiner, P., Sturdik, E., Improved selectivity of microbial biosensor using membrane coating: Application to the analysis of ethanol during fermentation. *Biosens. Bioelectron.*, 18, 1125, 2003.
45. Wang, J., Bian, C., Tong, J., Sun, J., Hong, W., Xia, S., Reduced carboxylic graphene/palladium nanoparticles composite modified ultramicroelectrode array and its application in biochemical oxygen demand microsensor. *Electrochim. Acta*, 145, 64, 2014.
46. Wang, J., Li, Y., Bian, C., Tong, J., Fang, Y., Xia, S., Ultramicroelectrode array modified with magnetically labelled *Bacillus subtilis*, palladium nanoparticles and reduced carboxy graphene for amperometric determination of biochemical oxygen demand. *Microchim. Acta*, 184, 763, 2017.
47. Shan, D., Li, Q.B., Ding, S.N., Xu, J.Q., Cosnier, S., Xue, H.G., Reagentless biosensor for hydrogen peroxide based on self-assembled films of horseradish peroxidase/laponite/chitosan and the primary investigation on the inhibitory effect by sulfide. *Biosens. Bioelectron.*, 26, 536, 2010.
48. Savizi, I.S.P., Kariminia, H.R., Ghadiri, M., Roosta-Azad, R., Amperometric sulfide detection using Coprinus cinereus peroxidase immobilized on screen printed electrode in an enzyme inhibition based biosensor. *Biosens. Bioelectron.*, 35, 297–301, 2012.
49. Qi, P., Zhang, D., Wan, Y., Development of an amperometric microbial biosensor based on Thiobacillus thioparus cells for sulfide and its application to detection of sulfate-reducing bacteria. *Electroanalysis*, 26, 1824–1830, 2014.
50. Janfada, B., Yazdian, F., Amoabediny, G., Rahaie, M., Use of sulfur-oxidizing bacteria as recognition elements in hydrogen sulfide biosensing system. *Biotechnol. Appl. Biochem.*, 62, 349, 2015.
51. Liu, Z., Ma, H., Sun, H., Gao, R., Liu, H., Wang, X., Xu, P., Xun, L., Nanoporous gold-based microbial biosensor for direct determination of sulfide. *Biosens. Bioelectron.*, 98, 29, 2017.
52. Yadav, A., Kharewal, T., Verma, N., Tehri, N., Gahlaut, A., Hooda, V., Electrochemical biosensors for the quantification of streptomycin in food systems: An overview. *J. Environ. Anal. Chem.*, 1–16, 1944620, 2021.
53. Liu, B., Zhang, B., Cui, Y., Chen, H., Gao, Z., Tang, D., Multifunctional gold-silica nanostructures for ultrasensitive electrochemical immunoassay of streptomycin residues. *ACS Appl. Mater. Interfaces*, 3, 12, 4668–76, 2011.
54. Abbasian, F., Ghafar-Zadeh, E., Magierowski, S., Microbiological sensing technologies: A review. *Bioengineering*, 5, 20, 2018.
55. Ahmed, A., Rushworth, J.V., Hirst, N.A., Millner, P.A., Biosensors for whole-cell bacterial detection. *Clin. Microbiol. Rev.*, 27, 63, 646, 2014.
56. Daniels, J.S. and Pourmand, N., Label-free impedance biosensors: Opportunities and challenges. *Electroanalysis*, 19, 1239–1257, 2007.
57. Fracchiolla, N.S., Artuso, S., Cortelezzi, A., Biosensors in clinical practice: Focus on oncohematology. *Sensors (Basel)*, 13, 6423, 2013.
58. Grabowska, I., Malecka, K., Jarocka, U., Radec ki, J., Radecka, H., Electrochemical biosensors for detection of avian influenza virus–current status and future trends. *Acta Biochim. Pol.*, 61, 471, 2014.

59. Mazzei, F., Antiochia, R., Botre, F., Favero, G., Tortolini, C., Affinity-based biosensors in sport medicine and doping control analysis. *Bioanalysis*, 6, 225–245, 2014.
60. Sun, J.Z., Peter, K.G., Si, R.W., Zhai, D.D., Liao, Z.H., Sun, D.Z. *et al.*, Microbial fuel cell-based biosensors for environmental monitoring: A review. *Water Sci. Technol.*, 71, 801–809, 2015.
61. Lei, Y., Chen, W., Mulchandani, A., Microbial biosensors. *Anal. Chim. Acta*, 568, 200–210, 2006.
62. Lim, J.W., Ha, D., Lee, J., Lee, S.K., Kim, T., Review of micro/nanotechnologies for microbial biosensors. *Front. Bioeng. Biotechnol.*, 3, 1–13, 2015.

# 9

# Plasmonic Nanomaterials in Sensors

**Noor Mohammadd[1], Ruhul Amin[1,2], Kawsar Ahmed[3,4]* and Francis M. Bui[3]**

[1]*Dept. of Electrical and Electronic Engineering, Ahsanullah University of Science and Technology, Dhaka, Bangladesh*
[2]*Department of Electrical and Electronic Engineering, Bangladesh University of Business and Technology (BUBT), Dhaka, Bangladesh*
[3]*Department of Electrical and Computer Engineering, University of Saskatchewan, Canada*
[4]*Group of Bio-Photomatiχ, Department of Information and Communication Technology, Mawlana Bhashani Science and Technology University, Santosh, Tangail, Bangladesh*

***Abstract***

Plasmonic nanomaterials exhibit distinctive optical properties when exposed to incident light with a proper wavelength. These noble materials have found widespread applications in optical sensing and sensors, energy transfer and conversion, biological analyte detection, chemical sensing, environmental control, nanomedicine, medical diagnostics, tumor and tissue imaging, food security, therapeutic and biomedical engineering, etc., thanks to their unique catalytic, mechanical and optical properties. As a result, the characteristics, synthesis and pertinent practical uses of plasmonic nanomaterials have piqued the attention of many researchers in recent decades, especially in the field of nanotechnology. Surface plasmon resonance (SPR) based sensors have a high sensitivity owing to the minimal deviation in sample refractive index (RI) in comparison to other optical sensing methods. Photonic crystal fiber (PCF)-based plasmonic sensors have recently gained unprecedented popularity owing to their compact architecture and novel light-control capabilities. We explore various types of plasmonic nanomaterials, their properties and recent applications in this book chapter. Moreover, we briefly compare and distinguish between different types of plasmonic nanomaterials, as well as discuss the main challenges and future prospects of plasmonic nanomaterials in sensors.

***Keywords*:** Nanomaterials, plasmon, surface plasmon resonance, plasmonic nanomaterials, optical sensors

## 9.1 Introduction

Research on nanomaterials has been bolstered by recent advancements in enhanced characterization and processing methodologies [1–3]. Nanoscale materials include a plethora of distinctive optical, acoustic, electrical, thermal and magnetic properties that make them

**Corresponding author*: kawsar.ict@mbstu.ac.bd; k.ahmed.bd@ieee.org; k.ahmed@usask.ca
ORCID ID: https://orcid.org/0000-0002-4034-9819

Inamuddin and Tariq Altalhi (eds.) Biosensors Nanotechnology, 2nd Edition, (185–200) © 2023 Scrivener Publishing LLC

potentially useful in a variety of pertinent applications, including energy transfer and conversion [4–6], nanomedicine and medical diagnostics [7, 8], etc. Recently, researchers have introduced a novel and unique type of nanoparticles—plasmonic nanomaterials—to boost the performance of nanomaterials and explore their applications [9, 10]. To keep up with the continual evolution of diverse scientific discoveries and developing technologies, the application area of plasmonic nanostructured-materials is expeditiously growing.

In the last few decades, optical sensor devices have been proposed as a viable substitute for traditional solid-state devices that are flat, fragile, barely bendable, stiff, etc. [11]. Optical sensors offer a number of notable advantages over electronic sensors, including reduced production costs, simple procedures, faster reaction times, and greater dependability. Emissions of electromagnetic noise or thermal interference also have a significant impact on electronic equipment [12]. When it comes to physical sensing and monitoring complex environments and their surroundings, optical platforms are increasingly being used. These platforms can be used to monitor a wide range of natural or ecological variables, namely temperature, torsion, moisture, pressure, stress and strain, maintaining influential practical utilizations in robotics and wearable sensors, as well as health and safety monitoring [13–17]. Accordingly, because of the many noteworthy advantages optical sensor devices possess, including minimal cost, superior sensitivity, reduced noise and interference, compactness, rapid response, dependability, etc., they have been proven to be a convenient option for chemical, gas and oil-sensing applications [18, 19].

It is now widely accepted that SPR is an essential technology for a wide range of sensing applications. As an influential and compelling tool, the SPR sensors are extensively applied in a variety of applications, including bioimaging [20] and biosensing [21, 22] as well as water testing [23], food security [24], chemical sensing [25–28], environmental control [29], aqueous specimen identification [30]. They are also used in gas [31] and biological analyte sensing [25], as shown in Figure 9.1. Numerous optoelectronic devices, namely SPR imaging [32], optical configurable filters [33], layer thickness control [34], optical oscillators and modulators [35] have all been included in SPR technology as a result of its continual improvement. Moreover, SPR technology allows for the combination of nanophotonic and nano-electronic segments, intending to produce ultra-solid optoelectronic devices [36]. In 1907, a German physicist named Zenneck initially came up with a theory about the origin of SPR [39]. In this theory, it was shown that the surface electromagnetic waves can be detected at the interface between two materials or mediums, where one of the mediums is lossless and the other is either metal or lossy dielectric. The first observation of surface waves was made by Sommerfeld in 1909, when he discovered that they are rapid and diminish exponentially with height above the contact [37]. Ritchie was responsible for the day-to-day advancement of SPR. The physical presence of surface waves was first illustrated in 1957 in a metallic dielectric contact and has since been confirmed many times [38].

The traditional prism-coupled SPR sensors can be categorized into two classes: Kretschmann [40] and Otto configurations [41]. The attenuated total reflection concept governs the operation of both of these sensors. Even though SPR sensors with Kretschmann-based configurations are commonly used because of their superior sensing capabilities, they have a number of drawbacks that must be considered. Specifically, these SPR-based sensors are of large and complex structure, and they include both optical and mechanical components that are constantly changing positions. Their inability to be transported and utilized for distant sensing applications is the consequence of this drawback [27]. Furthermore, the

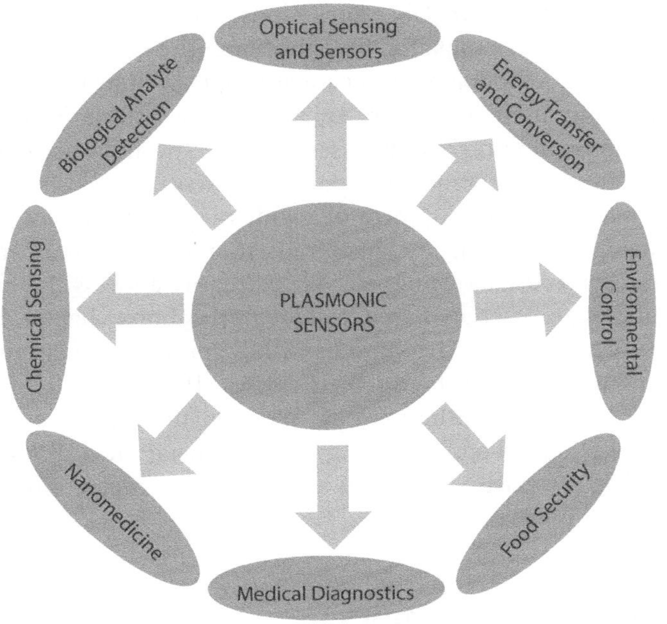


**Figure 9.1** Applications of plasmonic nanomaterials in sensors.

overall cost of implementing a spectral-based measurement in practice is high, and the ability to minimize the sensor size is limited. SPRs based on optical fibers have been developed to efficiently address such potential issues [42–45]. Plain and lightweight optical fibers are a good choice for their flexible architecture, as well as they are feasible to significantly decrease the sensor dimensions, which might be employed for distant sensing applications [45]. SPR sensors based on optical fibers have a greater dynamic range for identification, as well as a better resolution. Nevertheless, they are only useful for limited acceptance angles [46]. Numerous theoretical and practical studies have been conducted on optical fiber SPR sensors [42–45].

Photonic crystal fibers (PCFs) are a new kind of optical fiber that represents a much-improved and next-generational category of widely used and growing technology. In recent years, PCFs have moved to the top of the scientific priority list, garnering the attention of an increasing number of academics throughout the world. PCFs have ushered in an age of new potential possibilities across a broad variety of research and technological sectors, irrevocably breaking many of the weakened principles of old fiber optics theory and practice in the process. Due to its distinctive and attractive optical features over the existing traditional fiber optics, PCF-based SPR sensors have recently been widely researched and investigated by scientists [47–50]. PCFs hold numerous unique optical properties, among which birefringence, single-mode propagation, nonlinearity, confinement loss, etc. are mention-worthy. These special guiding traits can be readily modified, allowing the overall performance of the sensor to be controlled [47]. It is feasible to operate the sensor at its best by adjusting structural factors, such as pitch, number of air hole count, air hole size, etc. [49]. Current fabrication methods have advanced to the point where PCF sensors can be manufactured in a realistic manner. However, PCF-based SPR sensors have not yet been thoroughly characterized in the laboratory. Consequently, present PCF-reliant sensors are mostly shown by numerical simulation investigations. The finite element method (FEM) is a robust computational methodology for characterizing the optical characteristics of PCF sensors that are frequently used. To further enhance the overall performance of the PCF-reliant SPR sensors, thorough numerical analyses based on spectral wavelength and phase or amplitude intensity data are most frequently applied to assess their sensing capability [48].

Plasmonics has enabled the development of a wide range of pertinent practical applications, including sensing, bioimaging, metamaterials, theranostics, nonlinear optics, photocatalysis, energy transfer and harvesting, etc. We primarily address the breakthroughs in the area of plasmonic nanomaterials-based sensors in this book chapter, briefly depicting their basics and features in order to comprehend how plasmonic nanomaterials interact with light and distinguish between their diverse classes. This will be followed by a discussion of current developments in synthesizing plasmonic nanoparticles, as well as their optical characteristics and potential applications in various metals and alloys. Our discussion concludes with a look at the existing obstacles and future prospects for building plasmonic nanostructured-material sensors.

## 9.2 Fundamentals of Plasmonics

Plasmon is an optical phenomenon that occurs when a certain wavelength of an incoming light strikes a noble metallic surface at the dielectric interface, causing the conducted electrons to accrue energy and oscillate. The plasmon is a combined motion of unrestrained electrons on a metal surface restricted to particle dimensions of about 300 nanometers [51, 52]. When seen from a classical perspective, plasmons can be defined as an oscillation of electron density concerning the fixed positive ions present in a metal. In order to see a plasma oscillation, we can consider a cube of metal that has been put in an external electric field that is heading rightward. Electrons will travel to the left side of the metal, revealing positive ions on the right side, until the electrons balance out the field inside the metal. Upon removing the electric field, the electrons will gravitate to the right, repelled by one another and dragged to the positive ions that have been left uncovered on the right flank. For as long as there is a resistance or dampening, they oscillate at the plasma frequency. Plasmons quantize this kind of oscillation and interact significantly with light, eventuating in a polariton. Dielectric constants, such as air, vacuum, glass, etc. display a positive real portion of their relative permittivity, whereas metals and strongly doped semiconductors exhibit a negative real part of their permittivity at the given frequency of light. Real permittivity, in addition to its opposite sign, should generally have a greater magnitude in its negative permittivity region than its positive permittivity region. Otherwise, the light would not be limited to the surface. It is important to choose materials carefully since they may have an enormous impact on propagation distance and light confinement. Surface plasmons may also occur on non-flat interfaces including particles, cylinders, and other shapes. A wide variety of shapes have been studied by researchers since surface plasmons may restrict light below the diffraction limit.

Propagation SPR (PSPR) and Localized SPR (LSPR) are two distinctive forms of SPR based on the size of the noble metal [53–55]. It is common for PSPR to be formed on thin metal sheets, and the SPR may travel up to hundreds of micrometers or more along with the metal or dielectric surface in this manner. This characteristic has been exploited in sensing [56] due to the fact that it may be changed by biomolecules. SPR involves the three-dimensional (3D) propagation of plasmons in all three axes (X, Y, and Z) in conjunction with a thick dielectric or metal contact. SPR is a potent detecting method since it is very susceptible to changes in the surface coating, causing shifts in the SPR angle. Commercial SPR takes advantage of incoming light by employing an elevated RI glass prism. In reaction to the analyte, the RI of metal film coatings (such as silver and gold) varies. SPR imaging uses a charge-coupled device (CCD) camera to record reflected light for further investigation. SPR imaging is used to make measurements at a consistent wavelength and angle. The quantity of analyte attached is shown by the brightness of each flow cell.

LSPR is induced when the noble metals are generally lesser in size than the incident wavelength, where the SPR frequency extremely relies upon the nanoparticle separation distance and dielectric environment, as well as size, composition, geometry, etc. [57–60]. Specifically, LSPR is a technique for detecting electromagnetic waves trapped in a metal surface that have been locally generated (generally in the order of 30 nanometers). Variations in metal nanostructure type, dimension, shape, size, and geometry allow for a broad range of wavelengths to be tailored for LSPR features, such as peak extinction. Consequently, the extinction peak is dependent on several factors that may be exploited to construct colorimetric sensors with great sensitivity [61, 62]. In the visible spectrum (VS), silver, gold, copper, etc. plasmonic nanoparticles show LSPR effects. Consequently, they may be employed as colorimetric reporters to detect analytes in a variety of applications [63].

## 9.3 Optical Properties of Plasmonic Nanomaterials

Nowadays, most up-to-date plasmonic devices are solely based upon noble metals that reveal a moderately minimal optical loss in the VS. For instance, silver (Ag) is an excellent choice for nanoscale plasmonics from an optical standpoint, since it holds no interband transitions in the VS. Besides, silver possesses low optical damping and has a plasma wavelength of 137 nanometers deep in the ultraviolet (UV), thereby constituting a perfect Drude metal [64]. Regardless, silver has the significant drawback of being corrosive, resulting in the production of brittle oxide coatings very immediately even in aqueous solutions, which has precluded it from finding broad use. Aluminum (Al) would seem to be a tempting option owing to its intrinsically high electron density [65], but it, too, exhibits comparable corrosion concerns, has somewhat greater damping, and is difficult to fabricate into structures.

In many applications, gold (Au) is used as the primary plasmonic substance as it is innately stagnant, simple to fabricate, biocompatible and long-lasting, despite the fact that it has a somewhat high optical damping in the VS owing to significant interband transitions [64, 66]. It is not possible to lessen the thickness of gold layers because the growth of gold coatings is seed-initiated, resulting in intermittent layers when the thickness is less than ten nanometers [67, 68]. However, increasing the thickness of gold films is possible. A further issue is that gold holds weak adherence to glass substrates, which implies that the supplementary adhesion films (such as Cr or Ti) must be deposited between the metal and the dielectric, which may degrade the optical capabilities of the device, especially if the film thickness is less than 10 nm. Further, transferring the sample to a new deposition device might result in an oxide layer forming on the adhesion layer (particularly for Ti), which can drastically modify and eventually, degrade the optical characteristics of gold. A relatively new and novel strategy for improving gold layer characteristics depends on chemically developing single-crystal gold coatings rather than employing gas-based deposition procedures [69, 70], resulting in superior film quality when reaching the sub-ten nanometer extent. Copper (Cu), which exhibits nearly identical interband transition and optical damping as gold, is another promising plasmonic substance with wavelengths between 600 and 750 nm. Copper, however, is susceptible to oxidation, which restricts its widespread use as a plasmonic material.

The usage of graphene on top of Cu or Ag has recently gained popularity in order to avert the oxidation problem [72, 73, 98]. Due to its chemical inertness and mechanical strength, graphene is incapable of coming into touch with an aqueous solution [82]. PCF-reliant SPR sensors, based upon silver–graphene [74] and copper–graphene [72], exhibit interminable resilience as well as steady performance over an expansive range of temperatures. As a result, when the thickness of the layers is reduced to less than ten nanometers, many commonly employed plasmonic materials become impracticable due to the formation of unsteady films [67, 68].

Niobium (Nb) is a new plasmonic substance with excellent chemical resistance and mechanical resilience [75] and it has the potential to be used in many relevant practical applications in the future. The adherence of a niobium layer on silica glass is so substantial that no adhesive layer is needed. Also, unlike gold, niobium produces a persistent layer when the layer thickness is smaller than 10 nm [76, 77]. Besides, when a niobium layer is placed on a silica surface, multiple atomic protective coatings develop spontaneously in a natural way [78, 97].

## 9.4 Fiber Optic and PCF-Based Plasmonic Sensors

In comparison to the prism-coupled SPR sensors, fiber optic-reliant SPR sensors grant multiple benefits, including simplicity of design, compactness and the ability to perform distant sensing and measurements *in vivo* [83]. The prism-coupled structure of the traditional SPR technique is superseded by the core, which utilizes a total internal reflection (TIR) mechanism to steer light. During the manufacture of fiber optic-based SPR sensors, a small piece of the silica coating is cleared and replaced with a metal layer. Later on, a dielectric sensor layer is applied to the metal layer to provide more sensitivity. The light is projected from one end of the fiber by means of a light source, and it propagates through the fiber, thanks to the TIR mechanism. Surface plasmons are formed at the fiber core–metal contact as a result of the evanescent field created. Fiber optic-based plasmonic sensors are generally reliant upon the reflection or transmission characteristics of directed light [83], depending on the applications.

PCF-based SPR sensors have significant benefits over traditional optical fiber sensors. It is feasible to modify the guiding qualities of a PCF by changing the air hole diameter and number of rings in the PCF. Generally, PCFs are lightweight and can be produced at the micron size. PCF-based SPR sensors, unlike traditional fiber optics-based SPR sensors and prism-coupling, provide a possible platform for sensor downscaling. Furthermore, the distinctive and unheard-of properties of PCFs have unveiled a new avenue for the development of nano-sensors. The evanescent field may be modified using a variety of cladding structures, namely circular, octagonal, hexagonal, square, and so on, to acquire the best sensing performance [47–50].

## 9.5 Effects of Plasmonic Nanomaterials in PCF-Based SPR Sensors

Plasmonic nanomaterials have attracted considerable interest in photonics research as they enable unparalleled manipulation of light through designed oscillating charge distributions, referred to as plasmonic resonances. It is possible to control the reflection characteristics of optical coatings by using a mix of metallic and high RI films with thicknesses below 10 nanometers, which has interesting applications in green photonics and optoelectronics [87]. Nowadays, most plasmonic devices are made entirely of noble metals that display a low optical loss in the visible range. Gold, silver, copper, aluminum, niobium, etc. are some of the most popular and up-to-date plasmonic materials that are being extensively used in a plethora of diverse applications thanks to the notable properties they possess, as depicted in Figure 9.2.

### 9.5.1 Copper

In terms of conductivity, copper (Cu) is well-accepted as the second most conductive substance after silver, and it is considerably less expensive than both gold and silver. Notably, Cu

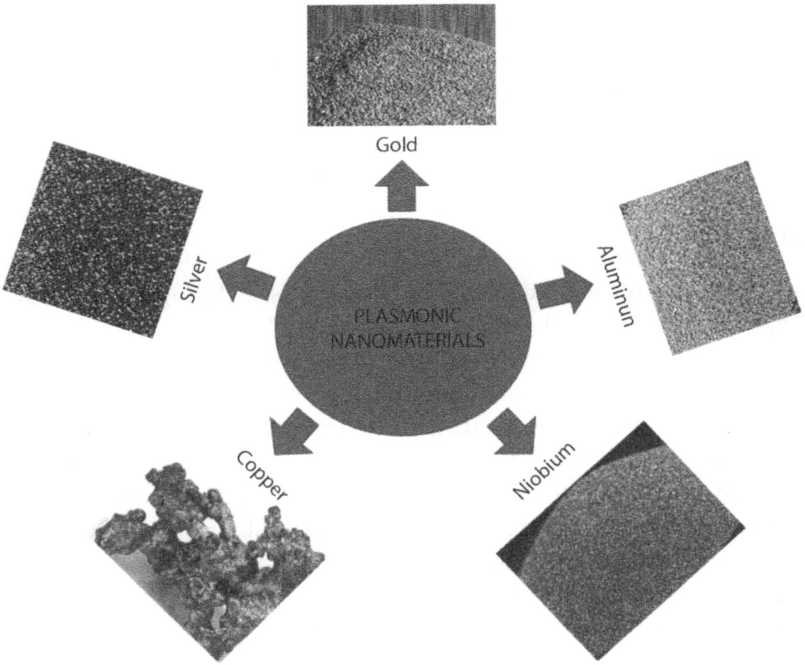


**Figure 9.2** Noble plasmonic materials.

has the same material damping rate and interband transition as Au [71]. For this reason, it has gotten less attention than Au and Ag because of its tendency to readily oxidize. With a graphene layer, however, the oxidation of copper might be avoided, since graphene is mechanically robust, chemically static, and has a hexagonal lattice arrangement that is impenetrable to gas molecules smaller than helium, and so it limits the infiltration of oxygen [73, 98]. Over the course of a year, copper-graphene layered plasmonic properties have shown to be more stable and long-lasting [72]. Recently, Rifat *et al.* came up with a Cu-filled PCF-based SPR sensor, where the thickness of Cu is 30 nm and the value of pitch is 2 μm, as illustrated in Figure 9.3(a). Their offered sensor exhibits a moderate wavelength sensitivity (WS) of 2000 nm/RIU over a RI range of 1.33 to 1.37 for analytes [84]. Copper possesses some significant drawbacks as well. To illustrate, it is susceptible to oxidation, which limits its across-the-board usage as a plasmonic material [84].

### 9.5.2 Silver

In comparison to the other plasmonic materials, silver (Ag) is proven to be the most conductive substance since it has a smaller loss and a stronger resonance peak. From an optical viewpoint, it is an ideal candidate for nano-scale plasmonics as it holds no interband transitions in the VS. Furthermore, silver has negligible optical damping and a plasma wavelength of 137 nm deep in the ultraviolet (UV), making it a suitable Drude metal [64]. Recently, Haque *et al.* designed a silver-based PCF sensor, where the thickness of Ag is 65 nm, which is much greater than the thickness of Cu, as depicted in Figure 9.3(b). In the sensing range of 1.29 to 1.39, their designed sensor demonstrates a relatively higher WS and amplitude sensitivity (AS) of 116000 nm/RIU and 2452 $RIU^{-1}$, sequentially [86]. Regardless, silver possesses the significant drawback of being corrosive. Additionally, it is not chemically unstable and may rapidly deteriorate. The oxidation issue might be solved using a silver film with a graphene layer [72, 85]. However, silver-graphene is not appropriate for durable plasmonic utilizations because of its quick oxidation and degradation features. It has recently been discovered that the characteristics of silver-graphene layered plasmonic substances decline with time.

### 9.5.3 Gold

Gold (Au) is one of the most used plasmonic materials. Although this noble substance has superior optical damping in the VS due to substantial interband transitions, it is often utilized as the backdrop plasmonic material in many applications as it is naturally inert, easy to manufacture, biocompatible, and long-lasting [84]. Recently, Anik *et al.* introduced a D-shaped PCF-based SPR sensor with gold as the backdrop plasmonic substance, as demonstrated in Figure 9.3(c). In the sensing range of 1.14 to 1.36, their designed PCF displays an elevated WS of 53,800 nm/RIU. However, their achieved AS (328 $RIU^{-1}$) is relatively lower than most other structures [96]. Notably, gold also holds numerous drawbacks. For starters, since seed-initiated development of gold coatings results in intermittent layers when the thickness is fewer than 10 nm, it is not feasible to reduce their thickness [67, 68]. It is, nevertheless, feasible to increase the thickness of gold coatings. Another concern is that gold has poor adhesion to glass substrates, necessitating the deposit of extra adhesion layers (such as titanium or chromium) between the metal and the dielectric, which may

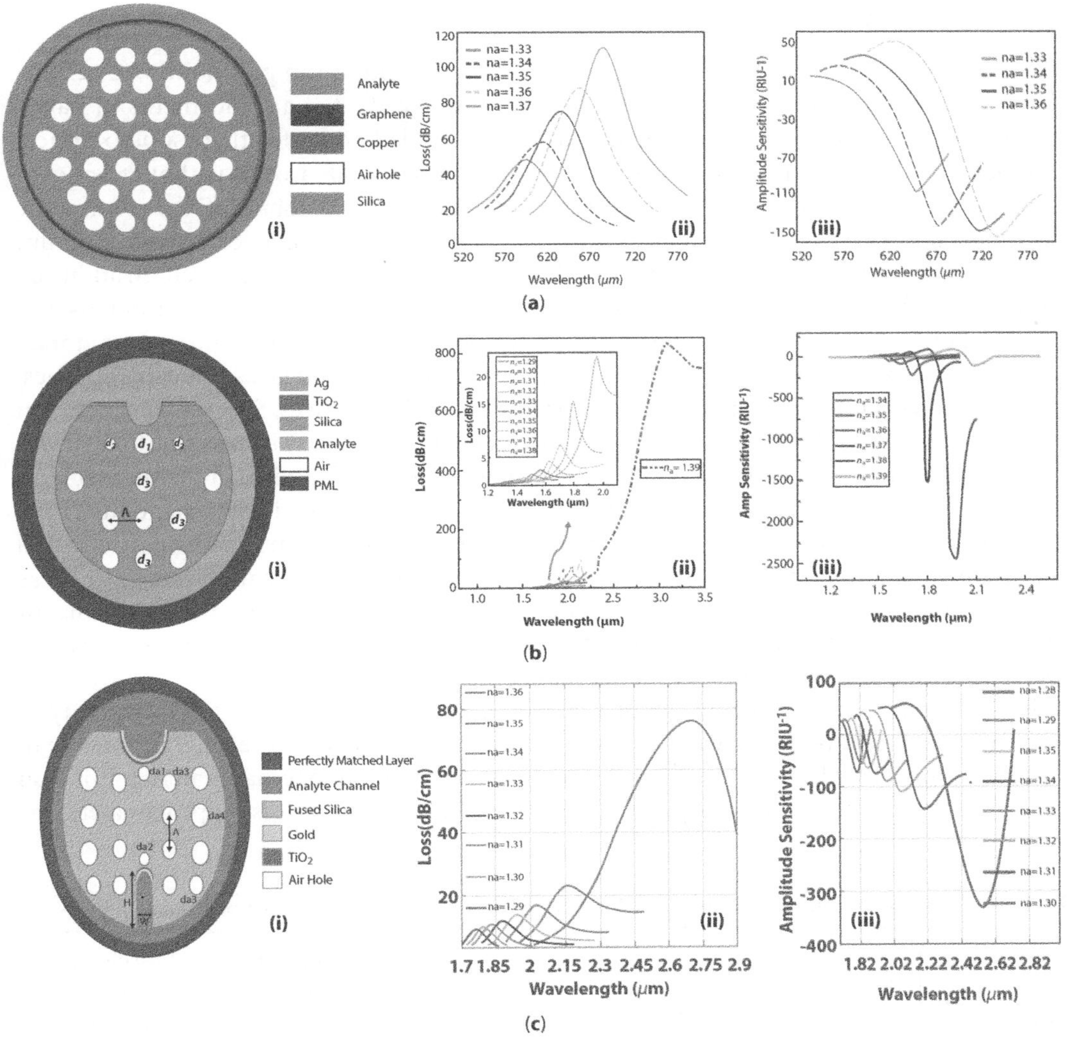

**Figure 9.3** (a) Copper-based plasmonic biosensor, a(i) Schematic illustration of their stated design, a(ii) and a(iii) Loss and amplitude sensitivity versus the wavelength in the sensing range of 1.33 to 1.37 [84]. (b) Silver-based dual core D-shaped PCF sensor, b(i) Geometry of their designed architecture, b(ii) and b(iii) Leakage loss and amplitude sensitivity concerning the wavelength in the range of 1.29 to 1.39 [86]. (c) Gold-based milled-microchannel D-timbered PCF sensor, c(i) Schematic illustration of their proposed structure, c(ii) and c(iii) Leakage loss and amplitude sensitivity versus wavelength in the sensing range of 1.14 to 1.36 [96].

decrease the optical capabilities of the device, notably if the nano-scale film thickness is less than 10 nm. Furthermore, moving the sample to a different deposition device may eventually lead to the emergence of an oxide layer on the adhesion layer (especially for Ti), which may significantly alter and even damage the optical traits of the substance [87]. Gold layer properties may be improved by chemically generating single-crystal gold coatings instead of applying gas-based deposition approaches [69, 70], resulting in enhanced film quality in the sub-ten nanometer range.

### 9.5.4 Niobium

Ultrathin metallic nanofilms have lately received a lot of interest in plasmonics, for example as building blocks of metasurfaces, thanks to unprecedented advancements in nanostructuring technology. Silver and gold, two noble metals known for their superb optical qualities, are often used in optical devices. However, these metals are not without their own set of drawbacks. There are several possible real-world uses for niobium (Nb) as a unique plasmonic substance due to its high mechanical and chemical resilience [75]. A niobium layer adheres so well to silica glass that no additional adhesive layer is required. In addition, unlike gold, when the film thickness is less than 10 nm, niobium generates a continuous film [76, 77]. Furthermore, when a niobium layer is formed on a silica surface, numerous atomic protective coatings form spontaneously [78]. Despite the fact that it lacks optical properties, this film protects the film from external environmental disturbances. SPR sensors based on indium tin oxide (ITO) have recently gained considerable interest because of their lower dimension plasma frequency [79]. Its optical dampening is also virtually comparable to gold and silver [80, 81]. Recently, Wieduwilt *et al.* reported that the niobium nanofilms have bulk optical characteristics, are continuous, homogeneous, and immune to environmental effects, and hence, have various advantages over existing noble metal nanolayers [87]. These findings show that ultrathin niobium nanofilms may be used to create novel platforms for superconducting photonics, biomedical diagnostics and ultrathin metasurfaces, as well as new sorts of optoelectronic devices.

It is evident from the below tabular representation that relatively low sensitivity is obtained when gold is employed as the sole plasmonic substance. However, it is possible to achieve significantly higher sensitivity by coating gold with an external material. It is also seen that it is feasible to design operational sensors with low thickness (approximately one

**Table 9.1** A comparison of various plasmonic materials and their sensing performance analysis.

| References | Plasmonic material | Additional material | Sensing range | Amplitude sensitivity $RIU^{-1}$ | Wavelength sensitivity nm/RIU |
|---|---|---|---|---|---|
| [84] | Copper | Graphene | 1.33–1.37 | 137 | 2000 |
| [86] | Silver | Titanium Dioxide | 1.29–1.39 | 2452 | 116000 |
| [88] | Gold | N/A | 1.33–1.38 | 371.5 | 4600 |
| [89] | Gold | N/A | 1.33–1.40 | 478 | 12000 |
| [90] | Gold | N/A | 1.33–1.43 | 1119 | 12800 |
| [91] | Gold | N/A | 1.34–1.42 | 427 | 18000 |
| [92] | Gold | Titanium Dioxide | 1.33–1.38 | 1411 | 25000 |
| [93] | Gold | Titanium Dioxide | 1.33–1.42 | 6829 | 28000 |
| [94] | Gold | Titanium Dioxide | 1.33–1.44 | 1405 | 121000 |
| [95] | Niobium | Aluminum Oxide | 1.36–1.41 | 1560 | 8000 |

third of gold) by using an ultrathin plasmonic substance named niobium. Moreover, it is apparent from the Table 9.1 that the maximum sensitivity can be achieved by using silver as the backdrop plasmonic substance.

## 9.6 Current Challenges and Future Directions

No plasmonic nanomaterial is flawless or suited for all purposes. Regardless, it is possible to pick materials with characteristics that are appropriate for particular applications sensibly and methodically. Active plasmonics research has widened the field of plasmonic nanomaterials, enabling the creation of ways for achieving advanced or unheard-of features. Hybridization or alloying of existing materials, in addition to the quest for novel materials, results in improved plasmonic characteristics, leading to new or synergistic effects not apparent in separate components. Even though a variety of other plasmonic materials beyond noble metals have been suggested and produced, their synthesis and structural flexibility remain difficult in most situations. A rigorous study on the expansive synthesis of cost-efficient non-noble metal-based plasmonic substances is vital for relevant real-world applications since majority of commercial-scale production requires a sustainable, high-throughput, and cost-effective method. By removing or reducing the existing barriers in the current challenges of controllability, synthesis, and chemical stability of non-noble metal-based nanostructures, it is possible to obtain complimentary and better functionality using non-noble metals compared to noble metals. SPR and other features of nanostructures are influenced by several factors, including size and shape, arrangement, chemical solidity, plasmonic coupling, and surface functionalization. As a result, developing techniques for acquiring various types of plasmonically-coupled nanostructures employing different non-noble plasmonic nanomaterials, reliably and controllably attaching ligands and operational groups to these plasmonic nanomaterials, and coupling the suitable practical applications to the pertinent plasmonic nanostructure with the appropriate property is critical for comprehending and acknowledging the unprecedented plasmonic characteristics of these plasmonic nanomaterials.

## 9.7 Conclusion

Plasmonic nanomaterials outperform most conventional nanomaterials in terms of overall sensing performance. Recently, these nanomaterials have gained widespread popularity among scientists as they hold influential pertinent applications in the fields of optical sensing and sensors, biological analyte detection, environmental control, chemical sensing, nanomedicine, biomedical engineering, and so on, owing to their unique optical characteristics. Consequently, this book chapter comes up with a compelling overview of plasmonic nanomaterials in sensors, as well as comprehensively addresses the pertinent contemporary literature and advancements of noble plasmonic nanomaterial-based sensors. Additionally, the performance characteristics of numerous sensors are thoroughly investigated, and it can be concluded that, while numerous simulated investigations have been conducted on various plasmonic nanomaterial-based sensors, additional experimental studies are a must to practically implement them in relevant applications, namely tumor and tissue imaging, as well as medical diagnostics, etc. Conclusively, the current challenges and their probable solutions for prospective advanced sensing applications are briefly discussed in this book chapter.

## Acknowledgment

This work was supported in part by funding from the Natural Sciences and Engineering Research Council of Canada (NSERC).

## References

1. Duan, H., Wang, T., Su, Z., Pang, H., Chen, C., Recent progress and challenges in plasmonic nanomaterials. *Nanotechnol. Rev.*, 11, 846, 2022.
2. Wei, Q. *et al.*, Porous one-dimensional nanomaterials: Design, fabrication and applications in electrochemical energy storage. *Adv. Mater.*, 29, 1602300, 2017.
3. Zhang, F., Wei, Y., Wu, X., Jiang, H., Wang, W., Li, H., Hollow zeolitic imidazolate framework nanospheres as highly efficient cooperative catalysts for [3 + 3] cycloaddition reactions. *J. Am. Chem. Soc.*, 136, 13963, 2014.
4. Sadeghi, S.M. and Gutha, R.R., Ultralong phase-correlated networks of plasmonic nanoantennas coherently driven by photonic modes. *Appl. Mater. Today*, 22, 100932, 2021.
5. Shan, Y., Li, Y., Pang, H., Applications of tin sulfide-based materials in lithium-ion batteries and sodium-ion batteries. *Adv. Funct. Mater.*, 30, 2001298, 2020.
6. Peng, L., Fang, Z., Zhu, Y., Yan, C., Yu, G., Holey 2D nanomaterials for electrochemical energy storage. *Adv. Energy Mater.*, 8, 1702179, 2018.
7. Nguyen, T.T., Mammeri, F., Ammar, S., Iron oxide and gold based magneto-plasmonic nanostructures for medical applications: A review. *Nanomaterials*, 8, 149, 2018.
8. Eatemadi, A. *et al.*, Carbon nanotubes: Properties, synthesis, purification, and medical applications. *Nanoscale Res. Lett.*, 9, 1, 2014.
9. Zhou, J., Yang, T., Chen, J., Wang, C., Zhang, H., Shao, Y., Two dimensional nanomaterial-based plasmonic sensing applications: Advances and challenges. *Coord. Chem. Rev.*, 410, 213218, 2020.
10. Zhou, C., Zou, H., Sun, C., Ren, D., Chen, J., Li, Y., Signal amplification strategies for DNA-based surface plasmon resonance biosensors. *Biosens. Bioelectron.*, 117, 678, 2018.
11. Do, T.N. and Visell, Y., Stretchable, twisted conductive microtubules for wearable computing, robotics, electronics, and healthcare. *Sci. Rep.*, 7, 1–12, 2017.
12. Ahuja, D. and Parande, D., Optical sensors and their applications. *J. Sci. Res. Rev.*, 1, 60, 2012.
13. Yeo, J.C. and Lim, C.T., Emerging flexible and wearable physical sensing platforms for healthcare and biomedical applications. *Microsyst. Nanoeng.*, 2, 1–19, 2016.
14. Morin, S.A., Shepherd, R.F., Kwok, S.W., Stokes, A.A., Nemiroski, A., Whitesides, G.M., Camouflage and display for soft machines. *Science*, 337, 828, 2012.
15. Gong, S., Schwalb, W., Wang, Y. *et al.*, A wearable and highly sensitive pressure sensor with ultrathin gold nanowires. *Nat. Commun.*, 5, 3132, 2014.
16. Xu, S., Zhang, Y., Jia, L., Mathewson, K.E., Jang, K.-I., Kim, J. *et al.*, Soft microfluidic assemblies of sensors, circuits, and radios for the skin. *Science*, 344, 70, 2014.
17. Araci, I.E., Su, B., Quake, S.R., Mandel, Y., An implantable microfluidic device for self-monitoring of intraocular pressure. *Nat. Med.*, 20, 1074, 2014.

18. Ding, W., Jiang, Y., Gao, R., Liu, Y., High-temperature fiber-optic Fabry-Perot interferometric sensors. *Rev. Sci. Instrum.*, 86, 055001, 2015.
19. Liu, Y., Wang, D., Chen, W., Crescent shaped Fabry-Perot fiber cavity for ultra-sensitive strain measurement. *Sci. Rep.*, 6, 1–9, 2016.
20. Fang, Y., Label-free cell-based assays with optical biosensors in drug discovery. *Assay Drug Dev. Technol.*, 4, 583, 2006.
21. Berger, C.E. and Greve, J., Differential SPR immunosensing. *Sens. Actuators B Chem.*, 63, 103, 2000.
22. Stemmler, I., Brecht, A., Gauglitz, G., Compact surface plasmon resonance-transducers with spectral readout for biosensing applications. *Sens. Actuators B Chem.*, 54, 98, 1999.
23. Mouvet, C., Harris, R., Maciag, C., Luff, B., Wilkinson, J., Piehler, J. *et al.*, Determination of simazine in water samples by waveguide surface plasmon resonance. *Anal. Chim. Acta*, 338, 109, 1997.
24. Homola, J., Dostálek, J., Chen, S., Rasooly, A., Jiang, S., Yee, S.S., Spectral surface plasmon resonance biosensor for detection of staphylococcal enterotoxin B in milk. *Int. J. Food Microbiol.*, 75, 61, 2002.
25. Homola, J., Surface plasmon resonance sensors for detection of chemical and biological species. *Chem. Rev.*, 108, 462, 2008.
26. Jorgenson, R. and Yee, S., A fiber-optic chemical sensor based on surface plasmon resonance. *Sens. Actuators B Chem.*, 12, 213, 1993.
27. Gupta, B.D. and Verma, R.K., Surface plasmon resonance-based fiber optic sensors: Principle, probe designs, and some applications. *J. Sens.*, 2009, 2009.
28. Lee, B., Roh, S., Park, J., Current status of micro-and nano-structured optical fiber sensors. *Opt. Fiber Technol.*, 15, 209, 2009.
29. Nooke, A., Beck, U., Hertwig, A., Krause, A., Krüger, H., Lohse, V. *et al.*, On the application of gold based SPR sensors for the detection of hazardous gases. *Sens. Actuators B Chem.*, 149, 194, 2010.
30. Cahill, C.P., Johnston, K.S., Yee, S.S., A surface plasmon resonance sensor probe based on retro-reflection. *Sens. Actuators B Chem.*, 45, 161, 1997.
31. Ashwell, G. and Roberts, M., Highly selective surface plasmon resonance sensor for NO2. *Electron. Lett.*, 32, 2089, 1996.
32. Su, Y.-D., Chen, S.-J., Yeh, T.-L., Common-path phase-shift interferometry surface plasmon resonance imaging system. *Opt. Lett.*, 30, 1488, 2005.
33. Kajenski, P.J., Tunable optical filter using long-range surface plasmons. *Opt. Eng.*, 36, 1537, 1997.
34. Akimoto, T., Sasaki, S., Ikebukuro, K., Karube, I., Refractive-index and thickness sensitivity in surface plasmon resonance spectroscopy. *Appl. Opt.*, 38, 4058, 1999.
35. Schildkraut, J.S., Long-range surface plasmon electrooptic modulator. *Appl. Opt.*, 27, 4587, 1988.
36. Maier, S.A., Plasmonics: The promise of highly integrated optical devices. *IEEE J. Sel. Top. Quantum Electron.*, 12, 1671, 2006.
37. Sommerfeld, A., Über die Ausbreitung der Wellen in der drahtlosen Telegraphie. *Ann. Phys.*, 333, 665, 1909.
38. Ritchie, R., Plasma losses by fast electrons in thin films. *Phys. Rev.*, 106, 874, 1957.
39. Zenneck, J., Über die Fortpflanzung ebener elektromagnetischer Wellen längs einer ebenen Leiterfläche und ihre Beziehung zur drahtlosen Telegraphie. *Ann. Phys.*, 328, 846, 1907.
40. Kretschmann, E. and Raether, H., Radiative decay of non radiative surface plasmons excited by light. *Z. Naturforsch. A*, 23, 2135, 1968.
41. Otto, A., Excitation of nonradiative surface plasma waves in silver by the method of frustrated total reflection. *Z. Phys.*, 216, 398, 1968.

42. Piliarik, M., Homola, J., Manıková, Z., Čtyroký, J., Surface plasmon resonance sensor based on a single-mode polarization-maintaining optical fiber. *Sens. Actuators B Chem.*, 90, 236, 2003.
43. Monzón-Hernández, D. and Villatoro, J., High-resolution refractive index sensing by means of a multiple-peak surface plasmon resonance optical fiber sensor. *Sens. Actuators B Chem.*, 115, 227, 2006.
44. Monzón-Hernández, D., Villatoro, J., Talavera, D., Luna-Moreno, D., Optical-fiber surface-plasmon resonance sensor with multiple resonance peaks. *Appl. Opt.*, 43, 1216, 2004.
45. Gupta, B. and Sharma, A.K., Sensitivity evaluation of a multi-layered surface plasmon resonance-based fiber optic sensor: A theoretical study. *Sens. Actuators B Chem.*, 107, 40, 2005.
46. Skorobogatiy, M. and Kabashin, A.V., Photon crystal waveguide-based surface plasmon resonance biosensor. *Appl. Phys. Lett.*, 89, 143518, 2006.
47. Amin, R., Abdulrazak, L.F., Mohammadd, N., Ahmed, K., Bui, F.M., Ibrahim, S.M., GaAs-filled elliptical core-based hexagonal PCF with excellent optical properties for nonlinear optical applications. *Ceram. Int.*, 48, 5617, 2022.
48. Mohammadd, N. *et al.*, GaP-filled PCF with ultra-high birefringence and nonlinearity for distinctive optical applications. *J. Ovonic Res.*, 18, 129, 2022.
49. Amin, R. *et al.*, Tellurite glass based optical fiber for the investigation of supercontinuum generation and nonlinear properties. *Phys. Scr.*, 97, 030007, 2022.
50. Amin, R., Khan, M.E., Abdulrazak, L.F., Al-Zahrani, F.A., Ahmed, K., Design of novel models for optical communication with ultra-high non-linearity, birefringence and low loss profile. *Phys. Scr.*, 96, 125107, 2021.
51. Lim, W.Q. and Gao, Z., Plasmonic nanoparticles in biomedicine. *Nano Today*, 11, 168, 2016.
52. Park, J.E., Kim, M., Hwang, J.H., Nam, J.M., Golden opportunities: Plasmonic gold nanostructures for biomedical applications based on the second near-infrared window. *Small Methods*, 1, 1600032, 2017.
53. Lee, S., Sun, Y., Cao, Y., Kang, S.H., Plasmonic nanostructure-based bioimaging and detection techniques at the single-cell level. *TrAC Trends Anal. Chem.*, 117, 58, 2019.
54. Kaushal, S., Nanda, S.S., Samal, S., Yi, D.K., Strategies for the development of metallic-nanoparticle-based label-free biosensors and their biomedical applications. *ChemBioChem*, 21, 576, 2020.
55. Nguyen, H.H., Park, J., Kang, S., Kim, M., Surface plasmon resonance: A versatile technique for biosensor applications. *Sensors*, 15, 10481, 2015.
56. Jatschka, J., Dathe, A., Csáki, A., Fritzsche, W., Stranik, O., Propagating and localized surface plasmon resonance sensing—A critical comparison based on measurements and theory. *Sens. Bio-Sens. Res.*, 7, 62, 2016.
57. Petryayeva, E. and Krull, U.J., Localized surface plasmon resonance: Nanostructures, bioassays and biosensing—A review. *Anal. Chim. Acta*, 706, 8, 2011.
58. Huang, X. and El-Sayed, M.A., Gold nanoparticles: Optical properties and implementations in cancer diagnosis and photothermal therapy. *J. Adv. Res.*, 1, 13, 2010.
59. Singh, P., LSPR biosensing: Recent advances and approaches, in: *Rev. Plasmonics 2016*, p. 211, 2017.
60. Boken, J., Khurana, P., Thatai, S., Kumar, D., Prasad, S., Plasmonic nanoparticles and their analytical applications: A review. *Appl. Spectrosc. Rev.*, 52, 774, 2017.
61. Liu, J., He, H., Xiao, D., Yin, S., Ji, W., Jiang, S. *et al.*, Recent advances of plasmonic nanoparticles and their applications. *Materials*, 11, 1833, 2018.
62. Borghei, Y.S., Hosseini, M., Dadmehr, M., Hosseinkhani, S., Ganjali, M.R., Sheikhnejad, R., Visual detection of cancer cells by colorimetric aptasensor based on aggregation of gold nanoparticles induced by DNA hybridization. *Anal. Chim. Acta*, 904, 92, 2016.

63. Borghei, Y.S. and Hosseinkhani, S., Colorimetric assay of apoptosis through *in-situ* biosynthesized gold nanoparticles inside living breast cancer cells. *Talanta*, 208, 120463, 2020.
64. Johnson, P.B. and Christy, R.W., Optical constants of noble metals. *Phys. Rev. B*, 6, 4370, 1972.
65. Ordal, M.A., Bell, R.J., Alexander, R.W., Long, L.L., Querry, M.R., Optical-properties of 14 metals in the infrared and far infrared - Al, Co, Cu, Au, Fe, Pb, Mo, Ni, Pd, Pt, Ag, Ti, V, and W. *Appl. Opt.*, 24, 4493, 1985.
66. Etchegoin, P.G., Le Ru, E.C., Meyer, M., An analytic model for the optical properties of gold. *J. Chem. Phys.*, 125, 164705, 2006.
67. Doron-Mor, I., Barkay, Z., Filip-Granit, N., Vaskevich, A., Rubinstein, I., Ultrathin gold island films on silanized glass. Morphology and optical properties. *Chem. Mater.*, 16, 3476, 2004.
68. Szunerits, S., Praig, V.G., Manesse, M., Boukherroub, R., Gold island films on indium tin oxide for localized surface plasmon sensing. *Nanotechnology*, 19, 195712, 2008.
69. Kossoy, A. *et al.*, Optical and structural properties of ultra-thin gold films. *Adv. Opt. Mater.*, 3, 71, 2015.
70. Stec, H.M., Williams, R.J., Jones, T.S., Hatton, R.A., Ultrathin transparent Au electrodes for organic photovoltaics fabricated using a mixed mono-molecular nucleation layer. *Adv. Funct. Mater.*, 21, 1709, 2011.
71. West, P.R., Ishii, S., Naik, G.V., Emani, N.K., Shalaev, V.M., Boltasseva, A., Searching for better plasmonic materials. *Laser Photonics Rev.*, 4, 795, 2010.
72. Kravets, V., Jalil, R., Kim, Y.-J., Ansell, D., Aznakayeva, D., Thackray, B. *et al.*, Graphene-protected copper and silver plasmonics. *Sci. Rep.*, 4, 72, 2014.
73. Chowdhury, M.M.A., Mohammadd, N., Hasan, M.K., Imam, S.-A., Thermo-optical effect dependent absorption enhancement in a graphene embedded LiNbO3 based cavity. *5th Int. Conf. Electr. Inf. Commun. Technol. (EICT)*, pp. 1–5, 2021.
74. Rifat, A.A., Mahdiraji, G.A., Chow, D.M., Shee, Y.G., Ahmed, R., Adikan, F.R.M., Photonic crystal fiber-based surface plasmon resonance sensor with selective analyte channels and graphene-silver deposited core. *Sensors*, 15, 11499, 2015.
75. Granata, C., Vettoliere, A., Russo, M., Ruggiero, B., Noise theory of dc nano-SQUIDs based on Dayem nanobridges. *Phys. Rev. B*, 84, 224516, 2011.
76. Troeman, A., van der Ploeg, S., Il'Ichev, E., Meyer, H.-G., Golubov, A.A., Kupriyanov, M.Y. *et al.*, Temperature dependence measurements of the supercurrent-phase relationship in niobium nanobridges. *Phys. Rev. B*, 77, 024509, 2008.
77. Schmelz, M., Matsui, Y., Stolz, R., Zakosarenko, V., Schönau, T., Anders, S. *et al.*, Investigation of all niobium nano-SQUIDs based on sub-micrometer cross-type Josephson junctions. *Supercond. Sci. Technol.*, 28, 015004, 2014.
78. Chowdhury, M.M.A., Priyam, A.G., Hasan, M.K., Mohammadd, N., Imam, S.-A., Design of MgO-doped-LiNbO3 based temperature tunable dual-narrowband absorber. *IEEE Int. Conf. Telecommun. Photonics (ICTP)*, pp. 1–5, 2021.
79. Dash, J.N. and Jha, R., SPR biosensor based on polymer PCF coated with conducting metal oxide. *IEEE Photonics Technol. Lett.*, 26, 595, 2014.
80. Franzen, S., Surface plasmon polaritons and screened plasma absorption in indium tin oxide compared to silver and gold. *J. Phys. Chem. C*, 112, 6027, 2008.
81. Rhodes, C., Cerruti, M., Efremenko, A., Losego, M., Aspnes, D., Maria, J.-P. *et al.*, Dependence of plasmon polaritons on the thickness of indium tin oxide thin films. *J. Appl. Phys.*, 103, 093108, 2008.
82. Huq, M.M., Hsieh, C.-T., Lin, Z.-W., Yuan, C.-Y., One-step electrophoretic fabrication of a graphene and carbon nanotube-based scaffold for manganese-based pseudocapacitors. *RSC Adv.*, 6, 87961, 2016.

83. Zhao, Y., Deng, Z.-Q., Li, J., Photonic crystal fiber based surface plasmon resonance chemical sensors. *Sens. Actuators B Chem.*, 202, 557, 2014.
84. Rifat, A.A. *et al.*, Copper-graphene-based photonic crystal fiber plasmonic biosensor. *IEEE Photonics J.*, 8, 1, 2016.
85. Dash, J. and Jha, R., Graphene based birefringent photonic crystal fiber sensor using surface plasmon resonance. *IEEE Photon. Technol. Lett.*, 26, 1092, 2014.
86. Haque, E., Mahmuda, S., Hossain, M.A., Hai, N.H., Namihira, Y., Ahmed, F., Highly sensitive dual-core PCF based plasmonic refractive index sensor for low refractive index detection. *IEEE Photonics J.*, 11, 1, 2019.
87. Wieduwilt, T., Tuniz, A., Linzen, S. *et al.*, Ultrathin niobium nanofilms on fiber optical tapers – a new route towards low-loss hybrid plasmonic modes. *Sci. Rep.*, 5, 17060, 2015.
88. Hasan, M.R. *et al.*, Spiral photonic crystal fiber-based dual-polarized surface plasmon resonance biosensor. *IEEE Sens. J.*, 18, 133, 2018.
89. Haider, F., Aoni, R.A., Ahmed, R. *et al.*, Alphabetic-core assisted microstructure fiber based plasmonic biosensor. *Plasmonics*, 15, 1949, 2020.
90. Islam, A., Haider, F., Aoni, R.A., Hossen, M., Begum, F., Ahmed, R., U-grooved dual-channel plasmonic sensor for simultaneous multi-analyte detection. *J. Opt. Soc. Am. B*, 38, 3055, 2021.
91. Haider, F., Aoni, R.A., Ahmed, R., Mahdiraji, G.A., Azman, M.F., Adikan, F.R.M., Mode-multiplex plasmonic sensor for multi-analyte detection. *Opt. Lett.*, 45, 3945, 2020.
92. Islam, M.S. *et al.*, A hi-bi ultra-sensitive surface plasmon resonance fiber sensor. *IEEE Access*, 7, 79085, 2019.
93. Mahfuz, M.A., Hossain, M.A., Haque, E., Hai, N.H., Namihira, Y., Ahmed, F., Dual-core photonic crystal fiber-based plasmonic RI sensor in the visible to near-IR operating band. *IEEE Sens. J.*, 20, 7692, 2020.
94. Islam, M.S. *et al.*, Dual-polarized highly sensitive plasmonic sensor in the visible to near-IR spectrum. *Opt. Express*, 26, 30347, 2018.
95. Hasan, M.R., Akter, S., Ahmed, K., Abbott, D., Plasmonic refractive index sensor employing niobium nanofilm on photonic crystal fiber. *IEEE Photon. Technol. Lett.*, 30, 315, 2018.
96. Anik, M.H.K. *et al.*, Milled microchannel-assisted open D-channel photonic crystal fiber plasmonic biosensor. *IEEE Access*, 9, 2924, 2021.
97. Sokhey, K., Rai, S., Lodha, G., Oxidation studies of niobium thin films at room temperature by X-ray reflectivity. *Appl. Surf. Sci.*, 257, 222, 2010.
98. Schriver, M., Regan, W., Gannett, W.J., Zaniewski, A.M., Crommie, M.F., Zettl, A., Graphene as a long-term metal oxidation barrier: Worse than nothing. *ACS Nano*, 7, 5763, 2013.

# 10

# Magnetic Biosensors

**Sumaiya Akhtar Mitu[1], Kawsar Ahmed[1,2*] and Francis M. Bui[2]**

[1]*Group of Bio-Photomatiχ, Department of Information and Communication Technology, Mawlana Bhashani Science and Technology University, Santosh, Tangail, Bangladesh*
[2]*Department of Electrical and Computer Engineering, University of Saskatchewan, Saskatoon, Canada*

***Abstract***
In the diversifying technologies, magnetic nanoparticles have the features of solid magnetic properties, as well as liquid characteristics. These outstanding combined properties are the concern in the sensing era. In the field of biosensing, various magnetic nanoparticles are used to sense biological objects, such as molecular detection. Recently, new diagnostic platforms have been developed for the efficient measurement of cells and biomolecules. Among most of the biological samples are magnetic susceptible. Using this property, different magnetic sensors are designed for disease detection or getting valuable information. In this chapter, the variations of different magnetic sensors and their corresponding sensitivity responses and effects will be discussed. Magnetic sensors with different magnetic fluids, such as $Fe_3O_4$, $MnFe_2O_4$, have been used to analyze the effects on different particles. Different structures of sensors are simulated to monitor biological interactions with the distinction of magnetic strength variation (Oe). In the photonic crystal fiber (PCF) technology, the magnetic sensors have unique properties, such that they can be oppressed for the rapid detection.

***Keywords:*** Magnetic sensor, photonic crystal fiber, biosensor, magnetic field strength

## 10.1 Introduction

One of the major challenges in the biological field is to detect any disease rapidly. Proper treatment mostly depends on effectively identifying the disease. Magnetic nanoparticles are a more versatile and powerful diagnostic instrument in biology. Based on the situation of antibodies, different structures, molecules, and microorganisms are used to design the magnetic biosensor [1]. The magnetic nanoparticles, such as Madison, Quantum Magnetics [2], CT, are the combination of two or three magnetite crystals whose thickness are 10 to 15 nm [3]. The applications of the biosensors and magnetic nanoparticles in different sectors are represented in Figure 10.1(a) and (b). Magnetic nanoparticles can be used as a biosensor, which will be described here. Besides, the applications of biosensors are mentioned in

**Corresponding author*: kawsar.ict@mbstu.ac.bd; k.ahmed@usask.ca; k.ahmed.bd@ieee.org
ORCID ID: 0000-0002-4034-9819

Inamuddin and Tariq Altalhi (eds.) Biosensors Nanotechnology, 2nd Edition, (201–214) © 2023 Scrivener Publishing LLC

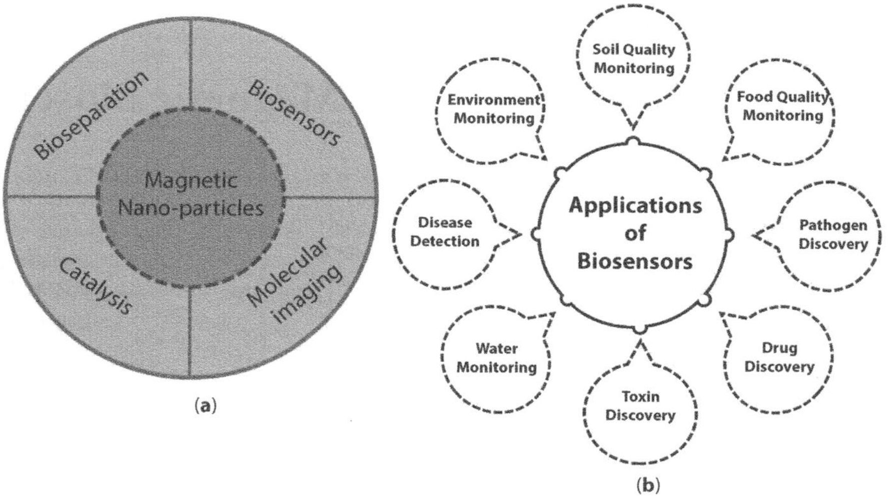


**Figure 10.1** Applications fields of magnetic nanoparticles.

Figure 10.1(b). The biosensors are mostly used in any kind of disease detection and environmental objects detection [4].

Magnetic nanofluids (MNF) are the stable colloidal suspensions where the paramagnetic nanoparticles are mixed in a carrier liquid (e.g., an organic solvent or oil or water) [5]. Magnetic nanofluids are also known as magnetic liquids or ferrofluids that contain the features of both magnetic materials and the liquids [6]. Moreover, these interesting features are used to increase its application in the field of heat transfer, optoelectronics, aerospace, solar sciences, and so on [7]. As a conventional sensor [8, 9], it may be affected by the electromagnetic field strength, optical devices that have appealed interest owing to outstanding advantages over conventional electromagnetic sensors those are small in size, high sensing response, and easily adaptable to the electromagnetic interference [10, 11].

Among the vast applications of biosensors, such as clinical diagnosis, environmental and water monitoring, food industry, pathogen detection has great impact in the food industry [12]. The topmost three bacteriological pathogens in United States are *Clostridium perfringens*, *Campylobacter jejuni*, and nontyphoidal *salmonellosis* those are the reasons for the foodborne illnesses [13]. The infective dose for the nontyphoidal *salmonellosis* is very low. The large number ($> 10^6$) of vegetative cells or $>10^6$ spores/g of food are the cause of the illness for *Clostridium perfringens*. The infective dosage for *Campylobacter jejuni* is typically greater than 10,000 cells; however, this varies depending on the type of contaminated food and a person's health status [14].

## 10.2 History

Bead array counter (BARC), named a magnetic biosensor, is revoluted by the Naval Research Laboratory (NRL) group. This ground-breaking sensor consisted of 64 identical GMR sensors and was designed to enable for the simultaneous detection of individual molecule interactions utilizing arrays of magneto resistive sensors [15]. The BARC sensor is a DNA

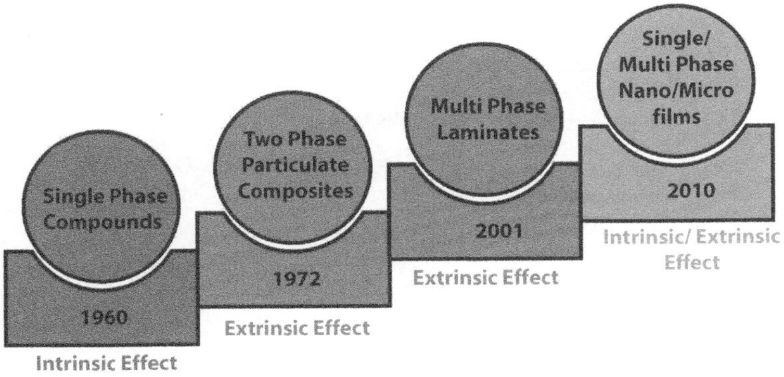


**Figure 10.2** History of development of the magnetic sensor.

hybridization sensor. Figure 10.2 depicts the historical advancements of magnetic sensors in detail.

i. In 2002, the performance of magneto biosensors was developed using the on chip current line structure for controlling the bimolecular movements and magnetic beads to detect the magnetic beads. This revolution was done by the Portugal's researchers of the INESC-MN, Lisbon [16]. The proposed line structure had the diameter of 400 nm and 2 μm.
ii. Later, the diameter of the spiral-shaped sensor was reduced as 70 μm for the magneto resistive sensor by the University of Bielefeld [17]. The magnetic beads from Laboratories had a signal of 0.86 m Bangs, which was proportional to their coverage. Besides, the detection limit was mentioned at 5% of surface coverage and 200 beads.
iii. After that, for detecting the single Dynabeads M-280 magnetic bead [18], the group of Stanford University established a new structure. The demonstrated structure used a spin valve material that had 10.3% MR ratio and 18.41 Ω sheet resistances. The sensor was designed using the photoresist layer of 1 μm.

## 10.3 Structural Design

Different analyses of structural design for magnetic biosensor have been exhibited from Figure 10.3 to Figure 10.6. The cross sectional views for the simulated structures have been designed using COMSOL Multiphysics 5.5(a). In the designed structures, silica is used as a ground material and perfectly matched layer (PML). The diameters of the simulated structures have been kept from 30 μm to 40 μm. The thinness of the PML layer has been kept 10% of the total diameter of the designed PCF based sensor to effectively calculate the loss profile [19]. The sol-gel process is used to fabricate the PCF sensor with a liquid filled core. The demonstrated structure was created using currently available nanotechnology, which can be used to create any type of complicated structure. Various magnetic fluids are used for the investigations of magnetic sensors. The simulated different magnetic biosensor

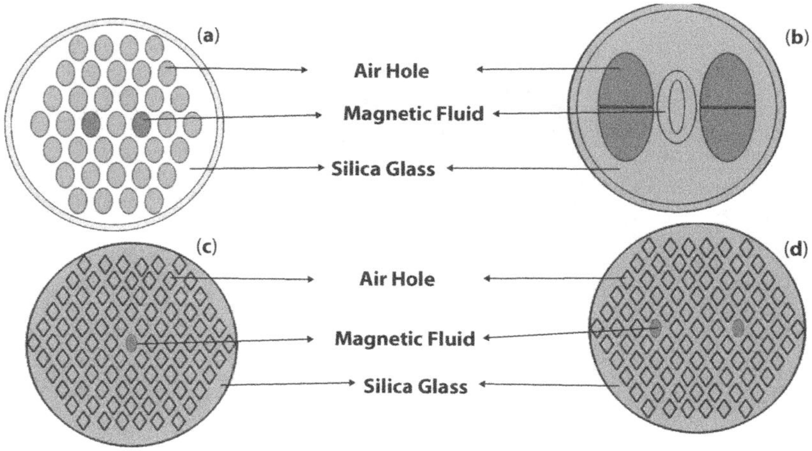


**Figure 10.3** Simulated structures for a magnetic biosensor, including ((a) double-core hexagonal elliptical (b) butterfly (c) single-core hexagonal diamond (d) double-core hexagonal diamond-shaped air holes) [25, 26, 28, 29]

structures are highlighted in Figure 10.3. It is noted that Figure 10.3 (a) and (d) are the dual core structure. And Figure 10.3 (b) and (c) are single core structures. In the single core structure light can be passed through the single core region as a result only core more can be noticed. Moreover, through the dual core region the light can be passed through two core regions as a result different polarization modes can be obtained from the dual core structure. Besides, the magnetic fluid can be changed to observe the sensing performances. At first $Fe_3O_4$ is used to observe the effect in the single core structure. After that $MnFe_2O_4$ is used to detect further changes. On the other hand, in the dual core structure, two cores can be filled with the same magnetic fluid or different fluids can be used in two cores. For individual structures different diameters are used for the air hole diameter and the core regions.

## 10.4 Numerical Analysis

The refractive index (RI) of the magnetic fluid can be measured at the room temperature (20°C) using the empirical equation of Langevin–function (1) [20].

$$n_{MF} = [n_m - n_o]\left[cothcoth\left(\alpha \frac{H - H_{c,n}}{T}\right) - \frac{T}{\alpha(H - H_{c,n})}\right] + n_o \tag{10.1}$$

Here, $n_m$ is used to represent the maximum RI and $n_o$ refers to the original RI of the magnetic field strength where $n_m$ is 1.4475 and $n_o$ is 1.4244. α is noted as the fitting parameter. H denotes the magnetic fluid strength that is greater than the critical magnetic fluid strength ($H_{c,n}$). The RI of the ground or base material, silica glass, can be measured using Eq. (10.2) [21].

$$n_{silica}(\lambda) = \sqrt{1 + \frac{A_1\lambda^2}{\lambda^2 - B_1} + \frac{A_2\lambda^2}{\lambda^2 - B_2} + \frac{A_3\lambda^2}{\lambda^2 - B_3}} \tag{10.2}$$

Here, the constant values of $A_1$, $A_2$, $A_3$, $B_1$, $B_2$, and $B_3$ are 0.6961663, 0.4079426, 0.8974794, 0.0684043, 0.1162414, and 98.96161 respectively.

The differences of the refractive indices of two distinct polarizations are known as birefringence (B). It can be defined as Eq. (10.3) [22]:

$$B = |n_x - n_y| \tag{10.3}$$

where $n_x$ and $n_y$ signify the effective RI for both polarizations. By keeping the birefringence greater than $10^{-4}$, the polarized modes can be easily distinguished. The low value of birefringence is a major contributor to the lengthening of beats and is also beneficial to excellent sensing performance.

The coupling length (CL) for the appropriate sensor can be calculated using the birefringence value. The CL for the simulated sensor structure can be defined as Eq. (10.4) [23].

$$CL = \frac{\lambda}{2 \times B} \tag{10.4}$$

Similarly, the CL of the planned model 2 may be determined. The RI difference is inversely proportional to the sensor's CL.

The beat length (BL) or CL, which is significantly related to the CL, can illuminate the sensor's power value. The Eq. (10.5) can be used to compute the power spectrum (PS) [23].

$$P(out) = sin^2\left(\frac{B \times \neq \times CL}{\lambda}\right) \tag{10.5}$$

The PS has an impact on the transmitted power spectrum (TPS). The Eq. (10.6) can be used to calculate the TPS [23].

$$T = 10 \times log\left(\frac{P(out)}{0.985945}\right) \tag{10.6}$$

where $P(out)$ indicates the output power spectrum and for applied the input power value of 0.985945.

The amplitude sensitivity (AS) of the simulated sensor can be obtained using Eq. (10.7) [24].

$$S(RIU^{-1}) = \frac{\Delta\lambda_{peak}}{\Delta n_{eff}} \tag{10.7}$$

where $\Delta\lambda_{peak}$ refers to the change of peak wavelength and $\Delta n_{eff}$ indicates the difference of RI. The TPS can be used to obtain these values. Eq. 10.8 is used to calculate the sensitivity of magnetic field sensors [24].

**Table 10.1** Different numerical parameters for the investigations of the sensor profile.

| SI | Parameter | Explanation | Unit |
|---|---|---|---|
| 1 | $n_{MF}$ | RI of magnetic fluid | RIU |
| 2 | $n_{silica}$ | RI of silica glass | RIU |
| 3 | $\lambda$ | Operating wavelength | μm |
| 4 | $B$ | Birefringence | RIU |
| 5 | $CL$ | Confinement loss | dB/m |
| 6 | $P(out)$ | Output power spectrum | dB |
| 7 | $T$ | Transmitted power spectrum | dB |
| 8 | $S$ | Amplitude sensitivity (AS) with respect to the operating wavelength | $RUI^{-1}$ |
| 9 | $S$ | AS in relation to the strength of the applied magnetic field | $Oe^{-1}$ |
| 10 | $H_{Oe}$ | Applied magnetic field strength | Oe |

$$S(Oe^{-1}) = \frac{\Delta\lambda_{peak}}{\Delta H_{Oe}} \tag{10.8}$$

Where, $\Delta\lambda_{peak}$ indicates the change of wavelength and $\Delta H_{Oe}$ denotes the variation of the magnetic strength.

In Table 10.1, the numerical parameters employed in this sensor design are listed together with an explanation and the corresponding unit value.

## 10.5 Outcome Analysis

### 10.5.1 Magnetic Fluid Sensor

The birefringence value for the different structures will be described here. The birefringence values using Eq. (10.3) for magnetic field sensors are given in Figure 10.4. In this case, the values are growing as the operating wavelength for the dual core structure increases.

CL for the magnetic sensor using Eq. (10.4) is drawn in Figure 10.5. It is noted that the values of CL is inversely proportional to the birefringence value. As a result, the CL values are reducing with the change of operating wavelength.

PS values can be measured using Eq. (10.5). The values of PS are exhibited in Figure 10.6. The power values are saturated from 0 to 1. These power values are needed to measure the transmission spectrum as well as sensing performance.

From the power spectrum values, transmission values can be easily measured. The sensing performances can be observed through the transmission values. These values can be attained using Eq. (10.6) and represented in Figure 10.7.

The sensitivity response is the most important part of a sensor. This sensitivity response depends on the transmission spectrum. From the peak values, the sensing performances can be achieved. The sensitivity response has a significant impact on the sensor's performance and efficacy. Figure 10.8 shows the sensitivity response of the modeled structure for both polarizations. The efficiency of a sensor can also be determined by the confinement loss spectrum, and these results are shown in Figure 10.9.

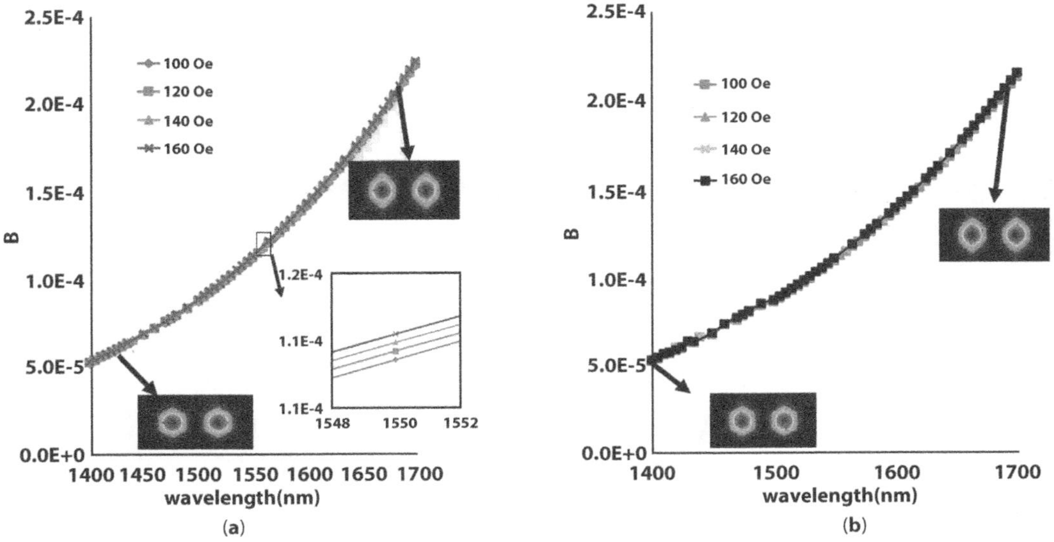


**Figure 10.4** Refractive index differences or birefringence value for different Oe values [28, 29].

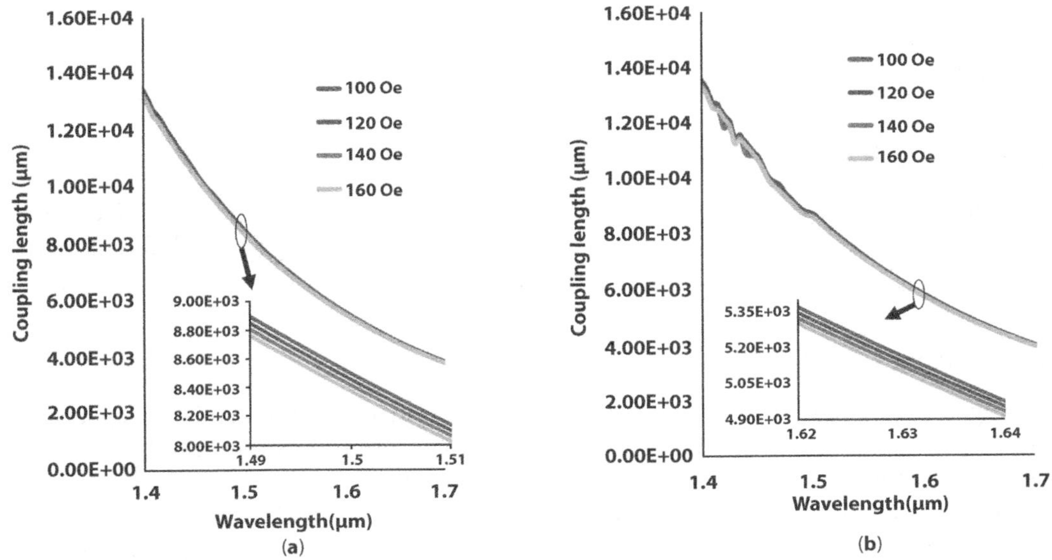


**Figure 10.5** The length of the coupling varies depending on the wavelength and Oe values [28, 29].

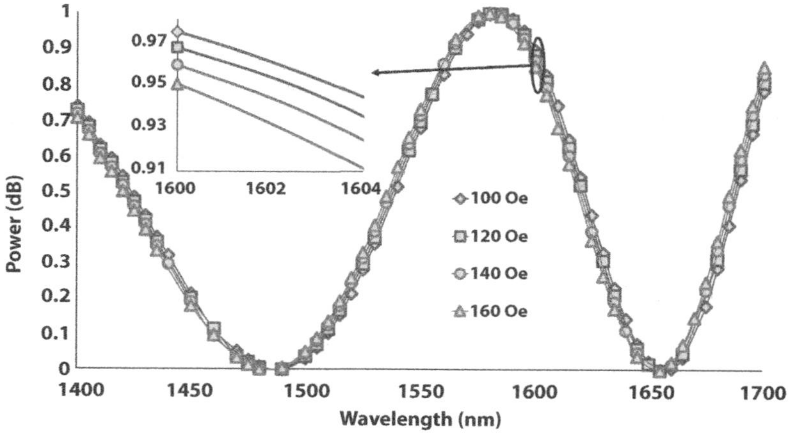


**Figure 10.6** Variations in Oe values and wavelength affect the PS [28, 29].

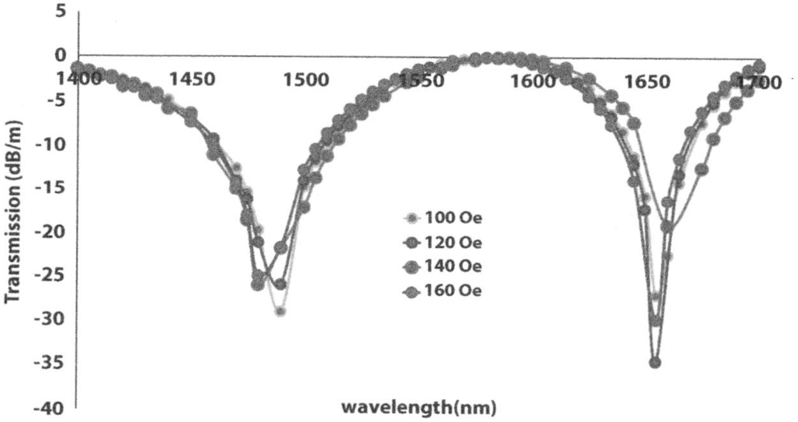


**Figure 10.7** Variations in Oe values affect the TPS [27–29].

### 10.5.2 Elliptical Hole-Assisted Magnetic Fluid Sensor

Now, the simulated structure is changed to observe the sensing performance. Here, the core portion is changed. Elliptical core is used to observe the variations of different mode distributions.

### 10.5.3 Ring Core Fiber

By using ring core fiber different outcomes have been noticed. The transmission spectrum analysis is given here for the ring core fiber structure. More peak values can be achieved from this investigation; those are represented in Figure 10.10.

Table 10.2 includes a summary of the outcome analysis for several parameters. The comparison study of the sensitivity responses in relation to the intensity of the applied magnetic field and the matching operating wavelength is shown in Table 10.2. It can be noted that

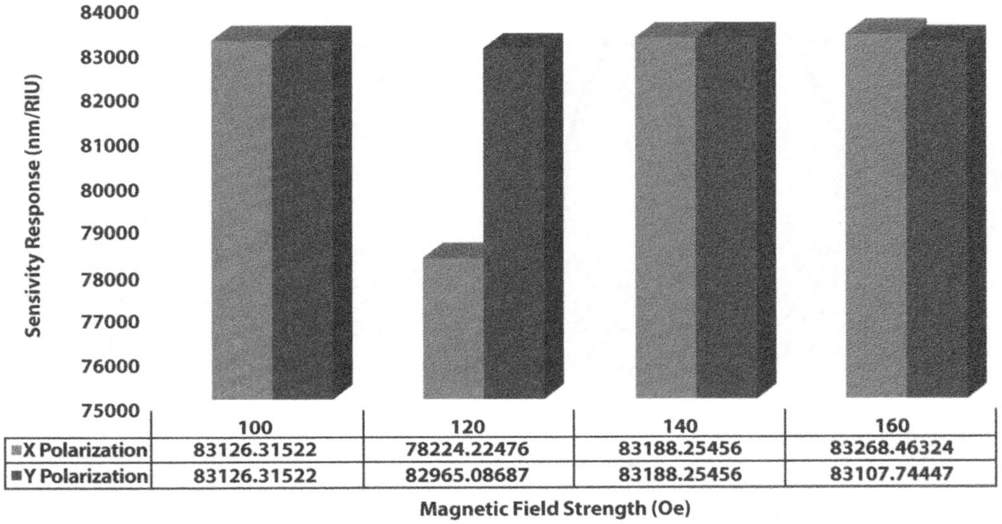


| | 100 | 120 | 140 | 160 |
|---|---|---|---|---|
| X Polarization | 83126.31522 | 78224.22476 | 83188.25456 | 83268.46324 |
| Y Polarization | 83126.31522 | 82965.08687 | 83188.25456 | 83107.74447 |

**Figure 10.8** Sensitivity response for different Oe values [27–29].

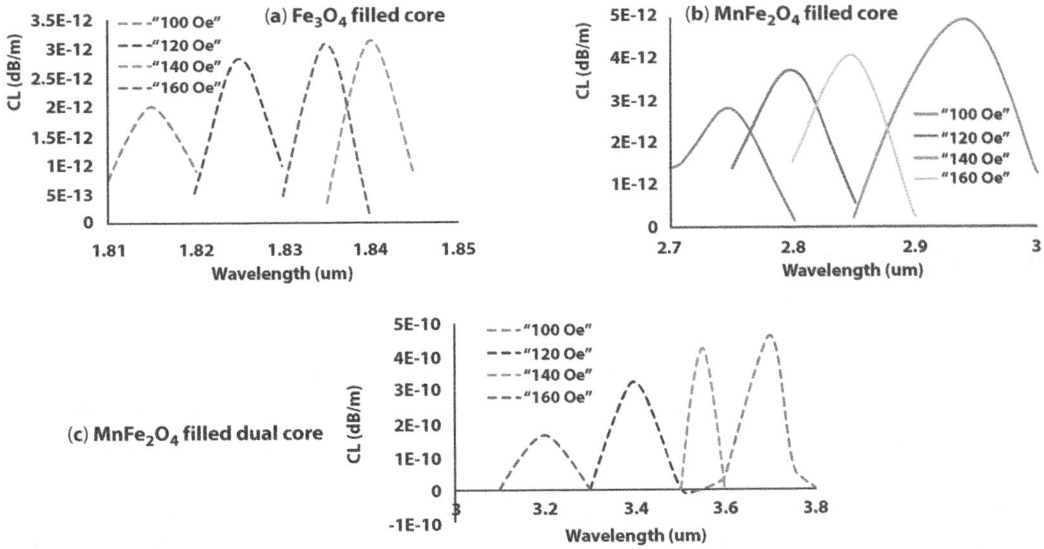


**Figure 10.9** Confinement loss spectrum for using different magnetic fluid [26, 28, 29].

more analysis had been performed to design a convenient sensor. Recent technologies such as cavity dielectric modulated [33], cancer cell detection using split ring resonator [34], three-dimensional metamaterial detection [35] and so on. As a result, a large number of applications have been widening for the magnetic biosensors [36].

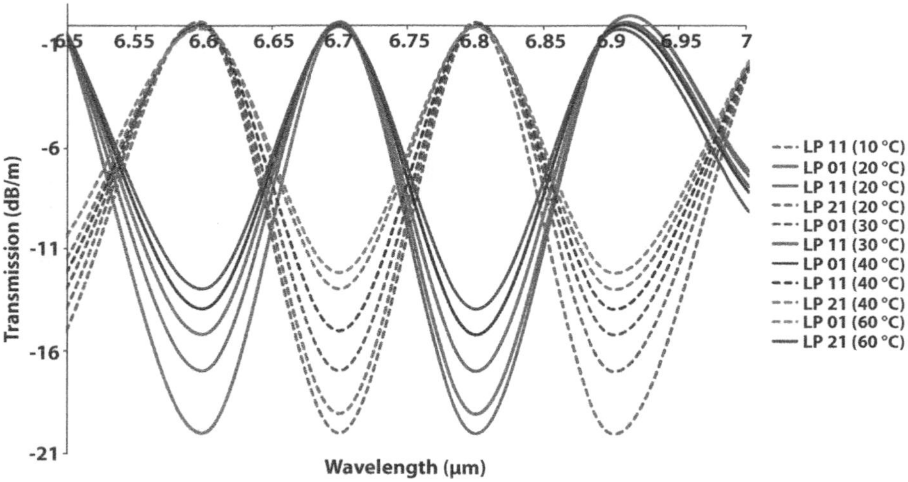


**Figure 10.10** Transmission spectrum for ring core fiber for different mode distributions [26].

**Table 10.2** Overall performance analysis.

| SI | $\lambda(\mu m)$ | Oe range | AS (nm/Oe) | Ref. |
|---|---|---|---|---|
| 1 | 0.6–1.6 | 0–250 | 1.56 | [30] |
| 2 | 1.8–1.92 | 410–600 | 384 | [31] |
| 3 | 1.56–1.8 | 30–130 | 700 | [32] |
| 4 | 1–4 | 50–200 | 40 | [25] |
| 5 | 1.4–1.7 | 100–160 | 5000 | [28] |

## 10.6 Conclusion

In this exposition, magnetic biosensor concept has been explored to improve the effectiveness and use of magnetic biosensors. Generally the magnetic biosensors are the more effective and easy to detect biological targets. The detection procedure is easy and efficient due to the simple and compact design of the sensor structure. Compared with other types of biosensors, the designed sensor structure has some advantages. First of all, the proposed magnetic biosensors have high sensing performance. Individual E. coli, for example, can be easily recognized using the proposed magnetic biosensors. Secondly, these sensors are more cost effective than the normal sensors. Moreover, these sensors have a simple structural design, they are small in size. Additionally, no external magnetic field source is required. Magnetic biosensors produce better output responses than optical biosensors.

## Acknowledgment

This work was supported by funding from the Natural Sciences and Engineering Research Council of Canada (NSERC).


## References


1. Häfeli, U., Schütt, W., Teller, J., Zborowski, M. (Eds.), *Scientific and clinical applications of magnetic carriers*, Springer Science & Business Media, Springer New York, NY, 2013.
2. Chemla, Y.R., Grossman, H.L., Poon, Y., McDermott, R., Stevens, R., Alper, M.D., Clarke, J., Ultrasensitive magnetic biosensor for homogeneous immunoassay. *Proc. Natl. Acad. Sci. U.S.A.*, 97, 14268, 2000.
3. Ananthi, V. and Arun, A., Pathogen identification through surface marker recognition methods, in: *Handbook of Microbial Nanotechnology*, pp. 355–373, 2022.
4. Ul-Islam, M., Ullah, M.W., Khan, S., Manan, S., Khattak, W.A., Ahmad, W., Shah, N., Park, J.K., Current advancements of magnetic nanoparticles in adsorption and degradation of organic pollutants. *Environ. Sci. Pollut. Res.*, 24, 12713, 2017.
5. Liu, Q., Li, S.G., Wang, X., Sensing characteristics of a MF-filled photonic crystal fiber Sagnac interferometer for magnetic field detecting. *Sens. Actuators B: Chem.*, 242, 949, 2017.
6. Vekas, L., Avdeev, M.V., Bica, D., Magnetic nanofluids: Synthesis and structure, in: *J. Biomed. Nanotechnol*, pp. 650–728, Springer, Berlin, Heidelberg, 2009.
7. Kumar, A. and Subudhi, S., Preparation, characteristics, convection and applications of magnetic nanofluids: A review. *Int. J. Heat Mass Transf.*, 54, 241, 2018.
8. Kawahito, S., Cerman, A., Aramaki, K., Tadokoro, Y., A weak magnetic field measurement system using micro-fluxgate sensors and delta-sigma interface. *IEEE Trans. Instrum. Meas.*, 52, 103, 2003.
9. Alipieva, E., Gateva, S.V., Taskova, E., Potential of the single-frequency CPT resonances for magnetic field measurement. *IEEE Trans. Instrum. Meas.*, 54, 738, 2005.
10. Cecelja, F. and Balachandran, W. (Eds.), Optimized CdTe sensors for measurement of electric and magnetic fields in the near field region, in: *IMTC/99. Proceedings of the 16th IEEE Instrumentation and Measurement Technology Conference*, May, IEEE, pp. 279–284, 1999.
11. Orr, P. and Niewczas, P., Polarization-switching FBG interrogator for distributed point measurement of magnetic field strength and temperature. *IEEE Sens. J.*, 11, 1220, 2010.
12. Lazcka, O., Del Campo, F.J., Munoz, F.X., Pathogen detection: A perspective of traditional methods and biosensors. *Biosens. Bioelectron.*, 22, 1205, 2007.
13. Scallan, E., Griffin, P.M., Angulo, F.J., Tauxe, R.V., Hoekstra, R.M., Foodborne illness acquired in the United States—Unspecified agents. *Emerg. Infect. Dis.*, 17, 16, 2011.
14. Barry, J.F., Turner, M.J., Schloss, J.M., Glenn, D.R., Song, Y., Lukin, M.D., Park, H., Walsworth, R.L., Optical magnetic detection of single-neuron action potentials using quantum defects in diamond. *Proc. Natl. Acad. Sci. U.S.A.*, 113, 14133, 2016.
15. Baselt, D.R., Lee, G.U., Natesan, M., Metzger, S.W., Sheehan, P.E., Colton, R.J., A biosensor based on magnetoresistance technology. *Biosens. Bioelectron.*, 13, 731, 1998.
16. Buchatip, S., Ananthanawat, C., Sithigorngul, P., Sangvanich, P., Rengpipat, S., Hoven, V.P., Detection of the shrimp pathogenic bacteria, Vibrio harveyi, by a quartz crystal microbalance-specific antibody based sensor. *Actuators B: Chem.*, 145, 259, 2010.

17. Schotter, J., Kamp, P.B., Becker, A., Puhler, A., Brinkmann, D., Schepper, W., Bruckl, H., Reiss, G., A biochip based on magnetoresistive sensors. *IEEE Trans. Magn.*, 38, 3365, 2002.
18. Li, G., Joshi, V., White, R.L., Wang, S.X., Kemp, J.T., Webb, C., Davis, R.W., Sun, S., Detection of single micron-sized magnetic beads and magnetic nanoparticles using spin valve sensors for biological applications. *J. Appl. Phys.*, 93, 7557, 2003.
19. Arif, M., Huq, F., Ahmed, K., Asaduzzaman, S., Azad, M., Kalam, A., Design and optimization of photonic crystal fiber for liquid sensing applications. *Photonic Sens.*, 6, 279, 2016.
20. Nallusamy, N., Zu, P., Raja, R.V.J., Arzate, N., Vigneswaran, D., Degenerate four-wave mixing for measurement of magnetic field using nanoparticles-doped highly nonlinear photonic crystal fiber. *Appl. Opt.*, 58, 333, 2019.
21. Chakma, S., Khalek, M.A., Paul, B.K., Ahmed, K., Hasan, M.R., Bahar, A.N., Gold-coated photonic crystal fiber biosensor based on surface plasmon resonance: Design and analysis. *Sens. Bio-Sens. Res.*, 18, 7, 2018.
22. Thakur, H.V., Nalawade, S.M., Gupta, S., Kitture, R., Kale, S.N., Photonic crystal fiber injected with Fe3O4 nanofluid for magnetic field detection. *Appl. Phys. Lett.*, 99, 161101, 2011.
23. Liu, Q., Li, S.G., Wang, X., Sensing characteristics of a MF-filled photonic crystal fiber Sagnac interferometer for magnetic field detecting. *Actuators B: Chem.*, 242, 949, 2017.
24. Chen, H., Li, S., Li, J., Fan, Z., Magnetic field sensor based on magnetic fluid selectively infilling photonic crystal fibers. *IEEE Photon. Technol. Lett.*, 27, 717, 2015.
25. Mitu, S.A., Ahmed, K., Al Zahrani, F.A., Abdullah, H., Hossain, M., Paul, B.K., Micro-structure ring fiber–based novel magnetic sensor with high birefringence and high sensitivity response in broad waveband. *Plasmonics*, 16, 905, 2021.
26. Abdullah, H., Mitu, S.A., Ahmed, K., Magnetic fluid-injected ring-core-based micro-structured optical fiber for temperature sensing in a broad wavelength spectrum. *J. Electron. Mater.*, 49, 4969, 2020.
27. Mitu, S.A., Ahmed, K., Bui, F.M., Nithya, P., Al-Zahrani, F.A., Mollah, M.A., Rajan, M.M., Novel nested anti-resonant fiber based magnetic fluids sensor: Performance and bending effects inspection. *J. Magn. Magn. Mater.*, 538, 168230, 2021.
28. Mitu, S.A., Ahmed, K., Hossain, M., Paul, B.K., Nguyen, T.K., Dhasarathan, V., Design of magnetic fluid sensor using elliptically hole assisted photonic crystal fiber (PCF). *J. Supercond. Nov. Magn.*, 33, 2189, 2020.
29. Mitu, S.A., Dey, D.K., Ahmed, K., Paul, B.K., Luo, Y., Zakaria, R., Dhasarathan, V., Fe3O4 nanofluid injected photonic crystal fiber for magnetic field sensing applications. *J. Magn. Magn. Mater.*, 494, 165831, 2020.
30. Zu, P., Chiu Chan, C., Gong, T., Jin, Y., Chang Wong, W., Dong, X., Magneto-optical fiber sensor based on bandgap effect of photonic crystal fiber infiltrated with magnetic fluid. *Appl. Phys. Lett.*, 101, 241118, 2012.
31. Arif, M., Huq, F., Ahmed, K., Asaduzzaman, S., Azad, M., Kalam, A., Design and optimization of photonic crystal fiber for liquid sensing applications. *Photonic Sens.*, 6, 279, 2016.
32. Han, B., Zhang, Y.N., Siyu, E., Wang, X., Yang, D., Wang, T., Lu, K., Wang, F., Simultaneous measurement of temperature and strain based on dual SPR effect in PCF. *Opt. Laser Technol.*, 113, 46, 2019.
33. Zhou, X., Mai, E., Sveiven, M., Pochet, C., Jiang, H., Huang, C.C., Hall, D.A., A 9.7-$nT_r$, 704-ms Magnetic biosensor front-end for detecting magneto-relaxation. *IEEE J. Solid-State Circuits*, 56, 2171, 2021.
34. Oueslati, A., Hlali, A., Zairi, H., Modeling of a metamaterial biosensor based on split ring resonators for cancer cells detection, in: *2021 18th International Multi-Conference on Systems, Signals & Devices (SSD)*, March, IEEE, pp. 392–396, 2021.

35. Chen, J., Yang, C., Gu, P., Kuang, Y., Tang, C., Chen, S., Liu, Z., High sensing properties of magnetic plasmon resonance by strong coupling in three-dimensional metamaterials. *J. Light. Technol.*, 39, 562, 2021.
36. Wu, K., Tonini, D., Liang, S., Saha, R., Chugh, V.K., Wang, J.P., Giant magnetoresistance biosensors in biomedical applications. *ACS Appl. Mater. Interfaces*, 14, 9945, 2022.

# 11

# Biosensors for Salivary Biomarker Detection of Cancer and Neurodegenerative Diseases

**Bhama Sajeevan, Gopika M.G., Sreelekshmi, Rejithammol R., Santhy Antherjanam and Beena Saraswathyamma***

*Department of Chemistry, Amrita Vishwa Vidyapeetham, Amritapuri, Kollam, India*

***Abstract***

Biosensors are devices that make use of biorecognition elements and produce measurable signals proportional to the concentration of the target biomolecules for the prognosis and diagnosis of cancer and neurological diseases. Since the prognosis and diagnosis of these diseases are of utmost concern, the development of diagnostic tools for them is a focus area of research. When a person is affected with cancer or neurological diseases, the patients' body secretes some molecules that act as biomarkers which is a warning sign that confirms that a person is suffering from these diseases and proper tests have to be done to confirm the disease. These biomarkers can detect diseases even in the primary stages and it can be pivotal for timely treatment and cure. Biomarker-based expression techniques that are currently used in diagnosis require invasive approaches to accumulate tissue samples of individuals which are expensive, time consuming and cumbersome, making them impractical for use. Salivary samples have been used in analysis of a variety of diseases, including cancers and, neurological and neuropsychiatric diseases which will be discussed in this chapter.

***Keywords*:** Biosensor, salivary biomarker, neurodegenerative disease, cancer

## 11.1 Introduction

Prognosis and diagnosis of cancer and neurological diseases are one of major areas of concern and the development of diagnostic tools for these diseases has been a focus area of research for a long time. The clinical analysis for the detection of the neurological disorders and certain tumor biomarkers are performed in blood, cerebrospinal fluid (CSF), urine, saliva etc. [1–6]. The analysis of CSF or lumbar puncture and blood tests involves discomfort, and it is more invasive for the patients [7, 8]. Saliva testing includes various advantages over other biofluid analysis such as easy and safe collection and sampling procedures, cost-effective, reduced risk of infections, no clotting of samples as in the case of blood and urine and simple diagnostic procedures, etc. [9–11]. Hence, research on the development of noninvasive clinical analysis leads to salivary fluid diagnosis for the detection of biomarkers

**Corresponding author*: beenas@am.amrita.edu

Inamuddin and Tariq Altalhi (eds.) Biosensors Nanotechnology, 2nd Edition, (215–244) © 2023 Scrivener Publishing LLC

for neurological disorders, tumors and other disorders [12–14]. Saliva is a complex biofluid, which contains many proteins, biomolecules, microbial organisms, electrolytes etc. [15]. Human saliva contains about 2000 proteins and approximately one fourth of these proteins are present in blood serum too [16]. Salivary gland is highly pervious and covered by capillaries which will enable the free transport of molecules from bloodstream to acinar cells of salivary glands [17, 18]. Thus saliva can be believed equivalent to that of blood serum in physiological functions as they have a lot of the same components [19]. Literature reports suggested that saliva contains a large amount of information related to health conditions of the human body and hence salivary fluids may be a possible medium for the study of various disorders in molecular diagnostics [20].

Saliva contains biomarkers for various diseases including cardiovascular diseases, renal problems, oral diseases, diabetes, carcinomas, autoimmune diseases, neurological diseases etc. [21, 22]. In this chapter, we are highlighting the detection and analysis of saliva-based biomarkers for neurological and most fatal cancer diseases. Analysis is done employing varied tools such as chromatographic techniques, immunoassays, spectrophotometric techniques, proteomic analysis by mass spectrometry, microfluidics and electrochemical biosensors [23–29]. Among the other analytical techniques electrochemical biosensors offer more flexibility for the development of portable devices [30], detect very low concentration of analytes with high sensitivity and these techniques are less expensive compared to other techniques [31].

The concept of electrochemical biosensing explored the way for the development of smart, flexible and point-of-care diagnostic devices for healthcare monitoring [32]. Electroanalytical techniques focus on analysis of an analyte through potential and/or current study by using the electrochemical cell. The study gives every important, even minute information about the analyte like its chemical nature and behavior, about its quantity etc. The acting principle of electroanalytical technique is that there should be some chemical changes happening to the analyte molecule that is, the analyte molecule has to directly react or indirectly react through various coupled reaction. Such electroanalytical techniques, due to economic feasibility, sensitivity, specificity, precision, accuracy, miniaturization and low detection limit are found to have wide range of uses and applications in various fields like clinical, pharmaceutical, biomedical, environment, industry etc. There are different categories of electroanalytical techniques based on which aspect of the electrochemical cell is controlled and which aspect is measured.

These devices are operated by the specific interaction of a recognition element like nucleic acids, enzymes, antibodies etc. with the analyte species by different electrochemical techniques and convert the electrochemical signals produced to electrical signals [33, 34]. There are several reports based on the fabrication of electrochemical biosensors for non-invasive monitoring of diseases from the biological fluids, such as sweat, saliva, etc. [35, 36]. Biosensing devices incorporated with smart phone technologies can revolutionize the area of healthcare monitoring to a large extent [37].

Biosensors produce fast results, are very economical, less time consuming and are very much feasible for miniaturization [38]. High accuracy and selectivity of these sensors are known which are owed to their biosignaling mechanism is based on biorecognition. The popular biosensors typically contain two components which are:

(a) Transducer elements that act at the recognition probe for the detection of the target analyte molecules from the samples. This component of electrochemical biosensors is available with two parts that are; the bioreceptors and the interface that is the transducer part. Bioreceptors are the core part of the electrochemical biosensors, which are the recognition part for sensing the target analytes from the sample, that is, they are the core recognition elements that bind with the target analyte molecules to generate signals, promising sensitivity and selectivity. These bioreceptors that are commonly employed in electrochemical biosensors are antibody enzymes, DNA molecules, cells etc. While the interface (transducer) gives a substrate part for the bioreceptors to be immobilized on them which broadly include nanoparticle surfaces, nanowire array, and field effect transistor (FET). There is a necessity of a redox-active tag in this transducer part for getting signaling output. These redox active molecules are either covalently bound to the bioreceptors molecules or are free in the electrolytic solution.

(b) An electronic system that contains an amplifier for amplifying the signals and a displayer. There are biological or chemical reactions occurring in the electrochemical system that is, mainly the redox reactions are happening between the target analyte molecules and the bioreceptor molecules and the signals are then analyzed, and quantitative results can be obtained from the signals shown so as to make accurate measurements of the molecule of interest [39].

Redox-active electrochemical reactions are found to occur either directly on the surface of the substrate or on the close proximity of the substrate surface. Many factors like the surface modification, the substrate material, and even its dimensions are found to have much impact on these redox-active electrochemical reactions [40]. Modifications on these sensors using nanomaterials is common due to their varied properties being advantageous such as high reactivity and selectivity. These modifications guarantee the assembly of the bioreceptors molecules, biomaterials like the DNA, antibodies, etc., and as a result, this improves the performance of the biosensor that is, reduces the background current, improves the ratio of signal-to-noise, as well as accelerates the speed of electron transfer [41]. Many nanomaterials, mostly precious metals gold, silver, and platinum nanoparticles, graphene oxide, carbon nanotubes, carbon spheres, etc. [42, 43].

Electrochemical biosensors [44] can be mainly classified as amperometric biosensors, potentiometric biosensors, conductometric biosensors, impedance biosensors, and FET-based biosensors (Figure 11.1).

(a) Voltammetric biosensors: they are the most commonly used electrochemical biosensors because of their ease of operation and flexibility in signal interpretation. They work when potential is applied at the working electrode as the input signal and the resulting current signal is measured as the output signal. A redox reaction occurs during the experiment producing a current signal. This is due to the analyte molecule transfer that happens from the bulk to the surface of the electrode [44, 45].

(b) Impedance biosensors: these biosensors are highly sensitive, selective, and of low cost. They are typically employed to determine the characteristics like the capacitance along with the resistance of the monomer material that is

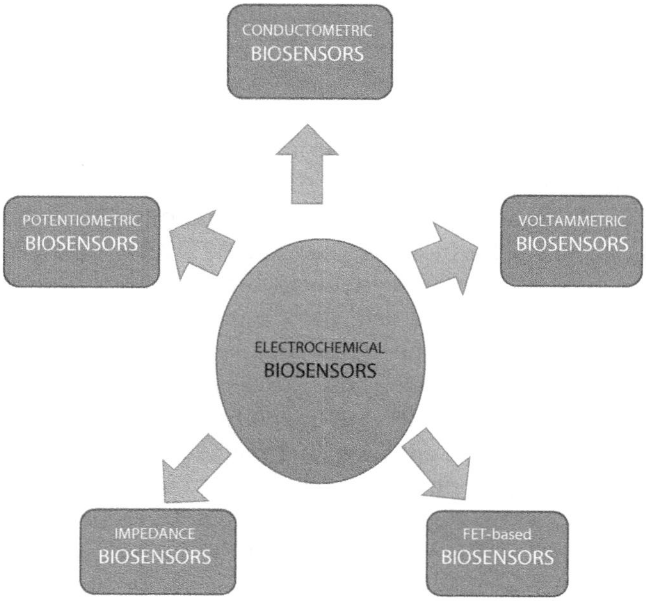


**Figure 11.1** Main classification of electrochemical biosensors.

adsorbed on the surface of the working electrode. EIS or impedance spectroscopy employs an AC excitation signal to study the capacitance along with the resistance of the material.

(c) Potentiometric biosensors: these biosensors are found very useful in measuring the charge potential, which accumulates on the working electrode of the setup when there is no net current flow occurring in the system. These biosensors are typically employed to get information regarding the electrochemical reactions happening within the electrochemical cell between the target analyte molecule and the bioreceptors molecule.

(d) Conductometric biosensors: they are an important category of biosensors which typically measures the conductivity of a solution during the electrochemical experiment as there is a change occurring to the components of the solution when there is an electrochemical reaction happening in the solution.

(e) FET-based biosensors: these biosensors measure conductivity. When the bioreceptors molecules on the substrate surface of the electrode bind to the studied analyte molecule, there is an alteration in the electric field as a result, it produces a change that is measurable in the source-drain conductivity. This conductivity produced between the drain and the source is measured in FET-based biosensors.

## 11.2 Biosensors for Neurodegenerative Diseases

Most of the neurodegenerative disorders are characterized by gradual loss of specific neuronal cell populations and are associated with aggregation of protein. These disorders are found to occur even in late stages of other major diseases, and they are multiple sclerosis (MS), which are found to begin with demyelination and then lead to loss of neurons progressively. Mostly, these disorders are found to be associated with extensive oxidative stress which results in the malfunctioning and finally the death of neuronal cells that are responsible for disease pathogenesis. These neurodegenerative disorders can cause brain tissue damage as they are prone to oxidative damage due to high oxygen intake due to presence of reactive oxygen species, regenerative capacity and presence of low levels of antioxidants.

ALS is the third most prominent among said disorders targeting people in the age group 50 to 70 years with a high risk for individuals at the age of late 60s or early 70s. The disease symptoms vary in the early stages based on the progressive dysfunctioning of upper and lower nerve cells. Multiple sclerosis (MS) is an autoimmune disease of the central nervous system that appears to be the most common reason for neurological disability shown by young adults. There is uniqueness in the clinical result of each patient suffering from neurodegenerative disorders, which is primarily due to the combination of different mechanisms of gliosis, inflammation-demyelination, remyelination-repair, and axonal damage neurodegeneration in a vast extend also influenced by idiosyncratic factors.

It is a well-known fact that within the next five years a set of new potential biomarkers, especially salivary biomarkers will be recognized and confirmed for the proper diagnosis of various neurodegenerative disorders. Such potential salivary biomarkers could easily aid a physician or a specialized doctor to easily detect potential neurodegenerative disease cases. Surrogate end point biomarkers will also be available by the time so as to help the physician to monitor the progression in the treatment. Figure 11.2 shows the main types of neurodegenerative diseases that are discussed below. Table 11.1 gives the list of biomarkers observed for main neurological disorders.

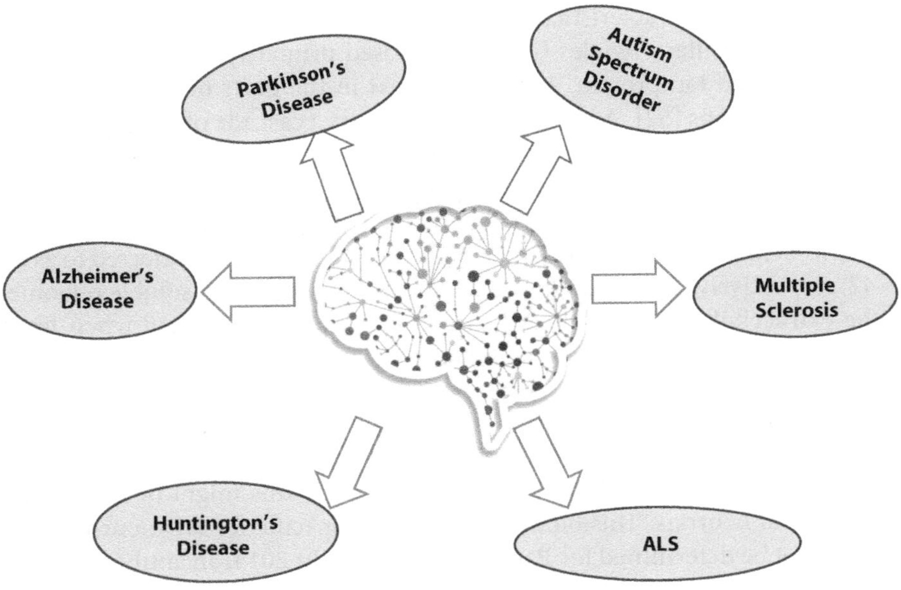


**Figure 11.2** Main types of neurodegenerative diseases.

**Table 11.1** Biosensors for salivary biomarkers of neurodegenerative disorders.

| Disease | Biomarker | Sensor type | LOD | |
|---|---|---|---|---|
| **Parkinson's disease** | 1. α-syn | DNA aptamer | 10 pM | [76] |
| | 2. DJ1 | Neurobiosensor | 0.5 fg $ml^{-1}$ | [83] |
| **Amyotrophic lateral sclerosis** | Cortisol | 1. Immunosensor | 0.1–10 ng $ml^{-1}$ | [96] |
| | | 2. Immunosensor | 1 pM | [99] |
| | | 3. Immunosensor | 1 pM | [100] |
| | | 4. Optic Sensor | 25 fg $ml^{-1}$ | [102] |

### 11.2.1 Alzheimer's Disease

Alzheimer's disease (AD) is a continuing condition that causes the shrinkage of the brain and gradual loss of brain cells. It can lead to dementia, a decrease in the thinking process, behavioral changes, and can even affect the patient's ability to function on their own. Severe memory loss in day-to-day events can be seen as a result of AD. Slow and nonlinear progression, gradual cognitive impairment, and memory loss characterize this devastating neurodegenerative condition of the brain [46–48]. Most biomarkers are taken by using Cerebrospinal fluid (CSF) as a sample which is invasive. Thereby many research studies ended up in a biological fluid called saliva as a diagnosis tool which results in salivary biomarkers. Saliva is easy to handle rather than blood. The collection of saliva is easy, noninvasive, does not take much time, no need of clinical training and its large availability [49–51]. Recent studies have discovered a wide range of biomolecules in saliva that can be used to identify possible biomarkers.

The key pathological characteristics of AD comprise plaques composed of beta-amyloid (Aβ) peptide, neurofibrillary tangles (NFT) composed primarily of hyperphosphorylated microtubule-associated tau protein, and neuron loss in the parts of brain supporting the complementary functions [52]. A build-up of Aβ triggers a cascade of pathogenic processes that cause neuronal injury, according to several studies [53]. The precursor enzyme of amyloid precursor protein (APP) is known for making Aβ peptide [54]. The protein is cleaved by β-secretase, resulting in a 99-amino-acid part. The γ-secretase then divides the fragment into a 40-amino-acid peptide and a 42-amino-acid peptide, which are referred to as Aβ1–40 and Aβ1–42, respectively. These peptides have a tendency to form insoluble oligomers and fibrils in the extracellular region [55, 56]. Amyloid plaques are formed when fibrils join together to create plaques, creating toxic build up that eventually lead to damages in neurons [57]. An early study revealed that increased Aβ1-42 levels determined in salivary samples amounted higher in patients with moderate AD than normal people. But the Aβ1-40 level has no significant difference between two categories of people [58]. But this biomarker fails to show higher Aβ1-42 levels for severe AD patients. This might be due to excessive loss of intracerebral neurons. This biomarker is highly specific for AD neurodegenerative disease and cannot be determined for Parkinson's disease. In 2014, an antibody-based magnet nanoparticles immunoassay for detection of Aβ1-42 was discovered [59]. The advantage of this report revealed more accurate results than previous study. By employing this

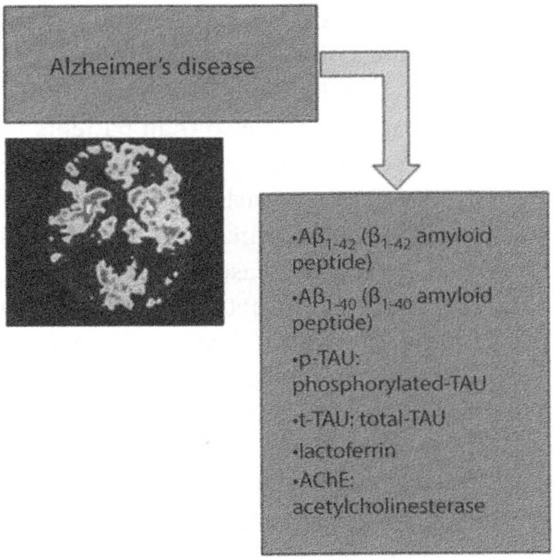


**Figure 11.3** Biomarkers for AD.

technology to identify Aβ1-42, Lee *et al.* was able to distinguish between healthy people and people who were at risk of developing AD. His findings can predict the disease risk in people with family history [60]. But the study cannot distinguish between mild, moderate and severe AD patients.

Tau protein is a microtubule related protein that binds to tubulin and helps to keep microtubules stable and flexible [61]. They are crucial for maintaining neuronal structure and plasticity as well as ensuring axonal transport. TAU's proper functioning is regulated appropriate levels of phosphorylation and dephosphorylation in healthy environments. Whereas in AD cases, the protein sequence experiences a mutation that leads to increased phosphorylation of the protein leading to a build-up of NFT and eventually culminates in the neuron loss [62, 63]. Phosphorylated tau and total tau, and the ratio between them serves as a biomarker for AD. The neurons that innervate the salivary glands and the acinar epithelial cells are likely sources of TAU proteins in saliva [64]. Shi *et al.* made a study on this biomarker by quantifying p-tau and t-tau using mass spectrometry by finding the ratio of p-tau/t-tau per individual shows higher levels in AD patients compared to healthy people [65]. This biomarker has high sensitivity and high specificity. This indicates that there is a significant production of p-tau by salivary glands in AD patients. Pekeles *et al.* made a study in another way by quantifying the ratio of t- and p- tau proteins by using western blot analysis. This ratio is seen in higher concentrations in people suffering with AD than in unaffected people [66]. Ashton *et al.* evaluated only t-tau level using Ultrasensitive-single array molecule technology but didn't get substantial change in t-tau level of people with Alzheimer's in comparison to healthy people [67].

In 2017, an ultrasensitive assay was fabricated for the biomarkers related with AD, amyloid beta peptide, pau and t-pau. Varied samples including blood, saliva, urine etc were analyzed for the same. The test is dependent on the target protein interaction with their pair of antibodies [68]. This developed versatile detection test is highly promising in quantification of AD biomarkers in a noninvasive way.

Another important marker for the disease is lactoferrin, a peptide which has antimicrobial properties and is present in the saliva [69]. Using tools like mass spectrometry and ELISA, it was shown by Carro *et al.* that patients with the disorder have decreased levels of the peptide than normal, healthy people and also than patients with Parkinson's disease [70]. Reduced salivary lactoferrin levels are AD-specific biomarkers, according to the findings, and can assist discriminate and diagnose early clinical stages of the disease.

Acetylcholinesterase is a neurotransmitter which is an excellent biomarker for neurodegenerative diseases such as AD. There is a decreased concentration in ACh in AD patients compared to healthy people, and thus we can detect it. This drop in acetylcholine (ACh) concentration is induced by cholinergic neuron death and the severe cholinergic conductivity impairment seen even in the early stages of AD [71]. Bakhtiari and coworkers using the Ellman colorimetric method in his study discovered that salivary AChE activity is lower for AD patients [72]. But this study did not show any correlation between AChE, age and disease progression. Thus, all the above discussed noninvasive salivary biomarkers are a promising tool for detecting the disease out early. Most of the studies associated with biosensors reported till date mostly used serum, CSF or human blood as real samples. Therefore, salivary biosensors are highly promising in future as it can be used in a noninvasive way as the real sample used is saliva.

### 11.2.2 Parkinson's Disease

Parkinson's disease (PD) is a brain disease which is commonly reported, and it affects both the peripheral and central nervous system. The gradual loss of neurons that synthesize neurons which leads to a decrease in the levels of dopamine in the brain is the main mechanism of PD, along with the presence of Lewy bodies [65, 73]. These are formed by the aggregation of alpha-synuclein proteins. It was much more difficult to find a salivary marker for idiopathic PD compared to that of AD.

The prominent salivary biomarkers, shown in Figure 11.4 which can be potentially used for the diagnosis of PD are α-synuclein and DJ-1 proteins. The protein, α-synuclein is acidic, soluble and heat resistant which is found in the neuron's presynaptic terminals [47, 48]. It affects the integrity of the neuronal membrane as well as membrane trafficking via vesicular transport [64]. In healthy people, α-synuclein is monomeric, but in Parkinson's disease sufferers, it is predominantly oligomeric. Those oligomers are further transformed and eventually form Lewy bodies [74]. A study using ELISA kit was used to quantify the levels of a-synuclein in salivary samples which lead to the conclusion that the protein level was significantly higher in the case of people with PD than healthy people proving a-synuclein can be a biomarker for the disease [75]. In 2021, Xiuxiu Yang, Xiaofang Zhao, Fengwei Liu fabricated a sensor for the detection of α-syn. The system shows high specificity and rapid response which makes it more advantageous in detecting α-syn [76]. This constructed biosensor thus will become an encouraging analytical tool for future noninvasive type of collection of α-syn detection too.

Another biomarker discovered to detect early PD is DJ-1, yet another protein thought to be a pleiotropic neuroprotective protein with antioxidant and anti-mitochondrial dysfunction properties [77, 78]. Parkinson's disease (PD) is caused by the alterations in the salivary protein, and oxidized DJ-1 is discovered in the brains of people with idiopathic PD [79].

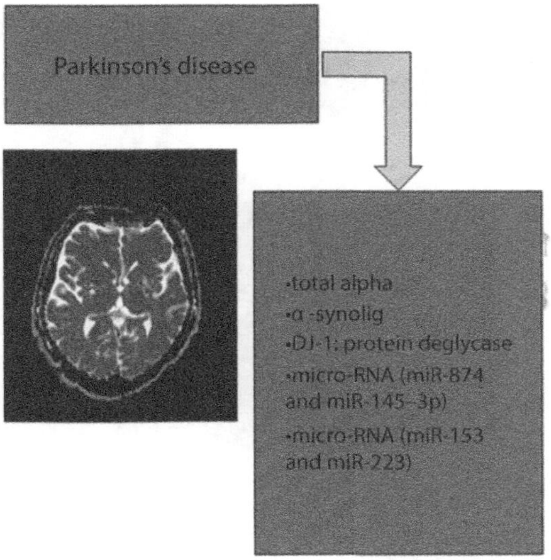


**Figure 11.4** Biomarkers for PD.

Thus, the researchers also quantified DJ 1 protein as a promising tool to be used as a biomarker for PD. DJ-1 is primarily found in the cytoplasm of dopaminergic neurons, with minor amounts in the mitochondria and nuclei [47, 80]. In a study done by quantifying DJ 1 and α-syntotal, Devic *et al.* replicated the results of α-syntotal as in previous research [81]. But on the other hand, there is a modest increase in DJ 1 in people with Parkinson's compared to people who haven't been diagnosed with PD, but the results were not statistically significant. Kang *et al.* in his study reported that DJ 1 can distinguish values between illness stages and subtypes [82]. Thus, we can make sure that these above discussed biomarkers make a promising tool for the early diagnosis of PD. In 2020, Münteha Nur Sonuç Karabog, Mustafa Kemal Sezgintürk developed an electrochemical neurobiosensor for the biomarker DJ 1 detection using saliva and CSF as real samples. In this study the researchers used multiwalled carbon nanotube (MWCNT) and gold nanocomposite (Au-NP) doped, 11-amino-1-undecanethiol modified polyethylene terephthalate coated indium tin oxide electrodes. The sensor had an LOD of 0.5 fg $ml^{-1}$. Thus, the developed sensor can be best used for the selective determination of DJ 1 [83].

### 11.2.3 Huntington's Disease

It is an autosomal dominant disease in the brain, which is caused by Huntington gene (Htt) expansion. In Huntington's disease (HD), proteases break down the mutant huntingtin protein unleashing N-terminal polyglutamine sequence which interferes with all the functions and causes neurodegeneration [47]. So, for finding HD one has to look for gene mutations in Htt protein. Bloom et al, using ELISA kit made study on the HD by taking concentration of Htt protein in saliva. Some of the important biomarkers for HD are listed in Figure 11.5.

From the study he found out that patients have higher total Htt protein content than compared to healthy people [84].

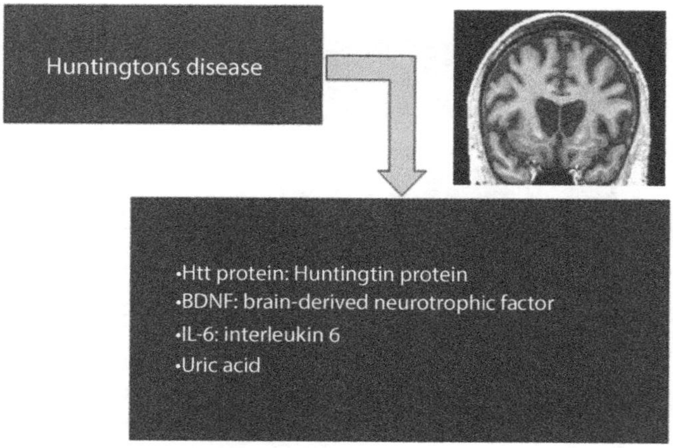


**Figure 11.5** Biomarkers for HD.

### 11.2.4 Amyotrophic Lateral Sclerosis

Amyotrophic lateral sclerosis (ALS) is one of the very dangerous, incurable neurodegenerative disorders commonly observed in adults and is distinguished by behavioral disorders [85]. The condition is marked by the cumulative loss of motor neurons, both upper and lower neurons in the spinal cord, motor cortex, and brain stem which results in the cramps, hyporeflexia, malfunctioning of tissues, spasticity, fasciculations, progressive muscle weakness, and can finally lead to the death of the paralyzed patient within 2 to 5 years after the detection of this disorder by respiratory related issues [86]. It may take an average of one-year delay in the detection of this severe disorder due to lack of proper detection method with available biomarkers. ALS is mainly reported with people in the age group 50–70 years with a high risk for individuals at the age of late 60s or early 70s. Most commonly ALS is detected in males rather than in females. The disease symptoms vary in the early stages based on the progressive dysfunctioning of upper motor neurons (UMN) or lower motor neurons (LMN) [87, 88].

Many different mechanisms were proposed for the ALS disorder, mainly like viral, genetic, immunological, epidemiological, and environmental factors, but still the actual mechanism is unknown. It will take a long lag-time of approximately a year for the proper diagnosis of ALS disorder from the time symptoms are shown by the patient; this has obstructed the recognition of the potential biomarkers for the specific detection of ALS. Now it is possible to detect the common motoneuron and similar neurodegenerative diseases, but no specific diagnostic test can detect and discriminate ALS from these similar neurodegenerative diseases. As a result of this delayed detection, patients are drawn to the extreme side of the disease within the long lag-time due to lack of proper treatment and therapies. Recently, the detection of ALS is found possible through tedious processes by combining the neurophysiological evidence and clinical data together along with the proper valuation of the symptoms; this has limited the time for quick action and the correct choice of an individualized therapy. Thus, it is the need of this era to come up with a new potential biomarker for ALS which should be quickly and easily detectable and easily accessible for the untimely diagnosis, classification into different groups, and evaluation for an appropriate rehabilitation therapy [89, 90].

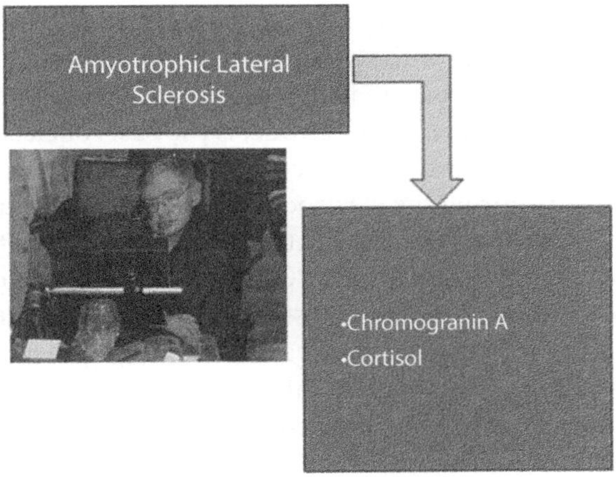


**Figure 11.6** Biomarkers for ALS.

Plasma, Saliva and CSF which are the protein biomarkers in biofluids are thought to be useful surrogate end point biomarkers for ALS detection as they might substitute for a clinically significant end point, and they could be even used for early phase clinical trials to make rapid decisions in drug development process. One of the major complex biofluids, saliva is composed of different molecules in an aqueous environment including proteins, nucleic acids, carbohydrates, metabolites, and hormones which are marked as potential biomarkers as they undertake many active and passive processes of transport. The presence and concentration of the specific salivary molecules are an indication of a particular disease as well as it shows how the rehabilitation and pharmacological treatment for a particular disease progressed. Recent advancement in the area has brought about the roles of salivary biomarkers as a potential diagnostic tool for many diseases like virus and bacterial infections, AD and PD, diabetes, lung cancer and also for detecting the existence of certain drugs in the body [85, 89, 90]. Some recent studies showed that there are certain salivary molecules that can act as potential biomarker for ALS detection, especially, cortisol and chromogranin A, shown in Figure 11.6 [91] (ChA) but more intense studies have to be performed to ensure that they can be useful surrogate end point biomarkers. It has been found that salivary chromogranin A has been increased tremendously in people diagnosed with last-stage ALS, increased two times in moderately affected patients. ChA (Chromogranin A) inhibits the release of glucose-induced insulin from pancreas which causes the decreased levels of insulin in serum and CSF of ALS patients. ChA is observed to be a much valuable indicator of ALS as they are suitable for indicating the disease progression. Though ChA level increase cause decreased levels of insulin in body as well as neuronal glucose utilization observed in connection with ALS, it is not necessary that ALS should involve diabetes mellitus as a result, it is still a hypothetical finding and cannot validate whether ChA is a potential biomarker for ALS. The mean salivary CgA level in last-stage ALS patients was found to be much higher than the level in moderate ALS patients, then in tube-fed vascular dementia and then in unaffected people [92, 93].

It is a well-known fact that within the next 5 years a set of new biomarkers, especially salivary biomarkers will be identified and validated for the proper diagnosis of ALS. Such potential salivary biomarkers could easily aid a physician or a specialized doctor to easily detect potential ALS cases. Surrogate end point biomarkers will also be available by the time so as to help the physician to monitor the progression in the treatment.

The chemical reaction between our analyte molecule, which is to be detected with the biomolecules that are immobilized in the electrode, is the basis of electrochemical biosensing. This chemical reaction either produces or consumes electrons, that is, either oxidation or reduction reaction occurs during the electrochemical detection of target analytes. These reactions affect the electrical properties like electric potential or current of the solution [94].

For the evaluation of cortisol from saliva at POC was demonstrated on a biosensor based on a portable immune-chromatographic test strip [95]. For the evaluation of cortisol in saliva, a chromatographic test-strip composing a cortisol conjugate glucose oxidase (GOD) of dimensions 5 × 1.5 × 50 mm is prepared. This cortisol biosensor which is based on calorimetric protocol can detect cortisol of concentration 1 to 10 ng $ml^{-1}$ within 25 minutes [96]. Sun *et al.* developed a method for the recognition of cortisol in saliva which used gold electrodes [97]. During the measurements there is production of p-nitrophenol (pNP) which is considered as an indicator and measured during cortisol detection. This system could not get continuous monitoring of cortisol in saliva, but it could show a lower limit of cortisol detection. Arya *et al.*, detected cortisol from saliva by electrochemical impedance (EIS) method and they used interdigitated microelectrodes to modify SAM of Dithiobis (succinimidyl propionate) [98].

Reports on development of electrochemical immunosensor that used zinc oxide nanoflakes and nanorods which are synthesized on gold coated substrates were found attractive [99]. In this work, the used 1 D nanorod had high surface area and 2 D nanoflakes supplied high surface charge for more loading of Anti-Cmab. These 1D nanorods and 2D nanoflakes were synthesized by sonochemical approach and are deposited using an electron beam evaporator onto a film of gold working electrode. The sensor showed high sensitivity for the selective detection of cortisol from saliva. This sensor is found to have less popularity due to complex and high cost of fabrication process, so not used for POC applications and evaluations.

An immunosensor was developed for the determination of cortisol in saliva samples by using the chemiresistor principle and an immunosensor was synthesized using carbon nanotubes (CNTs). This immunosensor which is based on CNTs was found to give rapid and label-free quantification of cortisol from saliva samples for POC [100].

Later, an ultrasensitive system was developed by Arun Kumar *et al.* for the same, which was developed using gold nanowires which are functionalized [101]. In 2017 Usha *et al.*, introduced an optic sensor for the determination of cortisol from saliva which was based on resonance loss mode. This optic sensor showed the LOD of 25.9 fg $ml^{-1}$ [102].

A faster POC salivary cortisol detecting biosensor was developed by Fernandez *et al.*, which was developed on a paper substrate. This sensor was found to be more ecological as it was inkjet-printed on a paper substrate which was flexible. The paper substrate was with an ink which was a metalloporphyrin-based macrocyclic catalyst which interacts with cortisol electrochemically and is captured by aptamer-functionalized magnetic microbeads [103].

### 11.2.5 Multiple Sclerosis

Multiple sclerosis (MS) is an immune-mediated, autoimmune neurodegenerative which appears to be the most common reason for neurological disability shown by young adults [104]. Visible symptoms linked with MS are visual problems, pain, fatigue, motor deficits, sexual dysfunction depression, and bowel and bladder [105, 106]. It is distinguished by the intrusion of immune system cells, high diverseness in the clinical, pathologic, radiologic features, and with a sophisticated clinical course distinguished by demyelination, inflammation, and axonal degeneration. Still there are confusions related to the origin of the disease, but it is vaguely clear that both environmental triggers mainly, vitamin D or ultraviolet Blight (UVB) exposure, and genetic predisposition that is, many genes increase disease proneness are found to be accountable for the cause of MS [107]. It is found that the disease mostly begins in young adulthood and majorly affects women more often than men. The disease causes excessive immune-response which results in the malfunction of the blood–brain barrier in turn brings about the ingress of B lymphocytes and T helper lymphocytes into the central nervous system. Thus, there is a need for a demyelination process so the gush of events in CNS set off the activation of an inflammatory reaction.

There is uniqueness in the clinical result of each patient which is primarily due to the combination of different mechanisms of gliosis, inflammation-demyelination, remyelination-repair, and axonal damage-neurodegeneration in a vast extend also influenced by idiosyncratic factors [104, 108]. For arriving at a proper treatment strategy, it is very necessary to identify these idiosyncratic factors along with understanding the correct mechanism in each case. As a result, current researchers focus on discovering reliable potential biomarkers for each and every MS pathogenic factor. A potential biomarker is believed to ideally behave as a surrogate endpoint biomarker as they might substitute for a clinically significant end point, and they could be even used for early phase clinical trials to make rapid decisions in the drug development process. Since no existing biomarkers can fully contemplate the immensity of the complex MS pathogenic [109] mechanisms to create a unique result, at this time the idea of a surrogate end point biomarker exists only theoretically.

Different types of biochemical molecules, like DNA, RNA, or proteins belong to the category of biomarkers for MS. Choice of a potential biomarker for MS detection depends on the clinical technique that is used for the disease detection. Since there is a direct relationship between the salivary glands and the central nervous system, salivary molecules can act as potential mediums of biomarkers to detect MS.

A study conducted on people with MS to determine the different levels of salivary biomarkers concluded that materials that are responsive to thiobarbituric acid and advanced glycation end-products were shown in increased levels in the test specimen [110]. Such studies helped to discover many biomarkers for MS but, the results are still not sufficient to generalize these biomarkers as potential ones. Further studies have to be carried out to arrive at a well-defined conclusion about these salivary biomarkers.

Many potential biomarkers present in salivary samples have been discovered for MS (in Figure 11.7), but the sensor development for the quantification of these markers requires more research to be carried out carefully so as to know the exact electrochemical reactions that have to happen during the electrochemical sensing experiments. Thus, fabrication of such sensors for the salivary biomarkers for the determination of MS is a growing field with many research possibilities.

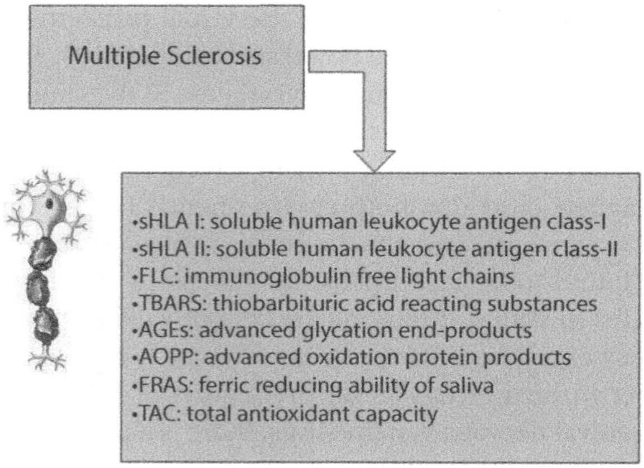


**Figure 11.7** Biomarkers of MS.

### 11.2.6 Neuropsychiatric Disorder

Neuropsychiatric disorders enclose a broad spectrum of medical disorders including seizures, attention deficit disorders, palsies, schizophrenia, cognitive, attention-deficit disorders, etc. The gravity of these diseases is different in children and adults, and from person to person. Symptoms include a wide range of memory and learning problems, emotional and mood problems, depression, and many more. The use of biomarkers can be vital in the diagnosis of such disorders as their symptoms present themselves in different ways and stronger conclusions can at times be difficult. Salivary biomarkers present an easy, noninvasive method of diagnosis for the same as patients with neuropsychiatric disorders can be difficult to treat as limited cooperation is expected from the patient's side [111].

In the past few years, application of saliva to monitor changing levels of hormones has gained popularity. Easily available steroid hormone monitoring equipment is currently available [112]. The levels of cortisol, dehydroepiandrosterone, estradiol, estriol, progesterone, testosterone, and other hormones, with great precision, can be analyzed in saliva, which can be helpful in assessing mood and cognitive-emotional behavior, diagnosing premenstrual depression, assessing ovarian function, evaluating, studying child health and development, and predicting sexual activity in adolescent hormones of proteins [113].

Emotional problems, anxiety and various other psychiatric problems are depicted by the marker cortisol. The level of this marker varies when the subject is exposed to both stress and emotional support. There is a very strong relationship between the level of the marker and saliva secretion which makes saliva a very helpful sample for the detection of the marker [114–116]. Alpha-amylase is also another marker that comes in handy when it comes to diagnosing neurological disorders. This particular marker is also more accurate than cortisol when measuring stress levels [117, 118].

Alcoholism is a very important social issue and is known to adversely affect the mental health of its consumers. A major trend seen among adults now is binge drinking [119]. Here, increased levels of hexosaminidase and immunoglobulin A has been proved in subjects that binge drink. People with alcohol dependency also have abnormalities in salivary immunity. Furthermore, even a single drop of alcohol was shown to raise salivary immunoglobulin A levels [120] and Inherited salivary immunity was disrupted because of binge drinking.

A biosensor to determine the levels of cortisol as a point-of-care diagnosis tool was fabricated using immobilized 2-D tin disulphide flakes and the antibody of the marker on a carbon electrode. It worked based on the interaction between antibody and antigen and showed comparable results with parallel methods [121]. An aptamer-based biosensor was also developed for the marker. The sensor had conjugated silver nanoparticles and the results obtained using this method corroborated with that of the ELISA results [122].

## 11.3 Biosensor for Cancer

Cancer is a fatal disease where there is an atypical growth of cells which usually cannot be controlled and can spread to any part of the body [123]. The mechanism where the abnormal cell growth initiates any tissue or organ in the human body and invades to any nearby parts of the body and spreads to other tissues or organs is called metastasizing. By 2018, it was reported that about 9.6 million people were newly diagnosed with cancer and almost 8.2 million patients were reported to be dead suffering from cancer. Among these, patient's lung cancer patients are about 2.09 million, those with colorectal cancer are about 1.8 million, with breast cancer are about 2.09 million, skin cancer is about 1.04 million, prostrate are about 1.28 million and so on [124]. Cancer is one of the most frequently reported causes of death especially when it is trailing behind a heart disease [125]. Among men, the most common cancer is lung cancer which is almost 16.7% then the prostate which is almost 15.0% then the intestinal cancer which is almost 10.0%, then the stomach cancer that is almost 8.5%, and the liver cancer which is almost 7.5%. Though there is an advent of more and more patients in developed countries, the death rate is found to be the highest at almost 80% in developing countries [126]. When a person is affected with cancer, the patients' body secretes some molecules that act as biomarkers which is the warning sign that confirms that the person is suffering from cancer and proper tests have to be done to confirm the disease. These cancer biomarkers are able to detect the early onset of the disease for better diagnosis and cure in people who are affected. Recent advancements in the area have introduced many biomarkers which are found to be potent for the early-stage detection of different cancers [127].

Since the symptoms of the different tumors are usually not so specific or may be absent like, the lung cancer which is usually detected during its final stages, that is, mostly the cancers are not detected until they have already metastasized. As a result, it is very important to introduce a rapid and highly sensitive, selective and accurate method for the early detection, screening, diagnostics, prognostics and staging of different cancers. Different body fluids like the blood, saliva, urine etc. are taken as samples for the analysis which are expected to have certain molecules that can act as potential biomarkers for diagnosing the cancer. Among these biofluids, salivary samples are the recent research area which is a complex fluid with large number of constituents molecules and has secretions from many major as well as minor salivary glands. The known biomarkers molecules like the DNAs, RNAs, metabolites, proteins, microbiota which are present in the blood are also found in the saliva so, saliva can also be a sample that can provide detectable biomarkers for cancer detection. Many salivary biomarkers have been found to act as potential biomarkers for the detection and evaluation of various cancers in the human body like lung cancer, breast cancer, pancreatic cancers etc. (Figure 11.8).

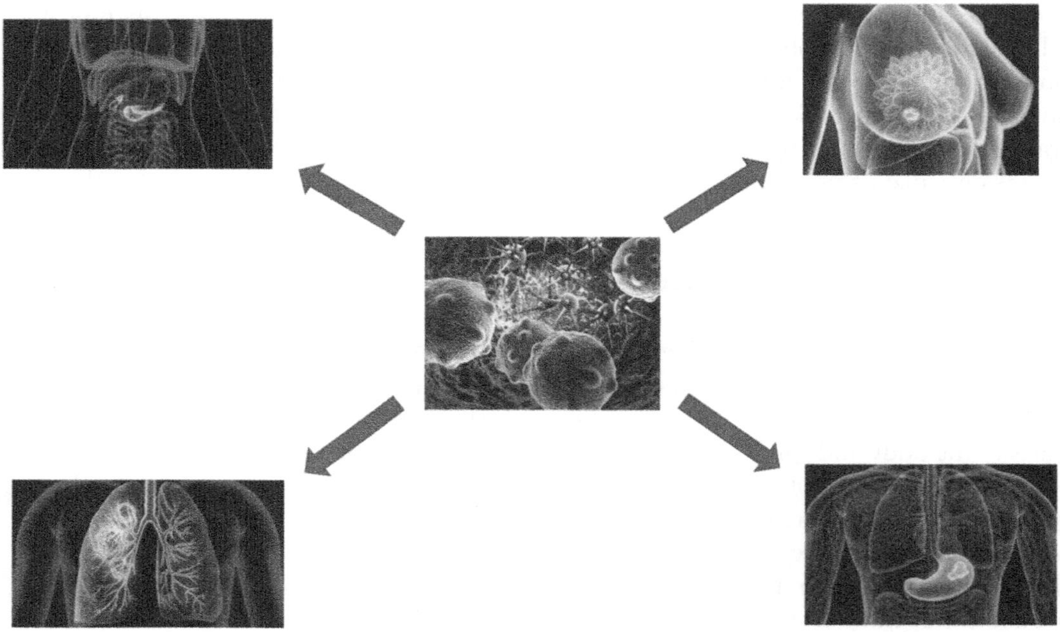

**Figure 11.8** Four major cancer types that are fatal.

### 11.3.1 Breast Cancer

Breast cancer (BC) is the single most leading cause of cancer related deaths in women [128]. It is often defined as cancers arising from within the breast tissues, usually from the inner lining of milk ducts. Breast cancers are also found in men but on a lesser rate, although the prognosis in males is poorer as it is not often diagnosed at the right time. According to latest statistics by About 2.6 million fresh cases of BC have been reported in 2020 and 0.6 million deaths occurred which corresponds to about one-fourth of all cancer cases.

Diagnosis in the early stages of BC can reduce the long-term mortality rate. Techniques which are currently in use for the BC diagnosis include mammography, biopsy, magnetic resonance imaging, ultrasound. Diagnosis via biomarker detection is achieved by various methods such as ELISA, radioimmunoassay (RIA) etc. These methods, however, are invasive in nature and methods of sample collection are not often convenient, economically feasible or easy. Hence, development of highly sensitive noninvasive methods is highly demanded by society.

Electrochemical biosensors offer highly sensitive, fast, real-time detection, reliable, specific and economical platforms for the BC diagnosis which rely on direct monitoring of cancer biomarkers from various samples collected from the body. They have an important role in the bedside-testing diagnostics as they are fast, economically feasible, selective, specific and can be miniaturized.

The most commonly used method of screening BC is mammography, which is proven to have average sensitivity from 55-77 % varying with the type of the procedure [129]. Therefore, many women who have tested negative for BC might actually have it. Presenting false-positive results is also a drawback for standard screening methods which makes the patient exposed to unwanted chemical and radiation therapy and other invasive procedures [130]. Owing to these complications in the standards, a need for finding alternative methods of screening is evident.

Presently, scientific advances have aided biomarker development to the position that saliva is today identified as a superb screening tool that can be obtained with ease. Diagnosis based on biomarkers extracted from saliva are giving way for a new and innovative method for characterizing particular diseases.

CA15-3, transmembrane glycoprotein, is known for its role it plays in cell adhesion [131]. A study by Agha-Hosseini *et al.* compared the amount of the protein in women with and without breast cancer in both serum and saliva. It was found that the levels of CA15-3 had significant correlation in women who had stage 2 BC [132]. Another study by Steckfus *et al.* suggested the protein was able to detect 65% of malignant BC cases in patients [133]. More studies validating the presence of biomarkers such as c-erbB-2, carcinoma antigen 15-3, EGFR, Cathepsin-D and p53 were conducted and they were proven to have communication with salivary glands [134, 135]. It was shown in a study that there is a notable rise in the concentration of erbB-2, 15-3 and p53 in the salivary samples of people affected with BC than unaffected people, proving that all three biomarkers can be used to potentially screen BC in its initial diagnosis. Her-2 in saliva is another biomarker for BC which showed considerable differences between patients before and after receiving therapy for BC suggesting they can be used for monitoring BC for recurrences during follow-up testing [136].

EGF of Epidermal growth factor is yet another biomarker for BC. In 1997, a study showed that EGF levels in salivary samples of women with primary or recurrent BC are elevated compared to that of women without them. The rise in levels were shown in maximum where there were local recurrences which suggested such a biomarker can be employed when it comes to follow-up checking of post-operative BC patients [137].

A number of biomarkers for the detection of the early onset and post-operative monitoring of BC patients. A lot of them are yet to be approved by the FDA so that they can be used as a reliable method for the detection of cancer. Biomarkers for the qualitative detection of these markers are very limited in numbers and aren't very sensitive as the levels of biomarkers in saliva are very low compared to that of other biological fluids.

A sandwich type electrochemical immunosensor was developed for the detection of CA15-3 in 2022 showing an extremely lower LOD of 0.08fg $mL^{-1}$. Screen printed electrode had a layer-by-layer film of gold nanoparticles and reduced graphene oxide deposited together and was selective for the biomarker from human saliva and serum [138]. A Silver/Zinc Oxide based surface plasmon resonance biosensor could also successfully detect the levels of the glycoprotein from salivary samples [139].

A low-cost biosensor for salivary HER-2 was developed immobilizing anti-HER2 antibodies on a screen-printed electrode on top of an electrodeposited layer of diazonium salt. The sensor worked based on electrochemical impedance spectroscopy and showed good results with a high linear range of detection [140]. Three biosensors for the detection of biomarkers, EGFR, HER-2 and p53 have been developed by Imad Abrao Nemeir. They showcase good sensitivity and were significantly accurate in determining the biomarkers from salivary samples [141].

### 11.3.2 Lung Cancer

Lung cancer (LC) is one of the forerunners in cancer-related deaths across the globe, occurring in a higher percentage in people who smoke. The two main categories of LC are Small-cell LC (SCLC) and non–small-cell LC (NSCLC) [142]. Various causes of lung cancers are found which includes inhalation of asbestos, second-hand smoke, arsenic and various other toxic metals and exposure to complex hydrocarbons [143]. Even with all these, the prime cause of lung cancer is proven to be smoking of tobacco and tobacco products.

Non–small-cell lung cancer is usually identified in its later stages with several distant metastases which results in a poor survival rate of 25%. One type of NSCLC is Adenocarcinoma, which usually materializes on the outer parts of the lungs which makes it easier to diagnose before further metastasis. Another type of NSCLC, squamous cell carcinomas. Patients usually are smokers, and it is found in the middle part of the lungs. The last type is large cell carcinoma, which can be seen anywhere and develops and extends quickly resulting in late diagnosis.

Small cell lung cancers or SCLCs are the second category of lung cancer. They tend to grow and spread very fast so that at the time of diagnosis, considerable spreading would have happened. They react well to chemotherapy and radiation but are prone to come back at a different point of time.

SCLC has a very low survival rate owing to its rapid and aggressive growth and metastasis [144]. NSCLC has a better rate of survival albeit being small as it is often diagnosed at a later stage making treatment difficult.

The tools mainly used for the diagnosis of both types of cancers are Low-Dosed Computerized Tomography and chest radiographs. Studies have shown that these methods are very limited in their detection and tend to show false positives which can lead to dangerous and unnecessary radiation treatment and further testing including invasive procedures. They also tend to give false negatives which restricts the patient from getting the treatment in the right time. Because of this, detection of biomarkers can be considered a highly efficient method for lung cancer diagnosis which helps in collecting precise data and assigning patients to targeted treatments in primary stages.

Therefore, several biomarkers, shown in Table 11.2 are collected from samples such as blood, urine, sputum etc. extensively for pre-diagnosis and diagnosis of lung cancer. But only limited data exists on liquid biopsy samples collected from saliva for the same. Some of the important biomarkers for lung cancer are shown in Figure 11.9.

Diagnosis and prognosis of LC using salivary biomarkers are a limited but growing area of research. Accurate detection of such markers is much needed, and use of biosensors is apt for it even though their numbers are limited at this point.

A disposable biosensor for α-amylase was synthesized using molecular imprinting (MI) technology from samples. It was proved to be cost effective and portable which is apt for point-of-care diagnosis. The template biomolecule used in the synthesis was immobilized over the working area which was activated prior to the process using a self-assembled monolayer (SAM) of cysteamine. Sensor was highly selective to the target molecule and showed a very wide linear range $3.0 \times 10^{-4}$ to $3.0 \times 10^{-2}$ in human saliva [156]. In 2010, a tri-enzymatic biosensor was developed for the above biomarker in samples that monitored the amounts of maltos that is a product of hydrolysis of maltopentose mediated by the enzyme of interest. The sensor was co-immobilized with α-glucosidase with Prussian Blue. The sensor was disposable and stable [157].

| | Markers | Category |
|---|---|---|
| Metabolic | Catalase activity, triene conjugates, Schiff bases, pH, sialic acid, alkaline phosphatase, chloride | Amino acids, biochemical |
| | ALT, AST/ALT, ALP, GGT, α-amylase | |
| Inflammatory | IL-1β, ILIRN, IL7, IL 10, CCL11, CXCL10, PDGF-BB, TNF | Protein |
| Proteomic | Haptoglobin, zinc-α-2-glycoprotein, calprotectin | Protein |
| Transcriptomic | *CCNI, EGFR, FGF19, FRS2, GREB1* | mRNA |
| | *EGFR* 19-del<br>*EGFR* 21-L858R | DNA |
| | *EGFR* 19-del<br>*EGFR* 21-L858R | DNA |
| | *EGFR* 19-del<br>*EGFR* 21-L858R | scfDNA† |
| Microbial | *Veillonella, Streptococcus* | 16S rRNA |
| | *Capnocytophaga, Veillonella* | 16S rRNA |
| | *Sphingomonas, Blastomonas, Acinetobacter, Streptococcus* | 16S rRNA |

**Figure 11.9** Salivary biomarkers for lung cancer [145–155].

Aydin and Sezgintürk developed an immunosensor based on the immobilization 6-Phosphono Hexanoic acid and further immobilization of anti-IL-1β as a biorecognition tool on the carboxylic group of the former. It was sensitive, repeatable and reproducible, showing high recovery from human serum and saliva and was deemed fit for routine analyses owing to its efficiency and cost-effectiveness. [158]. An impedimetric biosensor based on polythiophene polymer with a wide array of carboxylic groups working as binding sites for IL-1β was developed in 2018 with a wide linear range and had a low detection limit [159].

Immunosensor using electro-click technology designed by Guerrero *et al.* provides a greener tool for determination of the biomarker and it is significantly faster than the more traditional methods of detection [160]. Another important method for the pre-diagnosis and diagnosis of varied cancers including lung cancer is a microfluidic chip that has a microelectrode and an electrochemical biosensor that detects the presence of NSCLC cytokeratin fragment 21-1 (CYFRA 21-1). It showed promising results with a very low limit of detection [161]. Another non-invasive strategy for detecting NSCLC is a microfluidic paper-based biosensor which uses target DNA as sample to determine EGFR mutations for the successful detection of LC providing a low-cost, highly reproducible option [162].

### 11.3.3 Pancreatic Cancer

Pancreatic cancer (PC), a greatly threatening and forceful tumor, is the fourth principal cause of death among cancer related diseases, with an incident rate nearly equal to the

**Table 11.2** Biosensors for salivary biomarkers of cancer.

| Disease | Biomarker | Biosensor type | LOD | |
|---|---|---|---|---|
| **Breast cancer** | CA15-3 | Electrochemical immunosensor | 0.08fg $mL^{-1}$ | [138] |
| | | Surface plasmon resonance | - | [139] |
| | HER-2 | SPE | - | [140] |
| | EGFR, HER-2, p53 | - | - | [141] |
| **Lung cancer** | A-amylase | MIP | $< 3.0 \times 10^{-4}$ mg $mL^{-1}$ | [156] |
| | | | | [157] |
| | IL-1β | Immunosensor | 5 units $mL^{1}$ | [158] |
| | | Impedimetric biosensor | 3 fg $mL^{-1}$ | [159] |
| **Pancreatic cancer** | CA 19-9 | HEMT | - | [170] |
| **Gastric cancer** | | Microfluidic optoelectronic sensor | - | [174] |

mortality rate [163]. PC has poor prognosis and poor survival rate mostly because of it being diagnosed at late stages where treatment is difficult. PC does not show many symptoms in patients during the earlier stages and the tumor biology showcases that it adds to faster relapses and metastasis and does not respond well to cancer treatment plans [164]. To have better prognosis, their need for a better screening process is evident. Family history of PC [165], Cigarette smoking [166], chronic pancreatitis [167], diabetes mellitus [168], etc. are all proven to be risk factors for the disease but the screening of these high-risk candidates for PC are very limited.

The carbohydrate antigen 19-9 (CA 19-9) is the most used and studied biomarker when it comes to PC detection and prognosis, but it is nowhere near perfect owing to its many limitations, including bad specificity and the fact that it shows higher false positives when in the presence of obstructive jaundice [169]. This gives way for the need to have a better, nay, ideal biomarker for the diagnosis and prognosis of PC.

The detection of CA 19-9 is facilitated by an antibody functionalized AlGaN/GaN high electron mobility transistor (HEMT) sensor which employs self-assembled monolayer technique. It can potentially be made portable and can be used for PC screening [170].

A study done by Zhang *et al.* confirmed a group of four messenger RNA biomarkers identified together can be used to effectively tell apart PC cancer patients. The four markers are K-RAS, MBD3L2, ACRV1, and DPM1 [171].

### 11.3.4 Gastric Cancer

This is an aggressive disease, which is physiologically and genetically diverse, and its causes are not yet understood clearly [172]. The disease is caused by the growth of cancer cells in the stomach lining. The disease does not show early symptoms thus the disease grows inside the stomach unknowingly. This makes the disease diagnosis difficult and thus we need appropriate methods to overcome this death rate. There are many surgical and clinical ways of approaching the disease after it is known but this again will not make any difference in the patient's condition permanently. In this case, the upcoming world seeks something that is most suitable called biosensors for the early detection of gastric cancer (GC).

In 2016, HuaXiao, Yan Zhang *et al.* developed a biomarker in a very different approach that is, quantitative proteomic approach for the diagnosis of GC. They used a technique called Tandem Mass Tag (TMT) technology by collecting human salivary protein to study the difference between patients and healthy controls [173]. The researchers studied many proteins overall and found three of them; cystatin B, triosephosphate isomerase and malignant brain tumors 1 showed variation in gastric patients and healthy controls in a significant manner. This combination of biomarkers shows high specificity and sensitivity and is highly encouraged as it can be used in a non-invasive way. Zilberman *et al.* constructed a microfluidic optoelectronic sensor for the non-invasive detection of GC. This sensor detects the amount of $CO_2$ and $NH_3$ because stomach cancer causing bacteria Helicobacter pylori. produce urease enzymes which get converted to $CO_2$ and $NH_3$ (in saliva). The results of this sensor are highly encouraging, having potential of early diagnosis of GC in a non-invasive way [174]. The most commonly accessed biomarkers of the GC are Carcinoembryonic Antigen (CEA), Carbohydrate Antigen 19-9 (CA19-9), and Serum Protein p53. In 2020, a study took place in which these three biomarkers were determined simultaneously when a microsensor was developed based on different graphene materials modified with a Protoporphyrin IX solution which is utilized as the identification molecule. The sensor shows a broad linear range so that one can diagnose the different stages of cancer [175].

## 11.4 Conclusion

Salivary samples collected from living beings are explored in the study and detection of a variety of diseases, including fatal cancers, neurological disorders and neuropsychiatric diseases (alcoholism, steroid conception). Saliva's utility as a biomarker has been applied to neurological and cancer diseases as well. Saliva is indicated as a great resource for the diagnostics of and systemic disorders, as well as for detection of neuropsychiatric disorders, which is currently an active area of research.

## References

1. Farah, R., Haraty, H., Salame, Z., Fares, Y., Ojcius, D.M., Said Sadier, N., Salivary biomarkers for the diagnosis and monitoring of neurological diseases. *Biomed. J.*, 41, 63–87, 2018.

2. Espíndola, O.M., Brandão, C.O., Gomes, Y.C.P., Siqueira, M., Soares, C.N., Lima, M.A.S.D. *et al.*, Cerebrospinal fluid findings in neurological diseases associated with COVID-19 and insights into mechanisms of disease development. *Int. J. Infect. Dis.*, 102, 155–62, 2021.
3. Xiao, F., Lv, S., Zong, Z. *et al.*, Cerebrospinal fluid biomarkers for brain tumor detection: Clinical roles and current progress. *Am. J. Transl. Res.*, 12, 4, 1379–1396, 2020.
4. Ahn, S.B., Sharma, S., Mohamedali, A., Mahboob, S., Redmond, W.J., Pascovici, D. *et al.*, Potential early clinical stage colorectal cancer diagnosis using a proteomics blood test panel. *Clin. Proteomics*, 16, 1–20, 2020.
5. Yokota, K., Uchida, H., Sakairi, M., Abe, M., Tanaka, Y., Tainaka, T. *et al.*, Identification of novel neuroblastoma biomarkers in urine samples. *Sci. Rep.*, 11, 1–9, 2021.
6. Malon, R.S.P., Sadir, S., Balakrishnan, M., Córcoles, E.P., Saliva-based biosensors: Non-invasive monitoring tool for clinical diagnostics. *Biomed. Res. Int.*, 2014, 962903, 20, 2014.
7. Williamson, S., Munro, C., Pickler, R., Grap, M.J., Jr, R.K.E., Comparison of biomarkers in blood and saliva in healthy adults. *Nurs. Res. Pract.*, 2012, 246178, 4, 2012.
8. Tumani, H., Petereit, H.F., Gerritzen, A., Gross, C.C., Huss, A., Isenmann, S., Jesse, S., Khalil, M., Lewczuk, P., Lewerenz, J., Leypoldt, F., S1 guidelines "lumbar puncture and cerebrospinal fluid analysis" (abridged and translated version). *Neurol. Res. Pract.*, 2, 1, 1–28, 2020.
9. Javaid, M.A., Ahmed, A.S., Durand, R., Tran, S.D., Saliva as a diagnostic tool for oral and systemic diseases. *J. Oral. Biol. Craniofac. Res.*, 6, 67–76, 2016.
10. Lasisi, T.J. and Lawal, F.B., Preference of saliva over other body fluids as samples for clinical and laboratory investigations among healthcare workers in Ibadan, Nigeria. *Pan Afr. Med. J.*, 34, 191, 2019.
11. D'Amone, L., Matzeu, G., Omenetto, F.G., Stabilization of salivary biomarkers. *ACS Biomater. Sci. Eng.*, 7, 12, 5451–5473, 2021.
12. Podzimek, S., Vondrackova, L., Duskova, J., Janatova, T., Broukal, Z., Salivary markers for periodontal and general diseases. *Dis. Markers*, 2016, 9179632, 8, 2016.
13. Kaczor-Urbanowicz, K.E., Wei, F., Rao, S.L., Kim, J., Shin, H., Cheng, J. *et al.*, Clinical validity of saliva and novel technology for cancer detection. *Biochim. Biophys. Acta – Rev. Cancer*, 1872, 49–59, 2019.
14. Walton, E.L., Saliva biomarkers in neurological disorders: A "spitting image" of brain health? *Biomed. J.*, 41, 59–62, 2018.
15. Humphrey, and Williamson, R.T., A review of saliva normal composition, flow, and function. *J. Prosthet. Dent.*, 85, 162–9, 2001.
16. Gug, I.T., Tertis, M., Hosu, O., Cristea, C., Salivary biomarkers detection: Analytical and immunological methods overview. *Trends Anal. Chem. - TrAC*, 113, 301–316, 2019.
17. Punj, A., Secretions of human salivary gland, in: *Salivary glands - new approaches in diagnostics and treatment*, IntechOpen, 2019.
18. Yoshizawa, J.M., Schafer, C.A., Schafer, J.J., Farrell, J.J., Paster, B.J., Wong, D.T.W., Salivary biomarkers: Toward future clinical and diagnostic utilities. *Clin. Microbiol. Rev.*, 26 4, 781–791, 2013.
19. Jasim, H., Carlsson, A., Hedenberg-Magnusson, B., Ghafouri, B., Ernberg, M., Saliva as a medium to detect and measure biomarkers related to pain. *Sci. Rep.*, 8, 3220, 2018.
20. Vârlan, C., Dimitriu, B., Vârlan, V., Bodnar, D., Suciu, I., Current opinions concerning the restoration of endodontically treated teeth: Basic principles. *J. Med. Life*, 2, 2,165–172, 2009.
21. Streckfus, C.F. and Bigler, L.R., Saliva as a diagnostic fluid. *Oral. Dis.*, 8, 69–76, 2002.
22. Pawlik, P. and Błochowiak, K., The role of salivary biomarkers in the early diagnosis of Alzheimer's disease and Parkinson's disease. *Diagnostics*, 11, 2, 371, 2021.
23. Sener, A., Jurysta, C., Bulur, N., Oguzhan, B., Satman, I., Yilmaz, T.M. *et al.*, Salivary glucose concentration and excretion in normal and diabetic subjects. *Biomed. Biotechnol.*, 2009, 430426, 6, 2009.

24. Campuzano, S., Yánez-Sedeño, P., Pingarrón, J.M., Electrochemical bioaffinity sensors for salivary biomarkers detection. *Trends Anal. Chem. - TrAC*, 86, 14–24, 2017.
25. Christodoulides, N., Floriano, P.N., Miller, C.S., Ebersole, J.L., Mohanty, S., Dharshan, P. *et al.*, Lab-on-a-chip methods for point-of-care measurements of salivary biomarkers of periodontitis. *Ann. N. Y. Acad. Sci.*, 1098, 1, 411–428, 2007.
26. Griffin, S.M., Converse, R.R., Leon, J.S., Wade, T.J., Jiang, X., Moe, C.L. *et al.*, Application of salivary antibody immunoassays for the detection of incident infections with Norwalk virus in a group of volunteers. *J. Immunol. Methods*, 424, 53–63, 2015.
27. Wang, Q., Gao, P., Cheng, F., Wang, X., Duan, Y., Measurement of salivary metabolite biomarkers for early monitoring of oral cancer with ultra-performance liquid chromatography-mass spectrometry. *Talanta*, 119, 299–305, 2014.
28. Bigler, L.R., Streckfus, C.F., Dubinsky, W.P., Salivary biomarkers for the detection of malignant tumors that are remote from the oral cavity. *Clin. Lab. Med.*, 29, 1, 71–85, 2009.
29. Zhang, W., Li, P., Geng, Q., Duan, Y., Guo, M., Cao, Y., Simultaneous determination of glutathione, cysteine, homocysteine, and cysteinylglycine in biological fluids by ion-pairing high-performance liquid chromatography coupled with precolumn derivatization. *J. Agric. Food Chem.*, 62, 25, 5845–5852, 2014.
30. Yoon, J., Cho, H.Y., Shin, M., Choi, H.K., Lee, T., Choi, J.W., Flexible electrochemical biosensors for healthcare monitoring. *J. Mater. Chem. B*, 8, 7303–7318, 2020.
31. Maheshwaran, S., Akilarasan, M., Chen, S.-M., Chen, T.-W., Tamilalagan, E., Tzu, C.Y. *et al.*, An ultra-sensitive electrochemical sensor for the detection of oxidative stress biomarker 3-Nitro-l-tyrosine in human blood serum and saliva samples based on reduced graphene oxide entrapped zirconium (IV) oxide. *J. Electrochem. Soc.*, 167, 6, 066517, 2020.
32. Yang, A. and Yan, F., Flexible electrochemical biosensors for health monitoring. *ACS Appl. Electron. Mater.*, 3, 1, 53–67, 2021.
33. Mathew, M., Radhakrishnan, S., Vaidyanathan, A., Chakraborty, B., Rout, C.S., Flexible and wearable electrochemical biosensors based on two-dimensional materials: Recent developments. *Anal. Bioanal. Chem.*, 413, 727–762, 2021.
34. Yang, X. and Cheng, H., Recent developments of flexible and stretchable electrochemical biosensors. *Micromachines*, 11, 3, 243, 2020.
35. Xu, J., Fang, Y., Chen, J., Wearable biosensors for non-invasive sweat diagnostics. *Biosensors*, 11, 8, 245, 2021.
36. Kadian, S., Arya, B.D., Kumar, S., Sharma, S.N., Chauhan, R.P., Srivastava, A. *et al.*, Synthesis and application of PHT-TiO2 nanohybrid for amperometric glucose detection in human saliva sample. *Electroanalysis*, 30, 11, 2793–2802, 2018.
37. Shin Low, S., Pan, Y., Ji, D., Li, Y., Lu, Y., He, Y. *et al.*, Smartphone-based portable electrochemical biosensing system for detection of circulating microRNA-21 in saliva as a proof-of-concept. *Sens. Actuators B Chem.*, 308, 127718, 2020.
38. J. Say, M.F. Tomasco, A. Heller, Y. Gal, B. Aria, E. Heller, P.J. Plante, M.S. Vreeke, Process for producing an electrochemical biosensor. U.S. Patent 6,103,033, 2000.
39. Mehrvar, M. and Abdi, M., Recent developments, characteristics, and potential applications of electrochemical biosensors. *Anal. Sci.*, 20, 1113–1126, 2004.
40. Wang, J., Nanomaterial-based electrochemical biosensors. *Analyst*, 130, 421–426, 2005.
41. Zhu, C., Yang, G. *et al.*, Electrochemical sensors and biosensors based on nanomaterials and nanostructures. *Anal. Chem.*, 130, 421–425, 2005.
42. Ge, X., Asiri, A.M. *et al.*, Electrochemical sensors and biosensors based on nanomaterials and nanostructures. *Trends Anal. Chem.*, 58, 31–39, 2014.
43. Sotiropoulou, S., Gavalas, V. *et al.*, Electrochemical biosensing systems based on carbon nanotubes and carbon nanofibers. *Biosens. Bioelectron.*, 18, 211–215, 2005.

44. Jing, L., Xie, C., Li, Q., Yang, M., Li, S., Li, H., Xia, F., Electrochemical biosensor for the analysis of breast cancer biomarkers: From design to application. *Anal. Chem.*, 2022, 94, 269–296, 2022.
45. Talsness, C.E., Andrade, A.J.M., B.E. *et al.*, Components of plastic: Experimental studies in animals and relevance for human health. *Anal. Technol. Sci.*, 4, 364–422, 2002.
46. Tvarijonaviciute, A., Zamora, C., Ceron, J.J. *et al.*, Salivary biomarkers in Alzheimer's disease. *Clin. Oral. Investig.*, 24, 3437–3444, 2020.
47. Farah, R., Haraty, H., Salame, Z. *et al.*, Salivary biomarkers for the diagnosis and monitoring of neurological diseases. *Biomed. J.*, 41, 63–87, 2018.
48. Schepici, G., Silvestro, S., Trubiani, O. *et al.*, Salivary biomarkers: Future approaches for early diagnosis of neurodegenerative diseases. *Brain Sci.*, 10, 245, 2020.
49. Javaid, M.A., Ahmed, A.S., Durand, R., Tran, S.D., Saliva as a diagnostic tool for oral and systemic diseases. *J. Oral. Biol. Craniofac. Res.*, 6, 67–76, 2016.
50. Lasisi, T.J. and Lawal, F.B., Preference of saliva over other body fluids as samples for clinical and laboratory investigations among healthcare workers in Ibadan, Nigeria. *Pan Afr. Med. J.*, 34–191, 2019.
51. D'Amone, L., Matzeu, G., Omenetto, F.G., Stabilization of salivary biomarkers. *ACS Biomater. Sci. Eng.*, 7, 12, 5451–5473, 2021.
52. Lane, C.A., Hardy, J., Schott, J.M., Alzheimer's disease. *Eur. J. Neurol.*, 25, 59–70, 2018.
53. Hardy, J., Alzheimer's disease: The amyloid cascade hypothesis: An update and reappraisal. *J. Alzheimer's Dis.*, 9, 151–153, 2006.
54. Zhang, Y.W., Thompson, R., Zhang, H., Xu, H.X., App processing in Alzheimer's disease. *Mol. Brain*, 4, 3, 2011.
55. Parsons, R.B. and Austen, B.M., Protein-protein interactions in the assembly and subcellular trafficking of the base (beta-site amyloid precursor protein-cleaving enzyme) complex of Alzheimer's disease. *Biochem. Soc. Trans.*, 35, 974–979, 2007.
56. Zhang, H., Ma, Q.L., Zhang, Y.W., Xu, H.X., Proteolytic processing of Alzheimer's ss-amyloid precursor protein. *J. Neurochem.*, 120, 9–21, 2012.
57. Naslund, J., Haroutunian, V., Mohs, R., Davis, K.L., Davies, P., Greengard, P., Buxbaum, J.D., Correlation between elevated levels of amyloid beta-peptide in the brain and cognitive decline. *J. Am. Med. Assoc.*, 283, 1571–1577, 2000.
58. Bermejo-Pareja, F., Antequera, D., Vargas, T., Molina, J.A., Carro, E., Saliva levels of abeta1–42 as potential biomarkers of Alzheimer's disease: A pilot study. *BMC Neurol.*, 10, 108, 2010.
59. Kim, C.B., Choi, Y.Y., Song, W.K., Song, K.B., Antibody-based magnetic nanoparticle immunoassay for quantification of Alzheimer's disease pathogenic factor. *J. Biomed. Opt.*, 19, 051205, 2014.
60. Lee, M., Guo, J.P., Kennedy, K., McGeer, E.G., McGeer, P.L., A method for diagnosing Alzheimer's disease based on salivary amyloid-beta protein 42 levels. *J. Alzheimers Dis.*, 55, 1175–1182, 2017.
61. Drubin, D.G. and Kirschner, M.W., Tau protein function in living cells. *J. Cell Biol.*, 103, 2739–46, 1986.
62. Lindwall, G. and Cole, R.D., Phosphorylation affects the ability of tau protein to promote microtubule assembly. *J. Biol. Chem.*, 259, 5301–5305, 1984.
63. Iqbal, K., Alonso Adel, C., Chen, S., Chohan, M.O., El-Akkad, E., Gong, C.X., Khatoon, S., Li, B., Liu, F., Rahman, A. *et al.*, Tau pathology in Alzheimer disease and other tauopathies. *Biochim. Biophys. Acta*, 1739, 198–210, 2005.
64. Pawlik, P. and Błochowiak, K., The role of salivary biomarkers in the early diagnosis of Alzheimer's disease and parkinson's disease. *Diagnostics*, 11, 371, 2021.

65. Shi, M., Sui, Y.T., Peskind, E.R., Li, G., Hwang, H., Devic, I., Ginghina, C., Edgar, J.S., Pan, C., Goodlett, D.R. *et al.*, Salivary tau species are potential biomarkers of Alzheimer's disease. *J. Alzheimers Dis.*, 27, 299–305, 2011.
66. Pekeles, H., Qureshi, H.Y., Paudel, H.K., Schipper, H.M., Gornistky, M., Chertkow, H., Development and validation of a salivary tau biomarker in Alzheimer's disease. *Alzheimer's Dement.*, 11, 53–60, 2019.
67. Ashton, N.J., Ide, M., Scholl, M., Blennow, K., Lovestone, S., Hye, A., Zetterberg, H., No association of salivary total tau concentration with Alzheimer's disease. *Neurobiol. Aging*, 70, 125–127, 2018.
68. Chan, H.-N., Xu, D., Ho, S.-L. *et al.*, Ultra-sensitive detection of protein biomarkers for diagnosis of Alzheimer's disease. *Chem. Sci.*, 8, 4012–4018, 2017.
69. Carro, E., Bartolom, F. *et al.*, Early diagnosis of mild cognitive impairment and Alzheimer's disease based on salivary lactoferrin. *Alzheimer's Dement.: Diagn. Assess. Dis. Monit.*, 8, 131–138, 2017.
70. Carro, E., Bartolome, F., Bermejo-Pareja, F., Villarejo-Galende, A., Molina, J.A., Ortiz, P. *et al.*, Early diagnosis of mild cognitive impairment and Alzheimer's disease based on salivary lactoferrin. *Alzheimer's Dement.: Diagnosis Assess. Dis. Monit.*, 8, 131–8, 2017.
71. Jann, M.W., Rivastigmine, a new-generation cholinesterase inhibitor for the treatment of Alzheimer's disease. *Pharmacother: J. Hum. Pharmacol. Drug Ther.*, 20, 1–12, 2000.
72. Bakhtiari, S., Beladi Moghadam, N., Ehsani, M., Mortazavi, H., Sabour, S., Bakhshi, M., Can salivary acetylcholinesterase be a diagnostic biomarker for alzheimer. *J. Clin. Diagn. Res.*, 11, ZC58–ZC60, 2017.
73. Reichmann, H., Diagnosis and treatment of Parkinson's disease. *MMW Fortschr. Med.*, 159, 63–72, 2017.
74. Al-Nimer, M.S., Mshatat, S.F., Abdulla, H.I., Saliva alpha-synuclein and a high extinction coefficient protein: A novel approach in assessment biomarkers of Parkinson's disease. *J. Med. Sci.*, 6, 633–637, 2014.
75. Vivacqua, G., Latorre, A., Suppa, A., Nardi, M., Pietracupa, S., Mancinelli, R., Fabbrini, G., Colosimo, C., Gaudio, E., Berardelli, A., Abnormal salivary total and oligomeric alpha-synuclein in Parkinson's disease. *PloS One*, 11, 0151156, 2016.
76. Yang, X., Zhao, X., Liu, F., Li, H., Zhang, C.X., Yang, Z., Simple, rapid and sensitive detection of Parkinson's disease related alpha-synuclein using a DNA aptamer assisted liquid crystal biosensor. *Soft Matter*, 17, 4842, 2021.
77. Chen, J., Li, L., Chin, L.S., Parkinson disease protein DJ-1converts from a zymogen to a protease by carboxyl terminal cleavage. *Hum. Mol. Genet.*, 19, 12, 2395–2408, 2010.
78. Tang, B., Xiong, H., Sun, P., Zhang, Y., Wang, D., Hu, Z. *et al.*, Association of PINK1 and DJ-1 confers digenic inheritance of early-onset Parkinson's disease. *Hum. Mol. Genet.*, 15, 11, 1816–1825, 2006.
79. Chen, J., Li, L., Chin, L.S., Parkinson disease protein DJ-1 converts from a zymogen to a protease by carboxyl-terminal cleavage. *Hum. Mol. Genet.*, 19, 2395–408, 2010.
80. Junn, E., Jang, W.H., Zhao, X., Jeong, B.S., Mouradian, M.M., Mitochondrial localization of DJ-1 leads to enhanced neuroprotection. *J. Neurosci. Res.*, 87, 123–9, 2009.
81. Devic, I., Hwang, H., Edgar, J.S., Izutsu, K., Presland, R., Pan, C. *et al.*, Salivary a-synuclein and DJ-1: Potential biomarkers for Parkinson's disease. *Brain*, 134, 7, e178–e178, 2011.
82. Kang, W.Y., Yang, Q., Jiang, X.F., Chen, W., Zhang, L.Y., Wang, X.Y. *et al.*, Salivary DJ-1 could be an indicator of Parkinson's disease progression. *Front. Aging Neurosci.*, 6, 102, 2014.
83. Karaboga, M.N.S. and Sezgintürk, M.K., A nano-composite based regenerative neuro biosensor sensitive to Parkinsonism-associated protein DJ-1/Park7 in cerebrospinal fluid and saliva. *Bioelectrochemistry*, 138, 107734, 2020.

84. Corey-Bloom, J., Aikin, A., Garza, M., Haque, A., Park, S., Herndon, A. *et al.*, Salivary Huntington protein as a peripheral biomarker for Huntington's disease (P1.053). *Neurology*, 86, 2016.
85. Ryberg, H. and Bowser, R., Protein biomarkers for amyotrophic lateral sclerosis. *Expert Rev. Proteom.*, 5, 2, 249–262, 2008.
86. Obayashi, K., Sato, K., Shimazaki, R. *et al.*, Salivary chromogranin a: Useful and quantitative biochemical marker of affective state in patients with amyotrophic lateral sclerosis. *Intern. Med.*, 2008, 47, 1875–1879, 2008.
87. Krüger, T., Lautenschläger, J., Grosskreutz, J., Rhode, H., Proteome analysis of body fluids for amyotrophic lateral sclerosis biomarker discovery. *Proteomics Clin. Appl.*, 7, 1–2, 123–135, 2013.
88. Cloutier, F. and Marrero, A., MicroRNAs as potential circulating biomarkers for amyotrophic lateral sclerosis. *J. Mol. Neurosci.*, 56, 102–112, 2015.
89. Xu, Z., Henderson, R.D. *et al.*, Neurofilaments as biomarkers for amyotrophic lateral sclerosis: A systematic review and meta-analysis. *PLoS One*, 11, 10, e0164625, 2016.
90. Turner, M.R., Bowser, R. *et al.*, Mechanisms, models and biomarkers in amyotrophic lateral sclerosis. *Amyotroph. Lateral Scler. Frontotemporal Degener.*, 14, sup1, 19–32, 2013.
91. Obayashi, K., Sato, K. *et al.*, Salivary chromogranin A: Useful and quantitative biochemical marker of affective state in patients with amyotrophic lateral sclerosis. *Intern. Med.*, 47, 21, 1875–9, 2008.
92. Vejux, A., Namsi, A. *et al.*, Biomarkers of amyotrophic lateral sclerosis: Current status and interest of oxysterols and phytosterols front. *Mol. Neurosci.*, 11, 12, 2018.
93. Roozendaal, B., Kim, S. *et al.*, The cortisol awakening response in amyotrophic lateral sclerosis is blunted and correlates with clinical status and depressive mood. *Psychoneuroendocrinology*, 37, 1, 20–26, 2012.
94. Monosik, R., Streďanský, M., Šturdík, E., Biosensors - classification, characterization and new trends. *Acta Chim. Slov.*, 5, 1, 109–120, 2012.
95. Yamaguchi, M., Matsuda, Y., Sasaki, S., Sasaki, M., Kadoma, Y., Imai, Y., Niwa, D., Shetty, V., Immunosensor with fluid control mechanism for salivary cortisol analysis. *Biosens. Bioelectron.*, 41, 186, 2013.
96. Kaushik, A., Vasudev, A., Arya, S.K., Pasha, S.K., Bhansali, S., Recent advances in cortisol sensing technologies for point-of-care application. *Biosens. Bioelectron.*, 53, 499, 2014.
97. Sun, K., Ramgir, N., Bhansali, S., The superior performance of the electrochemically grown ZnO thin films as methane sensor. *Sens. Actuator B-Chem.*, 133, 533, 2008.
98. Arya, S.K., Dey, A., Bhansali, S., Electrochemical sensing of cortisol: A recent update. *Biosens. Bioelectron.*, 28, 166, 2011.
99. Vabbina, P.K., Kaushik, A., Pokhrel, N., Bhansali, S., Pala, N., Electrochemical cortisol immunosensors based on sonochemically synthesized zinc oxide 1D nanorods and 2D nanoflakes. *Biosens. Bioelectron.*, 15, 63, 124–130, 2015.
100. Vabbina, P.K., Kaushik, A., Pokhrel, N., Bhansali, S., Pala, N., Interaction of graphene electrolyte gate field-effect transistor for detection of cortisol biomarker. *Biosens. Bioelectron.*, 63, 124, 2015.
101. Tlili, C., Myung, N.V., Shetty, V., Mulchandani, A., Carbon nanotubes and graphene nano field-effect transistor-based biosensors. *Biosens. Bioelectron.*, 26, 4382, 2011.
102. Kumar, A., Aravamudhan, S., Gordic, M., Bhansali, S., Mohapatra, S.S., Immunosensor with fluid control mechanism for salivary cortisol analysis. *Biosens. Bioelectron.*, 22, 2138, 2007.
103. Usha, S.P., Shrivastav, A.M., Gupta, B.D., Saliva, a magic biofluid available for multilevel assessment and a mirror of general health—A systematic review. *Biosens. Bioelectron.*, 87, 178, 2017.

104. Fernandez, R.E., Umasankar, Y., Manickam, P., Nickel, J.C., Iwasaki, L.R., Kawamoto, B.K., Todoki, K.C., Scott, J.M., Bhansali, S., Disposable aptamer-sensor aided by magnetic nanoparticle enrichment for detection of salivary cortisol variations in obstructive sleep apnea patients. *Sci. Rep.*, 7, 17992, 2017.
105. Loreficec, L., Coccoc, E., Castagnolab, M., Messanab, I., Olianas, A., Top-down proteomic profiling of human saliva in multiple sclerosis patients. *J. Proteomics*, 187, 212–222, 2018 Sep 15.
106. Frohman, E.M., Racke, M.K., Raine, C.S., Multiple sclerosis–the plaque and its pathogenesis. *N. Engl. J. Med.*, 354, 942–955, 2006.
107. Gugliandolo, A., Longo, F. *et al.*, A multicentric pharmacovigilance study: Collection and analysis of adverse drug reactions in relapsing-remitting multiple sclerosis patients. *Ther. Clin. Risk Manage.*, 14, 1765–1788, 2018.
108. Comabella, M. and Montalban, X., Body fluid biomarkers in multiple sclerosis. *Lancet Neurol.*, 13, 1, 113–26, 2014.
109. Mirzaii-Dizgaha, M.-H., Mirzaii-Dizgahb, M.-R., Mirzaii-Dizgahc. Serum, I., and saliva total tau protein as a marker for relapsing-remitting multiple sclerosis. *Med. Hypotheses*, 135, 109476, 2020.
110. Ngounou Wetie, A.G., Wormwood, K.L. *et al.*, A pilot proteomic analysis of salivary biomarkers in autism spectrum disorder. *Autism Res.*, 8, 3, 338–50, 2015.
111. Kułak-Bejda, A., Waszkiewicz, N., Bejda, G., Zalewska, A., Maciejczyk, M., Diagnostic value of salivary markers in neuropsychiatric disorders. *Dis. Markers*, 2019, 4360612, 6, 2019.
112. Streckfus, C.F. and Bigler, L.R., Saliva as a diagnostic fluid. *Oral. Dis.*, 8, 2, 69–76, 2002.
113. Richards, J.A.S., Russell, D.L., Ochsner, S., Espey, L.L., Ovulation: New dimensions and new regulators of the inflammatory-like response. *Annu. Rev. Physiol.*, 64, 1, 69–92, 2002.
114. Miller, G.E., Chen, E., Zhou, E.S., If it goes up, must it come down? Chronic stress and the hypothalamic-pituitaryadrenocortical axis in humans. *Psychol. Bull.*, 133, 1, 25–45, 2007.
115. Carrasco, G.A. and Van de Kar, L.D., Neuroendocrine pharmacology of stress. *Eur. J. Pharmacol.*, 463, 1-3, 235–272, 2003.
116. La Marca-Ghaemmaghami, P., La Marca, R., Dainese, S.M., Haller, M., Zimmermann, R., Ehlert, U., The association between perceived emotional support, maternal mood, salivary cortisol, salivary cortisone, and the ratio between the two compounds in response to acute stress in second trimester pregnant women. *J. Psychosom.*, 75, 4, 314–320, 2013.
117. Rashkova, M.R., Ribagin, L.S., Toneva, N.G., Correlation between salivary alpha-amylase and stress-related anxiety. *Folia Med.*, 54, 2, 46–51, 2012.
118. Lim, I.S., Correlation between salivary alpha-amylase, anxiety, and game records in the archery competition. *J. Nutr. Biochem.*, 20, 4, 44–47, 2016.
119. Waszkiewicz, N., Szajda, S.D., Jankowska, A. *et al.*, The effect of the binge drinking session on the activity of salivary, serum and urinary β-hexosaminidase: Preliminary data. *Alcohol Alcohol.*, 43, 4, 446–450, 2008.
120. Waszkiewicz, N., Galińska-Skok, B., Zalewska, A. *et al.*, Salivary immune proteins monitoring can help detection of binge and chronic alcohol drinkers: Preliminary findings. *Drug Alcohol Depend.*, 183, 13–18, 2018.
121. Liu, X., Hsu, S.P.C. *et al.*, Salivary electrochemical cortisol biosensor based on tin disulfide nanoflakes. *Nanoscale Res. Lett.*, 14, 1, 189, 2019.
122. Dalirirad, S., Han, D. *et al.*, Aptamer-based lateral flow biosensor for rapid detection of salivary cortisol. *ACS Omega*, *5*, 51, 32890–32898, 2020.
123. WHO, Cancer, 2020, https://who.int/health-topics/cancer.
124. Ferlay, J., Soerjomataram, I. *et al.*, Cancer incidence and mortality worldwide: Sources, methods and major patterns in GLOBOCAN 2012. *Int. J. Cancer*, 136, E–359, 2015.

125. Wang, X., Kaczor-Urbanowicz, K.E. *et al.*, Salivary biomarkers in cancer detection. *Med. Oncol.*, 34, 7, 1–8, 2017.
126. WHO, Cancer, 2022, https://who.int/news-room/fact-sheets/detail/cancer.
127. Crulhas, B.P., Basso, C.R. *et al.*, Review—Recent advances based on a sensor for cancer biomarker detection. *ECS J. Solid State Sci. Technol.*, 10, 4, 047004, 2021.
128. Siegel, R., Naishadham, D., Jemal, A., Cancer statistics, 2013. *CA: Cancer J. Clin.*, 63, 11–30, 2013.
129. Berg, W.A., Gutierrez, L., NessAiver, M.S., Carter, W.B., Bhargavan, M., Lewis, R.S. *et al.*, Diagnostic accuracy of mammography, clinical examination, US, and MR imaging in preoperative assessment of breast cancer. *Radiology*, 233, 830–849, 2003.
130. Drukteinis, J.S., Mooney, B.P., Flowers, C.I., Gatenby, R.A., Beyond mammography: New frontiers in breast cancer screening. *Am. J. Med.*, 126, 472–479, M.J., 2013.
131. Duffy, S., Sherry, F., McDermott, E., O'Higgins, N., CA 15-3: A prognostic marker in breast cancer. *Int. J. Biol. Markers*, 15, 330–333, 2000.
132. Agha-Hosseini, F., Mirzaii-Dizgah, I., Rahimi, A., Correlation of serum and salivary CA15-3 levels in patients with breast cancer. *Med. Oral. Patol. Oral. Cir. Bucal*, 10, E521–e524, 2009.
133. Streckfus, C., Bigler, L., Dellinger, T., Dai, X., Kingman Thigpen Jt, A., The presence of soluble c-erbB-2 in saliva and serum among women with breast carcinoma: A preliminary study. *Clin. Cancer Res.*, 6, 2363–2370, 2000.
134. Streckfus, C., Bigler, L., Dellinger, T., Dai, X., Kingman, A., Thigpen, J.T., A preliminary study of CA 15-3, c-erbB-2, epidermal growth factor receptor, Cathepsin-D, and p53 in saliva among women with breast carcinoma. *Cancer Invest.*, 18, 101–109, 2000.
135. Bigler, L.R., Streckfus, C.F., Copeland, L., Burns, R., Dai, X., Kuhn, M. *et al.*, The potential use of saliva to detect recurrence of disease in high-risk breast cancer patients. *Oral. Pathol. Med.*, 31, 421–431, 2002.
136. Streckfus, C.F. and Bigler, L., The use of soluble, salivary c-erbB-2 for the detection and post-operative follow-up of breast cancer in women: The results of a five-year translational study. *Adv. Dent. Res.*, 18, 17–24, 2005.
137. Navarro, A.M., Mesía, R., Díez-Gibert, O., Rueda, A., Ojeda, B., Alonso, M.C., Epidermal growth factor in plasma and saliva of patients with active breast cancer and breast cancer patients in follow-up compared with healthy women. *Breast Cancer Res. Treat.*, 42, 83–86, 1997.
138. Martins, T.S., Bott-Neto, J.L., Oliveira, O.N. *et al.*, A sandwich-type electrochemical immunosensor based on Au-rGO composite for CA15-3 tumor marker detection. *Microchim. Acta*, 189, 38, 2022.
139. Liang, Y.-H., Chang, C.-C., Chen, C.-C., Chu-Su, Y., Lin, C.-W., Development of an Au/ZnO thin film surface plasmon resonance-based biosensor immunoassay for the detection of carbohydrate antigen 15-3 in human saliva. *Clin. Biochem.*, 45, 18, 1689–1693, 2012.
140. Nemeir, I.A., Mouawad, L., Saab, J., Hleihel, W., Errachid, A., Zine, N., Electrochemical impedance spectroscopy characterization of label-free biosensors for the detection of HER2 in saliva. *MDPI*, 60, 17, 2020.
141. Nemeir, I.A., Development of a new bioanalytical tool for the detection and monitoring of breast cancer biomarkers via saliva, Doctoral Dissertation, Université de Lyon; Université Saint-Esprit, Kaslik, Liban, 2020.
142. Lu, C., Onn, A., Vaporciyan, A.A. *et al.*, Chapter 84: Cancer of the lung, in: *Holland-Frei Cancer Medicine*, 9th ed., Wiley Blackwell, 2017.
143. Basumallik, N. and Agarwal, M., Small cell lung cancer, in: *StatPearls [internet]*, StatPearls Publishing, Treasure Island (FL), 2021.
144. Byers, L.A. and Rudin, C.M., Small cell lung cancer: Where do we go from here? *Cancer*, 121, 5, 664–672, 2015.

145. Bel'skaya, L.V., Sarf, E.A., Kosenok, V.K., Gundyrev, I.A., Biochemical markers of saliva in lung cancer: Diagnostic and prognostic perspectives. *Diagnostics*, 10, 4, 186, 2020.
146. Bel'skaya, L. and Kosenok, V., The activity of metabolic enzymes in the saliva of lung cancer patients. *Natl. J. Physiol. Pharm. Pharmacol.*, 7, 6, 1, 2017.
147. Goh, Y.M., Antonowicz, S.S., Boshier, P., Hanna, G.B., Metabolic biomarkers of squamous cell carcinoma of the aerodigestive tract: A systematic review and quality assessment. *Oxid. Med. Cell. Longev.*, 2020, Article ID 2930347, 13, 2020.
148. Pernemalm, M., de Petris, L., Eriksson, H. *et al.*, Use of narrow-range peptide IEF to improve detection of lung adenocarcinoma markers in plasma and pleural effusion. *Proteomics*, 9, 13, 3414–3424, 2009.
149. Xiao, H., Zhang, L., Zhou, H., Lee, J.M., Garon, E.B., Wong, D.T.W., Proteomic analysis of human saliva from lung cancer patients using two- dimensional difference gel electrophoresis and mass spectrometry. *Mol. Cell Proteomics: MCP*, 11, 2, M111.012112–M111.012112, 2012.
150. Zhou, W. and Christiani, D.C., East meets West: Ethnic differences in epidemiology and clinical behaviors of lung cancer between East Asians and Caucasians. *Chin. J. Cancer*, 30, 5, 287–292, 2011.
151. Liao, B.-C., Lin, C.-C., Yang, J.C.-H., Second and third-generation epidermal growth factor receptor tyrosine kinase inhibitors in advanced nonsmall cell lung cancer. *Curr. Opin. Oncol.*, 27, 2, 94–101, 2015.
152. Lin, C.-C., Huang, W.L., Wei, F., Su, W.C., Wong, D.T., Emerging platforms using liquid biopsy to detect EGFR mutations in lung cancer. *Expert Rev. Mol. Diagn.*, 15, 11, 1427–1440, 2015.
153. Krishnan, S. and Eslick, G., Streptococcus bovis infection and colorectal neoplasia: A meta-analysis. *Int. J. Colorectal Dis.*, 16, 9, 672–680, 2014.
154. Zhang, W., Luo, J., Dong, X. *et al.*, Salivary microbial dysbiosis is associated with systemic inflammatory markers and predicted oral metabolites in non-small cell lung cancer patients. *J. Cancer*, 10, 7, 1651–1662, 2019.
155. Skallevold, H.E., Vallenari, E.M., Sapkota, D., Salivary biomarkers in lung cancer. *Mediators Inflamm.*, 2021, Article ID 6019791, 10, 2021.
156. Rebelo, T.S.C.R., Miranda, I.M., Brandão, A.T.S.C., Sousa, L.I.G., Ribeiro, J.A., Silva, A.F., Pereira, C.M., A disposable saliva electrochemical MIP-based biosensor for detection of the stress biomarker α-amylase in point-of-care applications, pp 427–438. *Electrochem.*, *2*, 427–438, 2021.
157. Mahosenaho, M., Caprio, F., Micheli, L. *et al.*, A disposable biosensor for the determination of alpha-amylase in human saliva. *Microchim. Acta*, 170, 243–249, 2010.
158. Aydın, E.B. and Sezgintürk, M.K., A disposable and ultrasensitive ITO based biosensor modified by 6-phosphonohexanoic acid for electrochemical sensing of IL-1β in human serum and saliva. *Anal. Chim. Acta*, 1039, 41–50, 2018.
159. Aydın, E.B., Aydın, M., Sezgintürk, M.K., Highly sensitive electrochemical immunosensor based on polythiophene polymer with densely populated carboxyl groups as immobilization matrix for detection of interleukin 1β in human serum and saliva. *Sens. Actuators B Chem.*, 270, 18–27, 2018.
160. Guerrero, S., Agüí, L., Yáñez-Sedeño, P., Pingarrón, J.M., Design of electrochemical immunosensors using electro-click chemistry. Application to the detection of IL-1β cytokine in saliva. *Bioelectrochemistry*, 133, 107484, 2020.
161. Feng, J., Wu, T., Cheng, Q., Ma, H., Ren, X., Wang, X., Lee, J., Wei, Q., Ju, H., A microfluidic cathodic photoelectrochemical biosensor chip for the targeted detection of cytokeratin 19 fragments 21-1. *Lab. Chip.*, 21, 2, 378–384, 2021.

162. Tian, Liu, H., Li, Yu, J., Ge, S., Song, X., Yan, M., Paper-based biosensor for non-invasive detection of epidermal growth factor receptor mutations in non-small cell lung cancer patients. *Sens. Actuators B Chem.*, 251, 440–445, 2017.
163. Siegel, R., Ma, J., Zou, Z. *et al.*, Cancer statistics, 2014. *CA Cancer J. Clin.*, 64, 9–29, 2014.
164. Kamisawa, T., Wood, L.D., Itoi, T., Takaori, K., Pancreatic cancer. *Lancet*, 388, 10039, 73–85, 2016.
165. Hruban, R.H., Canto, M.I., Goggins, M. *et al.*, Update on familial pancreatic cancer. *Adv. Surg.*, 2, 44, 293–311, 2010.
166. Iodice, S., Gandini, S., Maisonneuve, P. *et al.*, Tobacco and the risk of pancreatic cancer: A review and meta-analysis. *Langenbecks Arch. Surg.*, 393, 535–4, 2008.
167. Raimondi, S., Lowenfels, A.B., Morselli-Labate, A.M. *et al.*, Pancreatic cancer in chronic pancreatitis; aetiology, incidence, and early detection. *Best Pract. Res. Clin. Gastroenterol.*, 24, 349–58, 2010.
168. Bosetti, C., Rosato, V., Li, D. *et al.*, Diabetes, antidiabetic medications, and pancreatic cancer risk: An analysis from the international pancreatic cancer case-control consortium. *Ann. Oncol.*, 25, 2065–72, 2014.
169. Ballehaninna, U.K. and Chamberlain, R.S., The clinical utility of serum CA 19-9 in the diagnosis, prognosis and management of pancreatic adenocarcinoma: An evidence-based appraisal. *J. Gastrointest. Oncol.*, 3, 2, 105–19, 2012.
170. Chang, C.-W., Chen, P.-H., Wang, S.-H., Hsu, S.-Y., Hsu, W.-T., Tsai, C.-C., Wadekar, P.V., Puttaswamy, S., Cheng, K.-H., Hsieh, S., Wang, H.-Y.J., Kuo, K.-K., Sun, Y., Tu, L.-W., Fast detection of tumor marker CA 19-9 using AlGaN/GaN high electron mobility transistors. *Sens. Actuators B Chem.*, 267, 191–197, 2018.
171. Zhang, L., Farrell, J.J., Zhou, H. *et al.*, Salivary transcriptomic biomarkers for detection of resectable pancreatic cancer. *Gastroenterology*, 138, 39, 49–57, 2010.
172. Durães, C., Almeida, G.M., Seruca, R., Oliveira, C., Carneiro, F., Biomarkers for gastric cancer: Prognostic, predictive or targets of therapy. *Virchows. Arch.*, 464, 367–378, 2014.
173. Xiao, H., Zhang, Y., Kim, Y., Kim, S., Kim, J.J., Kim, K.M. *et al.*, Differential proteomic analysis of human saliva using tandem mass tags quantification for gastric cancer detection. *Sci. Rep.*, 6, 22165, 2016.
174. Zilberman, Y. and Sonkusale, S.R., Microfluidic optoelectronic sensor for salivary diagnostics of stomach cancer. *Biosens. Bioelectron.*, 67, 465–71, 2015.
175. Stefan-van Stadena, R.-I., Ilie-Mihaia, R.-M., Gurzu, S., Simultaneous determination of carcinoembryonic antigen (CEA), carbohydrate antigen 19-9 (CA19-9), and serum protein p53 in biological samples with protoporphyrin IX (PIX) used for recognition by stochastic microsensors. *Anal. Lett.*, 53, 16, 2545–2558, 2020.

12

# Design and Development of Fluorescent Chemosensors for the Recognition of Biological Amines and Their Cell Imaging Studies

**Nelson Malini, Sepperumal Murugesan and Ayyanar Siva***

*Supramolecular and Organometallic Chemistry Lab, Department of Inorganic Chemistry, School of Chemistry, Madurai Kamaraj University, Madurai, Tamil Nadu, India*

***Abstract***

In recent years, the perception of ecologically significant biological ions has arisen as a substantial area in the field of chemical sensors. Among the several sophisticated analytical methods, fluorescence chemosensors has been a prevailing tool due to its ease detection method, fast response, high detection limit and to bioimaging applications. Besides, the fluorescent biosensors should be low-cost, transportable, and accessible, require minimal preparation methods, and would be able to perceive evolving various types of biological molecules for widespread acceptance. Recognition moiety for selective binding of particular biological ions is the main crucial component of a novel biosensor. Eventually, the good biosensor is extremely dependent on its selectivity, sensitivity, reversibility, and extensive pertinence of the recognition ion. In this chapter, we discuss different molecules for biological amine detection by fluorescent chemosensors in detail with its specific emphasis on the use of bioimaging applications. As the field advances, the sustained improvements in fluorescent biosensors are anticipated to result in universal biosensors for essential biological amines which can be detected in real time and in all environmental surroundings.

***Keywords***: Biogenic amines, fluorescence, chemosensor, imaging applications

## 12.1 Introduction

In general, chemosensors had been regularly utilized in several fields for the selective detection of biologically significant amines for about 150 years ago. In that way, usage of efficient fluorescent receptors has developed a range of biological progressions and assessed several sensing techniques to discover toxic compounds that have been released into the ecosystem. In the fluorescent sensing platforms, the fluorescent probes rapidly respond to analytes with dramatic disparities in emission intensity or color. In addition, the diverse fluorescent mechanisms including photo-induced electron transfer (PET), fluorescent resonance energy transfer (FRET), and excited-state intramolecular proton transfer (ESIPT),

**Corresponding author*: drasiva@gmail.com

Inamuddin and Tariq Altalhi (eds.) Biosensors Nanotechnology, 2nd Edition, (245–266) © 2023 Scrivener Publishing LLC

intramolecular charge transfer (ICT) are described for the fluorescence changes. The advancement of chemosensors for the large array of analytes including anions, cations, organic compounds, and other important biological molecules interacts with chemosensors to change their respective structures and luminescent properties. They also expand to biosensing with outstanding applications in diagnosis.

Although it is possible to detect biological ions by more complex systems, there is a need for simple, inexpensive fluorescent and colorimetric chemosensors for the naked eye discernment due to their ease of use in solution as well as in the solid state. Consequently, the attention in developing optical chemosensors explicitly identifying a target biological amine has grown progressively and will still constitute a breakthrough. In this chapter, we dealt with some basics of fluorescent chemosensors and its applications for detecting biologically important amines with its importance in cell imaging studies.

## 12.2 Chemosensors

Molecular sensors are molecular systems whose physicochemical properties change when they are exposed to a chemical species. The chemical species may be cations, anions or neutral molecules. The components of fluorescent chemosensors were a fluorophore, a spacer and a receptor unit in which the analyte specifically coordinates to the receptor and changed into a signal by the signaling moiety. Covalent bonds or noncovalent reversible interactions are responsible for the binding of analytes. In a similar way, the receptor and analyte can interact noncovalently through hydrogen bonding, electrostatic interactions, π-π interactions or reversible covalent bond, etc. (Figure 12.1) [1]. Chemosensors are often used to detect and quantify the metal ions/anions in an easier manner. Compared with the other sophisticated techniques, fluorescent and colorimetric chemosensors have attracted specific consideration in the biological studies for the finding of negligible amounts of cations and anions on account of its adaptability, high selectivity/sensitivity, consistency and reversibility, very low detection limit (LOD), inexpensive, noninvasive and leads to many real time advantageous applications [2, 3]. This led to the advancement of extensive chemosensors for the determination of various metal ions with ease. Hence, due to its advantages and easy binding interaction with the analytes, the design and development of chemosensors has become great importance for the chemists in order to meet several environmental problems prevailing nowadays.

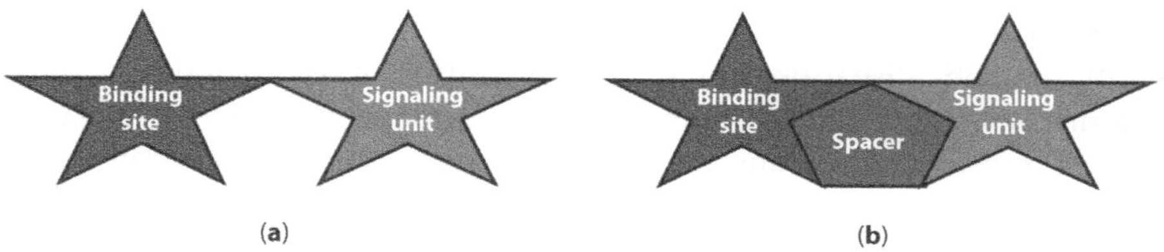


**Figure 12.1** Schematic representation for the design of chemosensors. (a) There is no spacer between the two units. (b) Binding site and signaling unit were connected by a spacer that leads to changes in absorption as well as in the emission spectra.

## 12.3 Importance of Biogenic Amines

Biogenic amines (**BAs**) are a group of bio-active amines with low molecular mass containing amino groups. They can be found in commonly used food items like fish, meat, in many dairy products, etc. [4, 5]. These amines are primarily conceived by the decarboxylation of certain amino acids by their microbial action. An adequate level of biogenic amines in the body can promote normal physical functions; however, the excessive consumption of biogenic amines produces many adverse effects [6]. On the basis of the number of amine groups present, biogenic amines are classified as monoamines, diamines and as polyamines. The most commonly found BAs were putrescine, cadaverine, spermine, spermidine, β-phenylethylamine, histamine, tyramine, trimethylamine, tryptamine, and agmatine (Figure 12.2) [7]. These biogenic amines have important biological functions such as regulation of body temperature, controlling hypertension, in the improvement of gene regulation, in the eventration of cell functions, in the progression of immune responses from the central nervous systems and also in the gastrointestinal emissions [8]. Nevertheless, the presence of BAs in nourishment is expressed as a measure for food decay and the concentration of their presence were prejudiced from the raw materials, microorganisms present in food, processing and during the storage of food [9]. The detection of biogenic amines by various methods is tabulated in Table 12.1.

In order to maintain the wellbeing of food, further accurate, hasty, movable, and ease handling techniques to perceive the presence of biogenic amines were essential. The methods extremely utilized for the determination of biogenic amines were chromatographic practices. Yet, these above-mentioned practices had some notable drawbacks like need of expert personnel, pre-preparation protocols for sample in multiple steps, time consuming etc.

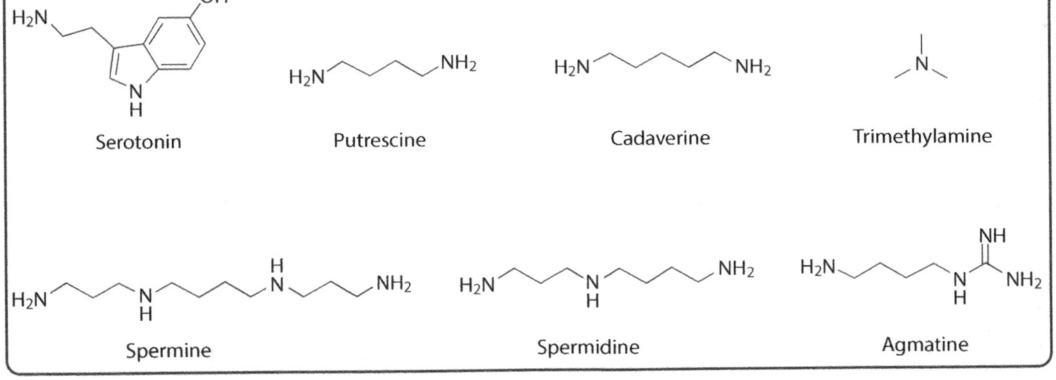


**Figure 12.2** Structures of several important biogenic amines [10].

**Table 12.1** Comparison of different transduction methods used for the detection of biologically significant amines.

| Sl. no. | Detection methods | Biogenic amine | Limit of detection | Applications | References |
|---|---|---|---|---|---|
| 1 | DNA- modified electrode | Aromatic amine | - | | *Dontha et al., Anal. Chem., 68, 4365,* ***1996.*** |
| 2 | Capillary electrophoresis | Primary amines | $10^{-18}$ M | - | *Novtony et al., Anal. Chem., 63, 408,* ***1998.*** |
| 3 | RP-HPLC | Histamine | 0.019 mg/L | Wine | *Moreno-Arribas et al., Food. Res. Int., 38, 387,* ***2005.*** |
| 4 | Copper oxide nano array | Hydrazine | - | - | *Fang et al., Electro. Commun. 11, 631,* ***2009.*** |
| 5 | Ion mobile spectrometry | Triethylamine | - | - | *Puton et al., Talanta, 84, 116,* ***2011.*** |
| 6 | Nano hybrid | Spermine | 115–854 nM | Urine sample | *Huang et al., Anal. Chim. Acta., 1009, 89,* ***2018.*** |
| 7 | Colorimetric approach | Spermine & Spermidine | 25.1 nM & 30.7 nM | - | *Aikawa et al., Tet. Lett., 30, 151302,* ***2019.*** |
| 8 | Fluorescence sensor | Tryptamine | $0.004 \times 10^{-8}$ M | Imaging application, Zebrafish embryos & in HeLa cells | *Siva et al., New J. Chem., 43, 9021,* ***2019.*** |
| 9 | Gas chromatography | Phenylethylamine analogues | 0.809- 4.70 mg/L | - | *Marginean et al., Foren. Chem., 21, 100281,* ***2020.*** |
| 10 | Solid state potentiometric sensor | Tyramine | 7.9 ppm | Food spoilage | *Saad et al., Food. Chem., 346, 128911,* ***2021.*** |

Different from those approaches, fluorescent sensors were advanced and automatic strategies that suggested fast, modest and gainful means for the biogenic amine recognition. With fluorescent biosensing, we list the biological amines that can be identified by fluorescent chemosensors and their specific integrations with bioimaging techniques.

### 12.3.1 Histamine-Based Biosensors

Histamine is a biologically active substance and is known to the intermediaries of antipathy, ulcer and other neurological disorders. Histamine was commonly deposited selectively in mast cells and basophils [11, 12]. An accurate understanding of histamine for *in vivo* and *in vitro* applications were substantial problems in genetic sciences. In addition, its presence in human urine is a vital clinical parameter for various diseases [13]. Therefore, it is important to find an exact, sensitive and novel technique for quantifying histamine in human beings also. In such a way, a simple approach with organic molecules having fluorescent nature should be suitable for histamine detection under mild conditions.

In 2002, Tong *et al.* [14] used Zinc (II)–protoporphyrin (ZnPP), an innovative and efficient monomer for the selective recognition of histamine. They found the association constant and the number of accessible sites by using Scatchard's plot and the histamine concentration is about 0.1–1 mmol/l. They chose methacrylic acid (MAA) **1** as the monomer to prepare the ZnPP-based polymer by multi-binding model. Such that in **1**, the monomer MAA acts as an explicit detection site for specific biomolecules compared to other molecularly imprinted polymers. Next, Seong and his co-workers [15] developed dendritic porphyrin-incorporated nanofibers 2 by electrospinning method for the histamine discovery from other interferences by fluorescent chemosensing technique. In this work, the survival and uniform arrangement of dendritic porphyrin on nanofibers surface were confirmed by EDX, ATR–FTIR and XPS. In addition, 2 showed the specific change with histamine alone, compared with other competitors in the emission spectra. The binding of their synthesized dendritic polymer with histamine was due to the adduct formed between 2 and histamine via coordination and quenching mechanism (Figure 12.3).

Thereafter, Seto *et al.* reported different histamine detection methods by means of a fluorescent compound (Figure 12.4) based on complex formation and ligand exchange mechanism. The emission spectral intensity of 3 had been abridged owing to the efficient quenching of $Ni^{2+}$ ion which on upsurges considerably by the addition of the biomolecule, histamine. Moreover, **3** showed high specificity towards the particular biogenic amine among the other interfering neurotransmitters in phosphate buffer solution (0.1 M, pH 7.4). They also successfully involved the bioimaging applications in RAW264 cells for histamine identification [16].

### 12.3.2 Tryptamine-Based Biosensors

Among the various biogenic amines, the major impact for tryptamine is played by the central nervous system, which originates in the lowest levels in mammals' brains [17, 18]. Consequently, an accurate and fast determination of tryptamine is highly recommended. Numerous sophisticated systematic methods had been exploited for the ascertainment of tryptamine. So, the chemosensing approach was the exclusive method because of its ease handling, inexpensive and larger sensitivity [19–22].

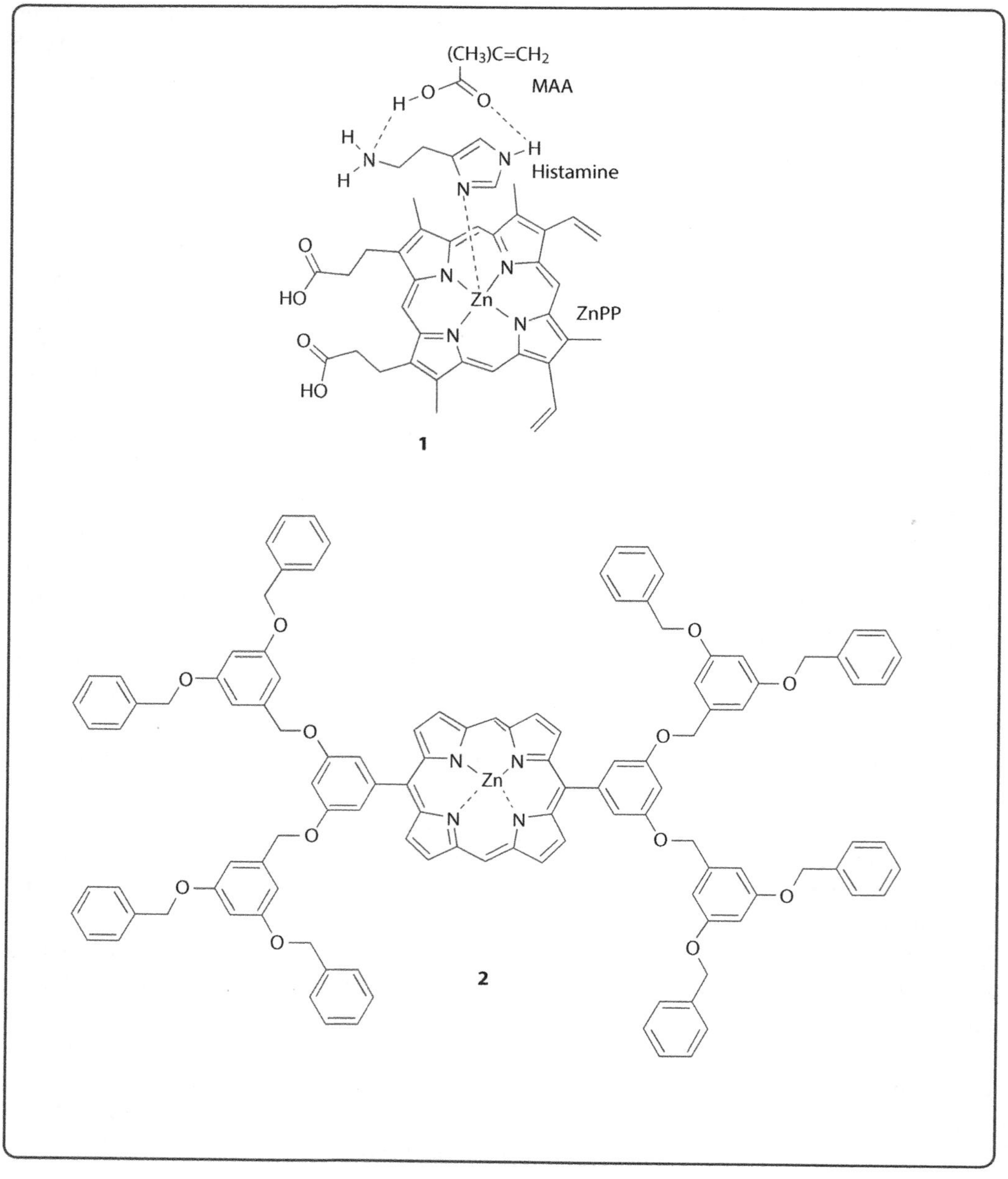


**Figure 12.3** Structures of sensors 1-2 for histamine detection.

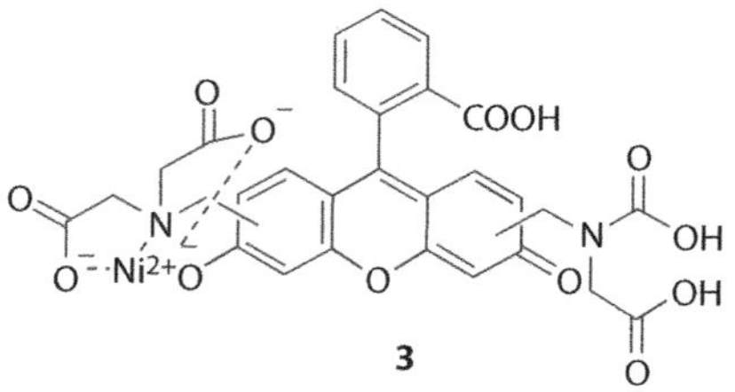


**Figure 12.4** Histamine detection by a nickel bound calcein derivative 3.

Our group Siva *et al.* [23] established the efficient chemosensors for the detection of tryptamine. In 2019, we reported naphthyl hydrazone attached Schiff base derivative **4** had been prepared and proficiently used for the perception of tryptamine and $F^-$ anions. **4** revealed high spectral changes in the absorbance of tryptamine. Besides, the emission spectra of **4** proposed the replication of excited state intramolecular proton transfer (ESIPT) by the enhancement in the emission spectra for tryptamine and $F^-$ anions. Binding stoichiometry of **4** with tryptamine and $F^-$ions had been determined as 1:1 by absorbance and emission spectral data using Job's method. Withal, we performed *in vivo* analysis of both tryptamine and fluoride ions in zebrafish embryos, which designates that our fluorescent chemosensor **4** had good cell permeability and biocompatibility nature.

Similarly, we profound a fluorescent chemosensor **5** for the discerning and selective sensing of tryptamine. **5** was designed for the quick determination of tryptamine in aqueous environment by using fluorescent chemosensor. **5** showed instantaneous and amazing specificity towards tryptamine and the lod accomplished is about 0.6 μM. Binding behavior of **5** with tryptamine features emission enhancement by the inhibition of ESIPT. Outstandingly, chemosensor **5** is applied to tryptamine detection by bio-imaging in HeLa cells for the first time and exploited to depicting logic circuits (Figure 12.5).

As mentioned in Figure 12.6a, amid the various amines tested with 5, tryptamine alone interrelates with 5 effectively and the blue fluorescence was clearly noted by our naked eye while the further enduring amines do not show any response under 365 nm (UV chamber). Also, the strong blue fluorescence changes of 5 can be easily examined by the emission spectra with several biogenic amines. And then, the tryptamine alone has distinct changes with 5 in the emission spectra with a histrionic increment of fluorescence at 448 nm. Simultaneously, there was desertion of fluorescence spectra at 564 nm on account of the inhibition of proton transfer mechanism in 5 representing change in emission from weak to

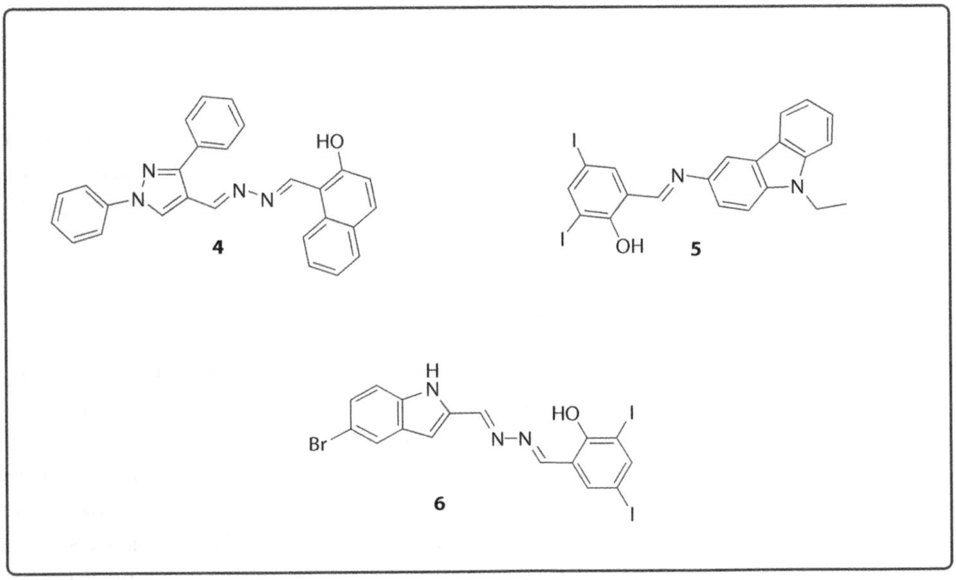


**Figure 12.5** Detection of tryptamine by sensors **4-6**.

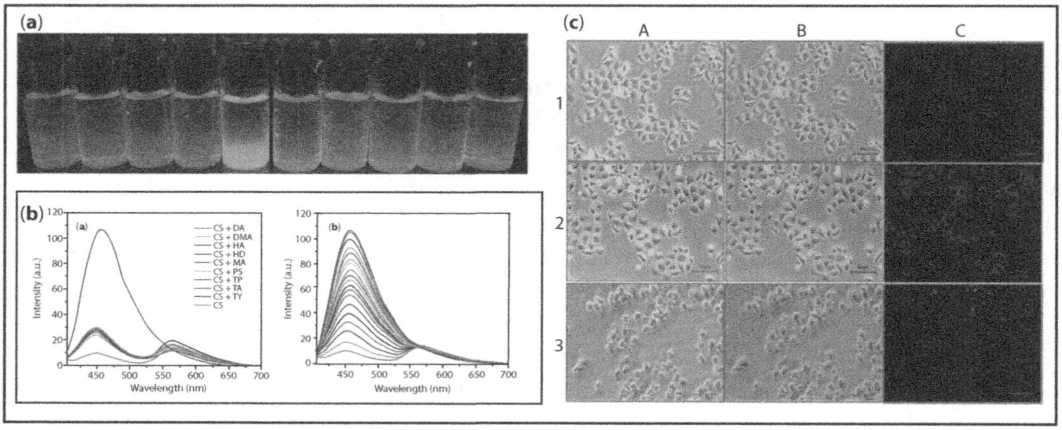


**Figure 12.6** (a) Fluorimetric photographs of 5 alone, 5 with various biogenic amines. (b) (a) Fluorescence spectra of 5 for explaining its selectivity response towards tryptamine. (b) Fluorescence spectra of 5 for notifying its high sensitivity behavior by the gradual addition of tryptamine. (c) Conception of emission enhancement properties of **5** with tryptamine in HeLa cells. [Journal of Molecular Liquids, 2021, 337, 116445]. Copyright 2022 Elsevier.

strong blue. Correspondingly, the titration study was done for 5, which causes progressive increase in emission band at 448 nm whereas the keto band vanishes with the sequential addition of tryptamine (Figure 12.6b).

Then, our synthesized chemosensor 5 was applied to the biological applications in HeLa cells. Initially, the cytotoxicity of 5 was determined by MTT assay and then the IC50 concentration of 5 was calculated as 29.16 μg. The development of 5 with HeLa cells alone resulted in negligible emission. In continuation to this, 5 and HeLa cells were again treated with tryptamine, blue fluorescence was perceived (Figure 12.6c) [24].

In the same way, in our next work we anticipated another multipurpose fluorescent chemosensor derived from combining indole derivative and salicylic hydrazone moiety (**6**) had been used for the discriminate detection of tryptamine and $F^-$ ions [25]. The binding between **6** towards $F^-$/tryptamine had been observed by absorbance and emission spectroscopy. The mechanistic nature behind the sensing was due to the inhibition of ESIPT on the probe **6** and followed by the deprotonation process. As well, the DFT studies were also executed as the evidence for the mechanism explained on **6** on interaction with tryptamine/fluoride anion.

### 12.3.3 Spermine-Based Biosensors

Spermine is a polyamine present naturally in all eukaryotic cells and body fluids. It paves a prominent part in cell growth, proliferation, differentiation, signal transduction and in protein synthesis. It is formed by the decarboxylation of amino acids by enzymatic processes [26–30]. Likewise, high levels of spermine presence in urine causes many abnormal physiological processes and even leads to cancer [31, 32]. Yet, spermine had little volatility, low molecular mass, and the absenteeism of chromophore suggestively hinder the specificity of spermine identification. So, optical methods are fascinating alternative methods as the responses are fast and easy to accomplish. In such aspects, Fukushima *et al.* [33], advanced an innovative strategy for the discovery of spermine by using a modified catechol and boronic acid derivative 7 in neutral medium. 7 resulted in insignificant spectral deviations and showed only trivial variations in color with the other competing biogenic polyamines. Likewise, when spermine is treated to **7**, the uv-vis spectra instigated a red shift with the color alteration from yellow to reddish-brown due to the inhibition of electrostatic repulsive interactions. Also, the lod was in the range of 6.24 micromolar for spermine. They contributed a simplistic method for spermine identification by colorimetric chemosensor tactics.

In another work, Fletcher and his co-workers [34] designed a dicarboxylated ethynylarene moiety 8 to behave as an excellent fluorophore for the low concentration detection of polyamines when mixed in the existence of various metals at micromolar concentrations. In that case, on relating with other diamines, particularly spermine perceived a high fold emission improvement by utilizing mediator as Pb (II) metal ion. Lod was also calculated as 25 μM by gradually enhancing the proportion of Pb (II) to 8. They also exploited the delicate effect towards the spermine identity by biosensing approach with metal chelation. Ko¨stereli *et al.* explored an amphiphile 9 as a chemosensor for spermine detection. Chemosensor 9 permits the spermine identification up to the very low nanomolar concentration range. They also applied sensor 9 which prompted us to examine the sensing of spermine in artificial urine (Figure 12.7) [35].

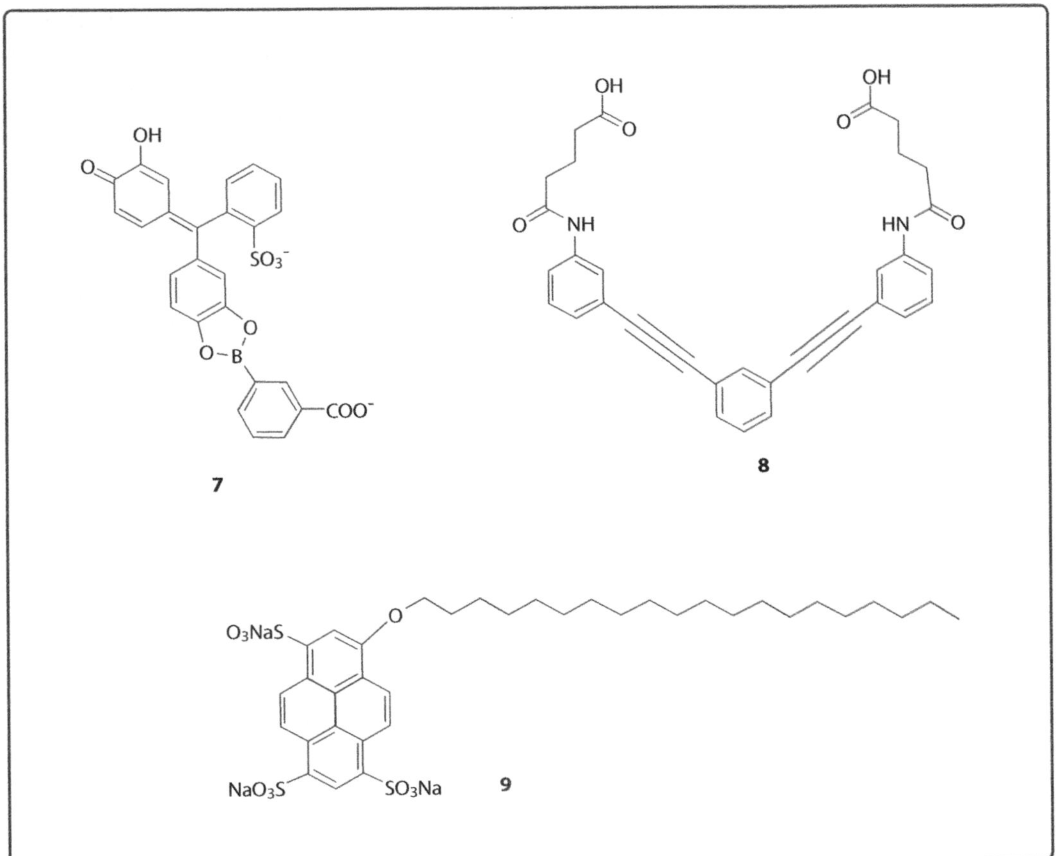


**Figure 12.7** Spermine recognition by the different type of compounds (7–9).

### 12.3.4 Tyramine-Based Chemosensor

Tyramine is one of the biogenic amines responsible for the migraines, pressor action, and hypertension issues and show noxious issues after consuming in large quantities. Whereas the milligram level of tyramine may lead to raised blood pressure, nausea and the higher concentration level accountable for hypertensive crisis. Accordingly, the harmful physical and toxic effects of tyramine presence in foods creates a potential risk to mankind [36, 37]. Henceforth, progress of hasty and consistent perception of tyramine using fluorescent chemosensor method is an essential one. Herein, Kaur and his coworkers [38], described a tetrapodal receptor **10** and then treated it with nanoparticles for the detection of biogenic amine tyramine in water. The cognizance performance of **10** with nanoparticles to diverse cations were primarily reviewed with emission spectroscopy, of these $Fe^{3+}$ ions displayed the specific quenching in the emission band. Next, the $Fe^{3+}$ ion with 10 and its nanoparticle assembly were furthermore treated with dissimilar biogenic amines and it showed a selective response for tyramine with the lod of 377 nM.

In the next approach, Gaganpreet Kaur and his co-workers [38] established a Biginelli based chemosensor **11** using a multicomponent reaction by zinc perchlorate as catalyst in a single pot reaction. Later, the sensor **11** was subjected to acquire organic nanoparticles and categorized by DLS and TEM imaging. The emission spectra in water exposed organic nanoparticles illustrate a fascinating probe for Ag (I) recognition with quenching in its fluorescence. Finally, the subsequent nanoparticles-Ag with chemosensor **11** adduct showed a characteristic response and spectral changes towards tyramine with the limit of detection in nanomolar range (3.91 nM) by turn-on fluorescence response (Figure 12.8).

### 12.3.5 Hydrazine-Based Chemosensor

Hydrazine ($N_2H_4$) is an inorganic amine. It is a highly responsive base and strong reducing agent. It is widely used in the manufacture of many compounds exerted in various fields like in pharmaceuticals, pesticides, photography, etc. [39–42]. Also, it displays extensive lethal

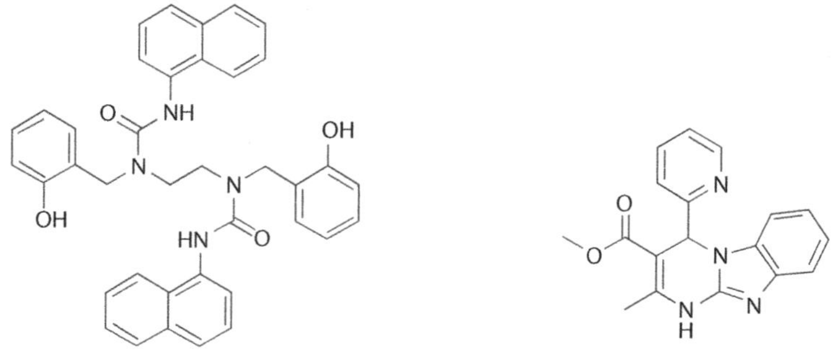


10

11

**Figure 12.8** Two distinct variants (**10-11**) for the identification of tyramine.

issues for mankind [43], and it had been considered as a carcinogenic substance with the limit of detection of about 10 ppb proclaimed by the U.S. Environmental Protection Agency (EPA) [44, 45] and affords impetus efforts on exploring novel simple methods for the proficient sensing of hydrazine even present in trace level.

As stated above, it is necessary to detect hydrazine by a modest method like fluorescent chemosensor approach. Thus, Xia *et al.* [46], described fluorescent chemosensor 12 by a naphthalimide scaffold for the discernment of hydrazine using the intramolecular charge transfer (ICT) mechanism. The probe 12 reveals a very low limit of detection for hydrazine noted as 9.40 ± 0.12 nM. They advanced the probe 12 with high selectivity and applied it to detect gaseous hydrazine in water samples. Besides, they also applied 12 to monitor the hydrazine presence in live cells.

In another way, He and his co-workers [47] found a dual-response chemosensor 13 for the perception of $ClO^-$ and $N_2H_4$ by turn-on emission performance. The chemosensor 13 experienced the off-on response in emission after the interaction between $ClO^-/N_2H_4$ by bond cleavage As well, the sensor 13 had specific features for monitoring hydrazine in fluorescence imaging with virtuous response. The owing distinguishing concerts of chemosensor 13 allowed $ClO^-$ and $N_2H_4$ sensing in Pseudomonas aeruginosa and zebrafish with turn on emission and tested in test strips using Whatman filter paper. Thus, they successfully demonstrated 13 to examine $ClO^-$ and $N_2H_4$ in living cells and in all ecological settings.

At another instance, Xu *et al.* [48], synthesized benzothiazole based fluorogenic receptor 14 for hydrazine. On treating the receptor 14 with hydrazine, emission behavior changed from green to blue. 14 might specifically identify the particular amine in 20 to 40 mM concentration with the lod at ppb range. Notably, the receptor 14 was effectively utilized to perceive $N_2H_4$ in living cells. Li *et al.* [49], designed an innovative chemosensor 15 for discriminating hydrazine from its corresponding interferences. The probe 15 were derived from nopinone, a natural product derivative. As expected, the fluorescence intensity of 15 showed that the $N_2H_4$ concentration levels used were 0 to 900 μM and the limit of detection was about 1.03 μM. In addition, the bioimaging applications evidenced the chemosensor 15 as a powerful ensemble to extricate $N_2H_4$ in live cells. Tan and his coworkers [50] intended innovative sensor 16 by the intramolecular charge transfer (ICT) effect. The probe 16 displayed a considerable enhancement in emission up to 120-fold and the lod calculated was as low as 0.11 ppb. This hydrazine detecting sensor 16 should be an auspicious receptor appropriate for environmental safety, water analysis and other safety examinations.

In continuation to this, Mahapatra *et al.* [51] deliberated a BODIPY dye 17 for the explicit discovery of a strong reducing agent, hydrazine. Specificity experiments proved that the probe 17 had outstanding discernment to hydrazine over other amine containing species. 17 were efficiently exploited for vapor hydrazine into a solid state by using silica gel TLC plates. As well, the probe 17 applied for the identification of hydrazine in Vero cells with high selectivity. Wu and his coworkers accomplished coumarin modified sensor 18 for the ridiculous sensing of hydrazine which resulted in protruding fluorescence off-on response toward hydrazine without any interferences. The lod of hydrazine by the chemosensor 18 was about 3.2 ppb. At last, they applied coumarin derivative 18 for observing hydrazine presence in U251 glioma cells (Figure 12.9) [52]. In addition to this, the hydrazine detection by different types of probes by chemosensing methodology were explained in Table 12.2.

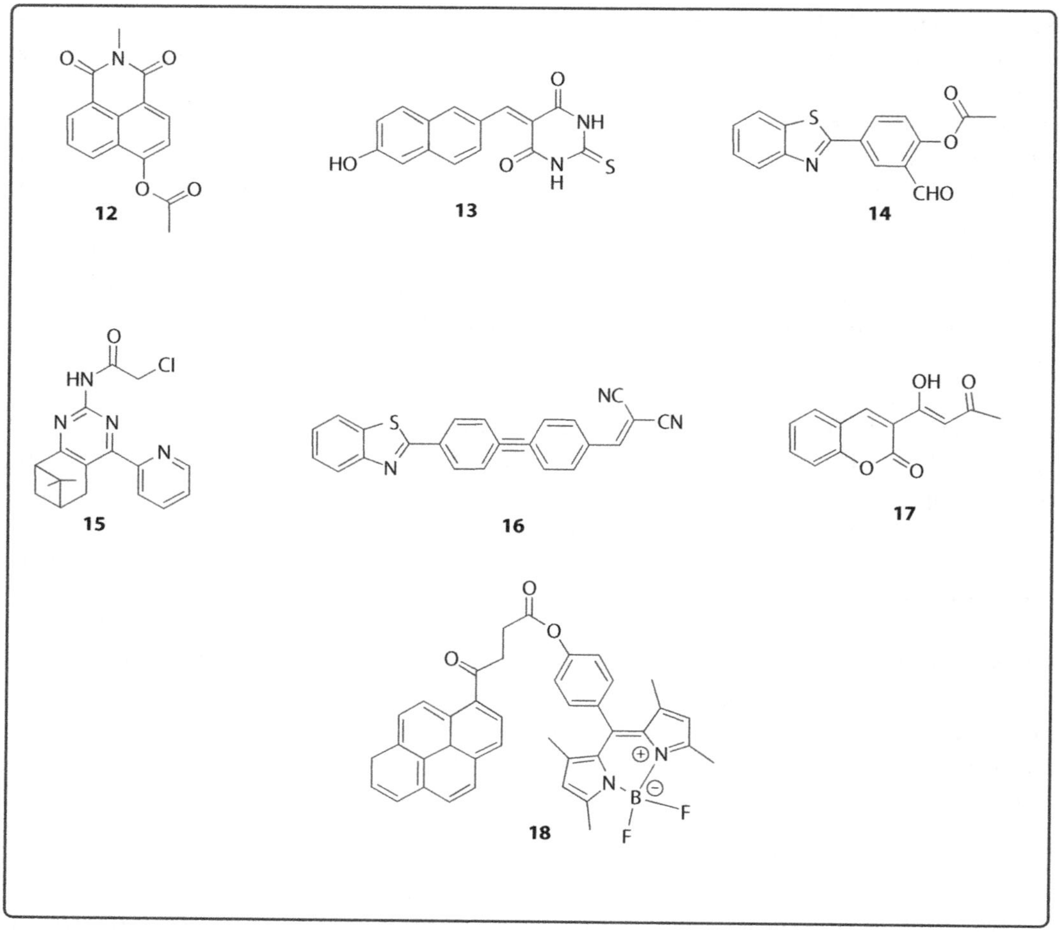


**Figure 12.9** Structures for the hydrazine recognition by the diverse synthesized sensors (12–18).

## 12.3.6 Polyamine-Based Chemosensor

Polyamines are organic molecules obtained from the amino acid enzymatic decarboxylation in nature. Polyamines present mostly in all living cells and execute indispensable purposes like cell development, differentiation, and further cellular progressions [53]. Significantly, irregular amounts of polyamines might be associated with many disorders and could assist as cancer indicators attributable to the polyamine presence in affected person's blood and urine models even in trace amounts [54–56]. Thus, the polyamines detection in physiological samples is measured as an appropriate and prevailing technique for appraising the incidence of cancer tumors. Alternatively, fluorogenic chemosensor provokes higher selectivity for the growth of instantaneous analysis of polyamines in real samples.

Kim *et al.* [57], fruitfully investigated a turn on sensor 19 to detect polyamines in buffered aqueous media through great spectral alteration. The probe 19 would associate with electrostatic interactions leading to aggregates which in turn activates emission in the excimer state. They also performed polyamine analysis in urine by using aqueous solvents.

**Table 12.2** Assessment of various novel fluorescent probes for the hydrazine detection.

| Sl. no. | Fluorescent receptor | LOD | Uses | References |
|---|---|---|---|---|
| 1 | 2-(4-((4-(Benzo[d]thiazol-2-yl) phenyl) ethynyl) benzylidene)- malononitrile | 0.11 ppb | - | *Qian et al., Dyes. Pigms, 99, 996,* ***2013***. |
| 2 | 2-(4-(N, N-diethylamino)-2-hydroxybenzoyl) benzoic acid | 9.6 ppb | Cell imaging in live HepG2 cells | *Kongsaeree et al., Spectrachim. Acta., 185, 228,* ***2017***. |
| 3 | 2-phenyl-benzothiazole based probe | 2.80 ppb | Live cells | *Li et al., J. Photochem. Photobiol., 356, 610,* ***2018***. |
| 4 | (E, E)-N, N'-bis(4 nitro-benzylidene) butane-1,4-diamine | $9.77 \times 10^{-8}$ M | Water sample | *Patra et al., Int. J. Environ. Anal. Chem., 98, 1160,* ***2018***. |
| 5 | [3-(3-methyl-1H-pyrazol-5-yl) coumarin] | 3.2 ppb | Glioma cell line U251 imaging | *Xu et al., Spectrachim. Acta. 188, 80,* ***2018***. |
| 6 | 2,5-bis(benzo[d]thiazol-2-yl) phenol | $1.22 \times 10^{-7}$ M | Detection in serum and gaseous medium | *Zheng et al., Tet. Lett. 60, 151219,* ***2019***. |

In another instance, Bao and his co-workers [58] developed functionalized thiadiazole moiety 20 to notice simultaneously diamine as well as for polyamines. After the interaction with respective amines, 20 became cross-linked by electrostatic interactions to induce interpolymer aggregation through the FRET mechanism. Then, it promotes a great change in emission from blue to orange. Furthermore, the fluorescence response of 20 towards polyamine may attain a detection range as low as 2 mM, representing the probe 20 as an effective fluorescent sensor to spot the minimum level of biogenic amines in an easy manner (Figure 12.10).

### 12.3.7 Aliphatic Amine-Based Chemosensors

In general, amines play a decisive role in all chemical industries mainly in the synthesis of organic compounds [59–61]. For example, ethylenediamine, methylamine and aniline had been broadly manipulated in dye industries [62]. However, in spite of their multipurpose use, these amines are harmful for humans as well as hazardous to the environment. Amongst them, fluorescent sensors display noteworthy advantages, as these amines possess modest, naked eye and quick recognition for the particular receptors.

In this regard, Liu *et al.* [63] designed oligothiophene substituents (21-23) for the revelation of aliphatic amines and hydrazine. But these derivatives (21-23) do not show any

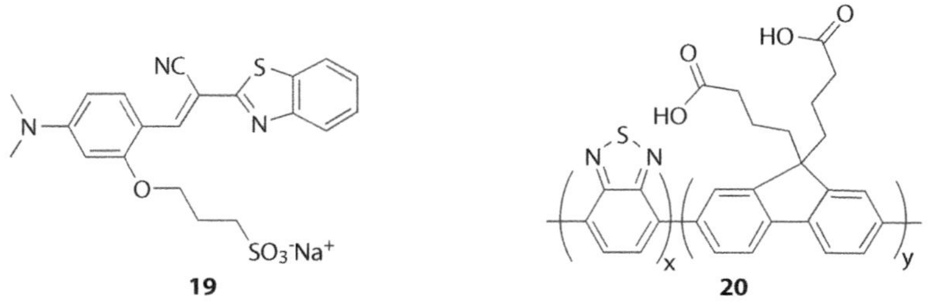


**Figure 12.10** Polyamines detection by the modulation of two discriminate sensors (**19, 20**).

obvious effect on the aromatic amines. Furthermore, they found smartphone related applications and they differentiate hydrazine and aliphatic amines with high satisfaction. Nawaz and his workers [64] created a cellulose containing multi-responsive fluorescent sensor 24 and functioned to the ultrasensitive sensing of numerous amines along with $PO_4^{3-}$, $CO_3^{2-}$, and $BO_3^{3-}$ ions. When paper strips were exposed to several amines in the gaseous and liquid state, 24 disclosed dissimilar emission responses to easily extricate the undesirable amines. Responsive behavior of 24 instigates the distinguished interactions amongst the chemosensor and amines (Figure 12.11).

In the similar method, Ruiu and his coworkers [65] reported the progression of pyrene compounds (25-27) as efficient chemosensors for the sensing of amines in water medium. The sensing ability of 25-27 was tested with useful amines by spectroscopic techniques and the detection limit is in the nanomolar concentration range of the amines. The high affinity

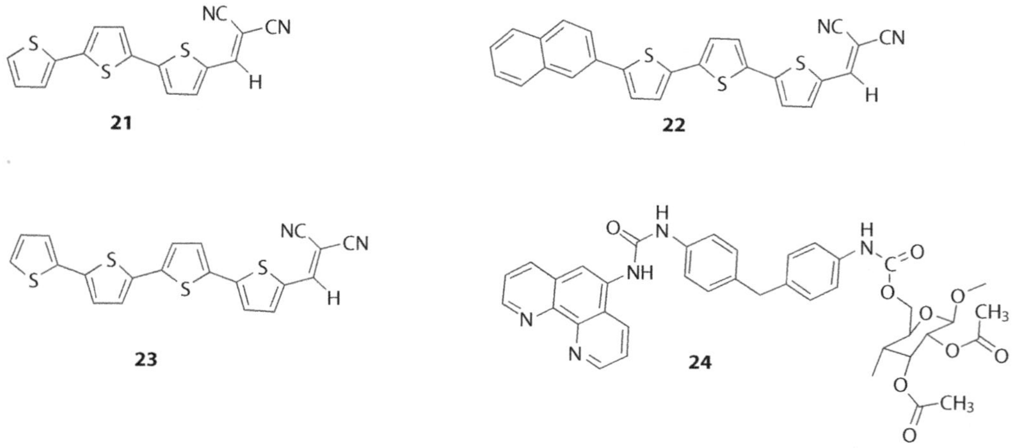


**Figure 12.11** Compounds (21-24) for functioning as fluorescent chemosensors for the aliphatic amine identification.

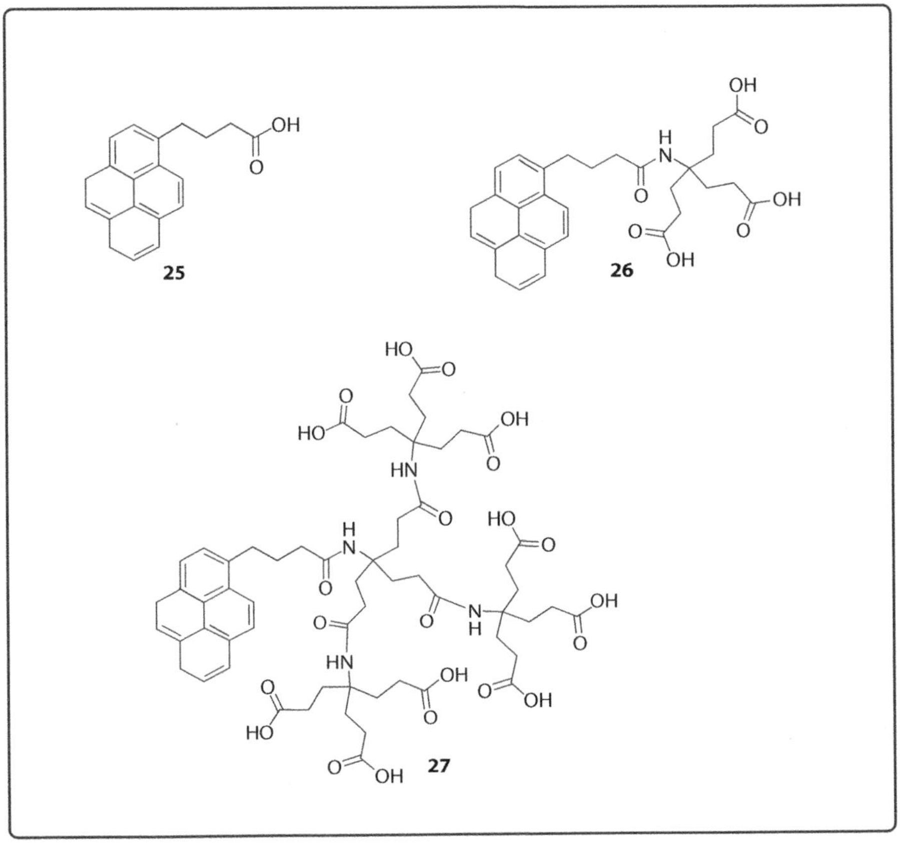

**Figure 12.12** The other set of various novel probes (25-27) for the aliphatic amine detection.

of these compounds with high selectivity for specific aliphatic amines were by turn off response as a result of the interactions in the periphery of the dendrimers (25-27) (Figure 12.12).

### 12.3.8 Norepinephrine-Based Chemosensor

Norepinephrine is a neurotransmitter, which controls several perilous functions, such as remembrance, reactions, and cardiovascular functions. It was evolved from chromaffin cells responsible for catecholamines production in neurosecretory vesicles [66, 67]. Presently, these amines could be analyzed through electrochemical and chromatographic methods [68, 69]. It is very hard to develop specific compounds having fluorescence nature for norepinephrine. However, Secor *et al.* [70] synthesized a boronic acid with coumarin derivative 28 for sensing dopamine and norepinephrine. The sensor 28 interacts with catecholamines, such as dopamine and norepinephrine by ion formation. In addition to this, intramolecular hydrogen bonding in 28 also gave spectacular visual color change to the specific amines over other simple amines.

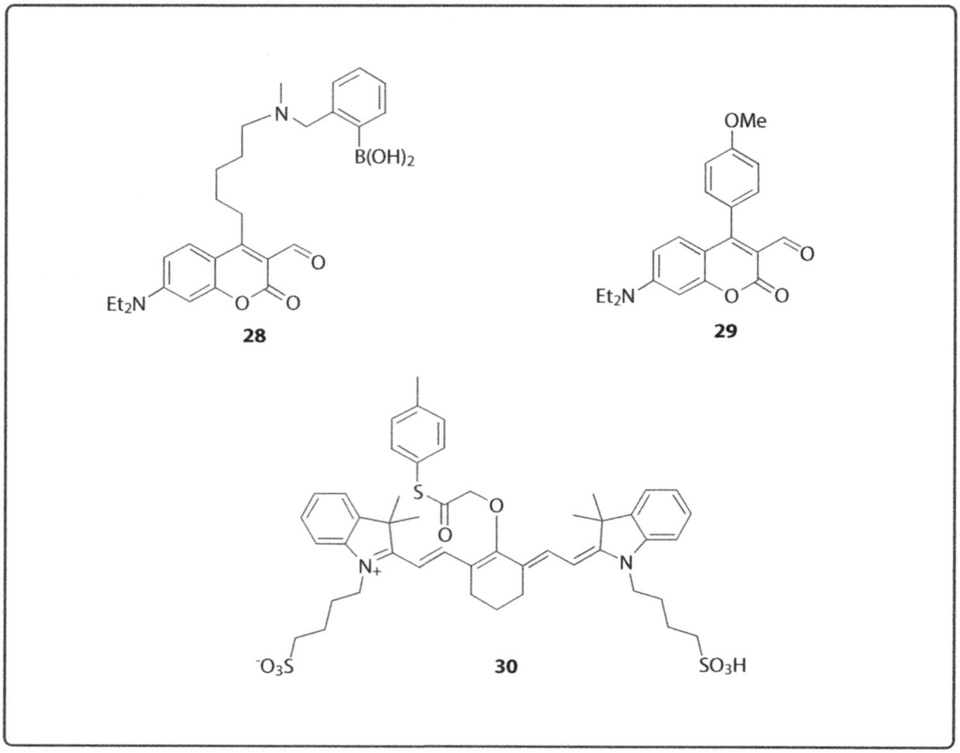


**Figure 12.13** Structural representation three different sensors (29-30) for the perceptive sensing of norepinephrine.

Next, Hettie and his coworkers [71] developed receptor 29 for norepinephrine and dopamine by discerning labeling and bioimaging in secretory cells which discriminate between other interferences in the chromaffin cells and then authorized with different antibodies. Zhou *et al.* [72] utilized longer emission wavelength cyanine protected thiophenol (30) for norepinephrine. The specific response towards norepinephrine is due to the nucleophilic substitution and intramolecular cyclization, which in turn induces the specific red fluorescence. More importantly for the first time, they performed bioimaging of nerve signals induced by $K^+$ ion and real-time imaging in rat brains (Figure 12.13).

### 12.3.9 Serotonin-Based Chemosensor

Serotonin is also known as 5-hydroxytryptamine, a monoamine neurotransmitter in the nervous system. It acts as a crucial part in adaptable emotion, mood, and sense of well-being [73]. Lack of serotonin levels implicated in serious disorders which include anxiety, bipolar issues, and depressive disorders [74, 75]. Consequently, the molecular imaging tools for detecting vesicular serotonin levels must modify their pool size and contain autofluorescence and radiolabeled ligands [76, 77]. Chemosensors with fluorescent compounds were convincing expertise to accomplish imaging applications in serotonin. Thence, Hettie and his co-workers [78] found the assessment of novel neurosensor 31. It showed the

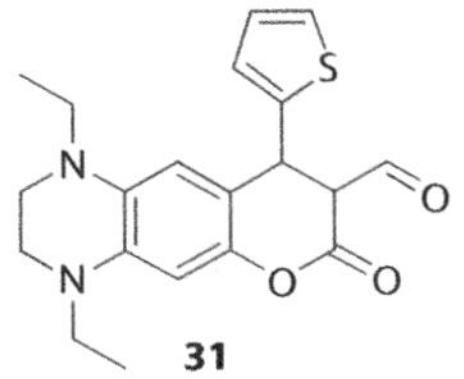


**Figure 12.14** Structure of synthesized neurosensor 31 for the detection of serotonin.

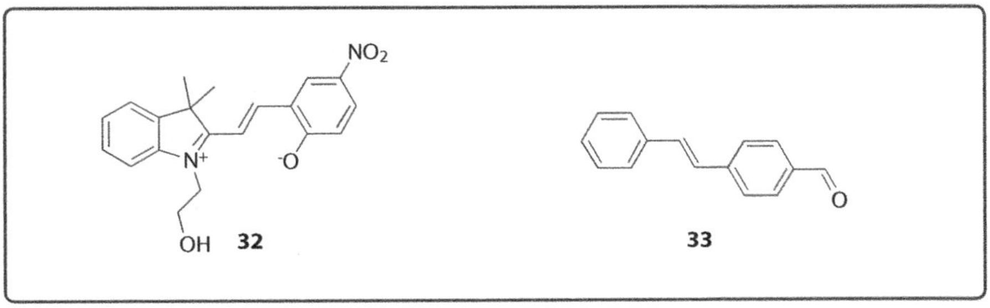


**Figure 12.15** Compounds (**32, 33**) developed for the specific sensing of disparate aromatic amines.

near-infrared fluorescence enhancement to serotonin. From the spectroscopic studies, the neurosensor 31 revealed an 8-fold fluorescence enhancement with large change in emission band at 715. Finally, the neurosensor **31** was applied to neuroimaging with the live chromaffin cells (Figure 12.14).

### 12.3.10 Aromatic Amine-Based Chemosensor

Xue and his coworkers [79] widely advanced spiropyran moiety **32** for detecting aromatic amines. They modified the spiropyran derivative **32** with hydroxyl group for the specific recognition of amines especially for aromatic amines. The sensing mechanisms for binding of aromatic amines was further evaluated by Absorbance spectroscopy, Fourier transform infrared (FT-IR) and Nuclear magnetic resonance (NMR) spectroscopy. At last, they used the synthesized sensor **32** for amine recognition in environments and biological systems.

In another tactic, Lee *et al.* [80] prepared trans-styrylbenzene based motif 33, an excellent receptor for aromatic amines. But the prepared sensor 33 showed large spectral deviations to 4-nitroaniline with high selectivity. In advance to that, they modified the synthesized chemosensor 33 with silica nanoparticles on its surface by improved Stober method. The enhanced sensor 33 works through fluorescence quenching in emission to sense the aniline substituents with high selectivity and pale green emission dissipates with increasing 4-nitroaniline concentration (Figure 12.15).

## 12.4 Conclusion

In conclusion, we simplified the importance of fluorescence chemosensor only for the recognition of some selective biogenic amines in a concise manner by using simple derivatives in numerous approaches. Interestingly, most of the sensors were applied to imaging applications as well as to several environmental and biological interests. It is also noteworthy that the fluorescent chemosensing methodology paves a significant role in detecting severe dreadful diseases like tumors and the growth in the field of research was at peak in various aspects now-a-days. We anticipate that this chapter will entice more attention towards biological amines by chemosensing techniques.

## References


1. Frick, A.A., Busetti, F., Cross, A., Lewis, S.W., Aqueous nile blue: A simple, versatile and safe reagent for the detection of latent fingermarks. *Chem. Commun.*, 50, 3341, 2014.
2. Costa-Fernández, J.M., Pereiro, R., Sanz-Medel, A., The use of luminescent quantum dots for optical sensing. *TrAC - Trends Anal. Chem.*, 25, 207, 2006.
3. Vendrell, M., Zhai, D., Er, J.C., Chang, Y., Combinatorial strategies in fluorescent probe development. *Chem. Rev.*, 112, 4391, 2012.
4. Vasconcelos, H., Coelho, C.C., Matias, A., Saraiva, C., Jorge, P.A.S., De Almeida, J.M.M.M., Biosensors for biogenic amines: A review. *Biosensors*, 11, 82, 2021.
5. Naila, A., Flint, S., Fletcher, G., Bremer, P., Meerdink, G., Control of biogenic amines in food: Existing and emerging approaches. *J. Food Sci.*, 75, 139, 2010.
6. Landete, J.M., De Rivas, B., Marcobal, A., Muñoz, R., Molecular methods for the detection of biogenic amine-producing bacteria on foods. *Int. J. Food Microbiol.*, 117, 258, 2007.
7. Biji, K.B., Ravishankar, C.N., Venkateswarlu, R., Mohan, C.O., Biogenic amines in seafood: A review. *J. Food Sci. Technol.*, 53, 2210, 2016.
8. Ruiz-capillas, C. and Herrero, A.M., Impact of biogenic amines on food quality and safety. *Foods*, 8, 62, 2019.
9. Wunderlichov, L. and Bu, L., Formation, degradation, and detoxification of putrescine by food-borne bacteria: A review. *Compr. Rev. Food Sci. Food Saf.*, 13, 1012, 2014.
10. Schirone, M., Esposito, L., D'Onofrio, F., Visciano, P., Martuscelli, M., Mastrocola, D., Paparella, A., Biogenic amines in meat and meat products: A review of the science and future perspectives. *Foods*, 11, 788, 2022.
11. Merétey, K. and Falus, K., Histamine: An early messenger.in inflammatory and immune reactions. *Immunol. Today*, 13, 154, 1992.
12. Schwartz, J., Pollard, H., Quach, T.T., Histamine as a neurotransmitter in mammalian brain: Neurochemical evidence. *J. Neurochem.*, 35, 26, 1980.
13. Nishiwaki, F., Kuroda, K., Inoue, Y., Endo, G., Determination of histamine, 1-methylhistamine and N-methylhistamine by capillary electrophoresis with micelles. *Biomed. Chromatogr.*, 187, 184, 2000.
14. Tong, A., Dong, H., Li, L., Molecular imprinting-based fluorescent chemosensor for histamine using zinc (II)– protoporphyrin as a functional monomer. *Anal. Chim. Acta*, 466, 31, 2002.
15. Young, D., Choi, M., Kim, Y., Fluorescent chemosensor for the detection of histamine based on dendritic porphyrin-incorporated nanofibers. *Eur. Polym. J.*, 48, 1988, 2012.
16. Seto, D., Soh, N., Nakano, K., Imato, T., Selective fluorescence detection of histamine based on ligand exchange mechanism and its application to biomonitoring. *Anal. Biochem.*, 404, 135, 2010.
17. Jones, R.S.G., Tryptamine: A neuromodulator or neurotransmitter in mammalian brain? *Prog. Neurobiol.*, 19, 117, 1982.
18. Burchett, S.A. and Hicks, T.P., The mysterious trace amines: Protean neuromodulators of synaptic transmission in mammalian brain. *Prog. Neurobiol.*, 79, 223, 2006.

19. Gao, M. and Tang, B.Z., Fluorescent sensors based on aggregation-induced emission: Recent advances and perspectives. *ACS Sens.*, 2, 1382, 2017.
20. Silva, A.P., Gunaratne, H.Q.N., Gunnlaugsson, T., Huxley, A.J.M., Mccoy, C.P., Rademacher, J.T., Rice, T.E., Signaling recognition events with fluorescent sensors and switches. *Chem. Rev.*, 97, 1515, 1997.
21. Chen, X., Tian, X., Yoon, J., Fluorescent and luminescent probes for detection of reactive oxygen and nitrogen species. *Chem. Soc. Rev.*, 40, 4783, 2011.
22. Gunnlaugsson, T., Akkaya, E.U., Yoon, J., James, T.D., Fluorescent chemosensors: The past, present and future. *Chem. Soc. Rev.*, 46, 23, 2017.
23. Krishnaveni, K., Murugesan, S., Siva, A., Dual-mode recognition of biogenic amine tryptamine and fluoride ions by a naphthyl hydrazone platform: Application in fluorescence imaging of HeLa cells and zebrafish embryos. *New J. Chem.*, 43, 9021, 2019.
24. Nelson, M., Muniyasamy, H., Kubendran, A.M., Balasubramaniem, A., Sepperumal, M., Ayyanar, S., Carbazole based fluorescent chemosensor for the meticulous detection of tryptamine in aqueous medium and its efficacy in cell-imaging and molecular logic gate. *J. Mol. Liq.*, 337, 116445, 2021.
25. Karuppiah, K., Nelson, M., Alam, M.M., Selvaraj, M., Sepperumal, M., Ayyanar, S., A new 5-bromoindolehydrazone anchored diiodosalicylaldehyde derivative as efficient fluoro and chromophore for selective and sensitive detection of tryptamine and $F^-$ ions: Applications in live cell imaging. *Spectrochim. Acta Part A Mol. Biomol. Spectrosc.*, 269, 120777, 2022.
26. Bachrach, U., Polyamines and cancer: Minireview article. *Amino Acids*, 26, 307, 2004.
27. Moinard, C., Cynober, L., De Bandt, J., Polyamines: Metabolism and implications in human diseases. *Clin. Nutr.*, 24, 184, 2005.
28. Agostinelli, E., Marques, M.P.M., Calheiros, R., Gil, F.P.S.C., Tempera, G., Viceconte, N., Battaglia, V., Grancara, S., Toninello, A., Polyamines: Fundamental characters in chemistry and biology. *Amino Acids*, 38, 393, 2010.
29. Stefanelli, C., Bonavita, F., Stanic, I., Mignani, M., Facchini, A., Pignatti, C., Flamigni, F., Caldarera, C.M., Spermine causes caspase activation in leukaemia cells. *FEBS Lett.*, 437, 233, 1998.
30. Teti, D., Visalli, M., Mcnair, H., Analysis of polyamines as markers of (patho) physiological conditions. *J. Chromatogr. B*, 781, 107, 2002.
31. Celanos, P., Baylin, B., Casero, R.A., Polyamines differentially modulate the transcription of growth- associated genes in human colon carcinoma cells. *J. Biol. Chem.*, 264, 8922, 1989.
32. Fan, J., Feng, Z., Chen, N., Spermidine as a target for cancer therapy. *Pharmacol. Res.*, 159, 104943, 2020.
33. Fukushima, Y. and Aikawa, S., Colorimetric chemosensor for spermine based on pyrocatechol violet and anionic phenylboronic acid in aqueous solution. *Microchem. J.*, 162, 105867, 2020.
34. Fletcher, J.T. and Bruck, B.S., Spermine detection via metal-mediated ethynylarene 'turn-on' fluorescence signaling. *Sens. Actuators B. Chem.*, 207, 843, 2015.
35. Severin, K. and Köstereli, Z., Fluorescence sensing of spermine with a frustrated amphiphile. *Chem. Commun.*, 48, 5841, 2012.
36. Ramon-Marquez, T., Medina-Castillo, A.L., Fernandez-Gutierrez, A., Fernandez-Sanchez, J.F., Novel optical sensing film based on a functional nonwoven nanofiber mat for an easy, fast and highly selective and sensitive detection of tryptamine in beer. *Biosens. Bioelectron.*, 79, 600, 2016.
37. Wang, W., Ren, S., Zhang, H., Yu, J., An, W., Hu, J., Yang, M., Occurrence of nine nitrosamines and secondary amines in source water and drinking water: Potential of secondary amines as nitrosamine precursors. *Water Res.*, 45, 4930, 2011.

38. Kaur, N., Kaur, M., Chopra, S., Singh, J., Kuwar, A., Singh, N., Fe (III) conjugated fluorescent organic nanoparticles for ratiometric detection of tyramine in aqueous medium: A novel method to determine food quality. *Food Chem.*, 245, 1257, 2017.
39. Nandi, S., Sahana, A., Mandal, S., Sengupta, A., Chatterjee, A., Safin, D.A., Babashinka, M.G., Tumonov, N.A., Filinchuk, Y., Das, D., Hydrazine selective dual signaling chemodosimetric probe in physiological conditions and its application in live cells. *Anal. Chim. Acta*, 893, 84, 2015.
40. Umar, A., Muzibur, M., Kim, S.H., Rusling, J.F., Schubert, U.S., Umar, A., Rahman, M.M., Kim, S.H., Hahn, Y., Zinc oxide nanonail based chemical sensor for hydrazine detection. *Chem. Commun.*, 7345, 1, 2008.
41. Garrod, S., Bollard, M.E., Nicholls, A.W., Connor, S.C., Connelly, J., Nicholson, J.K., Holmes, E., Integrated metabonomic analysis of the multiorgan effects of hydrazine toxicity in the rat. *Chem. Res. Toxicol.*, 18, 115, 2005.
42. Yamada, K., Yasuda, K., Fujiwara, N., Siroma, Z., Potential application of anion-exchange membrane for hydrazine fuel cell electrolyte. *Electrochem. Commun.*, 5, 892, 2003.
43. Choi, M.G., Hwang, J., Moon, J.O., Sung, J., Hydrazine-selective chromogenic and fluorogenic probe based on levulinated coumarin. *Org. Lett.*, 13, 5260, 2011.
44. Richardson, S.D. and Ternes, T.A., Water analysis: Emerging contaminants and current issues. *Anal. Chem.*, 90, 398, 2018.
45. Xiao, L., Tu, J., Sun, S., Pei, Z., Pei, Y., Pang, Y., Xu, Y., A fluorescent probe for hydrazine and it's *in vivo* applications. *RSC Adv.*, 4, 41807, 2014.
46. Xia, X., Zeng, F., Zhang, P., Lyu, J., Huang, Y., Wu, S., An ICT-based ratiometric fluorescent probe for hydrazine detection and its application in living cells and *in vivo*. *Sens. Actuators B. Chem.*, 227, 411, 2016.
47. He, X., Deng, Z., Xu, W., Li, Y., Xu, C., Chen, H., Shen, J., A novel dual-response chemosensor for bioimaging of Exogenous/Endogenous hypochlorite and hydrazine in living cells, Pseudomonas aeruginosa and zebrafish. *Sens. Actuators B. Chem.*, 321, 128450, 2020.
48. Xu, W., Liu, W., Zhou, T., Yang, Y., Li, W., A novel PBT-based fluorescent probe for hydrazine detection and its application in living cells. *J. Photochem. Photobiol. A. Chem.*, 356, 610, 2018.
49. Li, M., He, J., Wang, Z., Jiang, Q., Yang, H., Song, J., Yang, Y., Xu, X., Wang, S., Novel nopinone based turn on fluorescent probe for hydrazine in living cells with high selectivity. *Ind. Eng. Chem. Res.*, 58, 22754, 2019.
50. Tan, Y., Yu, J., Gao, J., Cui, Y., Yang, Y., Qian, G., A new fluorescent and colorimetric probe for trace hydrazine with a wide detection range in aqueous solution. *Dyes Pigm.*, 99, 966, 2013.
51. V.A., Maji, R., Maiti, K., Manna, S.K., Mondal, S., Ali, S.S., Manna, S., Sahoo, P., Mandal, S., Uddin, R., Mandal, D., A BODIPY/pyrene-based chemodosimetric fluorescent chemosensor for selective sensing of hydrazine in the gas and aqueous solution state and its imaging in living cells. *RSC Adv.*, 5, 58228, 2015.
52. Wu, W., Wu, H., Wang, Y., Mao, X., Zhao, X., A highly sensitive and selective off–on fluorescent chemosensor for hydrazine based on coumarin β-diketone. *Spectrochim. Acta Part A Mol. Biomol. Spectrosc.*, 188, 80, 2015.
53. Yamaguchi, K., Berberich, T., Takahashi, Y., Advances in polyamine research in 2007. *J. Plant Res.*, 120, 345, 2007.
54. Clarke, R.A., Schirra, H.J., Catto, J.W., Lavin, M.F., Gardiner, R.A., Markers for detection of prostate cancer. *Cancers*, 2, 1125, 2010.
55. Nuria, G., Rojas-benedicto, A., Albors-vaquer, A., López-Guerrero, J.A., Pineda-Lucena, A., Puchades-Carrasco, L., Metabolomics contributions to the discovery of prostate cancer biomarkers. *Metabolites*, 9, 48, 2019.

56. Tsoi, T., Chan, C., Chan, W., Chiu, K., Wong, W., Ng, C., Wong, K., Urinary polyamines: A pilot study on their roles as prostate cancer detection biomarkers. *PLoS One*, 11, 1, 2016.
57. Kim, T. and Kim, Y., Analyte-directed formation of emissive excimers for the selective detection of polyamines. *Chem. Commun.*, 52, 10648, 2016.
58. Bao, B., Yuwen, L., Zheng, X., Weng, L., Zhu, X., A fluorescent conjugated polymer for trace detection of diamines and biogenic polyamines. *J. Mater. Chem.*, 20, 9628, 2010.
59. Mohr, G.J., Demuth, C., Spichiger-keller, U.E., Application of chromogenic and fluorogenic reactands in the optical sensing of dissolved aliphatic amines. *Anal. Chem.*, 70, 3868, 1978.
60. Akceylana, E., Bahadir, M., Yılmaz, M., Removal efficiency of a calix [4] arene-based polymer for water-soluble carcinogenic direct azo dyes and aromatic amines. *J. Hazard. Mater.*, 162, 960, 2009.
61. Fan, J., Sun, W., Hu, M., Cao, J., Cheng, G., Dong, H., Song, K., Liu, Y., Sun, S., Peng, X., An ICT-based ratiometric probe for hydrazine and its application in live cells. *Chem. Commun.*, 48, 8117, 2012.
62. Cao, X., Ding, Q., Zhao, N., Gao, A., Jing, Q., Supramolecular self-assembly system based on naphthalimide boric acid ester derivative for detection of organic amine. *Sens. Actuators B. Chem.*, 256, 711, 2018.
63. Liu, T., Yang, L., Feng, W., Liu, K., Ran, Q., Wang, W., Liu, Q., Peng, H., Ding, L., Fang, Y., Dual-mode photonic sensor array for detecting and discriminating hydrazine and aliphatic amines. *ACS Appl. Mater. Interfaces*, 12, 11084, 2012.
64. Nawaz, H., Zhang, J., Tian, W., Jin, K., Jia, R., Yang, T., Zhang, J., Cellulose-based fluorescent sensor for visual and versatile detection of amines and anions. *J. Hazard. Mater.*, 387, 121719, 2020.
65. Ruiu, A., Vonlanthen, M., Morales-espinoza, E.G., Rojas-Montoya, S.M., Rivera, E., Gonz, I., Pyrene chemosensors for nanomolar detection of toxic and cancerogenic amines. *J. Mol. Struct.*, 1196, 1, 2019.
66. Zhang, K., Chen, Y., Wen, G., Mahata, M., Rao, F., Friese, R.S., Mahata, S.K., Hamilton, B.A., Connor, D.T.O., Catecholamine storage vesicles: Role of core protein genetic polymorphisms in hypertension. *Curr. Hypertens. Rep.*, 13, 36, 2011.
67. Borges, R., Pereda, D., Beltrán, B., Prunell, M., Rodríguez, M., Machado, J.D., Intravesicular factors controlling exocytosis in chromaffin cells. *Cell. Mol. Neurobiol.*, 30, 1359, 2010.
68. Carrera, V., Sabater, E., Vilanova, E., Sogorb, M.A., A simple and rapid HPLC – MS method for the simultaneous determination of epinephrine, norepinephrine, dopamine and 5-hydroxytryptamine: Application to the secretion of bovine chromaffin cell cultures. *J. Chromatogr. B*, 847, 88, 2007.
69. Robinson, D.L., Hermans, A., Seipel, A.T., Wightman, R.M., Monitoring rapid chemical communication in the brain. *Chem. Rev.*, 108, 2554, 2008.
70. Secor, K.E. and Glass, T.E., Selective amine recognition: Development of a chemosensor for dopamine and norepinephrine. *Org. Lett.*, 6, 3183, 2004.
71. Hettie, K.S., Liu, X., Gillis, K.D., Glass, T.E., Selective catecholamine recognition with NeuroSensor 521: A fluorescent sensor for the visualization of norepinephrine in fixed and live cells. *ACS Chem. Neurosci.*, 4, 918, 2013.
72. Zhou, N., Huo, F., Yue, Y., Yin, C., Specific fluorescent probe based on "protect–deprotect" to visualize the norepinephrine signaling pathway and drug intervention tracers. *J. Am. Chem. Soc.*, 142, 17751, 2020.
73. Kepser, L. and Homberg, J.R., The neurodevelopmental effects of serotonin: A behavioral perspective. *Behav. Brain Res.*, 277, 3, 2014.
74. Best, J., Nijhout, H.F., Reed, M., Bursts and the efficacy of selective serotonin reuptake inhibitors. *Pharmacopsychiatry*, 44, 76, 2011.

75. Blakely, R.D. and Edwards, R.H., Vesicular and plasma membrane transporters for neurotransmitters. *Cold. Spring. Harb. Perspect. Biol.*, 4, a005595, 2012.
76. Omiatek, D.M., Cans, A., Heien, M.L., Ewing, A.G., Analytical approaches to investigate transmitter content and release from single secretory vesicles. *Anal. Bioanal. Chem.*, 397, 3269, 2010.
77. Ravichandiran, P., Prabakaran, D.S., Bella, A.P., Boguszewska-Czubara, A., Masłyk, M., Dineshkumar, K., Johnson, P.M., Park, B.H., Han, M.K., Kim, H.G., Yoo, D.J., Naphthoquinone-dopamine linked colorimetric and fluorescence chemosensor for selective detection of $Sn^{2+}$Ion in aqueous medium and its bio-imaging applications. *ACS Sustain. Chem. Eng.*, 8, 10947, 2020.
78. Hettie, K.S. and Glass, T.E., Turn-on near-infrared fluorescent sensor for selectively imaging serotonin. *ACS Chem. Neurosci.*, 7, 21, 2016.
79. Xue, Y., Tian, J., Tian, W., Zhang, K., Xuan, J., Zhang, X., Spiropyran based recognitions of amines: UV–Vis spectra and mechanisms. *Spectrochim. Acta Part A Mol. Biomol. Spectrosc.*, 250, 119385, 2021.
80. Lee, Y.J., Kang, I.S., Shin, Y.S., Lee, S.W., Nanomaterials for detection of primary aromatic amine derivatives based on a fluorescent probe. *Mol. Cryst. Liq. Cryst.*, 704, 57, 2020.

# 13

# Application of Optical Nanoprobes for Supramolecular Biosensing: Recent Trends and Future Perspectives

**Riyanka Das[1,2], Rajeshwari Pal[1,2], Sourav Bej[1,2,3#], Moumita Mondal[1,2#] and Priyabrata Banerjee[1,2*]**

[1]*Electric Mobility & Tribology Research Group, CSIR-Central Mechanical Engineering Research Institute, Mahatma Gandhi Avenue, Durgapur, India*
[2]*Academy of Scientific & Innovative Research (AcSIR), Ghaziabad, Uttar Pradesh, India*
[3]*School of Science, Harbin Institute of Technology (Shenzhen), Shenzhen, China*

***Abstract***
Optical nanoprobes for biosensing applications have gained widespread attention for recognition of a variety of biorelevant ions or molecules present in human body. In the present chapter, the nanoprobes are classified into four categories: zero, one, two, and three-dimensional nanomaterials and herein the nanomaterials include carbon quantum dots (CQDs), graphene quantum dots (GQDs), semiconductor inorganic quantum dots (IQDs), polymer dots (PDs), gold nanoparticles (AuNPs), silver nanoparticles (AgNPs), upconversion nanoparticles (UCNPs), magnetic nanoparticles (MNPs) for 0-D, carbon nanotubes (CNTs), gold nanorods (AuNRs), silicon nanowires (SiNWs), nanofibers and nanoribbons for 1-D, $MnO_2$ nanosheets, 2-D MOF nanosheets, graphitic carbon nitride, graphene oxide (GO), reduced graphene oxide (rGO) for 2-D and hybrid nanoflowers (HNFs) and 3-D MOFs for 3-D nanomaterials. The chapter briefly covers diverse nanomaterials and nanocomposites and their applications in chromo-fluorgenic biosensing applications. Furthermore, the applications of the nanoprobes for targeted analytes recognition from complicated biological samples as well as in intracellular imaging have also been discussed. Finally, the loopholes of the present research and future directions of the research are also enlightened to keep an ample footstep in the domain of medical diagnosis.

***Keywords***: Nanomaterials, chromogenic and/or fluorogenic sensing, supramolecular interaction, bio-analyte sensing, enzymatic activity, disease diagnosis

## 13.1 Introduction

Nanomaterials can be normally described as "natural, incidental or manufactured materials that contain particles in the aggregate or nonaggregate state where at least 50% of these

**Corresponding author*: pr_banerjee@cmeri.res.in; priyabrata_banerjee@yahoo.co.in
#These authors contributed equally.

Inamuddin and Tariq Altalhi (eds.) Biosensors Nanotechnology, 2nd Edition, (267–326) © 2023 Scrivener Publishing LLC

particles have dimensions between 1 nm and 100 nm." [1]. In the recent past, nanomaterials have been introduced as potential candidates for the different domains of biosensing applications owing to their high surface area, unique electronic, superior thermal conductivity with physico-chemical properties [2]. Therefore, the nanomaterials can interact with the targeted biomolecules more efficiently than those of their bulky counterparts. The sensing performance of the nanomaterials is highly dependent on several factors, like the quality of the crystal structure, dimensions, morphology, orientation of the crystallographic axis etc. [3]. Nanomaterials-based sensors are generally composed of three components: a nanomaterial(s), a recognition element (like aptamer, antibody etc.) that undergoes target-specific interaction and a signal transduction unit, which generate optical, electrochemical or magnetic signals. Among different signal transductions, chromo-fluorogenic signal transduction is much more desirable because of its numerous advantages, like good specificity, high selectivity, ultra-sensitivity, fast response, cost-effectiveness, high photostability, visual change etc.

Nanomaterials can be broadly categorized into four classes: 0-D (carbon dots, quantum dots, nanoparticles etc.), 1-D (nanorods, nanotubes, nanowires etc.), 2-D (nanosheet, nanoplates etc.) and 3-D nanomaterials (nanoflowers, 3-D nanoMOFs etc.). Recently, 2-D transition metal oxide-based nanosheets are gaining much attention in the domain of biosensing. Herein, the atoms in the thin layer are covalently bonded whereas the layers are attached with weak interactions to provide flexibility in the entire system [4]. Graphene is an important class of two-dimensional nanomaterials, having flat sheet like structures. Due to the structural uniqueness, graphene and its derivatives, like GO or rGO exhibit interesting optoelectronic properties, which are advantageous for applications in the domain of biosensing [5]. Again, single walled or multi-walled carbon nanotubes (CNTs) are one-dimensional nanomaterials, which are also beneficial for biosensing applications.

Apart from graphene or CNTs, semiconductor quantum dots also possess notable luminescence properties. Carbon quantum dots and graphene quantum dots are sustainable and biocompatible congeners of the zero-dimensional nanomaterials as they do not contain any kind of metallic counterpart. Other metallic nanoparticles, like gold or silver nanoparticles possess unique properties, like good biocompatibility, localized surface plasmon resonance (LSPR) phenomenon etc., [6] based on which they can exhibit target-specific aggregation-disaggregation induced chromogenic responses. Moreover, these kinds of metallic nanoparticles can also be used as fluorescence quenchers to act as resonance energy acceptors from luminescent materials in FRET (fluorescence resonance energy transfer) process [7]. Thus 0-D nanomaterials can also be combined with other 1-D, 2-D or 3-D nanomaterials to improve the sensing specificity.

Metal-organic frameworks (MOFs) having 2-D or 3-D structures are an important class of nanomaterials. MOFs have attained burgeoning research attention in the last few decades owing to their high thermal stability, high surface area, large porosity etc. [8]. Based on the potential biosensing applications of the nanomaterials, various field-portable devices are being developed nowadays due to their cost-effectiveness, fast response, high sensitivity etc. Therefore, a wide variety of optical sensors are being developed for practical applications based on colorimetric or fluorescence signalling responses.

Supramolecular chemistry is termed as the chemistry of non-covalent interaction. Supramolecular sensing may be of two types: specific sensing [9] and differential sensing [10], where the former relies on the detection of a single analyte while the latter deals with the multiple analytes recognition present in a mixture. To design supramolecular nanobiosensors for target-specific analyte detection the receptor unit is to be chosen in such a way that it can undergo selective, sensitive and specific interactions with the targeted analytes not only in the ideal buffer condition but also in the complex biological matrices for real day applications. Currently, the progress of the literatures on the nanomaterials towards biosensing applications has been highlighted in many reviews. Again, some of those are particularly focused on a particular type of nanomaterials. However, the current knowledge of nanomaterials-based on optical biosensing applications by different nanomaterials is scantily updated in the existing literature.

In this perspective, nowadays emphasis is on the chromo-fluorogenic optical sensing of the biorelevant and clinically relevant molecules (e.g, ATP, $H_2O_2$, glucose, ascorbic acid, biothiols, dopamine, Vitamin C, lysine, intracellular pH etc.) by zero-dimensional (e.g., CQDs, GQDs, semiconducting IQDs, AgNPs, AuNPs, magnetic nanoparticles, polymer dots and UCNPs), one-dimensional (e.g, SiNWs, CNTs, AuNRs, nanofibers and nanoribbons), two-dimensional (e.g., GO, rGO, graphitic carbon nitride, 2-D nanoMOFs and $MnO_2$ nanosheets) and three-dimensional nanomaterials (e.g., 3-D nanoMOFs and hybrid nanoflowers) under one umbrella. The progress of the research endeavor from 2014 onwards has been included herein (Figure 13.1). The sensing performance of the bare nanomaterials, as well as functionalized nanomaterials, has also been described in the present chapter. The detection method, detection thresholds and linear dynamic ranges are also adeptly studied. Moreover, the applications of the described nanomaterials in intracellular imaging as well as applications in complex bio-matrices, like serum, urine have also been discussed herein for clinical *point-of-care* applications. Therefore, the information presented in the current chapter is expected to endow motivation of the researchers in near future towards designing target-specific nanosensor. The flow of the current research is not only analyzed herein, but

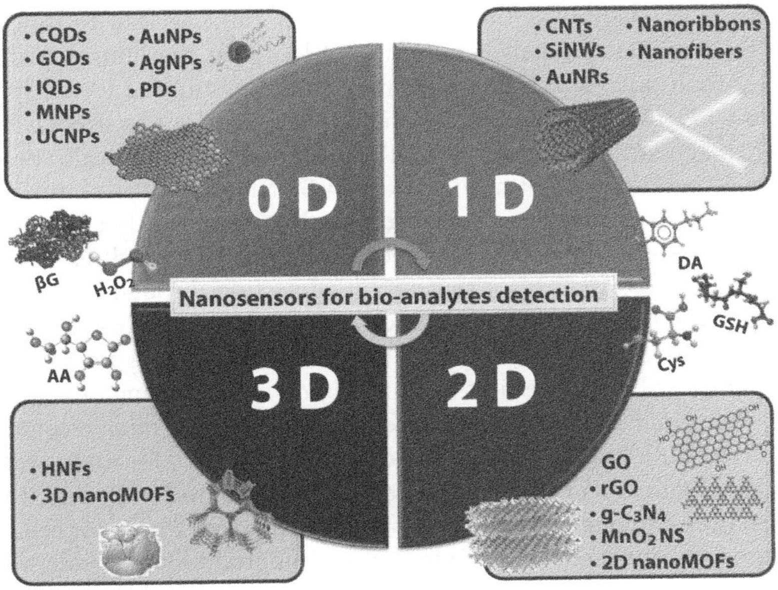


**Figure 13.1** Application of 0-D, 1-D, 2-D and 3-D nanomaterials for the target-specific detection of bio-analytes.

also the lacunas of the present research and future research endeavor for increasing selectivity, sensitivity, specificity, etc. are also discussed to overcome the current challenges of large-scale industrial applications.

## 13.2 Optical Nanoprobes for Biosensing Applications

### 13.2.1 Zero-Dimensional Nanoprobes for Optical Biosensing

#### *13.2.1.1 Carbon Quantum Dots*

Carbon quantum dots (CQDs) are carbon-containing nanomaterials, which are currently gaining burgeoning attention in biosensing [11, 12], drug delivery [13], optoelectronic applications [14], bioimaging etc. because of their good excitation-dependent optical properties, large quantum yield, high chemical stability, superior biocompatibility, good dispersibility etc. [15, 16]. By judicious selection of the precursors containing heteroatoms and proper reaction conditions, it is possible to synthesize CQDs with emission in the longer wavelength region for efficient bioimaging applications. Currently, CQDs are employed in various biosensing applications for the recognition of a variety of biologically relevant analytes like Vitamin B7, lysine, intracellular pH, β-glucuronidase, ascorbic acid, adenosine triphosphate, $Fe^{3+}$ etc. via variation of fluorescence response [11, 12, 17–19].

Lysine is an important biomolecule, of which abnormal level is related to various metabolic disorders [20, 21]. This urges the sensitive detection of lysine from different bio-matrices. Ren and Xiao *et al.* synthesized dual emissive carbon dots (dCDs) towards selective ratiometric fluorescence sensing of lysine as well as intracellular pH [17] to exhibit potential applicability toward the diagnosis of pH or lysine-related disorders (Figure 13.2). Biotin or Vitamin B7 is an important micronutrient, playing an imperative role in cellular growth and various metabolic pathways. The deficiency of biotin may lead to different inborn disorders, which necessitates timely monitoring of biotin in the human body. Lin *et al.* have reported carbon dots, which can selectively detect biotin due to the specific interaction of biotin with streptavidin and ultimately superparamagnetic streptavidin-coupled nanobeads were magnetically separated [11]. The proposed strategy has also been exploited for the effective recognition of biotin from Vitamin B7 tablets. β-glucuronidase (βG) is an important biomarker for various types of cancer like liver cancer [22], renal cancer [23] etc. Therefore, sensitive determination of βG is necessary for early-stage disease diagnosis. Liu *et al.* reported green emissive CDs for sensitive βG detection through IFE [12]. Moreover, the sensitivity has also been checked via intracellular detection of βG, exhibiting potential biomedical applications. Again, ascorbic acid (AA) and $Fe^{3+}$ sensing are also important as these are the important bioanalytes in the human body for regulating different physiological processes. Shamsipur *et al.* developed green emissive CDs towards recognition of $Fe^{3+}$ and AA respectively [18]. Initially, $Fe^{3+}$ quenched fluorescence of CDs but in presence of AA CDs again became luminescent owing to the reduction of $Fe^{3+}$ to $Fe^{2+}$ by AA. The proposed method has also been exploited for real urine sample analysis as well as for cellular imaging purposes. Adenosine triphosphate (ATP) is another important biomolecule, which acts as the energy source of our daily activities [24]. Tong *et al.* developed carbon dots doped with nitrogen (N-CDs) that can also recognize $Fe^{3+}$ and ATP in a similar way [19]. Here also,

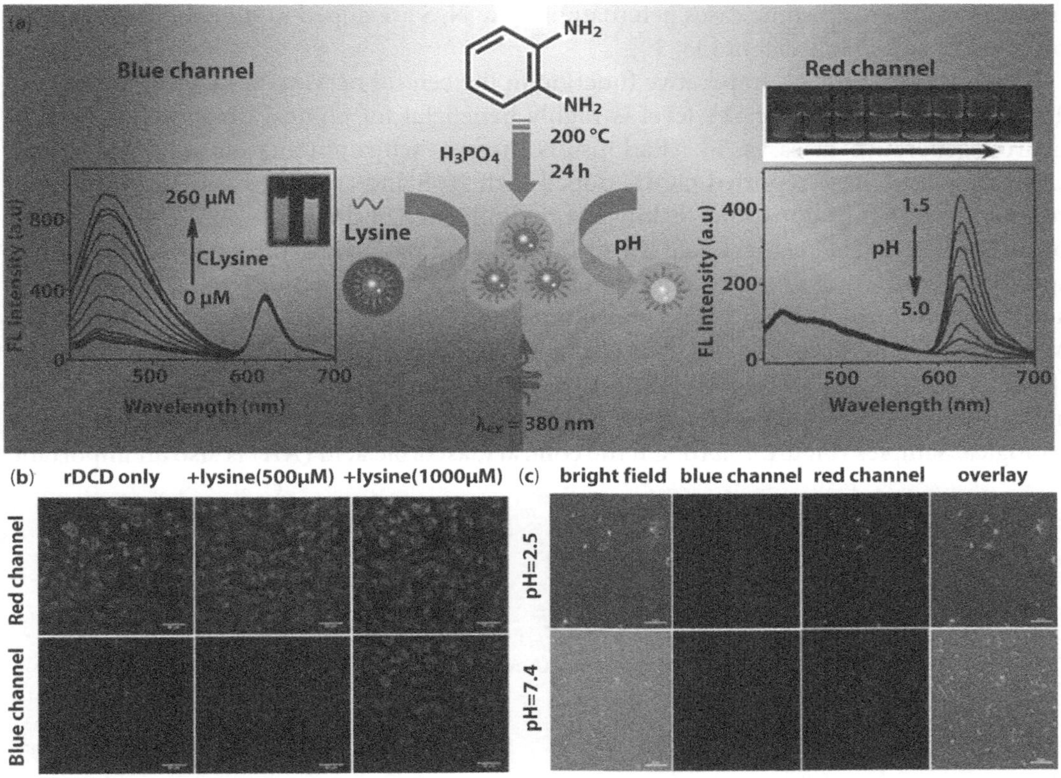


**Figure 13.2** (a) The schematic representation of dCDs synthesis towards fluorogenic determination of lysine and pH (Inset: Variation of fluorescence spectra of dCDs by lysine and at different pH), (b) Live cell imaging with HeLa cells for lysine detection in red and blue channels, (c) pH imaging in *E. coli*, Adapted with permission from Ref. 17.

ATP could regenerate the initially quenched fluorescence of the NCDs/$Fe^{3+}$ ensemble owing to Fe-O-P bond formation. The sensitivity of the probe has also been checked via intracellular imaging as well as via analysis of real serum samples also.

### *13.2.1.2 Graphene Quantum Dots*

Graphene quantum dots (GQDs) are quasi-zero-dimensional fluorescent nanomaterials having a particle size<100 nm, which are recently explored in the domain of bioimaging [25] as well as biosensing of a range of bioanalytes, like ascorbic acid, dopamine, $Fe^{3+}$ etc. [26–30] due to their various advantages, like good biocompatibility, aqueous phase dispersibility, high photostability, quantum confinement effects etc. The GQDs are generally prepared via two approaches: either the "top-down" approach or the "bottom-up" technique [31, 32]. Top-down approach is involved with the breakage of large size of graphene into smaller GQDs via various physico-chemical methods, like hydrothermal method [33], electrochemical method [34], chemical abalation [35] etc. while the latter is involved in preparing GQDs from smaller-sized carbon materials via thermal pyrolysis [36], solvothermal treatment [37] etc. In some cases heteroatoms, like N, S are doped to tune the fluorescence property to a desirable extent [38, 39].

Dopamine (DA) plays imperative function in the central nervous systems of mammals. Therefore, monitoring the DA level is highly beneficial for clinical diagnosis of various neurodegenerative diseases, like Parkinson's disease, schizophrenia etc. [40, 41]. Chen and Wang *et al.* have reported GQDs doped with N, S and co-doped with S, N for fluorogenic detection of dopamine, exhibiting the influence of doping toward the sensitivity of the GQDs [26]. DA selectively led to fluorescence quenching of N-GQDs due to the specific interaction of phenoxy anion of DA with pyridinic N sites of N-GQDs via surface adsorption to form a ground-state non-fluorescent complex (Figure 13.3a). Furthermore, M. El-Wekil *et al.* have also reported N-doped GQDs functionalized with β-cyclodextrin (β-CD) (β-CD/N@GQDs) towards selective chromo-fluorogenic sensing of DA due to catalytic peroxidase like activity via "inner filter effect" [27]. Moreover, it can also detect DA in biological samples (Figure 13.3b). On the contrary, ascorbic acid (AA) is also an important

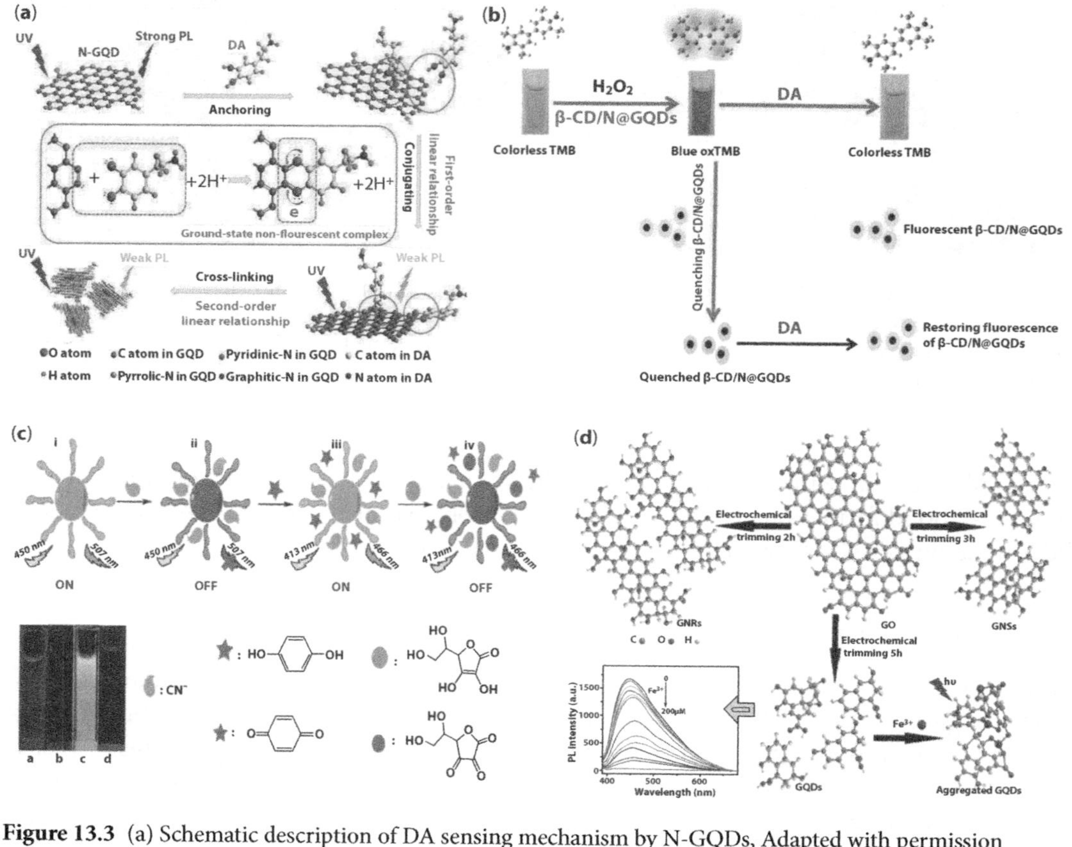


**Figure 13.3** (a) Schematic description of DA sensing mechanism by N-GQDs, Adapted with permission from Ref. 26, (b) Schematic diagram of peroxidase like activity of β-CDs/N@GQDs towards DA recognition, Adapted with permission from Ref. 27, (c) Plausible Mechanistic pathway towards sensing of $CN^-$, HQ and AA (Inset: (i) C-GQDs upon (ii) $CN^-$ addition, (iii) HQ and (iv) AA successively under UV Light), (d) Schematic Diagrams on the Electrochemical Trimming of GO Nanosheets and the Detection Mechanism of $Fe^{3+}$ by GQDs (Inset: the corresponding fluorescence spectral response of GQDs in presence of $Fe^{3+}$), Adapted with permission from Ref. 29.

biomolecule, of which abnormal levels in blood can lead to various diseases [42]. Therefore, monitoring AA in the human body is also important. Recently, Ma *et al.* reported coumarin conjugated C-GQDs for $CN^-$ detection, Hydroquinone (HQ) and C-GQD-$CN^-$-HQ ensemble simultaneously can monitor the level of AA via "turn-off" fluorescence response based on specific redox reaction of AA with benzoquinone (BQ), which is the oxidation product of HQ (Figure 13.3c) [28]. Iron is also an important bioanalyte of which regular monitoring in the human body is truly important as it is related to various physiological processes, like enzymatic catalysis, cellular metabolism etc. [43] Li and Wang *et al.* developed electrochemical trimming of graphene oxide to produce GQDs for "on-off" fluorescence sensing of $Fe^{3+}$ due to the aggregated GQDs formation by $Fe^{3+}$ with a low LOD of 0.23 μM [29]. The quenched fluorescence was again restored in presence of AA via AA-induced reduction of $Fe^{3+}$ to $Fe^{2+}$ (Figure 13.3d). In addition, Xi *et al.* reported boron-doped GQDs, which can selectively detect $Fe^{3+}$ or phosphate (Pi) along with cytochrome C (Cyt C), which are also important bioanalytes in the human body to control various physiological processes [30]. $Fe^{3+}$ quenched the fluorescence of B-GQDs via aggregation of B-GQDs while the fluorescence was again revived in presence of Pi via redispersion of aggregated B-GQDs. Again, Cyt C also led to fluorescence quenching of B-GQDs by absorption of excitation light of B-GQDs by Cyt C along with analyte-induced aggregation of B-GQDs.

#### *13.2.1.3 Inorganic Quantum Dots*

Inorganic quantum dots (IQDs) of several II-VI [44] or III-V semiconductors [45] with the light-emitting properties are particularly advantageous due to high fluorescence quantum yield, high resistance toward photobleaching, broad excitation spectra, good biocompatibility etc. [46, 47]. The surface modifications or conjugation of the semiconductor quantum dots with other metallic nanoparticles or other nanomaterials, like AuNPs [48], graphene [49], 2-D nanosheets ($WS_2$, $MnO_2$ etc.) [50] can make them efficient biosensor probes via FRET mechanism, where the QDs are the energy donor while the other nanomaterials act as an acceptor of energy. Recently, these types of QDs are widely used for biosensing of diverse bioanalytes, like ascorbic acid, biothiols, $H_2O_2$, etc. [46, 51–55].

Abnormal levels of inflammation biomarkers, like C-reactive protein (CRP), serum amyloid A (SAA) may lead to severe bacteriological infection, cardiovascular pathogenesis, etc. [56, 57]. Recently, Li, Wu and Guo *et al.* developed QDs-based immunosorbent assay for SAA and CRP sensing (Figure 13.4a, b) [51]. Moreover, the nanosensor has also been applied for the detection of the targeted SAA and CRP from serum samples. Su *et al.* reported CdTe QDs capped with thioglycolic acid (TGA), carbon dots (CDs) and MOF-based nanoprobe for specific ratiometric fluorogenic sensing of AA followed by monitoring of ascorbate oxidase (AAO) activity [52]. $Hg^{2+}$ quenched the fluorescence of nanoprobe via generation of size-quantized HgS on the QD surface via ion binding as well as electron transfer along with ~20 nm bathochromic shifting of the UV-Vis absorption spectra (Figure 13.4c). However, the presence of AA in the QDs/CDs@MOFs-$Hg^{2+}$ ensemble the fluorescence was again restored with the further blue shifting of the UV-Vis absorption spectra via a specific redox reaction of AA with $Hg^{2+}$. However, the presence of AAO in the same solution restricted the redox reaction between $Hg^{2+}$ and AA due to the oxidation of AA by AAO. Therefore, the proposed method provides "off-on" and "off-on-off" detection strategy for AA and AAO respectively. Wu and Geng *et al.* reported CdTe QDs, which can

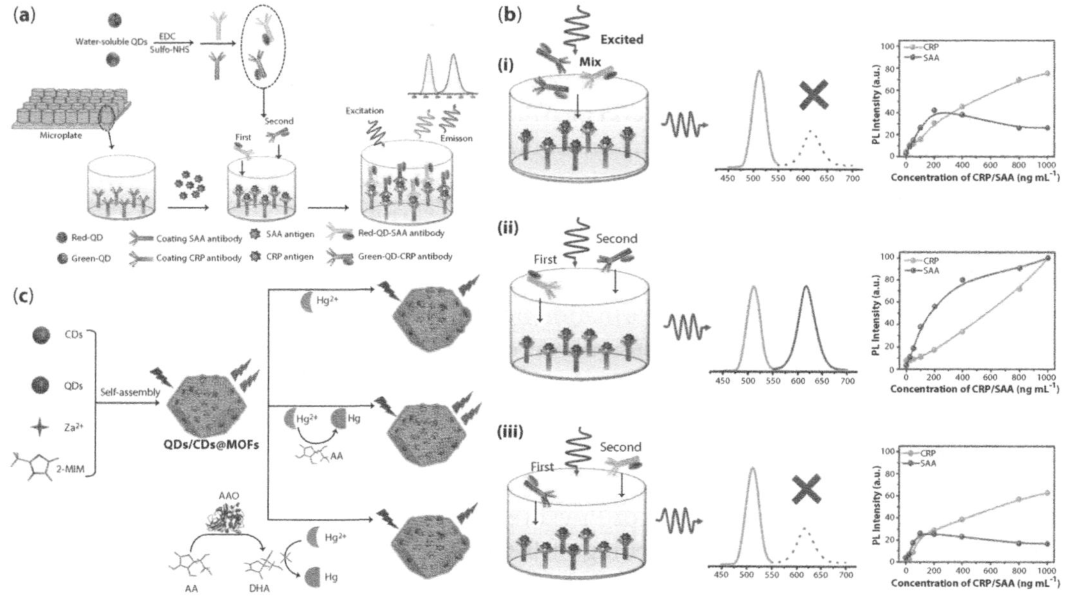


**Figure 13.4** (a) Schematic illustration of dQDs-FLISA procedure (b) The variable addition chronology of red-QD-SAA antibody and green QD-CRP antibody, Adapted with permission from Ref. 51, (c) Schematic illustration of QDs/CDs@MOFs nanoprobe synthesis and its sensing application, Adapted with permission from Ref. 52.

selectively detect AA and alkaline phosphatase (ALP), important enzyme and coenzyme for different metabolic processes [53]. The fluorescence recognition phenomenon is governed by the specific reaction of QDs with $Ag^+$. However, AA led to the reduction of $Ag^+$ to AgNPs. Therefore, fluorescence of the nanosensor remained in the ON state. Moreover, by enzymatic hydrolysis ALP is transformed to AA. Therefore, relying on this strategy monitoring of ALP is also possible by this approach. Santos *et al.* reported CdTe QDs for detection of different bioactive thiols, like L-cysteine, glutathione, captopril, thiomalic acid and coenzyme M-based on FRET between CdTe QDs with AuNPs, which acts as promising fluorescence quencher [54]. However, the fluorescence was recovered by the targeted analytes via restriction of the FRET process due to the aggregation of AuNPs. The sensitivity of the developed probe has also been checked in different complex sample matrices with a promising response. Again, Wang *et al.* have reported green emissive and red emissive CdTe quantum dots for ratiometric fluorescence sensing of $H_2O_2$ via electron transfer and target-induced formation of aggregated QDs [55].

### 13.2.1.4 *Noble Metal Nanoparticles*

Noble metal nanoparticles, like AgNPs, AuNPs, etc. have lured major consideration in the domain of biosensing application owing to inimitable physicochemical characteristics, like large molar extinction coefficient, ease of preparation, good biocompatibility etc. [58]. The basis of chromogenic sensing of the targeted analytes by the AuNPs and AgNPs is the different degrees of aggregation or disaggregation of the nanoparticles.

#### 13.2.1.4.1 Gold Nanoparticles

Among various nanoparticles, gold nanoparticle (AuNP)-based chromogenic biosensing is widely explored because of their inimitable magnetic and optoelectrical properties, biostability, easy preparation etc. [59]. In presence of the targeted analytes selective chromogenic responses are obtained by AuNPs via Surface Plasmon Resonance (SPR)-based sensing strategy, leading to chromogenic variation from blue to red or vice versa along with the SPR band shifting due to aggregation or disaggregation of the particles. In some cases, they are conjugated with carbon quantum dots to act as an efficient energy acceptor, thereby leading to obtain a target-specific fluorogenic response. Recently these are widely used for the chromogenic detection of various bioanalytes, like glutathione, histidine, DNA etc. [60–64].

Glutathione (GSH) is an important biothiol, which acts as an antioxidant in the human body [65]. However, inconsistent level of it may lead to cancer, aging, heart diseases etc. [66]. Liu *et al.* reported carbon dots co-doped with nitrogen and sulfur (N, S-CDs) and AuNPs-based nano-assembly for "off-on" fluorescence sensing of GSH via disaggregation of aggregated nanoprobe [60]. Similarly, Gong and Dong *et al.* developed N, S-CDs and AuNPs-based FRET pair towards fluorescence sensing of GSH via "turn-on" mode [62]. L-histidine is another important amino acid, the abrupt concentration of which may lead to renal disorder, Alzheimer's disease, cancer etc. [67–69]. Qiang and Chen *et al.* reported DNA duplex and AuNPs conjugated nano assembly for chromogenic recognition of L-histidine [61]. Initially, in the high salt environment, AuNPs remained disaggregated form due to interaction with the DNA duplex. However, in presence of the targeted analyte, DNA duplex was cleaved and AuNPs were absorbed on the single-stranded DNA and therefore AuNPs were aggregated along with visual alteration of color from red to blue. Cancer is the second deadliest disease worldwide. So, early-stage cancer detection is highly desirable for disease diagnosis. Recently, Sachdev and Bandyopadhyay *et al.* developed AuNPs-based nanosensor for colorimetric recognition of a cancer antigen, protein-phosphatase-1-gamma-2 (PP1γ2), which acts as a potential cervical cancer (CaCx) biomarker [63]. The immunosensor was prepared by the chemical modification of AuNPs for the attachment of the NPs with the biomarker specific antibody. However, in presence of the targeted biomarker, there occurs specific chromogenic alteration of the immunosensor due to particular epitope-paratope interaction. The method acts as a potential noninvasive tool for the detection of the targeted analyte from human urine with the fabrication of a POCT prototype for real-time applications (Figure 13.5a, b). Sanguinarine (SNG) is an important bioactive alkaloid, acting as a potential anticancer drug. Kaushik *et al.* reported citrate reduced AuNPs (CI-Au NPs) towards selective chromogenic detection of Calf thymus DNA (Ct-DNA), of which detection is one of the daunting tasks in the biomedical field [64]. Here SNG acted as an aggregation agent. Initially, due to the electrostatic interaction of SNG and CI-AuNPs, NPs were aggregated with a visual colorimetric alteration from red to blue. However, SNG-induced aggregation was masked in presence of Ct-DNA and the original color of the NPs was revived. The sensitivity of the nanosensor towards Ct-DNA was checked in human urine samples with qualified recovery (Figure 13.5c, d).

#### 13.2.1.4.2 Silver Nanoparticles

Silver nanoparticle (AgNP)-based sensing platforms are pertinent in the domain biosensing applications and are advantageous over AuNPs as they possess several unique

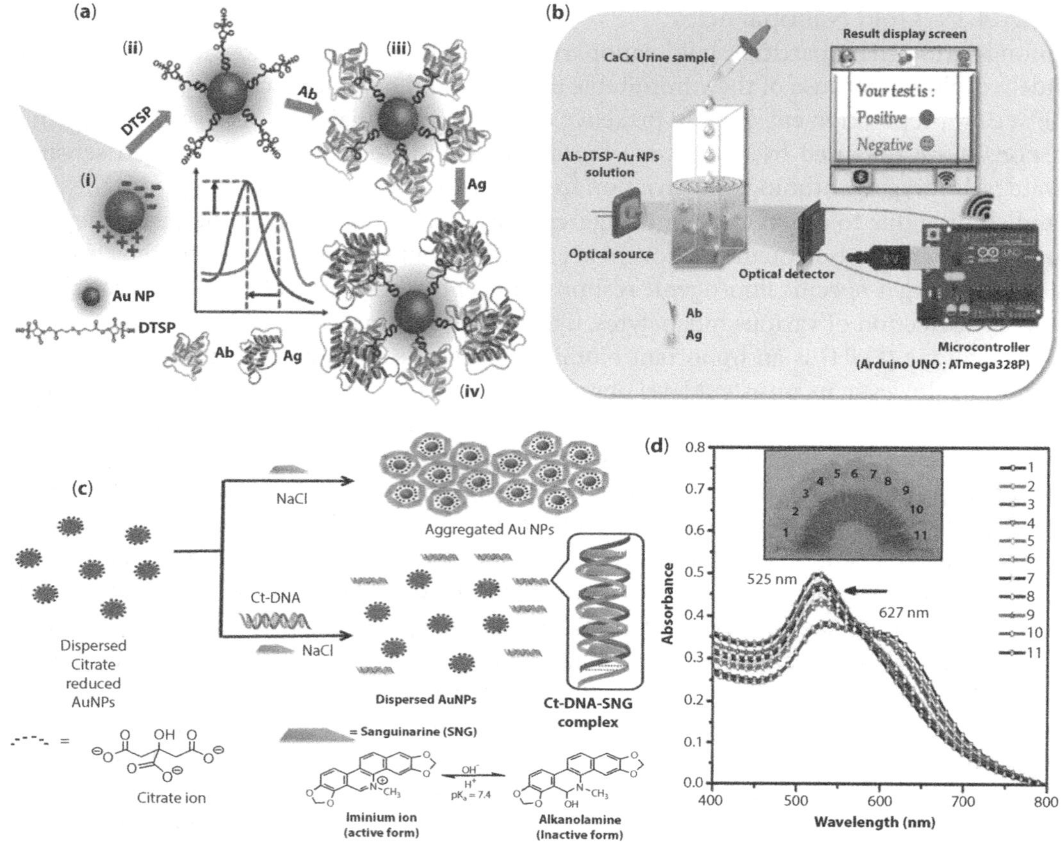


**Figure 13.5** (a) Stages of biomolecule attachment on AuNPs surface. (i) pristine AuNPs. (ii) attachment of 3,3'-dithiodipropionic acid di(n-hydroxysuccinimide ester) (DTSP) to AuNPs. (iii) attachment of Ab on the previous composite, (iv) final attachment of the selective protein-NCB-Ag on the surface of Ab-immobilized DTSP-linked Au NPs, (b) Schematic diagram of the portable POCT device for noninvasive cervical cancer specific biomarker antigen identification from urine samples, Adapted with permission from Ref. 63, (c) Plausible mechanistic course of interaction for detection of SNG and Ct-DNA using CI-AuNPs as chromogenic nanosensor, (d) Absorption spectra of CI-AuNPs/SNG in absence and presence of increasing concentration of Ct-DNA with the colorimetric images, Adapted with permission from Ref. 64.

characteristics like large molar extinction coefficient, sharp extinction bands, high selectivity etc. [70]. In recent times, AgNPs are broadly applied for sensing of a broad range of biomolecules, *like* adenosine, uric acid, cysteine, dopamine, etc. [71–75].

Adenosine is an important purine nucleoside, playing a vital role in the human body and can act as an important cancer biomarker [76]. Recently, Saraji *et al.* have reported an aptasensor using unfunctionalized-aptamer as the recognition probe and AgNPs as the chromogenic indicator for selective sensing of adenosine [71]. Adenosine helped the nanoparticles to be remained in dispersed state in the solution due to the strong coordination of aptamer with adenosine and based on the corresponding chromogenic changes of AgNPs and the UV-Vis spectral variation, quantification of adenosine was achieved. Moreover, the sensing specificity has also been examined for the detection of adenosine in urine samples with the recovery of 98–107%. Uric acid (UA) is also an important bioanalyte, of which abnormal level may lead to gout, renal disorder, hyperuricemia etc. [77, 78]. Li *et al.* reported a chromogenic recognition strategy via oxidative etching of AgNPs by chloroauric acid ($HAuCl_4$) for selective detection of uric acid [72]. Initially, Ag(0) was oxidized to Ag(I) while Au(III) was reduced to Au(0) forming honeycomb-shaped Au-Ag NPs. This led to the corresponding change of color from yellow to brown. However, UA could stop the oxidation of Ag(0), leading to fading of the brown color to yellow with the UV-Vis spectral shift from 477 to 428 nm. Ascorbic acid is another important biomolecule involved in cell development, immunity improvement etc. [79]. Abnormal levels of AA may act as a biomarker for diabetes mellitus, cancer, etc. [80]. Mehdinia *et al.* depicted selective sensing of AA via seed-mediated growth of AgNPs [73]. Actually, by AA, $Ag^+$ ions underwent a redox reaction to be reduced to Ag atoms, which led to yellow coloration from colorless with the emergence of a new band at 420 nm. The sensing efficacy has also been checked in commercial lemonade and pharmaceutical tablets. Rahim and Nishan *et al.* [74] reported Augmentin drug functionalized AgNPs-ionic liquid (IL) nano-conjugate toward selective colorimetric dopamine sensing, which is an important biomarker for Parkinson's disease [81]. In presence of dopamine, there occurs a chromogenic change of the nano-assembly from light grey to reddish-brown as dopamine was oxidized to dopamine quinone (Figure 13.6a, b). Ghosh *et al.* [75] depicted IL functionalized AgNPs, which can selectively detect cysteine, which is an important biothiol involved in various biological processes in the human body [82]. Upon interaction with the targeted analyte the electrostatic interaction was increased between ILs of AgNPs with Cys, which was monitored via zeta potential analysis and via a bathochromic shift of the SPR band with the corresponding colorimetric alteration from yellow to purple (Figure 13.6c).

#### *13.2.1.5 Others*

There are some other classes of zero-dimensional nanomaterials, like magnetic nanoparticles, upconversion nanoparticles, polymer dots etc. Among them, metal-based magnetic nanoparticles (MNPs), like spinel ferrites ($MFe_2O_4$, M= Zn, Co, Ni, Cd, Mn etc., e.g, $ZnFe_2O_4$ [83], $Fe_3O_4$ [84], $CoFe_2O_4$ [85], $MnFe_2O_4$ [86] etc.) are attracting escalating attention in the contemporary research domain because of their distinct optoelectronics and novel physiological properties. Nowadays these types of nanomaterials are extensively utilized for the chromogenic and/or fluorogenic biosensing of $H_2O_2$, glucose, ascorbic acid, etc. [84, 87–90] due to various advantages, like high sensitivity, specificity, ease of separation etc. [91].

Hunan *et al.* have reported $Fe_3O_4$@Au-based nanocomposite for selective chromogenic detection of GSH [84]. The developed nanocomposite exhibited superior peroxidase like activity and could produce hydroxyl radical (•OH), which was helpful for the corresponding chromogenic changes to quantify the $H_2O_2$ level. Moreover, the colorimetric sensing could also be visualized with the aid of a smartphone via RGB analysis. Ultimately, the detection specificity was evaluated in some complex sample matrices with satisfactory results. Furthermore, Lu, Meng and Zhang *et al.* reported peroxidase mimetic reduced $Co_3O_4$ nanoparticles for chromogenic recognition of $H_2O_2$ [87]. This is basically due to the fact that after reduction oxygen vacancies are introduced, which indirectly helped to increase the adsorption of $H_2O_2$. Similarly, Ponpandian *et al.* have synthesized $NiFe_2O_4$–PANI (PANI=Polyaniline)-based nanocomposite for chromogenic and electrochemical

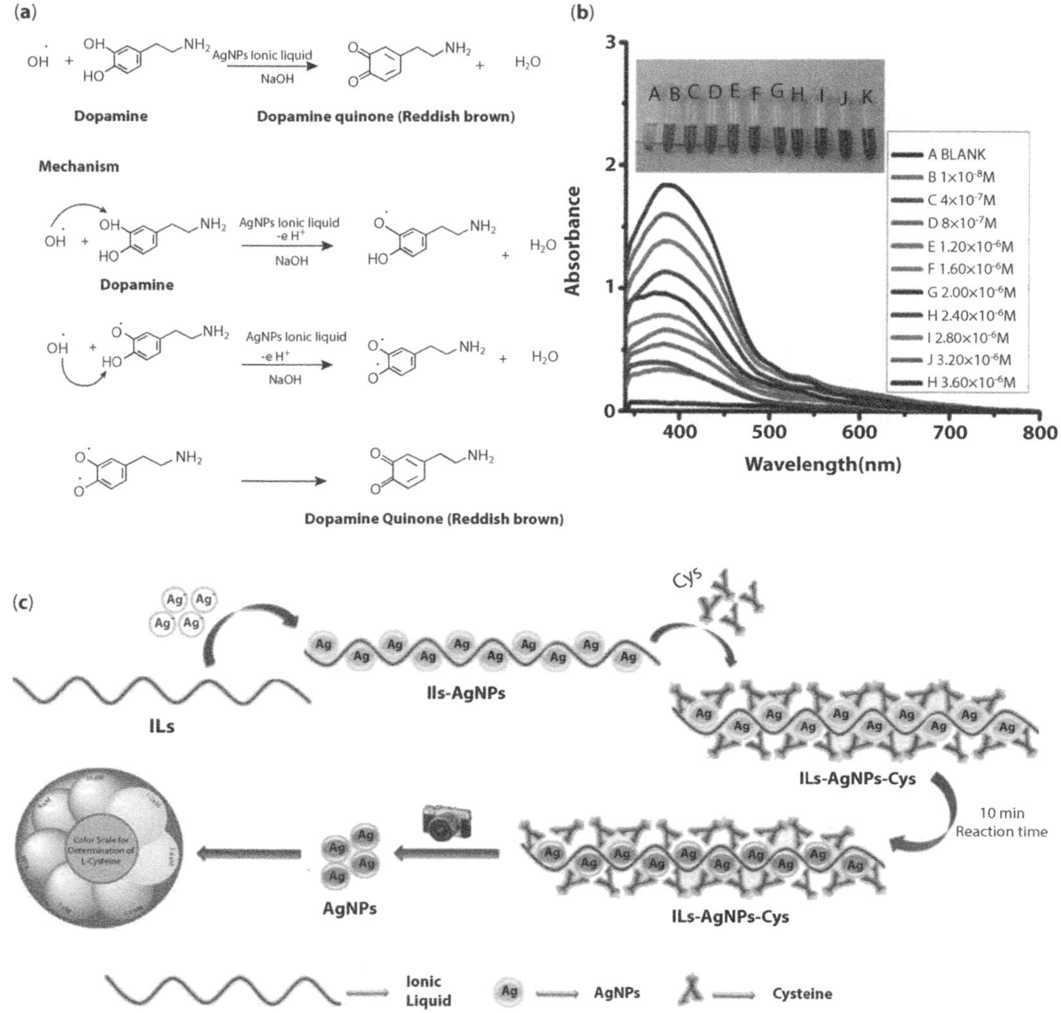


**Figure 13.6** (a) Proposed sensing mechanism of dopamine by IL capped AgNPs, (b) Variation of UV-Vis absorption spectra upon increasing dopamine concentration (Inset: observed colorimetric alterations), Adapted with permission from Ref. 74, (c) Proposed Cys sensing mechanism by IL functionalized AgNPs, Adapted with permission from Ref. 75.

detection of ascorbic acid via peroxidase like sensing of $H_2O_2$, where 3,5,3',5'-tetramethylbenzidine (TMB) acted as a colorimetric indicator [88]. Moreover, it is also responsive to some real fruit extracts, vitamin C tablets etc. Furthermore, Wang *et al.* have also reported fluorescent magnetic nanocomposite $Fe_3O_4@SiO_2@Au$ for $H_2O_2$ and glucose sensing [89]. Moreover, the detection strategy has also been exploited for the analysis of blood samples with good accuracy. Similar peroxidase like activity has been exhibited by Wu *et al.* for selective chromogenic sensing of glucose by mesoporous $MnFe_2O_4$ magnetic nanoparticles (Figure 13.7a) [90]. The sensor achieved the LOD of 0.7 mM with potential applicability in real urine sample analysis.

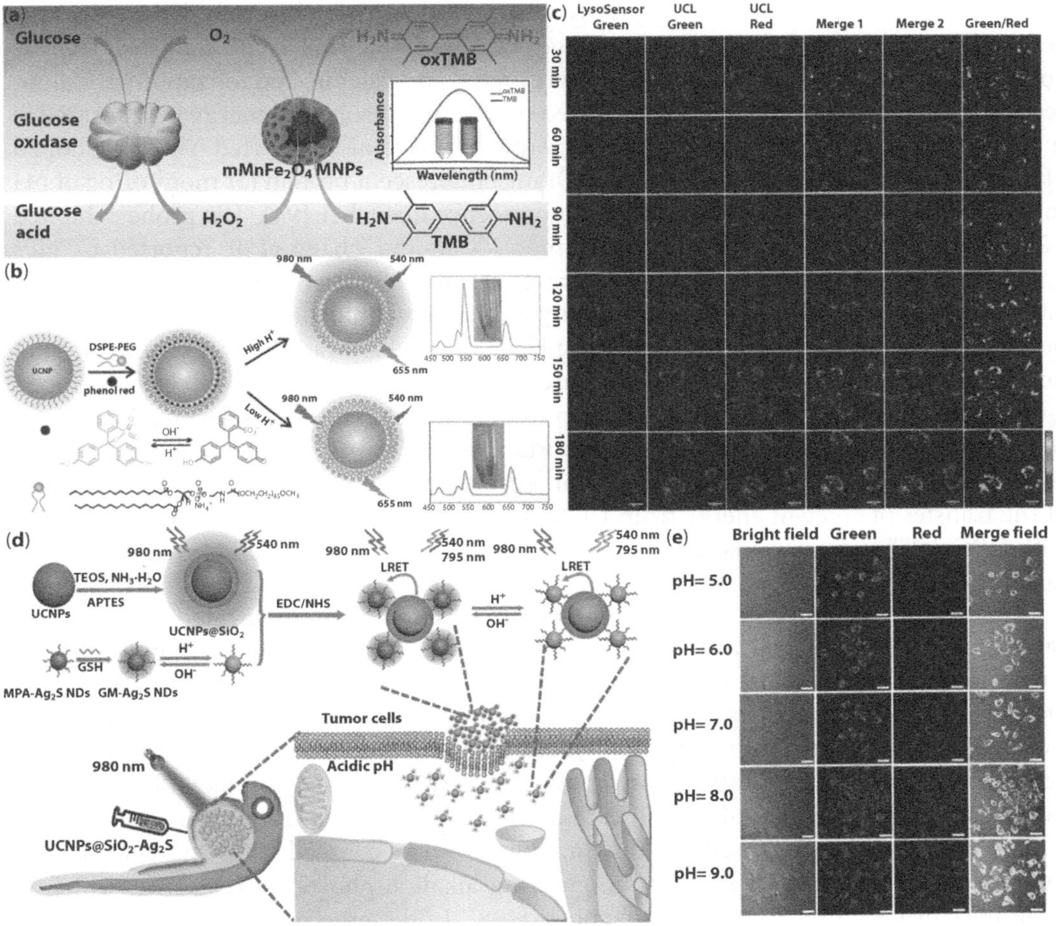


**Figure 13.7** (a) Schematic representation of glucose sensing by mesoporous $MnFe_2O_4$ based magnetic nanoparticles, (b) Schematic Diagram of the construction of UCNPs as a promising pH sensor *via* Inner Filtration Effect and (c) corresponding fluorescence imaging in HeLa cells for intracellular pH monitoring, Adapted with permission from Ref. 92, (d) Schematic representation of developed UCNPs-based nanoplatform as a ratiometric nanoprobe towards monitoring pH and (e) corresponding cellular imaging in HeLa cells, Adapted with permission from Ref. 103.

On the other hand, NIR light can be transformed into ultraviolet, visible or NIR light by lanthanide doped upconversion nanoparticles (UCNPs). UCNPs exhibit excellent biosensing of Myeloperoxidase, intracellular pH etc. [92–95] due to several advantages, like weak autofluorescence, high photostability, good biocompatibility etc. [96]. Generally, UCNPs are prepared depending on the luminescence energy transfer (LRET) mechanism [97]. Generally, 2-D nanomaterials (e.g, graphene [98], $MnO_2$ nanosheet [99] etc.), gold or silver nanoparticles [100], small organic dyes [101] etc. act as energy acceptor for increasing LRET efficiency of UCNPs.

Wu and Zhao *et al.* [92] reported self-assembly of upconversion nanoparticles (UCNPs) with phenol red and poly(ethylene glycol) (PEG)ylated phospholipid (DSPE-PEG) towards ratiometric pH monitoring in HeLa cells via IFE (Figure 13.7b, c) [102]. Based on the pH change, absorption property of phenol red fluctuated, which ultimately led to the corresponding ratiometric fluorescence change. Furthermore, Hall *et al.* reported UCNPs coupled with two anthraquinone dyes, alizarin red S and calcium red for ratiometric fluorescence monitoring of pH fluctuation based on IFE and resonance energy transfer (RET) mechanism [95]. Additionally, Chattopadhyay *et al.* have also reported $Yb^{3+}/Er^{3+}$-doped $NaGdF_4$-based UCNPs conjugated with mOrange fluorescent protein for monitoring of pH in HeLa cells, where later one acts as the fluorescence quencher [94]. The probe exhibited ~20% FRET efficiency at pH~7.0. Furthermore, Xian and Zhang *et al.* reported UCNPs coupled with $Ag_2S$ nanodots (NDs) for intracellular monitoring of pH over a pH range of 4.0 to 9.0 (Figure 13.7d) [103]. The sensitivity of the present biocompatible nanosensor has been employed for living cell imaging using HeLa cell and Zebra fish for discrimination of normal tissues with cancerous tissues (Figure 13.7e). Again, the same group has developed UCNPs-based NIR sensor integrated with phycocyanin (PC) towards selective fluorogenic sensing of Myeloperoxidase (MPO), an important pro-inflammatory enzyme in the human body [93]. Initially, the upconversion luminescence (UCL) remained in off state because of the transfer of efficient energy from UCNPs to PC. However, upon interaction of MPO with OPC, UCL was recovered and the sensing strategy has also been employed for cellular imaging during the inflammatory process.

Semiconducting polymer dots are obtained from fluorescent conjugated polymer and can respond to a particular analyte by specific functionalization with small molecules via covalent coupling or electrostatic assembly [104, 105]. These functionalized polymer dots display unique characteristics, like good photostability, signal amplification capability, good water dispersibility, good biocompatibility, etc. [106, 107], which are favorable for bioimaging and biosensing applications for monitoring various bioanalytes, like formaldehyde, sulfur dioxide, intracellular pH, alkaline phosphatase activity etc. [108–110].

Gao *et al.* have reported semiconducting polymer dots coupled with $Cu^{2+}$ coordinated rhodamine B hydrazide (RB-hy) for monitoring alkaline phosphatase activity, which is an important indicator for medical diagnosis [108]. In the present system, FRET is operated between polymer dots and $Cu^{2+}$-RB-hy (Figure 13.8a). However, in presence of PPi, a substrate of ALP, FRET is hindered due to the competitive interaction of $Cu^{2+}$ with Pi, produced from PPi by ALP. The probe can also detect targeted analytes in serum samples with good recoveries ranging from 95.9% to 102.0%. Jin and Huang *et al.* reported 3 polymer dots conjugated with 3 fluorescent dyes, for monitoring intracellular pH in HeLa cells (Figure 13.8b) [109]. Again, the group of Gao has also reported two-photon semiconducting polymer dot (BF@Pdots) for selective detection of $SO_2$ and formaldehyde based on the FRET mechanism [110]. Moreover, the sensing strategy was effectively applied for *in-vivo* imaging of $SO_2$ and formaldehyde (Figure 13.8c).

### 13.2.2 One-Dimensional Nanoprobes for Optical Biosensing

#### *13.2.2.1 Carbon Nanotubes*

The exceptional optical properties of carbon nanotubes (CNTs) have lured extraordinary research attention towards the development of different biosensors for detection of biomarkers followed by their *in vivo*/*in vitro* bioimaging. The extraordinary emission property of SWCNTs between 700-1400 nm allows deep penetration of the rays through layers of

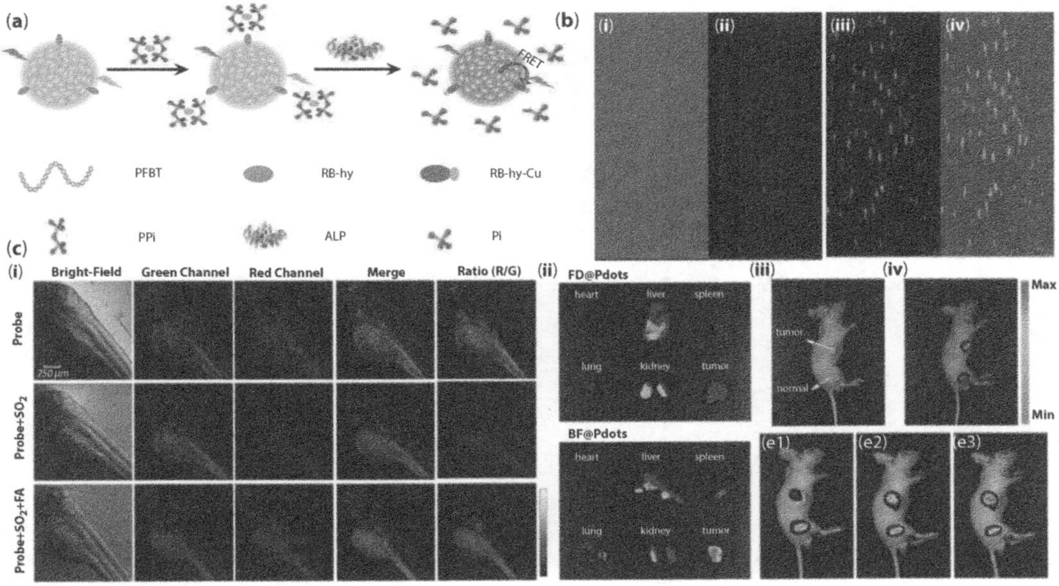


**Figure 13.8** (a) Schematic representation of ALP detection by polymer dots, Adapted with permission from Ref. 108, (b) Live cell imaging with HeLa cells by Pdots-PPF, (c) *In-vivo* imaging of $SO_2$ and formaldehyde in mitochondria of tumor cells by BF@Pdots, Adapted with permission from Ref. 110.

skin. It helps to *in vivo* tissue and organs imaging, which requires a dose limit of ~15 times lower, thereby enhancing its biocompatibility [111]. CNTs exhibit photoluminescence in the NIR region. Additionally, SWCNTs offer high photostability besides low autofluorescence which restricts from the false response. In this type of mechanism, different types of biomolecules may be exploited for tuning the emission property of CNTs which ultimately may be used for biosensing purposes [112–123]. Likewise, Strano *et al.* [112] developed luciferase enzyme grafted SWCNTs-based fluorescent sensor for detecting adenosine 5'-triphosphate (ATP) in living cells. Herein, in presence of ATP, enzyme catalyzed the transformation of D-luciferin to oxyluciferin took place which resulted in quenching in fluorescence of the developed CNTs with a LOD of 240 nM. The same group has developed functionalized CNTs by incorporating bombolitin peptides for recognition of nitroaromatics. The change in the secondary structure of the peptide by the nitroaromatics resulted change in emission profile [113]. Similar kind of approach was again chosen by this group of Strano, wherein, SWCNTs were functionalized with adenine (A)-thymine(T) base pair via weak interaction. The functionalised CNTs when interacted with nitric oxide (NO), the fluorescence was quenched with LOD of 300 nM [114]. Next, SWCNTs were dispersed in antibodies-integrated chitosan film. The developed chitosan film was exploited as a fluorescent sensor towards targeted protein troponin T—a cardiac-specific biomarker that shifted the fluorescence emission of SWCNTs after interaction with grafted antibodies [115].

Next, Safaee *et al.* [116] have developed single-stranded DNA (ssDNA) conjugated SWCNTs-based flexible microfibrous textiles-based fluorogenic probe (Figure 13.9). The ssDNA-SWCNTs nanosensor was exploited for detecting reactive oxygen species (specifically $H_2O_2$) and it has been utilized in wound dressing for monitoring of wound healing

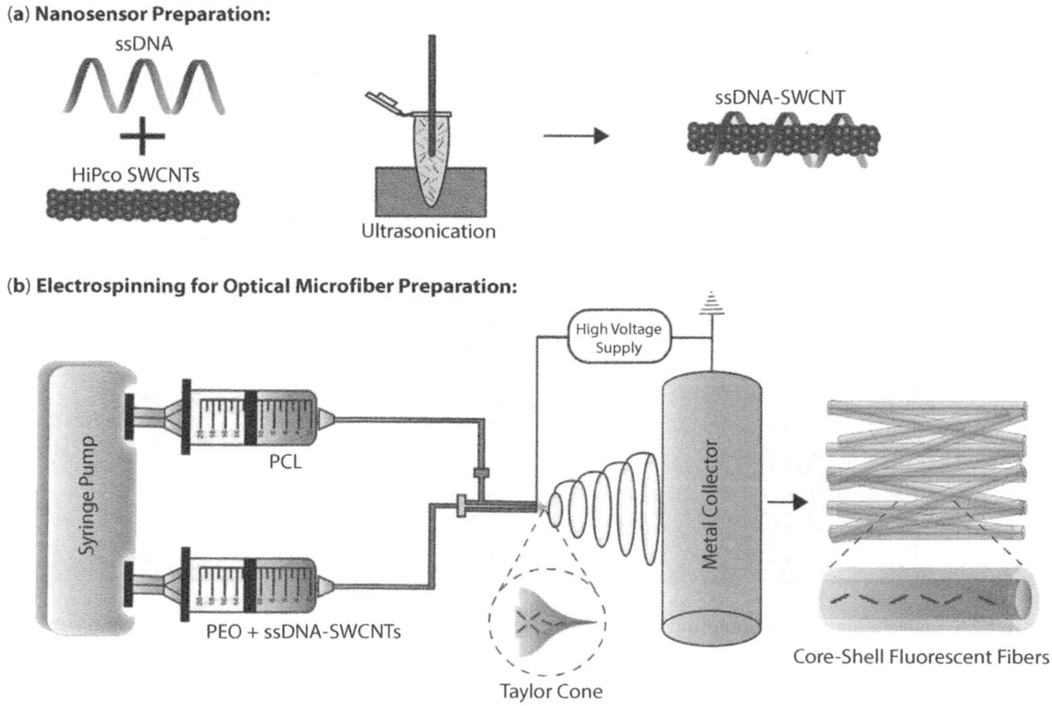


**Figure 13.9** Schematic diagram of fabricating optical core-shell microfibers. (a) Dispersion of SWCNTs was *via* sonication in presence of ssDNA, (b) The ssDNA-SWCNTs dispersion was mixed with poly (ethylene oxide) (PEO) and polycaprolactone (PCL), Adapted with permission from Ref. 116.

in real time. After that, Lisi *et al.* [117] has fabricated antibodies conjugated CNTs-based biosensors wherein CNTs have been exploited as signal amplifiers in surface plasmon resonance (SPR) to recognize Alzheimer's disease biomarker, Tau protein (LOD: 125 pM). Yao *et al.* have reported a lateral flow device using thymine-rich capture and label oligonucleotides for detecting mercury in naked eye.

Then, Nandeshwar *et al.* [118] have reported a colorimetric strategy, which relies on the degradation of SWCNTs. Herein, detection of myeloperoxidase enzyme has been emphasized with a detection limit of 327 ng/mL. In presence of $H_2O_2$ and $Cl^-$, the enzyme causes degradation of SWCNTs. the targeted enzyme may be detected by monitoring the change in optical contrast at the NIR wavelength.

Currently, the CNTs have also been applied for detecting COVID-19. Pinals *et al.* [119] have developed a detection strategy for the recognition of SARS-CoV-2 spike protein depending on the weak interaction with ACE2. By using this technique, virus-like particles and spike protein have been detected in the concentration range of 35 mg/L with a response time of only 5 s.

### *13.2.2.2 Silicon Nanowires*

Nanowires are another important type of one-dimensional nanostructures that can be easily prepared in suitable rigidity, straightness and variety of lengths, having potential applications in chemical and/or biological fields including sensing. Among the variety of nanowires, silicone (Si)-based nanowires (NWs) is considered as a potential contender because of their nontoxicity, increasing adhesion force [124], surface tailor ability, admirable mechanical/electronic/optical properties and biocompatibility [125]. Silicon nanowires (SiNWs) sensors are formed of three main parts, i.e., source, drain and gate electrodes [126]. Nowadays, SiNWs are used in biosensing applications following a variety of mechanistic pathways e.g., fluorescence and field-effect transistors (FET), electrochemical methods, etc. Among them fluorogenic sensing attracts more due to a bunch of advantages like the low detection limit, high sensitivity, immense selectivity etc. Therefore, it is a burgeoning research topic nowadays to explore the sensing mechanism of SiNWs as the anchoring substrate for biological detection as being an indirect band gap material silicon-based nanowire sensor array exhibit mature fabrication technologies for mass production [127]. Currently, these are being used for the detection of various bioanalytes like $ClO^-$, $H_2S$, $Ca^{2+}$, C-reactive protein etc. [128–131].

Intracellular ionic calcium ($Ca^{2+}$) acts as second important intracellular messenger especially in the neurons by helping in plasticity and synaptic transmission processes [132]. The aberrant concentration of $Ca^{2+}$ would lead to neurodegenerative disorders, chronic heart failure, skeletal muscle defects etc. [133] Mu *et al.* had designed a ratiometric nanobiosensor, which is comprised of single silicon nanowires (SiNWs) anchored with Ru-based dye, $Ru(bpy)_2(mcbpy\text{-}O\text{-}Su\text{-}ester)(PF_6)_2$ as the reference and Fluo-3 as a response molecule for analyzing intracellular $Ca^{2+}$ with high accuracy [128]. The present SiNWs-based nanosensor was successfully used in L929 cell. Notably, this nanowire sensor opens a new horizon to determine neuronal $Ca^{2+}$ at a high spatial resolution. Again, among reactive oxygen species (ROS), hypochlorous acid (HClO) acts as a pivotal function. It generates hypochlorite ion ($ClO^-$) in physiological pH solutions by partial dissociation [134]. The unusual level of this may cause the deviation in myeloperoxidase (MPO) level in leukocytes, monocytes, neutrophils and macrophages, resulting serious disease [135]. In this aspect, to monitor $ClO^-$ ion, Mu and Shi *et al.* had constructed a fluorescent nanosensor-based on a single SiNW by embellishing a near-infrared dye IR780 with the aid of confocal microscopy and micromanipulation [129]. In presence of $ClO^-$, there occurs one electron oxidation of IR780, which would cleave of the polymethine chain of IR780, resulting the quenching of fluorescence. This SiNW sensor has a distinguished response towards $ClO^-$ in HeLa Cells and RAW264.7 by showing the potential application to detect the intracellular and endogenous $ClO^-$ at the single-cell level. Hydrogen sulfide ($H_2S$), the third gaseous transmitter in human body, helps to regulate the intracellular redox process and other important physiological processes and its abnormal concentration would lead to serious health issues like Alzheimer's disease [136] etc. In this aspect, Mu and Shi *et al.* had fabricated a fluorescent "turn-on" SiNWs array-based sensor by covalently immobilizing the naphthalimide azide derivative for $H_2S$ detection due to the generation of donor-π-acceptor structure [130]. The as-designed SiNW sensor is highly selective for $H_2S$ having 7.13 μM LOD and was efficaciously engaged for live cell imaging with HeLa cells (Figure 13.10a, b). In the present era, cardiovascular diseases are serious global threat to death. An acute-phase protein named C-reactive protein (CRP) found in human saliva, is recognized as a biomarker for cardiovascular problems and heart attack incidence [137]. To reduce the risk of cardiovascular disorders, Irrera *et al.* proposed a SiNWs array-based room temperature light-emitting luminescent sensing platform to selectively detect CRP in the femtomolar level of detection

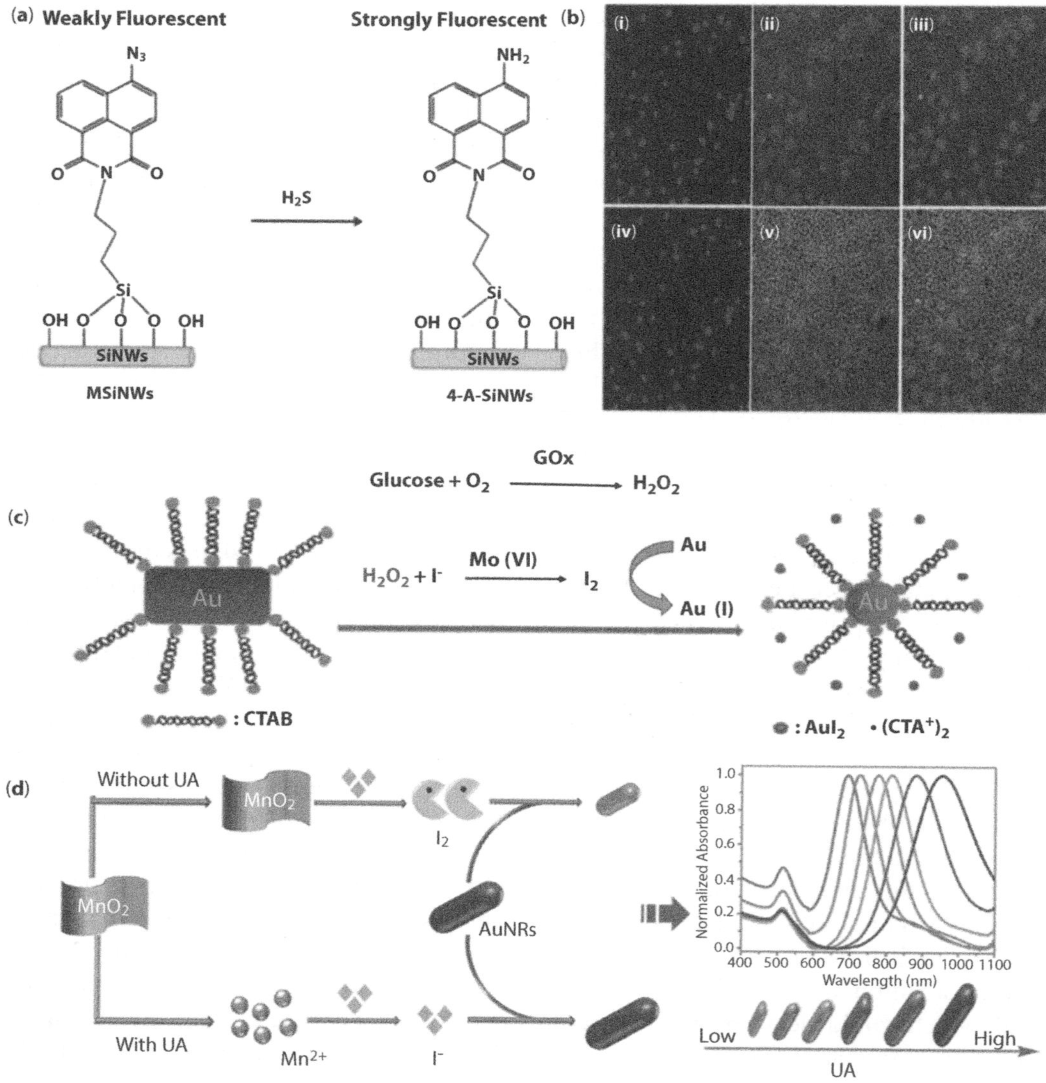


**Figure 13.10** (a) Schematic diagram of $H_2S$ response of MSiNWs, (b) Fluorescence images of *in situ* cell-captured MSiNW sensor incubation with extracellular $H_2S$ for live HeLa cells before (i-iii) and after (iv-vi). a and d: images of HeLa cells captured by substrate with Hoechst staining. (ii) and (v): the fluorescence images of cell-captured MSiNW arrays; (iii) and (vi): overlaid with (i), (iv) and (ii), (v); (c) Schematic illutration of glucose sensing analysis based on the GNRs-etching, Adapted with permission from Ref. 158, (d) Schematic diagram of colorimetric Uric Acid recognition, Adapted with permission from Ref. 166.

threshold (LOD) with $10^{-8}$-$10^{-2}$ μg/mL linear concentration range [131]. This nanoprobe achieves an economically and industrially viable methodology to detect CRP levels in saliva directly by an ultrasensitive and highly selective manner over a broad concentrations range. Markedly, this SiNW sensor is label-free, and highly selective towards CRP without the requirement of chemical treatment of the analyte. This as-prepared nanosensor is successfully applied in human serum to detect CRP and exhibited potential application towards preventing the risk of heart attack.

### 13.2.2.3 Gold Nanorods

The mesmerizing size-dependent properties of one-dimensional noble metal (e.g, Au, Ag, Cu etc.) nanomaterials are burgeoning research topic for their diverse applications including electronic, optical and biomedical applications. Among them, gold nanorods (GNRs) are an important captivating member [138] because of their unique characteristics, like adaptable plasmonic properties and anisotropic shape. These are pseudo-one-dimensional rod-shaped nanoparticles [139]. GNRs have two surface plasmon resonance absorptions, one in the visible region, and another in the NIR region [140, 141]. This optical behavior is justified by Gan's theory [142]. GNRs possess well established biocompatibility, good colloidal stability and exhibit advanced magnetic properties with high optothermal conversion efficiency [143]. The preparation as well as functionalization of GNRs is very facile, which can enhance the versatility of applications like disease diagnostics, sensing and therapeutics. Contextually, the three most popular strategies for modification of GNR surface are (1) ligand exchange, (2) self-assembly through layer by layer and (3) surface coating [144]. The formation of GNRs was involved mainly two methods, which are template method and electrochemical synthesis since the 1990s [145]. Nowadays there are a bunch of strategies to synthesize Au nanorods such as electrochemical synthesis [146], photochemical methods [147], seed-mediated growth method [148], template-assessed synthesis using porous aluminum template [149] and lithographic fabrication methods [150]. Among these, the seed mediated pathway plays a pivotal role to prepare Au nanorods as its simple and flexible scheme with numerous aspect ratios produces a relatively high yield of GNR [151]. It is worth mentioning that the synthesis of GNRs via green approach paves a new endeavor to surpass the challenges faced during the seed-mediated growth method. To establish the green synthesis, plants or microorganisms are required in the process where the designing of anisotropic Au nanorods is carried out via a metallic substrate (produced from several eukaryotic and prokaryotic species) reduction by enzymatic complexes [152]. Biosynthesis of Au nanorods imparts innumerable advantages like nonpure starting material usage, low working pressures and low temperatures for the reaction, eco-friendliness and economic viability [153]. The etching of GNRs produces tiny nanoparticles [154], which provide great importance in the fabrication of chemo- and/or bio-sensor due to its non-aggregation-based [155] shape-dependent localized surface plasmon resonance (LSPR) property [156], decreased aspect ratio, selectivity, sensitivity, response time and the feasible detection limit. Nowadays, biomolecules sensing through chromo and/or fluorometrically by GNRs are an alluring research topic.

As higher glucose level in urine or blood causes diabetes, it is necessary to monitor glucose on a regular basis [157]. Recently, Chen *et al.* developed an ultrasensitive colorimetric-based GNRs sensor to detect glucose through oxidase-catalyzed oxidation of glucose and molybdate-catalyzed etching of GNRs by $H_2O_2$ in presence of $I^-$ (Figure 13.10c) [158]. The probe is also tremendously sensitive towards real time monitoring of glucose in urine. On the other hand, dopamine is an essential neurotransmitter whose abnormal content in human body causes serious nerve disease [159]. To surpass this, Lin *et al.* had designed a chromogenic assay of dopamine (DA) detection by iodine mediated GNRs etching via hypsochromic shifting of LSPR peak of GNRs with a detection threshold of 0.62 μmol/L having 0.80-60.0 μmol/L concentration range [160]. Notably, this strategy is efficient for dopamine detection from real human urine sample.

Again, Rameshkumar and Huang *et al.* synthesized a selective chromogenic ophthalmic probe for the recognition of dopamine based on dopamine induced GNRs aggregation [161]. This developed sensor achieved 30 nM LOD with the concentration range 100 nM-10 mM and successfully used for various concentration of dopamine detection from human urine sample with excellent recoveries. Acid phosphatase (ACP) acts as catalyst in the acidic hydrolysis of phosphomonoesters in mammalian tissues [162] and unusual concentrations of ACP may cause even prostate cancer and other serious health disorders [163]. In this regard, Wu *et al.* established a rapid, low-cost and effective smartphone-assisted chromogenic visual detection of ACP through controlled GNR etching by iodine ($I_2$) [164]. This assay exhibits a good linear response 0-15.0 U/L having 0.97 U/L LOD. Interestingly, this method successfully achieved practical applicability by detecting the ACP in human serum sample.

Again, uric acid is the main antioxidant in blood and unexpected concentration causes alarming health issues like gout etc. [165]. Geng and Wang *et al.* had fabricated an enzyme-free chromogenic sensing assay for uric acid (UA) detection by the inhibition of AuNRs etching (Figure 13.10d) [166]. This method is sensitive and simple, achieved with naked eyes and UV-Vis absorption spectral response with 0.76 μM LOD. Markedly, this as-designed strategy is applicable in real sample determination from human urine and serum sample with good accuracy.

### *13.2.2.4 Nanoribbons*

Nanoribbons are another type of one-dimensional nanomaterials showing a great potential in a variety of applications like sensing, biomedical diagnostics etc. Nanoribbons may be categorized as carbon nanoribbons, phosphorene nanoribbons, silicene nanoribbons etc. Especially, carbon nanoribbons, elongated slim strips of $sp^2$ carbon atoms, have caught attention in this domain as it is easily modified or functionalized due to their high conductivity, good flexibility, good π-π stacking in basal planes, large length-to-width ratio, tunable fluorescence etc. [167]. Currently, nanoribbons have utilized as an appreciable sensor for a variety of bioanalytes, like citrate, quercetin, dopamine, α-glucosidase etc. via chromogenic and/or fluorogenic methodology [168–171].

An important metabolite of human body is citrate ($C_6H_5O_7^{3-}$) which has a variety of biological functions e.g., coagulation of blood, mitochondrial energy formation etc. [172] and the deficiency of citrate causes the serious dysfunction of kidney like nephrolithiasis and nephrocalcinosis [173]. Wang *et al.* depicted $C_3N_4$ nanoribbons-based *"off-on"* blue fluorogenic probe based on to recognize citrate anion ($C_6H_5O_7^{3-}$) [168]. This nanoprobe shows high selectivity and sensitivity towards the citrate anion showing the detection threshold of 0.78 μM. Initially, the fluorescence of the nanosensor was quenched in presence of $Cu^{2+}$ via photoinduced electron transfer (PET) [174] and in presence of $C_6H_5O_7^{3-}$ the fluorescence was again regenerated (Figure 13.11a). It is worth to mention that the group had also hypothesized about the inhibited interaction between $Cu^{2+}$ and $C_3N_4$ due to the strong coordination of $Cu^{2+}$ and ($C_6H_5O_7^{3-}$) [175]. Again, a natural flavonoid quercetin, is widely found in vegetables, fruits, flowers etc. [176]. The detection of quercetin is significant as it has a pivotal role in many pharmacological processes e.g., anticancer, antioxidant, etc. [177]. Wang and Fan *et al.* had fabricated carbon nanoribbons co-doped with sulfur and nitrogen (SNCNRs) polymer for quercetin detection [169] with 21.13 nM LOD and 50.0

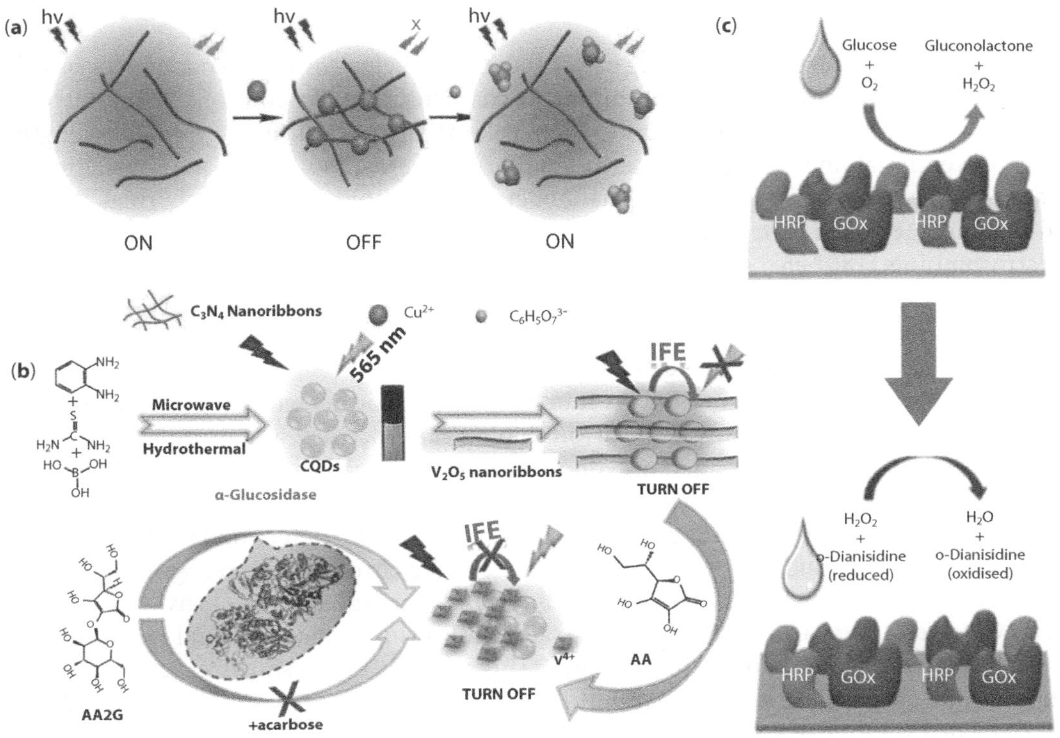


**Figure 13.11** (a) Schematic demonstration of citrate ion ($C_6H_5O_7^{3-}$) detection by $C_3N_4$ nanoribbon-based fluorescent sensor, (b) Schematic diagram of CQDs synthesis and mechanistic pathway for α-glucosidase detection based on CQDs-$V_2O_5$ nanosensor, Adapted with permission from Ref. 171, (c) Schematic mechanism of glucose recognition by utilizing CS/PVA nanofiber strips, Adapted with permission from Ref. 188.

nM-200 μM linear detection range. Notably, this nanosensor was used to detect quercetin in bovine serum sample to analyze real field applicability. On the other hand, an important catecholamine neurotransmitter dopamine (DA) has a significant role in the nervous system and the aberrant concentration of DA causes several neurological diseases [178]. Mehdinia and Jabbari *et al.* depicted a graphene nanoribbon-based label-free chromogenic nanobiosensor for DA sensing depending on peroxidase-like activity [170]. This nanoprobe exhibits the LOD of 0.035 μM with 0.1-50 μM linear concentration range. Markedly, the as-prepared peroxidase mimetic nanoribbon was successfully utilized for DA recognition in human serum samples, providing a sensitive sensing platform having potential advantages in diagnostic and biological applications. α-Glucosidase enzyme catalyse α-Glucose formation, which is the main reason for diabetes [179]. In this regard, Chen and Niu *et al.* had proposed an effective and sensitive "turn on" fluorogenic pathway based on carbon quantum dots (CQDs) co-doped with boron (B), sulfur (S) and nitrogen (N) heteroatoms and vanadium oxide ($V_2O_5$) nanoribbons towards detection of α-glucosidase and its inhibitors [171]. This as-fabricated nanoprobe exhibits fluorescence quenching by $V_2O_5$ nanoribbons (Figure 13.11b). Interestingly, this as-prepared nanoribbon-based nanosensor was also effective for the identification of targeted enzymes in real samples.

### 13.2.2.5 Nanofibers

Nanofibers are nanosized fibers having below 100 nm diameter. Nowadays nanofibers-based sensors are attractive because of their inimitable characteristics like controllable pore structure with remarkable interconnectivity, high volume of pore, stiffness, ductile strength than conventional fibers, large surface-to-volume ratio, etc. [180]. There are several techniques presents for the preparation of nanofibers like melt-blowing, hydrothermal, physical drawing [181], template synthesis [182], phase separation [183], self-assembling [184], electrospinning [185] etc. Among them electrospinning is the most effective pathway to get the dreamed nanofibers even in required dimensions [185]. Here, some chromogenic electrospun nanofibers are discussed for the detection of several biological molecules like L-cysteine, ascorbic acid, $H_2O_2$, glucose etc. [186–190].

Among various vitamins present in human body, Ascorbic acid (AA) is the soluble one, has a significant contribution in various biological processes, and its imbalance concentration may produce ROS and antioxidants, which may lead to tissue related disorders [191]. Lu *et al.* designed peroxidase mimetic $Fe_3O_4$/nitrogen-doped carbon ($Fe_3O_4$/N-C) hybrid nanofibers-based chromogenic probe for AA and $H_2O_2$ detection [186]. Again, Lu *et al.* have also fabricated another type of nanofiber for chromogenic recognition of AA and $H_2O_2$ which is comprised of carbon dots (CDs) impregnated $Fe_3O_4$ hybrid nanofibers (CDs/$Fe_3O_4$) via an electrospinning strategy following a hydrothermal reaction and a thermochemical reduction pathway [187]. This nanofiber also shows excellent peroxidase mimic catalytic activity for AA and $H_2O_2$ detection. These two types of nanofibers exhibit a broad range of applications including environmental monitoring, medical diagnosis, biosensors etc. Again, L-cysteine (Cys) is a vital amino acid for human body whose unusual level *in-vivo* may cause serious health problems, such as cerebrovascular disease, coronary heart disease etc. [192]. So, to detect Cys with a high performance nanoprobe, Lu *et al.* have synthesized Prussian blue (PB)/polypyrrole (PPy) hybrid nanofibers using α-$Fe_2O_3$ nanofibers [189]. This as-prepared nanofiber exhibited chromogenic detection of L-cysteine with 16 nM of LOD. Notably, this nanofiber sensor exhibits peroxidase mimic catalytic activity and it is highly selective toward L-cysteine over other biological influential components. This methodology opens a bunch of applications in biotechnology, environmental science, medicine and food industry. Again, Lu *et al.* also reported superparamagnetic $Fe_3O_4$-based nanofiber [190] for chromogenic determination of L-cysteine with LOD of 0.028 μM.

Again, glucose is the main source of energy in human body and helps in metabolic homeostasis and other physiological processes. The aberrant level of glucose in human body causes the very common health disorder, diabetes mellitus [193]. Filiz *et al.* designed a nanofiber-based chromogenic nanosensor towards monitoring glucose level [188]. They had fabricated the nanofiber by chitosan (CS)-poly(vinyl alcohol) (PVA) blended via electrospun method to get the bead free, uniformed structure of that CS/PVA nanofiber. This nanosensor showed 2.7 mM limit of detection (LOD) with 2.7-13.8 mM linear range of glucose concentration (Figure 13.11c). The applicability of this nanosensor was investigated with the human blood sample. This biobased CS/PVA nanofiber is economic viable and eco-friendly and designed as a testing strip for visual detection of glucose level present in human body.

### 13.2.3 Two-Dimensional Nanoprobes for Optical Biosensing

#### *13.2.3.1 Graphene*

Graphene has a single layer of honeycomb carbon lattice and similar to CNTs, graphene and its derivatives, GO, rGO exhibit unique electrochemical and optical properties due to their various advantages, like ultrathin nature, good biocompatibility, excellent electrical properties etc. [194]. Graphene can undergo target-specific interaction with biomolecules through electrostatic and/or π-π stacking interactions, which advantageous for biosensing applications.

##### 13.2.3.1.1 Graphene Oxide (GO)

GO is the two-dimensional oxidized from of graphite [195]. This is generally composed of carbon–carbon $sp^2$ domains along with various functionalities, like $-CO_2H$, -OH, epoxy group etc. [196]. Furthermore, GO can be used as a 2-D support material of various functionalized materials for effective biosensing applications. Recently, GO are found to be widely explored in the realm of biosensing of $H_2O_2$, glucose, $H_2S$ etc. [197–199], where GO acts as an efficient fluorescence quencher via FRET process [200].

Qi *et al.* have developed graphene oxide (GO)-AuNPs derived nanozyme for selective $H_2O_2$ detection based on peroxidase-mimetic activity, effectively catalyzing the oxidation of the enzymatic substrate, TMB to the corresponding oxidized product, TMB oxide, leading requisite colorimetric changes from colorless to blue [197]. The sensitivity of the present probe is 1~2 times greater than that of the single nanoparticles derived nanozymes. Wu and Zhao *et al.* developed graphene oxide-based assay towards fluorogenic monitoring of DNA exonuclease enzymatic activity [198]. Initially, fluorophore-tagged DNA and GO underwent FRET and π-π stacking interaction, due to which the fluorescence remained in the quenched state. However, in presence of Exo-III, fluorophore tagged DNA was digested to lead inefficient fluorescence quenching, based on which Exo-III activity could be quantified. Hao and Xing *et al.* have reported γ-$Al_2O_3$ nanorods anchored GO nanosheets conjugated with DNA/sulfide fluorophore for selective sensing of $H_2S$, a significant gas signaling agent in the human body [199]. Initially, the fluorescence of the nanocomposite remained in the quenched state. However, in presence of $H_2S$, DNA/sulfide adduct was detached from the surface of GO due to stronger interaction of $H_2S$ with the γ-$Al_2O_3$-GO, resulting the fluorescence enhancement corresponding to the degree of the present $H_2S$ (Figure 13.12a).

##### 13.2.3.1.2 Reduced Graphene Oxide (rGO)

rGO, another important class of graphene derivative, has appeared as promising tools for the detection of biomolecules because of their non-toxicity, photostability, cost-effectiveness etc. However, due to the absence of any functional groups, these kinds of nanomaterials are functionalized or conjugated with other organic dyes, metallic nanoparticles etc. via covalent or non-covalent supramolecular interaction to improve the sensitivity of rGO for chromo-fluorogenic biosensing applications.

Swain *et al.* reported phenylboronic acid integrated rGO for fluorogenic enzyme-free sensing of glucose [201]. Initially, the fluorescence of rGO-PBA nanoassembly remained in the quenched state due to the cyclic boronate ester formation. However, glucose could regenerate the initial fluorescence to provide quantitative information about the glucose

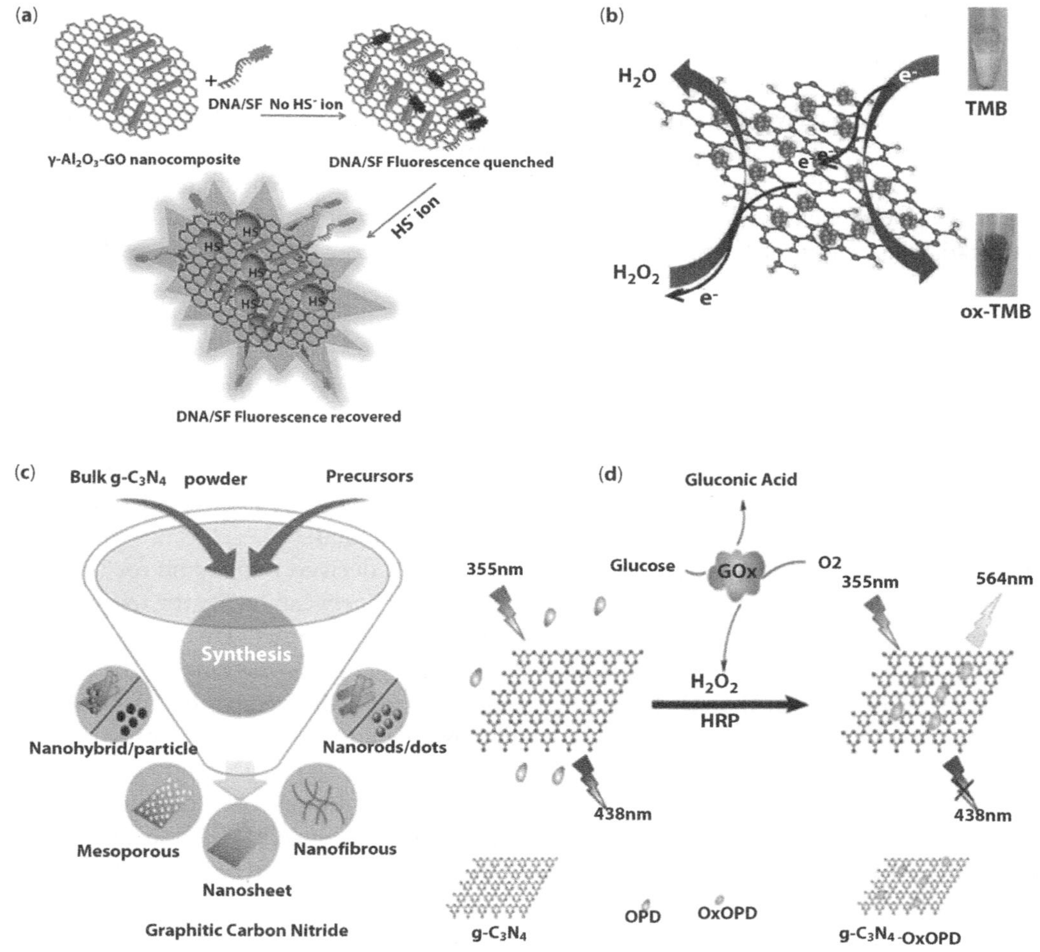


**Figure 13.12** (a) Sensing mechanism of γ-$Al_2O_3$–GO/DNA/SF nanobiosensor, Adapted with permission from Ref. 199, (b) schematic description of peroxidase-like activity of $IrO_2$/rGO, Adapted with permission from Ref. 204, (c) Graphitic carbon nitrides synthesis, (d) Schematic representation of g-$C_3N_4$ nanosheets for fluorogenic determination of $H_2O_2$ and glucose, Adapted with permission from Ref. 201.

concentration. Muthoosamy *et al.* have reported RNase A/AuNCs, folic acid and rGO-based "turn-on" fluorescence nanobiosensor for selective sensing of intracellular $Na^+$ and $K^+$ ions, which are the abundant ions in the human body [202]. The fluorescence enhancement in presence of $Na^+$ and $K^+$ ions may be due to the definite interactions of the metal ions with the anionic carboxylate functionalities present in the protein skeleton, ultimately recovering the fluorescence of RNase A/AuNCs, which might also be the result of specific oxidation/reduction reactions. The same group has reported folic acid-integrated rGO-modified BSA (bovine serum albumin) gold nanoclusters for selective fluorogenic detection of glutathione (GSH), which is an important cancer biomarker [203]. In presence of GSH, GSH underwent supramolecular interaction (H-bonding and van der Waals interaction) with BSA to form corresponding BSA-GSH complex, which destabilizes the FA-rGO-BSA/AuNCs and ultimately there was fluorescence quenching. The quenching of fluorescence may also be occured because of the GSH-$Au^+$ complex formation via target–specific etching reaction. This work paves the way towards detection of GSH for biomedical diagnosis of cancer cells. Yang *et al.* reported $IrO_2$/rGO nanocomposites for chromogenic sensing of small biothiols by peroxidase like enzymatic activity [204]. Herein, TMB was used as enzymatic substrate, which was adsorbed on rGO surface and transferred electrons to $IrO_2$ nanoparticles, which then transferred electrons to $H_2O_2$, by which $H_2O_2$ was reduced to water and TMB was oxidized to produce the colorimetric alteration from colorless to blue according to the concentration of $H_2O_2$. The inhibition of the enzymatic activity was employed for effectual detection of low molecular-weight biothiols, like Cys, Hcy, GSH from human serum samples (Figure 13.12b). Similar peroxidase like activity was reported by Ponpandian *et al.* towards synthesizing $WS_2$ nanosheets conjugated rGO for selective detection of $H_2O_2$ [205]. Here also TMB was used as the enzymatic substrate. Moreover, the assay could detect $H_2O_2$ in urine samples with good recovery, thereby exhibiting its importance in medical diagnosis.

### *13.2.3.2 Graphitic Carbon Nitride (g-$C_3N_4$)*

g-$C_3N_4$, exclusive graphene-like two-dimentional carbon-based nanomaterial having higher and stronger fluorescence response, made up of nitrogen containing cyanamide, melamine, thiourea dicynamide, ammonium thiocyanate etc. by polymerization technique (Figure 13.12c) [206]. In g-$C_3N_4$ nanosheets, the presence of tri-s-triazine with larger surface area and higher degree of condensation results stronger fluorescence useful for the evolution of highly selective and specific nanobioprobe or nanobiosening material towards recognition of a range of biomolecules, like glucose, Cysteine, $H_2O_2$, alkaline phosphatase (ALP), $Fe^{3+}$ etc. [207–211].

Cysteine is an important amino acid, consists of thiol group, which acts a pivotal function in various biological activities like synthesizing protein, metabolic processes. An excessive amount of Cysteine leads to such diseases like AIDS, Alzheimers's and Parkinson's diseases [212]. Hu and Hao *et al.* reported graphitic carbon nitride nanosheets (CNNS) for specific turn-off fluorescence sensing of $Hg^{2+}$, however, L-Cysteine (Cys) could revive the fluorescence of CNNs via specific interaction of $Hg^{2+}$ with the thiol group of L-Cys. Moreover, this fluorescent nanoprobe was applied to detect the targeted analytes from real field water sample analysis like with reasonable outcomes [208]. Again, alkaline phosphatase (ALP) enzyme plays important roles in many biological activities and acts as potential biomarker in clinical diagnosis [213]. In this aspect, Tang and Jiang *et al.* have also reported water-dispersible ultrathin g-$C_3N_4$ nanosheets as label-free biosensor for selective and sensitive detection for ALP with 0.08 U/L LOD value. This upgraded method is convenient because of its rapidness for selective detection of ALP in biomedical applications [209]. On the other hand, iron plays key role in various physiological processes like transport of oxygen, enzymatic electron transfer etc. The abnormal level of iron effects human health, leading to anaemia, liver and kidney damage etc. [214]. Recently, Zhang and Huang *et al.* have reported lanthanum (La) loaded g-$C_3N_4$ nanosheets (CNNS). $Fe^{3+}$ quenched its fluorescence via PET [210]. $H_2O_2$ is an important ROS related to numerous biological processes like cell division, cell proliferation, differentiation etc. Unusual amount of $H_2O_2$ results some diseases like cancer, nervous breakdown, mental disbalance etc. [215]. In this aspect, Jiang and Yu *et al.* reported a unique g-$C_3N_4$ nanosheets for specific sensing of $H_2O_2$ in presence of *o*-phenylenediamine (OPD) as the substrate and horseradish peroxidase (HRP) enzyme followed by selective detection of glucose by enzymatic catalysis of glucose oxidase (GOx). Moreover, the newly synthesized probe could also analyze real serum samples [211] (Figure 13.12d). Nano enzymes are functional nanomaterials with catalytic activity like enzymes. Peroxidase-mimicking nanozymes are used to formulate biosensors for diverse targeted analytes like NAs (nucleic acids), biomacromolecules etc. Phosphates play crucial functions in several physiological processes in the human body [216]. Wei *et al.* reported a peroxidase mimic ratiometric sensing platform, comprised of three nanozymes-based on g-$C_3N_4$ for sensitive recognition of $H_2O_2$ (LOD of 4.20μM and linear range of 10-1000μM), glucose (LOD of 4.73μM and linear range of 10-250μM) and discriminative detection of five phosphates, ADP, ATP, AMP, pyrophosphate (PPi) and phosphate (Pi) and the usefulness of the nanozymes was checked by examining the targeted analytes from unknown blind samples [207].

### *13.2.3.3 $MnO_2$ Nanosheets ($MnO_2$-NS)*

Manganese dioxides ($MnO_2$)-based 2-D nanosheets have exhibited important biosensing applications because of their novel optical characteristics, high specific capacitance, high surface area, good biocompatibility, cost-effectiveness etc. Moreover, they can efficiently quench the fluorescence of luminescent nanomaterials, which is having the capacity of fluorogenic sensing of the targeted analytes [217]. $MnO_2$ nanosheets exhibit unique oxidase-like activity and UV–Vis absorption, which permitted them to act as potential "on–off", "off–on" or "ratiometric" fluorescence sensors. Particularly, $MnO_2$ nanosheets exhibit the UV-Vis absorption peak in 250-500 nm, which coincides with the emission or excitation spectrum of various fluorescence nanomaterials. Therefore, $MnO_2$ nanosheets have the ability to reduce the emission property of fluorescence nanoparticles via IFE or FRET, where $MnO_2$ acts as the quencher or energy acceptor [218].

Unusual level of glutathione (GSH) in human body may cause various diseases, such as liver damage, leucocyte loss, psoriasis etc. [219]. Recently, Li and wang *et al.* have designed "turn-on" nanosensor for GSH, which is relied on based on the IFE of $MnO_2$ nanosheets on Boron Nitride Quantum Dots. Herein, redoxable $MnO_2$ effectively quenched the fluorescence of BNQDs via efficient light absorption capability of $MnO_2$ NS. However, GSH converted $MnO_2$ to $Mn^{2+}$, resulting the recovery of the partial fluorescence of BNQDs via weakening of IFE [220]. Dong *et al.* synthesized a ratiometric oxidase mimetic $MnO_2$ nanosheets-based fluorogenic nanosensor for GSH sensing. They have used Amplex Red (AR) and Scopoletin (SC) as fluorescent substrates of peroxidase. $MnO_2$ could not sufficiently quench the fluorescence of SC while led to fluorescence enhancement of non-fluorescent Amplex Red (AR) via oxidation. On the other hand, in presence of GSH, $MnO_2$ nanosheets were reduced to $Mn^{2+}$. Thus oxidase property of the nanosensor was lost. This results the enhancement of the fluorescence intensity of Scopoletin (SC) and quenching of Amplex Red (AR), which was helpful for ratiometric sensing of GSH with the LOD of 6.7 nM (Figure 13.13a) [221]. Likewise, Liu *et al.* designed oxidase mimetic $MnO_2$ nanosheets-based chromo-fluorogenic nanosensor for GSH. Here also, $MnO_2$ quenched the fluorescence of curcumin (CUR) whereas Amplex Red (AR), which has almost no fluorescence, exhibits fluorescence upon catalytic oxidation reaction. This led to the chromogenic alteration of the entire system from yellow to orange. Nevertheless, in presence of GSH, the oxidase activity was lost, which led to fluorescence enhancement of CUR with the fluorescence quenching of oxidized AR along with revival

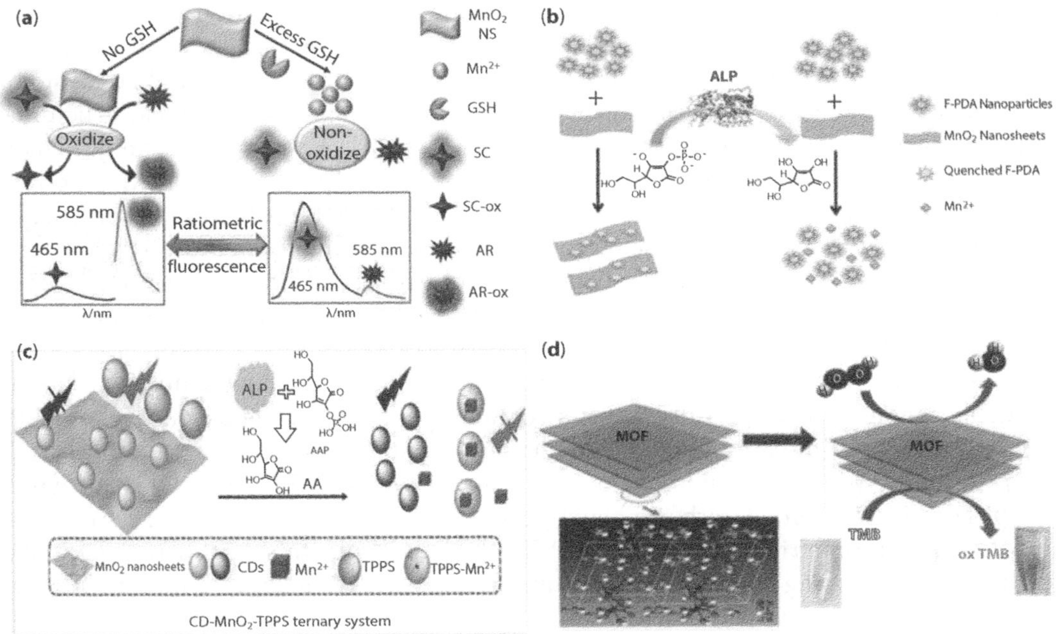


**Figure 13.13** (a) Schematic depiction of $MnO_2$ NS-based ratiometric fluorogenic detection of GSH, Adapted with permission from Ref. 221, (b) Schematic diagram of F-PDA–$MnO_2$ nanosensor for the FRET based recognition of ALP, Adapted with permission from Ref. 224, (c) CDs–$MnO_2$–TPPS based ternary system for AA and ALP sensing, (d) Schematic representation of peroxidase-like activity of 2-D MOF Nanosheets for chromogenic recognition of $H_2O_2$, Adapted with permission from Ref. 233.

of the chromogenic state from orange to yellow. Moreover, the sensory receptors were also applicable for monitoring of GSH from human serum samples [222]. Alkaline phosphatase (ALP) is an important protein present inside the body tissues and its higher level can damage liver and bone disorder resulting bone metastasis and liver related diseases etc. [223]. Yang *et al.* reported the fluorescent polydopamine (F-PDA) nanoparticles-based $MnO_2$ nanosheets for specific and ultrasensitive "off-on" detection of ALP activity through FRET process, where $MnO_2$ acts as the nano quencher and the F-PDA nanoparticles act as the donor. This method was effectively employed for bioimaging and monitoring of ALP activity in serum samples (Figure 13.13b) [224]. On the other hand, ascorbic Acid (AA) acts as nutritional factor, enzyme cofactor etc. [225]. Recently, Lu *et al.* depicted carbon dots (CDs), $MnO_2$ nanosheets and tetraphenyl porphyrin tetra sulfonic acid (TPPS)-based ternary system for fluorogenic detection of AA. Actually, $MnO_2$ NS at first reduced the fluorescence of CDs, keeping the fluorescence of TPPS as it was. Nevertheless, by AA $MnO_2$ NS was reduced to $Mn^{2+}$, which could help towards fluorescence recovery of CDs whereas coordination with $Mn^{2+}$ led to fluorescence quenching of TPPS. This method was also applied for monitoring of ALP via ALP-triggered evolution of AA (Figure 13.13c) [218].

### *13.2.3.4 2-D NanoMOFs*

2-D Metal-organic-frameworks (MOFs), a newly discovered member of 2-D nanomaterials, have attracted major interest in the current scenario. These classes of materials bear the special quality of both 2-D nanomaterials with the MOF crystals. MOFs are composed of responsive organic ligands and a wide array of metal ions or clusters as an emerging category of porous sensory materials. It is having a broad range of applications like gas storage, separation, catalysis, chemical sensing etc. because of its large surface area, structural tunability of pore size, functionalized nano-spaces etc. [226]. A variety of techniques like mass spectrometry, electrochemical techniques, atomic absorption spectrometry, atomic emission spectrometry, etc. are used for biosensing by 2-D-NanoMOFs. Nevertheless, chromo-fluorogenic sensing is advantageous over these techniques because of their fast response, ease to handle, detection capability in both solid and liquid phases etc. [227]. Over the years, MOFs have gained substantial consideration as fluorometric nanoprobes for selective sensing of environmentally relevant species [228].

Alkaline phosphatase (ALP) plays an important role in cellular transphosphorylation and dephosphorylation of phosphate esters-containing molecules. Abnormal level of ALP may lead to various diseases, like bone metastasis, hepatitic disease, prostate cancer, etc. [229]. Recently, Wei *et al.* reported peroxidase-like 2-D-Zn-TCPP(Fe) MOF nanozyme (TCPP(Fe)= Fe(III) tetra(4-carboxyphenyl)porphine) chloride) with sheet-like structure for selective colorimetric detection of ALP [230]. As a whole, this ALP detection strategy provides a stepping stone in the domain of nanozyme bioanalysis. Similarly, Huang *et al.* have reported peroxidase mimetic 2-D Fe-BTC MOF nanosheets (BTC=1,3,5-benzenetricarboxylic acid), a specific colorimetric biosensing strategy for glucose and hydrogen peroxide, of which proper level is related to signaling transduction and cellular functions [231]. The proposed assay was also employed for the selective determination of $H_2O_2$ and glucose. Interestingly, it has been observed that the enzymatic activity of Fe-BTC was 7 and 2.2 times higher than its template Cu(HBTC) nanosheet and 3-D MIL-100(Fe). Moreover, the sensing specificity has also been checked in human serum samples [232]. Furthermore, Shu and Hu *et al.* have also reported ultrathin 2-D Ni-MOFs nanosheets with enhanced peroxidase nanozyme activity towards chromogenic $H_2O_2$ sensing [233]. The sensitivity of the present method towards detection of $H_2O_2$ was monitored in disinfectant and human serum with promising results (Figure 13.13d). Li *et al.* have reported 2-D Co-MOFs nanosheets-based nanozyme with high peroxidase catalysis activity towards chromogenic recognition of AA and ALP via the oxidation of TMB to oxTMB along with chromogenic alteration from colorless to blue [234]. Moreover, ascorbic acid could be quantified by this assay with the aid of a smartphone. Ruan and Li *et al.* have reported luminescent 2-D-MOF nanosheets for fluorogenic sensing of $H_2O_2$ and glucose via multi-enzyme cascade catalysis along with glucose oxidase activity assay [235]. Furthermore, the sensitivity of the present method has also been evaluated in human serum samples with good recovery (97.5-98.5%).

### 13.2.4 Three-Dimensional Nanoprobes for Optical Biosensing

#### *13.2.4.1 Hybrid Nanoflowers*

Nanoflowers are an improved class of 3-D nanomaterials, which have broad applications, like antimicrobial activities, biosensors development for different bioanalytes detection. Synthesis process of hybrid nanoflowers (HNFs) is simply comprised of four-step processes —i) nucleation & coordination, ii) coprecipitation, iii) self-assembly and iv) anisotropic growth. In synthesis process, various metal ions and different biomolecules take part efficiently. Later on, the formation of protein-metal ion complex leads to the coordination bonding between various metal ions and -$NH_2$ group of protein (Figure 13.14a) [236, 237]. Next in the growth step in the production process, primary crystals and protein molecules

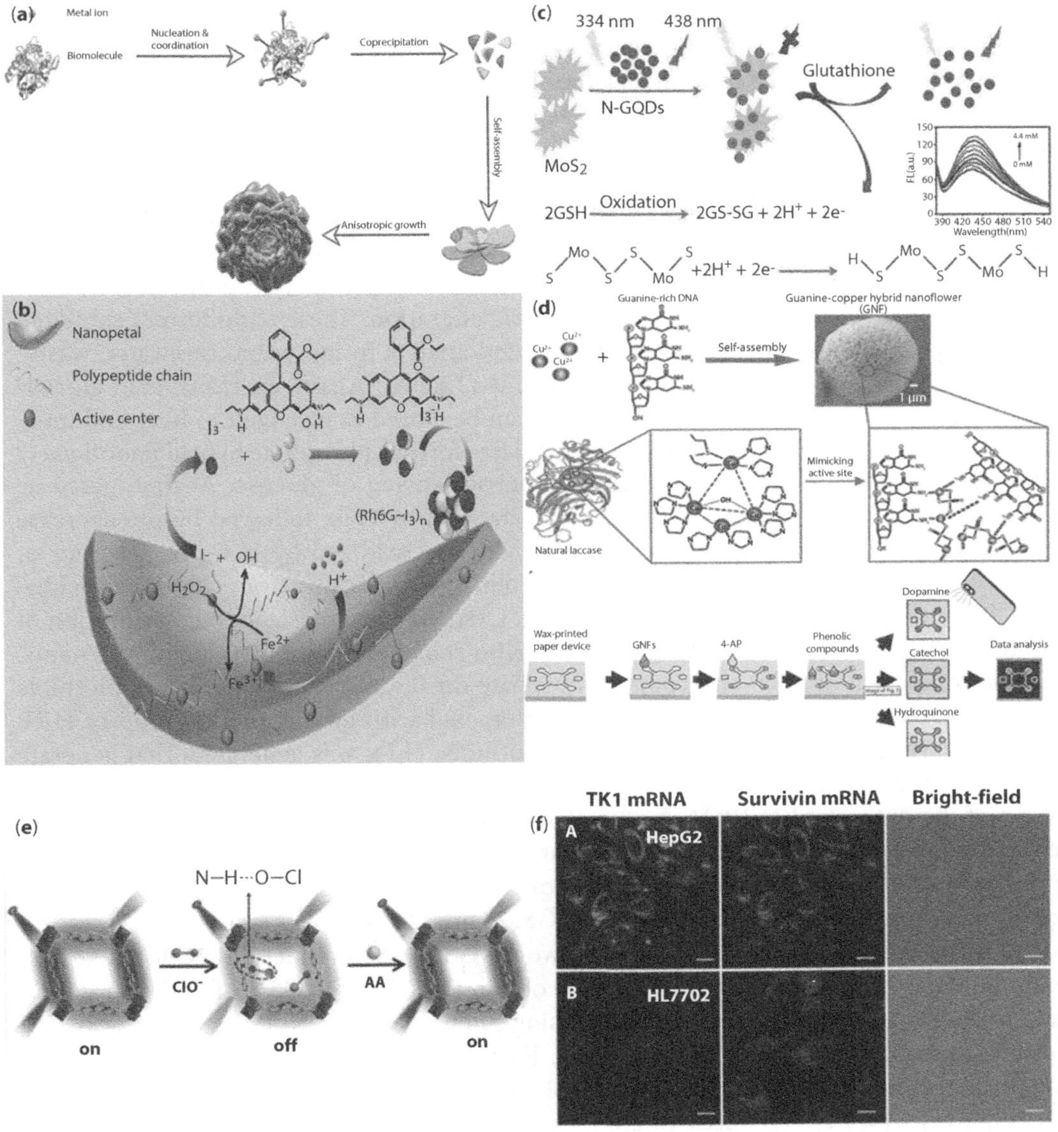


**Figure 13.14** (a) Synthesis of Hybrid Nanoflowers (HNFs) with the addition of several ions and biomolecules ($Co^{2+}$ Chloroperoxidase and $Zn^{2+}$ Papain *etc.*), Adapted with permission from Ref. 237, (b) Schematic depiction of biocatalytic reaction systems for $H_2O_2$ sensing, Adapted with permission from Ref. 242, (c) Schematic diagram of N-GQD-$MoS_2$ towards GSH sensing, inset: fluorescence spectra of N-GQDs-$MoS_2$ upon introduction of different GSH concentration, Adapted with permission from Ref. 239, (d) Colorimetric recognition of dopamine, catechol, hydroquinone by laccase-mimicking guanine-copper hybrid nanoflowers (GNFs), Adapted with permission from Ref. 240, (e) Proposed sensing mechanism of $ClO^-$ and AA by the 3-D MOF based nanosensor, Adapted with permission from Ref. 251, (f) Fluorescence imaging of TK1 mRNA and surviving mRNA in (A) HepG2 and (B) HL7702 cells, Adapted with permission from Ref. 252.

agglutinate together. Finally, separate petals are appeared at individual metal ion binding sites at the surface of agglomerates. After that, there occurs anisotropic growth, which ultimately led to the formulation of HNFs, which are recently used for sensing of various biomolecules like (GSH), cysteine (Cys), homocysteine (Hcy), dopamine, catechol, hydroquinone, Progesterone etc. [238–242].

Hydrogen peroxide ($H_2O_2$) is a regular oxidizing agent in human body. Sometimes, its higher level (75 ppm) may become dangerous for human health and causes many disorders [243]. Liu and Pang *et al.* have reported 3-D Hemoglobin (Hb)-$Cu_3(PO_4)_2$ organic/inorganic hybrid nanoflowers (Hb-$Cu_3(PO_4)_2$ HNFs) for the chromo-fluorogenic sensing of $H_2O_2$ with the LOD of 0.1 ppb and 0.01 ppb for chromogenic and fluorescent sensing method respectively (Figure 13.14b) [242]. The proposed HNFs helped to decompose $H_2O_2$ to the corresponding hydroxyl radical. This hydroxyl radical could lead to the oxidation of $I^-$ to $I_3^-$. This again combined with Rhodamine 6G for the production of the corresponding fluorescence quenching and the colorimetric alteration. These nanoflowers exhibited potential application for detection of the targeted analyte in tap water, rainwater, wastewater samples with the recovery of 102.82% and 112.36%. Again, excess level of glucose in bloodstream causes diabetes, sometimes it can be stored as glycogen in liver and muscles. Zhao *et al.* reported one novel glucose oxidase–ferrous phosphate hybrid nanoflowers (GOx–$Fe_3(PO_4)_2{\cdot}8H_2O$ HNFs) for the colorimetric sensing of glucose. At first, gluconic acid and $H_2O_2$ were produced from glucose catalyzed by GOx followed by the decrease of pH the reaction system to ~4, at which TMB was oxidized by $H_2O_2$ in presence of $Fe_3(PO_4)_2{\cdot}8H_2O$ catalyst, resulting prominent colorimetric change from colourless to blue [244]. The unusual level of bio-thiols, like GSH, Cys, and Hcy may lead to deterioration of immune and nervous systems [245]. Recently, Kim *et al.* have demonstrated hybrid nanoflowers, comprised of HRP and copper phosphate for fluorescence detection of biothiols via enzymatic activity. In this case, copper interacted with amine/amide group of HRP to form hybrid nanoflowers. In presence of biothiols, $H_2O_2$ was generated, that generate sharp fluorescence via oxidation of the employed Amplex UltraRed substrate due to activation of the entrapped HRP. Moreover, the newly synthesized nanoflowers have shown good results for biothiols detection in human serum samples [238]. Furthermore, Cheng and Cheng *et al.* described graphene quantum dots doped with nitrogen (N-GQDs)-$MoS_2$ nanoflowers towards fluorogenic sensing of GSH. Initially, N-GQDs were non-fluorescent due to the occurrence of FRET process. However, GSH led to reduction of $MoS_2$ to $Mo^{3+}$, leading to revival of the GQDs emission to obtain the LOD for GSH of 2.47 μM within the range of 0.4–4.4 mM. Based on the sensing phenomenon, XNOR logic gate was fabricated using GSH and $MoS_2$ as two inputs. Furthermore, the probe was also effectively employed for monitoring of intracellular GSH in retinal pigment epithelium cells (ARPE-19) (Figure 13.14c) [239]. Laccases are oxidase enzymes consisting of copper that catalyze the oxidation of variety of aromatic biomolecules, different types of phenols specially biologically active phenolic compounds like dopamine, hydroquinone and catechol with high toxicity level. Abnormal level of these compounds affects human health and ecosystems [246]. Kim *et al.* developed DNA-copper hybrid nanoflowers, which was synthesized via guanine-rich ssDNA and copper ions. This exhibited laccase-mimicking activity for chromogenic detection of several biologically important phenolic compounds such as catechol, hydroquinone and dopamine. In addition, this fresh nanomaterial found apposite for instrument-free microfluidics-based point-of-care testing (POCT) device fabrication towards the quantification of targeted analytes (Figure 13.14d) [240]. On the other hand, progesterone (P4) is a glucocorticoid hormone related to the reproductive system, which effectively harmful for both mental and physical health of female body. Abnormal level of P4 may cause some psychological diseases such as nervousness, manic depression, dysthymia disorder etc. [247]. Ding and Zhao *et al.* proposed a fresh idea of synthesizing a nanoflower from copper phosphate, streptavidin (SA), Pt/$IrO_2$ nanocomposites (Pt/$IrO_2$ NPs) and horseradish peroxidase (HRP) for specific colorimetric detection of progesterone via multiple catalytic pathways. The method acquired the LOD of P4 to be 0.076 ng $mL^{-1}$. Moreover, the reported immunoassay was also effective towards P4 detection in human saliva samples with recovery rate of 79.6–107% [241].

#### *13.2.4.2 3-D NanoMOFs*

3-D nanoMOFs are a special type of hybrid nanoMOFs with its unique structural diversity like uniform pore structure, atomic level structural uniformity, flexibility in dimension and chemical functionality [248]. These nanoMOFs are advantageous over 2-D nanoMOFs because of its higher stability, more robustness, porosity and higher surface area. 3-D nanoMOFs posses various applications in biological systems including enzyme immobilization, hydrogen generation, sensing, drug delivery, etc. [249]. Recently, these kinds of MOFs materials are being used in sensing of different bioanalytes, like $H_2O_2$, $ClO^-$, ascorbic acid etc. [250–252].

Overproduction of hydrogen peroxide ($H_2O_2$) may be responsible various inflammatory diseases [253]. Conversely, ascorbic acid (AA) plays a key role in maintaining intracellular connective tissues, osteoid, dentine, collagen etc. and its deficiency can cause scurvy and abnormal osteoid [254]. Liu *et al.* have accounted that in existence of o-phenylenediamine (OPD) acting as a substrate, PMA (phosphomolybdic acid) encapsulated metal–organic framework (MOFs) with metal carbene structure (Cu-MOFs) act as bifunctional enzymes for colorimetric sensing of $H_2O_2$ and AA [250]. Therefore, these MOFs have its practical applications in the field of medical diagnostics and biotechnology etc. Again, excessive level of $ClO^-$ may lead to various diseases like reproductive malfunction, cancer etc. [255]. On this basis, Cao and Lin *et al.* have proposed dual-emission multivariate 5-5-Eu/BPyDC@MOF-253-$NH_2$ acting as the ratiometric nano biosensor for "on–off" detection of hypochlorite ($ClO^-$) with red emission [251]. However, in presence of AA, it exhibited "on–off–on" sensing via redox reaction between $ClO^-$ and AA and accordingly blue emission was recovered (Figure 13.14e). Messenger RNA (mRNA) is a type of RNA, which helps in many biological activities and acts as a potential cancer biomarker [256]. Recently, Zhang and Li *et al.* described entropy-driven DNA amplifier-functionalized UIO-66 MOFs (DNA amplifier-MOFs)-based nanosensor towards fluorogenic sensing of mRNA and it was employed for intracellular imaging of mRNA in HepG2 and HL7702 cells (Figure 13.14f) [252].

## 13.3 Conclusions and Future Perspectives

The strategic development of nanomaterials has fueled the progression of the chromo-fluorogenic nanobiosensors towards sensing of the targeted biorelevant analytes, ubiq-

**Table 13.1** Summary of the performance of zero-dimensional optical nano-biosensors.

| Nanosensing materials | | Analyte detected | Biosensing method | LOD | Linear detection range | Ref. |
|---|---|---|---|---|---|---|
| Carbon Quantum Dots (CQDs) | | Vitamin B7 | Fluorogenic | 0.1 ng·mL$^{-1}$ | 0.5 to 100 ng·mL$^{-1}$ | 11 |
| | | β-glucuronidase (βG) | Fluorogenic | 0.3 U/L | 1-15 U/L | 12 |
| | | Lysine<br>pH | Fluorogenic | 94 nM | Lysine: 0.5-260 μM<br>pH: 1.5-5.0 | 17 |
| | | $Fe^{3+}$<br>AA | Fluorogenic | $Fe^{3+}$:13.7 nmol L$^{-1}$<br>AA: 82.0 nmol L$^{-1}$ | $Fe^{3+}$:0.05–10.0 μmol L$^{-1}$<br>AA: 0.2 to 11.0 μmol L$^{-1}$ | 18 |
| | | $Fe^{3+}$<br>ATP | Fluorogenic | $Fe^{3+}$:0.01 μM.<br>ATP: 0.005 μM. | $Fe^{3+}$: 0-40 μM, 40-350 μM<br>ATP: 0.01- 450 μM | 19 |
| Graphene Quantum dots (GQDs) | N-GQDs | Dopamine | Fluorogenic | 3.3 nM<br>611 nM | *10-3000 nM*<br>*3000-7000 nM* | 26 |
| | β-CD/N@GQDs | Dopamine | Both chromogenic and fluorogenic | Chromogenic: 0.04 μM<br>Fluorogenic: 0.009 μM. | Chromogenic: *0.12–7.5 mM*<br>Fluorogenic: 0.028–1.5 mM | 27 |
| | Coumarin functionalized C-GQD | Ascorbic acid | Fluorogenic | 2.2 nM | *0- 666.7 μM* | 28 |
| | GQDs | $Fe^{3+}$ | Fluorogenic | 0.23 μM | 0.1–20 μM | 29 |
| | B-GQDs | $Fe^{3+}$<br>Pi<br>Cyt C | Fluorogenic | $Fe^{3+}$: 31.2 nM<br>Pi: 340 nM<br>Cyt C: 5.9 μg/ml | $Fe^{3+}$: 50 nM-220 μM<br>Pi: 3-40 μM<br>Cyt C: 10-300 μg/ml | 30 |

*(Continued)*

**Table 13.1** Summary of the performance of zero-dimensional optical nano-biosensors. (*Continued*)

| Nanosensing materials | | Analyte detected | Biosensing method | LOD | Linear detection range | Ref. |
|---|---|---|---|---|---|---|
| Inorganic quantum dots (IQDs) | CdSe/ZnS QDs | serum amyloid A (SAA) and C-reactive protein (CRP) | Fluorescence-linked immunosorbent assay | SAA: 2.39 ng $mL^{-1}$<br>CRP: 6.37 ng $mL^{-1}$ | SAA: 10–1,000 ng $mL^{-1}$<br>CRP: 10–1,000 ng $mL^{-1}$ | 51 |
| | CdTe QDs/CDs@ MOFs | AA<br>AAO | Fluorogenic | AA: 4 nM<br>AAO: 0.02 $U{\cdot}L^{-1}$ | AA: 0.01-0.2 μM<br>AAO: 0.05–4 $U{\cdot}L^{-1}$ | 52 |
| | CdTe QDs | AA<br>ALP | Fluorogenic | AA: 3 μM<br>ALP: 0.25 $UL^{-1}$ | AA: 0-800 μM<br>ALP: 1-1000 U $L^{-1}$ | 53 |
| | CdTe QDs/Au NPs | Captopril, glutathione, L-cysteine, Mercaptosuccinic acid | Fluorogenic | - | At pH 8,<br>Captopril: 0.025-0.375 $mmol.L^{-1}$<br>GSH: 0.025-0.175 $mmol.L^{-1}$<br>L-cysteine: 0.025-0.375 $mmol.L^{-1}$ Mercaptosuccinic acid: 0.025-0.100 $mmol.L^{-1}$ | 54 |
| | CdTe QDs | $H_2O_2$ | Fluorogenic | 0.3 μM | 10-125 μM | 55 |

(*Continued*)

**Table 13.1** Summary of the performance of zero-dimensional optical nano-biosensors. (*Continued*)

| Nanosensing materials | | Analyte detected | Biosensing method | LOD | Linear detection range | Ref. |
|---|---|---|---|---|---|---|
| Gold nanoparticles | N, S-CDs/AuNPs | Glutathione | Fluorogenic | 3.6 nM | 0.01–5.0 μM | 60 |
| | DNA duplex- AuNPs | L-Histidine | Chromogenic | 3.6 nM | 0–400 nM | 61 |
| | N, S-CDs/AuNPs | Glutathione | Fluorogenic | 0.21 μM | 3.8- 415.1 μM | 62 |
| | AuNPs based immunosensor | protein-phosphatase-1-gamma-2 (PP1γ2) | Chromogenic | - | - | 63 |
| | AuNPs | DNA | Chromogenic | 0.36 μM | *0-5* μM | 64 |
| Silver nanoparticles | AgNPs | Adenosine | Chromogenic | 21 nM | 60–280 nM | 71 |
| | | Uric acid | Chromogenic | 0.03 nM | 0.1 nM-0.1 mM | 72 |
| | | Ascorbic acid | Chromogenic | 0.054 μM | 0.25–25 μM | 73 |
| | | Dopamine | Chromogenic | $1.18\times10^{-7}$ M | $1.0\times10^{-8}$–$3.6\times10^{-6}$ M | 74 |
| | | Cysteine | Chromogenic | 4.0 nM | 0-100 ng $mL^{-1}$ | 75 |

(*Continued*)

**Table 13.1** Summary of the performance of zero-dimensional optical nano-biosensors. (*Continued*)

| Nanosensing materials | | Analyte detected | Biosensing method | LOD | Linear detection range | Ref. |
|---|---|---|---|---|---|---|
| Magnetic nanoparticles (MNPs) | $Fe_3O_4$@Au | GSH | Chromogenic | 0.013 mol/L (Smartphone) | 0–0.25 mmol/L (Smartphone) | 84 |
| | Reduced $Co_3O_4$ nanopaticles | $H_2O_2$ Glucose | Chromogenic | $H_2O_2$: 4.3×10$^{-7}$ mol/L Glucose: 3.2 ×10$^{-7}$ mol/L | $H_2O_2$: 1–30 μM Glucose: 1-20 μM | 87 |
| | $NiFe_2O_4$–PANI | Ascorbic acid | Chromogenic | 232 nM | 10–100 μM | 88 |
| | $Fe_3O_4$@$SiO_2$@ Au magnetic nanoparticles (MNPs) | $H_2O_2$ Glucose | Chromogenic | $H_2O_2$: 3.0×10$^{-6}$ M Glucose: 3 μM | $H_2O_2$: 5×10$^{-6}$~3.5×10$^{-4}$ M Glucose: 5-350 μM | 89 |
| | mesoporous $MnFe_2O_4$ MNPs | Glucose | Chromogenic | 0.7 mM | 0.5-16.0 μM | 90 |
| Up-conversion-al nanoparti-cles (UCNPs) | Polyethyleni-mine-$NaGdF_4$:Yb, Er UCNPs- phycocyan-in (PC) conjugate | Myeloperoxidase (MPO) | Fluorogenic | $ClO^-$ activity: 66.88 nM $ONOO^-$ activity: 78.50 nM | $ClO^-$ activity: 0.1–5.0 μM $ONOO^-$ activity: 0.6- -7.2 μM | 93 |
| | $Yb^{3+}$/$Er^{3+}$-doped $NaGdF_4$ UCNPs-mOrange fluorescent protein conjugate | pH | Fluorogenic | - | - | 94 |

(*Continued*)

**Table 13.1** Summary of the performance of zero-dimensional optical nano-biosensors. (*Continued*)

| Nanosensing materials | | Analyte detected | Biosensing method | LOD | Linear detection range | Ref. |
|---|---|---|---|---|---|---|
| | Oleate-capped $NaYF_4$: 20% Yb, 2% Er@ $NaYF_4$ UCNPs coupled with Calcium Red and Alizarin Red S | pH | Fluorogenic | - | - | 95 |
| | $NaYF_4$: 18%Yb, 2% Er @$NaYF_4$) upconversion nanoparticles (UCNPs)- phenol red -poly(ethylene glycol) (PEG) ylated phospholipid (DSPE-PEG) self-assembly | pH | Fluorogenic | - | - | 102 |
| Polymer dots (PDs) | Pdots@RB-hy | Alkaline phosphatase (ALP) | Fluorogenic | 0.0018 $UL^{-1}$ | 0.005-15 $UL^{-1}$ | 108 |
| | Pdots-PF, Pdots-PP, and Pdots-PPF | pH | Fluorogenic | - | 3.0-8.0 | 109 |
| | BF@Pdots | $SO_2$ Formaldehyde | Fluorogenic | - | FA: 3 to 50 μM | 110 |

**Table 13.2** Summary of the performance of one-dimensional optical nano-biosensors.

| **Nanosensing materials** | **Analyte detected** | **Biosensing method** | **LOD** | **Linear detection range** | **Ref.** |
|---|---|---|---|---|---|
| Carbon nanotubes (CNTs) | ATP | Fluorogenic | $2.4\times10^{-7}$ M | - | 112 |
| | Nitric oxide | Fluorogenic | $3\times10^{-7}$ M | - | 114 |
| | Troponin T | Fluorogenic | $2.5\times10^{-9}$ M | - | 115 |
| | SARS-CoV-2 spike Protein | Fluorogenic | 35 mg/L | - | 119 |
| | Hydrogen Peroxide | Fluorogenic | - | $5\times10^{-6}$ - $5\times10^{-3}$ M | 120 |
| | MicroRNA 155 | Fluorogenic | $3.34\times10^{-14}$ M | $1\times10^{-13}$ - $1\times10^{-9}$ M | 121 |
| | Digoxin | Fluorogenic | $7.95\times10^{-12}$ M | $2.65\times10^{-11}$–$6.8\times10^{-10}$ M | 122 |
| | Cyclin A2 | Fluorogenic | $5\times10^{-9}$ M | - | 123 |
| Silicon nanowire (SiNW) | Ionic calcium ($Ca^{2+}$) | Fluorogenic | - | - | 128 |
| | Hypochlorite ion ($ClO^-$) | Fluorogenic | - | 0-$50\times10^{-6}$ M | 129 |
| | Hydrogen sulfide ($H_2S$) | Fluorogenic | 7.13 μM | 0-40 μM | 130 |
| | C-reactive protein (CRP) | Fluorogenic | $10^{-7}$ μg/mL | $10^{-8}$-$10^{-2}$ μg/mL | 131 |
| Gold nanorods (GNRs) | Glucose | Chromogenic | 0.1 μM | 1-10 μM | 158 |
| | Dopamine | Chromogenic | 0.62 μmol/L | 0.80-60.0 μmol/L | 160 |
| | Dopamine | Chromogenic | 30 nM | 100 nM-10 mM | 161 |
| | Acidphosphatase (ACP) | Chromogenic | 0.97 U/L | 0-15.0 U/L | 164 |
| | Uric acid (UA) | Chromogenic | 0.76 μM | 0.8-30μM and 30-300 μM | 166 |

*(Continued)*

**Table 13.2** Summary of the performance of one-dimensional optical nano-biosensors. (*Continued*)

| Nanosensing materials | | Analyte detected | Biosensing method | LOD | Linear detection range | Ref. |
|---|---|---|---|---|---|---|
| Nanoribbons | $C_3N_4$ nanoribbons | Citrate anion ($C_6H_5O_7^{3-}$) | Fluorogenic | 0.78 μM | 1-400 μM | 168 |
| | SNC nanoribbons | Quercetin | Fluorogenic | 21.13 nM | 50 nM-200 μM | 169 |
| | Graphene nanoribbons (GNR) | Dopamine (DA) | Chromogenic | 0.035 μM | 0.1-50 μM | 170 |
| | CQDs-$V_2O_5$ nanoribbon | α-Glucosidase | Fluorogenic | 0.003 U $mL^{-1}$ | 0.01-5 U $mL^{-1}$ | 171 |
| Nanofibres | $Fe_3O_4$/N-C | Ascorbic acid (AA) $H_2O_2$ | Chromogenic | AA: 0.04 μM $H_2O_2$:0.52 μM | 0-50 μM | 186 |
| | CDs/$Fe_3O_4$ | Ascorbic acid (AA) $H_2O_2$ | Chromogenic | AA: 0.285 μM $H_2O_2$: 0.917 μM | AA: 1-30 μM $H_2O_2$: 1-20 μM | 187 |
| | CS/PVA | Glucose | Chromogenic | 2.7 mM | 2.7-13.8 mM | 188 |
| | PB/PPy | L-cysteine | Chromogenic | 16 nM | 1-20 μM | 189 |
| | Superparamagnetic $Fe_3O_4$ | L-cysteine | Chromogenic | 0.028 μM | 2-10 μM | 190 |

**Table 13.3** Summary of the performance of two-dimensional optical nano-biosensors.

| Nanosensing materials | | Analyte detected | Biosensing method | LOD | Linear detection range | Ref. |
|---|---|---|---|---|---|---|
| Graphene Oxide (GO) | Graphene oxide-AuNPs nanocomposite | $H_2O_2$<br>Glucose | Chromogenic | $H_2O_2$: $4.2\times10^{-8}$ M<br>Glucose: $6.3\times10^{-7}$ M | $H_2O_2$: $3.8\times10^{-7}$–$5.5\times10^{-5}$ M<br>Glucose: $5.1\times10^{-6}$–$5.1\times10^{-4}$ M | 197 |
| | GO-fluorophore tagged DNA | Exonuclease Enzymatic Activity | Fluorogenic | 0.001 U/mL | 0.01-0.5 U/mL | 198 |
| | γ-$Al_2O_3$–GO/DNA/SF | $H_2S$ | Fluorogenic | 2.5 μM | 2.5 – 75 μM | 199 |
| Reduced Graphene Oxide (rGO) | rGO–PBA (PBA= phenylboronic acid) | Glucose | Fluorogenic | - | 2-75 mg/mL | 201 |
| | FA-rGO-RNase A/AuNCs | $Na^+$<br>$K^+$ | Fluorogenic | $Na^+$: RNase A/AuNCs: 49 mM<br>FA-rGO-RNase A/AuNCs: 110 mM<br>$K^+$:<br>RNase A/AuNCs: 74 mM<br>FA-rGO-RNase A/AuNCs: 15.7 mM | $Na^+$:<br>RNase A/AuNCs: 0-100 mM<br>FA-rGO-RNase A/AuNCs: 0-200 mM<br>$K^+$:<br>RNase A/AuNCs: 0-200 mM<br>FA-rGO-RNase A/AuNCs: 0-25 mM | 202 |

(*Continued*)

**Table 13.3** Summary of the performance of two-dimensional optical nano-biosensors. (*Continued*)

| Nanosensing materials | | Analyte detected | Biosensing method | LOD | Linear detection range | Ref. |
|---|---|---|---|---|---|---|
| | FA-rGO, BSA/AuNCs | GSH | Fluorogenic | 0.1 μM | 0- 1.75 μM | 203 |
| | IrO2/rGO nanocomposites | Biothiols | Chromogenic | Cys: 40 nM<br>Hcy: 57 nM<br>GSH: 83 nM | Biothiols: 0.1~50 μM | 204 |
| | $WS_2$/rGO nanocomposites | $H_2O_2$ | Chromogenic | 82 nM | - | 205 |
| Graphitic carbon nitride (g-$C_3N_4$) | Graphitic carbon nitride ($C_3N_4$)-based nanozymes | $H_2O_2$<br>Glucose<br>phosphates | Fluorogenic | $H_2O_2$: 4.20μM<br>Glucose: 4.73μM | $H_2O_2$:10-1000 μM<br>Glucose:10-250 μM | 207 |
| | Graphitic Carbon Nitride (g-$C_3N_4$) | L-Cysteine | Fluorogenic | - | - | 208 |
| | Graphitic Carbon Nitride (g-$C_3N_4$) | Alkaline phosphatase (ALP) | Fluorogenic | 0.08 U/L | 0.1 to 1000 U/L | 209 |
| | Graphitic Carbon Nitride (g-$C_3N_4$) | Iron ($Fe^{3+}$) | Fluorogenic | 0.0232μM | 0.1-20 μM | 210 |
| | Graphitic Carbon Nitride (g-$C_3N_4$) | $H_2O_2$<br>Glucose | Fluorogenic | $H_2O_2$:50 nM<br>Glucose:0.4 μM | $H_2O_2$: 0.1 to 2000 μM | 211 |

(*Continued*)

**Table 13.3** Summary of the performance of two-dimensional optical nano-biosensors. (*Continued*)

| Nanosensing materials | | Analyte detected | Biosensing method | LOD | Linear detection range | Ref. |
|---|---|---|---|---|---|---|
| $MnO_2$ nanosheets | CDs-$MnO_2$ sheets-tetraphenyl porphyrin tetra sulfonic acid (TPPS) | Ascorbic Acid (AA)<br>Alkaline phosphatase (ALP) | Fluorogenic | AA:0.13 μM<br>ALP:0.04 mU/mL | AA:0.5-40 μM<br>ALP:0.1–100 mU/mL | 218 |
| | $MnO_2$-NS | Glutathione (GSH) | Fluorogenic | 160 nM | 0.5-250 μM | 220 |
| | $MnO_2$-NS | Glutathione (GSH) | Fluorogenic | 6.7nM | - | 221 |
| | $MnO_2$-NS | Glutathione (GSH) | Both chromogenic and Fluorogenic | Chromogenic: 16 nM<br>Fluorogenic: 4.2 nM | Chromogenic: 50–15000 nM<br>Fluorogenic: 10-5000 nM | 222 |
| | Fluorescent polydopamine (F-PDA) nanoparticles–$MnO_2$ nanosheets | Alkaline phosphatase (ALP) | Fluorogenic | 0.34 mU/mL | 1-80 mU/mL | 224 |

(*Continued*)

**Table 13.3** Summary of the performance of two-dimensional optical nano-biosensors. (*Continued*)

| Nanosensing materials | | Analyte detected | Biosensing method | LOD | Linear detection range | Ref. |
|---|---|---|---|---|---|---|
| 2-D MOF nanosheets | Zn-TCPP(Fe) 2-D MOF nanosheets (TCPP(Fe)= Fe(III) tetra(4-carboxyphenyl) porphine) chloride) | ALP | Chromogenic | - | 2.5-20 U/L, 5-60 U/L, 50-200 U/L using PPi, ATP, and ADP as inhibitors | 230 |
| | Fe-BTC MOF nanosheets | $H_2O_2$<br>Glucose | Chromogenic | $H_2O_2$:36 nM<br>Glucose:39 nM | $H_2O_2$: 0.04-30 μM<br>Glucose: 0.04-20 μM | 232 |
| | Ni-MOFs nanosheets | $H_2O_2$ | Chromogenic | 0.008 μM | 0.04-160 μM | 233 |
| | Co-MOF nanosheets | Ascorbic acid (AA)<br>Alkaline phosphatase (ALP) | Chromogenic | AA: 0.47 μM<br>0.69 μM (Smartphone)<br>ALP: 0.33 U. $L^{-1}$ | 1-30 μM with the aid of Smartphone | 234 |
| | In-MOF nanosheets | $H_2O_2$<br>Glucose | Fluorogenic | $H_2O_2$: 0.87 μM<br>Glucose: 1.3 μM | $H_2O_2$: 0-160 μM | 235 |

**Table 13.4** Summary of the performance of three-dimensional optical nano-biosensors.

| Nanosensing materials | | Analyte detected | Biosensing method | LOD | Linear detection range | Ref. |
|---|---|---|---|---|---|---|
| Hybrid nano-flowers (HNFs) | HRP-copper phosphate NFs (HRP= horseradish peroxidase) | Glutathione (GSH) Cysteine (Cys) Homocysteine (Hcy) | Fluorogenic | GSH:13.4 nM Cys:4.5 nM Hcy:18.3 nM | 0.1–1μM | 238 |
| | N-GQDs-$MoS_2$ hybrid nanoflowers (HNFs) | Glutathione (GSH) | Fluorogenic | 2.47 μM | 0.4–4.4 mM | 239 |
| | Guanine-rich ssDNA-copper hybrid nanoflowers (GNFs) | Dopamine Catechol Hydroquinone | Chromogenic | Dopamine: 4.5 μg/mL Catechol: 3.0 μg/mL Hydroquinone:4.5 μg/mL | Upto 25 μg/mL | 240 |
| | Copper phosphate, Pt/$IrO_2$@SA@ HRP HNFs (SA=streptavidin) | Progesterone (P4) | Chromogenic | 0.076 ng. $mL^{-1}$ | 0.217- 7.934 ng $mL^{-1}$ | 241 |
| | Hb-$Cu_3(PO_4)_2$ HNFs (Hb= Hemoglobin) | Hydrogen peroxide ($H_2O_2$) | Both chromogenic and fluorogenic | Chromogenic: 0.1 ppb Fluorogenic: 0.01 ppb | 2-10 ppb 20-100 ppb | 242 |
| | GOx-$Fe_3(PO_4)_2{\cdot}8H_2O$ HNFs | Glucose | Chromogenic | 0.1 μM | 0.01-20 mM | 244 |
| 3-D nanoMOFs | PMA encapsulating-Cu-MOFs (PMA= Phosphomolybdic acid) | Hydrogen peroxide ($H_2O_2$) Ascorbic acid (AA) | Chromogenic | $H_2O_2$:0.222 μM AA: 0.0046 μM | $H_2O_2$:1-100 μM AA: 3-100 μM | 250 |
| | 5-5-Eu/BPyDC@ MOF-253-$NH_2$ | Hypochlorite ($ClO^-$) (AA) Ascorbic acid (AA) | Fluorogenic | $ClO^-$:0.094 μM AA: 0.73 μM | 0.1–30 μM | 251 |
| | DNA amplifier-MOF | Intracellular mRNA | Fluorogenic | 3.3 pM | | 252 |

uitously present in the human body. Due to large surface area, tunable structure, nanomaterials can be effectively used as sensory materials.

In most of the current review articles, the biosensing applications of only one kind of nanomaterial have been focused. On the contrary, in the current chapter, we have presented different dimensional nanomaterials and their applications in the domain of chromo-fluorogenic biosensing to obtain more sensitive and specific signals. Along with this, we have also emphasized the underlying mechanistic course of interaction. Some of the nanomaterials like IQDs, GQDs, CQDs, exhibit excellent photoluminescence behavior. They can be efficiently used as the "turn-off" or ratiometric fluorescence sensor development. Meanwhile, metallic nanoparticles, like AgNPs, AuNPs etc., GO or rGO can also be conjugated with the fluorescent nanomaterials, like CQDs, GQDs, IQDs etc. where the fluorescent materials act as donors while the former materials act as energy acceptor to produce effective FRET, which can be tuned during target-specific analyte recognition. Therefore, based on the electronic nature of the analytes of our interest, judicious designing of the nanomaterials or nanoassembly should be carried out to obtain a better response. Tables 13.1–13.4 highlights some of the present research works in the realm of chromo-fluorogenic biosensing with the corresponding detection limits and the linear detection ranges.

To further improve the selectivity, specificity, sensitivity, durability or reusability of the nanosensor, more economical, sustainable nanomaterial development should be attempted focusing on the signal amplification strategy. Moreover, multi-analyte sensing or the fabrication of sensor arrays for multiplex detection should also be prioritized to reduce the production cost as well as detection time. The toxicity of the developed nanomaterials should also be evaluated to ensure the fruitful intracellular applications. Another major thing is that, in most of the current literature, the sensitivity has been checked in spiked serum or urine samples. Therefore, for more reliable clinical applications, real sample analyses are highly desirable. Finally, the large-scale synthesis of the nanosensors and their industrial scale applications via portable device or wearable device fabrication should also be kept in mind for implementing nanomaterials as the next-generation disease diagnostic tool, especially in the developing countries. Therefore, in near future, the image processing integrated machine learning approach will be a fertile land for nanomaterials based clinical applications and only with a smartphone the status of the specific analyte will also be possible to monitor even at home. We hope that the present chapter will inspire the future researchers to not only prudently design the nanosensors towards the recognition of the biorelevant analytes for monitoring of the physiological balance in the human body, but also meet the industrial demands to improve our lifestyle on this planet.

## Acknowledgment

DST-INSPIRE fellowship (IF180252) is hereby acknowledged by RD. We duly acknowledge the project GAP-225612 (*vide* sanction order no. 78 (Sanc.)/ST/P/S&T/6G-1/2018) for financial support. DST SERB-CRG sponsored project, GAP-240712 (vide reference no. CRG/2022/001679) is also hereby acknowledged.

## References

1. Bleeker, E.A.J., Jong, W.H. d., Geertsma, R.E., Groenewold, M., Heugens, E.H.W., Koers-Jacquemijns, M., Meent, D. v. d., Popma, J.R., Rietveld, A.G., Wijnhoven, S.W.P., Cassee, F.R., Oomen, A.G., Considerations on the EU definition of a nanomaterial: Science to support policy making. *Regul. Toxicol. Pharmacol.*, 65, 119–125, 2013.
2. Kerman, K., Saito, M., Tamiya, E., Yamamura, S., Takamura, Y., Nanomaterial-based electrochemical biosensors for medical applications. *TrAC - Trends Anal. Chem.*, 27, 585–592, 2008.

3. Chen, A. and Chatterjee, S., Nanomaterials based electrochemical sensors for biomedical applications. *Chem. Soc. Rev.*, 42, 5425–5438, 2013.
4. Azadmanjiri, J., Srivastava, V.K., Kumar, P., Sofer, Z., Min, J., Gong, J., Graphene-supported 2D transition metal dichalcogenide van der waals heterostructures. *Appl. Mater. Today*, 19, 100600, 2020.
5. Della Pelle, F. and Compagnone, D., Nanomaterial-based sensing and biosensing of phenolic compounds and related antioxidant capacity in food. *Sensors*, 18, 462, 2018.
6. Kumari, Y., Kaur, G., Kumar, R., Singh, S.K., Gulati, M., Khursheed, R., Clarisse, A., Gowthamarajan, K., Karri, V.V.S.N.R., Mahalingam, R., Ghosh, D., Awasthi, A., Kumar, R., Yadav, A.K., Kapoor, B., Singh, P.K., Dua, K., Porwal, O., Gold nanoparticles: New routes across old boundaries. *Adv. Colloid Interface Sci.*, 274, 102037, 2019.
7. Zhan, Y., Yang, J., Guo, L., Luo, F., Qiu, B., Hong, G., Lin, Z., Targets regulated formation of boron nitride quantum dots – gold nanoparticles nanocomposites for ultrasensitive detection of acetylcholinesterase activity and its inhibitors. *Sens. Actuators B Chem.*, 279, 61–68, 2019.
8. Amini, A., Kazemi, S., Safarifard, V., Metal-organic framework-based nanocomposites for sensing applications – a review. *Polyhedron*, 177, 114260, 2020.
9. Pirondini, L. and Dalcanale, E., Molecular recognition at the gas–solid interface: A powerful tool for chemical sensing. *Chem. Soc. Rev.*, 36, 695–706, 2007.
10. Lavigne, J.J. and Anslyn, E.V., Sensing A paradigm shift in the field of molecular recognition: From selective to differential receptors. *Angew. Chem. Int. Ed.*, 40, 3118–3130, 2001.
11. Yao, W., Wu, N., Lin, Z., Chen, J., Li, S., Weng, S., Zhang, L., Liu, A., Lin, X., Fluorescent turn-off competitive immunoassay for biotin based on hydrothermally synthesized carbon dots. *Microhim. Acta*, 184, 907–914, 2017.
12. Gong, P., Sun, L., Wang, F., Liu, X., Yan, Z., Wang, M., Zhang, L., Tian, Z., Liu, Z., You, J., Highly fluorescent N-doped carbon dots with two-photon emission for ultrasensitive detection of tumor marker and visual monitor anticancer drug loading and delivery. *Chem. Eng. J.*, 356, 994–1002, 2019.
13. Chowdhuri, A.R., Singh, T., Ghosh, S.K., Sahu, S.K., Carbon dots embedded magnetic nanoparticles @chitosan @metal organic framework as a nanoprobe for pH sensitive targeted anticancer drug delivery. *ACS Appl. Mater. Interfaces*, 8, 16573–16583, 2016.
14. Li, X., Rui, M., Song, J., Shen, Z., Zeng, H., Carbon and graphene quantum dots for optoelectronic and energy devices: A review. *Adv. Funct. Mater.*, 25, 4929–4947, 2015.
15. Das, P., Ganguly, S., Bose, M., Mondal, S., Choudhary, S., Gangopadhyay, S., Das, A.K., Banerjee, S., Das, N.C., Zinc and nitrogen ornamented bluish white luminescent carbon dots for engrossing bacteriostatic activity and Fenton based bio-sensor. *Mater. Sci. Eng. C*, 88, 115–129, 2018.
16. Hassan, M., Gomes, V.G., Dehghani, A., Ardekani, S.M., Engineering carbon quantum dots for photomediated theranostics. *Nano Res.*, 11, 1–41, 2018.
17. Song, W., Duan, W., Liu, Y., Ye, Z., Chen, Y., Chen, H., Qi, S., Wu, J., Liu, D., Xiao, L., Ratiometric detection of intracellular lysine and pH with one-pot synthesized dual emissive carbon dots. *Anal. Chem.*, 89, 13626–13633, 2017.
18. Shamsipur, M., Molaei, K., Molaabasi, F., Alipour, M., Alizadeh, N., Hosseinkhani, S., Hosseini, M., Facile preparation and characterization of new green emitting carbon dots for Fe3+ sensitive and selective off/on detection of ion and ascorbic acid in water and urine samples and intracellular imaging in living cells. *Talanta*, 183, 122–130, 2018.
19. Huang, Q., Li, Q., Chen, Y., Tong, L., Lin, X., Zhu, J., Tong, Q., High quantum yield nitrogen-doped carbon dots: Green synthesis and $Fe^{3+}$ application as "off-on" fluorescent sensors for the determination of and adenosine triphosphate in biological samples. *Sens. Actuators B Chem.*, 276, 82–88, 2018.

20. Zhou, Y., Won, J., Lee, J., Yoon, Y., Studies leading to the development of a highly selective colorimetric and fluorescent chemosensor for lysine. *Chem. Commun.*, 47, 1997–1999, 2011.
21. Yang, M., Song, Y., Zhang, M., Lin, S., Hao, Z., Liang, Y., Zhang, D., Chen, P.R., Converting a solvatochromic fluorophore into a protein-based pH indicator for extreme acidity. *Angew Chem. Int. Ed.*, 51, 7674–7679, 2012.
22. Cheng, T.-C., Roffler, S.R., Tzou, S.-C., Chuang, K.-H., Su, Y.-C., Chuang, C.-H., Kao, C.-H., Chen, C.-S., Harn, I.H., Liu, K.-Y., Cheng, T.-L., Leu, Y.-L., An activity-based near-infrared glucuronide trapping probe for imaging ß-glucuronidase expression in deep tissues. *J. Am. Chem. Soc.*, 134, 3103–3110, 2012.
23. Bohnenstengel, F., Kroemer, H.K., Sperker, B., *In vitro* cleavage of paracetamol glucuronide by human liver and kidney ß-glucuronidase: determination of paracetamol by capillary electrophoresis. *J. Chromatogr. B*, 721, 295–299, 1999.
24. Zhu, J.H., Yu, C., Chen, Y., Shin, J., Cao, Q.Y., Kim, J.S., A self-assembled amphiphilic imidazolium based ATP probe. *Chem. Commun.*, 53, 4342–4345, 2017.
25. Zhu, S., Zhang, J., Tang, S., Qiao, C., Wang, L., Wang, H., Liu, X., Li, B., Li, Y., Yu, W., Surface chemistry routes to modulate the photoluminescence of graphene quantum dots: From fluorescence mechanism to up-conversion bioimaging applications. *Adv. Funct. Mater.*, 22, 4732–4740, 2012.
26. Ma, Y., Chen, A.Y., Xie, X.F., Wang, X.Y., Wang, D., Wang, P., Li, H.J., Yang, J.H., Li, Y., Doping effect and fluorescence quenching mechanism of N-doped graphene quantum dots in the detection of dopamine. *Talanta*, 196, 563–571, 2019.
27. Mahmoud, A.M., Mahnashi, M.H., Alhazzani, K., Alanazi, A.Z., Algahtani, M.M., Alaseem, A., Alyami, B.A., AlQarni, A.O., El-Wekil, M.M., Nitrogen doped graphene quantum dots based on host guest interaction for selective dual readout of dopamine. *Spectrochim. Acta - A: Mol. Biomol. Spectrosc.*, 252, 119516, 2021.
28. Yu, Z., Ma, W., Wu, T., Wen, J., Zhang, Y., Wang, L., He, Y., Chu, H., Hu, M., Coumarin-modified graphene quantum dots as a sensing platform for multicomponent detection and its applications in fruits and living cells. *ACS Omega*, 5, 7369–7378, 2020.
29. Qiang, R., Sun, W., Hou, K., Li, Z., Zhang, J., Ding, Y., Wang, J., Yang, S., Electrochemical trimming of graphene oxide affords graphene quantum dots for $Fe^{3+}$ detection. *ACS Appl. Nano Mater.*, 4, 5220–5229, 2021.
30. Ge, S., He, J., Ma, C., Liu, J., F., X., Dong, X., One-step synthesis of boron-doped graphene quantum dots for fluorescent sensors and biosensor. *Talanta*, 199, 581–589, 2019.
31. Dutta, C.A. and Doong, R.A., Highly sensitive and selective detection of Nanomolar ferric ions using dopamine functionalized graphene quantum dots. *ACS Appl. Mater. Interfaces*, 8, 21002–21010, 2016.
32. Xin, Y., Xiao, C., Liang-Shi, L., Synthesis of large, stable colloidal graphene quantum dots with tunable size. *J. Am. Chem. Soc.*, 132, 5944–5945, 2010.
33. Kim, S., Hwang, S.W., Kim, M.-K., Shin, D.Y., Kim, C.O. *et al.*, Anomalous behaviors of visible luminescence from graphene quantum dots: Interplay between size and shape. *ACS Nano*, 6, 8203–8208, 2012.
34. Li, Y., Hu, Y., Zhao, Y., Shi, G., Deng, L., Hou, Y., Qu, L., An electrochemical avenue to green-luminescent grapheme quantumdots as potential electron-acceptors for photovoltaics. *Adv. Mater.*, 23, 776–780, 2011.
35. Pan, D., Zhang, J., Li, Z., Wu, M., Hydrothermal route for cutting graphene sheets into blue-luminescent grapheme quantum dots. *Adv. Mater.*, 22, 734–738, 2010.
36. Liu, R., Wu, D., Feng, X., Müllen, K., Bottom-up fabrication of photoluminescent graphene quantum dots with uniform morphology. *J. Am. Chem. Soc.*, 133, 15221–15223, 2011.

37. Deng, D., Pan, X., Yu, L., Cui, Y., Jiang, Y., Qi, J., Li, W.X., Fu, Q., Ma, X., Xue, Q., Toward N-doped graphene via solvothermal synthesis. *Chem. Mater.*, 23, 1188–1193, 2011.
38. Liu, Y., Liu, Y., Park, S., Zhang, Y., Kim, T., Chae, S., Park, M., Kim, H., One-step synthesis of robust nitrogen-doped carbon dots: Acid evoked fluorescence enhancement and their application for $Fe^{3+}$ detection. *J. Mater. Chem. A*, 3, 17747–17754, 2015.
39. Xu, Q., Pu, P., Zhao, J., Dong, C., Gao, C., Chen, Y., Chen, J., Liu, Y., Zhou, H., Preparation of highly photoluminescent sulfur-doped carbon dots for Fe (III) detection. *J. Mater. Chem. A*, 3, 542–546, 2014.
40. Poewe, W., Seppi, K., Tanner, C.M., Halliday, G.M., Brundin, P., Volkmann, J., Schrag, A.E., Lang, A.E., Parkinson disease. *Nat. Rev. Dis. Primers*, 3, 17013–17033, 2017.
41. Grace, A.A., Dysregulation of the dopamine system in the pathophysiology of schizophrenia and depression. *Nat. Rev. Neurosci.*, 17, 524–532, 2016.
42. Liu, R., Yang, R., Qu, C., Mao, H., Hu, Y., Li, J., Qu, L., Synthesis of glycine-functionalized graphene quantum dots as highly sensitive and selective fluorescent sensor of ascorbic acid in human serum. *Sens. Actuators B: Chem.*, 241, 644–651, 2017.
43. Chowdhury, A.D. and Doong, R.-A., Highly sensitive and selective detection of Nanomolar ferric ions using dopamine functionalized graphene quantum dots. *ACS Appl. Mater. Interfaces*, 8, 21002–21010, 2016.
44. Talapin, D.V., Rogach, A.L., Kornowski, A., Haase, M., Weller, H., Highly luminescent monodisperse CdSe and CdSe/ZnS nanocrystals synthesized in a hexadecylamine trioctylphosphine oxide-trioctylphospine mixture. *Nano Lett.*, 1, 207–211, 2001.
45. Micic, O.L., Curtis, C.J., Jones, K.M., Sprague, J.R., Nozik, A.J., Synthesis and dots, characterization of InP quantum. *J. Phys. Chem.*, 98, 4966–4969, 1994.
46. Gill, R., Zayats, M., Willner, I., Semiconductor quantum dots for bioanalysis. *Angew. Chem. Int. Ed.*, 47, 7602–7625, 2008.
47. Wu, P., Li, Y., Yan, X.-P., CdTe quantum dots (QDs) based kinetic discrimination of $Fe^{2+}$ and $Fe^{3+}$, and CdTe QDs-fenton hybrid system for sensitive photoluminescent detection of $Fe^{2+}$. *Anal. Chem.*, 81, 6252–6257, 2009.
48. Zhu, J., Chang, H., Li, J.J., Li, X., Zhao, J.W., Dual-mode melamine detection based on gold nanoparticles aggregation-induced fluorescence turn-on and turn-off of CdTe quantum dots. *Sens. Actuators B: Chem.*, 239, 906–915, 2017.
49. Li, J., Lu, Q.C.H., Yao, Q.-H., Zhang, X.L., Liu, J.J., Yang, H.H., Chen, G.N., A graphene oxide platform for energy transfer-based detection of protease activity. *Biosens. Bioelectron.*, 26, 3894–3899, 2011.
50. Wang, Y., Jiang, K., Zhu, J.L., Zhang, L., Lin, H.W., A FRET-based carbon dot-$MnO_2$ nanosheet architecture for glutathione sensing in human whole blood samples. *Chem. Commun.*, 51, 12748–12751, 2015.
51. Lv, Y., Wang, F., Li, N., Wu, R., Li, J., Shen, H., Li, L.S., Guo, F., Development of dual quantum dots-based fluorescence-linked immunosorbent assay for simultaneous detection on inflammation biomarkers. *Sens. Actuators B: Chem.*, 301, 127118, 2019.
52. Chen, J., Jiang, S., Wang, M., Xie, X., Su, X., Self-assembled dual-emissive nanoprobe with metal–organic frameworks as scaffolds for enhanced ascorbic acid and ascorbate oxidase sensing. *Sens. Actuators B: Chem.*, 339, 129910, 2021.
53. Chen, P., Yan, S., Sawyer, E., Ying, B., Wei, X., Wu, Z., Geng, J., Rapid and simple detection of ascorbic acid and alkaline phosphatase via controlled generation of silver nanoparticles and selective recognition. *Analyst*, 144, 1147–1152, 2019.
54. Jiménez-López, J., Rodrigues, S.S.M., Ribeirob, D.S.M., Ortega-Barrales, P., Ruiz-Medina, A., Santos, J.L.M., Exploiting the fluorescence resonance energy transfer (FRET) between CdTe

quantum dots and Au nanoparticles for the determination of bioactive thiols. *Spectrochim. Acta - A: Mol. Biomol. Spectrosc.*, 212, 246–254, 2019.

55. Wang, Y., Yang, M., Ren, Y., Fan, J., Ratiometric determination of hydrogen peroxide based on the size-dependent green and red fluorescence of CdTe quantum dots capped with 3-mercaptopropionic acid. *Microchim. Acta*, 186, 277, 2019.
56. Kiernan, U.A., Tubbs, K.A., Nedelkov, D., Niederkoer, E.E., Nelson, R.W., Detection of novel truncated forms of human serum amyloid a protein in human plasma. *FEBS Lett.*, 537, 166–170, 2003.
57. Yeh, E.T.H. and Willerson, J.T., Coming of age of C-reactive protein: Using inflammation markers in cardiology. *Circulation*, 107, 370–371, 2003.
58. Sabela, M., Balme, S., Bechelany, M., Janot, J.M., Bisetty, K., A review of gold and silver nanoparticle-based colorimetric sensing assays. *Adv. Eng. Mater.*, 19, 1–24, 2017.
59. Li, Y., Schluesener, H.J., Xu, S., Gold nanoparticle-based biosensors. *Gold Bull.*, 43, 29–41, 2010.
60. Li, J., Rao, X., Xiang, F., Wei, J., Yuan, M., Liu, Z., A photoluminescence "switch-on" nanosensor composed of nitrogen and sulphur co-doped carbon dots and gold nanoparticles for discriminative detection of glutathione. *Analyst*, 143, 2083–2089, 2018.
61. Jiao, Y., Liu, Q., Qiang, H., Chen, Z., Colorimetric detection of L-histidine based on the target-triggered self-cleavage of swing-structured DNA duplex-induced aggregation of gold nanoparticles. *Microchim. Acta*, 185, 452, 2018.
62. Dong, W., Wang, R., Gong, X., Dong, C., An efficient turn-on fluorescence biosensor for the detection of glutathione based on FRET between N, S dual-doped carbon dots and gold nanoparticles. *Anal. Bioanal. Chem.*, 411, 6687–6695, 2019.
63. Basak, M., Mitra, S., Agnihotri, S.K., Jain, A., Vyas, A., Bhatt, M.L.B., Sachan, R., Sachdev, M., Nemade, H.B., Bandyopadhyay, D., Noninvasive point-of-care nanobiosensing of cervical cancer as an auxiliary to pap-smear test. *ACS Appl. Bio Mater.*, 4, 5378–5390, 2021.
64. Khurana, S., Kukreti, S., Kaushik, M., Designing a two-stage colorimetric sensing strategy based on citrate reduced gold nanoparticles: Sequential detection of Sanguinarine (anticancer drug) and visual sensing of DNA. *Spectrochim. Acta A Mol. Biomol. Spectrosc.*, 246, 119039, 2021.
65. Meister, A., Glutathione metabolism and its selective modification. *J. Biol. Chem.*, 263, 17205–17208, 1988.
66. Townsend, D.M., Tew, K.D., Tapiero, H., The importance of glutathione in human disease. *Biomed. Pharmacother.*, 57, 145–155, 2003.
67. Watanabe, M., Suliman, M.E., Qureshi, A.R., Garcia-Lopez, E., Barany, P., Heimburger, O., Stenvinkel, P., Lindholm, B., Consequences of low plasma histidine in chronic kidney disease patients: Associations with inflammation, oxidative stress, and mortality. *Am. J. Clin. Nutr.*, 87, 1860–1866, 2008.
68. Seshadri, S., Beiser, A., Selhub, J., Jacques, P.F., Rosenberg, I.H., D'Agostino, R.B., Wilson, P.W., Wolf, P.A., Plasma homocysteine as a risk factor for dementia and alzheimer's disease. *N. Engl. J. Med.*, 346, 476–483, 2002.
69. Verri, C., Roz, L., Conte, D., Liloglou, T., Livio, A., Vesin, A., Fabbri, A., Andriani, F., Brambilla, C., Tavecchio, L., Calarco, G., Calabro, E., Mancini, A., Tosi, D., Bossi, P., Field, J.K., Braimbilla, E., Sozzi, G., Fragile histidine triad gene inactivation in lung cancer: The European early lung cancer project. *Am. J. Respir. Crit. Care Med.*, 179, 396–401, 2009.
70. Amjadi, M., Fakhri, Z.A., Hallaj, T., Carbon dots-silver nanoparticles fluorescence resonance energy transfer system as a novel turn-on fluorescent probe for selective determination of cysteine. *J. Photochem. Photobiol. A*, 309, 8–14, 2015.
71. Yousefi, S. and Saraji, M., Optical aptasensor based on silver nanoparticles for the colorimetric detection of adenosine. *Spectrochim. Acta A Mol. Biomol. Spectrosc.*, 213, 1–5, 2019.

72. Li, L., Wang, J., Chen, Z., Colorimetric determination of uric acid based on the suppression of oxidative etching of silver nanoparticles by chloroauric acid. *Microchim. Acta*, 187, 18, 2020.
73. Rostami, S., Mehdinia, A., Jabbari, A., Seed-mediated grown silver nanoparticles as a colorimetric sensor for detection of ascorbic acid. *Spectrochim. Acta A Mol. Biomol. Spectrosc.*, 180, 204–210, 2017.
74. Nishan, U., Gul, R., Muhammad, N., Asad, M., Rahim, A., Shah, M., Iqbal, J., Uddin, J., Shah, A.-u.-H. A., Shujah, S., Colorimetric based sensing of dopamine using ionic liquid functionalized drug mediated silver nanostructures. *Microchem. J.*, 159, 105382, 2020.
75. Sahu, S., Sharma, S., Kant, T., Shrivas, K., Ghosh, K.K., Colorimetric determination of L-cysteine in milk samples with surface functionalized silver nanoparticles. *Spectrochim. Acta A Mol. Biomol. Spectrosc.*, 246, 118961, 2021.
76. Gomes, C.V., Kaster, M.P., Tomé, A.R., Agostinho, P.M., Cunha, R.A., Adenosine receptors and brain diseases: Neuroprotection and neurodegeneration. *Biochim. Biophys. Acta*, 1808, 1380–1399, 2011.
77. Chen, W.J., Wu, Y., Zhao, X., Liu, S., Song, F.R., Liu, Z.Y., Screening the anti-gout traditional herbs from TCM using an *in vitro* method. *Chin. Chem. Lett.*, 27, 1701–1707, 2016.
78. Johnson, R.J., Kang, D.H., Feig, D., Kivlighn, S., Kanellis, J., Watanabe, S., Tuttle, K.R., Rodriguez-Iturbe, B., Herrera-Acosta, J., Mazzali, M., Is there a pathogenetic role for uric acid in hypertension and cardiovascular and renal disease? *Hypertension*, 41, 1183–1190, 2003.
79. Ping, J., Wu, J., Wang, Y., Ying, Y., Simultaneous determination of ascorbic acid, dopamine and uric acid using high-performance screen-printed grapheme electrode. *Biosens. Bioelectron.*, 34, 70–76, 2012.
80. Paixao, T.R. and Bertotti, M., FIA determination of ascorbic acid at low potential using a ruthenium oxide hexacyanoferrate modified carbon electrode. *J. Pharm. Biomed. Anal.*, 46, 528–533, 2008.
81. Sitte, H.H., Pifl, C., Rajput, A.H., Hörtnagl, H., Tong, J., Lloyd, G.K., Kish, S.J., Hornykiewicz, O., Dopamine and noradrenaline, but not serotonin, in the human claustrum are greatly reduced in patients with Parkinson's disease: Possible functional implications. *Eur. J. Neurosci.*, 45, 192–197, 2017.
82. Palanisamy, S., Zhang, X., He, T., Simple colorimetric detection of dopamine using modified silver nanoparticles. *Sci. China Chem.*, 59, 387–393, 2016.
83. Navadeepthy, N.D., Rebekah, R.A., Viswanathan, V.C., Ponpandian, P.N., N-doped Graphene/ $ZnFe_2O_4$: A novel nanocomposite for intrinsic peroxidase based sensing of $H_2O_2$. *Mater. Res. Bull.*, 95, 1–8, 2017.
84. Huanan, G., Qiaoyan, W., Shuping, L., A smartphone-integrated dual-mode nanosensor based on $Fe_3O_4$@Au for rapid and highly selective detection of glutathione. *Spectrochim. Acta A Mol. Biomol. Spectrosc.*, 271, 120866, 2022.
85. Zhang, K., Zuo, W., Wang, Z., Liu, J., Li, T., Wang, B., Yang, Z., A simple route to $CoFe_2O_4$ nanoparticles with shape and size control and their tunable peroxidase like activity. *RSC Adv.*, 5, 10632–10640, 2015.
86. Peng, Y., Wang, Z., Liu, W., Zhang, H., Zuo, W., Tang, H., Chen, F., Wang, B., Size- and shape-dependent peroxidase-like catalytic activity of $MnFe_2O_4$ Nanoparticles and their applications in highly efficient colorimetric detection of target cancer cells. *Dalton Trans.*, 44, 12871–12877, 2015.
87. Lu, J., Zhang, H., Li, S., Guo, S., Shen, L., Zhou, T., Zhong, H., Wu, L., Meng, Q., Zhang, Y., Oxygen-vacancy-enhanced peroxidase-like activity of reduced $Co_3O_4$ nanocomposites for the colorimetric detection of $H_2O_2$ and glucose. *Inorg. Chem.*, 59, 3152–3159, 2020.

88. Navadeepthy, D., Thangapandian, M., Viswanathan, C., Ponpandian, N., A nanocomposite of $NiFe_2O_4$–PANI as a duo active electrocatalyst toward the sensitive colorimetric and electrochemical sensing of ascorbic acid. *Nanoscale Adv.*, 2, 3481–3493, 2020.
89. Luo, S., Liu, Y., Rao, H., Wang, Y., Wang, X., Fluorescence and magnetic nanocomposite $Fe_3O_4$@$SiO_2$@Au MNPs as peroxidase mimetics for glucose detection. *Anal. Biochem.*, 538, 26–33, 2017.
90. Liu, K., Su, J., Liang, J., Wu, Y., Mesoporous $MnFe_2O_4$ magnetic nanoparticles as a peroxidase mimic for the colorimetric detection of urine glucose. *RSC Adv.*, 11, 28375–28380, 2021.
91. (a) Justino, C.I.L., Rocha-Santos, T.A.P., Cardoso, S., Duarte, A.C., Strategies for enhancing the analytical performance of nanomaterial-based sensors. *Trends Anal. Chem.*, 47, 27–36, 2013, (b) Justino, C.I.L., Rocha-Santos, T.A.P., Duarte, A.C., Review of analytical figures of merit of sensors and biosensors in clinical applications. *Trends Anal. Chem.*, 29, 1172–1183, 2010.
92. Chen, E., Cai, K., Liu, X., Wu, S., Wu, Z., Ma, M., Chen, B., Zhao, Z., Label-free ratiometric upconversion nanoprobe for spatiotemporal pH mapping in living cells. *Anal. Chem.*, 93, 6895–6900, 2021.
93. You, Y., Cheng, S., Zhang, L., Zhu, Y., Zhang, C., Xian, Y., Rational modulation of the luminescence of upconversion nanomaterials with phycocyanin for the sensing and imaging of myeloperoxidase during an inflammatory process. *Anal. Chem.*, 92, 5091–5099, 2020.
94. Ghosh, S., Chang, Y.-F., Yang, D.-M., Chattopadhyay, S., Upconversion nanoparticle-mOrange protein FRET nanoprobes for self-ratiometric/ratiometric determination of intracellular pH, and single cell pH imaging. *Biosens. Bioelectron.*, 155, 112115, 2020.
95. Tsai, E.S., Himmelstoß, S.F., Wiesholler, L.M., Hirsch, T., Hall, E.A.H., Upconversion nanoparticles for sensing pH. *Analyst*, 144, 5547–5557, 2019.
96. Hudry, D., Howard, I.A., Popescu, R., Gerthsen, D., Richards, B.S., Structure–property relationships in lanthanide-doped upconverting nanocrystals: Recent advances in understanding core–shell structures. *Adv. Mater.*, 31, 1900623, 2019.
97. Su, Q., Feng, W., Yang, D., Li, F., Resonance energy transfer in upconversion nanoplatforms for selective biodetection. *Acc. Chem. Res.*, 50, 32–40, 2017.
98. Zhang, C.L., Yuan, Y.X., Zhang, S.M., Wang, Y.H., Liu, Z.H., Biosensing platform based on fluorescence resonance energy transfer from upconverting nanocrystals to graphene oxide. *Angew. Chem. Int. Ed.*, 50, 6851–6854, 2011.
99. Wu, Y., Li, D., Zhou, F., Liang, H., Liu, Y., Hou, W.J., Yuan, Q., Zhang, X.B., Tan, W.H., Versatile *in situ* synthesis of $MnO_2$ nanolayers on upconversion nanoparticles and their application in activatable fluorescence and MRI imaging. *Chem. Sci.*, 9, 5427–5434, 2018.
100. Wang, L., Yan, R., Huo, Z., Wang, L., Zeng, J., Bao, J., Wang, X., Peng, Q., Li, Y., Fluorescence resonant energy transfer biosensor based on upconversion-luminescent nanoparticles. *Angew. Chem. Int. Ed.*, 44, 6054–6057, 2005.
101. Hu, X., Wang, Y., Liu, H., Wang, J., Tan, Y., Wang, F., Yuan, Q., Tan, Naked eye detection of multiple tumor-related mRNAs from patients with photonic-crystal micropattern supported dual-modal upconversion bioprobes. *Chem. Sci.*, 8, 466–472, 2017.
102. Casey, J.R., Grinstein, S., Orlowski, J., Sensors and regulators of intracellular pH. *Nat. Rev. Mol. Cell Biol.*, 11, 50–61, 2010.
103. Ding, C., Cheng, S., Zhang, C., Xiong, Y., Ye, M., Xian, Y., Ratiometric upconversion luminescence nanoprobe with near-infrared $Ag_2S$ nanodots as the energy acceptor for sensing and imaging of pH *in vivo*. *Anal. Chem.*, 91, 7181–7188, 2019.
104. Pu, K., Shuhendler, A.J., Rao, J., Semiconducting polymer nanoprobe for *in vivo* imaging of reactive oxygen and nitrogen species. *Angew. Chem. Int. Ed.*, 52, 10325–10329, 2013.

105. Hou, W., Yuan, Y., Sun, Z., Guo, S., Dong, H., Wu, C., Ratiometric fluorescent detection of intracellular singlet oxygen by semiconducting polymer dots. *Anal. Chem.*, 90, 14629–14634, 2018.
106. Wu, C. and Chiu, D.T., Highly fluorescent semiconducting polymer dots for biology and medicine. *Angew. Chem. Int. Ed. Engl.*, 52, 3086–3109, 2013.
107. Huanan, G., Qiaoyan, W., Shuping, L., A smartphone-integrated dual-mode nanosensor based on $Fe_3O_4$@Au for rapid and highly selective detection of glutathione. *Spectrochim. Acta A Mol. Biomol. Spectrosc.*, 271, 120866, 2022.
108. Sun, J., Mei, H., Gao, F., Ratiometric detection of copper ions and alkaline phosphatase activity based on semiconducting polymer dots assembled with rhodamine B hydrazide. *Biosens. Bioelectron.*, 91, 70–75, 2017.
109. Chen, P., Ilyas, I., He, S., Xing, Y., Jin, Z., Huang, C., Ratiometric pH sensing and imaging in living cells with dual-emission semiconductor polymer dots. *Molecules*, 24, 2923, 2019.
110. Zhang, Q., Zhang, Z., Hu, X., Sun, J., Gao, F., Dual-targeting into the mitochondria of cancer cells for ratiometric investigation of the dynamic fluctuation of sulfur dioxide and formaldehyde with two-photon integrated semiconducting polymer dots. *ACS Appl. Mater. Interfaces*, 14, 179–190, 2022.
111. Smith, A.M., Mancini, M.C., Nie, S., Bioimaging:Second window for *in vivo* imaging. *Nat. Nanotechnol.*, 4, 710–711, 2009.
112. Kim, J.H., Ahn, J.-H., Barone, P.W., Jin, H., Zhang, J., Heller, D.A., Strano, M.S., A luciferase/single-walled carbon nanotubeconjugate for near-infrared fluorescent detection of cellular ATP. *Angew. Chem. Int. Ed.*, 48, 1456–1459, 2010.
113. Heller, D.A., Pratt, G.W., Zhang, J., Nair, N., Hansborough, A.J., Boghossian, A.A., Reuel, N.F., Barone, P.W., Strano, M.S., Peptidesecondary structure modulates single-walled carbon nanotube fluorescence as a chaperone sensor for nitroaromatics. *Proc. Natl. Acad. Sci. U.S.A.*, 108, 8544–8549, 2011.
114. Zhang, J., Boghossian, A.A., Barone, P.W., Rwei, A., Kim, J.-H., Lin, D., Heller, D.A., Hilmer, A.J., Nair, N., Reuel, N.F., Strano, M.S., Single molecule detection of nitric oxide enabled by d(AT)15 DNA adsorbed to near infrared fluorescent single-walled carbonnanotubes. *J. Am. Chem. Soc.*, 133, 567–581, 2011.
115. Zhang, J., Kruss, S., Hilmer, A.J., Shimizu, S., Schmois, Z., De La Cruz, F., Barone, P.W., Reuel, N.F., Heller, D.A., Strano, M.S., A rapid, direct, quantitative, and label-free detector of cardiac biomarker troponin T using near-infrared fluorescent single-walled carbon nanotube sensors. *Adv. Healthc. Mater.*, 14, 412–423, 2014.
116. Safee, M.M., Gravely, M., Roxbury, D., A wearable optical microfibrous biomaterial with encapsulated nanosensors enableswireless monitoring of oxidative stress. *Adv. Funct. Mater.*, 31, 2006254, 2021.
117. Lisi, S., Scaranoa, S., Fedelia, S., Pascalea, E., Cicchia, S., Raveletb, C., Peyrinb, E., Minunni, M., Toward sensitive immuno-baseddetection of tau protein by surface plasmon resonance coupled to carbon nanostructures as signal amplifiers. *Biosens. Bioelectron.*, 93, 289–292, 2017.
118. Nandeshwar, R. and Tallur, S., Integrated low cost optical biosensor for high resolution sensing of myeloperoxidase (MPO) activitythrough carbon nanotube degradation. *IEEE Sens. J.*, 21, 1236–1243, 2021.
119. Pinals, R.L., Ledesma, F., Yang, D., Navarro, N., Jeong, S., Pak, J.E., Kuo, L., Chuang, Y.-C., Cheng, Y.-W., Sun, H.-Y., Landry, M.P., Rapid SARS-CoV-2 spike protein detection by carbon nanotube-based near-infrared nanosensors. *Nano Lett.*, 21, 2272–2280, 2021.
120. Safee, M.M., Gravely, M., Roxbury, D., A wearable optical microfibrous biomaterial with encapsulated nanosensors enables wireless monitoring of oxidative stress. *Adv. Funct. Mater.*, 31, 2006254, 2021.

121. Ma, H., Xue, N., Li, Z., Xing, K., Miao, X., Ultrasensitive detection of miRNA-155 using multi-walled carbon nanotube-goldnanocomposites as a novel fluorescence quenching platform. *Sens. Actuators B: Chem.*, 266, 221–227, 2018.
122. Elmizadeh, H., Faridbod, F., Soleimani, M., Ganjali, M.R., Bardajee, G.R., Fluorescent apta-nanobiosensors for fast and sensitivedetection of digoxin in biological fluids using rGQDs: Comparison of two approaches for immobilization of apatamer. *Sens. Actuators B: Chem.*, 302, 127133, 2020.
123. Wang, X., Wang, C., Qu, K., Song, Y., Ren, J., Miyoshi, D., Sugimoto, N., Qu, X., Ultrasensitive detection of a prognostic indicatorin early-stage cancer using graphene oxide and carbon nanotubes. *Adv. Funct. Mater.*, 20, 3967–3971, 2010.
124. Shi, X., von dem Bussche, A., Hurt, R.H., Kane, A.B., Gao, H., Cell entry of one dimensional nanomaterials occurs by tip recognition and rotation. *Nat. Nanotechnol.*, 6, 714–719, 2011.
125. Irrera, A., Magazzù, A., Artoni, P., Simpson, S.H., Hanna, S., Jones, P.H., Priolo, F., Gucciardi, P.G., Maragò, O.M., Photonic torque microscopy of the nonconservative force field for optically trapped silicon nanowires. *Nano Lett.*, 16, 4181–4188, 2016.
126. Zhang, G.-J. and Ning, Y., Silicon nanowire biosensor and its applications in disease diagnostics: A review. *Anal. Chim. Acta*, 749, 1–15, 2012.
127. Patolsky, F., Zheng, G., Lieber, C.M., Nanowire-based biosensors. *Anal. Chem.*, 78, 13, 4260–4269, 2006.
128. Chen, M., Mu, L., Wang, S., Cao, X., Liang, S., Wang, Y., She, G., Yang, J., Wang, Y., Shi, W., A single silicon nanowire-based ratiometric biosensor for $Ca^{2+}$ at various locations in a neuron. *ACS Chem. Neurosci.*, 11, 9, 1283–1290, 2020.
129. Cao, X., Mu, L., Chen, M., Bu, C., Liang, S., She, G., Shi, W., Single silicon nanowire-based fluorescent sensor for endogenous hypochlorite in an individual cell. *Adv. Biosyst.*, 2, 12, 1800213, 2018.
130. Wang, H., Mu, L., She, G., Shi, W., Silicon nanowires-based fluorescent sensor for *in situ* detection of hydrogen sulfide in extracellular environment. *RSC Adv.*, 5, 65905–65908, 2015.
131. Leonardi, A.A., Faro, M.J.L., Franco, C.D., Palazzo, G., D'Andrea, C., Morganti, D., Manoli, K., Musumeci, P., Fazio, B., Lanza, M., Torsi, L., Priolo, F., Irrera, A., Silicon nanowire luminescent sensor for cardiovascular risk in saliva. *J. Mater. Sci.: Mater. Electron.*, 31, 1, 10–17, 2020.
132. Stringer, C. and Pachitariu, M., Computational processing of neural recordings from calcium imaging data. *Curr. Opin. Neurobiol.*, 55, 22–31, 2018.
133. Martin, V., Vale, C., Hirama, M., Yamashita, S., Rubiolo, J.A., Vieytes, M.R., Botana, L.M., Synthetic ciguatoxin CTX 3C induces a rapid imbalance in neuronal excitability. *Chem. Res. Toxicol.*, 28, 6, 1095–108, 2015.
134. Shepherd, J., Hilderbrand, S.A., Waterman, P., Heinecke, J.W., Weissleder, R., Libby, P., A fluorescent probe for the detection of myeloperoxidase activity in atherosclerosis-associated macrophages. *Chem. Biol.*, 14, 11, 1221–1231, 2007.
135. (a) Pan, B., Ren, H., Lv, X., Zhao, Y., Yu, B., He, Y., Ma, Y., Niu, C., Kong, J., Yu, F., Sun, W.-B., Zhang, Y., Willard, B., Zheng, L., Hypochlorite-induced oxidative stress elevates the capability of HDL in promoting breast cancer metastasis. *J. Transl. Med.*, 65, 10, 2012, (b) Sun, Z.-N., Liu, F.-Q., Chen, Y., Tam, P.K.H., Yang, D., A highly specific BODIPY-based fluorescent probe for the detection of hypochlorous acid. *Org. Lett.*, 10, 11, 2171–2174, 2008.
136. (a) Kabil, O. and Banerjee, R., Redox biochemistry of hydrogen sulfide. *J. Biol. Chem.*, 285, 29, 21903–21907, 2010, (b) Mancuso, C., Navarra, P., Preziosi, P., Roles of nitric oxide, carbon monoxide, and hydrogen sulfide in the regulation of the hypothalamic-pituitary-adrenal axis. *J. Neurochem.*, 113, 563–575, 2010, (c) Vandiver, M.S. and Snyder, S.,J., Hydrogen sulfide: A gasotransmitter of clinical relevance. *Mol. Med.*, 90, 255–263, 2012.

137. Salazar, J., Martínez, M.S., Chávez, M., Toledo, A., Añez, R., Torres, Y., Apruzzese, V., Silva, C., Rojas, J., Bermúdez, V., C-reactive protein: Clinical and epidemiological perspectives. *Cardiol. Res. Pract.*, 2014, 605810, 2014.
138. (a) Cioffi, N., Torsi, L., Ditaranto, N., Tantillo, G., Ghibelli, L., Sabbatini, L., Bleve-Zacheo, T., D'Alessio, M., Zambonin, P.G., Traversa, E., Copper nanoparticle/polymer composites with antifungal and bacteriostatic properties. *Chem. Mater.*, 17, 5255–5262, 2005, (b) DeVries, G.A., Brunnbauer, M., Hu, Y., Jackson, A.M., Long, B., Neltner, B.T., Uzun, O., Wunsch, B.H., Stellacci, F., Divalent metal nanoparticles. *Science*, 315, 358–361, 2007, (c) Jin, R.C., Cao, Y.W., Mirkin, C.A., Kelly, K.L., Schatz, G.C., Zheng, J.G., Photoinduced conversion of silver nanospheres to nanoprisms. *Science*, 294, 1901–1903, 2001.
139. Zheng, J., Cheng, X., Zhang, H., Bai, X., Ai, R., Shao, L., Wang, J., Gold nanorods: The most versatile plasmonic nanoparticles. *Chem. Rev.*, 121, 21, 13342–13453, 2021.
140. (a) El-Sayed, M.A., Some interesting properties of metals confined in time and nanometer space of different shapes. *Acc. Chem. Res.*, 34, 4, 257–264, 2001, (b) Perez-Juste, J., Pastoriza-Santos, I., Liz-Marzan, L.M., Mulvaney, P., Gold nanorods: Synthesis, characterization and applications. *Coord. Chem. Rev.*, 249, 17-18, 1870–1901, 2005, (c) Nikoobakhtand, B. and El-Sayed, M.A., Surface-enhanced Raman scattering studies on aggregated gold nanorods. *J. Phys. Chem. A*, 107, 18, 3372–3378, 2003.
141. (a) Smith, A.M., Mancini, M.C., Nie, S.M., Bioimaging second window for *in vivo* imaging. *Nat. Nanotechnol.*, 4, 710–711, 2009, (b) Perez-Juste, J., Correa-Duarte, M.A., Liz-Marzan, L.M., Silica gels with tailored, gold nanorod-driven optical functionalities. *Appl. Surf. Sci.*, 226, 1-3, 137–143, 2004.
142. Gans, R., Form of ultramicroscopic particles of silver. *Ann. Phys.*, 47, 270–284, 1915.
143. Jain, P.K., Lee, K.S., El-Sayed, I.H., El-Sayed, M.A., Calculated absorption and scattering properties of gold nanoparticlesof different size, shape, and composition: Applications in biological imaging and biomedicine. *J. Phys. Chem. B*, 110, 7238–7248, 2006.
144. Zhou, J., Cao, Z., Panwar, N., Hu, R., Wang, X., Qu, J., Tjin, S.C., Xu, G., Yong, K.-T., Functionalised gold nanorods for nanomedicine: Past, present and future. *Coord. Chem. Rev.*, 352, 15–66, 2017.
145. (a) Hulteen, J.C. and Martin, C.R., A general template-based method for the preparation of nanomaterials. *J. Mater. Chem.*, 7, 1075–1087, 1997, (b) Yu, Y.-Y., Chang, S.-S., Lee, C.-L., Wang, C.R.C., Gold nanorods: Electrochemical synthesis and optical properties. *J. Phys. Chem. B*, 101, 6661–6664, 1997.
146. Perez-Juste, J., Liz-Marzan, L.M., Carnie, S., Chan, D.Y.C., Mulvaney, P., Electric-field-directed growth forgold nanorods. *Adv. Funct. Mater.*, 14, 571–579, 2004.
147. Kim, F., Song, J.H., Yang, P.D., Photochemical synthesisof gold nanorods. *J. Am. Chem. Soc.*, 124, 14316–14317, 2002.
148. Murphy, C.J., Sau, T.K., Gole, A.M., Orendorff, C.J., Gao, J.X., Gou, L., Hunyadi, S.E., Li, T., Anisotropicmetal nanoparticles: Synthesis, assembly, and opticalapplications. *J. Phys. Chem. B*, 109, 13857–13870, 2005.
149. Huang, X., Neretina, S., El-Sayed, M.A., Gold nanorods: From synthesis and properties to biological and biomedicalapplications. *Adv. Mater.*, 21, 4880–4910, 2009.
150. Perez-Juste, J., Pastoriza-Santos, I., Liz-Marzan, L.M., Mulvaney, P., Gold nanorods: Synthesis, characterization and applications. *Coord. Chem. Rev.*, 249, 1870–1901, 2005.
151. Yang, D.P. and Cui, D.X., Advances and prospects of gold nanorods. *Chem. Asian J.*, 3, 2010–2022, 2008.
152. Hulkoti, N.I., and Taranath, T.C., Biosynthesis of nanoparticles using microbes – a review. *Colloids Surf. B Biointerfaces*, 121, 474–483, 2014.

153. Onaciu, A., Braicu, C., Zimta, A.-A., Moldovan, A., Stiufiuc, R., Buse, M., Ciocan, C., Buduru, S., Berindan-Neagoe, I., Gold nanorods: From anisotropy to opportunity. An evolution update. *Nanomedicine*, 14, 9, 1203–1226, 2019.
154. Kermanshahian, K., Yadegar, A., Ghourchian, H., Gold nanorods etching as a powerful signaling process for plasmonic multicolorimetric chemo-/biosensors: Strategies and applications. *Coord. Chem. Rev.*, 442, 213934, 2021.
155. Saa, L., Coronado-Puchau, M., Pavlov, V., Liz-Marzán, L.M., Enzymatic etching of gold nanorods by horseradish peroxidase and application to blood glucose detection. *Nanoscale*, 6, 13, 7405–7409, 2014.
156. Pérez-Juste, J., Pastoriza-Santos, I., Liz-Marzán, L.M., Mulvaney, P., Gold nanorods: Synthesis, characterization and applications. *Coord. Chem. Rev.*, 249, 17, 1870–1901, 2005.
157. Urakami, T., Suzuki, J., Yoshida, A., Saito, H., Mugishima, H., Incidence of children with slowly progressive form of type 1 diabetes detected by the urine glucose screening at schools in the Tokyo Metropolitan Area. *Diabetes Res. Clin. Pract.*, 80, 3, 473–476, 2008.
158. Zhang, Z., Chen, Z., Cheng, F., Zhang, Y., Chen, L., Highly sensitive on-site detection of glucose in human urine with naked eye based on enzymatic like reaction mediated etching of gold nanorods. *Biosens. Bioelectron.*, 89, 932–936, 2017.
159. (a) Wang, J., Du, R., Liu, W., Yao, L., Ding, F., Zou, P., Wang, Y.Y., Wang, X.X., Zhao, Q.B., Rao, H.B., Colorimetric and fluorometric dual-signal determination of dopamine by the use of Cu-Mn-O microcrystals and C-dots. *Sens. Actuators B*, 290, 125–132, 2019, (b) Maia, T.V. and Frank, M.J., An integrative perspective on the role of dopamine in schizophrenia. *Biol. Psychiatry*, 81, 1, 52–66, 2017.
160. Xiu, Q., Chun-Ling, Y., Rui, S., Shu-Zheng, W., Yi-Lin, W., Colorimetric detection of dopamine based on iodine-mediated etching of gold nanorods. *Chin. J. Anal. Chem.*, 49, 1, 60–67, 2021.
161. Teo, P.S., Rameshkumar, P., Pandikumar, A., Jiang, Z.-T., Altarawneh, M., Huang, N.M., Colorimetric and visual dopamine assay based on the use of gold nanorods. *Microchim. Acta*, 184, 4125–4132, 2017.
162. Bull, H., Murray, P.G., Thomas, D., Fraser, A.M., Nelson, P.N., Acid phosphatises. *Mol. Pathol.*, 55, 2, 65–72, 2002.
163. Leman, E.S. and Getzenberg, R.H., Biomarkers for prostate cancer. *J. Cell. Biochem.*, 108, 1, 3–9, 2009.
164. Liu, B.-W., Huang, P.-C., Wu, F.-Y., Smartphone colorimetric assay of acid phosphatase based on a controlled iodine-mediated etching of gold nanorods. *Anal. Bioanal. Chem.*, 412, 8051–8059, 2020.
165. (a) Huang, W., Deng, Y., He, Y., Visual colorimetric sensor array for discrimination of antioxidants in serum using $MnO_2$ nanosheets triggered multicolor chromogenic system. *Biosens. Bioelectron.*, 91, 89–94, 2017, (b) He, Y., Qi, F., Niu, X., Zhang, W., Zhang, X., Pan, J., Uricase-free on-demand colorimetric biosensing of uric acid enabled by integrated CoP nanosheet arrays as a monolithic peroxidase mimic. *Anal. Chim. Acta*, 1021, 113–120, 2018.
166. Qin, X., Yuan, C., Geng, G., Shi, R., Cheng, S., Wang, Y., Enzyme-free colorimetric determination of uric acid based on inhibition of gold nanorods etching. *Sens. Actuators: B. Chem.*, 333, 129638, 2021.
167. (a) Fukumori, M., Pandey, R.R., Fujiwara, T., TermehYousefi, A., Negishi, R., Kobayashi, Y., Tanaka, H., Ogawa, T., Diameter dependence of longitudinal unzipping of singlewalled carbon nanotube to obtain graphene nanoribbon. *Jpn. J. Appl. Phys.*, 56, 06GG12, 2017, (b) Sadeghi, S., Arjmand, M., Otero Navas, I., Zehtab Yazdi, A., Sundararaj, U., Effect of nanofiller geometry on network formation in polymeric nanocomposites: Comparison of rheological and electrical properties of multiwalled carbon nanotube and graphene nanoribbon. *Macromolecules*, 50, 3954–3967, 2017.

168. Hu, Y., Yang, D., Yang, C., Feng, N., Shao, Z., Zhang, L., Wang, X., Weng, L., Luo, Z., Wang, L., A novel "off-on" fluorescent probe based on carbon nitride nanoribbons for the detection of citrate anion and live cell imaging. *Sensors*, 18, 1163, 2018.
169. Wang, Z.-X., Gao, Y.-F., Jin, X., Yu, X.-H., Tao, X., Kong, F.-Y., Fan, D.-H., Wang, W., Excitation-independent emission carbon nanoribbons polymer as a ratiometric photoluminescent probe for highly selective and sensitive detection of quercetin. *Analyst*, 144, 2256–2263, 2019.
170. Rostamia, S., Mehdiniab, A., Jabbari, A., Intrinsic peroxidase-like activity of graphene nanoribbons for label-free colorimetric detection of dopamine. *Mater. Sci. Eng. C*, 114, 111034, 2020.
171. Yin, C., Wu, M., Liu, T., Fu, L., Sun, Q., Chen, L., Niu, N., Turn-on fluorescent inner filter effect-based B, S, N co-doped carbon quantum dots and vanadium oxide nanoribbons for α-glucosidase activity detection. *Microchem. J.*, 178, 107405, 2022.
172. Akram, M., Citric acid cycle and role of its intermediates in metabolism. *Cell Biochem. Biophys.*, 68, 475–478, 2014.
173. Liu, Z., Devaraj, S., Yang, C., Yen, Y., A new selective chromogenic and fluorogenic sensor for citrate ion. *Sens. Actuators B Chem.*, 174, 555–562, 2012.
174. (a) Lee, E.Z., Jun, Y.S., Hong, W.H., Thomas, A., Jin, M.M., Cubic mesoporous graphitic carbon(IV) nitride: An all-in-one chemosensor for selective optical sensing of metal ions. *Angew. Chem. Int. Ed.*, 49, 9706–9710, 2010, (b) Tian, J., Liu, Q., Asiri, A.M., Al-Youbi, A.O., Sun, X., Ultrathin graphitic carbon nitride nanosheet: A highly efficient fluorosensor for rapid, ultrasensitive detection of $Cu^{2+}$. *Anal. Chem.*, 85, 5595–5599, 2013.
175. (a) Field, T.B., McCourt, J.L., McBryde, W., Composition and stability of iron and copper citrate complexes in aqueous solution. *Can. J. Chem.*, 52, 3119–3124, 1974, (b) Still, E.R. and Wikberg, P., Solution studies of systems with polynuclear complex formation. 1. The copper(II) citrate system. *Inorg. Chim. Acta*, 46, 147–152, 1980, (c) Parry, R. and Dubois, F., Citrate complexes of copper in acid solutions. *J. Am. Chem. Soc.*, 74, 3749–3753, 1952.
176. Cai, X., Fang, Z., Dou, J., Yu, A., Zhai, G., Bioavailability of quercetin: Problems and promises. *Curr. Med. Chem.*, 20, 20, 2572–2582, 2013.
177. Loke, W.M., Proudfoot, J.M., Stewart, S., Mckinley, A.J., Needs, P.W., Kroon, P.A., Hodgson, J.M., Croft, K.D., Metabolic transformation has a profound effect on anti-inflammatory activity of flavonoids such as quercetin: Lack of association between antioxidant and lipoxygenase inhibitory activity. *Biochem. Pharmacol.*, 75, 1045–1053, 2008.
178. (a) El-Ghundi, M., O'Dowd, B.F., George, S.R., Insights into the role of dopamine receptor systems in learning and memory. *Rev. Neurosci.*, 18, 1, 37–66, 2007, (b) Robinson, D.L., Hermans, A., Seipel, A.T., Wightman, R.M., Monitoring rapid chemical communication in the brain. *Chem. Rev.*, 108, 7, 2554–2584, 2008.
179. Wang, M., Wang, M., Zhang, F., Su, X., A ratiometric fluorescent biosensor for sensitive determination of α-glucosidase activity and acarbose based on N-doped carbon dots. *Analyst*, 145, 5808–5815, 2020.
180. Rasouli, R., Barhoum, A., Bechelany, M., Dufresne, A., Nanofibers for biomedical and healthcare applications. *Macromol. Biosci.*, 19, 2, 1800256, 2019.
181. Ondarcuhu, T. and Joachim, C., Drawing a single nanofibre over hundreds of microns. *Europhys. Lett.*, 42, 2, 215, 1998.
182. Maiyalagan, T., Viswanathan, B., Varadaraju, U., Fabrication and characterization of uniform $TiO^2$ nanotube arrays by sol-gel template method. *Bull. Mater. Sci.*, 29, 7, 705, 2006.
183. Fujihara, K., Teo, W.E., Lim, T.C., Ma, Z., An introduction to electrospinning and nanofibers. *World Sci., Singapore*, 90, 288, 2005.
184. Liu, G., Ding, J., Qiao, L., Guo, A., Dymov, B.P., Gleeson, J.T., Hashimoto, T., Saijo, K., Polystyrene-block-poly (2-cinnamoylethyl methacrylate) nanofibers—Preparation, characterization, and liquid crystalline properties. *Chem. Eur. J.*, 5, 9, 2740–2749, 1999.

185. Hekmati, A.H., Rashidi, A., Ghazisaeidi, R., Drean, J.Y., Effect of needle length, electrospinning distance, and solution concentration on morphological properties of polyamide-6 electrospun nanowebs. *Text. Res. J.*, 83, 4, 1452–1466, 2013.
186. Jiang, Y., Song, N., Wang, C., Pinna, N., Lu, X., Facile synthesis of Fe3O4/nitrogen-doped carbon hybrid nanofibers as a robust peroxidase-like catalyst for sensitive colorimetric detection of ascorbic acid. *J. Mater. Chem. B*, 5, 5499–5505, 2017.
187. Chena, S., Chia, M., Yanga, Z., Gaoa, M., Wanga, C., Lua, X., Carbon dots/$Fe_3O_4$ hybrid nanofibers as efficient peroxidase mimics for sensitive detection of $H_2O_2$ and ascorbic acid. *Inorg. Chem. Front.*, 4, 1621–1627, 2017.
188. Filiz, B.C., Elalmis, Y.B., Bektas, I.S., Figen, A.K., Fabrication of stable electrospun blended chitosan-poly(vinyl alcohol) nanofibers for designing naked-eye colorimetric glucose biosensor based on GOx/HRP. *Int. J. Biol. Macromol.*, 192, 999–1012, 2021.
189. Song, N., Zhu, Y., Ma, F., Wang, C., Lu, X., Facile preparation of Prussian blue/polypyrrole hybrid nanofibers as robust peroxidase mimics for colorimetric detection of L-cysteine. *Mater. Chem. Front.*, 2, 768–774, 2018.
190. Chen, S., Chi, M., Zhu, Y., Gao, M., Wang, C., Lu, X., A facile synthesis of superparamagnetic $Fe_3O_4$ nanofibers with superior peroxidase-like catalytic activity for sensitive colorimetric detection of L-cysteine. *Appl. Surf. Sci.*, 440, 237–244, 2018.
191. Hsu, S.C., Cheng, H.T., Wu, P.X., Weng, C.J., Santiago, K.S., Yeh, J.M., Electrochemical sensor constructed using a carbon paste electrode modified with mesoporous silica encapsulating PANI chains decorated with GNPs for detection of ascorbic acid. *Electrochim. Acta*, 238, 246–256, 2017.
192. (a) Lim, S.Y. and Kim, H.J., Ratiometric detection of cysteine by a ferrocenyl Michael acceptor. *Tetrohedron Lett.*, 52, 25, 3189–3190, 2011, (b) Luo, Y., Shen, Z., Liu, P., Zhao, L., Wang, X., Facile fabrication and selective detection for cysteine of xylan/Au nanoparticles composite. *Carbohydr. Polym.*, 140, 122–128, 2016.
193. Lee, P.C., Li, N.S., Hsu, Y.P., Peng, C., Yang, H.W., Direct glucose detection in whole blood by colorimetric assay based on glucose oxidase-conjugated graphene oxide/$MnO_2$ nanozymes. *Analyst*, 144, 3038–3044, 2019.
194. Liu, X., Ye, C., Li, X., Cui, N., Wu, T., Du, S., Wei, Q., Fu, L., Yin, J., Lin, C.T., Highly sensitive and selective potassium ion detection based on graphene hall effect biosensors. *Materials*, 11, 399, 2018.
195. Wang, Y., Li, Z., Wang, J., Li, J., Lin, Y., Graphene and graphene oxide: Biofunctionalization and applications in biotechnology. *Trends Biotechnol.*, 29, 205–212, 2011.
196. Singh, R.K., Kumar, R., Singh, D.P., Graphene oxide: Strategies for synthesis, reduction and frontier applications. *RSC Adv.*, 6, 64993–65011, 2016.
197. Qi, Y., Chen, Y., He, J., Gao, X., Highly sensitive and simple colorimetric assay of hydrogen peroxide and glucose in human serum via the smart synergistic catalytic mechanism. *Spectrochim. Acta - A: Mol. Biomol. Spectrosc.*, 234, 118233, 2020.
198. Liu, X., Wu, Y., Wu, X., Zhao, J.X., Graphene oxide-based fluorescence assay for sensitive detection of DNA exonuclease enzymatic activity. *Analyst*, 144, 6231–6239, 2019.
199. Samak, N.A., Selim, M.S., Hao, Z., Xing, J., Controlled-synthesis of alumina grapheme oxide nanocomposite coupled with DNA/ sulfide fluorophore for eco-friendly "turn off/on" $H_2S$ nanobiosensor. *Talanta*, 211, 120665, 2020.
200. Liu, F., Choi, J.Y., Seo, T.S., Graphene oxide arrays for detecting specific DNA hybridization by fluorescence resonance energy transfer. *Biosens. Bioelectron.*, 25, 2361–2365, 2010.
201. Basiruddin, S.K. and Swain, S.K., Phenylboronic acid functionalized reduced graphene oxide based fluorescence nano sensor for glucose sensing. *Mater. Sci. Eng. C*, 58, 103–109, 2016.

202. Wong, X.Y., Quesada-González, D., Manickam, S., Muthoosamy, K., Fluorescence "turn-off/turn-on" biosensing of metal ions by gold nanoclusters, folic acid and reduced graphene oxide. *Anal. Chim. Acta*, 1175, 338745, 2021.
203. Wong, X.Y., Quesada-González, D., Manickam, S., New, S.Y., Muthoosamy, K., Merkoçi, A., Integrating gold nanoclusters, folic acid and reduced graphene oxide for nanosensing of glutathione based on "turn-off" fluorescence. *Sci. Rep.*, 11, 2375, 2021.
204. Liu, X., Wang, X., Han, Q., Qi, C., Wang, C., Yang, R., Facile synthesis of $IrO_2$/rGO nanocomposites with high peroxidase-like activity for sensitive colorimetric detection of low weight biothiols. *Talanta*, 203, 227–234, 2019.
205. Keerthana, S., Rajapriya, A., Viswanathan, C., Ponpandian, N., Enzyme like-colorimetric sensing of $H_2O_2$ based on intrinsic peroxidase mimic activity of $WS_2$ nanosheets anchored reduced graphene oxide. *J. Alloys Compd.*, 889, 161669, 2021.
206. Ahmad, R., Tripathy, N., Khosla, A., Khan, M., Mishra, P., Ansari, W.A., Syed, M.A., Hahn, Y.-B., Review—Recent advances in nanostructured graphitic carbon nitride as a sensing material for heavy metal ions. *J. Electrochem. Soc.*, 167, 037519, 2020.
207. Wang, X., Qin, L., Lin, M., Xing, H., Wei, H., Fluorescent graphitic carbon nitride-based nanozymes with peroxidase-like activities for ratiometric biosensing. *Anal. Chem.*, 91, 16, 10648–10656, 2019.
208. Zhang, H., Huang, Y., Hu, S., Huang, Q., Wei, C., Zhang, W., Kang, L., Huang, Z., Hao, A., Fluorescent probes for "off–on" sensitive and selective detection of mercury ions and L-cysteine based on graphitic carbon nitride nanosheets. *J. Mater. Chem. C*, 3, 2093–2100, 2015.
209. Xiang, M.-H., Liu, J.-W., Li, N., Tang, H., Yu, R.-Q., Jiang, J.-H., Fluorescent graphitic carbon nitride nanosheet biosensor for highly sensitive, label-free detection of alkaline phosphatase. *Nanoscale*, 8, 4727–4732, 2016.
210. Zhang, H., Huang, Y., Lin, X., Lu, F., Zhang, Z., Hu, Z., Lanthanum loaded graphitic carbon nitride nanosheets for highly sensitive and selective fluorescent detection of iron ions. *Sens. Actuators B: Chem.*, 255, 2218–2222, 2017.
211. Liu, J.-W., Luo, Y., Wang, Y.-M., Duan, L.-Y., Jiang, J.-H., Yu, R.-Q., Graphitic carbon nitride nanosheets-based ratiometric fluorescent probe for highly sensitive detection of $H_2O_2$ and glucose. *ACS Appl. Mater. Interfaces*, 8, 49, 33439–33445, 2016.
212. Luo, Y., Zhang, L., Liu, W., Yu, Y., Tian, Y., A single biosensor for evaluating the levels of copper ion and L-cysteine in a live rat brain with Alzheimer's disease. *Angew. Chem. Int. Ed. Engl.*, 54, 14053–14056, 2015.
213. Kuang, Y., Shi, J., Li, J., Yuan, D., Alberti, K.A., Xu, Q.B., Xu, B., Pericellular hydrogel/nanonets inhibit cancer cells. *Angew. Chem. Int. Ed. Engl.*, 53, 31, 8104–8107, 2014.
214. Zhang, C., Cui, Y., Song, L., Liu, X., Hu, Z., Microwave assisted one-pot synthesis of graphene quantum dots as highly sensitive fluorescent probes for detection of iron ions and pHvalue. *Talanta*, 150, 54–60, 2016.
215. Giorgio, M., Trinei, M., Migliaccio, E., Pelicci, P.G., Hydrogen peroxide: A metabolic by-product or a common mediator of ageing signals? *Nat. Rev. Mol. Cell Biol.*, 8, 9, 722–728, 2007.
216. Wang, Y.-M., Liu, J.-W., Adkins, G.B., Shen, W., Trinh, M.P., Duan, L.-Y., Jiang, J.-H., Zhong, W., Enhancement of the intrinsic peroxidase-like activity of graphitic carbon nitride nanosheets by ssDNAs and its application for detection of exosomes. *Anal. Chem.*, 89, 22, 12327–12333, 2017.
217. Jia, Y., Yi, X., Li, Z., Zhang, L., Yu, B., Zhang, J., Wang, X., Jia, X., Recent advance in biosensing applications based on two-dimensional transition metal oxide nanomaterials. *Talanta*, 219, 121308, 2020.
218. Lu, H. and Xu, S., CDs–$MnO_2$–TPPS ternary system for ratiometric fluorescence detection of ascorbic acid and alkaline phosphatase. *ACS Omega*, 6, 25, 16565–16572, 2021.

219. Yi, L., Li, H., Sun, L., Liu, L., Zhang, C., Xi, Z., A highly sensitive fluorescence probe for fast thiol-quantification assay of glutathione reductase. *Angew. Chem. Int. Ed. Engl.*, 48, 22, 4034–4037, 2009.
220. Peng, C., Xing, H., Fan, X., Xue, Y., Li, J., Wang, E., Glutathione regulated inner filter effect of MnO2 nanosheets on boron nitride quantum dots for sensitive assay. *Anal. Chem.*, 91, 9, 5762–5767, 2019.
221. Fan, D., Shang, C., Gu, W., Wang, E., Dong, S., Introducing ratiometric fluorescence to MnO2 nanosheet based biosensing: A simple, label-free ratiometric fluorescent sensor programmed by cascade logic circuit for ultrasensitive GSH detection. *ACS Appl. Mater. Interfaces*, 9, 31, 25870–25877, 2017.
222. Li, H., Wang, Z., Zhao, J., Guan, Y., Liu, Y., Dual colorimetric and ratiometric fluorescent responses for the determination of glutathione based on fluorescence quenching and oxidase-like activity of MnO2 nanosheets. *ACS Sustain. Chem. Eng.*, 8, 43, 16136–16142, 2020.
223. Kawaguchi, M., Hanaoka, K., Komatsu, T., Terai, T., Nagano, T., Development of a highly selective fluorescence probe for alkaline phosphatase. *Bioorg. Med. Chem. Lett.*, 21, 17, 5088–5091, 2011.
224. Xiao, T., Sun, J., Zhao, J., Wang, S., Liu, G., Yang, X., FRET effect between fluorescent polydopamine nanoparticles and MnO2 nanosheets and its application for sensitive sensing of alkaline phosphatase. *ACS Appl. Mater. Interfaces*, 10, 7, 6560–6569, 2018.
225. Liu, G., Zhao, J., Yan, M., Zhu, S., Dou, W., Sun, J., Yang, X., *In situ* formation of fluorescent silicon-containing polymer dots for alkaline phosphatase activity detection and immunoassay. *Sci. China Chem.*, 63, 554–560, 2020.
226. Kim, C.R., Uemura, T., Kitagawa, S., Inorganic nanoparticles in porous coordination polymers. *Chem. Soc. Rev.*, 45, 3828–3845, 2016.
227. Samanta, P., Desai, A.V., Sharma, S., Chandra, P., Ghosh, S.K., Selective recognition of $Hg^{2+}$ ion in water by a functionalized metal–organic framework (MOF) based chemo dosimeter. *Inorg. Chem.*, 57, 5, 2360–2364, 2018.
228. Yan, B., Lanthanide-functionalized metal–organic framework hybrid systems to create multiple luminescent centers for chemical sensing. *Acc. Chem. Res.*, 50, 11, 2789–2798, 2017.
229. Goggins, S., Naz, C., Marsh, B.J., Frost, C.G., Ratiometric electrochemical detection of alkaline phosphatase. *Chem. Commun.*, 51, 561–564, 2015.
230. Wang, X., Jiang, X., Wei, H., Phosphates-responsive 2D-metal—Organic-framework-nanozymes for colorimetric detection of alkaline phosphatase. *J. Mater. Chem. B*, 8, 6905–6911, 2020.
231. Jr., G.J.D., Carcamo, J.M., Bórquez-Ojeda, O., Shelton, C.C., Golde, D.W., Hydrogen peroxide generated extracellularly by receptor-ligand interaction facilitates cell signaling. *Proc. Natl. Acad. Sci. U.S.A.*, 102, 5044–5049, 2005.
232. Yuan, A., Lu, Y., Zhang, X., Chen, Q., Huang, Y., Two-dimensional iron MOF nanosheet as a highly efficient nanozyme for glucose biosensing. *J. Mater. Chem. B*, 8, 9295–9303, 2020.
233. Chen, J., Shu, Y., Li, H., Xu, Q., Hu, X., Nickel metal-organic framework 2D nanosheets with enhanced peroxidase nanozyme activity for colorimetric detection of $H_2O_2$. *Talanta*, 189, 254–261, 2018.
234. Wan, H., Wang, Y., Chen, J., Meng, H.-M., Li, Z., 2D Co-MOF nanosheet-based nanozyme with ultrahigh peroxidase catalytic activity for detection of biomolecules in human serum samples. *Microchim. Acta*, 188, 130, 2021.
235. Ning, D., Liu, Q., Wang, Q., Du, X.-M., Ruan, W.-J., Li, Y., Luminescent MOF nanosheets for enzyme assisted detection of $H_2O_2$ and glucose and activity assay of glucose oxidase. *Sens. Actuators B: Chem.*, 282, 443–448, 2019.

236. Shende, P., Kasture, P., Gaud, R.S., Nanoflowers: The future trend of nanotechnology for multi-applications. *Artif. Cells Nanomed. Biotechnol.*, 46, 413–422, 2018.
237. Dube, S. and Rawtani, D., Understanding intricacies of bioinspired organic-inorganic hybrid nanoflowers: A quest to achieve enhanced biomolecules immobilization for biocatalytic, biosensing and bioremediation applications. *Adv. Colloid Interface Sci.*, 295, 102484, 2021.
238. Le, X.A., Le, T.N., Kim, M., Dual-functional peroxidase-copper phosphate hybrid nanoflowers for sensitive detection of biological thiols. *Int. J. Mol. Sci.*, 23, 366, 2022.
239. Tang, S., Yu, C., Qian, L., Zhou, C., Zhen, Z., Liu, B., Cheng, X., Cheng, R., Nitrogen-doped graphene quantum dots-MoS2 nanoflowers as a fluorescence sensor with an off/on switch for intracellular glutathione detection and fabrication of molecular logic gates. *Microchem. J.*, 171, 106786, 2021.
240. Tran, T.D., Nguyen, P.T., Le, T.N., Kim, M. I.I., DNA-copper hybrid nanoflowers as efficient laccase mimics for colorimetric detection of phenolic compounds in paper microfluidic devices. *Biosens. Bioelectron.*, 182, 113187, 2021.
241. Lu, M., He, Q., Zhong, Y., Pan, J., Lao, Z., Lin, M., Wang, T., Cui, X., Ding, J., Zhao, S., An ultrasensitive colorimetric assay based on a multi-amplification strategy employing Pt/$IrO_2$@SA@HRP nanoflowers for the detection of progesterone in saliva samples. *Anal. Methods*, 13, 1164–1171, 2021.
242. Gao, J., Liu, H., Pang, L., Guo, K., Li, J., A biocatalyst and colorimetric/fluorescent dual biosensors of $H_2O_2$ constructed via hemoglobin- $Cu_3(PO_4)_2$ organic/inorganic hybrid nanoflowers. *ACS Appl. Mater. Interfaces*, 10, 30441–30450, 2018.
243. Sun, J., Li, C., Qi, Y., Guo, S., Xue, L., Optimizing colorimetric assay based on V2O5 nanozymes for sensitive detection of $H_2O_2$ and glucose. *Sensors*, 16, 584, 2016.
244. Guo, J., Wang, Y., Zhao, M., A self-activated nanobiocatalytic cascade system based on an enzyme-inorganic hybrid nanoflower for colorimetric and visual detection of glucose in human serum. *Sens. Actuators B: Chem.*, 284, 45–54, 2019.
245. Ballatori, N., Krance, S.M., Notenboom, S., Shi, S., Tieu, K., Hammond, C.L., Glutathione dysregulation and the etiology and progression of human diseases. *Biol. Chem.*, 390, 191–214, 2009.
246. Mate, D.M. and Alcalde, M., Laccase engineering: From rational design to directed evolution. *Biotechnol. Adv.*, 33, 1, 25–40, 2015.
247. Osler, M., Wium-Andersen, M.K., Wium-Andersen, I.K., Gronemann, F.H., Jorgensen, M.B., Rozing, M.P., Incidence of suicidal behaviour and violent crime following antidepressant medication: A Danish cohort study. *Acta Psychiatr. Scand.*, 140, 522–531, 2019.
248. Yuan, S., Feng, L., Wang, K., Pang, J., Bosch, M., Lollar, C., Sun, Y., Qin, J., Yang, X., Zhang, P., Wang, Q., Zou, L., Zhang, Y., Zhang, L., Fang, Y., Li, J., Zhou, H.-C., Stable metal–organic frameworks: Design, synthesis, and applications. *Adv. Mater.*, 37, 1704303, 2018.
249. Karmakar, A., Samanta, P., Dutta, S., Ghosh, S.K., Fluorescent "*turn-on*" sensing based on metal–organic frameworks (MOFs). *Chem. Asian J.*, 14, 4506–4519, 2019.
250. Wang, D., Li, Z., Zhao, Q., Zhang, J., Yang, G., Liu, H., Encapsulation of phosphomolybdate within metal–organic frameworks with dual enzyme-like activities for colorimetric detection of $H_2O_2$ and ascorbic acid. *J. Clust. Sci.*, 32, 1–9, 2021.
251. Zeng, Y.-N., Zheng, H.-Q., Gu, J.-F., Cao, G.-J., Zhuang, W.-E., Lin, J.-D., Cao, R., Lin, Z.-J., Dual-emissive metal–organic framework as a fluorescent "switch" for ratiometric sensing of hypochlorite and ascorbic acid. *Inorg. Chem.*, 58, 13360–13369, 2019.
252. Meng, H.-M., Shi, X., Chen, J., Gao, Y., Qu, L., Zhang, K., Zhang, X.-B., Li, Z., DNA amplifier-functionalized metal–organic frameworks for multiplexed detection and imaging of intracellular mRNA. *ACS Sens.*, 5, 103–109, 2020.

253. Wei, H. and Wang, E., $Fe_3O_4$ magnetic nanoparticles as peroxidase mimetics and their applications in $H_2O_2$ and glucose detection. *Anal. Chem.*, 80, 2250–2254, 2018.
254. Yang, H., Zhao, J., Qiu, M., Sun, P., Han, D., Niu, L., Cui, G., Hierarchical bi-continuous Pt decorated nano porous Au-Sn alloy on carbon fiber paper for ascorbic acid, dopamine and uric acid simultaneous sensing. *Biosens. Bioelectron.*, 124-125, 191–198, 2019.
255. Liu, L., Zhu, G., Zeng, W., Lv, B., Yi, Y., Highly sensitive and selective "off-on" fluorescent sensing platform for $ClO^-$ in water based on silicon quantum dots coupled with nanosilver. *Anal. Bioanal. Chem.*, 411, 1561–1568, 2019.
256. Schwarzenbach, H., Hoon, D.S., Pantel, K., Cell-free nucleic acids as biomarkers in cancer patients. *Nat. Rev. Cancer*, 11, 426–437, 2011.

# 14

# *In Vivo* Applications for Nanomaterials in Biosensors

**Abhinay Thakur[1] and Ashish Kumar[2]***

*[1]Department of Chemistry, Lovely Professional University, Phagwara, Punjab, India*
*[2]NCE, Bihar Engineering University, Department of Science and Technology, Government of Bihar, Bihar, India*

***Abstract***
Owing to the inherent link between human health and medical science, the rapid detection and monitoring of several lethal heavy metals, malignant cells, etc., requires more consideration. Currently, the integration of nanotechnology (such as functionalized nanoparticles (NPs), nanowires, nanotubes and nanofibers) with biosensors has greatly evolved from the *in vitro* studies offering rapid, simple, selective and sensitive detection methods for the analysis of heavy metals, toxins, and malignant cells. Nanomaterials (NMs) can enhance the surface of the sensors, transducing regions, increasing catalytic behaviors with the emergence of numerous innovative signal transduction technologies in biosensors. They offer a high surface-to-volume proportion, as well as a controlled geometry and architecture that encourages miniaturization, which is a useful feature when specimen volume is an issue. This chapter explores present and prospective breakthroughs in nanotechnology-based biosensors for real time assessment of targeted several analytes and toxicity mechanisms in living creatures, using primary data sets from 2018 onwards. Innovative biosensing technologies centered on unique sensing components and transduction concepts will receive special attention. Opportunities and future considerations for the utilization of NM-based biosensors for enhanced environmental and food-sensing devices will also be discussed.

***Keywords:*** Biosensors, nanomaterials, electrochemical sensor, cancer, detection, monitoring, treatment, *in vivo*

## 14.1 Introduction

As the median standard of living advances, human life expectancies will tend to climb. Several studies highlighting the large rise in the aging demographic, particularly in wealthy nations, provide support for advancements in medication and therapy. As expected, the worldwide community of persons aged 60 years and more might increase by 250% by 2050 as contrasted to 2013 [1–3]. Similarly, with the aging of society, enduring healthcare costs are expected to rise. The use of Assistive Health Technology (AHT) is increasing in an effort

**Corresponding author*: drashishchemlpu@gmail.com

Inamuddin and Tariq Altalhi (eds.) Biosensors Nanotechnology, 2nd Edition, (327–346) © 2023 Scrivener Publishing LLC

to ensure a strong aging population. As a result, significant attempts are being undertaken to achieve higher performance standards in healthcare organizations. Significant technology advancements will undoubtedly transform research in a range of subjects in the twenty-first century, including human wellbeing. Novel methods for tracking the health of individuals, behavior, and activities will be possible. Combination therapy is a critical element of health and well-being, with several researches demonstrating the significance of proper medication administration. Optimal aging is difficult to achieve and necessitates several critical methods. Effective medicine is one of these tactics that is primarily concerned with the individual's behavior. Furthermore, maximum ailments are treated mostly with pharmaceuticals, as it is universally acknowledged. As a result, it is necessary for the person to consume the prescription as prescribed by the medical professional. Combination therapy, on the other hand, continues to be a problem in the healthcare industry, particularly among the elderly. In fact, more than half of all elderly adults have numerous chronic conditions. Consequently, regular assessment and evaluation of an individual's compliance are critical to improving their health results. This must be done utilizing precise evaluation methodologies in order to be effective [4–7].

Numerous nanomaterials (NMs) have been manufactured for diverse purposes due to the tremendous growth of nanotechnology. Besides size, nanoparticles (NPs) are categorized according to their physical properties, such as electric charges; chemical properties, including the structure of the NP shell or core; form (cylinders, films, coils, etc.) [8–10]. The key obstacle restricting their application in the therapy and diagnostics of disorders is NP cytotoxicity for human beings. Currently, scientists are frequently confronted with the issue of toxicology and its adverse impacts. The selection of an appropriate experimental paradigm for predicting toxicity *in vitro* (cell lines) and *in vivo* (experimental animals) is critical in this regard. Just a thorough comprehension of the interconnections among all of the elements and pathways generating NP may lead to the development of benign, compatible NPs that could be employed for the detection and therapy of human diseases (Figure 14.1) [11].

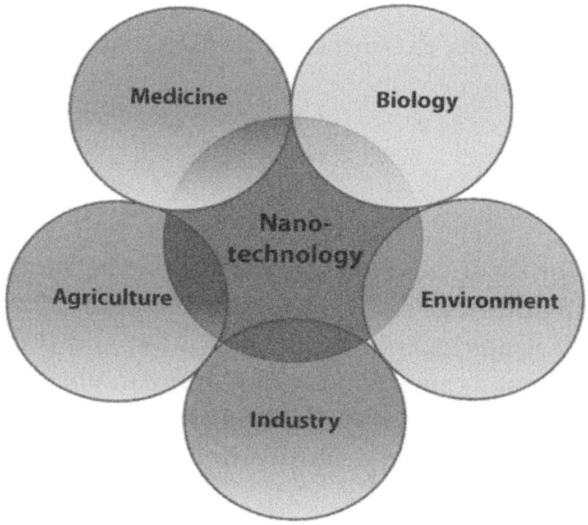


**Figure 14.1** Nanotechnology and NMs revolutionizing progress in agriculture, medicine, environment, industry, and biology. (Reprinted with authorization from Ref. [11] (MDPI) (2018) This work is licensed under a Creative Commons Attribution Based License 4.0.)

Although photoluminescence, which is often utilized in biological functional nanocomposites, is applied to certain approaches, like the luminescence optical emission of certain NMs. However, a UV-vis spectrophotometer was used to investigate drug binding and release activities for numerous NMs, the inorganic substance comprised of metals, and a specific spectrum of light is absorbed by metal-oxide nanostructures. As a result, this spectrum reaction aids the research into the biological sensing of a variety of nanobased elements. In contrast, the advent of biosensors and wearable electrochemical sensing systems for the identification of diverse biological fluids has unlocked ample opportunities for the treatment as well as prophylaxis of patients with chronic diseases in subsequent decades. An electrochemical biosensor, according to the International Union of Pure and Applied Chemistry (IUPAC), is a self-embedded interconnected machine that employs a biological recognition component (i.e., biochemical receptor) in a specific spatial interaction with an electrochemical transduction component to provide specific quantitative or semi-quantitative analytical data. With the growing need for improved stability and specificity in medical investigations, significant attention has subsequently been dedicated to investigating biologically reactive NMs for extremely accurate biosensors and bioimaging [12–15]. A biosensor is an instrument or monitor that generates measurable outputs by blending a biological element, including an antibody or enzyme, to an electric component that further records save and communicates data on physiological alterations and the existence of a variety of physiological and biochemical components in the surroundings. Biosensors come in various sizes, and then they could recognize and quantify extremely small concentrations of pathogens, harmful substances, and hydrogen potential (pH) values.

(a) Analyte: A fascinating substance whose constituents are being discovered or recognized.
(b) Bioreceptor: Bioreceptors are biomolecules or biological components that may recognize the targeted substrate. Biorecognition is the word used to describe the signal production that occurs during the interface between the analyte and the bioreceptor.
(c) Transducer: An energy convertible is a device that transforms one type of energy into the other. A biosensor's transducer is an essential component. It converts a biorecognition phenomenon into an electronic current that signals the presence of a biological and chemical substrate or reflects a quantity. Signalization is the name for this type of energy transformation. The quantity of analytes–bioreceptor associations is determined by the quantity of optical or electrical signals generated by transducers [16–19].
(d) Electronics: The attained signal is analyzed and modified for project once it has been transduced. The transducer's electric signals are amplified and converted to a digital representation. The signals that have been analyzed are quantified by the display unit.
(e) Display: A user interpretation equipment, this may comprise a computer or laptop, provides the output so that the individual may receive and comprehend the appropriate responses. Based on the user demand, the attained data may be in the format of numeric, topographical, data analysis, or visual (Figure 14.2) [20].

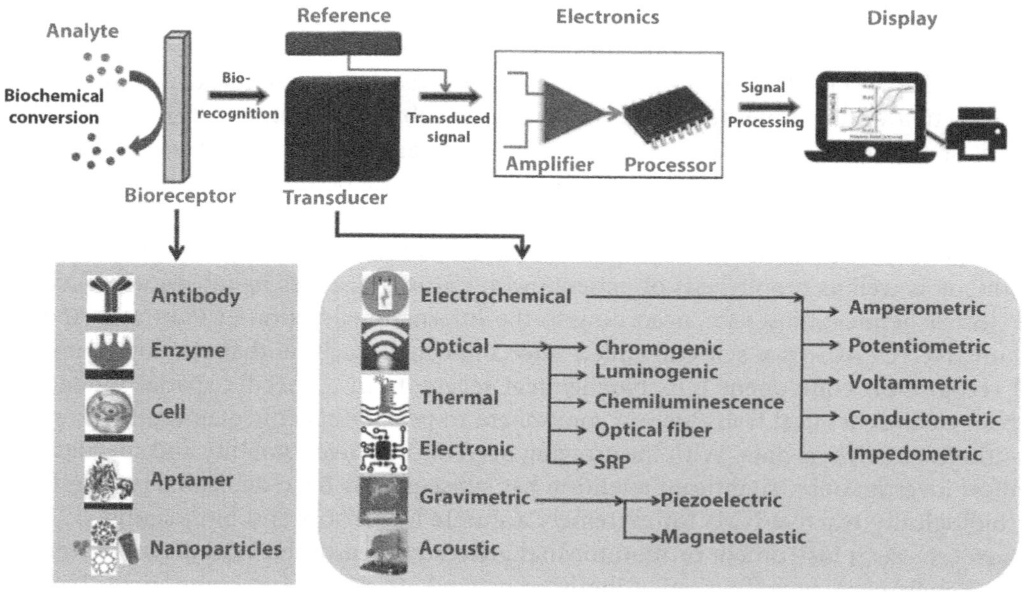


**Figure 14.2** Representation of a typical biosensor, which includes a bioreceptor, a transducer, an electrical circuit, and a screen, as well as numerous types of bioreceptors used in biosensors. (Reprinted with authorization from Ref. [20]) (MDPI) (2021). This work is licensed under a Creative Commons Attribution Based License 4.0.)

For instance, in an investigation, Wang *et al.* [21] evaluated the *in vivo* toxicity of SiNPs using *Caenorhabditis elegans* (*C. elegans*), a basic though physiologically and morphologically well-defined paradigm in living creatures. The investigation was carried out at the individual, sub-cellular, cytoplasmic, as well as molecular-scale levels. They glanced that SiNPs affected *C. elegans* aging, metal detoxification, apoptosis, and hypoxia reaction by ($N \geq 200$) whereas, mitochondrial strain, endoplasmic reticulum (ER) stress, autoimmune reactions and oxidative strain by ($N \geq 20$). Also it affects the body length ($N \geq 75$), and reproductive ability ($N \geq 10$). Even though the worms were administered a significant dosage of SiNPs (e.g., 50 mg/mL) during the developmental phases, the investigations revealed that SiNPs had no meaningful influence on growth, longevity, or reproduction capability ($p > 0.05$). The introduction of SiNPs did not disrupt the intracellular activities of the *C. elegans* gut ($p > 0.05$), according to the subcellular examination of the worms treated with SiNP. Toxicity tests showed that the SiNPs did not cause any detrimental or protective cellular processes, like ER strain, mitochondrial strain, or oxidative strain ($p > 0.05$). The results obtained revealed that SiNPs are biodegradable and minimal in cytotoxicity, implying that the substance is a suitable fluorescent nanoprobe for a broad series of physiological and medical purposes.

Similarly, Choi *et al.* [22] developed a biodegradable silk fibroin membrane with excellent permeability and gold (Au) micro-fractures as a very sensitive SERS sensor. Their silk-based SERS sensing took use of substantial local field amplification caused by Au nanostructures vaporized on a permeable silk substrates in nanoscale fracture areas. The suggested sensor's SERS efficiency was assessed in respect of limit of detection (LOD), sensitivities, and predictability. The surface topological characteristics of the manufactured

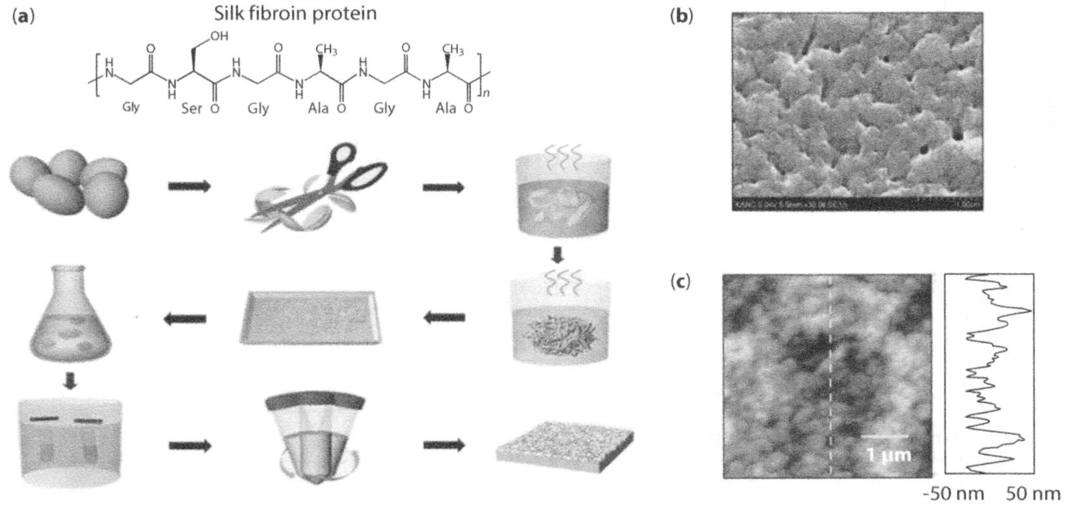


**Figure 14.3** SERS sensor implantation on an extremely permeable free-standing silk substrate. (a) Methods for making a permeable silk surface. (b) FE-SEM picture of the high-porosity manufactured silk specimen. (c) AFM picture of silk base and interface contour. (reprinted with authorization from Ref. [22] (MDPI) (2021) This work is licensed under a Creative Commons Attribution Based License 4.0.)

permeable silk surface were conformed using atomic force microscope (AFM) and field emission scanning electron microscopy (FE-SEM) pictures in Figure 14.3 [22]. A quite permeable surface pattern in Figure 14.3a was effectively generated via a simple yet efficient chemical approach shown in Figure 14.3b. The AFM results in Figure 14.3c vividly show the coarse and permeable silk texture. The silk surface was found to have extremely coarse surface characteristics, with a mean roughness of 38.55 nm, whereas the findings are not revealed the silicon wafer's surface irregularity that was as low as 0.15 nm. In the inclusion of Au nanostructures, coarse and permeable silk properties may lead to the stimulation of localized plasmons, resulting in increased Raman scattering.

SERS findings utilizing 4-ABT analytes show a considerable increase in the limit of detection and sensitivities in contrast of the efficiency of a corresponding SERS sensor having a Au foil sheet, as well as high predictability and a measurement range. More intriguingly, the overlapping was crucial, and a quantifiable assessment of the local range augmentation matches the empirical SERS increase quite well.

Similarly, Noh *et al.* [23] with the purpose of evolving an extremely delicate *in vivo* biosensor for glutathione disulfide (GSSG), GR (glutathione reductase) which is covalently immobilized, was used along with Au NPs and NADPH (b-nicotinamide adenine dinucleotide phosphate) accumulated over the poly[2-terthiophene-30-(p-benzoic acid)]. In aspects of electric voltage, pH, temperature and also, NADPH: GR ratio, analytical factors impacting biosensor efficiency were optimized. In the frequency response of 0.1 mM and 2.5 mM of GSSG, a standard curve graph was achieved utilizing chronoamperometry, having a LOD of 12.5 to 0.5 nM. The biosensor that was created was used to recognize GSSG in an actual plasma specimen. To measure the oxidative strain generated by diquat and t-butyl hydroperoxide, a micro biosensor was employed to measure the *in vivo* GSSG levels. The findings were consistent, suggesting that a GSSG biosensor could be useful in medical diagnosis

**Table 14.1** Several nanomaterials used in biomaterials.

| S. no. | Nanomaterial | Treatment/utilization | Reference |
|---|---|---|---|
| 1. | Fluorescent silicon nanoparticles | Immune response | [1] |
| 2. | Iron oxide NPs | Drug nanocarriers | [2] |
| 3. | Solid lipid nanocarriers | Neurodegenerative diseases and skin disorders | [3] |
| 4. | Gold nanoparticles | Heavy metal ions | [4] |
| 5. | Au DENPs-MPC | Cell-based immunotherapy | [5] |
| 6. | Fluorescent gold nanoclusters | Cancer | [6] |
| 7. | Iron oxide stealth nanoparticles | Medical imaging | [7] |
| 8. | Silver nanoparticles | Non-cytotoxic against fibroblast cells | [8] |
| 9. | Enzymeless porous nanomaterial (pNM)-modified microneedle electrode | Blood glucose monitoring | [9] |
| 10. | Gold-coated iron oxide nanoparticles capped with polyvinylpyrrolidone (Fe@Au) | Magnetic resonance imaging (MRI) and X-ray computed tomography (CT) | [10] |
| 11. | Phyto-synthesized silver nanoparticles | Anti-diabetic | [11] |
| 12. | Bioelectrochemically active infinite coordination polymer (ICP) nanoparticles | Electrochemical biosensors | [12] |

and treatment and oxidative stress surveillance. Table 14.1 shows the several nanomaterials used in biomaterials.

## 14.2 Types of NM-Based Biosensors

### 14.2.1 Fluorescent NM-Based Biosensors

Significant advancements in the use of fluorescent NMs in biological research have combined nanobiotechnology with cutting-edge imaging capabilities. Fluorescent NMs, like quantum dots (QDs), up-conversion nanophosphors, carbon, and silica, are being used in biological and medical research as fluorescent probes [24–26]. In an experiment, Bai *et al.* [27] discussed the physiochemical features of AuNCs, as well as the procedures for manufacturing them. They then began discussing AuNCs-based construction methodologies for detecting biomarkers such as tiny compounds, proteins and DNA. In addition, the usage of AuNCs for *in vivo* and *in vitro* tumor imaging were reviewed. Fluorescent AuNCs have a lot of possibilities for medical diagnostics like analytical testing and cell imaging because of their wide Stokes shifting, strong fluorescence, customizable outputs, outstanding durability, and excellent biocompatibility. Despite the fact that tremendous development has been made in the latest generations, there are still problems to be overcome and considerable attempts to be made. To begin with, some AuNCs possess poor fluorescence quantum outputs that exert a substantial impact on the analyte's LODs. Second, the architectures of AuNCs are often unknown, and novel techniques for determining their composition must be developed. Finally, the metabolism kinetics of AuNCs and their cytotoxicity should be investigated further into order to do bio-imaging *in vivo*. These fluorescent AuNCs have indeed demonstrated their possibilities as excellent luminescent probes for biosensors and cell imaging, but there are still a few difficulties to be addressed. Fluorescent SiNPs have been widely researched as potential possibilities for biodegradable fluorescent nanoparticles in recent times, the availing benefit of silicon's minimal cytotoxicity. They show significant potential for biosensors and bioimaging technologies. Fluorescence SiNPs, in particular, has been proven to be ideal for long-usage and cellular aiming, real-time imaging and transportation, fluorescence printing, bacterial infection diagnosis and photodynamic therapy, and cancer cell detection and treatment. Although progress has been made in comprehending the cytotoxicity of SiNPs in cultured cells and mammal experimental animals, this information remains scattered and contradictory. Additionally, Wang *et al.* [21] investigated the *in vivo* cytotoxicity of SiNPs of living creatures at various body functional levels using *Caenorhabditis elegans* (*C. elegans*), a well-determined paradigm. They glanced at how SiNPs affected *C. elegans* life expectancy ($N \geq 30$), body length ($N \geq 75$), reproductive ability ($N \geq 10$), endoplasmic reticulum (ER) stress ($N \geq 20$), endocytic grouping ($N \geq 20$), autoimmune reactions ($N \geq 20$), hypoxia reaction ($N \geq 200$), metal decontamination ($N \geq 200$) and ageing ($N \geq 200$). Investigations on *C. elegans* subjected to SiNPs followed carried out to learn more about their bioactivity, longevity, and clutch volume. Short-term (4 h) treatment to significant levels of SiNPs did not decrease longevity, as demonstrated in Figure 14.4a [21]. Figure 14.4b shows that worms subjected to SiNPs for 4 hours produced no more progeny than worms in the negative controls, showing that SiNPs had no toxic effect *in vivo*. Worms were fed 50 mg/mL SiNPs for the duration of the larval phase to examine the prolonged impact of SiNPs on lifespan and fertility. In comparison to the negative control group, they did not see any significant differences in longevity or progeny generation following SiNP administration (Figures 14.4c, d). The results show that the short-term or long-term retention of SiNPs in worm tissues had no effect on the species, demonstrating that SiNPs have minimal cytotoxicity and strong biocompatibility in live organisms.

Moreover, the manufacture of Au-coated Fe oxide NPs encapsulated with polyvinylpyrrolidone (Fe@Au NPs) was documented by Caro *et al.* [28]. The NPs were shown to be stable in aqueous conditions and to have outstanding properties due to contrast agents (CA) intended for X-ray computed tomography (X-ray CT). The NPs also displayed accessible "light-to-heat" conversions potential over the near-infrared range (NIR), a vital property of efficient photothermal treatments (PTT), owing to the inclusion of local surface plasmon resonances of Au. Biocompatibility and great efficacy in destroying glioblastoma cells through PTT were discovered *in vitro*. Zebrafish embryos were used as an intermediary stage across cells and rodent experiments to verify the NPs' *in vivo* nontoxicity. In mouse models, the intratumoral in addition to intravenous approaches were evaluated using MRI and CT to ensure that a successful treatment dosage was delivered inside the tumor.

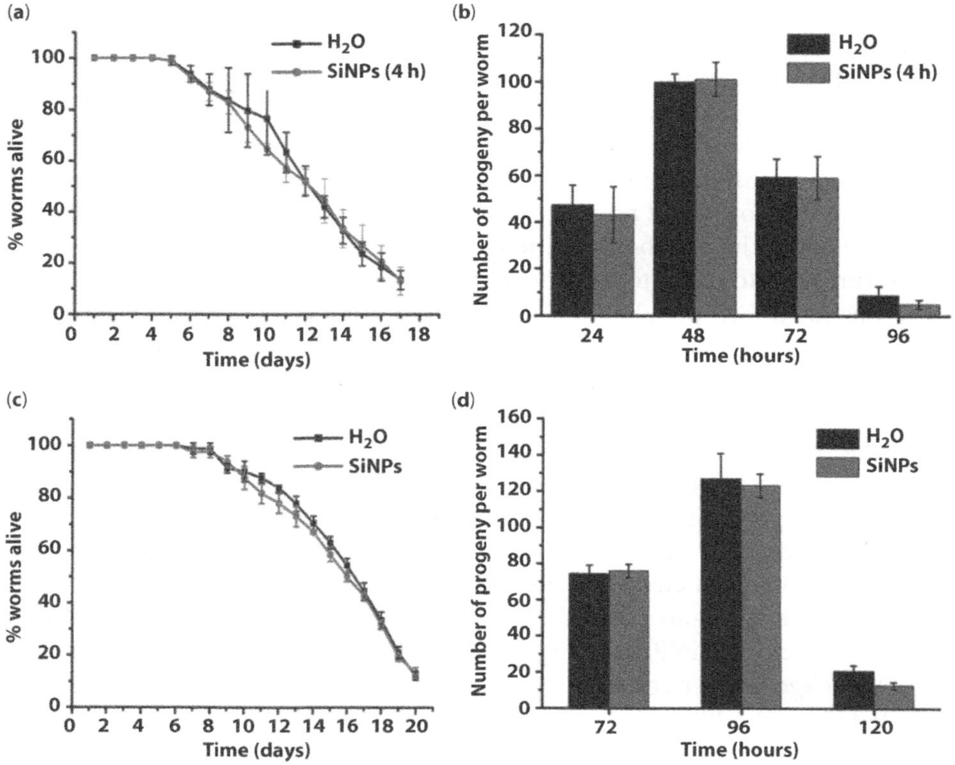


**Figure 14.4** Short- and long-term impacts of SiNPs on *C. elegans* longevity and fertility. *C. elegans* survival rates (a) and quantity of deposited eggs (b) after being given $H_2O$ or SiNPs for 4 hours from the L4 larval phase. From the L1 phase until the $H_2O$ control group attained the mature adult phase, the survival rates (c) and quantity of deposited eggs (d) of *C. elegans* were continually administered with $H_2O$ or SiNPs. All procedures were carried out three times separately (three biological replicates, N 30/condition for data in (a, c), N 10/condition for data in (d)) (b, d). (Reprinted with authorization from Ref. [21] (MDPI) (2022). This work is licensed under a Creative Commons Attribution Based License 4.0.)

The Fe@AuNPs' multimodal imaging CA abilities were verified by pharmacokinetics and bioavailability, which highlighted limitations of the intravenous administration for targeting the tumor, requiring the former approach for therapeutic potential. Furthermore, Fe@ Au NPs were effectively employed in a mouse prototype of glioblastoma multiforme for *in vivo* prototype of imaging targeted PTT.

Likewise, Liu *et al.* [29] presented the fabrication of Au nanoclusters that are stabilized using trypsin (try-AuNCs) having fluorescence near-infrared region to biosense folic acid (FA) and heparin altered try-AuNCs employed *in vivo* cancer bioimaging using surface plasmon-enhanced energy transfer (SPEET). The SPEET/try-AuNCs fluorescence biosensor was created by transferring energy across try-AuNCs and cysteamine altered Au NPs via heparin (cyst-AuNPs). Having a linear spectrum of 0.14.0 g $mL^{-1}$ and a LOD (3s) of 0.05 g $mL^{-1}$, the designed SPEET/try-AuNCs fluorescence biosensor permitted sensitivity and selectivity monitoring of heparin. The relative standard for eleven replicate detection techniques of 2.5 g $mL^{-1}$ heparin was 1.1%, while the spiking heparin recovery in human serum specimens varied from 96.9 to 100%. Additionally, folic acid was deposited over the interface of try-AuNCs for enhancing the selectivity of AuNCs for tumors, and near-infrared fluorescent FA-try-AuNCs were employed for *in vivo* cancer imaging of a HeLa tumor with high folate receptor expression (FR). An *in vivo* examination of the static activity and targeted capabilities of FA-try-AuNCs sensor in HeLa tumor-bearing mice and healthy hairless mice demonstrated the probe's comparatively high sensitivity for FR positive tumors. The results demonstrate that try-AuNCs generated in this method have a wide range of applications as multifunctional nanomaterials for bioimaging biomolecules in the SPEET method and *in vivo* malignancy imaging exhibiting high tailoring capacity.

Song *et al.* [30] proposed a generic meso-functionalization technique for soft substances. This was for the purpose of obtaining silk fibroin (SF) materials with additional functionalities, such as *in vivo* bioimaging and sensing. On the mesoscopic level, the upconversion fluorescence emission was achieved by 3D nanomaterials assembly of multifunctional components, lanthanide(Ln)-doped upconversion NPs (UCNPs). UCNPs' interfaces were modified by hydroxyl groups (-OH) from polyethylene glycol or $SiO_2$ coatings, which could interact to carbonyl groups (C-O) in SF frameworks to accomplish meso-functionalization. The functionalized silk scaffolds were then implanted beneath the skin of mice, allowing them to conduct *in vivo* bioimaging and similar biological activities. This method of molecular functionalization could open the way for the development of more general biomaterials.

### 14.2.2 Magnetic NM-Based Biosensors

Certain NPs and other particles exhibit magnetic properties, which is now well recognized. In fact, the magnetic properties of NMs are linked not only to their structure, although even to their size [17, 22, 31–34]. As a result of their shape, particles and NPs have magnetism and will respond differently in a magnetic field. In magnetic activities, bulk substances are particles having a mean diameter of up to 100 nm, while NPs are particles with an average diameter of less than 100 nm. For instance, Chen *et al.* [35] studied the impact of Au DENP-MPC embedded oligodeoxynucleotides (ODN) of unmethylated cytosine guanine dinucleotide on T cell-mediated immunotherapy (CPG). They generated Au DENPs-MPC first, then assessed their ability to compact and transmit CpG-ODN into formed BMDCs, as well as the prospect of employing T cells driven by developed BMDCs to limit tumor cell proliferation. As indicated by a 43.12% to 47.35% in different interface maturation indicators, the Au DENPs-MPC reduced Au DENP toxicity and improved CpG-ODN transmission to BMDCs for maturity. According to transwell assays, *ex vivo* stimulated T cells to exhibit remarkable anti-cancer efficacy and may greatly inhibit tumor cell proliferation. These findings show that Au DENPs-MPC can improve their antigen-presenting capability to stimulate T cells, meaning that T cells-based immunotherapy controlled by Au DENPs-MPC preloaded with CpG-ODN could represent a highly appealing malignancy treatment. Additionally, Bhat *et al.* [36] used the biological animal model *Drosophila melanogaster* to examine the cytotoxicity of mesoporous carbon nanospheres generated using scrap onion peel. After 25 days of treatment, the mortality experiments at various dosages of carbon NPs revealed that they were non-toxic. There were no development or behavioral issues found. The neurological toxicity, as measured by acetylcholinesterase activities, was also unaffected. Gastrointestinal ingestion of mesoporous carbon NPs for 25 days had no apparent adverse impact on *Drosophila melanogaster*, according to lifespan, behavior, and biochemical studies. As a result, mesoporous carbon NPs made from scrap onion peel could be employed as medication carriers in a variety of disease models.

Xu *et al.* [37] produced cysteine-functionalized Au NPs (D-/L-Au NPs) by depositing D-/L-cysteine over surfaces of Au NPs to have excellent suppression of *Escherichia coli* (*E. coli*) *in vivo/vitro*. Mice with *E. coli* diseased in the colon has been given D-Au NPs to examine antibacterial activity in a complicated physiological condition (Figure 14.5a) [37]. *E. coli* was not found in the healthy subjects. In the group compared, mice afflicted with *E. coli* were given PBS. The positive regulation was given kanamycin utilized to cure largely Gram-negative bacteria infestations in clinical contexts. Body weight (wt.) fluctuations in mice throughout therapy were crucial markers of therapeutic success. Figure 14.5b shows that the body wt. of mice processed with D-Au NPs decreased slightly on the first day, then gradually retrieved and increased 2 days later; the body wt. of mice processed using kanamycin regained following 4 days, and the physical wt. of mice processed to PBS does not change markedly. Following therapy, the bacterial CFU within the small intestine, cecum

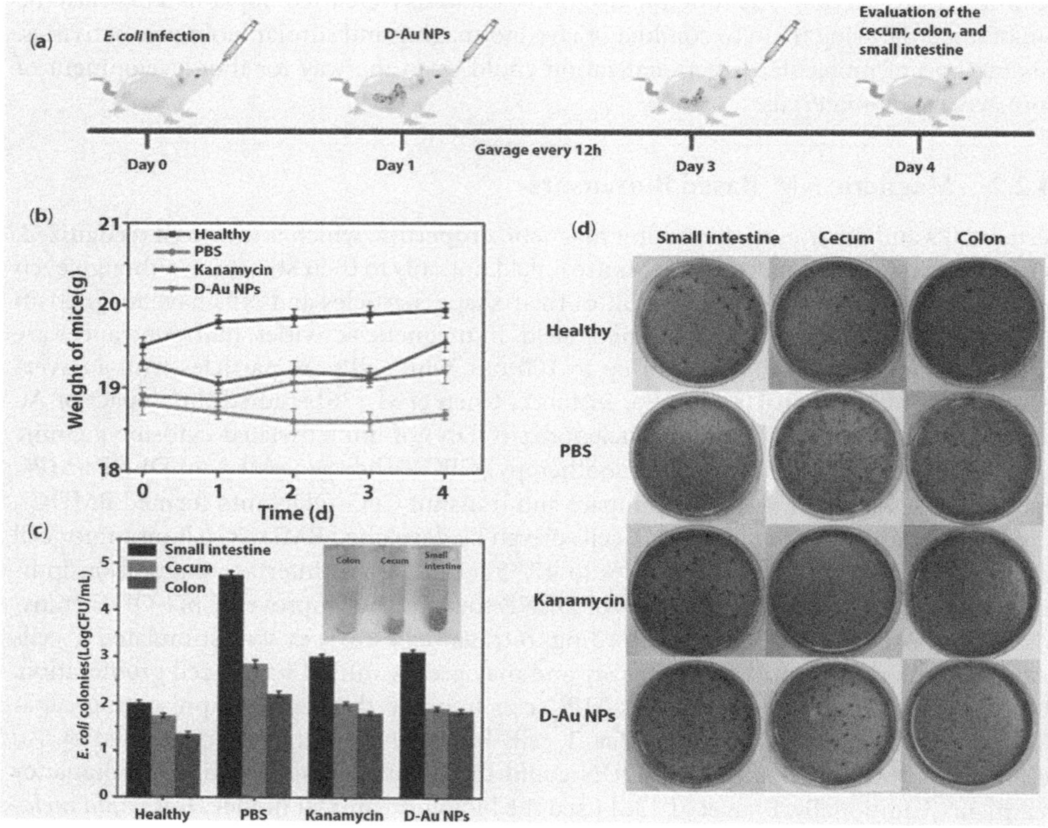


**Figure 14.5** (a) Conceptual representation of D-Au NPs-based in vivo treatment of *E. coli* infection. Mice were assessed following period of orogastric using PBS, D-Au NPs, and kanamycin. (b) Body wt. variations in mice following therapy. (c) Following various therapies, the *E. coli* remained in the intestines of diseased mice. (d) Following various therapies, agar plates exhibit bacterial colonies in the intestine (reprinted with authorization from Ref. [37] (MDPI) (2021). This work is licensed under a Creative Commons Attribution Based License 4.0.)

and colon in mice are shown in Figure 14.5c. According to the findings, treating *E. coli*–infected animals using D-Au NPs culminated in equivalent kanamycin impacts. Following diagnostic using D-Au NPs, the frequency of *E. coli* in minute specimens was reduced by much more about 1.5 magnitudes, comparable to the kanamycin class, and bacterial volume throughout the cecum and colon of mice was reduced to levels comparable to the healthful class (Figure 14.5d). In mice, D-Au NPs were efficient in treating *E. coli* infection.

They discovered that D-/L-Au NPs had antibacterial efficacy that was two to three times that of isolated L-cysteine, D-cysteine and Au NPs. The higher antibacterial efficacy was having by D-Au NPs than L-Au NPs, that was due to their distinctive steric structure. Chiral Au NPs had a greater damaging impact on cell membranes than the other groups, resulting in bacterial tissue demise and cytoplasm leakage. In the treatment of *E. coli* diseased mice, D-Au NPs displayed antibacterial activity comparable to kanamycin, but they had no effect on the proportion of intestinal microbiota. This research was critical in the creation of a viable chiral antibacterial weapon.

Haq *et al.* [38] intended to investigate if phyto-synthesized silver NPs (AgNPs) produced from *Phagnalon niveum* plant methanolic extract has anti-diabetic properties. Ultraviolet-Visible (UV-Vis) spectroscopy was used to examine the produced AgNPs derived using the methanolic extract of *P. niveum*, and Fourier Transform Infrared (FTIR) spectroscopy was utilized to recognize the functional moieties engaged in the diminution of the silver ions ($Ag^+$). XRD later on utilized to determine the dimensions and crystallinity. Uninfected healthy control rats were given simply distilled water ($H_2O$), diabetic rats were given glibenclamide 0.5 mg/kg physical wt. and diabetic regulated rats were given alloxan 200 mg/kg physical wt. DE: diabetic rats given 10 mg/kg body weight of methanolic extract of *P. niveum*; DAgNPs: diabetic rats administered 10 mg/kg body weight of *P. niveum* AgNPs. After dissecting at the completion of therapy (21 days), blood glucose quantity were checked over the periods 0, 7, and 14, and liver, lipid and renal characteristics were analyzed. On the completion of therapy (21 days), the participants were given an oral glucose tolerance test in which they were given 2 g/kg physical mass of glucose and their blood glucose levels were evaluated. A glucometer was used to determine the fasting glucose levels. Each animal's urine was obtained and examined for bilirubin, glucose, pH, nitrite and leukocytes among other things, using lab-made assay kits. A Dunnett's analysis and one-way ANOVA were utilized for quantitative investigation. Green-mediated AgNP production employing *P. niveum* methanolic extract resulted in mono-dispersed and spherical NPs varying in size from 14 to 26 nm (average, 21 nm). Considerable improvements in lipid, liver, and renal characteristics, and substantially reduced blood glucose quantities and an improvement in body wt., were seen. Biosynthesized AgNPs dramatically increased urine, body wt. and serum quantity, showing that it is a potential anti-diabetic drug.

### 14.2.3 Carbon Allotropes and Quantum Dots NM-Based Biosensors

Owing to its unique features, such as decent light endurance and $H_2O$ solubility, graphene quantum dots (GQDs) have gotten a lot of interest in recent years. GQDs are also notable for their ultra-small dimensions, extremely tunable photoluminescence, outstanding multiphoton excitation, excellent brightness, and chemical stability, among other things [2, 21, 26, 39–44]. Furthermore, different NMs, particularly GQDs and GQD-based NMs, have been shown to generate reactive oxygen species (ROS) in cells, suggesting that photodynamic therapy for cancer may be possible. In fluorescent imaging, image-guided surgery, and biosensing, GQDs have shown much more promising. When GQDs are used in biological applications, researchers pay particular consideration to their biocompatibility. When GQDs are co-incubated with diverse cells *in vitro*, they exhibit good bioactivity; *in vivo* investigations also reveal that GQDs are biocompatible. When quantities of N-doped graphene quantum dots are less than 200 g/mL, they have minimal toxicity. These findings point to GQDs and GQD-based materials having potential biological uses. In an investigation, Wang *et al.* [41] devised a simple one-pot approach for loading GQDs onto NPs, resulting in prolonged metabolic pathways in the blood and enhanced GQD transport to tumors. For their prospective usage in cancer therapy, optical and magneto ferric oxide@polypyrrole ($Fe_3O_4$@PPy) core-shell NPs which was considered. The *in vivo* data showed that adding GQDs allowed researchers to track NP dispersion and breakdown. This research revealed fresh information about the use of GQDs in lengthy-usage in specific *in vivo* monitoring. Before administration, a distinct circulatory framework was visible at the tumor region upon stimulation with 660 nm laser, as shown in Figure 14.6 [41]. Although certain compounds in the visible light range could evenly generate green fluorescence during visible-light excitation, a couple of green fluorescence lights were spotted in 488 nm. In order to monitor NPs *in vivo*, GQDs were used. Green fluorescence confirmed the existence of GQD-NPs 6 hours after injection, though the blood vessels around the tumor site remained visible (red). A huge quantity of GQD-NPs was generated within the tumor blood vessels, as seen in the merged image, while a minor number of GQD-NPs entered adjacent tissue across the tumor blood vessel membrane. The blood vessels in the tumor region remained vividly visible (red) after the GQD-NPs had been administered for 24 hours, and still showed numerous GQD-NPs in the tumor blood vessels (green). Because of the increased porosity and persistence (EPR) effect, a considerable quantity of GQD-NPs flowed by the tumor vascular wall and permeated inside the adjacent cells over time (from 6 to 24 hours), attaining enriched in the tumor tissues. The GQD-NPs fluorescence patterns with in tumor tissues and blood vessels were still present 48 hours after injection, however, the fluorescence patterns with in tumor tissues were greatly diminished (>24 h). This meant that a greater number of GQD-NPs had been digested. As a result, there was less GQD-NP elevation at the tumor location. The findings demonstrated that the dynamic changes of NPs might be followed *in vivo* by introducing GQDs into them.

Jin *et al.* [45] created a transdermal $H_2O_2$ electrochemical biosensor using MNs and a nanohybrid of reduced graphene oxide (RGO) and platinum NPs resulting in the formation of Pt/rGO. This blend increases the MN electrode's detecting capability, and the MNs were used as a noninvasive sublingual instrument to explore the *in vivo* surroundings. A $H_2O$-soluble polymer coating protects the Pt/rGO nanostructures from dynamic damage throughout the MN skin implantation procedure. The interstitial fluid easily dissolves the polymer layer, exposing the Pt/rGO over MNs enabling biosensing efficiently *in vivo*. Both pigskin and living mice were used to show the applicability of Pt/rGO-embedded MNs for real-time and in situ $H_2O_2$ screening *in vivo*. This research presents a novel real-time transdermal biosensing technology, that has the potential to be a useful technique for sensing *in vivo* with great sensitivities while remaining non-invasive.

Moreover, Lu *et al.* [46] established the development of a 3D conducting structure by combining single-walled carbon nanotubes (SWNTs) and infinite coordination polymer (ICP) NPs for *in vivo* electrochemical measurement and microdialysis. All biosensing components,

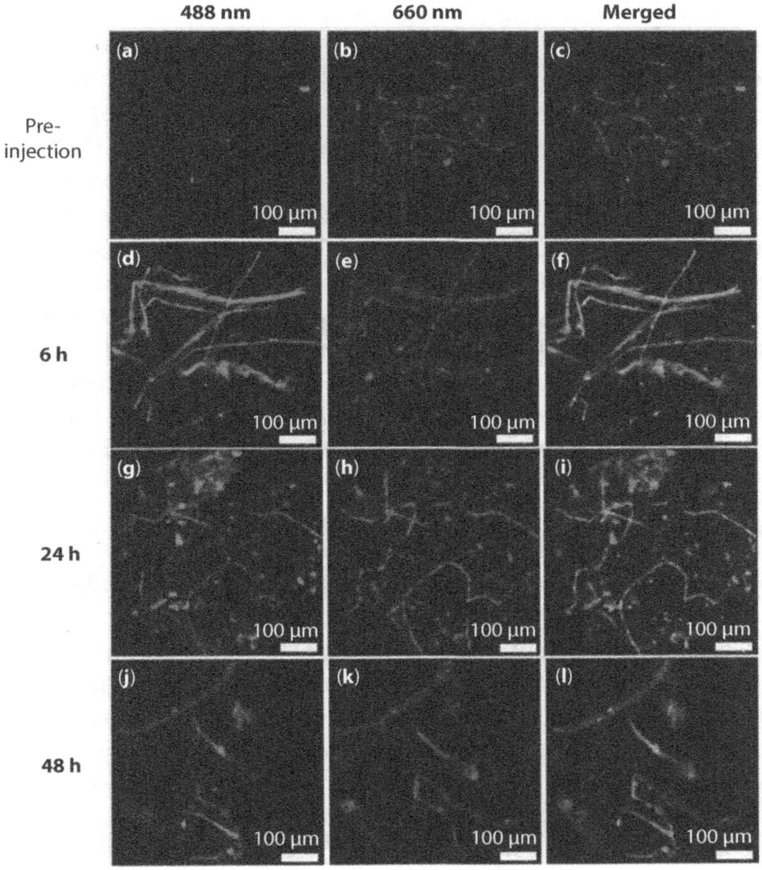


**Figure 14.6** GQD-NPs (photographed with 488 nm excitation), blood vessels (photographed by tail vein infusion of Evans blue at 660 nm excitation) in malignant tissues throughout the duration in fibered confocal fluorescence microscopy (FCFM) pictures (reprinted with authorization from Ref. [41] (MDPI) (2019) This work is licensed under a Creative Commons Attribution Based License 4.0.)

along with a cofactor (i.e., -nicotinamide adenine dinucleotide, $NAD^+$), enzyme (i.e., glucose dehydrogenase, GDH) and an electrocatalyst (i.e., methylene green, MG), were also flexibly encased in the bioelectrochemically active ICP NPs, that were produced via self-assembly. To produce ICP/SWNT-based biosensors, the as-synthesized ICP/SWNT nanocomposite was drop-coat over the vitreous carbon platform. Electrochemical experiments revealed that ICP/SWNT-based biosensors made merely have superior biosensing capabilities, including improved sensitivities and durability than ICP-based biosensors produced just using ICP NPs. They demonstrated a technologically simple but effective online electroanalytical framework for continuously monitoring glucose inside the brains of guinea pigs, using an ICP/SWNT-based nanosensor in a constant flow mechanism combined to an *in vivo* microdialysis, using a GDH-based electrochemical biosensor as an illustration. Under the conditions utilized here, the varied linear spectrum for glucose using the ICP/SWNT-biosensor ranged from 50 to 1000 μM. In addition, *in vivo* sensitivity investigations using GDH-free ICP-based biosensors revealed these biosensors to be especially sensitive for glucose detection in the brain region. In guinea pig striatal microdialysates, the baseline value of glucose was determined to be 0.31 0.03 μM (n = 3). The work described a straightforward method for manufacturing electrochemical biosensors, that could be beneficial for investigating the chemical processes occurring during several physiologic processes.

Li *et al.* [47], using the fluorescent "turn-on" aptasensor, built a Cy3-labeled ssDNA probe (P0-Cy3), pie-pie layered over the interface of oxidized mesoporous carbon nanospheres (OMCN). These OMCN-based aptasensor could not only discern mucin1 protein in the aqueous exhibiting a minimum LOD of 6.52 nmol/L, broad scope of 0.1 to 10.6 mol/L and high sensitivity, although it could also accurately measure malignant cells in the solvent having a LOD of 8500 cells/mL and linear range of $10^{-4}$ to $2\times10^{-6}$ cells/mL owing to the innate characteristics of OMCN. These aptasensor were employed to scan malignancy using solid tissues including cell, tissue sections, and *ex/in vivo* malignancies, highlighting the significant differences between malignancy and healthy tissues. It is a simple and reliable screening technology that can identify malignancy both *in vivo/vitro*.

### 14.2.4 Lipid NM-Based Biosensors

The benefits of lipid nanoparticles (LN) as medication delivery methods have been extensively studied. Elevated bioactivity, the capacity to pack hydrophilic and lipophilic substances, regulated drug release, and drug aiming are just a few of the characteristics that have urged scientists to look into the possibility of utilizing LN to produce a variety of active substances via various pathways of administering. Owing to their capacity to shield these compounds from breakdown and enhance their external bioavailability, SLN has received a lot of interest within the last year as transporters for cutaneous delivery of antioxidants. For instance, Montenegro *et al.* [44] used a molecule having well hydrating properties, pyroglutamic acid, to generate an IDE ester (IDEPCA); (2) IDEPCA was packed into lipid nanocarriers (LN). They compared *in vitro* antioxidant in addition to anti-glycation activities as well as *in vivo* hydration benefits of gel vehicles of IDEPCA LN to IDE SLN following topical administration in human volunteers. All of the SLN had excellent technical qualities (mean particle size less than 25 nm, polydispersity index less than 0.300, and high stability). IDEPCA LN and IDE LN displayed equal antioxidant capacity in the oxygen radical absorbance capability experiment, but IDEPCA LN was more efficient in the *in vitro* NO scavenge experiment. IDE LN and IDEPCA inhibited the production of intermediate glycation end products to the same extent. *In vivo* studies revealed that IDEPCA LN had a higher moisturizing impact than IDE LN. The outcomes stated that the method under investigation could be a possible option for developing external compositions with improved moisturizing properties. Fang *et al.* [48] presented an L-dopa biosensor that is minimally intrusive and is predicated on a flexible differential microneedle array (FDMA). L-Dopa and interference chemicals were detected by one working electrode, whereas electroactive external interference was detected by another. By removing common-mode interference, the distinct current responsiveness of listed electrodes had been correlated to the amount of L-Dopa. L-Dopa was processed by tyrosinase over the outermost surface of working elctrode and further oxidized to the dopaquinone upon *in vivo* monitoring. As a result, only electroactive interfering compounds elicited a response from working elctrode. L-Dopa and interferents may permeate via the outermost layer of working elctrode and enter the Au nanodendrites catalytic layer since there was no tyrosinase present on the interface. Under the given working potential, the working electrode may react to L-Dopa, also peripheral sites by electrochemically oxidizing peripheral sites and L-Dopa. The varied current sensitivity of such electrodes were contrasted to the quantity of L-Dopa through reducing basic-mode distortion. Underneath the presence of high-concentration interfering compounds, this differential technique efficiently handled the issues of low-concentration material identification as illustrated in Figure 14.7 [48].

The asymmetrical arrangement enhanced the sensor's precision while also providing exceptional anti-interference capabilities. The analytical efficacy of this innovative flexible microneedle sensor was good, with a broad continuous dynamic spectrum (0–20 M),

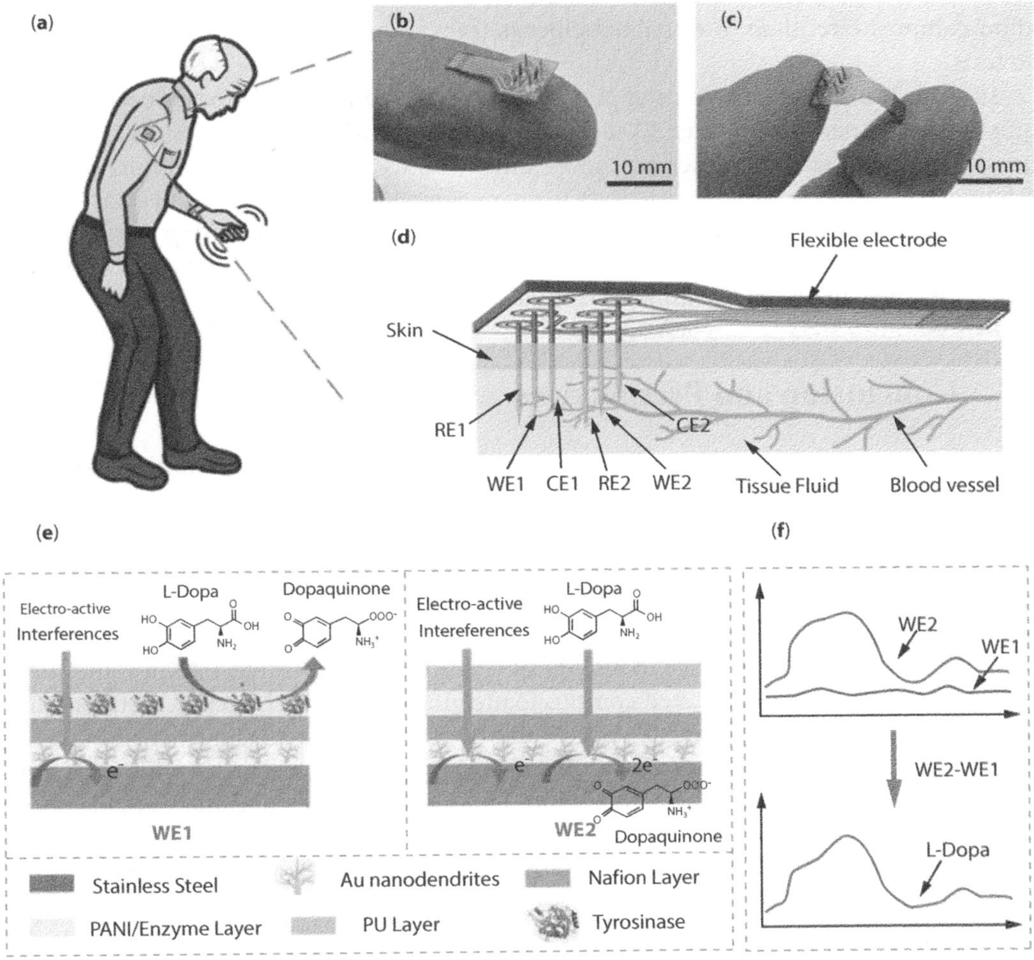


**Figure 14.7** (a) Schematic representation of the minimally intrusive L-Dopa sensor utilized to measure L-Dopa levels in Parkinson's sufferers on a continuous basis. Image of FDMA (b, c). (d) Conceptual illustration of FDMA subcutaneous injection with minimum invasiveness. (e) Conceptual diagram of the working electrodes WE1 and WE2 surface alteration architecture, (f) The L-Dopa differential detecting concept and the current response signal of WE1 and WE2. (Reprinted with authorization from Ref. [48] (MDPI) (2022). This work is licensed under a Creative Commons Attribution Based License 4.0.)

excellent sensitivity (10 nA $M^{-1}$ $cm^{-2}$), and long-lasting durability (14 days). Finally, the L-dopa detector displayed dynamic behavior in reaction to *in vivo* L-dopa, as well as significant anti-interruption capability. Altogether the impressive results suggested that this FDMA might be used to evaluate the dynamic concentration of L-dopa for Parkinson's disease in a less intrusive and continual manner.

Moreover, Chen *et al.* [49] showed that microporous PVDF membranes layered amid numerous levels of NMs could be employed *in vivo* for continual glucose measurement. This was accomplished by layering glucose oxidase enzyme, polyaniline nanofiber, Pt NPs, and porous films onto needle electrodes, which were effectively created using layer-by-layer accumulation. The incorporation of NPs into Polyaniline nanofibers that are conducting in nature results in a significant surface to volume proportion and glucose enzyme electrochemical performance. A membrane coated with permeable PVDF and nanosphere Nafion composite resulted in restricted glucose transfer and extended the longevity of *in vivo* tests.

A sub-microamperometric output current was exhibited by the glucose biosensor with a quick reaction period of under 30 seconds, and 0.23 A/mM of sensitivity. In regards to glucose level, the logarithmic sensing spectrum was 0 to 20 mM. Investigations on mice modeling revealed a remarkable reactivity to changes in glucose levels of blood even though preserving biocompatibility with the tissues in its vicinity. Following 7 days of continual observation, the sensitivity remained around 10% of the original sensitivity, and within 21 days, it remained at 70% of the initial sensitivity.

## 14.3 Conclusion and Perspectives

The relevance of nanomaterials and nanoscience's opens up more opportunities for NPs and their variants to be used for ecological, economical, and medical purposes. When engaging with protein, cells, and tissues, biological sensitive nanomaterials (NMs) have the ability to activate a biochemical response. A multitude of parameters, notably substance physical architecture and interface texture, affect the bioactivities of biologically responsive NMs. These features exert a significant effect on biological systems' encounters with NMs, culminating in a wide range of biological reactions. In medical technology and therapeutic investigations, sophisticated composite metamaterials, which include both organic and inorganic elements, are crucial. As a result, the tremendous advancement in human wellbeing and research is addressed by integrating NMs and biosensors with appropriate surface-modified architectures. To explore biomedical sensors and their applications, semiconductors metal-oxides, metals, and organic ingredients are thoroughly described. Reliable research is used to describe the architectural, optic, physical, and spectral features of nanoranged nanomaterials. As a result, the challenges of integrating nanotechnology, notably in the pharmaceutical industry's development of innovative pharmaceutical drugs and the resolution of complex health issues, are also reviewed in this chapter. Physical properties of nanostructures could cause changes in pharmacokinetics, pharmacological, and metabolism activities, that raises issues regarding their use. Their capacity to transgress through biomembranes toxic resources, and endure in the ecosystem and physiology with ease is a remarkable breakthrough.

## References


1. Jiang, Z., Han, X., Zhao, C., Wang, S., Tang, X., Recent advance in biological responsive nanomaterials for biosensing and molecular imaging application. *Int. J. Mol. Sci.*, **23**, 3, 1923, 2022.
2. Modi, S., Prajapati, R., Inwati, G.K., Deepa, N., Tirth, V., Yadav, V.K., Yadav, K.K., Islam, S., Gupta, P., Kim, D.H., Jeon, B.H., Recent trends in fascinating applications of nanotechnology in allied health sciences. *Crystals*, **12**, 1, 39, 2022.
3. Wallyn, J., Anton, N., Vandamme, T.F., Synthesis, principles, and properties of magnetite nanoparticles for *in vivo* imaging applications—A review. *Pharmaceutics*, **11**, 11, 1–29, 2019.
4. Bashir, S., Thakur, A., Lgaz, H., Chung, I.M., Kumar, A., Corrosion inhibition efficiency of bronopol on aluminium in 0.5 M HCl solution: Insights from experimental and quantum chemical studies. *Surf. Interfaces*, **20**, April, 100542, 2020.
5. Thakur, A., Kaya, S., Abousalem, A.S., Sharma, S., Ganjoo, R., Assad, H., Kumar, A., Computational and experimental studies on the corrosion inhibition performance of an aerial extract of Cnicus Benedictus weed on the acidic corrosion of mild steel. *Process Saf. Environ. Prot.*, **161**, 801–818, 2022.
6. Thakur, A., Kumar, A., Kaya, S., Vo, D.V.N., Sharma, A., Suppressing inhibitory compounds by nanomaterials for highly efficient biofuel production: A review. *Fuel*, **312**, September 2021, 122934, 2022.
7. Bashir, S., Thakur, A., Lgaz, H., Chung, I.-M., Kumar, A., Corrosion inhibition performance of acarbose on mild steel corrosion in acidic medium: An experimental and computational study. *Arab. J. Sci. Eng.*, **45**, 6, 4773–4783, 2020.
8. Shabbir, U., Rubab, M., Tyagi, A., Oh, D.H., Curcumin and its derivatives as theranostic agents in alzheimer's disease: The implication of nanotechnology. *Int. J. Mol. Sci.*, **22**, 1, 1–23, 2021.
9. Harish, V., Tewari, D., Gaur, M., Yadav, A.B., Swaroop, S., Bechelany, M., Barhoum, A., Review on nanoparticles and nanostructured materials: Bioimaging, biosensing, drug delivery, tissue engineering, antimicrobial, and agro-food applications. *Nanomaterials*, **12**, 3, 457, 2022.
10. Arshad, R., Barani, M., Rahdar, A., Sargazi, S., Cucchiarini, M., Pandey, S., Kang, M., Multi-functionalized nanomaterials and nanoparticles for diagnosis and treatment of retinoblastoma. *Biosensors*, **11**, 4, 1–19, 2021.
11. Ajdary, M., Moosavi, M.A., Rahmati, M., Falahati, M., Mahboubi, M., Mandegary, A., Jangjoo, S., Mohammadinejad, R., Varma, R.S., Health concerns of various nanoparticles: A review of their *in vitro* and *in vivo* toxicity. *Nanomaterials*, **8**, 9, 1–28, 2018.
12. Parveen, G., Bashir, S., Thakur, A., Saha, S.K., Banerjee, P., Kumar, A., Experimental and computational studies of imidazolium based ionic liquid 1-methyl- 3-propylimidazolium iodide on mild steel corrosion in acidic solution Experimental and computational studies of imidazolium based ionic liquid 1-methyl- 3-propylimidazolium. *Mater. Res. Express*, **7**, 1, 016510, 2020.
13. Bashir, S., Thakur, A., Lgaz, H., Chung, I.-M., Kumar, A., Computational and experimental studies on Phenylephrine as anti-corrosion substance of mild steel in acidic medium. *J. Mol. Liq.*, **293**, 111539, 2019.
14. Thakur, A. and Kumar, A., Corrosion inhibition activity of Pyrazole derivatives on metals and alloys in acidic environment: A review. *J. Emerg. Technol. Innov. Res.*, **6**, 1, 339–343, 2019.
15. Thakur, A. and Kumar, A., Sustainable inhibitors for corrosion mitigation in aggressive corrosive media: A comprehensive study. *J. Bio- Tribo-Corros.*, **7**, 2, 1–48, 2021.
16. Suhito, I.R., Koo, K.M., Kim, T.H., Recent advances in electrochemical sensors for the detection of biomolecules and whole cells. *Biomedicines*, **9**, 1, 1–20, 2021.
17. Li, Y.C.E. and Chi Lee, I., The current trends of biosensors in tissue engineering. *Biosensors*, **10**, 8, 88, 2020.
18. Bollella, P., Biosensors – recent advances and future challenges. *Biosensors*, 23, 1242, 2021.

19. Li, H., Qiao, R., Davis, T.P., Tang, S.Y., Biomedical applications of liquid metal nanoparticles: A critical review. *Biosensors*, **10**, 12, 1–21, 2020.
20. Naresh, V. and Lee, N., A review on biosensors and recent development of nanostructured materials-enabled biosensors. *Sensors (Switzerland)*, **21**, 4, 1–35, 2021.
21. Wang, Q., Zhu, Y., Song, B., Fu, R., Zhou, Y., The *in vivo* toxicity assessments of water-dispersed fluorescent silicon nanoparticles in caenorhabditis elegans. *Int. J. Environ. Res. Public Health*, **19**, 4101, 2022.
22. Choi, J.H., Choi, M., Kang, T., Ho, T.S., Choi, S.H., Byun, K.M., Combination of porous silk fibroin substrate and gold nanocracks as a novel sers platform for a high-sensitivity biosensor. *Biosensors*, **11**, 11, 441, 2021.
23. Noh, H.B., Chandra, P., Moon, J.O., Shim, Y.B., *In vivo* detection of glutathione disulfide and oxidative stress monitoring using a biosensor. *Biomaterials*, **33**, 9, 2600–2607, 2012.
24. Wu, J., Yu, Y., Su, G., Safety assessment of 2D MXenes: *In vitro* and *in vivo*. *Nanomaterials*, **12**, 5, 828, 2022.
25. Xu, W., Wang, D., Li, D., Liu, C.C., Recent developments of electrochemical and optical biosensors for antibody detection. *Int. J. Mol. Sci.*, **21**, 1, 134, 2020.
26. Saddow, S.E., Silicon carbide technology for advanced human healthcare applications. *Micromachines*, **13**, 3, 346, 2022.
27. Bai, Y., Shu, T., Su, L., Zhang, X., Fluorescent gold nanoclusters for biosensor and bioimaging application. *Crystals*, **10**, 5, 357, 2020.
28. Caro, C., Gámez, F., Quaresma, P., Páez-Muñoz, J.M., Domínguez, A., Pearson, J.R., Pernía Leal, M., Beltrán, A.M., Fernandez-Afonso, Y., De La Fuente, J.M., Franco, R., Pereira, E., García-Martín, M.L., Fe3O4-Au core-shell nanoparticles as a multimodal platform for *in vivo* imaging and focused photothermal therapy. *Pharmaceutics*, **13**, 3, 1–22, 2021.
29. Liu, J.M., Chen, J.T., Yan, X.P., Near infrared fluorescent trypsin stabilized gold nanoclusters as surface plasmon enhanced energy transfer biosensor and *in vivo* cancer imaging bioprobe. *Anal. Chem.*, **85**, 6, 3238–3245, 2013.
30. Song, Y., Lin, Z., Kong, L., Xing, Y., Lin, N., Zhang, Z., Chen, B.H., Liu, X.Y., Meso-functionalization of silk fibroin by upconversion fluorescence and near infrared *in vivo* biosensing. *Adv. Funct. Mater.*, **27**, 26, 1–10, 2017.
31. Juanola-Feliu, E., Miribel-Català, P.L., Avilés, C.P., Colomer-Farrarons, J., González-Piñero, M., Samitier, J., Design of a customized multipurpose nano-enabled implantable system for *in-vivo* theranostics. *Sensors*, 24, 12432, 2014.
32. Bhuckory, S., Kays, J.C., Dennis, A.M., *In vivo* biosensing using resonance energy transfer. *Biosensors*, **9**, 2, 76, 2019.
33. Campos, B., Almeida, T., Wong, A., Ribovski, L., Campanhã, F., Taboada, P., Fatibello-filho, O., The application of graphene for *in vitro* and *in vivo* electrochemica biosensing. *Biosens. Bioelectron.*, **89**, 224–233, 2017.
34. Sondhi, P., Maruf, M.H.U., Stine, K.J., Nanomaterials for biosensing lipopolysaccharide. *Biosensors*, **10**, 1, 2, 2020.
35. Chen, H., Zhang, Y., Li, L., Guo, R., Shi, X., Cao, X., Effective CpG delivery using zwitterion-functionalized dendrimer-entrapped gold nanoparticles to promote t cell-mediated immunotherapy of cancer cells. *Biosensors*, **12**, 2, 71, 2022.
36. Bhat, V.S., Kudva, A.K., Naik, H.V., Reshmi, G., Raghu, S.V., De Padova, P., Hegde, G., Toxicological profiling of onion-peel-derived mesoporous carbon nanospheres using *in vivo* drosophila melanogaster model. *Appl. Sci.*, **12**, 3, 1–16, 2022.
37. Xu, Y., Wang, H., Zhang, M., Zhang, J., Yan, W., Plasmon-enhanced antibacterial activity of chiral gold nanoparticles and *in vivo* therapeutic effect. *Nanomaterials*, **11**, 6, 2021, 2021.

38. Nisar, M., Haq, U., Shah, G.M., Gul, A., Foudah, A.I., Alqarni, M.H., Yusufoglu, H.S., Hussain, M., Alkreathy, H.M., Ullah, I., Khan, A.M., Jamil, S., Ahmed, M., Khan, R.A., Biogenic synthesis of silver nanoparticles using Phagnalon niveum and its *in vivo* anti-diabetic effect against alloxan-induced diabetic Wistar rats. *Nanomaterials*, **12**, 1–18, 2022.
39. Wang, M., Hu, C., Su, Q., Luminescent lifetime regulation of lanthanide-doped nanoparticles for biosensing. *Biosensors*, **12**, 2, 131, 2022.
40. Tang, X., Wang, Z., Wei, F., Mu, W., Han, X., Recent progress of lung cancer diagnosis using nanomaterials. *Crystals*, **11**, 1, 1–13, 2021.
41. Wang, Y., Xu, N., He, Y., Wang, J., Wang, D., Gao, Q., Xie, S., Li, Y., Zhang, R., Cai, Q., Loading graphene quantum dots into optical-magneto nanoparticles for real-time tracking *in vivo*. *Mater. (Basel)*, **12**, 13, 2191, 2019.
42. Grasso, G., Zane, D., Dragone, R., Microbial nanotechnology: Challenges and prospects for green biocatalytic synthesis of nanoscale materials for sensoristic and biomedical applications. *Nanomaterials*, **10**, 1, 11, 2020.
43. Dayem, A.A., Hossain, M.K., Lee, S.B., Kim, K., Saha, S.K., Yang, G.M., Choi, H.Y., Cho, S.G., The role of reactive oxygen species (ROS) in the biological activities of metallic nanoparticles. *Int. J. Mol. Sci.*, **18**, 1, 1–21, 2017.
44. Montenegro, L., Panico, A.M., Santagati, L.M., Siciliano, E.A., Intagliata, S., Modica, M.N., Solid lipid nanoparticles loading idebenone ester with pyroglutamic acid: *In vitro* antioxidant activity and *in vivo* topical efficacy. *Nanomaterials*, **9**, 1, 43, 2019.
45. Jin, Q., Chen, H.J., Li, X., Huang, X., Wu, Q., He, G., Hang, T., Yang, C., Jiang, Z., Li, E., Zhang, A., Lin, Z., Liu, F., Xie, X., Reduced graphene oxide nanohybrid–assembled microneedles as mini-invasive electrodes for real-time transdermal biosensing. *Small*, **15**, 6, 1–10, 2019.
46. Lu, X., Cheng, H., Huang, P., Yang, L., Yu, P., Mao, L., Hybridization of bioelectrochemically functional infinite coordination polymer nanoparticles with carbon nanotubes for highly sensitive and selective *in vivo* electrochemical monitoring. *Anal. Chem.*, **85**, 8, 4007–4013, 2013.
47. Li, C., Meng, Y., Wang, S., Qian, M., Wang, J., Lu, W., Huang, R., Mesoporous carbon nanospheres featured fluorescent aptasensor for multiple diagnosis of cancer *in vitro* and *in vivo*. *ACS Nano*, **9**, 12, 12096–12103, 2015.
48. Fang, L., Ren, H., Mao, X., Zhang, S., Cai, Y., Xu, S., Zhang, Y., Li, L., Ye, X., Liang, B., Differential amperometric microneedle biosensor for wearable levodopa monitoring of Parkinson's disease. *Biosensors*, **12**, 2, 102, 2022.
49. Chen, D., Wang, C., Chen, W., Chen, Y., Zhang, J.X.J., PVDF-Nafion nanomembranes coated microneedles for *in vivo* transcutaneous implantable glucose sensing. *Biosens. Bioelectron.*, **74**, 1047–1052, 2015.

# 15

# Biosensor and Nanotechnology for Diagnosis of Breast Cancer

**Kavitha Sharanappa Gudadur[1,2], Aiswarya Manammal[1,2] and Pandiyarasan Veluswamy[1,2,3*]**

*[1]SMart and Innovative Laboratory for Energy Devices (SMILE), Indian Institute of Information Technology Design and Manufacturing (IIITDM) Kancheepuram, Chennai, India*
*[2]School of Interdisciplinary Design and Innovation (SIDI), Indian Institute of Information Technology Design and Manufacturing (IIITDM) Kancheepuram, Chennai, India*
*[3]Department of Electronics and Communication Engineering, Indian Institute of Information Technology Design and Manufacturing (IIITDM) Kancheepuram, Chennai, India*

***Abstract***
A biosensor is a receptor-transducer device with integrated processing power that can turn a biological reaction into an electrical signal. An extensive range of biosensor applications, involving environmental monitoring, medication delivery, healthcare, and illness detection, have pushed the design of biosensors to the forefront of research in the last ten years. The key objection involved in biosensor growth are (i) the capture of biorecognition signals and their conversion into electrical, optical, electrochemical, and gravimetric signals (transduction process), (ii) improving transducer performance (i.e., reducing response time, enhancing reproducibility, sensitivity, and lowering recognition confines even to detect specific molecules), and (iii) biological sensing devices miniaturization using nanosynthesis or microsynthesis technologies. Those objections can be changed through the combination of sensing technologies with nanoscale materials (zero to three dimensional, have a strong conductivity, color tunability, and surface-to-volume ratio. Nanomaterials employed in the synthesis and nanobiosensors include nanoparticles (NPs) (organic or inorganic), polymeric NPs (large surface area, high thermal and electrical conductivity). This study delivers an outline of the development of biosensors, current methods used in biosensors for breast cancer diagnosis employing NPs, and their recent breakthrough in biomedical sensing technologies with the growth of nanotechnology.

***Keywords:*** Sensor, nanomaterials, nanobiosensor, breast cancer diagnosis

## 15.1 Introduction

### 15.1.1 Sensors

A device called a sensor is used to detect and convert physical quantities like pressure, humidity, heat, and force as well as electrical quantity such as current, into signals that can

**Corresponding author*: pandiyarasan@yahoo.co.in

Inamuddin and Tariq Altalhi (eds.) Biosensors Nanotechnology, 2nd Edition, (347–370) © 2023 Scrivener Publishing LLC

be tracked and examined [1, 2]. A gadget that can transform one form of energy into another is known as a transducer. A measurement system's beating heart is the sensor. Range, high resolution, drift, selectivity, calibration, repeatability, sensitivity, linearity, reproducibility, and reaction time are all characteristics of a great sensor [3, 4]. Environmental observation, medical diagnostics, industrial manufacturing, and space and defense have all benefited from sensor technology.

### 15.1.2 Biosensors

#### *15.1.2.1 Design and Principle*

A biosensor combines biological components like antibodies or enzymes with an electrical constituent to produce a detectable signal. The electronic element senses and communicates data about physiological changes as well as chemical and biological components in the environment. Biosensors are available in a variety of sizes and shapes, and they are capable of detecting and quantifying even minute concentrations of illnesses, harmful substances, and pH levels. The conventional biosensor includes an analyte, bioreceptor, transducer, and display as shown in Figure 15.1.

(a) Analyte: A substance that recognizes the constituents like alcohol, ammonia, glucose, and lactose.
(b) A bioreceptor is a biomolecule that detects the target substrate like aptamers, antibodies, cells, enzymes, DNA, or RNA. Signals are created in different forms i.e., heat, mass change, light, pH, and charge during the contact between the analyte and bioreceptor.
(c) Transducer: A transducer that converts energy into another form. A biosensor's transducer is a critical component. It transforms a biorecognition

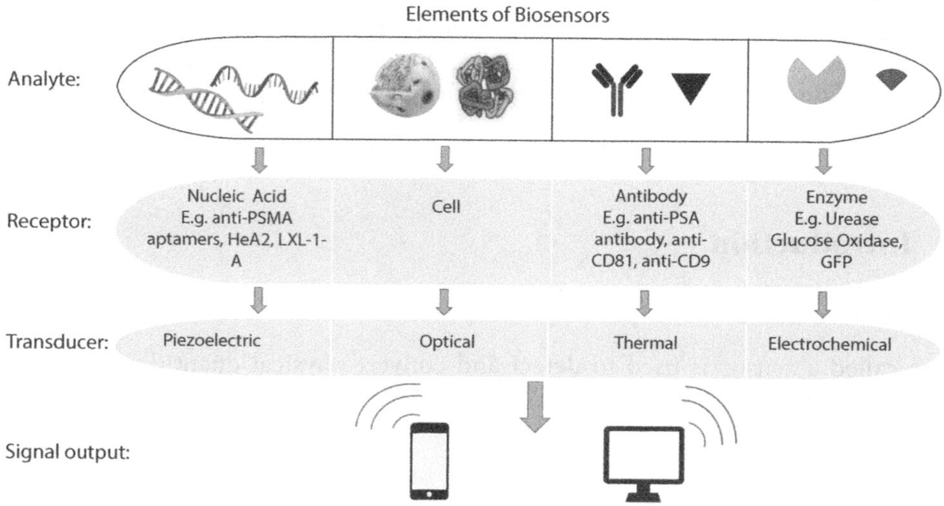


**Figure 15.1** A schematic design of a standard biosensor shows the transducer, bioreceptor, electronic system (CPU and amplifier), and displays (computer or printer) [5].

event into an electrical signal that indicates the presence of a biological and chemical target. Signalization is the name for this energy conversion process.

(d) Electronics: The signal is processed and the transducer's electrical impulses are transformed and amplified into numeral form. The display unit quantifies the processed signals.

(e) Display: It consists of a computer, which creates the output for the user to read and understand.

### *15.1.2.2 Roadmap of Biosensors*

The progresses of biosensors starting from the discovery of electric potential between fluid parts by M. Cramer in 1906 till the development of a nerve-on-chip biosensor developed by S. Girbi *et al.* for nerve evaluation is consolidated in Table 15.1. Moreover, the evaluation time of biosensors is represented in Figure 15.2.

**Table 15.1** Roadmap of biosensors.

| Year | Roadmap of biosensor |
|---|---|
| 1906 | M. Cramer [6] discovered the electric potential difference between fluid parts. |
| 1909 | Soren Sorensen was the first to propose the pH concept and the pH measure [7]. |
| 1909–1922 | Griffin and Nelson [8, 9], demonstrated the invertase enzyme that can be immobilized using charcoal and aluminum hydroxide. |
| 1922 | W.S. Hughes [10] invented a pH-measuring electrode. |
| 1956 | Leland C. Clark, Jr. [11] proposed an oxygen electrode. |
| 1962 | Leland C. Clark, Jr *et al.* [12] created an amperometry enzyme electrode for sensing glucose in the lab. |
| 1967 | Using glucose oxidase mounted atop an oxygen sensor, Updike and Hicks constructed the functioning enzyme electrode [13]. |
| 1969 | Guilbault and Montalvo [14] demonstrated a potentiometric enzyme electrode-based sensor for detecting urea. |
| 1971 | Bergveld proposed a field effect ion-sensitive transistor [25]. |
| 1973 | Guilbault and Lubrano developed hydrogen peroxide-based lactate and glucose enzyme sensors with the help of electrodes namely platinum [12]. |
| 1974 | B. Danielsson and K. Mosbach [13] created the enzyme thermistor. |
| 1975 | D.W. Lubbers and N. Opitz [14] proposed fiber-optic biosensors for oxygen and $CO_2$ detection. |
| 1975 | YSI developed a commercial biosensor for the detection of glucose [15, 16]. |
| 1975 | The microbe-based immune sensor was demonstrated by Suzuki *et al.* [17]. |
| 1976 | The artificial bedside pancreas was shown by Clemens *et al.* [18]. |
| 1980 | The *in vivo* blood gases and fiber-optic pH sensor were demonstrated by Peterson [19]. |

(*Continued*)

**Table 15.1** Roadmap of biosensors. (*Continued*)

| Year | Roadmap of biosensor |
|---|---|
| 1982 | Schultz [20] created a glucose-sensing fiber-optic biosensor. |
| 1983 | Liedberg *et al.* [21] discovered the surface plasmon resonance (SPR) immunosensor. |
| 1983 | Roederer and Bastiaans developed the first immunosensor based on piezoelectric detection [22] |
| 1984 | Ferrocene, which was paired with a glucose oxidase to detect glucose, was the first mediated amperometric biosensor [23]. |
| 1990 | SPR-based biosensors have been developed by Pharmacia Biacore [24]. |
| 1992 | i-SAT created a portable blood biosensor [24]. |
| 1999 | Poncharal *et al.* [25] proposed the first nanobiosensor. |
| 2018 | A nerve-on-chip biosensor was presented by S. Girbi *et al.* for nerve evaluation. The transmission of impulses [26]. |

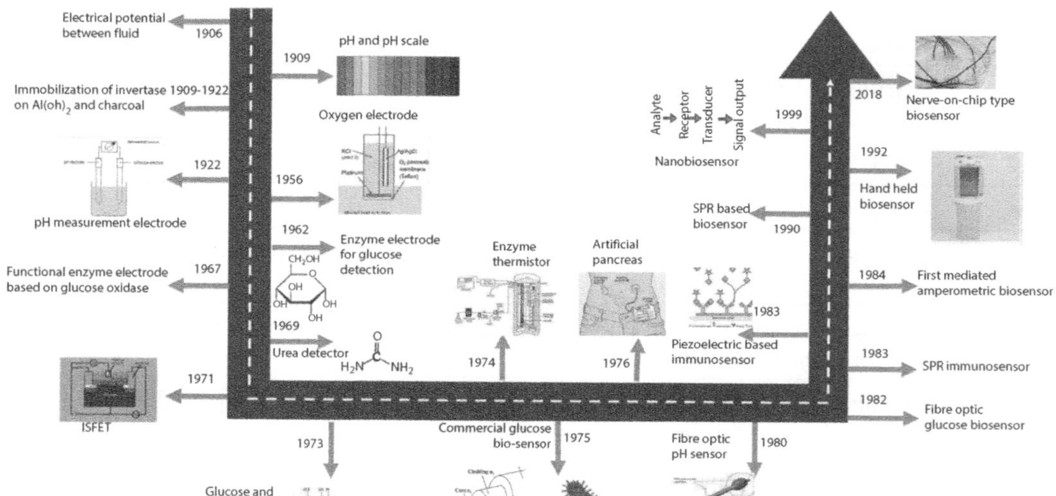


**Figure 15.2** Biosensor evolution timeline.

## 15.2 Characteristics of Biosensors

Certain static and dynamic qualities are required to produce a highly active biosensor system. These characteristics can be exploited to improve the performance of commercial biosensors [3, 4, 27].

(a) Selectivity: This is an important parameter to be considered during the bioreceptor selection for a biosensor. A bioreceptor identifies the target analyte in a sample including mixed and undesired pollutants.

(b) Sensitivity: The least quantity of analyte which is detected at very low dilatations (ng/mL) to validate the presence of trace analyte in a sample.
(c) Linearity: Assures the sensed results are accurate. The better the linearity, the more accurate the detection of substrate concentration (straight line).
(d) Response time: Time is taken by the material to achieve 95 percent of the findings.
(e) Reproducibility: When the same sample is measured multiple times, the biosensor can yield identical findings.
(f) Stability: One of the most important qualities that requires continuous monitoring. The degree of susceptibility to environmental perturbations both inside or outside the biosensing equipment is referred to as stability. Two elements that determine stability are the bioreceptor's affinity.

### 15.2.1 Cancer Treatment Using Nanotechnology

Traditional bio-sensing methods have been significantly improved by nanotechnology, resulting in a cost-effective and convenient molecular-level analytical diagnostic platform for breast cancer. The term "nanotechnology" was invented by Taniguchi in 1974, but Nobel Laureate Richard Feynman's lecture "There is Plenty of Room at the Bottom" gave it intellectual respectability. The study of enormously small matters is called nanotechnology, whereas the study of extremely small objects is called nanoscience. It entails very small-scale material manipulation and application. Atoms and molecules act differently at this scale and can be exploited for a variety of unexpected and fascinating purposes. Nanotechnology is well-defined as the manipulation, reduction, and fabrication of nanoscale materials with dimensions ranging from 1 to 100 nm with features such as high strength, biocompatibility, enhanced stability, precision targeting, and cheap cost [28].

Nanotechnology has been examined for the identification of aberrant breast cancer cells and *in vivo* imaging in biomedical application which is represented in Figure 15.3.

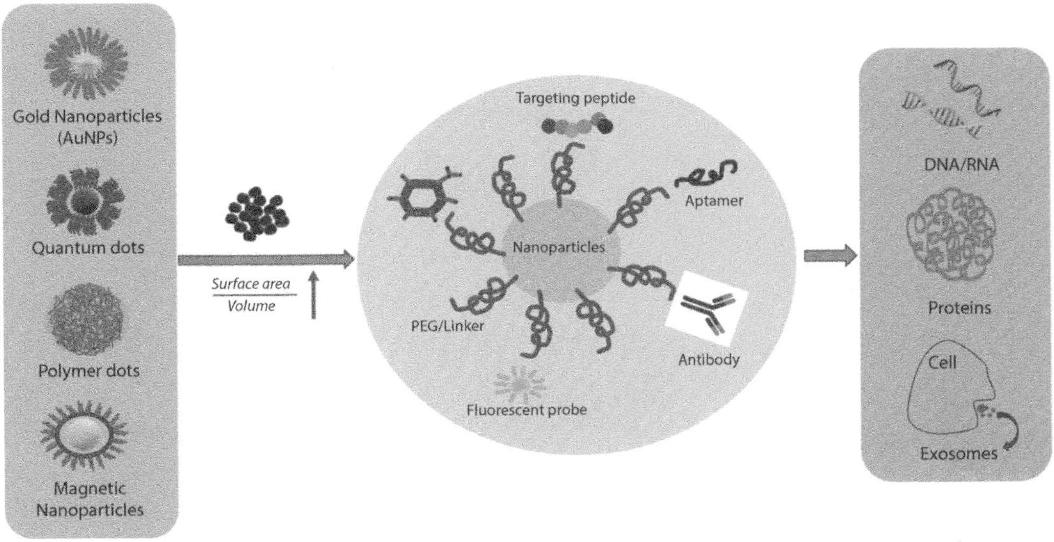


**Figure 15.3** Nanotechnology aids in the identification and diagnosis of cancer.

Moreover, it improves sensitivity, and specificity, and enables continuous monitoring of temperature changes in the breasts via multiplexed measurement capacity. As a result, nanotechnology is on the verge of incorporating wearable electronics.

## 15.3 Cancer Therapy with Nanomaterials

### 15.3.1 Biosensors with Nanomaterials

Nanomaterials are materials with nanometer-scale dimensions. They are materials with a $10^{-9}$ m dimension. Metal-based nanomaterials, dendrimers, carbon-based nanoparticles (NPs), and composites are the most common types of nanomaterials as depicted in Figure 15.4 (a). They are constructed in both a top-down and bottom-up manner as depicted in Figure 15.4 (b). To decrease original macroscopic structures to nanoscale structures, top-down strategies are applied. Top-down processes include laser machining, etching, micromachining, lithography, photolithography, and ball milling. Inert-gas expansion, layer deposition, sol-gel procedures, electrodeposition, ultrasonic dispersion, vapor condensation, inert-gas condensation, plasma-based synthesis, and molecular beam epitaxy are some of the techniques that have been employed.

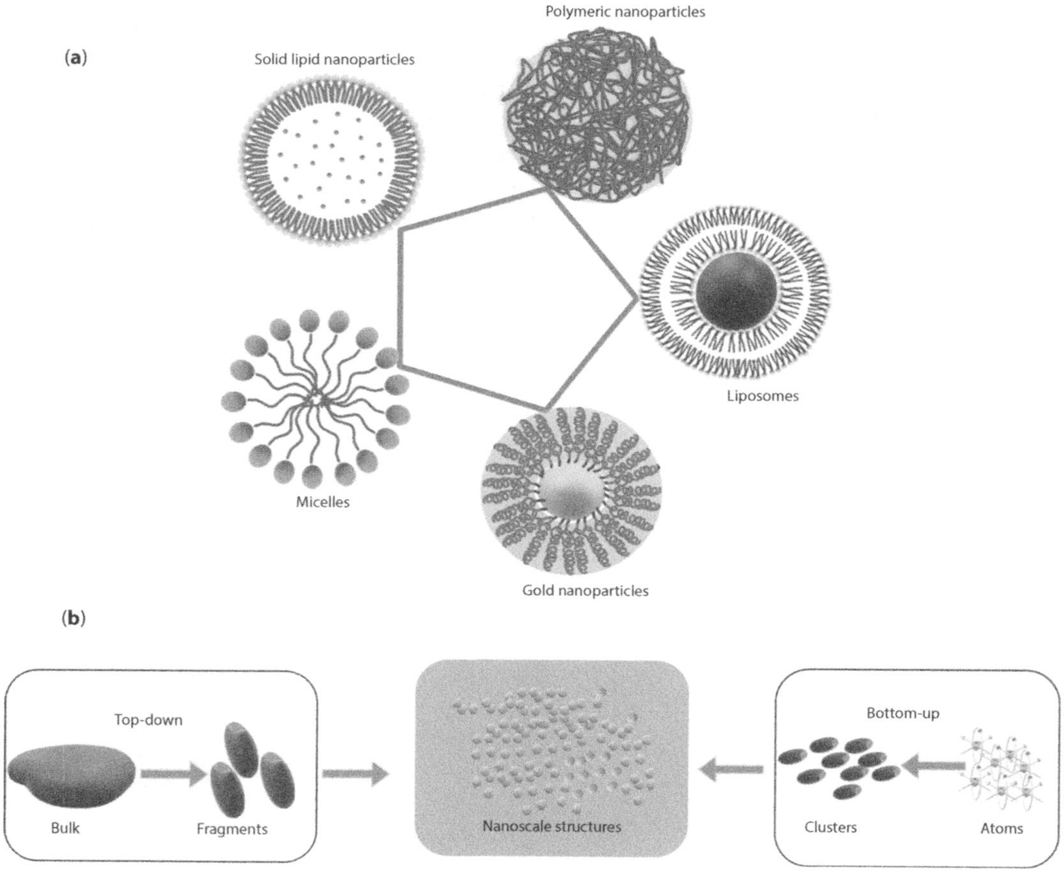


**Figure 15.4** (a) Biosensors made of various nanomaterials. (b) Various NP synthesis techniques.

### 15.3.2 Nanomaterials' Properties

In light of the quantum impacts, volume, and surface of NPs, nanomaterials offer uncommon mechanical properties. Metals' mechanical properties, similar to those of traditional materials, are comprised of 10 parts: inflexibility, durability, strength, pliability, versatility, weariness strength, flexibility, fragility, hardness, and yield pressure. Most inorganic non-metallic materials are lacking qualities like flexibility, durability, pliancy, and pliability. Moreover, a few natural materials are adaptable and miss the mark on characteristics of fragility and inflexibility. When NPs are added to a typical material, they enhance the grain to some extent, producing an intragranular structure, which improves the material mechanical properties [29, 30].

NPs have dissimilar physical or chemical properties than macroscale counterparts, such as bulk materials, due to their small size [31]. Because they have a greater surface area to volume ratio, they have stronger thermal, electric, catalytic, mechanical, and magnetic properties than macrostructures. They have a quantum confinement effect, which means that the bandgap widens as particle size decreases [32]. They could be employed in a diversity of applications, including solar energy recycling, catalysis, drug delivery, environmental remediation imaging, antimicrobials, fuel additives, fuel cells, optoelectronics, and data storage.

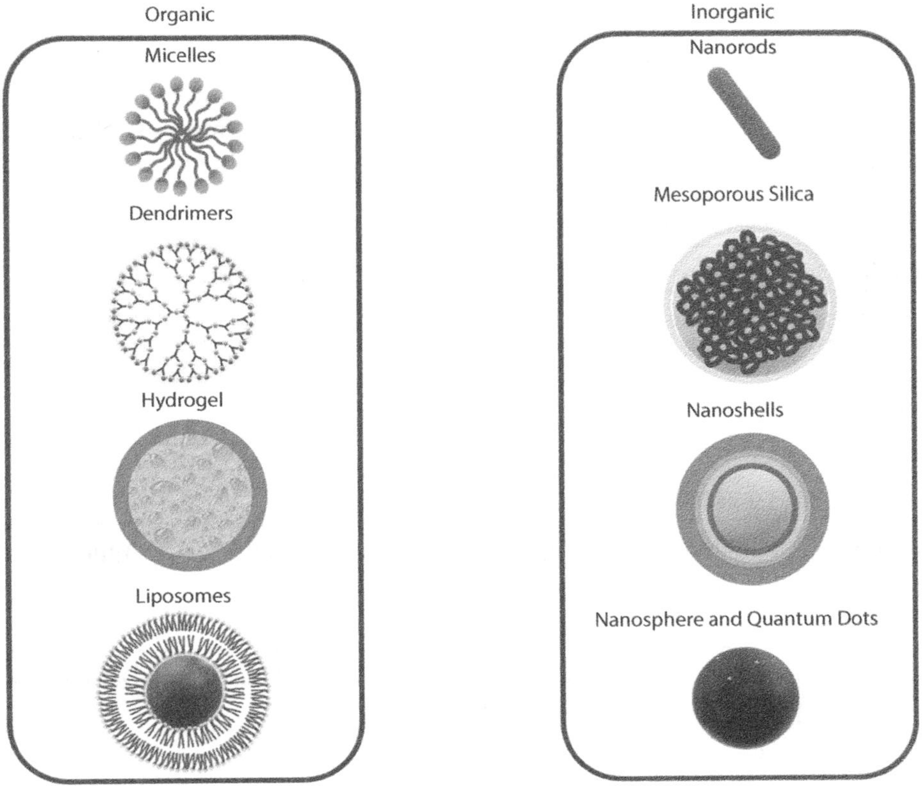


**Figure 15.5** Breast cancer detection and therapeutics based on nanotechnology have made substantial progress.

### 15.3.3 Organic and Inorganic Nanomaterials

Traditional chemotherapy medicines have a large volume of action that affects both normal and malignant cells [33]. Given that there is no complete cure for cancer in the twenty-first century, the researchers determined that using NPs (organic or inorganic) is the best cancer treatment [34, 35] as shown in Figure 15.5.

NPs offer a variety of benefits, including customized pharmaceutical release, stability, and solubility. NPs have a high permeability at tumor vasculature, allowing tumor medicines to enter more easily than healthy channels [36]. This is a key benefit in cancer treatment. Additionally, some complexes used in cancer treatment are shown in Table 15.2.

#### *15.3.3.1 Organic NPs*

Organic systems for cancer treatment have been devised by researchers, and each one has advantages and limitations shown in Table 15.3.

##### 15.3.3.1.1 Polymeric NPs

The primary advantages of polymeric NPs include biodegradability, biocompatibility, hydrophilicity, and non-toxicity. These can be used to treat cancer with great effectiveness [37, 38].

##### 15.3.3.1.2 Polymeric Micelles

L-lysine, aspartic acid, propylene oxide, and D, L-lactic acid form micelles with a hydrophilic shell and a hydrophobic core using amphiphilic block copolymers such PEG and poly (vinyl alcohol) [43, 44]. The hydrophilic shell stabilizes the aquaphobic core, which aids as

**Table 15.2** NP-based drugs are used to treat different stages of breast cancer.

| Device | NPs | Functions | Developmental phase | References |
|---|---|---|---|---|
| Combidex | Iron oxide NPs | tumors | Not available | [39] |
| NK105 | Micellar NPs | Breast cancer | Ph. 2 | [40] |
| CYT-6091 | Gold NPs | Breast cancer | Ph. 1/2 | [41] |

**Table 15.3** Types of organic NPs for cancer therapy [42].

| Types of NPs | Size (nm) | Advantage | Disadvantage | Application |
|---|---|---|---|---|
| Polymers | 10–1000 | Drug delivery, biodegradability | Delivery efficiency is inefficient. | Components delivery |
| Quantum dots | < 10 | Modification of the surface | UV makes it unstable. | Detection of cancer |
| Dendrimers | 43952 | Carriage of a lot of medications | Cytotoxic | Target delivery |
| Liposomes | 50–100 | Tractability | Irritation | Gene delivery |

a reservoir for aquaphobic medications. To load pharmaceuticals into a micelle for cancer treatment, physical encapsulation and chemical covalent coupling can be used [45, 46].

#### 15.3.3.1.3 Liposomes NPs

Liposomes, like cholesterol and phospholipids, are formed up of lipid bilayers with a spherical shape and a hydrophilic or hydrophobic center. The single lipid bilayer of liposomes has an aqueous core that entraps water-soluble medicines, whereas multi-bilayer liposomes entrap lipid-soluble drugs. Doxil®, Myocet®, and DaunoXome® are liposomal analogs of the anthracycline's doxorubicin and daunorubicin that have been licensed for the treatment of metastatic breast cancer [47, 48]. Liposomes have also been discovered to be good at encapsulating plasmid DNA and siRNA in their hydrophilic center [49]. In another trial, patients with taxane-resistant breast cancer were given PEGylated liposomal DOX instead of vinorelbine or mitomycin C + vinblastine, and their survival rates were reported to be greater [50]. According to Sriraman *et al.*, HeLa cell *in vitro* studies demonstrated that liposomes customized to folic acid and transferrin had higher efficacy and cell infiltration in delivering DOX than other methods [51]. The succeeding generation of liposomal medications includes immunoliposomes, which selectively transport medicine to the intended sites of action. One study revealed an *in vivo* mechanism for monoclonal antibody (MAb)-mediated NP (immunoliposome) directing solid tumors. To produce immunoliposomes specific to HER2 was coupled to liposome-grafted PEG chains, and *in vivo* tests demonstrated that this technique greatly increased NP absorption in HER2-expressing breast tumors. The system's efficacy was further tested in the absence of the anti-HER2 antibody showed reduced tumor [52]. The present research focuses on the progress of liposomes that could be used as MRI contrast agents [53].

### *15.3.3.2 Inorganic NPs*

Organic (NPs made up of organic materials) and inorganic (NPs made up of inorganic elements) make up the vast majority of NPs. The issue of "inorganic NPs" is new, as are its biomedical applications because this category was formed around the turn of the century. The two parts of inorganic NPs are the core (metals such as quantum dots (QDs), gold) and shell (organic polymers that shield the main from chemical correlations) [54].

#### 15.3.3.2.1 Gold NPs

Gold NPs used in tumor imaging and the design of multimodal cancer-targeting drug delivery systems because of their photosensitive and electric properties, low toxicity, ease of production, and the existence of undesirable sensitive groups on the surface [55, 56]. The gold atom core of these NPs can be functionalized by layering a monolayer of moieties on top of them. At the cellular level, several human cell lines have been established which are non-toxic to gold NPs, and gold NPs have been proven to be potentially biodegradable in vivo in numerous studies [57–60]. Very-small gold NPs (whose diameter may be modified by modifying physical and chemical variables) have a homogeneous distribution inside tumor tissues because of the ability to permeate across tissues, however, absorption is limited [61, 62]. Gold NPs (sizes ranging from 5 to 35 nm) were synthesized and tested against human cervical cancer cells in one study (HeLa). In HeLa cells *in vitro*, the NPS was reported to exhibit anticancer activity via generating DNA damage in the

G2/M phase [63]. In a separate study, gold NPs containing DOX were found to have an improving anticancer effect on HeLa cells than free DOX [64]. These NPs also have been used as drug delivery trajectories for tumor necrosis factors in a developing tumor in mice, to destroy tumor cells in animals [65, 66]. Gold NPs are predicted to be employed as CT contrast agents and antibacterial agents in the future due to their high CT attenuation efficacy and low cytotoxicity [67].

##### 15.3.3.2.2 QDs NPs

QDs are semiconducting material NPs has exclusive optical assets owing to their size effect and quantum effect [5]. An excellent QD should have the following characteristics for drug delivery applications:

- great capacity for drug loading—no drug response
- low toxicity and excellent biocompatibility
- *in vivo* residence time is longer—*in vivo* residence time is longer
- a specific amount of mechanical toughness [68].

QDs are characterized by inorganic NPs (e.g., CdS and PbS) but the utmost common inner core semiconductor QD system of CdSe coated with the ZnS outer shell [69]. Due to their exclusive properties, strong and stable fluorescence, and photobleaching resistance, for long periods, QDs are a new form of revolutionary biosensors used in cancer diagnostics [70]. Using QDs and emission spectrum scanning multiphoton imaging, Voura *et al.* [71] established a way to track tumor cell extravasation in an alive animal. Intravenously injected QD-labeled cells were followed as they were extravasated into lung tissue in mice. Using QDs and spectrum imaging, multiphoton laser stimulation researchers were able to detect five different populations of cells at the same time [71]. QDs were also associated with streptavidin and immunoglobulin G to detect nuclear antigens present inside the nucleus, microtubule fibers and stain actin in the cytoplasm, and identify the HER2 (breast cancer marker) on the living and fixed cancer cell surface. Two cellular targets were discovered using a single excitation wavelength and QDs coupled to streptavidin and IgG with different emission spectra [72].

### 15.3.4 Nanobiosensors

A variety of nanomaterials, including copper, silicon, silver, and gold NPs, as well as organic materials like graphene, graphite, and carbon nanotubes, are used to immobilize biosensors [73–81]. Furthermore, electrochemical and other types of biosensor development-based materials provide great specificity and sensitivity. Because of their oxidation stability [82] and near-zero toxicity, gold NPs have potential applicability among metallic NPs, whereas other NPs, such as silver, oxidize and have dangerous symptoms when used internally in medicine, such as drug administration. Using nanomaterials for biosensors in general presents challenges that must be overcome if they are to be used in biomedicine [83]. Furthermore, NP-based signal magnification approaches have both advantages and disadvantages [84].

Nanomaterials, on the other hand, are considered demanding components in bioanalytical instruments simply because they improve single-molecule detection sensitivity and detection limits [85]. In this context, platinum-based NPs with a single-label response are used for electrochemical magnification for the recognition of low DNA quantities [86]. Similarly, tumor-targeting ligands including peptides, monoclonal antibodies, and small compounds can be efficiently linked with optical and magnetic semiconductor QDs to target tumor antigens with high specificity and affinity [87]. QDs can be employed for a better understanding of the tumor microenvironment and to provide nanomedicine [88]. The use of biosensors with various cantilever diameters (millicantilevers, microcantilevers, and nanocantilevers) is also investigated due to their wide range of applications.

Biosensor research and development have become more open and diverse as a result of advances in nanotechnology. It is feasible to improve biosensor performance and detection power by altering the size and morphology of nanomaterials, such as NPS (oxide and metal-based), nanowires (NWs), and carbon nanotubes (CNTs) for various qualities. A variety of nanomaterial-based biosensors are shown in Figure 15.6 . Breast cancer detection with nanobiosensor is shown in Figure 15.7 and the use of nanobiosensors in biomedical applications is depicted in Figure 15.8.

Nanobiosensors use the same basic operating concept as macro and micro biosensors, but for signal or data translation, they use nanoscale components [91]. Nanobiosensors have high application diversity due to their dimensionality. Nanobiosensors are useful in the field of nanotechnology since they can detect viruses and bacteria.

a) Observing physical and chemical events in inaccessible regions.
b) Medical diagnostics and biochemical detection in cellular organelles.
c) Nanoparticle monitoring in the workplace and the environment.
d) Detecting potentially harmful substances at extremely low concentrations [91].

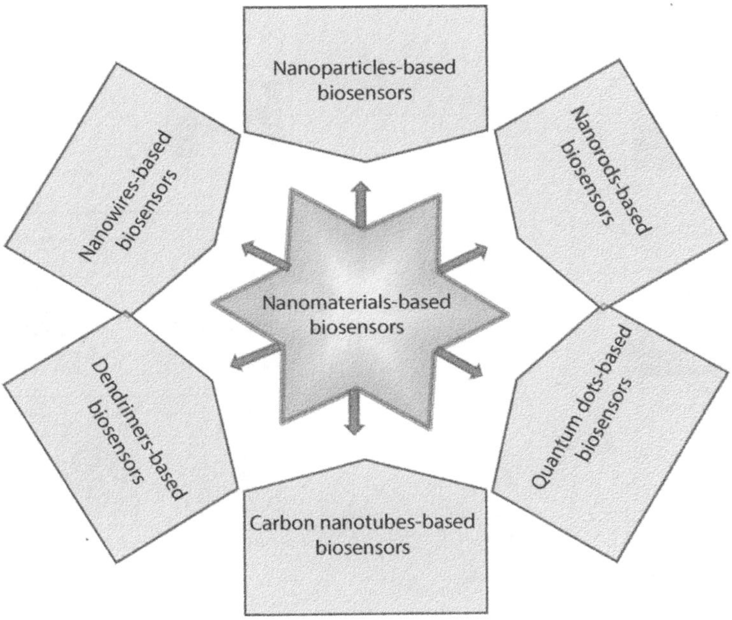


**Figure 15.6** Biosensors based on nanomaterials (nanobiosensors) [89].

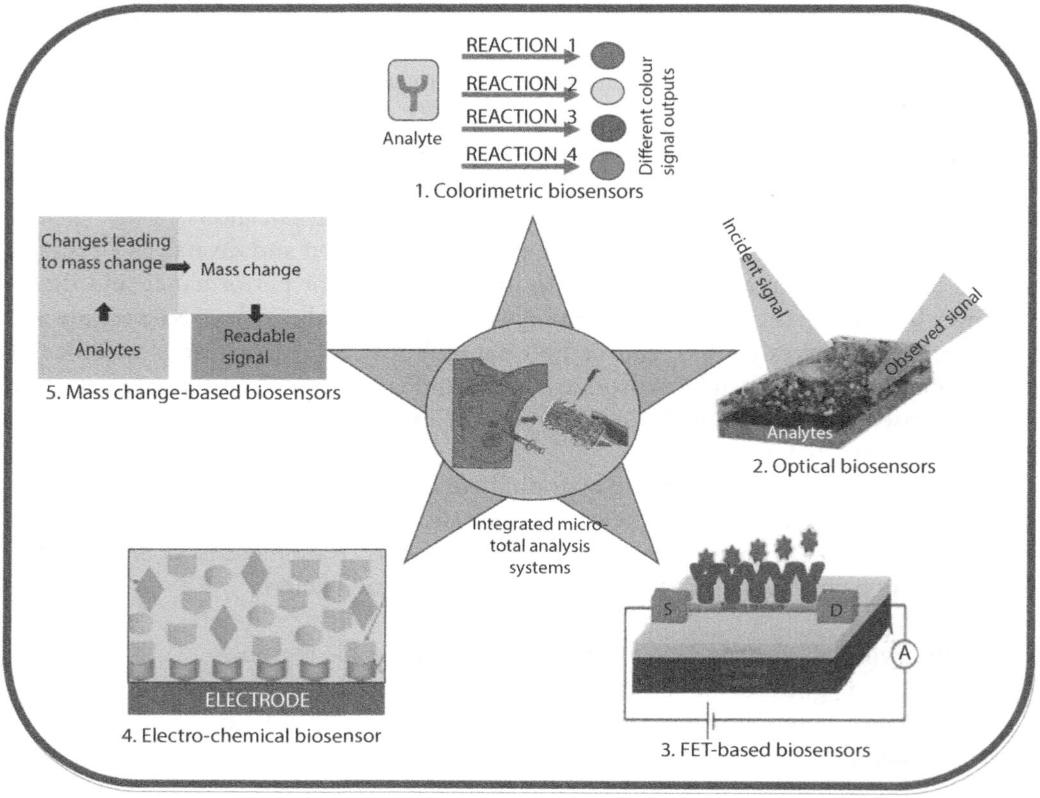


**Figure 15.7** Breast cancer detection with biosensors and nanobiosensors.

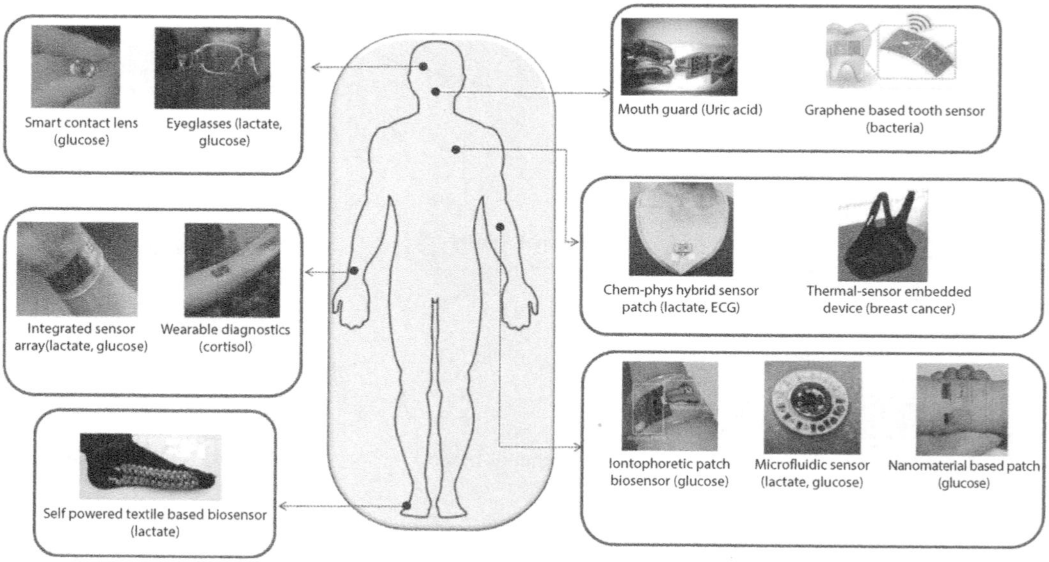


**Figure 15.8** Biomedical technology makes use of nanobiosensors [90].

## 15.4 Diagnosis of Breast Cancer

### 15.4.1 Breast Cancer

Breast cancer will be the highest frequent malignancy among women in 2020 [92], with 626 700 deaths and 2 million new cases [93]. Early detection and treatment have cut mortality in half over the last few decades. A reliable and efficient diagnosis is required for effective therapy. Lowering the cost of diagnosis is critical in resource-constrained health systems because it allows many more women to be identified and treated properly.

### 15.4.2 Diagnosis

#### *15.4.2.1 Analysis at the Point of Care (POC)*

Clinical examination, mammography, ultrasound, and biopsy are all contemporary state-of-the-art diagnostic techniques as shown in Figure 15.9. In some circumstances, positron emission tomography (PET) and magnetic resonance imaging (MRI) is required [94, 95]. The efficacy of tissue biopsy examination for everyday investigation analysis for early cancer indications is restricted since it requires complex instrumentation and specialist staff [96]. The recognition of novel diverse genetic and epigenetic patterns, potential molecular signatures, and alterations and the acceptance of actual FDA-approved biomarkers for early breast cancer detection has derived from the bottlenecks associated with current diagnostic techniques [97].

In recent years, developed new biosensing methods or assays for breast cancer leverage the differences in optical, electrochemical, colorimetric, quartz crystal microbalance, piezoelectric, electrical impedance, and acoustic between disease and normal tissue. Technology for signal amplification and bio-receptors has also advanced tremendously [98]. Researchers are continuing to focus on improving analytical performance in biological fluids for specific breast cancer biomarkers to widen the scope of the non-invasive type of liquid biopsy [99].

Only a few laboratory-based biosensing devices are widely utilized and deployed at the point-of-care (POC) [100], despite technical breakthroughs. There have been reports of breast cancer detection biomarkers, nanomaterial-mediated signal amplification techniques, bio-transducers, bioreceptors and nanobiosensing modalities [98, 101–103].

#### *15.4.2.2 Wearable Analysis*

##### 15.4.2.2.1 Wearable Temperature Sensor

Body temperature can be utilized to gather information for medical diagnosis and is a prominent indication of sleepiness, depression, fever, and metabolic process failure [104]. Although the most frequent way of determining body temperature is with a patchable (skin-based) sensor, a mercury thermometer [105]. Rapid reactions, better dependability, low weight, higher sensitivity, and a larger temperature measurement range, are the supreme required features of a flexible temperature sensor. Resistance changes are used to operate the skin-patchable temperature sensor as shown in Figure 15.10, which is accomplished by using fillers (conductive) on the polymer matrix. Single-walled CNTs with hydrogen-bond-based and carbonyl groups were utilized to build a soft heat sensor with

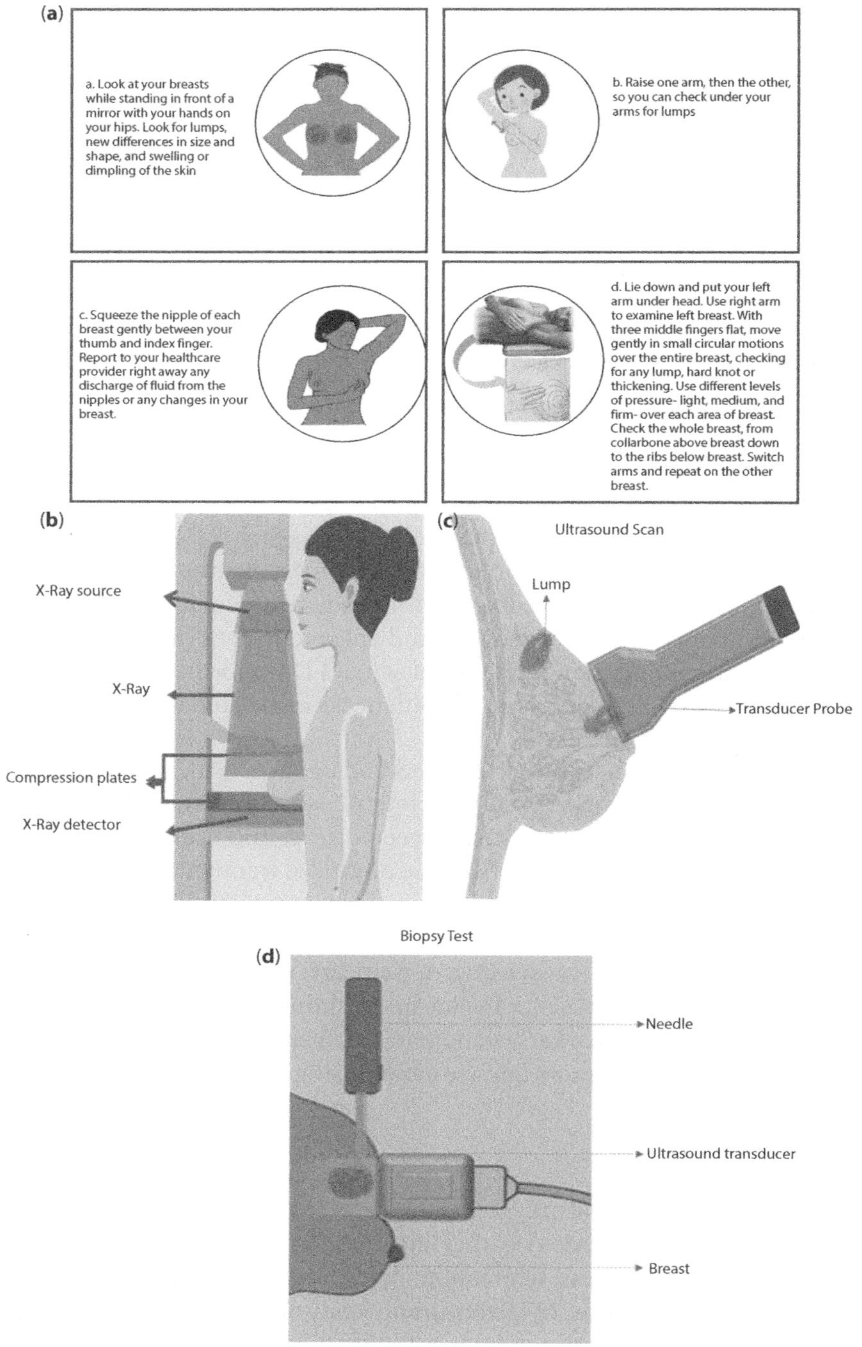


**Figure 15.9** (a) clinical examination, (b) mammography, (c) ultrasonography, and (d) biopsy are the current diagnostic procedures.

considerable mechanical flexibility [106]. Positive temperature coefficient materials have been discovered to be the best applicant for producing temperature-based flexible skin-based patchable sensors with greater response time and sensitivity [107]. Oh *et al.* [108] recently developed an extremely flexible, wearable and resistor-sensitive form temperature sensor with help of an octopus biomimetic adhesive. This sensor was made of a temperature-sensitive, poly(3,4-ethylene dioxythiophene), poly(N-isopropyl acrylamide) (pNI-PAM) hydrogel and carbon nanotubes (CNT); it had a sensitivity related to the temperature of about 2.6% between 25°C and 40°C, allowing it to detect body temperature changes of about 0.5 °C. The device was developed by coating an octopus-like rim assembly of cohesive polydimethylsiloxane (PDMS) layer with help of pNI-PAM in one mold using an undercutting photolithography method. The performance of sensors is unaffected afterward by many detachment cycles on the epidermis of the skin, and there was no long-term pain.

Zhu *et al.* [109] released a report that described numerous circuit design approaches that increased sensor accuracy and resilience using stretchy CNT-based transistors. Trung *et al.* developed a sensor using self-supporting single reduced graphene oxide (rGO), which monitors the temperature [110]. Fabrics that may be utilized as shock absorbers or undershirts were embedded with this fiber-based sensor. The sensor has a reaction time and a recovery time of about 7 sec. and 20 sec. Its performance was unaffected even when mechanical distortion was added.

Traditional skin-based patchable temperature sensors can be used to measure the temperature of the skin, which differs substantially from the main body temperature. So, to overwhelmed this problem, Ota *et al.* [111] proposed a wearable 3D printed "wearable" temperature sensor that can be worn in the ear and used to measure the main body temperature using an infrared sensor based on the tympanic membrane or eardrum. This device can be appropriately connected to a cellular module for optimal monitoring. The device's 3D printed technique enables an easy transition for personalized treatment.

#### 15.4.2.2.2 Optical-Based Skin-Patchable Sensors

Optical-based sensors can detect a variability of optical signals, including polarization, intensity, wavelength, and frequency. The performance was assessed using selectivity, sensitivity, and response time [112]. Additionally, the pulse oximeter contains two LEDs with different wavelengths that are positioned on the human body, and the photodetector detects the light incident from the inner tissue [113, 114], the photodetector is a significant constituent in most optical sensors. Conventional oximeters are bulky, which limits their real-world applications.

These devices were able to demonstrate more precise and rapid on or off-switching behavior with a greater device profit because of careful adjustment of the physical characteristics of the lively layer. The sensitivity of the sensor in the near-infrared (IR) spectrum was improved due to an excellent mix of mechanical conformability and responsiveness. In a current study, Xu *et al.* [115] established a wearable photoplethysmography sensor that provides dimensions to assess regular health monitoring.

A near-IR photoplethysmogram flexible (NIR PPG) sensor was paired with less power an extremely organic phototransistor (OPT) and an effective inorganic LED. It was established that flexible skin-based patch PPG sensors were proficient in assessing heart rate

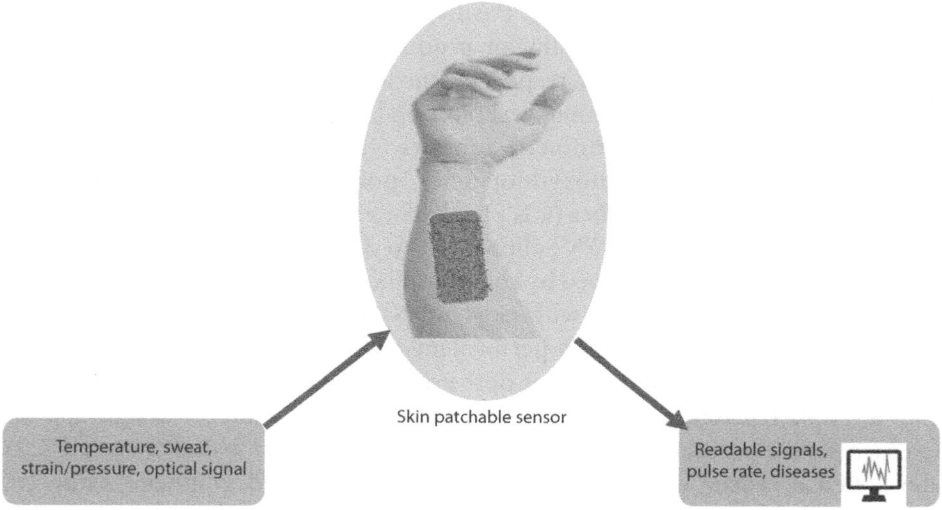


**Figure 15.10** Skin patchable sensor.

dissimilarity and accurate pulse pressure chasing in various human postures and that they performed better and consumed less power than commercially available PPG sensors.

## 15.5 Conclusion

This chapter looked at the fundamental design and mechanisms of biosensors. Biosensors have applications in medicine, food safety monitoring, engineering and technology, drug delivery, ecotoxicology, biomedical science, toxicology, and disease progression. Because of the usage of nanomaterials in biosensors, biosensor technology has grown dramatically in the previous decade. This is owing to the use of novel biorecognition components and transducers, as well as developments in micro-scale shrinking, design, and fabrication of nanostructured devices, as well as new NM synthesis techniques. The introduction of nanomaterials in measuring technology has introduced more diversification, resilience, and dynamics in current years. The transduction process has been substantially enhanced by using a variation of nanomaterials (such as CNTs and NPs) with different properties within biosensors (such as increased faster detection, sensitivity, reproducibility, and shorter reaction time). Though the use of nanoconstituents in biosensor applications is improving, there are still several limitations that prohibit these applications from moving further. CNT-based gas sensors, for example, suffer from a lack of selectivity, which limits their usage in CNT-based systems. By combining CNTs with other materials, however, this barrier can be circumvented. Other problems with these sensors include:

(i) A lack of research on nanostructures' long-term viability in sensing applications.
(ii) Nanostructure fabrication; and the presence of nanomaterials with a high toxicity level.

(iii) Integration and miniaturization of microfluidic systems.
(iv) The lack of security influence of nanobiosensors in terms of risks.
(v) To a lesser extent, biocompatible and biostable.

Most nanobiosensor used in bio-medical applications requires a greater sample size for detection, which can result in false-negative or false-positive results. Only a rare biosensor has reached global profitable success aside from lateral flow pregnancy tests and electrochemical glucose sensors. As a result, low-cost nanomaterial-based biosensors that produce correct data rapidly and are simple to use are in high demand.

The invention of patchable skin-based sensors is a significant step frontward in the advancement of diagnostic technology and health monitoring. A diversity of health-related metrics such as heartbeat, body temperature, breathing rate, motions of various body parts, and perspiration composition can be recorded after skin-based patchable sensors are interfaced with the skin. A combined two more sensors are placed in a solitary substrate for the progress of multisensory devices, as well as everyday discoveries in the arena of flexible electronics and thin film, which have aided the advance of skin-based patchable sensors greatly.

In comparison to standard diagnostic techniques, skin-based patchable devices hold promise for the rapid and easy identification of critical disease signs as well as the monitoring of routine health-related indicators including heartbeat, pulse rate, blood pressure, and body temperature. Additionally, the materials utilized in device manufacturing and existing production methods appear to enhance the overall price. The overall gadget price can be effectively decreased by using sensing materials, such as CNTs, polymers, graphene, and carbon-based substrates, as well as simple fabrication techniques. Overall, skin-based patch sensing nanoconstituents can be used in the upcoming for fast detection and health monitoring including blood pressure. Diabetic individuals might considerably benefit from early diagnosis in terms of blood sugar monitoring. This can also benefit medical scientists because it can lead to the development of new medications to treat a variety of diseases.

## References

1. Ensafi, A.A., *An introduction to sensors and biosensors*, pp. 1–10, Elsevier, Cambridge, 2019.
2. Ensafi, A.A., Karimi-Maleh, H., Ghiaci, M., Arshadi, M., Characterization of Mn-nanoparticles decorated organo-functionalized $SiO_2$-$Al_2O_3$ mixed-oxide as a novel electrochemical sensor: Application for voltametric determination of captopril. *J. Mater. Chem.*, 21, 15022–15030, 2011.
3. Theavenot, D.R., Toth, K., Durst, R.A., Wilson, G.S., Electrochemical biosensors: Recommended definitions and classification. *Biosens. Bioelectron.*, 16, 121–131, 2001.
4. Dincer, C., Bruch, R., Costa-Rama, E., Fernández-Abedul, M.T., Merkoçi, A., Manz, A., Urban, G.A., Güder, F., Disposable sensors in diagnostics, food, and environmental monitoring. *Adv. Mater.*, 31, 1806739, 2019.
5. Danliang, *et al.*, Recent advances on aptamer-based biosensors for detection of pathogenic bacteria. *World J. Microbiol. Biotechnol.*, 37, 3, 45, 2021.
6. Cremer, M., Über die Ursache der elektromotorischen Eigenschaften der Gewebe, zugleich ein Beitrag zur Lehre von denpolyphasischen Elektrolytketten. *Z. Biol.*, 47, 562–608, 1906.

7. Sörensen, S.P.L., Enzymstudien. II. Mitteilung. Über die Messung und die Bedeutung der Wasserstoffionenkoncentration beienzymatischen Prozessen [Enzyme studies. 2nd Report. On the measurement and the importance of hydrogen ion concentration during enzymatic processes]. *Biochem. Z.*, 21, 131–304, 1909.
8. Griffin, E.G. and Nelson, J.M., The influence of certain substances on the activity of invertase. *J. Am. Chem. Soc.*, 38, 722–730, 1916.
9. Nelson, J.M. and Griffin, E.G., Adsorption of invertase. *J. Am. Chem. Soc.*, 38, 1109–1115, 1916.
10. Hughes, W.S., The potential difference between glass and electrolytes in contact with the glass. *J. Am. Chem. Soc.*, 44, 2860–2867, 1922.
11. Heineman, W.R., Jensen, W.B., Leland C. Clark Jr., Nanostructured materials-enabled biosensors., *Biosens. Bioelectron.*, 21, 1403–1404, 2006.
12. Clark, L.C. and Lyons, C., Electrode systems for continuous monitoring in cardiovascular surgery. *Ann. N. Y. Acad. Sci.*, 102, 29–45, 1962.
13. Updike, S.J. and Hicks, G.P., The enzyme electrode. *Nature*, 214, 986–988, 1967.
14. Guilbault, G.G. and Montalvo, J.G., Jr., Urea-specific enzyme electrode. *J. Am. Chem. Soc.*, 91, 2164–2165, 1969.
15. Newman, J.D. and Turner, A.P.F., Home blood glucose biosensors: A commercial perspective. *Biosens. Bioelectron.*, 20, 2435–2453, 2005.
16. D'Orazio, P., Biosensors in clinical chemistry. *Clin. Chim. Acta*, 334, 41–69, 2003.
17. Suzuki, S., Takahashi, F., Satoh, I., Sonobe, N., Ethanol and lactic acid sensors using electrodes coated with dehydrogenase–collagen membranes. *Bull. Chem. Soc. Jpn.*, 48, 3246–3249, 1975.
18. Clemens, A.H., Chang, P.H., Myers, R.W., Development of an automatic system of insulin infusion controlled by blood sugar, its system for the determination of glucose and control algorithms. *Journ. Annu. Diabétol*, 269–278, 1976.
19. Yoo, E.-H. and Lee, S.-Y., Glucose biosensors: An overview of use in clinical practice. *Sensors*, 10, 4558–4576, 2010.
20. Roederer, J.E. and Bastiaans, G.J., Microgravimetric immunoassay with piezoelectric crystals. *Anal. Chem.*, 55, 2333–2336, 1983.
21. Cass, A.E., Davis, G., Francis, G.D., Hill, H.A., Aston, W.J., Higgins, I.J., Plotkin, E.V., Scott, L.D., Turner, A.P., Ferrocene-mediated enzyme electrode for amperometric determination of glucose. *Anal. Chem.*, 56, 667–671, 1984.
22. J.S. Schultz, Oxygen sensor of plasma constituents. U.S. Patent 4344438A, 1982.
23. Liedberg, W., Nylander, C., Lundstrm, I., Surface plasmon resonance for gas detection and biosensing. *Sens. Actuators A Phys.*, 4, 299–304, 1983.
24. Mun'delanji, C.V. and Tamiya, E., Nanobiosensors and nanobioanalyses: A review, in: *Nanobiosensors and Nanobioanalyses*, pp. 3–20, Springer, Tokyo, Japan, 2015.
25. Poncharal, P., Wang, Z.L., Ugarte, D., De Heer, W.A., Electrostatic deflections and electromechanical resonances of carbon nanotubes. *Science*, 283, 1513–1516, 1999.
26. Gribi, S., De Dunilac, S.B., Ghezzi, D., Lacour, S.P., A microfabricated nerve-on-a-chip platform for rapid assessment of neural conduction in explanted peripheral nerve fibers. *Nat. Commun.*, 9, 4403, 1–10, 2018.
27. Turner, A.P.F., Biosensors: Sense and sensibility. *Chem. Soc. Rev.*, 42, 3184–3196, 2013.
28. Chen, X.J., Zhang, X.Q., Liu, Q., Zhang, J., Zhou, G., Nanotechnology: A promising method for oral cancer detection and diagnosis. *J. Nanobiotechnology*, 16, 1, 1–17, 2018.
29. Zou, B., Huang, C.Z., Wang, J., Liu, B.Q., Effect of nano-scale TiN on the mechanical properties and microstructure of $Si_3N_4$ based ceramic tool materials. *Key Eng. Mater.*, 315, 154–158, 2006.
30. Wang, X.H., Xu, C.H., Yi, M.D., Zhang, H.F., Effects of Nano-$ZrO_2$ on the microstructure and mechanical properties of Ti (C, N)-based cermet die materials. *Adv. Mater. Res.*, 154, 1319–1323, 2011.

31. Ullah, M., Naz, A., Mahmood, T., Siddiq, M., Bano, M., Biochemical synthesis of nickel and cobalt oxide nanoparticles by using biomass waste. *Int. J. Enhanc. Res. Sci. Technol. Eng.*, 3, 4, 415–422, 2014.
32. Manzoor, U., Islam, M., Tabassam, L., Rahman, S.U., Quantum confinement effect in ZnO nanoparticles synthesized by co-precipitation method. *Phys. E*, 41, 9, 1669–1672, 2009.
33. Brigger, I., Dubernet, C., Couvreur, P., Nanoparticles in cancer therapy and diagnosis. *Adv. Drug Deliv. Rev.*, 54, 631–651, 2002.
34. Krishnan, S.R. and George, S.K., Nanotherapeuticsincancerprevention, diagnosis, and treatment, in: *Pharmacology and Therapeutics*, S.J. Thatha (Ed.), pp. 953–978, Gowder, 2014.
35. Ahn, S., Seo, E., Kim, K., Lee, S.J., Controlled cellular uptake and drug efficacy of nanotherapeutics. *Sci. Rep.*, 3, 1997, 1–10, 2013.
36. Lytton-Jean, A.K., Kauffman, K.J., Kaczmarek, J.C., Langer, R., Cancer nanotherapeutics in clinical trials. *Cancer Treat. Res.*, 166, 293–322, 2015.
37. Masood, F., Polymeric nanoparticles for targeted drug delivery system for cancer therapy. *Mater Sci. Eng. C*, 60, 569–578, 2016.
38. Parveen, S. and Sahoo, S.K., Polymeric nanoparticles for cancer therapy. *J. Drug Targeting*, 16, 108–123, 2008.
39. Bhattacharyya, S., Kudgus, R.A., Bhattacharya, R., Mukherjee, P., Inorganic nanoparticles in cancer therapy. *Pharm. Res.*, 28, 237–259, 2011.
40. Kato, K., Chin, K., Yoshikawa, T., Yamaguchi, K., Tsuji, Y. *et al.*, Phase II study of NK105, a paclitaxel-incorporating micellar nanoparticle, for previously treated advanced or recurrent gastric cancer. *Invest. New Drugs*, 30, 1621–1627, 2012.
41. Libutti, S.K., Paciotti, G.F., Byrnes, A.A., Alexander, H.R., Gannon, W.E. *et al.*, Phase I and pharmacokinetic studies of CYT-6091, a novel PEGylated colloidal gold-rhTNF nanomedicine. *Clin. Cancer Res.*, 16, 6139–6149, 2010.
42. Bakhtiary, Z., Saei, A.A., Hajipour, M.J., Raoufi, M., Vermesh, O. *et al.*, Targeted superparamagnetic iron oxide nanoparticles for early detection of cancer: Possibilities and challenges. *Nanomedicine*, 12, 287–307, 2016.
43. Cho, K., Wang, X., Nie, S., Shin, D.M., Therapeutic nanoparticles for drug delivery in cancer. *Clin. Cancer Res.*, 14, 1310–1316, 2008.
44. Felber, A.E., Dufresne, M.H., Leroux, J.C., pH-sensitive vesicles, polymericmicelles, and nanospheres prepared with polycarboxylates. *Adv. Drug Deliv. Rev.*, 64, 979–992, 2012.
45. Adams, M.L., Lavasanifar, A., Kwon, G.S., Amphiphilic block copolymers for drug delivery. *J. Pharm. Sci.*, 92, 1343–1355, 2003.
46. Raavé, R., de Vries, R.B., Massuger, L.F., van Kuppevelt, T.H., Daamen, W.F., Drug delivery systems for ovarian cancer treatment: A systematic review and meta-analysis of animal studies. *Peer J.*, 3, e1489, 2015.
47. Anajwala, C.C., Jani, G.K., Swamy, S.V., Current trends of nanotechnology for cancer therapy. *Int. J. Pharm. Sci. Nanotechnol.*, 3, 1043–1056, 2010.
48. Saengkrit, N., Saesoo, S., Srinuanchai, W., Phunpee, S., Ruktanonchai, U.R., Influence of curcumin-loaded cationic liposome on anticancer activity for cervical cancer therapy. *Colloids Surf. B*, 114, 349–356, 2014.
49. Lin, Y.Y., Kao, H.W., Li, J.J., Hwang, J.J., Tseng, Y.L. *et al.*, Tumor burden talks in cancer treatment with PEGylated liposomal drugs. *PLoS One*, 8, e63078, 2013.
50. Keller, A.M., Mennel, R.G., Georgoulias, V.A., Nabholtz, J.M., Erazo, A. *et al.*, Randomized phase III trial of pegylated liposomal doxorubicin versus vinorelbine or mitomycin C plus vinblastine in women with taxane-refractory advanced breast cancer. *J. Clin. Oncol.*, 22, 3893–3901, 2004.

51. Sriraman, S.K., Salzano, G., Sarisozen, C., Torchilin, V., Anti-cancer activity of doxorubicin-loaded liposomes co-modified with transferrin and folic acid. *Eur. J. Pharm. Biopharm.*, 105, 40–49, 2016.
52. Kirpotin, D.B., Drummond, D.C., Shao, Y., Shalaby, M.R., Hong, K. *et al.*, Antibody targeting of long-circulating lipidic nanoparticles does not increase tumor localization but does increase internalization in animal models. *Cancer Res.*, 66, 6732–6740, 2006.
53. German, S., Navolokin, N., Kuznetsova, N., Zuev, V., Inozemtseva, O. *et al.*, Liposomes loaded with hydrophilic magnetite nanoparticles: Preparation and application as contrast agents for magnetic resonance imaging. *Colloids Surf. B*, 135, 109–115, 2015.
54. Yezhelyev, M., Yacoub, R., O'Regan, R., Inorganic nanoparticles for predictive oncology of breast cancer. *Nanomedicine*, 4, 83–103, 2009.
55. Kim, D. and Jon, S., Gold nanoparticles in image-guided cancer therapy. *Inorg. Chim. Acta*, 393, 154–164, 2012.
56. Patra, C.R., Bhattacharya, R., Mukhopadhyay, D., Mukherjee, P., Fabrication of gold nanoparticles for targeted therapy in pancreatic cancer. *Adv. Drug Deliv. Rev.*, 62, 346–361, 2012.
57. Sun, T., Zhang, Y.S., Pang, B., Hyun, D.C., Yang, M. *et al.*, Engineered nanoparticles for drug delivery in cancer therapy. *Chem. Int. Ed.*, 53, 12320–12364, 2014.
58. Connor, E.E., Mwamuka, J., Gole, A., Murphy, C.J., Wyatt, M.D., Gold nanoparticles are taken up by human cells but do not cause acute cytotoxicity. *Small*, 1, 325–327, 2005.
59. Abadeer, N.S. and Murphy, C.J., Recent progress in cancer thermal therapy using gold nanoparticles. *J. Phys. Chem. C*, 120, 4691–4716, 2016.
60. Bhowmik, T., Saha, P.P., Sarkar, A., Gomes, A., Evaluation of cytotoxicity of a purified venom protein from Naja kaouthia (NKCT1) using gold nanoparticles for targeted delivery to cancer cell. *Chem. Biol. Int.*, 261, 35–49, 2017.
61. Haley, B. and Frenkel, E., Nanoparticles for drug delivery in cancer treatment. *Urol. Oncol.*, 26, 57–64, 2008.
62. Pang, B., Yang, X., Xia, Y., Putting gold nanocagesto work for optical imaging, controlled release, and cancer theranostics. *Nanomedicine*, 11, 1715–1728, 2016.
63. Jeyaraj, M., Arun, R., Sathishkumar, G., MubarakAli, D., Rajesh, M. *et al.*, An evidence on G2/M arrest, DNA damage and caspase mediated apoptotic effect of biosynthesized gold nanoparticles on human cervical carcinoma cells (HeLa). *Mater. Res. Bull.*, 52, 15–24, 2014.
64. Tomoaia, G., Horovitz, O., Mocanu, A., Nita, A., Avram, A. *et al.*, Effects of doxorubicin mediated by gold nanoparticles and resveratrol in two human cervical tumor cell lines. *Colloids Surf. B*, 135, 726–734, 2015.
65. Miller, G., Colloid gold nanoparticles deliver cancer-fighting drugs. *Pharm. Technol.*, 7, 17–19, 2003.
66. Mahmood, M., Karmakar, A., Fejleh, A., Mocan, T., Iancu, C. *et al.*, Synergistic enhancement of cancer therapy using a combination of carbon nanotubes and anti-tumor drug. *Nanomedicine*, 4, 883–893, 2009.
67. Huo, D., He, J., Li, H., Yu, H., Shi, T. *et al.*, Fabrication of Au@Ag coreshell NPs as enhanced CT contrast agents with broad antibacterial properties. *Colloids Surf. B*, 117, 29–35, 2014.
68. Zhao, M.X. and Zhu, B.J., The research and applications of quantum dots as nano-carriers for targeted drug delivery and cancer therapy. *Nanoscale Res. Lett.*, 11, 1, 2016.
69. Vashist, S.K., Tewari, R., Bajpai, R.P., Bharadwaj, L.M., Raiteri, R., Review of quantum dot technologies for cancer detection and treatment. *A. Zojono J. Nanotechnol. Online*, 2, 1–14, 2006.
70. Malik, P., Gulia, S., Kakkar, R., Quantum dots for diagnosis of cancers. *Adv. Mater. Lett.*, 4, 811–822, 2013.

71. Voura, E.B., Jaiswal, J.K., Mattoussi, H., Simon, S.M., Tracking metastatic tumor cell extravasation with quantum dot nanocrystals and fluorescence emission-scanning microscopy. *Nat. Med.*, 10, 993–998, 2004.
72. Wu, X., Liu, H., Liu, J., Haley, K.N., Treadway, J.A. *et al.*, Immunofluorescent labeling of cancer marker Her2 and other cellular targets with semiconductor quantum dots. *Nat. Biotechnol.*, 21, 41–46, 2003.
73. Li, M., Li, R., Li, C.M., Wu, N., Electrochemical and optical biosensors based on nanomaterials and nanostructures: A review. *Front. Biosci. (Schol. Ed.)*, 3, 1308–1331, 2011.
74. Zhou, Y., Chiu, C.W., Liang, H., Interfacial structures and properties of organic materials for biosensors: An overview. *Sensors (Basel)*, 12, 15036–15062, 2012.
75. Guo, X., Single-molecule electrical biosensors based on single-walled carbon nanotubes. *Adv. Mater.*, 25, 3397–3408, 2013.
76. Ko, P.J., Ishikawa, R., Sohn, H., Sandhu, A., Porous silicon platform for optical detection of functionalized magnetic particles bio-sensing. *J. Nanosci. Nanotechnol.*, 13, 2451–2460, 2013.
77. Senveli, S.U. and Tigli, O., Biosensors in the small scale: Methods and technology trends. *IET Nanobiotechnol.*, 7, 7–21, 2013.
78. Valentini, F., Galache, F.L., Tamburri, E., Palleschi, G., Single walled carbon nanotubes/polypyrrole-GOx composite films to modify gold microelectrodes for glucose biosensors: Study of the extended linearity. *Biosens. Bioelectron.*, 43, 75–78, 2013.
79. Lamprecht, C., Hinterdorfer, P., Ebner, A., Applications of biosensing atomic force microscopy in monitoring drug and nanoparticle delivery. *Expert Opin. Drug Deliv.*, 11, 1237–1253, 2014.
80. Shen, M.Y., Li, B.R., Li, Y.K., Silicon nanowire field-effect-transistor based bio-sensors: From sensitive to ultra-sensitive. *Biosens. Bioelectron.*, 60, 101–111, 2014.
81. Sang, S., Wang, Y., Feng, Q., Wei, Y., Ji, J., Zhang, W., Progress of new label-free techniques for biosensors: A review. *Crit. Rev. Biotechnol.*, 15, 1–17, 2015.
82. Hutter, E. and Maysinger, D., Gold-nanoparticle-based biosensors for detection of enzyme activity. *Trends Pharmacol. Sci.*, 34, 497–507, 2013.
83. Su, L., Jia, W., Hou, C., Lei, Y., Microbial biosensors: A review. *Biosens. Bioelectron.*, 26, 1788–1799, 2011.
84. Ding, L., Bond, A.M., Zhai, J., Zhang, J., Utilization of nanoparticle labels for signal amplification in ultrasensitive electrochemical affinity biosensors: A review. *Anal. Chim. Acta*, 797, 1–12, 2013.
85. Turner, A.P., Biosensors: Sense and sensibility. *Chem. Soc. Rev.*, 42, 3184–3196, 2013.
86. Kwon, S.J. and Bard, A.J., DNA analysis by application of Pt nanoparticle electrochemical amplification with single label response. *J. Am. Chem. Soc.*, 134, 10777–10779, 2012.
87. Nie, S., Xing, Y., Kim, G.J., Simons, J.W., Nanotechnology applications in cancer. *Annu. Rev. Biomed. Eng.*, 9, 257–288, 2007.
88. Jain, R.K., Normalizing tumor microenvironment to treat cancer: Bench to bedside to biomarkers. *J. Clin. Oncol.*, 31, 2205–2218, 2013.
89. Malik, P., Katyal, V., Malik, V., Asatkar, A., Inwati, G., Mukherjee, T.K., Nanobiosensors: Concepts and variations. *ISRN Nanomater.*, 2013, 9, 2013.
90. Kim, J., Campbell, A.S., De Avila, B.E., Wang, J., Wearable biosensors for healthcare monitoring. *Nat. Biotechnol.*, 37, 4, 389–406, 2019.
91. Karim, R.A., Reda, Y., Fattah, A.A., Review—Nanostructured materials-based nanosensors. *J. Electrochem. Soc.*, 167, 3, 1–11, 2020.
92. Abrao Nemeir, I., Saab, J., Hleihel, W., Errachid, A., Jafferzic-Renault, N., Zine, N., The advent of salivary breast cancer biomarker detection using affinity sensors. *Sensors*, 19, 10, 2373, 2019.

93. Sung, H., Ferlay, J., Siegel, R.L., Laversanne, M., Soerjomataram, I., Jemal, A., Bray, F., Global cancer statistics 2020: GLOBOCAN estimates of incidence and mortality worldwide for 36 cancers in 185 countries. *CA: Cancer J. Clin.*, 71, 3, 209–249, 2021.
94. Michaelson, J., Satija, S., Moore, R., Weber, G., Halpern, E., Garland, A., Kopans, D.B., The pattern of breast cancer screening utilization and its consequences. *Cancer*, 94, 1, 37–43, 2002.
95. Yahalom, G., Weiss, D., Novikov, I., Bevers, T.B., Radvanyi, L.G., Liu, M., Rosenberg, M.M., An antibody-based blood test utilizing a panel of biomarkers as a new method for improved breast cancer diagnosis. *Biomark. Cancer*, 5, BIC–S13236, 2013.
96. Ranjan, R., Esimbekova, E.N., Kratasyuk, V.A., Rapid biosensing tools for cancer biomarkers. *Biosens. Bioelectron.*, 87, 918–930, 2017.
97. Soper, S.A., Brown, K., Ellington, A., Frazier, B., Garcia-Manero, G., Gau, V., Wilson, D., Point-of-care biosensor systems for cancer diagnostics/ prognostics. *Biosens. Bioelectron.*, 21, 10, 1932–1942, 2006.
98. Mittal, S., Kaur, H., Gautam, N., Mantha, A.K., Biosensors for breast cancer diagnosis: A review of bio receptors, bio transducers and signal amplification strategies. *Biosens. Bioelectron.*, 88, 217–231, 2017.
99. Labib, M., Mohamadi, R.M., Poudineh, M., Ahmed, S.U., Ivanov, I., Huang, C.L., Kelley, S.O., Single-cell mRNA cytometry via sequence-specific nanoparticle clustering and trapping. *Nat. Chem.*, 10, 5, 489–495, 2018.
100. Hayes, B., Murphy, C., Crawley, A., O'Kennedy, R., Developments in point-of-care diagnostic technology for cancer detection. *Diagnostics*, 8, 2, 39, 2018.
101. Sharifi, M., Hasan, A., Attar, F., Taghizadeh, A., Falahati, M., Development of point-of-care nanobiosensors for breast cancers diagnosis. *Talanta*, 217, 121091, 2020.
102. Wang, L., Early diagnosis of breast cancer. *Sensors*, 17, 7, 1572, 2017.
103. Ranjan, P., Parihar, A., Jain, S., Kumar, N., Dhand, C., Murali, S., Khan, R., Biosensor-based diagnostic approaches for various cellular biomarkers of breast cancer: A comprehensive review. *Anal. Biochem.*, 610, 113996, 2020.
104. Xu, K., Lu, Y., Takei, K., Multifunctional skin-inspired flexible sensor systems for wearable electronics. *Adv. Mater. Technol.*, 4, 1800628, 2019.
105. Honda, W., Harada, S., Ishida, S., Arie, T., Akita, S., Takei, K., High-performance, mechanically flexible, and vertically integrated 3D carbon nanotube and InGaZnO complementary circuits with a temperature sensor. *Adv. Mater.*, 27, 32, 4674–4680, 2015.
106. Yang, H., Qi, D., Liu, Z., Chandran, B.K., Wang, T., Yu, J., Chen, X., Soft thermal sensor with mechanical adaptability. *Adv. Mater.*, 28, 41, 9175–9181, 2016.
107. Yokota, T., Inoue, Y., Terakawa, Y., Reeder, J., Kaltenbrunner, M., Ware, T., Yang, K., Mabuchi, K., Murakawa, T., Sekino, M. *et al.*, Ultraflexible, large-area, physiological temperature sensors for multi-point measurements. *Proc. Natl. Acad. Sci. U.S.A.*, 112, 47, 14533–14538, 2015.
108. Oh, J.H., Hong, S.Y., Park, H., Jin, S.W., Jeong, Y.R., Oh, S.Y., Yun, J., Lee, H., Kim, J.W., Ha, J.S., Fabrication of high-sensitivity skin attachable temperature sensors with bioinspired micro structured adhesive. *ACS Appl. Mater. Interfaces*, 10, 8, 7263–7270, 2018.
109. Zhu, C., Chortos, A., Wang, Y., Pfattner, R., Lei, T., Hinckley, A.C., Pochorovski, I., Yan, X., To, J.W.F., Oh, J.Y. *et al.*, Stretchable temperature-sensing circuits with strain suppression based on carbon nanotube transistors. *Nat. Electron.*, 1, 3, 183, 2018.
110. Trung, T.Q., Le, H.S., Dang, T.M.L., Ju, S., Park, S.Y., Lee, N.E., Freestanding, fiber-based, wearable temperature sensor with tunable thermal index for healthcare monitoring. *Adv. Healthcare Mater.*, 7, 12, 1800074, 2018.
111. Ota, H., Chao, M., Gao, Y., Wu, E., Tai, L.C., Chen, K., Matsuoka, Y., Iwai, K., Fahad, H.M., Gao, W. *et al.*, 3d printed "earable" smart devices for real-time detection of core body temperature. *ACS Sens.*, 2, 7, 990–997, 2017.

112. Dobrescu, A.I., Ardelean, L., Matei, C., Tampa, M., Puiu, M., Mihaicusa, S., Polysomnography test and sleep disordered breathing in prader-willi syndrome. *Mater. Plast.*, 51, 331–335, 2014.
113. Lee, W., Liu, Y., Lee, Y., Sharma, B.K., Shinde, S.M., Kim, S.D., Nan, K., Yan, Z., Han, M., Huang, Y. *et al.*, Two-dimensional materials in functional three-dimensional architectures with applications in photodetection and imaging. *Nat. Commun.*, 9, 1, 1417, 2018.
114. Park, S., Fukuda, K., Wang, M., Lee, C., Yokota, T., Jin, H., Jinno, H., Kimura, H., Zalar, P., Matsuhisa, N. *et al.*, Ultraflexible near-infrared organic photodetectors for conformal photoplethysmogram sensors. *Adv. Mater.*, 30, 34, 1802359, 2018.
115. Xu, H., Liu, J., Zhang, J., Zhou, G., Luo, N., Zhao, N., Flexible organic/inorganic hybrid near-infrared photoplethysmogram sensor for cardiovascular monitoring. *Adv. Mater.*, 29, 31, 1700975, 2017.

# 16

# Bioreceptors for Antigen–Antibody Interactions

**Vipul Prajapati[1]* and Princy Shrivastav[2]**

*[1]Department of Pharmaceutics, SSR College of Pharmacy (Permanently Affiliated to Savitribai Phule Pune University), Sayli-Silvassa Road, Sayli, Silvassa, Union Territory of Dadra Nagar Haveli & Daman Diu, India*

*[2]Department of Pharmaceutical Chemistry, SSR College of Pharmacy (Permanently Affiliated to Savitribai Phule Pune University), Sayli-Silvassa Road, Sayli, Silvassa, Union Territory of Dadra Nagar Haveli & Daman Diu, India*

***Abstract***

The analyte is specifically detected and bound by a bioreceptor in a biosensor that further interacts with the analyte through a process known as biorecognition. Recent developments in biosensor technology have largely focused on the immune system. The immune system works by producing antibodies after identifying exotic species called antigens. Antibodies are too specific in their ability to biorejugate, making them a suitable choice for incorporation into biosensors. This chapter provides a brief overview of antibody and antigen–antibody interactions. In addition, antibody-based biosensors (also known as immunosensors) were discussed along with their application of immunosensors in diagnostics, drug safety, food safety, and environmental control. Finally, some new approaches to antibody modifications were also discussed, such as antibody mimetics, camelid nanobodies, and revised nanobodies. These modifications offer several advantages over classical antigen–antibody receptors. In short, immunosensors are a key component in the food and healthcare industry due to their high selectivity and sensitivity, where they are used for a variety of purposes. Despite recent advances in immunosensors field, new methods are still required to enhance the sensitivity, selectivity and ease of use of such devices so that they can fulfill the stringent wants of clinical diagnostics or industry.

***Keywords*:** Bioreceptors, antigens, antibodies, interactions, immunosensors, biorecognition

## 16.1 Introduction

Biosensors are inorganic and organic devices that assess analytes in samples and provide qualitative and quantitative information. This analyte is specifically detected and bound by a bioreceptor (the organic component) in a biosensor and then a signal is generated in the sensor with the help of a transducer (the inorganic component) [1, 25]. Bioreceptors are organic molecular species that include enzymes, cells, proteins, and antibodies, to name a few. When bioreceptors interact with the analyte, changes in heat, charge, mass, pH, and other signals occur. This is called biorecognition (or biochemical recognition). To create a

*Corresponding author*: vippra2000@yahoo.com

Inamuddin and Tariq Altalhi (eds.) Biosensors Nanotechnology, 2nd Edition, (371–394) © 2023 Scrivener Publishing LLC

biosensor sensitive enough for application, the receptor must be selective for the biomarker of interest and generate adequate signal [1, 26, 28]. Because biosensors are designed to allow for fast and reliable analysis by target analysts, a critical step in optimizing a biosensor is the choice of bioreceptor. Four of the most important biorecognition elements are antigen binding units of whole fragments of monoclonal antibodies (mAb), Fab', aptamers and scFv fragments [31]. Recent advances in biosensor technology have mostly targeted the immune system, which is capable of distinguishing the self from the non-self. Antibodies, which are responsible for biorecognition of foreign species, such as infections and poisons, are synthesized in adequate amounts by the immune system. Antigens (Ag) are foreign organisms like this [2, 8]. Due to their high sensitivity and selectivity, as well as the availability of antibodies, immunochemical methods are helpful for a wide range of chemicals in a wide range of applications. Clinical research is one such application. Immunological sensors are biosensors consisting of antibodies [5, 9]. Polyclonal, monoclonal, or recombinant antibodies are utilized in a number of applications, including immunodiagnosis and biomarker detection [11]. Therefore, antibody–antigen interactions must be characterized in order to create antibody-based treatments [7, 12].

## 16.2 Antibodies: A Brief Overview

Due to their excellent affinity and specificity for targets, antibodies that are among the most carefully designed and naturally produced molecules play an important role in a variety of detection devices. This explains why, after aptamers and, more recently, mitochondrial RNA, they are the most widely used bioreceptors in the development of electrochemical immunological sensors, where they are normally immobilized on a substrate that acts as the sensor's surface [2, 26].

Around 3500 antibody fragment structures and a small number of complete antibody structures are currently available in the Protein Data Bank (PDB) [13].

### 16.2.1 What Are Antibodies?

Abdominals are highly soluble serum glycoproteins that are produced in animals in response to the presence of an Ag and are specific for this substance. Biochemical pathways can be used to identify these antibodies and are purified for biosensor applications [29, 33]. Immunoglobulins are another name for them [2, 3]. B lymphocytes, which come from the bone marrow of animals, form them [24].

Based on the sequences of their constant heavy chain ranges, they can be divided into five categories: IgM, IgD, IgG, IgE, and IgA. Of the five classes, IgG are mainly used for immunological testing [2]. All are structurally similar, with the only differences being glycosylation and the number of disulfide bridge locations. The elementary structure of an Ab is shown in Figure 16.1, which has two parts: the antigen binding fragment (Fab) and the constant fragment (Fc). The antibody structure contains an amine ($-NH_3^+$) and a carboxyl group (–COOH) at the upper and bottom end, respectively. These help in antibody immobilization and communication with other compounds. An Ab consists of four chains made up of polypeptide, out of which two are heavy chains and the other two are light chains, all connected by disulfide bonds. A variable region (VL) and a constant region (CR) form the

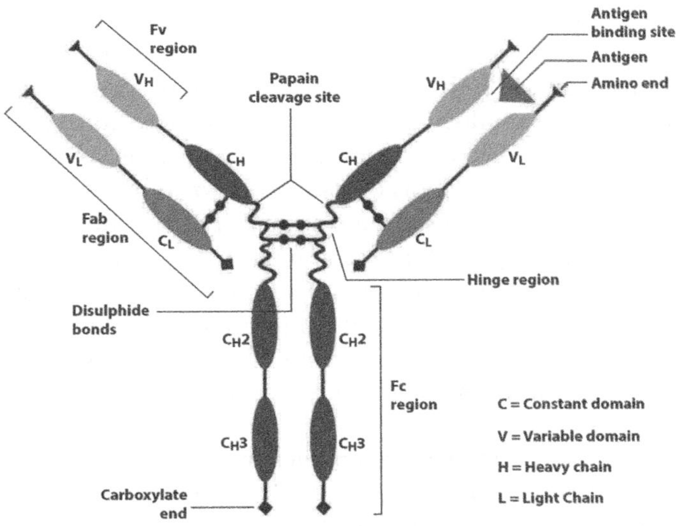


**Figure 16.1** A general structure of an antibody.

light chain (LC). One variable (V) and three constant regions (C) are found in IgD, IgG, and IgA. IgE and IgM are distinguished by one variable and four constant intervals. The additional J chain produces dimers in IgA and pentamer in IgM isotypes. The remaining isotypes are monomeric. In the area of fragment variables (Fv) are the regions that determine the complementarity (CDR) of the Fab component of Ab. The Binding Sites to the Ab antigen are formed by CDR that confer specificity to the antigen. The Fc region plays a role in cell-mediated dependent cytotoxicity, antibody-dependent cell phagocytosis (ADCC), complement-mediated lysis, degranulation, cell activation and proliferation, and antigen presentation to the immune system [4, 11, 13, 29, 30].

## 16.2.2 Types of Antibodies

Antibodies are divided into two categories depending on how they are produced by lymphocytes (monoclonal and polyclonal antibodies). Each of them is important for the immune system, diagnostic tests and treatment (see Figure 16.2).

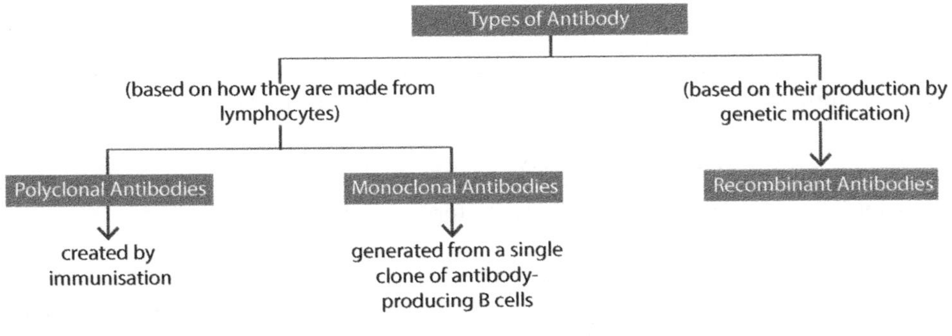


**Figure 16.2** A flowchart depicting the types of antibodies.

Polyclonal antibodies (pAbs) are produced by immunization and bind to numerous epitopes in a given antigen. The procedure is quite inexpensive, and a single extraction can produce large amounts of an antibody. This produces inhomogeneous antibodies.

A single clone of antibody-producing B cells produces monoclonal antibodies (mAbs). Following immunization, the spleen is removed, and the B cells combine with immortal myeloma cells. Fused B cells and myeloma cells (also known as hybridomas) are tested with an ELISA before being injected into a second mammal. The hybridoma's myeloma component promotes localized tumor growth in the animal, and ascites, an antibody-rich fluid, may be collected and separated from the tumor using spine chromatography. Monoclonal antibodies are particularly valuable because the immortal nature of the hybridoma allows for continuous culture, resulting in a constant supply of monoclonal antibodies. Monoclonal antibodies are often chosen in immunoassay procedures because this approach produces homogeneous antibodies.

In addition to these two, recombinant antibodies (rAbs) have become popular recently due to genetic modification and the use of a different producing cell. They are made of modified microbes. Many of these recombinant antibodies are isolated in large quantities and must be folded again. Recombinant antibodies give researchers more control over the antibodies they produce. This opens up the possibility of incorporating amino acids into the sensor scaffold at the back of the antibody [5, 26].

Immunosensors can be built from recombinant Ab fragments, which are useful and durable building blocks. Unlike typical Abs, rAb fragments are stable, small and have extremely focused immobilization ability. Despite the advantages mentioned above, rAbs are currently not used enough in the field of immunosensors. rAbs are expected to have much greater potential for the development of Ab-based sensors with a variety of new biomedical diagnostic applications [4]. The pros and cons of the three antibody types are summarized in Table 16.1.

**Table 16.1** Pros and cons of the three antibody types [4].

| Type of antibody | Pros | Cons |
|---|---|---|
| Polyclonal antibody | Low | Lack of reproducibility |
| | | Limited quantity |
| | | Specificity is too high |
| | | Intrinsic instability |
| Monoclonal antibody | Strong affinity<br>Wide specificity<br>High reproducibility | High costs |
| Recombinant antibody | Avoid the need for laboratory animals<br>High-speed production<br>Subsequent molecular increase in binder yield | High costs<br>Hour<br>Stability adjusted in terms of pH, ionic strength and temperature |

### 16.2.3 Production and Purification of Antibodies

The immune system of cleftines, mice, birds or sheep is used to produce monoclonal, polyclonal and recombinant antibodies. To produce bacterial pathogenic antibodies, these hosts can be immunized with heat-treated or unheated antigens. These antigens are usually administered with an adjuvant, and the host-triggered immune response after a series of vaccines can be evaluated with an ELISA to detect serial serum dilutions for antigen recognition [5, 11].

Many purification methods for the production of raw antibodies have been developed over the years. Since immunoglobulin molecules are typical proteins, conventional protein purification techniques can also be used to purify immunoglobulins [11].

#### *16.2.3.1 Polyclonal Antibodies*

pAbs occur regularly in rabbits, goats or sheep and their widespread use in immunosensors to detect pathogens is evidence of their popularity [11]. It is worth noting that pAbs can detect a variety of epitopes in a single cell due to their basic nature. In general, three processes are involved in the production of pAbs: (1) injection of an antigen into a mammal, (2) regular bleeding from the animal, and (3) collection of antibodies directly from serum [17].

Tolerance or hypersensitivity to many antigens occur rather spontaneously. For this reason, adjuvants are essential in most vaccination programs. The Titer-max and Freund adjuvants, in particular Freund's incomplete adjuvant (FIA) and Freund's complete adjuvant (FCA), are two well-known options [16].

Choosing a suitable animal is a key consideration in the development of a polyclonal antibody. Purebred animals are less expensive; however, their reproducibility is worse than that of inbred animals. Specific pathogen-free (SPF) animals help to improve specificity of pAbs. Animals with SPF, on the other hand, may be more susceptible to death due to a weakened immune system. For the production of polyclonal antibodies, animals like lambs, rats, goats, horses, hamsters, mice, and pigs should be used; however, rabbits seem to be the most comfortable. Chickens or eggs can be used, although the structure of bird

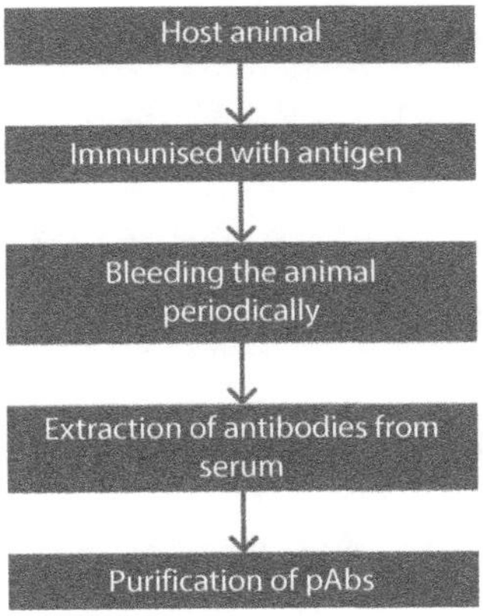


**Figure 16.3** The steps of production of polyclonal antibodies.

antibodies is somewhat different from that of human antibodies. Other major concerns are the number of antibodies needed and the availability of a collection of animals. Polyclonal antibodies can be used to produce antibodies against an entire microorganism or some small compounds such as organophosphate [11, 16]. The whole process is described in Figure 16.3.

### *16.2.3.2 Monoclonal Antibodies*

Hybridoma technology is used to produce monoclonal antibodies, and mouse hosts are often selected for immunization [11]. As shown in Figure 16.4, the experimental animal is immunized at the beginning of the antibody production process. Antibody-producing B lymphocytes are isolated from primary lymph nodes, bone marrow, or, most commonly, the spleen before fusing with mutant myeloma cells. The selective hypoxanthine-aminopterin-thymidine (HAT) medium is used to distinguish fused cells from unfused cells. Normal cells feature both the primary mechanism and an alternate method based on the enzyme hypoxanthine-guanine phosphoribosyl transferase (HGPRT) for nucleotide synthesis. Because the HGPRT genes in the myeloma cells were defective, 8-azaguanine killed the cells that translated the HGPRT enzyme. When aminopterin inhibits the major pathway for nucleotide synthesis, only cells fused (hybridomas) with HGPRT genes can arise from B lymphocytes and metabolize hypoxanthine. The resulting hybridomas produce some antibodies. Hybridoma cultures are diluted in a single cell per source of microtitration plates, with colonies forming from each cell. A single clone is responsible for the production of antibodies; therefore, they are monospecific. Only colonies that produce antibodies of the right quality should be grown, allowing for large-scale antibody production. Cell line stock can be stored permanently by freezing. Live animals can also be used to create monoclonal antibodies. When hybridoma cells are put into animals, they produce tumors that contain ascites fluid, an

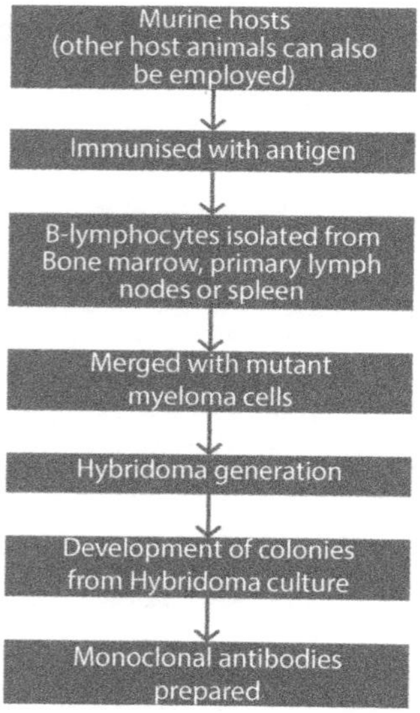


**Figure 16.4** The steps of production of monoclonal antibodies.

antibody-rich solution. Monoclonal antibodies are currently produced in fermentation chambers because this process is extremely animals and is sometimes considered unethical [11, 16].

Monoclonal antibodies have been created to combat many antigens, including tiny chemicals, bigger structures, and even whole cells. Human mAbs of the proper isotype can be produced against specific antigens using mice and standard hybridoma technologies. Humans are used to recreate the original Loci Ig in mice. Antibodies made in this manner are well tolerated by the human body, allowing for new treatments and diagnosis [16].

### *16.2.3.3 Recombinant Antibodies*

Recombinant antibodies (rAbs) for detection of a variety of structurally varied antigens were identified using biopanning antibody libraries and phage imaging technology against a target of interest. Naive, synthetic, and immune libraries are the three types of libraries that can be used to produce antibody clusters. Antibody fragments are classified into two types: Fab and single-chain variable fragment (scFv). How they are produced

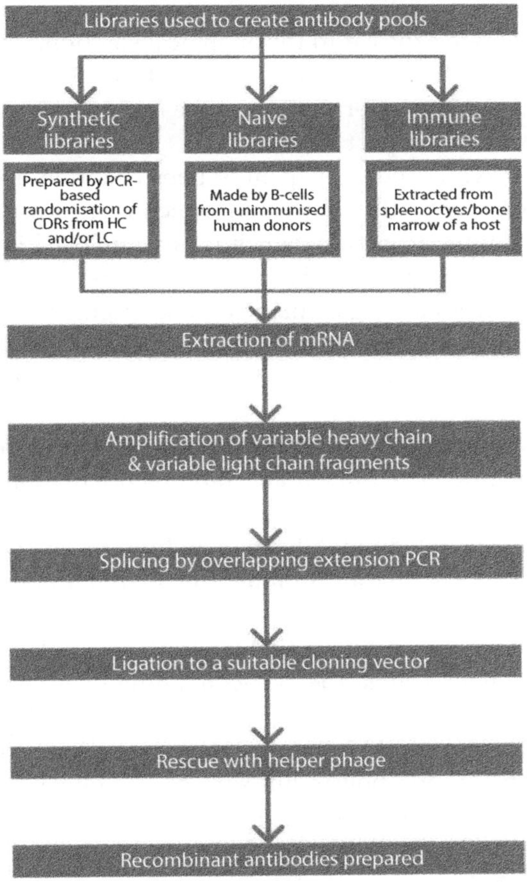


**Figure 16.5** The steps of production of recombinant antibodies.

is influenced by the vector in which the library is stored. Bio-panning is a method for selecting binders from an antibody library that contains many antibody-coding gene sequences. To do this, the antigen is either immobilized in the solid phase (immune tubes) or conjugated, and the antibody pool is subjected to a series of antigen selection cycles with stricter binding capacity criteria. The chosen binders are saved and tested further to increase their target specificity. Figure 16.5 summarizes this technique. The sensitivity and specificity of rAbs can be significantly improved by mutagenesis targeting the position or mixing of chains. Other advantages include the possibility to directly fuse other labels (such as proteins or green fluorescent enzymes) with the antibody fragment, making detection easier and simpler, and the availability of a range of antibody forms. The avian hosts used in the manufacture of high-affinity recombinant antibodies has been proven particularly advantageous [11].

### 16.2.4 Antibodies as Bioreceptors

Antibodies were introduced into the biosensors, allowing the development of biosensors *in vivo* [5]. A plasma immunoassay is a blood test used to detect antibodies. Insulin was the first to show that immunological components can be used as sensitive agents in humans. Antibody bioreceptors are immobilized in typical antibody biosensors on the surface of the transducer. After that, you are exposed to the analytical solution. The transducer converts biorecognition into a signal when the analyte binds to the antibody bioreceptor. The home pregnancy test (which detects hCG hormone) was the first commercially accessible immunological test designed with this aspect in mind.

With the proteolytic enzymes such as papain or pepsin, antibodies can be broken down into many constituent fragments. The Fab region, which is responsible for antigen specific binding, is segregated from the Fc region, which is not employed in immunosensors. The disulfide bridge in the Fab region may need to be reduced, and dithiothreitol is employed to remove non-specific interactions. Single-chain variable fragments (scFv) are smaller than Fab fragments, but have the same selectivity. They are manufactured by fusing the VL and VH chains. An extensive library of antibody fragments has been created for a variety of antigen sites.

Immunoglobulin macromolecules often bind asymmetrically to the sensor platforms on which they function, resulting in an unfavorable binding orientation and decreased sensor specificity and repeatability. Because the antibody fragments are tiny and symmetrical in structure, they enable for accurate immobilization and improved analyte selectivity due to less steric impairment. To construct an immunosensor containing insulin, the Fab fragment of a monoclonal antibody against bovine insulin was employed as the foundation for an SPR sensor. Another group employed scFv selection and cloning procedures to fuse an antibody's scFv component against the cyst-containing light chain residue at the light chain's C-terminus, resulting in an immunosensor that fixed the fragments evenly on the surface of an SPR chip. Unlabeled EIS immunosensors were created by immobilizing antibodies and antibody fragments on a polyaniline surface. The internal B antigen (as a marker of the foodborne pathogen Listeria monocytogenes) can be recognized by linking the polymer's redox state to charge transfers that occur during the formation of the Ab/Ag complex via impedance changes [15, 18].

## 16.3 Antigen–Antibody Reactions

The most critical role of antibody molecules in an immunological approach is to attach to their specific antigen to create an antibody–antigen complex [2]. Antigens react solely with self-produced antibodies or certain antigens that are closely linked to them in extremely specific reactions. Abs recognize antigen molecular structures (epitopes). The better the geometry and chemical character of the epitope that matches to the antibody combination site, the better the antibody–antigen interactions and the stronger the affinity of the antibody with the antigen. The affinity of the antibody for the antigen is one of the most essential factors in determining the efficacy of antibodies *in vivo*. An antigen–antibody interaction is defined as an irreversible bimolecular link between the antigen and the antibody. The antigen–antibody interaction is altered by non-covalent contacts between the epitope and the VH/VL region of the antibody. The non-covalent interactions between the Binding Site of Ab and the epitope are similar to how proteins connect to their cellular receptors or bind enzymes to their substrates. High ionic strength or excessive pH can break or break the bond. Hydrogen bonding, hydrophobic contacts, stearic repulsion, electrostatic forces, and Van der Waals forces are intermolecular forces that help stabilize the antibody–antigen complex. Since all of these non-covalent contacts occur at such a short distance (typically less than 1 Å), Ag–Ab interactions require an exceptional level of antigen–antibody similarity [14, 19].

This phenomenon serves a variety of purposes, one of which is the development of sensors. An immunosensor (also called an immune sensor) is a sensor that uses an antibody (as a bioreceptor) for accurate antigen recognition and thus produces a stable immune complex [3]. There are several types of Ag–Ab reactions, including:

1. Agglutination
2. Precipitation
3. Fixing the complement
4. Enzyme-linked immunosorbent assay
5. Radioimmunological assay
6. Western Blotting

### 16.3.1 Agglutination

The apparent aggregation that occurs when an antibody interacts with a particle antigen is called agglutination. Agglutinins are antibodies that cause such reactions. IgM antibodies bind more effectively than IgG antibodies. The prozoning phenomenon occurs when an excess of an antibody or antigen blocks the agglutination response. Incomplete or monovalent antibodies do not trigger agglutination, even if they interact with the antigen. They could act as blocking antibodies and prevent the entire antibody from agglutinating [19].

### 16.3.2 Precipitation

The antigen is soluble in this type of antigen–antibody interaction. The antigen–antibody complex forms an insoluble precipitate in the presence of electrolytes when a soluble antigen is combined with the respective antibody at the correct temperature and pH. A precipitation reaction is the term for this. A network between antigens and antibodies is formed, which in some cases may be visible as an insoluble precipitate. Precipitins are antibodies that bind to soluble antigens and group them together.

The development of an antigen–antibody complex is influenced by the valence of both the antibody and the antigen. The antibody must be bivalent, as monovalent Fab fragments do not precipitate. The antigen must be bivalent or multivalent and contain at least two copies of the same epitope or many epitopes that bind to separate antibodies in polyclonal sera. In relation to each other, both the antigen and the antibody must have the correct concentration [19].

### 16.3.3 Complement Fixation

It is a test that can be performed to assess the presence of antibodies in a patient's serum. The test is split into two parts. First, a system composed of sheep RBCs, complement-fixing antibodies, and an external complement source, commonly serum of guinea pig. When these ingredients are appropriately combined, an anti-sheep antibody attaches to the surface of RBCs. The complement system produces an antigen–antibody complex that attaches to and liquefies RBCs.

In addition to supplementation, a known antigen and a patient serum are added to a suspension of sheep RBCs. The two components of this method are investigated one after the other. The solution is then blended with the supplement after the patient's serum has been mixed with the known antigen. If there are antigen-specific antibodies in the serum, the antigen–antibody complexes bind to all complement components. Later, anti-sheep antibodies and sheep red blood cells are introduced. If the complement has not been bound by an antigen–antibody combination formed by the patient's blood and known antigens, it can be attached to the sheep cell and anti-sheep antibody signaling system. The absence of antibodies in the patient's serum, as well as a failed complement fixation test, is shown by the indicator of lysis of sheep red blood cells. The absence of red blood cell lysis implies a good prognosis if the patient's serum has an antibody that fixes the complement [19].

### 16.3.4 Radiomunoassay

The concept of isotopic dilution and the binding of a specific antibody to a part of the material to be analyzed are used in the radioimmunological testing process. When an antigen combines with a specific antibody against a chemically different material from the antigen or antibody, an antigen–antibody complex is formed. If there are not enough antibodies to completely complicate the entire antigen, the unlabeled antigen can be determined by combining the antibody with a known amount of labeled isotope antigen and an unknown amount of labeled antigen. The affinity of the antigen–antibody response and the highly specific binding sites in the antibodies used determine the sensitivity and specificity of radiomunoassay (RIA).

In RIA, sufficient amounts of labeled unknown standards, antigens, and antibodies are mixed into a buffer solution, and the reaction can reach equilibrium. Most competitive bond tests are equilibrium tests, in which all reactants are aggregated at the same time and the reaction can continue until equilibrium is reached. It may be important to add tracers later to improve sensitivity under certain circumstances. A sequential saturation test is the term for this type of exam. Free and bound fractures are separated using a suitable technique at the end of the incubation period. Counting the free or bonded radioactivity, or both, with the appropriate counting devices determines the distribution of radioactivity in each sample. The inhibition detected in the unknown is compared to that caused by standard solutions of known antigen concentration to estimate the antigen concentration. Dose-response curves are created using standard data, and the unknown concentration of antigens is read directly from the graph for this purpose [20].

### 16.3.5 Enzyme-Linked Immunosorbent Assay

Due to dangers of using radiation, RIA assays were modified to use an enzyme instead of the radioisotope, leading to modern-day enzyme-linked immunosorbent assay (ELISA). ELISA uses antigens and antibodies labeled with enzymes (glucose oxidase and alkaline phosphatase) to detect biological molecules. In 96-well microtitration plates, the fluid-phase antigen is often immobilized. Allowing the antigen to bind to a specific antibody, which is then detected by a secondary antibody coupled to the enzyme. A noticeable change in color or fluorescence caused by a chromogenic substrate for the enzyme indicates the presence of antigen. Colorimetric readings can be used to investigate both quantitative and qualitative information. Fluorogenic substrates have a higher sensitivity and can detect antigen concentrations in samples more accurately. In the ELISA experiment, the antigen is detected directly or indirectly by attaching or immobilizing the antigen-specific entrapment antigen or antibody directly to the well surface.

ELISA is a versatile technology that can be used for both scientific research and diagnostic purposes. It has the ability to detect a wide range of biological compounds at very low concentrations and amounts [21]. This method has recently been used to detect the COVID-19 antigen or its antibody in host body [25].

### 16.3.6 Western Blotting

Blotting is the process of transferring proteins or nucleic acids to microporous membranes [22]. Western Blot is a technique for identifying, quantifying and sizing proteins. Western Blotting evolved from Northern Blotting, which recognizes, quantifies, and estimates RNA size, using gel electrophoresis to separate RNA and Southern Blotting, which detects specific DNA sequences separated by gel electrophoresis. Protein blotting, commonly known as immunoblotting, is a technique for analyzing proteins that has recently gained popularity. SDS-PAGE (Sodium Dodecyl Sulfate Polyacrylamide Gel Electrophoresis) is used to separate native and denatured proteins. The proteins are transported to a membrane, where they are identified with the help of target-specific antibodies [23].

## 16.4 Antibody-Based Biosensors (Immunosensors)

Fast, simple and cost-effective analysis devices are required for the detection of traces of clinically and environmentally relevant substances, also for the food and pharmaceutical industries. Owing to their high sensitivity and selectivity, immunosensors are an optimal choice for detecting and monitoring the above domains. Recent progresses in immunosensors, such as optimal abdominal orientation, the inclusion of nanomaterials to improve multiplexing and sensitivity, and microfluidic-based devices, could have significant economic and therapeutic implications [2].

### 16.4.1 What are Immunosensors?

Immunosensors are solid-state devices connected to a transducer through an immunochemical reaction. They are one of the most frequent forms of affinity biosensors because, like the immunological test, they rely on antibodies for their biorecognition ability of antigens. Modern transducer technology, unlike immunological assays, enables for the identification and measurement of the immune complex without the need of a label. To avoid misinterpretations, tests based on immune responses are called "immunological assays," while complete instruments (biosensors) based on the immune response are called "immunosensors." Immunosensors greatly simplify analysis compared to conventional immunochemical techniques, making it fast and accurate [2, 44].

Immunosensors are divided into four categories depending on the type of transducer: optical, electrochemical, thermometric and microgravimetric. With these immunosensors, direct or indirect immunosensors can be used. Electrochemical immunosensors have become increasingly important recently due to their inexpensive, high sensitivity, low power consumption, and ability to reduce and automate [2, 43].

### 16.4.2 Selection of Antibodies Suitable for Immunosensors

This is the most important step in the immunosensor design process. Since Abs requires only a small fraction of the molecule to detect Ag, choosing an Abs based on stability and selectivity is crucial.

The IgG molecule has a complex glycosylation model based on disulfide bonds. This explains both the low thermal stability and the complicated and expensive production process. When comparing pAbs with mAbs, pAbs are better in situations where "very high" specificity is not required, such as. B simultaneous detection of several bacterial serotypes. However, when a single antibody is used to detect many bacterial serotypes or viral variants, mAbs provide extremely high specificity that can limit their application. Therefore, Mab cocktails should be used [24]. Compared to antibodies (which have a size of ~10–15 nm), aptamers (with a size of ~1–2 nm) can achieve greater surface coverage with more binding sites available per biosensor surface [29].

Antibodies directed against other antibodies are called secondary antibodies which are also used more recently in addition to primary antibodies. Secondary antibodies are typically coupled to a labeling molecule, allowing for better detection and quantification, as well as use in immunological assays for indirect target detection. An unmodified anti-target antibody first binds to the target trapped on the surface, then recognizes a labeled secondary antibody. Despite the fact that indirect detection takes longer than direct detection, detection limits are typically greater [24].

Camels and shark antibodies are also being investigated, as they are small in size and comparatively more stable than ordinary antibodies. The small antibody changes in development of sensor are designed to improve surface coverage while facilitating integration and contraction. This has been discussed in more detail in the relevant parts of the chapter [24, 29].

To verify that antibody to be used in the immunosensor is successful, a number of methodologies are employed to examine essential binding parameters. ELISA, optical methods like SPR and biolayer interferometry, and radioisotope tagging can all be beneficial [8].

### 16.4.3 Application of Immunosensors in Diagnostics

Immunosensors are interesting candidates for the development of diagnostic tools because of their most important property, the specificity of Ab-Ag interactions. Immunosensors may also provide the following benefits: lower medical expenditures, miniaturization, and faster response. Nucleic acids, proteins, hormones, metabolites, entire cells, and a variety of pathogenic microorganisms are bioanalytical objectives for the manufacture of diagnostic equipment [2].

#### *16.4.3.1 Antibodies for Detection of Proteins*

Clinical diagnosis is one of the most essential applications for sensors [3]. Antibodies can be used to detect protein biomarkers in conditions such as cancer, heart disease, and Alzheimer's. In this section, the presumed protein indicators of anti-canthion (CA125), prostate-specific antigen (PSA), mucin 1 (MUC1, CA 15-3), and carcinoembryonic antigen (CEA) [32] are first examined.

CA125 is a glycoprotein that can be used as a diagnostic and therapeutic surveillance biomarker for breast, stomach, liver and ovarian cancer. It has been exploited for the developing several optical sensors [2].

PSA is a prostatic protein that is overexpressed in prostate cancer. An unlabeled white light reflectance spectroscopy-based immunosensor was utilized to detect it in human serum. A biotinylated antibody was combined with streptavidin in a two-site immunoassay. Interferences detected by sensors included vitamin C, alpha-fetoprotein, bovine serum albumin, uric acid, and glucose [2, 37]. Mucins are glycoproteins that are released or attached to the membrane and have a unique expression pattern that varies as a tumor grows. They have recently been studied as possible epithelial biomarkers of cancer. MUC1 has received the utmost attention because it is related to tumor proliferation, metastases, chemoresistance, and changes in drug metabolism. Breast, ovarian, ovarian, hepatic, colon, and lung cancers have been linked to overexpression of MUC1 [2, 45]. Another tumor-associated molecule being studied as a potential target for optical or electrochemical sensors is CEA. It is used to diagnose colon, ovarian, and breast cancer as biomarkers [2].

In addition to malignancy, a variety of markers show higher blood levels in the presence of heart disease, making them useful diagnostic tools with significant prognostic implications. Real-time monitoring is essential for accurate results as levels of cardiac markers change. Therefore, immunosensors offer a quick and painless technique for diagnosing cardiovascular disease [2]. C-reactive protein (CRP) is a clinical biomarker commonly used to assess inflammation in the liver or heart. The presence of CRP in human blood plasma can be used to diagnose diseases such as diabetes and cancer [3]. Myoglobin (in combination with Creatinine kinase, Ck-MB, or Troponin) is a protein released from damaged muscle tissue that can be used to diagnose myocardial infarction. Ck-MB is a creatinine

**Table 16.2** Antibody biomarkers for detection of proteins.

| Sr. no. | Purpose | Antibody biomarkers | Application | Reference no. |
|---|---|---|---|---|
| (a) | Cancer Biomarkers | Carcinoembryonic Antigen (CEA) | Diagnosis of colon, ovarian, and breast cancer | [2] |
| | | Anticanthion (CA125) | Diagnosis and therapeutic surveillance biomarker for breast, stomach, liver and ovarian cancer | [2] |
| | | Prostate-specific Antigen (PSA) | Diagnosis of prostate cancer | [37] |
| | | Mucin 1 (MUC1) | Diagnosis of breast, ovarian, ovarian, hepatic, colon, and lung cancers | [45] |
| (b) | Cardiac Biomarkers | C-reactive protein (CRP) | Diagnosis of diabetes and cancer | [3] |
| | | Ck-MB | Indicator of severity of heart attack | [8, 9] |
| | | Troponin I | | |
| (c) | | Tumor necrosis factor alpha (TNF-alpha) | Associated with Alzheimer's, cancer and inflammatory diseases | [3, 10] |

phosphokinase enzyme found only in heart cells and when the heart muscle is injured, its levels increase. Troponin I or T can be used to detect heart injuries, as they enter the bloodstream in large amounts when heart muscle cells die. The serum levels of these markers rise rapidly after a heart attack and peak a few hours later, depending on the severity of the heart attack [8, 9, 35].

Tumor necrosis factor alpha (TNF-alpha) is a protein associated with Alzheimer's, cancer and inflammatory diseases, detection of which in another study demonstrates the potential of the thermometric immunosensor [3, 10]. Antibody biomarkers for detection of protein biomarkers have been summed up in Table 16.2.

#### *16.4.3.2 Antibodies for Detection of Metabolites*

The detection of metabolites for the diagnosis of various diseases has aroused interest in recent years, as metabolites have been associated with a variety of disorders. This approach was used to determine concentrations of metabolites such as creatinine and low-density lipoproteins (LDL). Creatinine (a metabolite of creatine) that is excreted in the urine after

**Table 16.3** Antibody biomarkers for detection of metabolites [2].

| Sr. no. | Antibody biomarkers | Application |
|---|---|---|
| (a) | Creatinine | Diagnosis of Renal Filtration Problems |
| (b) | Apolipoprotein-100-specific antibodies | LDL cholesterol detection |

the kidneys eliminate it. It is used as a biomarker to detect problems related to renal filtration. There are several commercially available point-of-care (POC) devices for detection of creatinine.

A promising approach to LDL cholesterol detection using apolipoprotein-100-specific antibodies has been reported: the immobilization of a large number of Abs in a film composed of gold-coated nanoparticles (AuNP) of silver chloride/polyaniline. An unlabeled impedance sensor was developed using a similar technique in which antiapolipoprotein B was immobilized on the electrode surface using reduced aminated graphene oxides [2]. Refer to Table 16.3 for antibodies that are used for detection of metabolites.

#### *16.4.3.3 Antibodies for Detection of Pathogens*

Infectious disorders produced by infectious microorganisms have increased rapidly in recent years, resulting in increased mortality rate owing to misdiagnosis, treatment delays, and other reasons. Most illnesses are highly infectious and can swiftly spread, resulting in an epidemic or pandemic. It is important to have pathogen detection methods that are fast and portable. For the diagnosis and treatment of diseases, in addition to limiting consent to antibiotics and minimizing antibiotic resistance, rapid detection of the distinction between bacterial and viral infections is crucial. Immunological tests, including immunochromatographic, agglutination, and immunofiltration tests, make up the majority of modern POC tests. Based on a self-organizing network for single-walled carbon nanotubes (SWCNTs) modified with antibodies against swine flu virus (SIV), an electrochemical immunosensor without a SIV label was developed [2].

Pontiac fever and Legionnaires' disease are deadly infections with a high mortality rate caused by *Legionella pneumophila*, a bacterium. A ZnO nanorod (based on electrochemical immunosensor technology) is reported to detect this. The nanobars were produced using a low-temperature hydrothermal technique that used electrostatic interactions to immobilize the primary abdominals. Ag was collected as soon as its peptidoglycan-associated lipoprotein was detected. The electroactivity of the enzymatic reaction product was investigated in a sandwich ELISA test by adding HRP-labeled secondary antibodies in the next step [2].

AIDS is caused by the human immunodeficiency virus (HIV), which is transmitted primarily through unprotected sexual contact or the use of contaminated needles. An unlabeled and fast capacitive immunosensor was created for the real-time detection of an HIV-1 capsid protein (p24 Ag) that can be found in serum at the start of infection. A straightforward ELISA sensor for detecting HIV-1 and HIV-2 antibodies was also announced. Anti-HIV sit-ups and phosphatase-labeled alkaline IgG were administered after immobilizing HIV-1 gp41 and HIV-2 gp36 Ags on the surface of the transducers [2, 6]. Several other immunosensors have been developed, including an immunosensor to detect the hepatitis C virus (based on optical immunosensor technology). For the detection of *Streptococci* using a photo mobilization approach, a disposable impedometry sensor based on gold SPE modified with polythyramine foil was described [2].

The gram-negative bacterium *Pseudomonas aeruginosa* is typically found in contaminated water. Infections with this pathogen in humans can cause lung infections, systemic infections, and urinary tract infections. Immobilization of polyclonal antibodies against *P. aeruginosa* in a screen-printed carbon electrode (SPCE) led to an immune impedance sensor for *P. aeruginosa*. Scarlet fever (rash), pharyngitis, impetigo, erysipelas and cellulite are caused by the gram-positive bacterium *Streptococcus pyogenes*, which has an impedimetric biosensor [9].

Lactone N-3-oxo-dodecanoyl-l-homoserine (HSL) is present in pathogenic wound infections, and activation of receptor-1 expression (TREM-1) is a biomarker that signals a response to bacterial sepsis. As a result, early detection of these compounds can be crucial for a rapid response to wound infection. The use of gold SPE in the development of an immune impedance sensor for these biomarkers has been shown to be effective [9].

**Table 16.4** Antibody biomarkers for detection of pathogens.

| Sr. no. | Disease type | Purpose | Immunosensors employed | Reference no. |
|---|---|---|---|---|
| (a) | Viral Diseases | Detection of Swine flu virus | Electrochemical Immunosensor based on single-walled carbon nanotubes | [2] |
| | | Detection of Hepatitis C virus | Optical Immunosensor | [2] |
| | | Detection of AIDS caused by HIV | Simple electrochemical ELISA sensor to detect HIV-1 and HIV-2 Abs | [6] |
| (b) | Bacterial Diseases | Detection of Pontiac fever and Legionnaires' disease caused by *Legionella pneumophila* | ZnO nanorod based on electrochemical Immunosensor | [2] |
| | | Detection of lung infections, systemic infections, and urinary tract infections caused by *Pseudomonas aeruginosa* | Impedance sensor based on SPCE | [9] |
| | | Detection of bacterial sepsis caused by pathogenic wound infections | Impedance sensor based on gold SPE to detect TREM-1 | [9] |

Some other examples of immunosensors for pathogen detection are a surface acoustic wave (SAW) immunosensor to measure dust mite allergens in the air: *Dermatophagoides farinae* (Derf1) on a regular basis [3] and the for a concanaline A (ConA) lectin immobilized agglutination test on an immunosensor that prevents the detection of sulfate-reducing bacteria (SRBs). A biosensor integrated into a microfluidic device detected samples of human and bovine tuberculosis at very low concentrations [9].

Finally, antibody-based microarray approaches were used to detect fundamentally different diseases in a sensitive and parallel manner. Bacterial strains that pose a threat to bioterrorism (such as *Francisella tularensis, Yersinia pestis*, and *Burkholderia mallei*) are as much a part of it as foodborne pathogens (*Campylobacter jejuni*) and mycotoxins, the bacterial cells that form spores (*B. globigii*) and toxins and viral particles (such as West Nile virus) [11, 42]. Antibodies that are used to detect pathogens responsible for bacterial and viral diseases have been given in Table 16.4.

### *16.4.3.4 Antibodies for Detection of Allergic Biomarkers*

For diagnosis of food allergies *in vitro* using blood sample, immunoanalytical examination of the particular concentration of IgE in a given allergen is done. Food allergies are divided into classes 0 to VI depending on the amount of IgE found in plasma or serum. The main commercially available *in vitro* tests for allergy detection are ImmunoCAP and the Universal Allergy Kit 3g [27]. These techniques include the very accurate and specific PCR and ELISA. When doing routine POC analyses, however, these present detection technologies provide no advantage in terms of simplicity or cost-effectiveness [36]. Furthermore, *in vitro* allergy testing has poorer clinical sensitivity than skin tests, thereby making these tests routinely excluded from established clinical procedures [27]. As a result, the creation of new techniques as alternative analysis tools should be reconsidered in order to meet the cost-effective, cost-effective, and simple detection requirements [36]. Figure 16.6 depicts various biosensing technologies used for detecting allergy biomarkers.

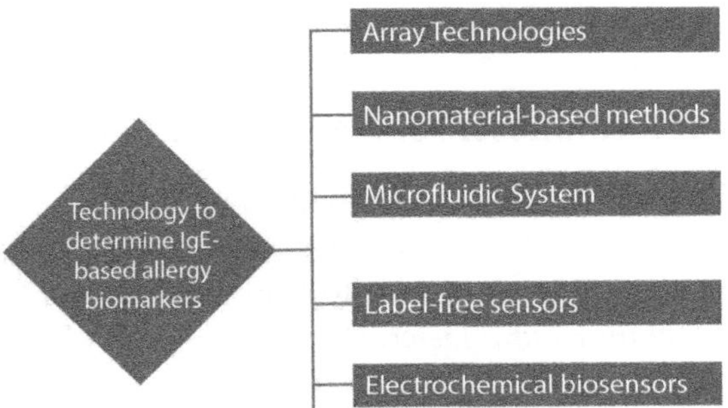


**Figure 16.6** Biosensing technologies used to determine Ig E-based allergy biomarkers.

### 16.4.4 Application of Immunosensors in the Safety of Medicines

Pharmacological adaptation requires therapeutic monitoring to achieve maximum efficacy with the least toxicity. Drug concentrations should ideally be assessed at the site of action, the receptor, but since this is not practical, they are tested in body fluids. Therefore, new minimally intrusive drug detection devices are urgently needed for use in the home, in laboratories and clinics. Special caution should be given when focusing on medications with a narrow therapeutic range that are slightly overdosed and underdosed.

Clenbuterol, a drug used to treat respiratory diseases, was detected with a basic ECL sensor containing gold-plated nanoparticles (AuNP) for high signals. EIS has also been used to detect ciprofloxacin, which is commonly used to treat infections of the lungs, urinary system, and gastrointestinal tract with an unlabeled immunosensor. Anti-ciprofloxacin abdominals were immobilized using electrogenerated NHS polypyrrole films [2].

The use of self-assembled monolayers (SAM) in 3-mercaptopropionic acid as an impedimetric sensor for the selective and sensitive detection of an anesthetic, ketamine, has been demonstrated. Load transfer resistance was used to assess Abs immobilization via SAM and target Ag identification, which can be used to detect ketamine levels. In addition, an unlabeled graphite oxide immobilization-based immunosensor platform has been demonstrated for the measurement of paracetamol in pharmaceutical formulations [2].

Drug abuse and doping are another serious social and health problem. A few of the most widely utilized medications for sports doping include steroid hormones, β-blockers, theophylline and its derivatives, growth hormones, methamphetamines, and peptides. The development of rapid, sensitive, and precise detection methods, particularly in sports, is becoming increasingly significant, with a focus on portable methods of detecting doping drugs on-site. Such chemicals can already be detected in saliva using commercially available methods. An electrochemical sensor for the simultaneous detection of methamphetamine and morphine has been developed [2].

An unlabeled SPR immunosensor has been proposed to identify and quantify enrofloxacin. Enrofloxacin belongs to the class of quinolone antibiotics used to treat infectious diseases in humans and animals. Large doses of enrofloxacin can cause bacterial resistance and a buildup of drugs in edible animal tissue, both of which can be harmful to human health. In a recent study, researchers developed a thermometric enzyme-linked immunosorbent assay (TELISA) for fast detection of diazepam [3].

### 16.4.5 Application of Immunosensors in the Food Safety Industry

Detecting food contamination is a key concern, as these impurities can have stern implications on human health, food safety, and medical and financial costs. To avoid poisoning and ensure proper treatment, it is important to immediately and accurately detect food impurities. *Salmonella typhimurium, Listeria monocytogenes, Escherichia coli* (*E. coli*), *Campylobacter jejuni, Enterobacter sakazakii (E. sakazakii), Clostridium perfringens, Staphylococcus aureus,* and *Yersinia enterocolitica* are the most common food pathogens [2, 38].

An electrochemical immunosensor was developed for the concurrent detection of *E. sakazakii* and *E. coli* using low-density matrices for carbon screen printing [3]. *Salmonella* is another foodborne pathogen that can cause everything from fever, abdominal cramps, and diarrhea to more serious health problems like Reiter's syndrome [2].

Infected foods such as meat, smoked fish, dairy products, ready-to-eat foods, and poultry spread Listeria monocytogenes (*L. monocytogenes*) to humans. TiO2 nanowire beam microelectrodes were used as an abdominal immobilization platform in an immunosensor for the detection of *L. monocytogenes*. The commercially available RAPTOR biosensor was used to detect *L. monocytogenes* in Frankfurterfleisch, *S. typhimurium* in depleted water samples of fresh alfalfa milk and seed samples and *Enterococcus faecalis* in recreational water samples [2, 11]. Furthermore, a recent study demonstrates an SPR-based immunosensor for the detection of extremely sensitive *S. typhimurium* by regulating the orientation of antibody molecules bound on the SPR surface using self-assembled G protein [39]. Another study revealed the creation and deployment of an LSPR-based immunosensor based on AuNP substrate to detect casein allergens in raw milk samples, which is a viable contender for developing massively parallel detectability for food safety in a miniaturized device [40].

### 16.4.6 Application of Immunosensors in Environmental Safety and Control

Monitoring of environmental air and water pollution has been the subject of intense scientific research due to its significant impact on public health. Many toxins enter the soil and eventually end up in wastewater and drinking water. Hormones, endocrine disruptors, antibiotics, surfactants, pesticides and their metabolites, drugs for human and veterinary use and explosives are just some of the contaminants that can be detected in different concentration ranges [2].

Antibodies of the immunoglobulin G (IgG) class are preferentially used in the development of such immunosensors. It has been demonstrated that it can detect mycotoxins such as DON, ochratoxin A, cyanotoxins, and certain aflatoxins [28]. One study addressed the development of immunosensors based on magnetic beads (MB) for detecting estradiol and ethinyl diestradiol in wastewater. Both of these act as endocrine disruptors at nanomolar concentrations, and cause breast and testicular cancer. The test used an ALP-labeled main AB that competed with free estrogen and immobilized estrogen in the sample [2, 3]. In addition to this, an immunosensor has been reported for the simultaneous detection of insecticides, paraoxon, and endosulfan [2].

### 16.4.7 Application of Immunosensors to Detect the COVID-19 Virus

SARS-CoV-2 belongs to a class of enveloped viruses that contain single-stranded RNA (coronaviruses) that was detected in 2019 and is known to cause severe acute respiratory syndrome (COVID-19) in all age groups worldwide. Coronaviruses use a variety of detecting groups to infect their hosts. In addition to receptor binding, spike protein machinery performs a variety of viral physiological activities. The spike protein's unique structure facilitates interaction between the S1 protein's receptor binding pattern (RBP) and the angiotensin-converting enzyme-2-ectodomain (ACE2).

When data from structural and genomic studies utilizing the microscope and sequencing became accessible, various molecular methods such as PCR and serological testing, as well as COVID-19 screening, were created and optimized. Using appropriate primers and probes based on the genomic sequence accessible in the database, multiple virus sequences were targeted using PCR. Antigens or antibodies obtained from blood samples were utilized to diagnose the virus in serological test kits. Most commercially available serological kits work on the principle of Lateral Flow Assay (LFA) or ELISA. There are around 100 assays available on the market for detecting COVID-19–specific antibodies [47].

The COVID-19 testing methods and tools are based on two factors, the target molecule and the type of sample. The viral genome can be identified using PCR, and oropharyngeal and nasopharyngeal smear is tested for the viral antigen using ELISA and LFA. In a blood sample, blood cell count, viral antigen, and antiviral antibodies are detected. Although RT-PCR (reverse transcription polymerase chain reaction) is widely used as COVID-19 diagnostics, other conventional methods that is being used include thoracic CT (computed tomography), immunological screening methods such as ELISA, etc. [25, 34]. The biomarkers used to detect COVID-19 can be both quantitative and qualitative which include inflammatory markers (CRP and Interleukins), hematological biomarkers (number of blood cells), and biochemical biomarkers (cardiac troponin, creatinine, total bilirubin, creatine kinase, LDH, aminotransferase) [25, 41].

## 16.5 Modified Antibodies as Bioreceptors: A Novel Approach

Antibodies have advantages, but they can have disadvantages. They can show batch to batch variations, require the use of animals, and can be unstable. As a result, several investigations have altered the antibodies before applying them to the surface of the transducer. As a result, the modified antibodies could be used more accurately, which improved the overall performance of the biosensor [8, 25].

Modification of intact antibodies can occur in one of two ways: chemical cleavage or enzymatic digestion. Disulfide bonds can be selectively reduced with moderate reducing agents like 2-mercaptoethylamine (2-MEA) or tris(2-carboxyethyl)-phosphine (TCEP) in the chemical fission approach, so only the disulfide bonds between the Fc domains are broken. Papain or pepsin are used to break down antibodies when enzymatic digestion is used. In the advancement of successful biosensors, the use of modifications of antibodies, semi-antibodies and antibody fragments has been demonstrated [8, 28, 46].

### 16.5.1 Antibody Mimetics

Many synthetic proteins have been produced as replacements to animal proteins in order to lessen reliance on animal sources. Typically, these proteins are built on a framework of preserved proteins with a complementary determination region (CDR) generated separately. Standard rotational methods can also be used to swiftly create them and analyze their bonding capacity. The resemblance of synthetic binding proteins to classical antibodies can be used to classify them.

ScFvs were first utilized as a replacement for antibodies. Monoclonal technology is used to unite the two different binding domains of the antibody's heavy and light chains to a short peptide. They are based on the immunoglobulin fold. ScFvs are less stable and may be more difficult to express than other antibody mimics due to their two Ig domains.

Various forms of antibody mimics use different protein motifs. In some or all of their designs, they use α propellers. Adnectin is a plant phytocystatin consisting of a four-stranded beta leaf and a single helix alpha; Adhirons is a scaffolding based on a plant phytocystatin consisting of a four-stranded beta leaf and a single helix alpha; and anticaline is a scaffold based on the lipocaline structure, consisting of two to four braided beta leaves, DARPin is an example of a synthetic binding protein whose structure is based exclusively on the helix alpha [8, 31].

### 16.5.2 Camelid Nanobodies

These are a very stable tiny recognition protein based on a single leaf with three CDR rings, created from animal vaccination. This is a sort of antibody derivative obtained after an animal has been immunized. The groups Chondrichthyes (rays and sharks) and Camillidae (alpacas, camels, dromedaries, llamas, and so on) produced these nanobodies in addition to standard mammalian isotypes as a result of convergent evolution. These heavy-chain antibodies have greater molecular stability and a wider operating temperature range. Nanobody-based technologies have recently emerged as a result of the isolation of their binding areas [8, 28]. They have more flexibility at antigen binding sites, so they can bind to epitopes that antibodies cannot reach. Another significant advantage of nanobodies is their ability to penetrate tissue, which can be used to identify intracellular and extracellular proteins in bacterial cells [30, 48].

### 16.5.3 Reengineered Nanobodies

Nanobodies with spatially defined anchor groups and an amino acid spacer of variable length can be generated using recombinant DNA techniques. The technique of effector function, humanization of antibodies, modulation of affinity, and improvement of stability are some of the techniques used to increase the efficacy and fabricability of antibodies, antibody fragments, antibodies, and antibody fusion products [8, 13].

## 16.6 Conclusion

Immunosensors are widely used diagnostic tools for practical applications in the finding of various analytes owing to their high specificity, sensitivity, low cost, and speed. Immunosensors have achieved incredible success in building a number of protocols using cutting-edge technologies and state-of-the-art tools, which has led to significant improvements in the functions of immunosensors. The associated Ab immobilization technique and the type of Ab used to make the immunosensor have a huge impact on sensor performance. Despite recent advances in immunosensors, new methods are still needed to expand sensitivity, selectivity, and conveniency of such sensors so that they can encounter the stringent requirements of industry or clinical diagnostics. The most difficult aspect of developing immunosensors is to extend their use beyond the laboratory. Mass production of reliable sensors with high sensitivity, specificity, and speed is critical to the successful commercialization of any immunosensor. Another barrier in this crucial area of research is the requirement to scale and integrate into an electronic platform capable of identifying biomarkers or viruses in real time. Numerous unique processes, different designs, and robust immunosensors have emerged as a result of recent advances in nanomaterials, particularly for use in field environments.

## References


1. Nikhil, B., Pawan, J., Nello, F., Pedro, E., Introduction to biosensors. *Essays Biochem.*, 60, 1, 1–8, 2016.
2. Cristea, C., Florea, A., Tertiş, M., Săndulescu, R., Chapter 6: Immunosensors, in: *Biosensors-Micro and Nanoscale Applications*, pp. 165–202, IntechOpen, United Kingdom, 2015.
3. Lima, S.A. and Ahmed, M.U., Introduction to immunosensors, in: *Immunosensors*, pp. 1–16, The Royal Society of Chemistry, United Kingdom, 2019.
4. Sharma, S., Byrne, H., O'Kennedy, R.J., Antibodies and antibody-derived analytical biosensors. *Essays Biochem.*, 60, 1, 9–18, 2016.
5. Richmond, A., Antibodies use in biosensors, AZoLifeSciences, 2021, https://www.azolifesciences.com/article/Antibodies-Use-in-Biosensors.aspx.
6. Regenmortel, M.H.V., Altschuh, D., Chatellier, J., Christensen, L., Rauffer-Bruyère, N., Richalet-Secordel, P., Witz, J., Zeder-Lutz, G., Measurement of antigen–antibody interactions with biosensors. *J. Mol. Recognit.: Interdiscip. J.*, 11, 1-6, 163–167, 1998.
7. Yang, D., Singh, A., Wu, H., Kroe-Barrett, R., Determination of high-affinity antibody-antigen binding kinetics using four biosensor platforms. *J. Vis. Exp.*, 122, e55659, 1–16, 2017.
8. Goode, J.A., Development of biosensors using novel bioreceptors; *Investigation and Optimisation of Fundamental Parameters at the Nanoscale*, Doctoral dissertation, University of Leeds, 2015.
9. Leva-Bueno, J., Peyman, S.A., Millner, P.A., A review on impedimetric immunosensors for pathogen and biomarker detection. *Med. Microbiol. Immunol.*, 209, 3, 343–362, 2020.
10. Karki, H.P., Jang, Y., Jung, J., Oh, J., Advances in the development paradigm of biosample-based biosensors for early ultrasensitive detection of Alzheimer's disease. *J. Nanobiotechnology*, 19, 1, 1–24, 2021.
11. Byrne, B., Stack, E., Gilmartin, N., O'Kennedy, R., Antibody-based sensors: Principles, problems and potential for detection of pathogens and associated toxins. *Sensors*, 9, 6, 4407–4445, 2009.
12. Yang, D., Singh, A., Wu, H., Kroe-Barrett, R., Comparison of biosensor platforms in the evaluation of high affinity antibody-antigen binding kinetics. *Anal. Biochem.*, 508, 78–96, 2016.
13. Chiu, M.L., Goulet, D.R., Teplyakov, A., Gilliland, G.L., Antibody structure and function: The basis for engineering therapeutics. *Antibodies*, 8, 4, 55, 2019.
14. Wilson, I.A. and Stanfield, R.L., Antibody-antigen interactions: New structures and new conformational changes. *Curr. Opin. Struct. Biol.*, 4, 6, 857–867, 1994.
15. Davies, D.R. and Cohen, G.H., Interactions of protein antigens with antibodies. *Proc. Natl. Acad. Sci.*, 93, 1, 7–12, 1996.
16. Pohanka, M., Monoclonal and polyclonal antibodies production-preparation of potent biorecognition element. *J. Appl. Biomed.*, 7, 3, 115–121, 2009.
17. Stills, H.F., Polyclonal antibody production, in: *The Laboratory Rabbit, Guinea Pig, Hamster, and Other Rodents*, pp. 259–274, 2012.
18. Holford, T.R., Davis, F., Higson, S.P., Recent trends in antibody based sensors. *Biosens. Bioelectron.*, 34, 1, 12–24, 2012.
19. Caponi, L., and Migliorini, P., *Antibody usage in the lab.*, pp. 19–23, Springer Science & Business Media, Berlin, Heidelberg, 2013.

20. Zaidi, P. and S.K., Raciloimmunoassay: Principle and technique. *J. Pak. Med. Assoc.*, 43, 12, 264–267, 1993.
21. Gan, S.D. and Patel, K.R., Enzyme immunoassay and enzyme-linked immunosorbent assay. *J. Invest. Dermatol.*, 133, 9, 1–3, 2013.
22. Kurien, B.T. and Scofield, R.H., Western blotting: An introduction. *Methods Mol Biol.*, 1312, 17–30, 2015.
23. Jensen, E.C., The basics of western blotting. *Anat. Rec.: Adv. Integr. Anat. Evolutionary Biol.*, 295, 3, 369–371, 2012.
24. Laczka, O.F., Immunosensors: Using antibodies to develop biosensors for detecting pathogens and their toxins, in: *Biological Toxins and Bioterrorism*, pp. 273–294, 2015.
25. Singh, B., Datta, B., Ashish, A., Dutta, G., A comprehensive review on current COVID-19 detection methods: From lab care to point of care diagnosis. *SI, 2*, 100119, 1–21, 2021.
26. Gaudin, V., Advances in biosensor development for the screening of antibiotic residues in food products of animal origin–a comprehensive review. *Biosens. Bioelectron.*, 90, 363–377, 2017.
27. Morais, S., Tortajada-Genaro, L.A., Maquieira, A., Martinez, M.A.G., Biosensors for food allergy detection according to specific IgE levels in serum. *TrAC, Trends Anal. Chem.*, 127, 115904, 1–12, 2020.
28. Bazin, I., Tria, S.A., Hayat, A., Marty, J.L., New biorecognition molecules in biosensors for the detection of toxins. *Biosens. Bioelectron.*, 87, 285–298, 2017.
29. Morales, M.A. and Halpern, J.M., Guide to selecting a biorecognition element for biosensors. *Bioconjugate Chem.*, 29, 10, 3231–3239, 2018.
30. Jayan, H., Pu, H., Sun, D.W., Recent development in rapid detection techniques for microorganism activities in food matrices using bio-recognition: A review. *Trends Food Sci. Technol.*, 95, 233–246, 2020.
31. Crivianu-Gaita, V. and Thompson, M., Aptamers, antibody scFv, and antibody Fab'fragments: An overview and comparison of three of the most versatile biosensor biorecognition elements. *Biosens. Bioelectron.*, 85, 32–45, 2016.
32. Carneiro, M.C., Sousa-Castillo, A., Correa-Duarte, M.A., Sales, M.G.F., Dual biorecognition by combining molecularly-imprinted polymer and antibody in SERS detection. Application to carcinoembryonic antigen. *Biosens. Bioelectron.*, 146, 1–9, 2019.
33. Anusha, J.R., Kim, B.C., Yu, K.H., Raj, C.J., Electrochemical biosensing of mosquito-borne viral disease, dengue: A review. *Biosens. Bioelectron.*, 142, 1–15, 2019.
34. Goud, K.Y., Reddy, K.K., Khorshed, A., Kumar, V.S., Mishra, R.K., Oraby, M., Ibrahim, A.H., Kim, H., Gobi, K.V., Electrochemical diagnostics of infectious viral diseases: Trends and challenges. *Biosens. Bioelectron.*, 180, 1–21, 2021.
35. Karimi-Maleh, H., Orooji, Y., Karimi, F., Alizadeh, M., Baghayeri, M., Rouhi, J., Tajik, S., Beitollahi, H., Agarwal, S., Gupta, V.K., Rajendran, S., A critical review on the use of potentiometric based biosensors for biomarkers detection. *Biosens. Bioelectron.*, 184, 1–16, 2021.
36. Sheng, K., Jiang, H., Fang, Y., Wang, L., Jiang, D., Emerging electrochemical biosensing approaches for detection of allergen in food samples: A review. *Trends Food Sci. Technol.*, 121, 93–104, 2022.
37. Dowlatshahi, S. and Abdekhodaie, M.J., Electrochemical prostate-specific antigen biosensors based on electroconductive nanomaterials and polymers. *Clin. Chim. Acta*, 516, 111–135, 2021.
38. Rowe, C.A., Scruggs, S.B., Feldstein, M.J., Golden, J.P., Ligler, F.S., An array immunosensor for simultaneous detection of clinical analytes. *Anal. Chem.*, 71, 2, 433–439, 1999.

39. Oh, B.K., Kim, Y.K., Park, K.W., Lee, W.H., Choi, J.W., Surface plasmon resonance immunosensor for the detection of Salmonella typhimurium. *Biosens. Bioelectron.*, 19, 11, 1497–1504, 2004.
40. Hiep, H.M., Endo, T., Kerman, K., Chikae, M., Kim, D.K., Yamamura, S., Takamura, Y., Tamiya, E., A localized surface plasmon resonance based immunosensor for the detection of casein in milk. *Sci. Technol. Adv. Mater.*, 8, 4, 331–338, 2007.
41. Eissa, S. and Zourob, M., Development of a low-cost cotton-tipped electrochemical immunosensor for the detection of SARS-CoV-2. *Anal. Chem.*, 93, 3, 1826–1833, 2020.
42. Byrne, B., Stack, E., Gilmartin, N., O'Kennedy, R., Antibody based sensors: Principles, problems, and potential for detection of pathogens and associated toxins. *Sensors*, 9, 6, 4407–4445, 2009.
43. Gmoshinski, I.V., Khotimchenko, S.A.E., Popov, V.O., Dzantiev, B.B., Zherdev, A.V., Demin, V.F., Buzulukov, Y.P., Nanomaterials and nanotechnologies: Methods of analysis and control. *Russ. Chem. Rev.*, *82*, 1, 48, 2013.
44. Sharma, S., Byrne, H., O'Kennedy, R.J., Antibodies and antibody-derived analytical biosensors. *Essays Biochem.*, 60, 1, 9–18, 2016.
45. Liang, S.L. and Chan, D.W., Enzymes and related proteins as cancer biomarkers: A proteomic approach. *Clin. Chim. Acta*, 381, 1, 93–97, 2007.
46. Singh, A.K., Kilpatrick, P.K., Carbonell, R.G., Application of antibody and fluorophore-derivatized liposomes to heterogeneous immunoassays for d-dimer. *Biotechnol. Prog.*, 12, 2, 272–280, 1996.
47. Wu, J., Liu, J., Li, S., Peng, Z., Xiao, Z., Wang, X., Luo, J., Detection and analysis of nucleic acid in various biological samples of COVID-19 patients. *Travel Med. Infect. Dis.*, 37, 101673, 2020.
48. Brazaca, L.C., Ribovski, L., Janegitz, B.C., Zucolotto, V., Chapter 10: Nanostructured materials and nanoparticles for point of care (POC) medical biosensors, in: *Medical Biosensors for Point of Care (POC) Applications*, pp. 229–254, Woodhead Publishing, Imprint of Elsevier, New York, United States, 2017.

17

# Biosensors for Paint and Pigment Analysis

**Sonal Desai[1], Priyal Desai[1] and Vipul Prajapati[2]***

*[1]Department of Pharmaceutical Quality Assurance, SSR College of Pharmacy (Permanently Affiliated to Savitribai Phule Pune University), Sayli-Silvassa Road, Sayli, Silvassa, Union Territory of Dadra Nagar Haveli & Daman Diu, India*

*[2]Department of Pharmaceutics, SSR College of Pharmacy (Permanently Affiliated to Savitribai Phule Pune University), Sayli-Silvassa Road, Sayli, Silvassa, Union Territory of Dadra Nagar Haveli & Daman Diu, India*

***Abstract***

Biosensors have emerged as an innovative analytical tool in the field of medicines, biomedical diagnosis and treatments, analysis of foods as well as for the care of environments. These devices use biological recognition systems for detection of target analyte by generating measurable signals. This chapter describes the current trends of biosensors for paint and pigment analysis. The history, components, and working principle of biosensors are also highlighted. Additionally, pigment-based biosensors, biosensors for various natural and synthetic pigments used in food and paints are discussed.

***Keywords:*** Biosensors, micro-biosensors, paints, pigments

## Abbreviations

| | |
|---|---|
| ADI | Acceptable Daily Intake |
| β-CD–rGO | Beta-cyclodextrin-reduced graphene oxide |
| CPE | Carbon Paste Electrode |
| CUR-Tz-TCO | Curcumin-1,2,4,5-tetrazine-trans-cyclooctene |
| Covid-19 | Coronavirus Disease 2019 |
| EDX | Energy Dispersive X-ray analysis |
| ERT | Erythrosine |
| FT-IR | Fourier Transform-Infrared Spectroscopy |
| GC-MS | Gas Chromatography-Mass Spectrometry |
| GCE | Glassy Carbon Electrode |

**Corresponding author*: vippra2000@yahoo.com

Inamuddin and Tariq Altalhi (eds.) Biosensors Nanotechnology, 2nd Edition, (395–418) © 2023 Scrivener Publishing LLC

| | |
|---|---|
| Gr-TPyP | Graphene–Porphyrin composite Paste |
| hBN | Hexagonal Boron Nitride |
| IUPAC | International Union of Pure and Applied Chemistry |
| L-DOPA | Levodopa |
| LC-MS | Liquid Chromatography-Mass Spectroscopy |
| NGQDs | Nitrogen-doped graphene quantum dots |
| PGMCPE | Modified carbon paste electrode |
| PDDA-Gr-Ni | Poly diallyl dimethylammonium chloride -Graphene Nickel |
| PGMCPE | Poly (glycine) modified carbon paste electrode |
| PLM | Polarized - Light Microscopy |
| Py-GC/MS | Pyrolysis-Gas Chromatography/Mass spectrometry |
| RGO | Reduced Graphene Oxide |
| SEM | Scanning Electron Microscopy |
| SPR | Surface plasmon resonance |
| $TiO_2$ | Titanium Dioxide |
| TPI | Terahertz pulsed imaging |
| USA | United States of America |
| UV | Ultra Violet |
| XRD | X-ray Diffraction |
| XRF | X-ray Fluorescence |
| YSI | Yellow Spring Instruments |

## 17.1 Paint and Pigments

The application of paints or coating to various surfaces such as plastic, concrete, metal and wood is widely adopted to provide protection against corrosion and damage [1]. Paint is a mixture primarily containing binder, pigments and solvent [2]; available as a liquid or as liquefiable form which is applied on any surface as a thin film used for identification, decoration, coloring or protection purposes. Some paints are also used as filler or concealer to hide imperfections [3]. Paints are used for coating or coloring of houses, articles, furniture, walls, surfaces, etc. [2]. Pigment and dye are both used to impart color or change color. Dyes are colored materials, dissolve in vehicles while pigments are colored, colorless or fluorescent compounds, organic and inorganic in nature, and are not soluble in vehicles. Dyes impart brighter color than the pigments, but they are light sensitive and are non-permanent. Dyes give color by adsorption, whereas dyes adsorb color both by adsorption or by scattering of light [3, 4]. Pigments are classified based on their nature (organic/inorganic), origin (natural/synthetic) and chemical structures. Figure 17.1 represents classifications of pigments/dyes with suitable examples.

As per the definition given by the American Coating Association, paints are the group of emulsions made up of pigments suspended in a liquid vehicle to be used for decoration or for protection. Paints have a history of more than 30,000 years. In ancient times, crude paints were used by Cave dwellers to leave behind footprints of their lives in the form of graphics. As per record, in 1700, the first paint mill in America was started in Boston by Thomas Child. In 1867, patent on first "ready mixed" paints were obtained by D.R. Averill in the United States. In the mid-1880s, there was a significant increase in paint industries in the USA [5]. Initially, the painter was the only manufacturer of paint except for the manufacturing of pigment white lead, which started much earlier [6]. The manufacturing of paint as a commodity began during the Civil war and after that use of various mills for manufacturing of paints was started. In early 1900, the industrial revolution led to huge development in new markets for paints and coatings [5]. Even pigments have a long history. Minerals have been used as colorants since ancient times. The first synthetic pigment was Egyptian blue which was extensively utilized by the 4th Dynasty. After that, new synthetic pigments continued to emerge. Table 17.1 depicts development of different pigments used in the paint industry.

Ever since, with fast developments in science and technology, new pigments/colors are being developed. Recently, Mas Subramanian *et al.* accidentally discovered the new blue pigment "YInMn Blue" when they were studying the electronic behavior of manganese

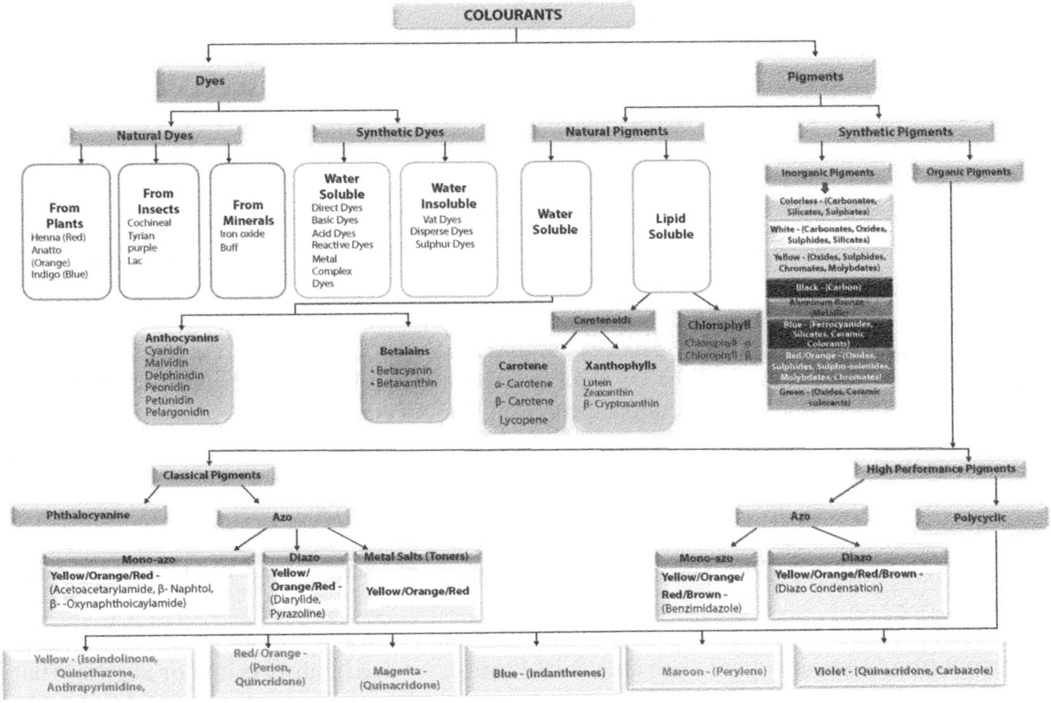


**Figure 17.1** Flowchart for classification of pigments/dyes.

**Table 17.1** Major breakthrough in pigment development [7, 8]

| Year | Breakthrough in pigment development |
|---|---|
| Egyptian Antiquity (4th Dynasty ~ 2500) | Development of the first synthetic pigment Egyptian Blue. Use of cobalt to colour glass |
| Mayan Antiquity, (4th-8th Century A.D.) | Maya Blue, indigo introduced with palygorskite clay |
| ~ 1704 | First synthetic pigment Prussian Blue was developed |
| 1804 | First plant of white lead in America |
| 1905 | Ultramarine Blue was developed from Red β – naphthol |
| 1907 | Toluidine red was introduced |
| 1909 | Hansa Yellow was firstly introduced in Germany |
| 1916 | First type of $TiO_2$ was prepared |
| 1919 | Production of $TiO_2$ started |
| 1923 | Anatase Pigments was developed in France |
| ~ 1930 | Fluorescent Paints were developed from Anthracene dyes in shellac |
| 1931 | $TiO_2$ pigment was prepared by sulphate process by DuPont |
| 1935 | Diarylide Yellow pigment was commercialized |
| 1938 | Commercialization of Phthalo green |
| 1948 | Commercialization of chloride process for preparation of rutile $TiO_2$ Pigments by DuPont |
| 1954 | Diazo condensation pigments were introduced |
| 1955 | Introduction of Quinacridones |
| 1964 | Development of Benzimidazoles pigments |
| 1963 | Patent on Metal oxide ($TiO_2$) coated mica flakes by DuPont |
| 1971 | Aluminum flakes pigments were used in metallic effect coatings |
| 1972 | The US paint industry had titanium dioxide pigments in the slurry form from Dupont and others |
| 1978 | White lead pigments were banned in US |

oxide. This pigment was approved for industrial use in September 2017 and for commercial use in May 2020.

Household paints are classified into two classes: a) Oil paints which are diluted with mineral turpentine or other organic solvents & b) Emulsion based paints (vinyl or acrylic) that are diluted by water [9]. Paint consists of a binder, pigments and solvent [2]. The binder is a polymeric in nature which is used to adhere the pigment to the surface to be coated. Binders are either solubilized or dispersed in paint. Binders for paints include alky resins, vinyl & acrylic emulsions, epoxy resins, & polyurethanes. Pigments are organic or inorganic in nature and are used to impart color to the paint. Additionally, pigments protect the surface from corrosion. Solvent used for paints can be water or any organic solvent depending on the solubility of the binder.

## 17.2 Characteristics of Pigments for Paints

The pigments used for paints can be organic or inorganic in nature. Pigments are used to provide appearance, color, and opacity. The color of the pigment depends on the chemical structure of the pigment. For organic pigments, and the color of pigments is because of the presence of chromophores whereas transfer spectra and d-d transition spectra are responsible for color of inorganic compounds. Also, the presence of metals in two valency states is also accountable for the color of inorganic pigments. Particle size of the pigments also affects its color. Generally, smaller particles are brighter in shades. Tinctorial strength of the pigments is also an important parameter to be considered while choosing the pigment as less quantity of pigment is required to get standard shade if tinctorial strength is high. The pigment must be insoluble in the vehicle or solvent in which it is incorporated. It should not chemically react with other components of paints. Opacity of pigment depends on its refractive index. Inorganic pigments have higher refractive index than organic pigments. Hence, inorganic pigments are relatively opaque than organic pigments. For this reason, titanium dioxide with a refractive index of more than 2.5 is extensively used as a white pigment in paints. The pigments used for paints must be durable with reference to light and weather fastness [10]. Figure 17.2 highlights important features of pigments used in paints.

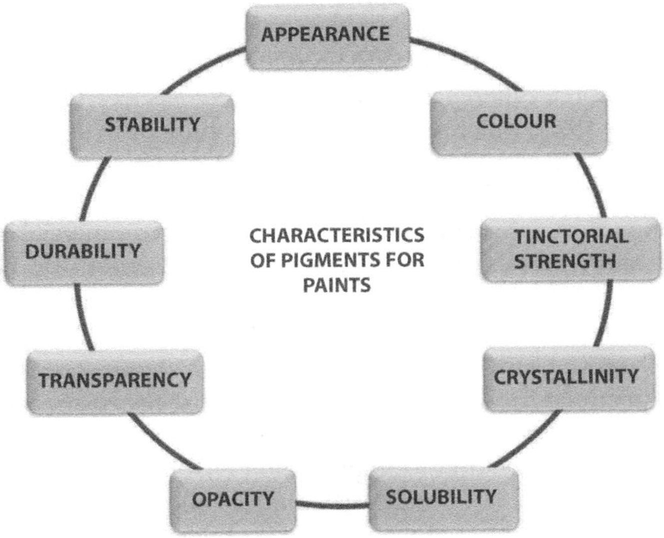


**Figure 17.2** Essential characteristics of pigments used in paints.

## 17.3 Analysis of Paints and Pigments

Many analytical procedures have been used for the identification and characterization of pigments namely EDX, FT-IR, Fluorescence spectroscopy, GC-MS, LC-MS, SEM, UV-Visible spectroscopy and XRD [11]. Visible reflectance spectroscopy which is based on the principle of selective light absorption is used to evaluate pigment composition [12]. Mass spectrometry/Pyrolysis-gas chromatography is employed for identification of synthetic organic dyes and resins in paints [13–15]. LC-MS techniques are also widely applicable to analyze various components of paints such as zinc pyrithione which is used as biocide in antifouling paints [16]. Reports are available for use of micro X-ray diffraction technique used for paint and pigment analysis [17]. FTIR [18] and Raman Spectroscopy [19] are used for identification of pigments in paint. In spite of various conventional and modern analytical techniques available for paint and pigment analysis, use of biosensors in this field is increasing due to its compact size, accuracy and quick response.

## 17.4 Biosensors and Their Background

Ever since its discovery, biosensors have been the center of attention of researchers. Biosensor is an analytical tool which measures the concentration of target analyte by converting biological response into detectable and processable electrical signal [20, 21]. Biosensors are small, self-contained analytical devices which utilize immobilized biomolecules combined with transducers to detect specific signals of an analyte [22]. Antibodies, enzymes, nucleic acids, and other biological sensing components are in proximity with a transducer (e.g., optical, electrochemical, piezoelectric, etc.) that converts the biorecognition into a signal. The analyte understudy concentration is directly related to the output signal intensity. Finally, utilizing electronic and software technologies, the output is processed and presented on a reader device [23].

Biosensors have a history of over 100 years. In 1906, M. Cremer correlated the conc of an acid in a fluid to the electric potential. In 1909, the concept of pH as a convenient way of expressing hydrogen ion concentration was given by a Danish biochemist, Sorensen [24]. The study of potential difference between glass and electrolytes in contact with the glass was presented in 1922, which was an extension study of that given by Haber [25]. In 1956, Leland C. Clark, Jr. developed an electrode for detection of oxygen. Later on, in 1962, he also developed a simple device for fast measurement of glucose [26]. In 1975, US based company Yellow Spring Instruments (YSI) fabricated a commercial biosensor for the first time [27]. Literature reveals that in 1977 the word 'Biosensor' was been first coined by Karl Cammann where he presented a detailed review on ion-selective electrodes-based biosensors [28]. Definitions of biosensor are provided by IUPAC [29–31]. The "Father of Biosensors" was Leland C. Clark Jr. at 'World Congress on Biosensors' in Geneva in 1992 [26]. Development of miniaturization of biosensors or micro biosensors was another breakthrough in the arena of biosensors. Micro biosensors have many advantages such as small sample requirement, easy adaptation to implantation enabling *in-vivo* measurement and possibility of bulk fabrication [32]. Miniaturization also allows integration of several sensors into arrays, allowing screening of natural and synthetic compounds accompanied by combinatorial libraries [33].

## 17.5 Components, Principle and Working of Biosensors

As per definition given by IUPAC, *an electrochemical biosensor is a self-reliant incorporated device that uses a biological recognition element (biochemical receptor) that is kept in direct longitudinal contact with an electrochemical transduction element to provide a particular quantitative or semi-quantitative analytical information* [29].

Major constituents of biosensor are bioreceptor, transducer, electronics & display [34].

**Bioreceptor:** It is a biomolecule used to identify the analyte. The relations between bioreceptor and analyte are known as biorecognition.

**Transducer:** It is the main component of biosensor, which is responsible for signalization; converts biorecognition into electrical signals. Different types of transducer include electrochemical, magnetic, optical, mass-based and thermal [25].

The electrochemical transducers measure change in either electric potential or electric current or any other property that generates during biorecognition and includes electrochemical (Conductometric-Measurement of conductance, amperometric-Measurement of current, Field effect transistors-Measurement of voltage, Ion selective electrode-Measurement of voltage, Gas sensing electrode-Measurement of voltage, Impedimetric-Measurement of impedance), piezoelectric crystal-Measurement of mass change, optoelectronic-Optical measurement and Thermistor- Measurement of heat [35, 36].

**Electronics and display:** To amplify transduced signal and convert into digital form to display, display can be in the form of a computer or printer to display response in the form of a number, graph, or figure.

Bioreceptor which is in connection with the transducer interacts with an analyte to complete biorecognition. The signal generated by the transducer due to biorecognition is amplified by the electronics component of the biosensor and measured [37]. When a biological system (biological molecule/element) is linked with a transducer, the transducer changes a biological retort to an electrical signal. Based on the type of biological system, either voltage or current is generated as an output. In case of later, the current is converted to voltage. Generally, the output signal voltage is low which is amplified and filtered by the indication processing element.

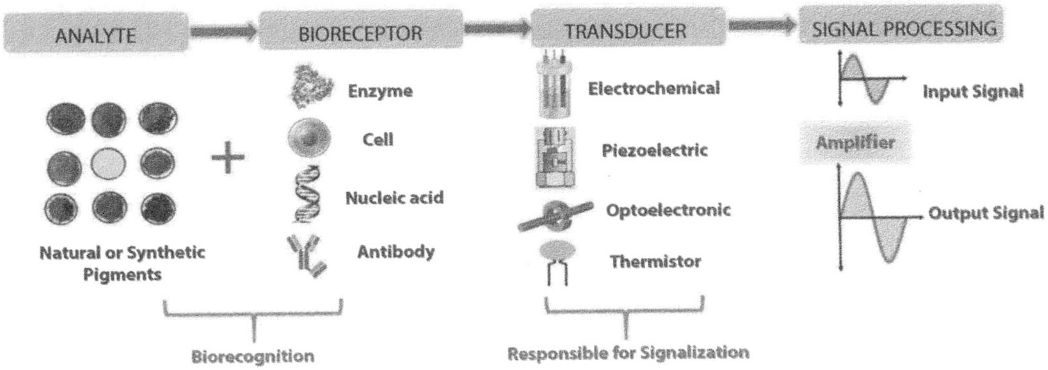


**Figure 17.3** Diagram representing working principle of biosensors.

The output of the signal processing part is an analog signal which is relative to the strength of analyte under study. The analog signal is changed into a digital signal by a microcontroller and displayed [38]. Figure 17.3 illustrates the working principle of biosensors.

## 17.6 Applications of Biosensors

Biosensors have been established to be useful in the medical as well as diagnostic fields. Their applications are also extended to food, pharma, beverages, farming, environmental monitoring, microbiology and biotechnological industries [39, 40]. Use of biosensors for diagnosis, progression and treatment of diseases is a major breakthrough in the medical field [41]. The role of biosensors in agriculture technology has also been extended such as in pre-harvest, in post-harvest process, for detection of artificial ripening agents and evaluation of quality of soil [42]. Biosensors have also emerged as an analytical tool for quality control of food [43]. In the food industry, biosensors are used to analyze nutrients, natural toxins and to check food processing [44]. Additionally, they are also utilized to evaluate freshness of food [45]. In the food industry, biosensors are used to analyze nutrients, natural toxins and to check food processing [46]. Also new developments are being carried out in biosensing technologies for diagnosis of COVID-19 virus [47–49]. Figure 17.4 outlines the various applications of biosensors.

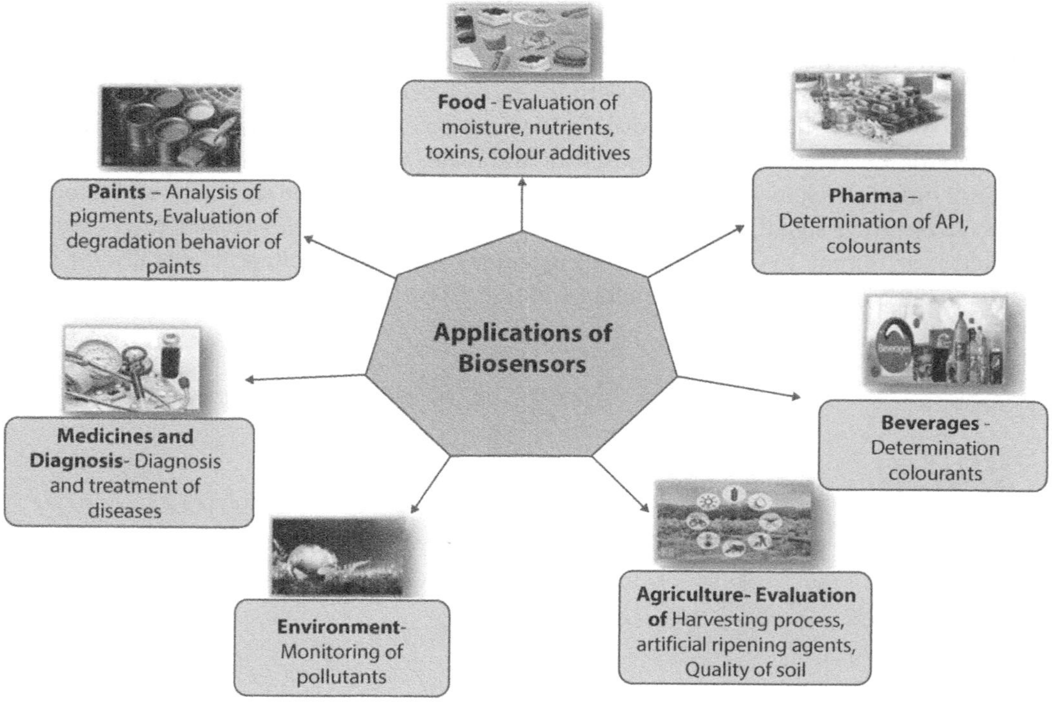


**Figure 17.4** Applications of biosensors in various fields.

### 17.6.1 Pigment-Based Biosensors

Pigment based biosensors revealed phenomenal is potential for the recognition of several analytes. The development of pigment-based biosensors, various pigments has been widely studied. Among these pigments, chlorophyll being the highest [50]. Chlorophyll's fluorescence activity makes it a suitable candidate to be used to develop optical biosensors [51]. Similarly, beetroot pigments "Betalains" have natural fluorescence rendering them useful for development of biosensors to produce consistent and strong signals, which is easily identified by conventional fluorescence techniques [52]. Preliminary investigation on Annatto dye found from seeds of the shrub Bixa orellana was carried out as possible pH-biosensor probes for adulterated milk [53]. Table 17.2 presents reported pigment-based biosensors for detection of various metals, contaminants and biomolecules.

Carotenoids synthesis is stimulated by the presence of heavy metals [54, 55]. This concept was utilized to design biosensors for recognition of heavy metals [50, 54, 56]. Biosensor based on the change in shade of chromoionophore was fabricated to estimate the concentration of formaldehyde. Oxidation of formaldehyde by alcohol oxidase enzyme resulted in increase in concentration of hydrogen ions which led to change in color of chromoionophore [57]. Another study used fluorescence characteristics of betaxanthins as output signal for biosensor for determination of copper ions. Copper incubated cells were reacted with L-DOPA, which is a precursor for betaxanthins, and the concentration of betaxanthins was measured in terms of fluorescence intensity [58]. Micro biosensor was developed using color (Deinoxanthin) producing extremophilic bacteria, *Deinococcus radiodurans* for determination of cadmium based on colorimetry [59]. Biosensor for recognition of pathogenic bacteria was established based on pigment lycopene through expression of biosynthetic genes namely geranylgeranyl pyrophosphate synthase and phytoene synthase. Isopentenyl pyrophosphate isomerase was added to increase precursor, dimethylallyl diphosphate to produce lycopene which was measured by optical detection [60]. Curcumin was used as a signal reporter to prepare enzyme free biosensor to detect *Salmonella typhimurium* using bovine serum albumin-curcumin nanoparticles which were converted to Quartz-CO bacteria. Curcumin released in the existence of an alkaline condition was optically measured and was an indicator of presence of *S. typhimurium* [61]. The presence of $\beta$-Diketones in the structure of curcumin forms metal complexes to form suitable system for detection. Curcumin-chitosan polymer coated glass carbon electrode was fabricated for detection of bilirubin. Curcumin being antioxidant in nature, it generates current which can be measured [62].

Dyes are added to food products and pharmaceutical formulations to increase their visual appearance and consumer compliance. However, their use is strictly controlled by legislation because of their toxicity. The legislation and acceptable daily intake (ADI) for these colored additives varies from country to country [63]. Hence, it becomes mandatory to analyze the exact concentration of these colorants. Many techniques are currently used for estimation of various dyes and pigments. But these techniques require extensive sample preparation, are very expensive and utilize specific equipment (HPLC, LC-MS, and GC-MS). On the other hand, biosensors have emerged as an important analytical tool which can replace these techniques, since they are fast, efficient and accurate and allow a much more widespread use since they are very low cost and easy to use. Table 17.3 depicts

**Table 17.2** Biosensors based on pigments with their applications.

| **Bioreceptor** | **Pigment used/Bioreporter** | **Transducer platform** | **Application** | **Year** | **References** |
|---|---|---|---|---|---|
| *Daucus carota* | β-Carotene | Spectrophotometry | Detection of heavy metals | 2015 | [50] |
| *Rhodopseudomonas palustris* | β -Carotene | Colorimetry | Detection of Arsenite | 2008 | [54] |
| *Escherichia coli* | Lycopene and β-carotene | Fluorescence | Evaluation of zinc levels in human | 2018 | [56] |
| Alcohol oxidase enzyme | Chromoionophore I | UV Spectrophotometry | Detection of formaldehyde | 2018 | [57] |
| *Cupriavidus metallidurans* | Betaxanthin | Fluorescence | Analysis of environmental copper | 2018 | [58] |
| Genetically engineered *Deinococcus radiodurans* | Deinoxanthin | Colorimetry | Detection of cadmium | 2012 | [59] |
| Engineered *Escherichia coli* | Lycopene | Optical detection | Detection of water contaminants namely *Pseudomonas aeruginosa* and *Burkholderia pseudomallei* | 2021 | [60] |
| Bovine serum albumin | Curcumin | Optical detection | Detection of *Salmonella typhimurium* | 2018 | [61] |
| Chitosan | Curcumin | Voltammetry | Quantification of bilirubin | 2019 | [62] |

**Table 17.3** Applications of biosensors/sensors for different dyes and pigments.

| Name of pigment/Dye | Bio sensor/Sensor | Matrix | Detection level | Linear range | Reference |
|---|---|---|---|---|---|
| Allura red dye | Optical sensor | Candies | - | 22.9-78.8 mg $kg^{-1}$ | [64] |
| | Electrochemical sensor based on poly diallyl dimethylammonium chloride, functionalized graphene and nickel nanoparticles composite (PDDA-Gr-Ni) | Real sample | 8.0 nmol $L^{-1}$ | 0.05-10.0 µmol $L^{-1}$ | [65] |
| | Voltammetry using Titania/Electro-Reduced Graphene Oxide Nanohybrid Electrochemical Sensor | Milk drink samples | 0.05 µM | 0.5-5.0 µM | [66] |
| Amaranth | Screen-printed electrode was modified by Pd/GO nanomaterial | Water, orange juice, and apple juice | 30.0 nM | 0.08 mM–360.0 mM | [67] |
| | Electrochemical sensor-Graphene/$TiO_2$ -Ag Based Composites | - | - | $1 \times 10^{-7}$ M | [68] |
| | Fluorescence sensor-Nitrogen-Doped Graphene Quantum Dots | Candy | 0.15 µg/L | 0.4 to 3.5 µg/L | [69] |
| β-carotene | Stochastic sensor based on a graphene–porphyrin composite paste | Soft drink | $1.00 \times 10^{-15}$ mol $L^{-1}$ | $1.00 \times 10^{-15}$ mol $L^{-1}$ and $1.00 \times 10^{-3}$ mol $L^{-1}$ | [70] |
| | Hexagonal boron nitride (hBN) modified glassy carbon electrode (hBN/GCE) | Simulated blood serum | 1.25 pM | 5 pM to 5 µM | [71] |

*(Continued)*

**Table 17.3** Applications of biosensors/sensors for different dyes and pigments. (*Continued*)

| Name of pigment/Dye | Bio sensor/Sensor | Matrix | Detection level | Linear range | Reference |
|---|---|---|---|---|---|
| Curcumin | The magneto-reactance effect of metallic glass is used to create a magnetic biosensor (Metglas 2714A) | Curcumin-loaded $Fe_3O_4$ nanoparticles | - | 0–50 ng/ml | [72] |
| | Surface plasmon resonance nano sensors based on Poly (2-hydroxyethyl methacrylate-N-met acryloyl-Ltryptophan) (poly (HEMA-MATrp)) nanoparticle | - | 0.0012 mg/L | 0.01–150 mg/L | [73] |
| | Electrochemical sensor based on beta-cyclodextrin-reduced graphene oxide nanocomposite | - | 33 nM | 0.05–10 mM | [74] |
| Cyanidin | Voltammetry using Glassy carbon electrode functionalized multiwalled nanotube | Fruit and food samples | 0.18 μg/l | 10 to 560 μg/l | [75] |
| Delphinidin | Voltammetry using Glassy carbon electrode (f-CNTs/GCE) | Fruit and food samples | 0.13 μg/l | 100 to 1400 μg/l | [75] |
| Erythrosine | Differential pulse voltammetric using $TiO_2$ nanoparticles modified gold electrode surface | Tablets, ERT Lake and fruit juice | 2.6 nM | 0.1 μM to 10.0 μM | [76] |
| | Voltammetry using Glucose modified carbon paste sensor | Tablets, ERT Lake and fruit juice | 21.6 nM | $1.0 \times 10^{-4}$ M to $1.0 \times 10^{-7}$ M | [77] |

(*Continued*)

**Table 17.3** Applications of biosensors/sensors for different dyes and pigments. (*Continued*)

| Name of pigment/Dye | Bio sensor/Sensor | Matrix | Detection level | Linear range | Reference |
|---|---|---|---|---|---|
| Eriochrome Black T | Amperometric biosensor with cellophane membranes | In Effluent | 8 ppm | 20-70 ppm | [78] |
| | Amperometric biosensor with nylon membrane | In Effluent | 3 ppm | 8-25 ppm | [78] |
| | Amperometric biosensor with chitosan/TEOS/EG membranes | In Effluent | 5 ppm | 15-40 ppm | [78] |
| Indigo carmine | Voltammetry using Modified carbon paste electrode (PGMCPE) | Indigo carmine injection | $11 \times 10^{-8}$ M | $2 \times 10^{-6}$ to $1 \times 10^{-5}$ M and $1.5 \times 10^{-5}$ M to $6 \times 10^{-5}$ M | [79] |
| | Voltammetry using Polyarginine modified carbon-based Electrode | Water and food samples | $2.53\text{x}10^{-8}$ | $2 \times 10^{-7}$ to $10^{-6}$ M, and $1.5 \times 10^{-6}$ to $3.5 \times 10^{-6}$ M | [80] |
| Sunset Yellow | Amperometric sensor based on Molecularly Imprinted polypyrrole | Wine samples | - | 0.4 to 2 μM and 2 to 8 μM | [81] |
| | Amperometric biosensor using laccase as biorecognition element | - | - | $5\times10^{-5}$ to $8\times10^{-3}$ M | [82] |
| | Au-Pd and reduced graphene oxide nanocomposites modify the glassy carbon electrode (Au-Pd-RGO/GCE) with the help of electrochemical sensor | Soft drinks | 1.5 nM | 0.686–331.686 mM | [83] |
| | Electrochemical biosensor based on Photocured Polyacrylamide Membrane | Soft drinks | 0.02 M | 0.08 to 10.00 M | [84] |

(*Continued*)

**Table 17.3** Applications of biosensors/sensors for different dyes and pigments. (*Continued*)

| Name of pigment/Dye | Bio sensor/Sensor | Matrix | Detection level | Linear range | Reference |
|---|---|---|---|---|---|
| | Electrochemical device based on glassy carbon electrode/ electrochemically reduced graphene oxide (GCE/ERGO) electrode | Soft drinks | 19.2 nM | 50 to 1000 nM | [85] |
| | Voltammetry using Clay modified carbon paste electrode (clay-CPE) | Tablets and urine samples | 2.6 nM | - | [86] |
| | Electrochemiluminescence sensor based on graphene quantum dots | - | 7.6 nM | 2.5 nM to 25μM | [87] |
| Tartrazine | Modified glassy carbon electrode electrochemical sensor with chitosan, gold nanoparticles, multiwalled carbon nanotubes and graphene oxide | Soft drinks, candies, and jellies | 1.45 mg $mL^{-1}$ | 0 - 90 mg mL-1 | [88] |
| | Voltammetry using Poly (glycine) modified carbon paste electrode | Candy and soft drink | 2.83 x $10^{-7}$ mol L-1 | $1 \times 10^{-6}$ to $2.7 \times 10^{-5}$ mol $L^{-1}$ and $3.5 \times 10^{-5}$ to $8.7 \times 10^{-5}$ mol $L^{-1}$ | [89] |
| | Electrochemiluminescence sensor using Nitrogen-doped carbon nanodots | Cookies | 0.18 μM | 0.5–30.0 μM | [90] |

reported applications of biosensors/sensors for different pigments/dyes used in food products and formulations with their detection level and linear working range.

The presence of certain groups in the pigments makes them suitable to be analyzed by different types of transducer systems. A group of scientists from Brazil developed an ocular sensor for determination of Allura red [64]. Allura Red contains a diazo group (–N=N–), which also undergoes oxidation making it electrochemically active and appropriate to be detected by voltammetry [65, 66]. Amaranth also undergoes oxidation and can be detected electrochemically [67, 68]. Additionally, the sulfonic group of amaranth reacts with the carboxyl group leading to slaking of fluorescence of nitrogen-doped graphene significant dots and can be detected by fluorimetry [69]. Azo dyes, namely sunset yellow and Eriochrome Black T, were subjected to an oxidative decolorization using laccase enzymes and the decrease in color was evaluated by amperometric biosensing [78, 82]. A magnetic biosensor was fabricated to detect curcumin in Fe3O4–alginate–curcumin nanoparticles based on change in magneto-reactance with reference to different concentrations of nanoparticles [72]. Surface plasmon resonance (SPR) nanosensor, optical based sensor, which measures difference in refractive index near gold or silver surface for detection of analyte. Curcumin imprinted nanoparticles were attached to the SPR chip for selective and rapid detection of curcumin in real samples [73]. Curcumin being polyphenol in nature, may be oxidized quickly on the electrode surface. This idea was used to create a beta-cyclodextrin–reduced graphene oxide electrode of curcumin for the voltammetric detection [74]. Similarly, cyanidin and delphinidin being polyphenolic compounds can be easily oxidized and detected by voltammetry [75]. Erythrosine was also oxidized by TiO2 nanoparticles modified gold electrode [76] and by glucose modified carbon paste electrode [77] to be evaluated electrochemically. Conversion of indigo carmine to dehydro indigo carmine using modified electrodes was utilized to detect indigo carmine in injection, food and water samples [79, 80]. Likewise, sunset yellow was detected by electrochemical sensors [81–86] based on its property of being oxidized and electrochemiluminescence sensor [87]. Many electrochemical sensors [88, 89] and an electrochemiluminescence sensor [90] for tartrazine are reported based on its property of being oxidized and ability to quench fluorescence produced by carbon dots, respectively.

### 17.6.2 Sensor-Based Paint and Pigment Analysis

Pigments have a long history of usage. However, being chemical substances, they have to meet with the safety assessment criteria provided by chemical legislation of the country where they are being manufactured and sold. The regulation of paint pigments is not uniform across the globe [91]. Many developed countries like the USA have banned the usage of lead pigments for household use due to health hazards associated with them [92]. Several paint industries throughout the world have taken initiatives to completely phase out the usage of lead pigments in the dyes [93].

Sensors play a vital role in the study of the paint and pigments. Infrared photo diode sensor was demonstrated for detection of white blue and red in painting using voltammetry [94]. Terahertz sensor which is based on terahertz pulsed imaging (TPI) is another technology which provides contactless measurement of paint thickness. This method is based on modification in the refractive index and/or extinction coefficient of the material with change in microstructure. The reflected pulses are well separated as a function of time [95].

**Table 17.4** Applications of biosensors for estimation of organic and inorganic pigments used in paints.

| Name of pigment/Dye | Bio sensor/Sensor | Matrix | Detection level | Linear range | Reference |
|---|---|---|---|---|---|
| Acid Green 25 | Voltammetry sensor using modified glassy carbon electrode | Hair dyes | −0.47 V | $1.0 \times 10^{-7}$ to $7.0 \times 10^{-6}$ mol·L$^{-1}$ $2.7 \times 10^{-9}$ mol·L$^{-1}$ | [96] |
| $\beta$-Naphthol | Florescence sensor using rhodamine 6G incorporated SH-$\beta$-CD functionalized gold nanoparticles | In Aqueous solution of naphthol mixtures | 8 nM | 0.1–20 M | [97] |
| Iron oxide | Voltammetry | Microsamples of primers from two baroque paintings | - | 0.3–30 mVs$^{-1}$ | [98] |
| Chromium Green | Microbial Sensor | Freshwater pond and tap water | 1 mM | 0–250 mM | [99] |
| Indigo and Prussian blue pigments | Voltammetry using Sample--modified graphite electrodes | Oil paint film specimens | - | - | [100] |
| Egyptian Blue Pigment | Voltammetry using graphite electrodes | Wall Paintings | - | - | [101] |
| Azurite, Cadmium red, Cadmium yellow, Chrome orange, Chrome yellow, Copper resonate, Malachite, Lead tin yellow, Lead white, Naples yellow, White zinc | Stripping Differential Pulse Voltammetry using immobilized into polymer film electrodes | Painting and polychromed sculptures | - | - | [102] |

*(Continued)*

**Table 17.4** Applications of biosensors for estimation of organic and inorganic pigments used in paints. (*Continued*)

| Name of pigment/Dye | Bio sensor/Sensor | Matrix | Detection level | Linear range | Reference |
|---|---|---|---|---|---|
| Copper pigments | Sample -modified Elva cite 2044 film electrodes were used by cyclic voltammetry | Baroque wall paintings | - | - | [103] |
| Lead White | Voltammetry | Nanoparticles from ancient wall paintings and polychromed sculptures | - | 80 ± 10 mV | [104] |
| Lead pigments (lead white, Naples yellow, minion) | Voltammetry using paraffin-impregnated graphite electrode | Synthetic paint specimens | - | - | [105] |

Table 17.4 represents reported applications of biosensors for examination of organic and inorganic stains used in paints.

Anthraquinone groups of dyes were reduced using glassy carbon electrode cyclic and voltammetry was performed to determine the concentration of hair dye [96]. $\beta$-naphthol was evaluated on the bases of quenching of dye Rhodamine which was incorporated in SH-$\beta$-CD functionalized gold nanoparticles [97]. This principle can be applied to detect $\beta$-naphthol in paint analysis. Voltammetric analysis of iron pigments namely yellow goethite, red hematite, black magnetite and brown maghemite in two microsamples of baroque paintings was carried out based on reduction of free ferric oxides [98]. Electrochemical behavior of indigo and Prussian blue in oil pain was evaluated by Doménech-Carbó *et al.* Microbial Sensor was fabricated for estimation of pigment chromium green in water samples [99]. The deprivation behavior of indigo and Prussian blue dyes on oil paint film specimens was studied. In the degradation study, indigo produced two peaks because of dropping of indigo to leucoindigo and oxidation of indigo to dehydroindigo, respectively while isatin gave peak because of hydroxylated derivatives by proton-assisted reduction. The chromatic shift of paint was found to be due to indigo oxidation to red crystalline compound (isatin) and Prussian blue oxidation to Berlin green [100]. Voltammetric analysis of Egyptian Blue Pigment using graphite electrodes was carried out [101]. Doménech-Carbó *et al.* identified inorganic pigments present in paintings and sculptures with several different colors by using differential pulse voltammetry to immobilize polymer film electrodes [102], copper pigments using Elva cite 2044 film electrodes with samples were modified by cyclic voltammetry [103], voltammetry of nanoparticles combined with atomic force microscopy were used to extract lead pigments from ancient paintings and polychromed sculptures [104] in paint samples using paraffin-impregnated graphite electrode by voltammetry [105].

## 17.7 Conclusion

Biosensors are specific, quick, low weight, and compact in contrast to the traditional analytical methods, which are time-consuming, noncompact, and expensive. They can be also miniaturized which in turn leads to being less expensive and portable. The sensitivity of biosensors is comparable to modern analytical techniques and show higher sensitivity and stability compared to traditional methods. As per literature review, many reports on paint and pigment/dye analysis by different biosensors/sensors are available. Most of them were electrochemical sensors based on oxidation–reduction reactions taking place on the surface of electrodes. Furthermore, biosensors based on fluorescence quenching by pigment/dye were also described.

## References

1. Brandt, A., Milne, A., Weyers, H., Paints and coatings. 9 Applications. *Ullmann's Encycl. Ind. Chem.*, 26, 105, 2012.
2. Pande, P. and Kiran, U.V., Solvent based paint and its impact on environment and human beings, in: *Environment and Society*, p. 198, 2020.

3. Gürses, A., Açıkyıldız, M., Güneş, K., Gürses, M.S., Dyes and pigments: Their structure and properties, in: *Dyes and Pigments*, pp. 13–29, Springer, Cham, 2016.
4. Kumar, A., Dixit, U., Singh, K., Gupta, S.P., Beg, M.S.J., Structure and properties of dyes and pigments, in: *Dyes and Pigments-Novel Applications and Waste Treatment*, R. Papadakis (Ed.), pp. 1–19, Intech Open, London, 2021.
5. Barnett, J.R., Miller, S., Pearce, E., Colour and Art: A brief history of pigments. *Opt. Laser Technol.*, 38, 445–453, 2006.
6. Heckel, G.B., A century of progress in the paint industry. *J. Chem. Educ.*, 11, 487, 1934.
7. Herbst, W. and Hunger, K., *Industrial organic pigments production, properties, applications*, p. 6, John Wiley & Sons, Weinheim, Germany, 2006.
8. Croll, S., *History of paint science and technology*, 1, 1–19, North Dakota State University, Fargo, North Dakota, 2009.
9. Waldie John, M., (Chairman, textbook editing committee), *Surface coatings volume 1 - raw materials and their usage, oil and color.* The oil and colour chemists' Association of Austalia, p. 1-3, 288-313, Springer Netherlands, Netherlands, 2013.
10. Abel, A.G., Pigments for paint, in: *Paint and Surface Coatings: Theory and Practice*, p. 111, 1999.
11. Burgio, L. and Clark, R.J., Library of FT-Raman spectra of pigments, minerals, pigment media and varnishes, and supplement to existing library of Raman spectra of pigments with visible excitation. *Spectrochim. Acta A Mol. Biomol. Spectrosc. Spectrochim. Acta A*, 57, 1491, 2001.
12. Cavaleri, T., Giovagnoli, A., Nervo, M., Pigments and mixtures identification by visible reflectance spectroscopy. *Proc. Chem.*, 8, 45, 2013.
13. Ghelardi, E., Degano, I., Colombini, M.P., Mazurek, J., Schilling, M., Learner, T., Py-GC/MS applied to the analysis of synthetic organic pigments: Characterization and identification in paint samples. *Anal. Bioanal. Chem.*, 407, 1415, 2015.
14. Redígolo, M.M., Amaral, P.O., Leão, C., Crepaldi, C., Russo, T., Mendonca, V.D., Munita, C.S., Bustillos, O.V., PY-GC-MS applied to the identification of synthetic resins in Brazilian painting. *International Nuclear Atlantic Conference – INAC*, p. 11431, 2015.
15. Prati, S., Fuentes, D., Sciutto, G., Mazzeo, R., The use of laser pyrolysis–GC–MS for the analysis of paint cross sections. *Anal. Appl. Pyrolysis*, 105, 327, 2014.
16. Yamaguchi, Y., Kumakura, A., Sugasawa, S., Harino, H., Yamada, Y., Shibata, K., Senda, T., Direct analysis of zinc pyrithione using LC-MS. *Int. J. Environ. Anal. Chem.*, 86, 83, 2006.
17. Lau, D., Hay, D., Wright, N., Micro X-ray diffraction for painting and pigment analysis. *AICCM Bull.*, 30, 38, 2006.
18. Harkins, T.R., Harris, J.T., Shreve, O.D., Identification of pigments in paint products by infrared spectroscopy. *Anal. Chem.*, 31, 541, 1959.
19. Prinsloo, L.C., Tournié, A., Colomban, P., Paris, C., Bassett, S.T., In search of the optimum Raman/IR signatures of potential ingredients used in San/Bushman rock art paint. *J. Archaeal. Sci.*, 40, 2981, 2013.
20. Bhalla, N., Jolly, P., Formisano, N., Estrela, P., Introduction to biosensors. *Essays Biochem.*, 60, 1, 2016.
21. Mehrotra, P., Biosensors and their applications–a review. *J. Oral. Biol. Craniofac. Res.*, 1, 6, 153, 2016.
22. Scheller, F., Schubert, F., Pfeiffer, D., Hintsche, R., Dransfeld, I., Renneberg, R., Wollenberger, U., Riedel, K., Pavlova, M., Kuhn, M., Muller, H.-G., Tan, P., Hoffmann, W., Moritz, W., Research and development of biosensors. A review. *Analyst*, 114, 653, 1989.
23. Malon, R.S., Sadir, S., Balakrishnan, M., Córcoles, E.P., Saliva-based biosensors: Noninvasive monitoring tool for clinical diagnostics. *BioMed. Res. Int.*, 2014, 1–21, 962903, 2014.
24. Camões, M.F., A century of pH measurements. *Chem. Int.-Newsmagazine for IUPAC*, 32, 3, 2010.

25. Hughes, W.S., The potential difference between glass and electrolytes in contact with the glass. *J. Am. Chem. Soc.*, 44, 2860, 1922.
26. Heineman, W.R., Jensen, W.B. Leland C. Clark Jr., Obituary Leland C. Clark Jr. (1918–2005) *Biosens. Bioelectron.*, 21, 1403–1404, 2006.
27. Yoo, E.H. and Lee, S.Y., Glucose biosensors: An overview of use in clinical practice. *Sensors*, 10, 4558, 2010.
28. Cammann, K., Biosensors based on ion-selective electrodes. *Fresen. Z. Anal. Chem.*, 287, 1, 1977.
29. Thevenot, D.R., Toth, K., Durst, R.A., Wilson, G.S., Electrochemical biosensors: Recommended definitions and classification. *Pure Appl. Chem.*, 71, 2333, 1999.
30. Thevenot, D.R., Toth, K., Durst, R.A., Wilson, G.S., Electrochemical biosensors: Recommended definitions and classification. *Biosens. Bioelectron.*, 16, 121, 2001.
31. Thevenot, D.R., Toth, K., Durst, R.A., Wilson, G.S., Electrochemical biosensors: Recommended definitions and classification. *Anal. Lett.*, 34, 635, 2001.
32. Karube, I., Sode, K., Tamiya, E., Micro biosensors. *J. Biotechnol.*, 15, 267, 1990.
33. Keusgen, M., Biosensors: New approaches in drug discovery. *Sci. Nat.*, 89, 433, 2002.
34. Naresh, V. and Lee, N., A review on biosensors and recent development of nanostructured materials-enabled biosensors. *Sensors*, 21, 1109, 2021.
35. Martinkova, P., Kostelnik, A., Válek, T., Pohanka, M., Main streams in the construction of biosensors and their applications. *Int. J. Electrochem. Sci.*, 12, 7386–7403, 2017.
36. Sethi, R.S., Transducer aspects of biosensors. *Biosens. Bioelectron.*, 9, 243, 1994.
37. Malhotra, S., Verma, A., Tyagi, N., Kumar, V., Biosensors: Principle, types and applications. *Int. J. Adv. Res. Innov. Ideas Educ.*, 3, 3639, 2017.
38. Lowe, C.R., An introduction to concept and technology of biosensor. *Biosensor*, 1, 3, 1985.
39. Rodriguez-Mozaz, S., Marco, M.-P., Lopez de Alda, M.J., Barceló, D., Biosensors for environmental applications: Future development trends. *Pure Appl. Chem.*, 76, 723, 2004.
40. Patel, S., Nanda, R., Sahoo, S., Mohapatra, E., Biosensors in healthcare: The milestones achieved in their development towards lab-on-chip-analysis. *Biochem. Res. Int.*, 2016, 1–12, 3130469, 2016.
41. Metkar, S.K. and Girigoswami, K., Diagnostic biosensors in medicine–a review. *Biocatal. Agric. Biotechnol.*, 1, 17, 271, 2019.
42. Kundu, M., Krishnan, P., Kotnala, R.K., Sumana, G., Recent developments in biosensors to combat agricultural challenges and their future prospects. *Trends Food Sci. Technol.*, 88, 157, 2019.
43. Murugaboopathi, G., Parthasarathy, V., Chellaram, C., Anand, T.P., Vinurajkumar, S., Applications of biosensors in food industry. *Biosci. Biotechnol. Res. Asia*, 10, 711, 2013.
44. Lim, S.A. and Ahmed, M.U., Chapter 1: Introduction to food biosensors, in: *Food Biosensors*, pp. 1–21, 2016.
45. Khalid, W.E. and Jais, N.I., A mini review on sensor and biosensor for food freshness detection. *Malaysian J. Anal. Sci.*, 25, 153, 2021.
46. Bahl, S., Javaid, M., Bagha, A.K., Singh, R.P., Haleem, A., Vaishya, R., Suman, R., Biosensors applications in fighting COVID-19 pandemic. *Apollo Med.*, 17, 221, 2020.
47. Chen, L., Zhang, G., Liu, L., Li, Z., Emerging biosensing technologies for improved diagnostics of COVID-19 and future pandemics. *Talanta*, 225, 121986, 2021.
48. Samson, R., Navale, G.R., Dharne, M.S., Biosensors: Frontiers in rapid detection of COVID-19. *3 Biotech.*, 10, 1, 2020.
49. Luong, A.D., Buzid, A., Vashist, S.K., Luong, J.H., Perspectives on electrochemical biosensing of COVID-19. *Curr. Opin. Electrochem.*, 30, 100794, 2021.

50. Wong, L.S., Lee, B.R., Koh, C.E., Ong, Y.Q., Choong, C.W., A novel *in vivo* β carotene biosensor for heavy metals detection. *J. Environ. Biol.*, 36, 1277, 2015.
51. Mandal, R. and Dutta, G., From photosynthesis to biosensing: Chlorophyll proves to be a versatile molecule. *SI*, 1, 100058, 2020.
52. Guerrero-Rubio, M.A., Escribano, J., García-Carmona, F., Gandía-Herrero, F., Light emission in betalains: From fluorescent flowers to biotechnological applications. *Trends Plant Sci.*, 25, 159, 2020.
53. Santos, T.T., Lourenço, L.R., de Lima, S.R., Goulart, L.R., Messias, D.N., Andrade, A.A., Pilla, V., Fluorescence quantum yields and lifetimes of annatto aqueous solutions dependent on hydrogen potential: Applications in adulterated milk. *J. Photochem. Photobiol. A: Chem.*, 8, 100080, 2021.
54. Yoshida, K., Inoue, K., Takahashi, Y., Ueda, S., Isoda, K., Yagi, K., Maeda, I., Novel carotenoid-based biosensor for simple visual detection of arsenite: Characterization and preliminary evaluation for environmental application. *Appl. Environ. Microbiol.*, 74, 6730, 2008.
55. Wong, L.S. and Choong, C.W., Rapid detection of heavy metals with the response of carotenoids in Daucus carota. *Int. J. Environ. Sci. Dev.*, 5, 270, 2014.
56. Watstein, D.M. and Styczynski, M.P., Development of a pigment-based whole-cell zinc biosensor for human serum. *ACS Synth. Biol.*, 19, 7, 267, 2018.
57. Fauzia, V., Imawan, C., Narayani, N.M., Putri, A.E., A localized surface plasmon resonance enhanced dye-based biosensor for formaldehyde detection. *Sens. Actuators B: Chem.*, 1, 257, 1128, 2018.
58. Chen, P.H., Lin, C., Guo, K.H., Yeh, Y.C., Development of a pigment-based whole-cell biosensor for the analysis of environmental copper. *RSC Adv.*, 7, 29302, 2017.
59. Joe, M.H., Lee, K.H., Lim, S.Y., Im, S.H., Song, H.P., Lee, I.S., Kim, D.H., Pigment-based whole-cell biosensor system for cadmium detection using genetically engineered *Deinococcus radiodurans. Bioprocess Biosyst. Eng.*, 35, 265–272, 2012.
60. Wu, Y., Wang, C.-W., Wang, D., Wei, N., A whole-cell biosensor for point-of-care detection of waterborne bacterial pathogens. *ACS Synth. Biol.*, 10, 333, 2021.
61. Sivalingam, T., Devasena, T., Dey, N., Maheswari, U., Curcumin-lHuang, F., Xue, L., Zhang, H., Guo, R., Li, Y., Liao, M., Wang, M., Lin, J., An enzyme-free biosensor for sensitive detection of *Salmonella* using curcumin as signal reporter and click chemistry for signal amplification. *Theragnostic*, 8, 6263, 2018.
62. Sivalingam, T., Devasena, T., Dey, N., Maheswari, U., Curcumin-loaded chitosan sensing system for electrochemical detection of bilirubin. *Sens. Lett.*, 17, 228, 2019.
63. Oplatowska-Stachowiak, M. and Elliott, C.T., Food colors: Existing and emerging food safety concerns. *Crit. Rev. Food Sci. Nutr.*, 57, 524, 2017.
64. Botelho, B.G., Dantas, K.C., Sena, M.M., Determination of Allura red dye in hard candies by using digital images obtained with a mobile phone and N-PLS. *Chemom. Intell. Lab. Syst.*, 167, 44, 2017.
65. Yu, L., Shi, M., Yue, X., Qu, L., Detection of Allura red based on the composite of poly (diallyl dimethylammonium chloride) functionalized graphene and nickel nanoparticles modified electrode. *Sens. Actuators B: Chem.*, 225, 398, 2016.
66. Li, G., Wu, J., Jin, H., Xia, Y., Liu, J., He, Q., Chen, D., Titania/electro-reduced graphene oxide nanohybrid as an efficient electrochemical sensor for the determination of Allura red. *Nanomaterials*, 10, 307, 2020.
67. Tajik, S., Beitollahi, H., Jang, H.W., Shokouhimehr, M., A simple and sensitive approach for the electrochemical determination of amaranth by a Pd/GO nanomaterial-modified screen-printed electrode. *RSC Adv.*, 11, 278, 2021.

68. Pogacean, F., Rosu, M.C., Coros, M., Magerusan, L., Moldovan, M., Sarosi, C., Porav, A.S., Stefan-van Staden, R.I., Pruneanu, S., Graphene/TiO2-Ag based composites used as sensitive electrode materials for amaranth electrochemical detection and degradation. *J. Electrochem. Soc.*, 165, B3054, 2018.
69. Li, Y., Luo, S., Sun, L., Kong, D., Sheng, J., Wang, K., Dong, C., A green, simple, and rapid recognition for amaranth in candy samples based on the fluorescence slaking of nitrogen-doped graphene quantum dots. *Food Anal. Methods*, 12, 1658, 2019.
70. Stefan-van Staden, R.I., Moscalu-Lungu, A., van Staden, J.F., Determination of β-carotene in soft drinks using a stochastic sensor based on a graphene–porphyrin composite. *Electrochem. Commun.*, 109, 106581, 2019.
71. Durai, L., Yadav, P., Pant, H., Srikanth, V.V., Badhulika, S., Label-free wide range electrochemical detection of β-carotene using solid state assisted synthesis of hexagonal boron nitride nanosheets. *New J. Chem.*, 44, 15919, 2020.
72. Devkota, J., Wingo, J., Mai, T.T., Nguyen, X.P., Huong, N.T., Mukherjee, P., Srikanth, H., Phan, M.H., A highly sensitive magnetic biosensor for detection and quantification of anticancer drugs tagged to superparamagnetic nanoparticles. *J. Appl. Phys.*, 115, 17B503, 2014.
73. Cikrik, S., Cimen, D., Denizli, N.B., Denizli, A., Preparation of surface plasmon resonance-based nanosensor for curcumin detection. *Turk. J. Chem.*, 46, 14, 2022.
74. Mirzaei, B., Zarrabi, A., Noorbakhsh, A., Amini, A., Makvandi, P., A reduced graphene oxide-β-cyclodextrin nanocomposite-based electrode for electrochemical detection of curcumin. *RSC Adv.*, 11, 7862, 2021.
75. Yang, L., Yang, T., Li, G., xi Ma, J., Yu, Y., Electrochemical studies of polyphenols, anthocyanins, and flavonoids extracted from blueberry fruit. *Int. J. Electrochem. Sci.*, 17, 2, 2022.
76. Shetti, N.P., Nayak, D.S., Kuchinad, G.T., Electrochemical oxidation of erythrosine at $TiO_2$ nanoparticles modified gold electrode—An environmental application. *J. Environ. Chem. Eng.*, 5, 2083, 2017.
77. Nayak, D.S. and Shetti, N.P., A novel sensor for a food dye erythrosine at glucose modified electrode. *Sens. Actuators B: Chem.*, 230, 140, 2016.
78. Sarika, C. and Rekha, K., A simple laccase based amperometric biosensor for detection of phenolic azo dyes-a comparative study on different membranes as immobilization supports. *Curr. Trends Biotechnol. Pharm.*, 15, 408, 2021.
79. Manjunatha, J.G., A novel poly (glycine) biosensor towards the detection of indigo carmine: A voltammetric study. *J. Food Drug Anal.*, 26, 292, 2018.
80. Edwin, D.S., Manjunatha, J.G., Raril, C., Girish, T., Ravishankar, D.K., Arpitha, H.J., Electrochemical analysis of indigo carmine using polyarginine modified carbon paste electrode. *J. Electrochem. Sci.*, 11, 87, 2021.
81. Xu, J., Zhang, Y., Zhou, H., Wang, M., Xu, P., Zhang, J., An amperometric sensor for sunset yellow for detection based on molecularly imprinted polypyrrole. *Sci. Res. Eng.*, 5, 159, 2012.
82. Sridevi, G., Gopal, M., Gopkumar, P., Amperometric bio-based sensor for azo compound detection using laccase as biorecognition element. *J. Chem. Pharm. Res.*, 1, 44, 2008.
83. Wang, J., Yang, B., Zhang, K., Bin, D., Shiraishi, Y., Yang, P., Du, Y., Highly sensitive electrochemical determination of Sunset Yellow based on the ultrafine Au-Pd and reduced graphene oxide nanocomposites. *J. Colloid Interface Sci.*, 481, 229, 2016.
84. Rozi, N., Ahmad, A., Yook Heng, L., Shyuan, L.K., Hanifah, S.A., Electrochemical sunset yellow biosensor based on photocured polyacrylamide membrane for food dye monitoring. *Sensors*, 18, 101, 2018.
85. Tran, Q.T., Phung, T.T., Nguyen, Q.T., Le, T.G., Lagrost, C., Highly sensitive and rapid determination of sunset yellow in drinks using a low-cost carbon material-based electrochemical sensor. *Anal. Bioanal. Chem.*, 411, 7539, 2019.

86. Moolya, C.V., Shetti, N.P., Nayak, D.S., Clay coated carbon electrode sensor for a food dye sunset yellow. *Mater. Today: Proc.*, 18, 1116, 2019.
87. Niu, H., Yang, X., Wang, Y., Li, M., Zhang, G., Pan, P., Qi, Y., Yang, Z., Wang, J., Liao, Z., Electrochemiluminescence detection of Sunset Yellow by graphene quantum dots. *Front. Chem.*, 8, 505, 2020.
88. Rovina, K., Siddiquee, S., Shaarani, S.M., Selective electrochemical sensor for the determination of tartrazine in food products. *Trans. Sci. Technol.*, 6, 54–59, 2019.
89. Manjunatha, J.G., A novel voltammetric method for the enhanced detection of the food additive tartrazine using an electrochemical sensor. *Heliyon*, 4, e00986, 2018.
90. Gümrükçüoğlu, A., Başoğlu, A., Kolayli, S., Dinç, S., Kara, M., Ocak, M., Ocak, Ü., Highly sensitive fluorometric method based on nitrogen-doped carbon dot clusters for tartrazine determination in cookies samples. *Turk. J. Chem.*, 44, 99, 2020.
91. Sauer, U.G. and Kreiling, R., The Grouping and Assessment Strategy for Organic Pigments (GRAPE): Scientific evidence to facilitate regulatory decision-making. *Regul. Toxicol. Pharmacol.*, 109, 104501, 2019.
92. Gottesfeld, P., Time to ban lead in industrial paints and coatings. *Public Health Front.*, 3, 144, 2015.
93. Anonymous, Global lead paint elimination report, 2017.
94. Ly, S.Y., Lee, C.H., Yoo, H.S., Lee, S.M., Pigment analysis using infrared photo diode working sensor. *J. Korean Soc. Sci. Art*, 15, 365, 2014.
95. Su, K., Shen, Y.C., Zeitler, J.A., Terahertz sensor for non-contact thickness and quality measurement of automobile paints of varying complexity. *IEEE Trans. Terahertz Sci. Technol.*, 4, 432–439, 2014.
96. De Oliveira, R., Hudari, F., Franco, J., Zanoni, M.V., Carbon nanotube-based electrochemical sensor for the determination of anthraquinone hair dyes in wastewaters. *Chemosensors*, 3, 22, 2015.
97. Li, X., Liu, D., Wang, Z., Highly selective recognition of naphthol isomers based on the fluorescence dye-incorporated SH-β-cyclodextrin functionalized gold nanoparticles. *Biosens. Bioelectron.*, 26, 2329, 2011.
98. Grygar, T., Bezdička, P., Hradil, D., Doménech-Carbó, A., Marken, F., Pikna, L., Cepriá, G., Voltammetric analysis of iron oxide pigments. *Analyst*, 127, 1100, 2002.
99. Chen, P.H., Lin, C., Guo, K.H., Yeh, Y.C., Development of a pigment-based whole-cell biosensor for the analysis of environmental copper. *RSC Adv.*, 7, 29302, 2017.
100. Doménech-Carbó, A., Doménech-Carbó, M.T., Osete-Cortina, L., Donnici, M., Guasch-Ferré, N., Gasol-Fargas, R.M., Iglesias-Campos, M.Á., Electrochemical assessment of pigments-binding medium interactions in oil paint deterioration: A case study on indigo and Prussian blue. *Herit. Sci.*, 8, 1, 2020.
101. Doménech-Carbó, A., Doménech-Carbó, M.T., López-López, F., Valle-Algarra, F.M., Osete-Cortina, L., Haartman, E.A., Electrochemical characterization of Egyptian blue pigment in wall paintings using the voltammetry of microparticles methodology. *Electroanalysis*, 25, 2621, 2013.
102. Doménech-Carbó, A., Doménech-Carbó, M.T., Moya-Moreno, M., Gimeno-Adelantado, J.V., Bosch-Reig, F., Identification of inorganic pigments from paintings and polychromed sculptures immobilized into polymer film electrodes by stripping differential pulse voltammetry. *Anal. Chim. Acta*, 407, 275, 2000.
103. Doménech-Carbó, A., Doménech-Carbó, M.T., Gimeno-Adelantado, J.V., Bosch-Reig, F., Saurí-Peris, M.C., Casas-Catalán, M.J., Electrochemical analysis of the alterations in copper pigments using charge transfer coefficient/peak potential diagrams. Application to microsamples

of baroque wall paintings attached to polymer film electrodes. *Fresenius J. Anal. Chem.*, 369, 576, 2001.

104. Doménech-Carbó, A., Doménech-Carbó, M.T., Mas-Barberá, X., Identification of lead pigments in nano samples from ancient paintings and polychromed sculptures using voltammetry of nanoparticles/atomic force microscopy. *Talanta*, 71, 1569, 2007.
105. Domenech Carbo, A., Domenech Carbo, M., Mas Barbera, X., Ciarrocci, J., Simultaneous identification of lead pigments and binding media in paint samples using voltammetry of microparticles. *Arche*, 2, 121, 2007.

# 18

# Bioreceptors for Tissue

**Vipul Prajapati[1]*, Jenifer Ferreir[2], Riya Patel[2], Shivani Patel[2] and Pragati Joshi[2]**

[1]*Department of Pharmaceutics, SSR College of Pharmacy (Permanently Affiliated to Savitribai Phule Pune University), Sayli-Silvassa Road, Sayli, Silvassa, Union Territory of Dadra Nagar Haveli & Daman Diu, India*

[2]*Department of Pharmacology, SSR College of Pharmacy (Permanently Affiliated to Savitribai Phule Pune University), Sayli-Silvassa Road, Sayli, Silvassa, Union Territory of Dadra Nagar Haveli & Daman Diu, India*

***Abstract***

Biosensors have attracted much more interest in the field of medicines and nanotechnology, and serve a great deal of interest in various fields. They have evolved into effective tools for analysing biological materials providing numerous advantages. The unprecedanted evolution of tissue engineering necessitates the use of the best analytical techniques for precise, thorough, and authentic genuine surveillance of the major attributes, improving the overall quality of health indicators. Biosensors are thus most useful instruments for the diagnosis of disease as they are sensitive, discriminatory and expeditious where they help in monitoring with the help of cutting edge biosensing tools. They are incorporated into transducer that are placed inside the body in terms of chips for detecting analytes such as drugs, neurotransmitters, toxic compounds and other compounds. The chapter describes the application of biosensors based on tissue. A detailed classification on different bioreceptors, transducers and different principles, such as optical, gravimetry, and thermal based is described. Its profuse applications for various purposes, such as diagnosis of disease and in disparate other prospects of tissue engineering is provided. Lastly the generalised areas which is encompassed by the tissue biosensor, its challenges and its future aspects are detailed out.

***Keywords:*** Biosensor, tissues, chemiluminescence, fluorescence, bioimaging, gene therapy, drug delivery, bioluminescence

## Abbreviations

| | |
|---|---|
| IUPAC | International Union of Pure and Applied Chemistry |
| DARPA | Defence Advance Research Projects Agency |
| JACS | Journal of the American Society |
| BRET | Bioluminescence Resonance Energy Transfer |

**Corresponding author*: vippra2000@yahoo.com

Inamuddin and Tariq Altalhi (eds.) Biosensors Nanotechnology, 2nd Edition, (419–442) © 2023 Scrivener Publishing LLC

| | |
|---|---|
| RET | Resonance energy transfer |
| 3D | Three-Dimensional |
| 2D | Two-Dmensional |
| SELEX | Systemic Evolution of Ligands by Experimental Enrichment |
| nAChR | Nicotinic Acetylcholine Receptor |
| SPR | Surface Plasmon Resonance |
| EW | Evanescent wave |
| FRET | Fluorescence Resonance Energy Transfer |
| LSPR | Localised Surface Plasmon Resonance |
| hMSCs | Human Mesenchymal Stem Cells |
| MEA | Microelectrode array |
| ECG | Electrocardiogram |
| NIR | Near-infrared |
| PPO | Polyphenol oxidase |
| AAO | Ascorbic acid oxidase |

## 18.1 Introduction

Biosensor, elucidated by the IUPAC, is a device that uses biological reactions, that detects particularly by using enzymes in isolated form, organelles or the cell in whole for the chemical substances having effects with the help of signals obtained thermally, electrically or optically [1].

A biosensor, as per a contemporary definition, is primarily composed of a transducer and an immobilised biological sensing constituent. The integration of these with biological system's exclusive capacity to comprehend attempts differentiates chemical sensors from biosensors (Figure 18.1) [2].

Figure 18.1 shows an outlook of the few basic components of a biosensor.

Biosensors are receiving attention as the next-generation sensing technologies, focusing on diverse and various form of information and innovations, spanning chemistry, physics, and biology. Biosensors offer a great comprehensive foundation for analyzing in applications of biomedical science, food, public health status, microbiology, and environmenatal issues by amalgamating disciplinary technologies. A biosensor system is the combination of biological sample receptors, transducers, and display systems that emply the use of electrical, chemical, or visual functionalities to accurately identify the results and then transform the signal which is used as recognition into signal which can be thus then measured.

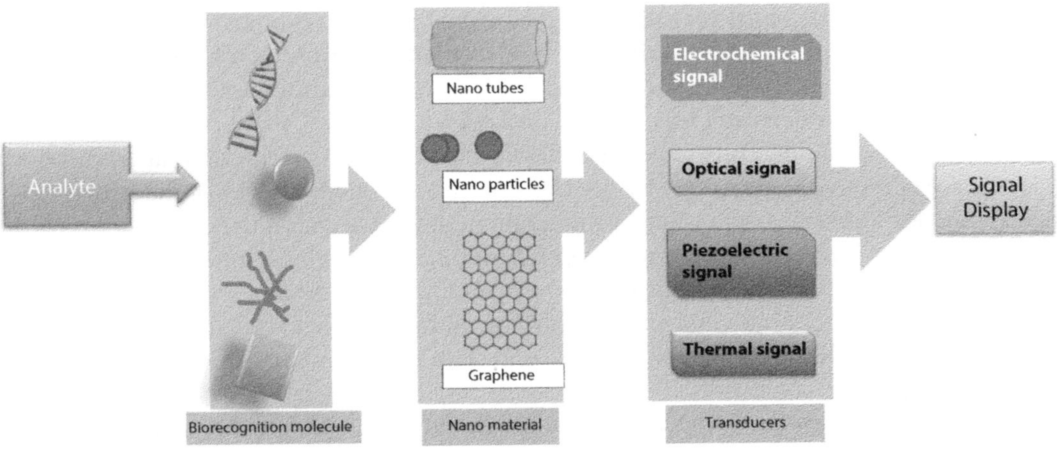


**Figure 18.1** Basic components of a biosensor including an analyte, biorecognitions molecule which includes DNA, antibody, tissue, cell, nano materials, transducer, signal display.

Biosensors focus on analyzing infinitesimal signals that is obtained from a finite or limited samples and use these components, delivering a helpful approach with great sensitivity and stability for effective detection and quantification without any high skilled laboratory training and expertise [3].

Biosensors have recently demonstrated tremendous opportunities for its use in tissue engineering and medicaments which are used in regeneration. Both fields are constantly developing in rapid rate in bioengineering that hold enormous promise for the progression of engineered tissue constructs for repairing the destroyed capabilities of diseased or damaged body cells and tissues [4].

Pathogenic organisms, proteolytic enzymes, cells, and tissue activities have now been recognized using bioreceptor components in biosensors [5].

Distinct biomolecular reactions which are bridled by the antigens, DNA, antibodies, isolated enzymes, and ion channels are used as biorecognition element in molecule-labeled biosensors. Perceiving cellular secretions in microfluidic biodevices, which is chip-specific, necessitates the biological recognition marker's placement which is either antibodies or aptamers that have close vicinity to cells [6]. Coupling of the biological components to biomolecules that have affinity in high, facilitates detecting various diverse analytes with excellent sensing and selectivity properties, which is the key strength of molecular-based biosensors [7].

pH and conductivity analysis have become common conventional analytical procedures, but their usability is hampered due to low specificity and sensitivity. Precise or error-free estimation in these methods is exortinate, necessitating usage of avant-garde equipment in most cases [8]. Because of its undemanding manageable design, sensitivity, and capacity to deliver real time information, biosensors could therefore surpass these restrictions. Biosensors are being designed for a broad range of purposes, including food safety and quality inspection, environmental and bioprocess control, agrobusiness, medicine and pharmaceuticals, clinical assessments, and diagnosis and treatment [9].

Tissue engineering and regenerative medicine are two promosingly burgeoning global topics in biomedical engineering that hold tremendous promise for the development of artificial tissue constructions that can restore the smooth running of tissues or organs which are diseased or defaced [10].

Biosensors have also demonstrated the ability to detect disease-specific biomarkers *in-vivo* [11]. In case of *in-vivo* scenario, the device can closely track wide varieties of the strong biological signals like protein or antibody production in defensive consequences to myocardial infraction, tissue injury, inflammatory events, muscular dystrophy or infectous events. As a corollary, biosensors have a considerable advantage in which they can quickly reveal health-related medical problems, making them an indispensable excellent mechanism for identification of preliminary disease and its respective therapy management in healthcare situations [12].

Biosensors, based on the living animal tissues, have the role in detection and measuring drugs, toxins and hormones. The specific usage of biosensors specific to tissue has exhibited in the field of biomedicals such as pharmacology, biodefense and in physiology. They are thus utilized by incorporating cells which are genetically modified directly introducing them in to the tissues of the animals or transplanting them into the tissues. Biosensor based systems provide a simple approach for detecting diverse signals arising from tissue, meaning that biosensor based systems could be used to diagnose diseases [13].

Specific chips are thus prepared for the functioning of the speific organs. The example of such chips for specific organs are brain on chip sysytems, heart on chip systems and also lung-liver on chip system which are thus being in used [14].

The Defense Advance Research Projects Agency (DARPA) has thus launched projects titled Tissue-Based Biosensors, which will fund the development of two- and three-dimensional tissue-based biosensors by research and business institutions [15].

The potential use of biosensors based on tissues specifically for human subjects will necessitate the resolution of various issues, including the mechanism of output produced detection and proof of the feasibility of the genetic alterations required for implementing biosensor proteins into cells [16].

## 18.2 History

Biosensors had been originally initiated in the early 1960s by Clark and Leland, who used the enzyme called glucose oxidase to amperometrically detect glucose. Scientist Leland and Clark invented the methodology for the detection of oxygen and was thus named as 'Father of Biosensors'. Even though it being the first functional as well as succesful coupling of an electrode along with an enzymatic reaction, Clark and Lyons' system is rarely regarded as the first biosensor since glucose oxidase operates in solution. Updike and Hicks thus illustrated the first approach with an immobilized enzyme in 1967. In 1969, George Guilbault published a distinct model on the basis of the use of potentio-metric electrode made of urea which in particular uses urease which was incapicicated along a sensor which was highly sensitive to pH in JACS. In 1974, it was then emphasized to use thermal transducers as biosensors. Klaus Mosbach and Bengt Daniellsson officially named the devices thermal-enzyme thermistors or probes of enzymes, correspondingly [17].

Thus several more publications began to proven that the notion of Biosensors, the combination of an enzyme and electrolytic sensors, starting in 1970. This was principally actual origin of a Biosensor, a peculiar scientific project that blended biological materials with electrochemical sensors.

## 18.3 Tissue-Based Biosensors

Generally, this type of biosensors can be produced via altered cells or directly modified genes to insert biosensor protein into an animal's tissue. In the last decades, tissue-based biosensors have paved the way for the advancement of *in-situ* monitoring cell methods [18]. These label-free biosensors have the effect of providing investigators to detect cellular and tissue reactions from a biological system by evaluating phenotype responses with high resolution.

Organelles are frequently utilized as nanostructured construction for native tissues in tissue-based multicellular culturing [19]. The usage of tissues obtained from animals as well as plants have been exclusively gained much attention in the past few decades. The main reason being its natural source of the enzyme for the development of the tissue based biosensors.

Thus several biomaterials such as plant tissues, animals tissue and bacterial cells have been used [20]. The plant tissue gained much interest after the initiation of the use of the banana tissue biosensor, proposed in the year 1985, was commomnly called as 'bananatrode'. Various plant tissues have also been used for the biosensing purpose [21].

Various developments were also made for the use of animal tissues and exclusively considered its interest for the detection of lactate from the enzyme lactate dehydrogenase and formic acid from the enzyme cytochrome C, respectively. Many other such tissues have been used for the detection of other compounds using discrete enzymes associated with it (Table 18.1) [22].

Table 18.1 shows the examples of plant tissues and animal tissues used along with their substrate, enzyme or analyte.

### 18.3.1 Tissue-Based Biosensor in Experimental Animals

The design of tissue-based biosensor for genetically modified cells that was based on living animals has been the focus of research in this aspect. Among the other areas of biomedicine, tissue based biosensors have application in physiology, pharmacology and biosensor as;

#### *18.3.1.1 Applications in Physiology*

Hormones and other physiological development substances are commonly examined in blood sample. Blood sample in small animals, such as mice, cannot be repeated since the blood volume will be depleted unless indwelling catheters are used.

Stressing the animals during blood sampling can have a negative impact on the experiment. The used of tissue-based biosensors in research involving hormonal changes, as well as the impact of energy restriction on ageing in rats, rabbits, would be extremely beneficial.

**Table 18.1** Plant and animal tissues with respect to their enzymes and substrates reported for biosensors.

| | Types of tissues | Enzyme | Substrate/analyte | References |
|---|---|---|---|---|
| Plant based tissues | Banana | PPO | Flavonoids | [23] |
| | Mushroom | Alcohol oxidase | Alcohol | [24] |
| | Potato | Polyphenol oxidase (PPO) | Atrazine | [25] |
| | Spinach | Glycolate oxidase | Glycolic acid | [26] |
| | Cucumber | AAO (Ascorbic acid oxidase) | Glutathione | [27] |
| | Sweet potato | Peroxidase | hydroquinone | [28] |
| | Avocado | PPO | Paracetamol | [29] |
| | Corn kernels | Pyruvate decarboxylase | Pyruvate | [30] |
| Animal based tissues | Bovine-erythrocytes | Acetyl choline | Choline | [31] |
| | Bovine-heart | Cytochrome C | Formic acid | [32] |
| | Heart of chicken | Cytochrome C | Formic acid | [33] |
| | Yeast | Alcohol dehydrogenase | Alcohol | [34] |
| | Porcine-pancreas | Cholesterol esterase | Cholesterol | [35] |
| | Pigeon-breast | Cytochrome C | Formic acid | [36] |
| | Human-erythrocytes | Lactate dehydrogenase | Lactate | [37] |
| | Equine-liver | Alcohol dehydrogenase | Alcohol | [38] |

#### *18.3.1.2 Drug Discovery and Testing*

Tissue-based biosensors also have a specific use that can monitor how new medications operate in an *in vivo* case study, allowing researchers to determine how specifically a drug potently affects a specific biochemical pathway as well as analyze it's pharmacokinetics.

#### *18.3.1.3 Biosensor*

By breathing the air, a human and a mouse are compared who breathes the same air respectively. The lungs are an effective delivery route for bringing compounds which determine them from the surrounding when in contact with the tissue biosensors, like the pathogen and then which converts the signals arising fom the pathogens into a physical signal.

### 18.3.2 Incorporating Biosensor Molecules Into Tissue

Organs on a chip, also known as micro-physiological systems, are specialized *in-vitro* technologies that can mimic the functioning of human organs. They are potential methods for tissue engineering and drug screening because of their unparalleled capacity to recreate key aspects of natural cellular conditions. The essential need for non-invasive, real-time monitoring of tissue structures necessitated biases being directly included [39].

### 18.3.3 Bioluminescence-Based Biosensor Tissues in Living Animals

Bioluminescence relies on a biological reaction involving the light produced by an enzyme after the substrate undergoes oxidation. Bioluminescence has been a popular study topic used by biomedical researchers for the variety of purposes which include non-invasive imaging, protien-protien interaction, molecular imaging, gene regulation and cell based assays.

### 18.3.4 Based on Bioluminescence Resonance Energy Transfer (BRET)

Resonance energy transfer (RET) is a photophysical phenomena in which an exciting donor chromophore transfers energy to a receiver chromophore through the non-radiative dipole–dipole coupling mechanism. BRET is a valuable approach for detecting small distance changes between the one donating and the one who accepts it, and thus has a vast applications in functional imaging, biosensing and protein–protein interactions, respectively [40].

## 18.4 Classification

Tisssue based biosensor has thus been distinguished based on its principles and its componental parts. Its been classified based on its receptor recognition princple and transducer. It has also been classified on its different principle namely optically, gravimetrically and thermally (Figure 18.2).

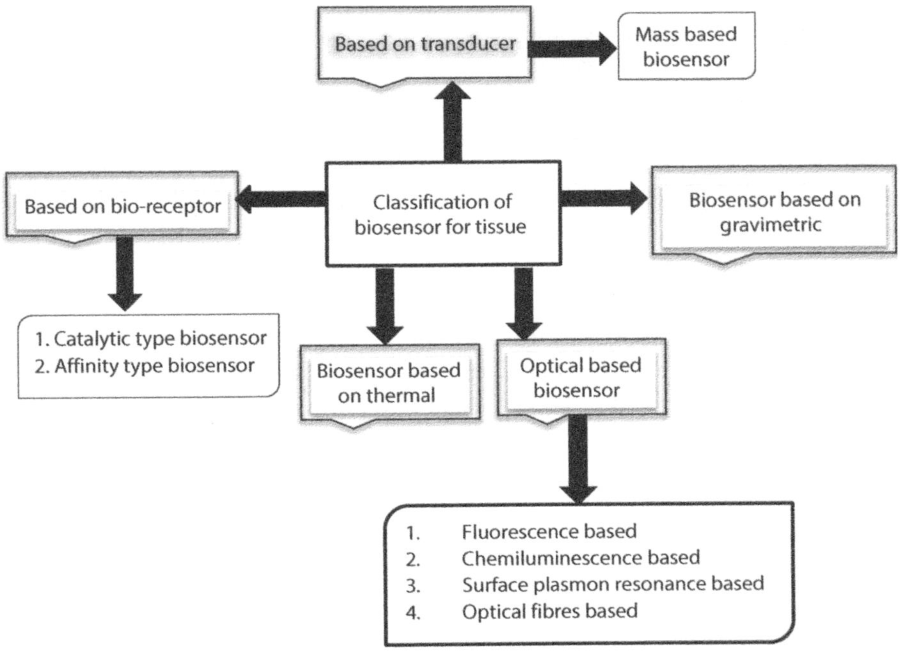


**Figure 18.2** A schematic representation of the classification of biosensor for tissue based on different principles.

### 18.4.1 Based on Bioreceptor

Bioreceptors are categorized as the catalytic, affinity or non-catalytic biosensors based on the biological recognition principle [41]. They are such as;

#### *18.4.1.1 Catalytic Type Biosensor*

Biomolecule interaction leads in the production of an unique biochemical compound in a catalytic biosensor.

#### *18.4.1.2 Enzyme-Based Biosensor*

Enzymatic biosensors are sensor that response on the basis of the binding among the enzyme and the binding of its substrate based on the desired sample. This kind of biosensor utilizes two major mechanisms namely such as the substrate identification and the enzyme inactivation. The basis for the substrate detection method is formed by the embeddment of the tranformation of substrate through the enzyme in the biosensor.

The capacity of the target molecule to limit enzymatic reactions is the operating idea of enzymatic biosensors based inhibition. Inhibition of enzyme approach involves measuring enzyme grown in presence and the removal of the substances that inhibits. The reduction in product level allows for the identification of inhibitory sites that block particular enzymes from working [42].

#### *18.4.1.3 Microbe-Based Biosensor*

An efficient method that identifies the specific analytes by immobilising microbes on a transducer is known as the biosensor based on whole cell microbials. The recognition components in the biosensors for accurately identifying particles or the overall state of the external environment might be bacteria and fungus as the examples of microorganisms. The majority of microorganism biosensors are under the size range of 0.5 and 5 micrometres. They are in different groups based on the kind of cell employed in them. They are the types of either nanobiosensors or microbiosensors. Among the benefit of biosensors based on microbes is that they are non-invasive that not necessitates filtration, which is a laborious and costly procedure. Proteins as well as a few other intrinsic components may be utilized to identify targets as biological recognition component analytes. Microorganisms can be combined with diverse chemicals, and biological processes to create sensitive biosensors. Optical and mass-based transducers are two types of transducers [43].

#### *18.4.1.4 Aptamer-Based Biosensor*

Aptamers are one stranded synthesized nucleotides ( sequences of DNA or RNA) that may be twisted into three-dimensional (3D) and two-dimensional (2D) shapes to preferentially target certain molecules. Because of origins of the nucliec acid, the aptamers are physically as well as functionally fixed across a vast range of storage conditions and temperature.

Unlike antibody, which needs the establishment of biological systems, aptamers may be made chemically, and are stable in the range of pH 2-12, and can tolerate the refolding of the thermal. The aptamers can chemically be altered to meet the recognition requirements of the desired molecule. Aptamers are single-stranded RNA and DNA snippets which are previously been chosen to bind to certain targets using *in-vitro* evaluation method commonly know as SELEX [44].

### *18.4.1.5 Affinity Type Biosensor*

The analyte is securely attached to the sensor in an affinity (non-catalytic) biosensor, and no new biological substance is formed during the interaction.

#### 18.4.1.5.1 Antibody-Based Sensor

Due to their strong binding capacity to the target, these distinct proteins can recognize analytes in other disrupting agents. Immunoglobin (Ig), which is structured like a 'Y' and is grouped into two subunits (H) and two light strands, works as a counteracting agent (L). Furthermore, some human antibodies utilize disulphide bonds and an antigen to construct dimeric or pentameric structures. J-chain (joining protein) is an extra protein. IgA, IgD, IgA, IgG, IgM and IgE are the five types of antibodies based on heavy chain differences. Immunosensors are the biosensors that have an implanted antibodies as a ligand or rely on the antibodies–antigen interaction. There are two types of immuno-sensors such as non-labeled type and labelled type. The antigen–antibody interaction is identified using the non-labeled immunosensors that evaluate the physiological changes that occurs as the complex evolves. A marker that is easily detectable is used in the context of labelled immunosensors. Label measurement is used to examine the antigen–antibody complex in a sensitive manner [45].

#### 18.4.1.5.2 Receptor-Based Biosensor

Biosensor signals for the single chemicals that ranges from low molecular weight to larger poly-peptides are neurotransmitters and hormone receptors. The transductions and the signal amplification in the biosystems take place via the range of processes, that ranges from the depolarization of neuronal member to the G-protein linked creation of the secondary messengers, as well as to activate or to inhibit the target gene expression. The idea of combination of the sensitive and particular protein receptors which senses with the help of optical, electrochemical and acoustical technologies to create analytical devices is thus inspiring. The medical, diagnostics, food, military, and environmental fields have greatly benefited from these biosensors which are receptor based. Utilizing the acetylcholine receptor, major histocompatibility complex related receptor, interleukin 6 receptors and the amino sensitive receptors that thereby contains the crab-antennules has thus been reported in the recent paper for the receptor-based biosensor. However, the nicotinic acetylcholine receptor (nAChR) from fish electric organs have been used in most of these studies [46].

#### 18.4.1.5.3 Nanoparticle

Nanomaterials has attracted attention in the areas of biosensors as a result of recent developments in nanotechnology and their hyper-sensitivity. Both chemical and biological biosensors have used a variety of nanoparticles, which includes oxide nanoparticles semiconductor nanoparticles, metal nanoparticles and also nanodimensional conducting polymers. Numerous labs have investigated using silver and gold nanoparticles and silver-silica blended nanostructures, for biosensor substrates. Metal nanoparticles are very often applied as the parts of 'electronic wires'. Oxide nanoparticles are also the types of nanoparticles. Biomolecules are frequently bounded with this technique, while semi-conductor nanoparticles are extensively employed in a variety of applications in tracers or labels [47].

## 18.4.2 Based on Transducers

According to the working mechanisms, transducers are classified as mass based, electrochemical, optical, thermal and gravimetric based.

### *18.4.2.1 Mass-Based Biosensor*

#### 18.4.2.1.1 Magnetoelastic-Based Biosensor

Amorphous ferromagnetic film strands are used to make magnetoelastic sensors. The ribbons are usually made of strong magneto-elastic coupled coefficient and mechanical durability iron-rich alloys. The magnetostrictive feature of ribbons cause their form to alter in the existence of a magnetic field. This instrument works in a similar way to that of material like quartz crystal micro-balance approach. The distinction of the magnetoelastic biosensors employ magnetoelasticity in place of piezoelectricity [48].

#### 18.4.2.1.2 Piezoelectric-Based Biosensor

The term "piezo" refers to the act of applying pressure or compressing. A potential distinction is created by mechanical stress on some materials which is known as piezoelectricity. The sensors are classified into two types such as the bulk acoustic wave and the surface-acoustic wave based piezoelectric sensor. In biosensors, based on peizoelectric, the bioreceptors are connected with piezoelectric materials like quartz and ceramics, which generates a detectable signal and detecting the piezoelectric crystal bounded to mass-induced oscillation changes [49].

#### 18.4.2.1.3 Electrochemical-Based Biosensor

Electrochemical biosensors monitor the current generated by the electrolytes' oxidation and reduction process. Electrochemical transducers use electrical factors including current, impedance, and potential, as well as a surface containing specialised enzymes, to translate biological interactions into an electrochemical signal [50].

#### 18.4.2.1.4 Potentiometric-Based Biosensor

When there is no current, potentiometric biosensors determine the charge that has accumulated at the working electrode because of the analyte-bioreceptor interaction related to the standard electrode. The types of trusistors such as the electrodes selective to ions and the field sensitive to the ions are used to change a biological process into a potential signal [51].

#### 18.4.2.1.5 Amperometric-Based Biosensor

Amperometric biosensors employ two or three electrode combinations. Amperometric can monitor the electric current produced by the electro-chemical oxidation or by the reduction of the electroactive species. They are at the working electrode when the voltage supply is constant and in relation to the standard electrode. The working electrode thus thereby generates an electric current which is on the surface, thereby proportional to the analyte concentration in the solution. Unlike the potentiometric biology, this method produces a sensitive, quick linear response making it more beneficial amperometric biosensors for the mass manufacturing. The disadvantage of these are that they have low sensitivity and interferance when compared to the other electo-active substances [52].

#### 18.4.2.1.6 Conductometric-Based Biosensors

Conductometric based biosensors evaluate the dissimilarity of the conductance in between two electrodes which ends to an electro-chemical process (i.e the change in the analyte's conductivity property). To detect metabolic activity in live organisms, biosensors that measure conductivity and impedance are routinely used [53].

#### 18.4.2.1.7 Impedimetric-Based Biosensors

When a momentory oscillatory excited pulse is supplied, an electrical block at the electrolye or the electrode is generated which is thus detected by the impedimetric based biosensors. When the voltage of low amplitude AC is provided to sensor electrode, the utilization of an electric-block analyzer is done which measures the response of phase-current to and fro, for which the fequency serves as the main function [54].

#### 18.4.2.1.8 Voltammetric-Based Sensor

Voltammetric biosensors measure the current of a controlled fluctuation for the given voltage to detect an analyte. The great benefits of the voltammetric sensors are the coetaneous detection of multiple analytes and its high sensitive readings [55].

### 18.4.3 Optical-Based Biosensor

Optical-based biosensors are analytical devices that blends a biological recognition element with an optical transducer system to create a biosensor. The basic concept of the operation of the optical based biosensor is that it generates the signals proportional to that of the analyte concentration which thereby allows the real time and label free parallel detection. Antibodies, enzymes and other biomarkers are used in optical biosensors. As biorecognition elements, aptamers, the entire cells, and tissues were used. When it comes to optical biosensors, the absorption, transmission, reflection, and refraction of light are all affected by the transduction process. These are classified as label-free or based, depending on this premise. The measured signal in label-free sensing is generated by the analyte's contact with the transducer. The optical signals are created by the methods such as luminescent, fluorescence or colorimetric in label-based sensing, alternatively. SPR (Surface Plasmon Resonance), chemiluminescence, refractive index, optical wave-guide interferometry, EW fluorescence and surface enhanced fluorescence are all acronyms for surfaced-enhanced fluorescence. Optical biosensors may be made using a variety of optical ideas, including Raman scattering. Optical biosensors which are fluorescence based, SPR-based, chemiluminescence based and optical fibre based are the examples of optical biosensors [56].

#### *18.4.3.1 Optical Biosensor Based on Fluorescence*

The basic phenomena of the optical based biosensors includes labelling for the detection of the molecule or the analyte. The optical biosensors based on the fluorescence when developed, received a lot of attention in consequence of this occurrence. This kind of the biosensor is the most thoroughly researched for medicinal treatment and environmental monitoring. Because of its excellent selectivity, sensitivity, and precision, it's ideal for food quality monitoring and a fast response time. Fluorescent dyes of various types, such as QDs, dyes, and this biosensor makes use of fluorescent proteins. Biosensors based on fluorescence include strategies of types as ‚Fluorescent quenching (turning off) and fluorescent enhancement(fluorescence resonance energy transfer)'. Optical biosensors, based on FRET, have recently gained popularity. Because of their increased sensitivity, they were used in the investigation of the intercellular process. The FRET is a group of people that work together to the procedure entails the transfer of nonradiative energy from an energised source [57].

#### *18.4.3.2 Optical Biosensors Based on Chemiluminescence*

The phenomenon of chemiluminescence occurs when light energy is released as a result of a chemical process. Because of its ease of use, low detection limit, wide calibration range, and inexpensive cost of instruments, Chemiluminescence-based biosensors have attracted a lot of attention. In recent years, in order to better chemiluminescence research, they have been broadened to nanomaterials intrinsic sensitivity, as well as new detection applications. The method has been developed for graphene which are accustomed in detecting the DNA oxide, within a linear gamut of 0.1–3 nm with great sensitivity and selectivity [58].

#### *18.4.3.3 Optical Biosensors Based Surface Plasmon Resonance (SPR)*

Surface plasmon waves are detected by Surface Plasmon Resonance-based biosensors, which measure the alteration in the refractive index induced at the surface of the metal by interaction at molecular level. This biosensor belongs to the label-free biosensor technologies category based on the SPR principle. When polarised light is combined with a magnetic field, the SPR phenomenon occurs, that lights a metal surface at the point where two mediums with differing refractive indices meet. At a particular angle, it forms plasmons, which are electron charge density waves. The SPR phenomenon is caused by the width of the layer at the metal surface at a certain location, and the power of the reflected light has decreased in comparison to the incident light.The resonance angle is a measurement of the angle between two points. The drop in intensity is proportionate to the decrease in intensity.

On the surface, there is a lot of mass, Furthermore, the SPR approach is based on variations in refractive index caused by the analyte's binding to the biorecognition component on the SPR or the transducer sensor. The phenomenon of SPR has a wide range of uses; Diagnosis of diseases, as well as monitoring of the quality of food and environment. Aftre the activation of the SPR, the nanomaterials of metal based like silver and gold, are also affected by this phenomena, which have resulted in the emergence of a novel phenomenon known as localised surface plasmon resonance (LSPR). The primary distinction between the phenomenon of the SPR and LSPR is of the plasma that is involved in both. The locally present total internal reflection which is on the surface of the object governs oscillations place of the metal surface, the nanostructure here is a lot of bulk on the surface. In addition, the SPR method is based on changes in refractive index generated by the analyte's binding on the SPR or the transducer sensor to the biorecognition component. The phenomenon of the SPR have several applications. Monitoring of the quality of food and environment, as well as illness diagnosis, and metal based nanomaterials are likewise impacted by this phenomenon when the SPR is active. As a result, a new phenomena known as localised surface plasmon resonance has emerged (LSPR) [59].

#### *18.4.3.4 Optical Biosensors Based on Optical Fibers*

An optical fibre biosensor is a sensor system made of optical fibres that uses an optical area for the measurement of biological species such as the aptamers, proteins, and an entire cell. Biosensors based on optical fibres are being touted as a possible alternative to standard biomolecule analysis methods. The Evanescent field-sense of the optical fibre, which can be seen in situation, is a reliable optical fibre technology. An evanescent wave is produced, when light is flown beacuse of the total internal reflection through the optical fibre at the interface of the sample. With increasing distance from the interface, this field decays exponentially. In the vicinity of a detecting surface, the evanescent wave can be employed to trigger fluorescence [60].

### 18.4.4 Biosensor Based on Gravimetric

Biosensors based on gravimetric are dependent on the mass biosensors that provide a measurable signal in action to a slight change based on the molecular weight of the binding materials on the surface, such as proteins or antibodies. Thin piezoelectric quartz is used in biosensors based on gravimetric crystals that vibrates at a given frequency. Based on the current which is applied, the applied voltage bulk of the material discovered [61].

### 18.4.5 Biosensor Based on Thermal

Thermal biosensors use the key aspects of biological processes (exothermic or endothermic) to detect the thermal energy received or emitted during ongoing reaction. The heat emitted during oxidation is measured by this thermal biosensor. Its studies provide a viable techhnique for the formation of a thermal self-regulating sensitive biosensor which has improved detection capabilites for the variety of biochemical compounds. MEMS thermal biosensors which are another type of biosensor enables the intergration and batch production of tiny devices, resulting in low-cost bioanalytical instrument [62].

## 18.5 Applications of Tissue-Based Biosensors

Apart from its distinctive use and importance, tissue based biosensors have varied advantages in various fields of medical as well as other prospects such as in treatment of various disease, in drug delivering, in biodimensions printing, in tissue engineering and also in genetic therapy. The detail applications of tissue based biosensors is summarized in Figure 18.3.

### 18.5.1 Biosensor in Cancer Treatment

Because of the cancer's prevalent frequency, large fatality rate, and reappearance after therapy, tumor diagnostic tests require a great deal of attention. Lung, prostate, breast, ovarian, hematologic, skin, and colon cancers, as well as leukemia, could assume over 200 different types, including both genetic and environmental factors, which are linked to an increased chance of acquiring malignancy. Both pathogens are strongly linked to somewhat different cancers [63, 64].

### 18.5.2 Biosensor in Diabetics

Chronic disease is a complex, chronic, and progressive illness that affects over 400 million people across the globe. Unregulated chronically high blood glucose levels damage and destroy different parts of the body, resulting in substantial morbidity and mortality. Sugar level control is very important to minimize the occurrence of these problems. Glucometer measurement, which is still being presented by syringe enzymatic electrical glucose biosensors for subcutaneous injection, is a key success in glycaemic control [65].

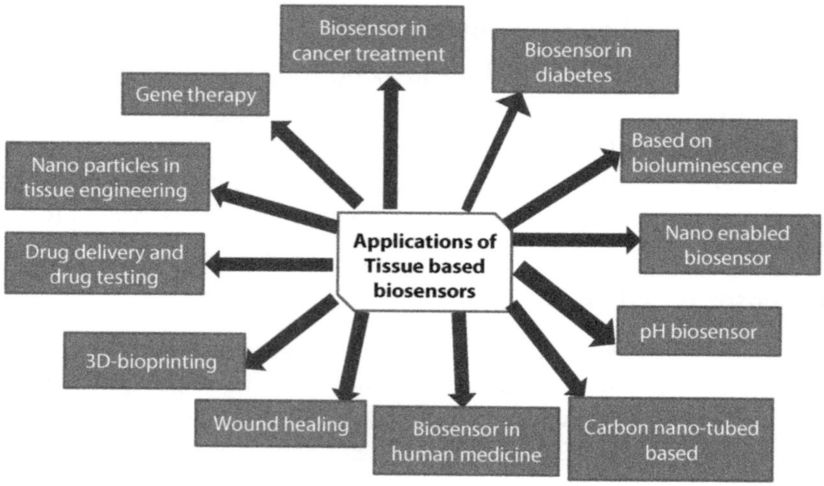


**Figure 18.3** A flowchart showing an outlook of applications of tissue based biosensors in different area.

### 18.5.3 Measurement of the Light Output in the Tissues of Living Animals From the Bioluminescence-Based Biosensor

Increasing the effectiveness of tissue-based biosensing that uses bioluminescence presents the number of tissues. Poor light transmission via tissues, like the dermis, that absorb photons substantially at frequencies below 600 nm requires new approaches. Its creation of luminescence enzymes that emits at higher wavelength, its use of highly sensitive detection equipment, through the use of epidermis border lets the amount of light reaching the sensor without ever being impeded by intermediate tissues all seems to be feasible ways to circumvent this constraint. Precursors would've been delivered directly and constantly to biosensing cells in a living creature under the suspect's command in the perfect situation [66].

### 18.5.4 Drug Delivery and Drug Testing

Tissue-based biosensors are used to track whether new medications behave like invigilators. These could be used to determine how dynamically a medicine effects a certain biochemical pathway, as well as to track its kinetics [67]. Because of its great drug administration capability to the specific tissues devoid of any adverse effects, the implementation of efficient dosage forms has risen in popularity in the last several generations. To increase medication bioavailability and pharmacological quality, dosage form systems are developed to deliver pharmaceuticals in a controlled manner to the target place. Scientists have mostly concentrated on nanotechnology for biomedical applications to improve treatment effects while minimizing negative effects with consideration of tissue based biosensors and bioreceptors [68].

### 18.5.5 Tissue-Based Biosensors in Human Medicine

Tissue-based biosensors have the potential to be a valuable tool for adult diagnostic purposes if the few major issues can be resolved. A clinical adoption of such a technique will most likely use the participant's individual cells. While it is possible to picturise that a transplant like the one previously described in this section may be customized to offer a wide range of biological monitoring systems, this must be done with little difficulty and without involving the patient at threat [69].

### 18.5.6 In the 3D Bioprinting

Biomarkers or biomedical elements can be manufactured using the 3D printing technologies, which is a potential advancement. Biorecognition and the 3D prints the manufacture of biosensor or biometric elements that exhibit the promising features. The 3D printer technique is a workfull to modified biosensors or the structure of biosensor, which has the requirements for the printers, that are significantly used [70].

### 18.5.7 Wound Healing

When injury occurs, epithelial cells are damaged, rapidly reducing the wound's efficiency and causing a difference among the injured and undestroyed tissue. Cation flows among the environment tissue to injury is induced by the potential gradients which can cause directional cell migration [71].

### 18.5.8 Nanoparticles in Tissue Engineering (TE)

Nanotechnology is used in TE for the wide range of applications including an improvement of the genetic, electrical, and mechanical performance. Also used in case of the gene delivery, DNA translation, virus transmission, and cell mapping as well as in promoting tissue regeneration, small molecule recognition and delivering of biomaterials [72].

### 18.5.9 Gene Therapy

Utilizing genome engineering technology to engage developed tissues or stem cells seems to have been a hot issue in TE. Human mesenchymal stem cells (hMSCs) are hematopoietic cells that can differentiate into a multicellular organism, including oligodendrocytes, bone formation, myofibrils, and adipose tissues [73].

### 18.5.10 As a pH Biosensor

Any variety of key intracellular signalling events are controlled by the pH level inside the extracellular matrix. A changed pH is becoming a new characteristic of cancer progression. An acidified pH (6.2–6.9) enhances, in correlation to the balanced cellular membrane pH of immune tissues (7.2–7.4). Three primary characteristics should be taken into consideration when planning technology to measure pH values in sophisticated tissue engineering modelling techniques: (i) its ultimate service is biocompatible and biodegradable; (ii) this same awareness in a wide pH range of reactions together with 3D tissue engineering; and (iii) system's non-invasiveness [74].

### 18.5.11 Nano-Enabled Sensors

Throughout the last generation, there seems to have been a surge of attention to the use of nanotechnology (NTs), nanoscale, and nanostructured materials in detectors. This is due to the wide diversity of original sources, shapes, and based on the increasing options available, all of which contribute to the adaptability of choosing the right nanoparticle for each study's sensing approach. Materials with as miniscule size of the pore, serves the better image resolution and utilisation of existing resources [75].

### 18.5.12 Carbon Nanotube-Based Sensor

Several of these innovative and enhanced embedded sensors have been made possible by carbon nanoparticle's morphology and surface capabilities. Early tumor monitoring in an *in-vitro* experiment is amongst the most current, interesting, and revolutionary ideas using nanocomposite membrane-based nanotube sensor [76].

## 18.6 Generalized Areas Encompassing Biosensors

### 18.6.1 Biosensors in Models of Neurological Disease

Biosensor-based systems provide a simple approach for detecting diverse signals from tissue, indicating that these systems could be used to diagnose diseases. Conventional neurological disease diagnosis is ardous as well as time-wasting along with being complicated because it requires the usage of a microscope. When a physician investigates a patient's symptoms, there is a chance that roughly 40% of them won't be able to help. Some ailments, for instance the Parkinson disease, can be predicted in its preliminary phase/period. MEA is also the most common method for determining neural circuits, disorders and physiology. The approach of the MEA has a perk of non-invasive electro-physiological policing, high through-put screening, multi-site recording and long term culture of the neurons. Pharmaceutical drugs were tested using the neural network's collective electro-physiological behaviour with the help of MEA in relation to the burst activity. Furthermore, Lourenco *et al.* devised a novel approach of multimodal for electric, metabolic as well as magnetic depolarization analysis in order to investigate spreading depolarization and neural activity using hemodynamic measures [77].

### 18.6.2 Biosensors in Models of Cardiac Disease

The preceding applications clearly demonstrate that one of the main properties of biosensors is the monitoring of physiological electrical signals. Bioelectrical activity, on the other hand, is a crucial cardiac function that readily retorts to the tissues of heart in aspects to its health. Usually, the cardiomyocytes that are activated by the change of pH could generate bioelectrical activity,and the action potential of cellular membrane. The structured electrical propagation allows for synchronised pumping activity. According to a recent study, changes in tissue oxygen levels dramatically disrupts the regular ordered length of action potentials. As a result, continuous electrocardiogram (ECG) monitoring is recommended.The methodology provides a clinically accepted way for detecting and diagnosing cardiac rhythm signals. For Heart-related illnesses, Lee *et al.* created a tiny device to readily monitor heart rhythm signals. A cardiac sensor which is flexible, inclusive of a chip of high-field communication-chip, an electrode, an electrochemical cell and a poly-urethane substrate which has a battery. An instantaneous visible signal is been provided by the compact cardiac biosensor, so that individuals may see changes in cardiac rhythm signals on their smartphones.

Feiner and his colleagues also manufactured a electronic scaffold which was degradeable for use as a cardiac patch. As a passivation layer, electrodes made of gold are assembled on an scaffold made up of albumin-fibre which was electrospuned. These flexible patches were sensitized to unconstrained signals obtained from the cardiac cells and which therby apply external electrical stimulation to regularize the contraction of these cells caused by the design [78].

### 18.6.3 Biosensors in Disease Models of the Liver/Lung and Immune Systems

Other types of tissue-based models, such as lungs and the liver are modelled with bosensors which are embedded to decry the functions of tissues-like living constructs, that has similarity to the brain, and heart-on-chip systems. Mitochondrial dysfunction, for example, is involved in the production of chemical or pharmacological products toxicity. Balvi *et al.* employed liver organoids made from HepG2 cells used as a model of tissue and then using microfluidic device, cultured on them.

It provides the microdevice as a biosensor, featuring to monitor the dynamic of malfunction of the mitrochondria by concuringly monitoring the metabolic activity of organoids by the liver. The liver thereby detects the glutaminolysis or oxidative phosphorylation of glycolysis via detecting oxidative phosphorylation.The use of an organoid-based system allows researchers to assess medication safety and the impact of the concentration of drug on the functional damange of the mitochondria.

Another important area is the diagnosis and prognosis of immunosystem diseases in the realm of biosensors. Autoimmune disease is usually caused by a malfunction of the immune system's regulatory system. Genetic factors whose expression is influenced by a variety of method [79].

### 18.6.4 Biosensor in the Model of Cancer Disease

Studies on cancer have been conducted for several decades. Typically, cancer research has focused on the treatment of cancer disorders, resulting in the development of numerous novel effective therapeutic strategies. Patients with cancer disorders, on the other hand, are frequently discovered at the end of their lives, resulting in death because patients did not receive treatment at the optimal time. As a result, how to quickly and correctly identify or prognose cancer disorders has recently been a hot topic in cancer research. Models based on biomimetic cancer have been extensively developed to investigate the genesis, metastasis, and treatment of cancer. Cancer cell's mutations and other mechanisms associate, for instance assembled a cancer-on-a-chip model, where the researchers used a liver cancerous cells, a heart and a microfluidic chip [80].

### 18.6.5 Biosensors in Bioimaging

Biosensors provide an *in vivo* focusing characteristic for boosting the ability to sense in bioimaging with addition to the applicability of diagonosis of *in-vitro* because biosensors have a speedy accurate labelling function. High surface areas, adjustable features, and other concepts are based on this concept. Nanomaterials have a wide range of use due to their small size and characterisation, which makes them a contender for diverse of purposes.

For bioimaging that depends on the type of diagnostic equipment used, consisting of CT imaging, MRI and many types of fluorescence microscope pictures silicas, carbon dots polymers and other nanomaterials have been developed for application in several functions.

Palantavida *et al.* created nanoparticles made of mesoporous silica encased in a polymer matrix. A chemical, specifically targeting LS277, is a near-infrared (NIR) fluorescent dye. When compared to using only LS277, the nanoparticle of mesoporous silica gives five times the resolution of using only LS277 under a brief exposure time for cell labelling bio-images. Moreover, the NIR-trigger technique was also used by coworkers to create carbon dot-based nanoparticles.

The NIR having a carbon dot particularly of S, N-doped was created by using citric acid, urea, and DMSO in the manufacturing of carbon dots. Fluorescence perchance could be

**Table 18.2** Types of biosensor and its application.

| Transduction | Biosensor type | Application | Reference |
|---|---|---|---|
| Electrochemical | Acetylcholinesterase inhibition biosensors | Pesticide determination | [83] |
| Electrochemical | Piezoelectric biosensors | Carbamate and organophosphate identification | [84] |
| Optical | Micro fabricated biosensor | In the field of innovative medication delivery (e.g., in optical corrections) | [85] |
| DNA biosensor | Gold NPs and graphene | Support for immobilisation | [86] |
| Enzymatic biosensor | Carboxy-graphene | Improve the electrical characteristics | [87] |
| Immunosensor | Gold NPs | Signal amplification | [88] |

generated and promptly removed from important organs like the kidney or the liver. After 24-hour period, the intravenous injection was given. Surprisingly, the NIR fluorescent pictures appeared after intravenous injection found that within three hours, S, N-doped carbon dots significantly aggregate in tumours, indicating S, N-doped carbon dots allows a molecule to be labelled for tumours [81].

### 18.6.6 Biosensor in Evaluation of Food

Antibody-coated optics are often used in the food business to detect food contaminants and infections. Because the signal is amplified by a fluorescence-based optical measuring instrument. As a result, for the light source in the biosensors, fluorescence can thus be utilized. Small molecules are detected and investigated using a variety of ligand-binding and immunological techniques. SPR-based sensor systems have been used to prepare vitamins that are water soluble and the residues of the drugs such as sulfonamines and agonists are frequently updated from existing ELISA or different immunilogical assays. Biosensors thus offers a practical, appealing, and cost-effective alternative to a variety of different procedures. Because it is dependable and fast to reply, it showed a lot of promise in food business for regulation of its quality and the safety, as well as in bioprocessor sectors [82].

The biosensor based on its transducer allong with applications is summarized in Table 18.2 including the specific enzymes involvement in it.

## 18.7 Conclusion

Over a long period of time, the application of the biosensors in the fields of tissue recognition has been tremendous. It has made great advancement, eventually lead to its own path, and undoubtedly has been is not only beneficial but helpful and resourceful for mankind. It has made a significant role in the research industry as well as in the medical industry. Even though the past advancement of the biosensors have been made in great achievment, there are still a whole lot of areas where the role of biosensors has not researched or is still incompetent. The scale-up approach and the stability in the long time run of the commercial goods are the two important issues for biosensors. The future challenges and perspective of biosensors has still a long way to pave. With the advancement of nanobiotechnology in conjugation of compact size of biosensors, the complex challanges associated with the tissue based biosensors for plants and animals can be dissolved. With the help of advanced tissue sensors, the targeted biomarkers can also be identified that aids in the clinical understanding of various diseases.

## References


1. Fetz, V., Knauer, S.K., Bier, C., Von Kries, J.P., Stauber, R.H., Translocation biosensors–cellular system integrators to dissect CRM1-dependent nuclear export by chemicogenomics. *J. Sens.*, 9, 7, 5423–5445, 2009.
2. Thevenot, D.R., Toth, K., Durst, R.A., Wilson., G.S., Electrochemical biosensors, recommended definitions and classification. *Pure Appl. Chem.*, 71, 12, 2333–2348, 1999.
3. Bhalla, N., Jolly, P., Formisoano, N., Estrela, P., Introduction to biosensors. *Essays Biochem.*, 60, 1–8, 2016.
4. Hasan, A., Memic, A., Annabi, N., Hossain, M., Paul, A., Dokmeci, M.R., Dehghani, F., Khademhosseini, A., Electrospun scaffolds for tissue engineering of vascular grafts. *Acta Biomater.*, 10, 1, 11–25, 2014.
5. Zhang, W., Chen, C., Yang, D., Dong, G., Jia, S., Zhao, B., Yan, L., Yao, Q., Sunna, A., Liu, Y., Optical biosensors based on nitrogen-doped graphene functionalized with magnetic nanoparticles. *Adv. Mater. Interfaces*, 3, 20, 1600590, 2016.
6. Ma, C., Fan, R., Ahmad, H., Shi, Q., Comin-Anduix, B., Chodon, T., Koya, R.C., Liu, C.C., Kwong, G.A., Radu, C.G., Ribas, A., A clinical microchip for evaluation of single immune cells reveals high functional heterogeneity in phenotypically similar T cells. *Nat. Med.*, 17, 6, 738–743, 2011.
7. Contreras-Naranjo, J.E. and Aguilar, O., Suppressing non-specific binding of proteins onto electrode surfaces in the development of electrochemical immunosensors. *Biosensors*, 9, 1, 15, 2019.
8. Kaur, H. and Sharma, A., Biosensors: Recent advancements in tissue engineering and cancer diagnosis. *Biosens. J.*, 4, 1000131, 1–3, 2015, 2015.
9. Hasan, A., Memic, A., Annabi, N., Hossain, M., Paul, A., Dokmeci, M.R., Dehghani, F., Khademhosseini, A., Electrospun scaffolds for tissue engineering of vascular grafts. *Acta Biomater.*, 10, 1, 11–25, 2014.
10. Hasan, A., Ragaert, K., Swieszkowski, W., Selimovic, S., Paul, A., Camci-Unal, G., Mofrad, M.R., Khademhosseini, A., Biomechanical properties of native and tissue AS engineered heart valve constructs. *J. Biomech.*, 47, 9, 1949–1963, 2014.
11. Malima, A., Siavoshi, S., Musacchio, T., Upponi, J., Yilmaz, C., Somu, S., Hartner, W., Torchilin, V., Busnaina, A., Highly sensitive microscale *in vivo* sensor enabled by electrophoretic assembly of nanoparticles for multiple biomarker detection. *Lab. Chip*, 12, 22, 4748–4754, 2012.
12. Giepmans, B.N., Adams, S.R., Ellisman, M.H., Tsien, R.Y., The fluorescent toolbox for assessing protein location and function. *Science*, 312, 5771, 217–224, 2006.
13. Raknim, P. and Lan, K.C., Gait monitoring for early neurological disorder detection using sensors in a smartphone, validation and a case study of parkinsonism. *Telemed. J. e-Health*, 22, 1, 75–81, 2016.

14. Giana, G., Romano, E., Porfirio, M.C., D'Ambrosio, R., Giovinazzo, S., Troianiello, M., Barlocci, E., Travaglini, D., Ranstrem, O., Pascale, E., Detection of auto-antibodies to DAT in the serum: Interactions with DAT genotype and psycho-stimulant therapy for ADHD. *J. Neuroimmunol.*, 278, 212–222, 2015.
15. Low, L.A., Tagle, D.A., Organs-on-chips: Progress, challenges, and future directions. *Exp. Biol. Med. (Maywood),* 242, 1573-1578, 2017.
16. Acha, V., Andrews, T., Huang, Q., Sardar, D.K., Hornsby, P.J., Tissue-based biosensors. *J. Mol. Recognit.*, pp. 365–281, Springer, New York, 2010.
17. Mohanty, S.P. and Kougianos, E., Biosensors, a tutorial review. *IEEE Potentials*, 25, 2, 35–40, 2006.
18. Acha, V., Andrews, T., Huang, Q., Sardar, D.K., Hornsby, P.J., Tissue-based biosensors, in: *Recognition Receptors in Biosensors*, M. Zourob, (Ed.), pp. 365–381, Springer, New York, 2010.
19. Li, Y.C. and Lee, I., The current trends of biosensors in tissue engineering. *Biosensors*, 10, 8, 88, 2020.
20. Schroth, P., Luth, H., Hummel, H.E., Schutz, S., Schoning, M.J., Characterising an insect antenna as a receptor for a biosensor by means of impedance spectroscopy. *Electrochim. Acta*, 47, 293–297, 2001.
21. Bezerra, V.S., de Lima Filho, J.L., Montenegro, MCBSM, Araujo, A.N., da Silva, V.L., Flow-injection amperometric determination of dopamine in pharmaceuticals using a polyphenol oxidase biosensor obtained from soursop pulp. *J. Pharm. Biomed. Anal.*, 33, 1025–1031, 2003.
22. Rekha, K., Gouda, M.D., Thakur, M.S., Karanth, N.G., *Biosens. Bioelectron.*, 15, 9, 499–502, 2000.
23. Griffiths, L.A., Detection and identification of the polyphenoloxidase substrate of the banana. *Nature*, 184, 4679, 58–59, 1959.
24. Zhao, H., Lan, Y., Liu, H., Zhu, Y., Liu, W., Zhang, J., Jia, L., Antioxidant and hepatoprotective activities of polysaccharides from spent mushroom substrates (Laetiporus sulphureus) in acute alcohol-induced mice. *Oxid. Med. Cell. Longev.*, 2017, 5863523, 1–12, 2017.
25. Edwards, R. and Owen, W.J., The comparative metabolism of the s-triazine herbicides atrazine and terbutryne in suspension cultures of potato and wheat. *Pestic. Biochem. Physiol.*, 34, 3, 246–254, 1989.
26. Richardson, K.E. and Tolbert, N.E., Oxidation of glyoxylic acid to oxalic acid by glycolic acid oxidase. *Jf*, 236, 5, 1280–1284, 1961.
27. Semida, W.M., Hemida, K.A., Rady, M.M., Sequenced ascorbate-proline-glutathione seed treatment elevates cadmium tolerance in cucumber transplants. *Ecotoxicol. Environ. Saf.*, 154, 171–179, 2018.
28. Vieira, I.C. and Fatibello-Filho., O., Biosensor based on paraffin/graphite modified with sweet potato tissue for the determination of hydroquinone in cosmetic cream in organic phase. *Talanta*, 52, 4, 681–689, 2000.
29. Garcia, L.F., Benjamin, S.R., Antunes, R.S., Lopes, F.M., Somerset, V.S., Gil, E.D.S., Solanum melongena polyphenol oxidase biosensor for the electrochemical analysis of paracetamol. *Prep. Biochem. Biotechnol.*, 46, 8, 850–855, 2016.
30. He, X. and Rechnitz, G.A., Plant tissue-based fiber-optic pyruvate sensor. *Anal. Chim. Acta*, 316, 1, 57–63, 1995.
31. Krupka, R.M. and Deves, R., The electrostatic contribution to binding in the choline transport system of erythrocytes. *J. Biol. Chem.*, 255, 18, 8546–8549, 1980.
32. Ballantine, D.S., White, R.M., Martin, S.J., Tissue-based. *Chem.*, 71, 2205–2214, 2005.
33. Sharma, S.K., Sehgal, N., Kumar, A., Biomolecules for development of biosensors and their applications. *Curr. Appl. Phys.*, *3*, 2-3, 307–316, 2003.
34. Lutwak-Mann, C., Alcohol dehydrogenase of animal tissues. *Biochem. J.*, 32, 8, 1364, 1938.

35. Breithaupt, D.E., Bamedi, A., Wirt, U., Carotenol fatty acid esters: Easy substrates for digestive enzymes? *Comp. Biochem. Physiol. B, Biochem.*, 132, 721–728, 2002.
36. Krebs, H.A. and Hems, R., Some reactions of adenosine and inosine phosphates in animal tissues. *Biochem. Biophys. Acta Biomembr.*, 12, 1-2, 172–180, 1953.
37. Schurr, A., Lactate, the ultimate cerebral oxidative energy substrate? *J. Cereb. Blood Flow Metab.*, 26, 1, 142–152, 2006.
38. Shaw, C.R. and Koen, A.L., Response, galactose dehydrogenase, nothing dehydrogenase, and alcohol dehydrogenase: Interrelation. *Science*, 156, 3781, 1517–1518, 1967.
39. Ferrari, E., Palma, C., Vesentini, S., Occhetta, P., Rasponi, M., Integrating biosensors in organs-on-chip devices, a perspective on current strategies to monitor microphysiological systems. *Biosensors*, 10, 9, 110, 2020.
40. Yeh, H.W. and Ai, H.W., Development and applications of bioluminescent and chemiluminescent reporters and biosensors. *Annu. Rev. Anal. Chem.*, 12, 129–150, 2019.
41. Nguyen, H.H., Lee, S.H., Lee, U.J., Fermin, C.D., Kim, M., Immobilized enzymes in biosensor applications. *Mater.*, 12, 1, 121, 2019.
42. Asal, M., Ozen, O., Şahinler, M., Polatoglu, I., Recent developments in enzyme, DNA and immuno-based biosensors. *J. Sens.*, 18, 6, 1924, 2018.
43. Gui, Q., Lawson, T., Shan, S., Yan, L., Liu, Y., The application of whole cell-based biosensors for use in environmental analysis and in medical diagnostics. *J. Sens.*, 17, 7, 1623, 2017.
44. Ali, M.H., Elsherbiny, M.E., Emara, M., Updates on aptamer research. *Int. J. Mol. Sci.*, 20, 10, 2511, 2019.
45. Lim, S.A. and Ahmed, M.U., Electrochemical immunosensors and their recent nanomaterial-based signal amplification strategies. *RSV. Adv.*, 6, 30, 4995–5014, 2016.
46. Rogers, K.R., Valdes, J.J., Eldefrawi, M.E., Acetylcholine receptor fiber-optic evanescent fluorosensor. *Anal. Biochem.*, 182, 2, 353–359, 1989.
47. Luo, X., Morrin, A., Killard, A.J., Smyth, M.R., *Electroanalysis*, 18, 319, 2006.
48. Schierhorn, M., Lee, S.J., Boettcher, S.W., Stucky, G.D., Moskovits, M., Metal–silica hybrid nanostructures for surface-enhanced Raman spectroscopy. *Adv. Mater.*, 18, 21, 2829–2832, 2006.
49. Cai, H., Xu, Y., Zhu, N., He, P., Fang, Y., An electrochemical DNA hybridization detection assay based on a silver nanoparticle label. *Analyst*, 127, 6, 803–808, 2002.
50. Nassarawa, S.S., Luo, Z., Lu, Y., Conventional and emerging techniques for detection of food-borne pathogens in horticulture crops, a leap to food safety. *Food Bioprocess. Technol.*, 15, 1–20, 2022.
51. Pohanka, M., Overview of piezoelectric biosensors, immunosensors and DNA sensors and their applications. *Mater.*, 11, 3, 448, 2018.
52. Malhotra, B.D. and Ali, M.A., *Chapter 1: Nanomaterials in biosensors: Fundamentals and applications. J. Nanomater , in: Nanomaterials for Biosensors Fundamentals and Applications,* A volume in micro and nano technologies, pp. 1–74, Elsevier Inc., Cambridge, United States, 2018.
53. Naresh, V. and Lee, N., A review on biosensors and recent development of nanostructured materials-enabled biosensors. *J. Sens.*, 21, 4, 1109, 2021.
54. Radhakrishnan, R., Suni, I.I., Bever, C.S., Hammock, B.D., Impedance biosensors, applications to sustainability and remaining technical challenges. *ACS Sustain. Chem. Eng.*, 2, 7, 1649–1655, 2014.
55. Martinkova, P., Kostelnik, A., Valek, T., Pohanka, M., Main streams in the construction of biosensors and their applications. *Int. J. Electrochem. Sci.*, 12, 7386–7403, 2017.
56. Liu, G., Feng, D.Q., Qian, Y., Wang, W., Zhu, J.J., Construction of FRET biosensor for off-on detection of lead ions based on carbon dots and gold nanorods. *Talanta*, 201, 90–95, 2019.

57. Dippel, A.B., Anderson, W.A., Evans, R.S., Deutsch, S., Hammond, M.C., Chemiluminescent biosensors for detection of second messenger cyclic di-GMP. *ACS Chem. Biol.*, 13, 7, 1872–1879, 2018.
58. Solaimuthu, A., Vijayan, A.N., Murali, P., Korrapati, P.S., Nano-biosensors and their relevance in tissue engineering. *Curr. Opin. Biomed. Eng.*, 13, 84–93, 2020.
59. Srivastava, K.R., Awasthi, S., Mishra, P.K., Srivastava, P.K., Biosensors/molecular tools for detection of waterborne pathogens, in: *Waterborne Pathogens*, pp. 237–277, 2020.
60. Cali, K., Tuccori, E., Persaud, K.C., Gravimetric biosensors. *Meth. Enzymol.*, 98, 435–468, 2020.
61. Xie, B. and Danielsson, B., Thermal biosensor and microbiosensor techniques. *Biochip J.*, 2, 1–9, 2007.
62. Sun, Z., Liu, H., Wang, X., Thermal self-regulatory intelligent biosensor based on carbon nanotubes-decorated phase change microcapsules for the enhancement of glucose detection. *Biosens. Bioelectron.*, 195, 113586, 1–11, 2022.
63. Choi, Y.E., Kwak, J.W., Park, J.W., Nanotechnology for early cancer detection. *J. Sens.*, 10, 1, 428–55, 2010.
64. Bohunicky, B. and Mousa, S.A., Biosensors, the new wave in cancer diagnosis. *Nanotechnol. Sci. Appl.*, 4, 1–10, 2011.
65. Roglic, G., WHO Global report on diabetes, a summary. *Int. J. Non-Commun. Dis.*, 1, 1, 3, 2016.
66. Tamborlane., W.V., Beck, R.W., Bode, B.W., Buckingham, B., Chase, H.P., Clemons, R., Fiallo-Scharer, R., Fox, L.A., Gilliam, L.K., Hirsch, I.B., Huang, E.S., Juvenile diabetes research foundation continuous glucose monitoring study group continuous glucose monitoring and intensive treatment of type 1 diabetes. *N. Engl. J. Med.*, 359, 14, 1464–1476, 2008.
67. Nelson, J.F., Karelus, K., Bergman, M.D., Felicio, L.S., Neuroendocrine involvement in aging, evidence from studies of reproductive aging and caloric restriction. *Neurobiol. Aging*, 16, 5, 837–843, 1995.
68. Gourley, P.L., Brief overview of BioMicroNano technologies. *Biotechnol. Prog.*, 21, 1, 2–10, 2005.
69. Peltomaa, R., Glahn-Martínez, B., Benito-Pena, E., Moreno-Bondi, M.C., Optical biosensors for label-free detection of small molecules. *J. Sens.*, 18, 12, 4126, 2018.
70. Remaggi, G., Zaccarelli, A., Elviri, L., 3D printing technologies in biosensors production: Recent developments. *Chemosensors*, 10, 2, 65, 2022.
71. Tyler, S.E.B., Nature's electric potential, a systematic review of the role of bioelectricity in wound healing and regenerative processes in animals, humans, and plants. *Front. Physiol.*, 8, 627, 2017.
72. Memic, A., Alhadrami, H.A., Hussain, M.A., Aldhahri, M., Al Nowaiser, F., Al-Hazmi, F., Oklu, R., Khademhosseini, A., Hydrogels 2.0, improved properties with nanomaterial composites for biomedical applications. *Biomed. Mater.*, 11, 1, 014104, 2015.
73. Wang, S., Castro, R., An, X., Song, C., Luo, Y., Shen, M., Tomas, H., Zhu, M., Shi, X., Electrospun laponite-doped poly (lactic-co-glycolic acid) nanofibers for osteogenic differentiation of human mesenchymal stem cells. *J. Mater. Chem. A*, 22, 44, 23357–23367, 2012.
74. Chitnis, T., Glanz, B.I., Gonzalez, C., Healy, B.C., Saraceno, T.J., Sattarnezhad, N., Diaz-Cruz, C., Polgar-Turcsanyi, M., Tummala, S., Bakshi, R., Bajaj, V.S., Quantifying neurologic disease using biosensor measurements in-clinic and in free-living settings in multiple sclerosis. *NPJ Digit. Med.*, 2, 1, 1–8, 2019.
75. Zhang, X., Yuxiang, L., Gillies, R.J., Tumor pH and its measurement. *J. Nucl. Med.*, 51, 8, 1167–1170, 2010.
76. Naresh, V. and Lee, N., A review on biosensors and recent development of nanostructured materials-enabled biosensors. *J. Sens.*, 21, 4, 1109, 2021.

77. Hasan, A., Nurunnabi, M., Morshed, M., Paul, A., Polini, A., Kuila, T., Al Hariri, M., Lee, Y.K., Jaffa, A.A., Recent advances in application of biosensors in tissue engineering. *Biomed. Res. Int.*, 2014, 307519, 1–18, 2014.
78. Li, Y.C. and Lee, I., The current trends of biosensors in tissue engineering. *Biosensors*, 10, 8, 88, 2020.
79. Mollarasouli, F., Kurbanoglu, S., Ozkan, S.A., The role of electrochemical immunosensors in clinical analysis. *Biosensors*, 9, 3, 86, 2019.
80. Zhou, J., Tian, G., Zeng, L., Song, X., Bian, X.W., Nanoscaled metal-organic frameworks for biosensing, imaging, and cancer therapy. *Adv. Healthc. Mater.*, 7, 10, 1800022, 2018.
81. Pirsaheb, M., Mohammadi, S., Salimi, A., Current advances of carbon dots based biosensors for tumor marker detection, cancer cells analysis and bioimaging. *TrAC - Trends Anal. Chem.*, 1, 115, 83–99, 2019.
82. Lu, Y., Shi, Z., Liu, Q., Smartphone-based biosensors for portable food evaluation. *Curr. Opin. Food Sci.*, 28, 74–81, 2019.
83. Singh, A.P., Balayan, S., Hooda, V., Sarin, R.K., Chauhan, N., Nano-interface driven electrochemical sensor for pesticides detection based on the acetylcholinesterase enzyme inhibition. *Int. J. Biol. Macromol.*, 1, 164, 3943–3952, 2020.
84. Ukhurebor, K.E. and Adetunji, C.O., Relevance of biosensor in climate smart organic agriculture and their role in environmental sustainability, what has been done and what we need to do?, in: *Biosensors in Agriculture: Recent Trends and Future Perspectives*, pp. 115–136, Springer, Cham, 2021.
85. Shumyantseva, V.V., Kuzikov, A.V., Masamrekh, R.A., Bulko, T.V., Archakov, A.I., From electrochemistry to enzyme kinetics of cytochrome P450. *Biosens. Bioelectron.*, 121, 192–204, 2018.
86. Shi, A., Wang, J., Han, X., Fang, X., Zhang, Y., A sensitive electrochemical DNA biosensor based on gold nanomaterial and graphene amplified signal. *Sens. Actuators B Chem.*, 200, 206–212, 2014.
87. Liang, B., Fang, L., Yang, G., Hu, Y., Guo, X., Ye, X., Direct electron transfer glucose biosensor based on glucose oxidase self-assembled on electrochemically reduced carboxyl graphene. *Biosens. Bioelectron.*, 15, 43, 131–136, 2013.
88. Xiong, P., Gan, N., Cui, H., Zhou, J., Cao, Y., Hu, F., Li, T., Incubation-free electrochemical immunoassay for diethylstilbestrol in milk using gold nanoparticle-antibody conjugates for signal amplification. *MCA*, 181, 3, 453–462, 2014.

# 19

# Biosensors for Pesticide Detection

**Hoang Vinh Tran**

*School of Chemical Engineering, Hanoi University of Science and Technology, Hanoi, Vietnam*

***Abstract***

Pesticides (PTCs) are toxic organic compounds, including insecticides, herbicides, nematicides, acaricides, fungicides, and rodenticides, and are used in increasing yield in agricultural production. Most PTCs are neurotoxic compounds and have adverse effects on plants and foods when released into the environment (water, atmosphere, and soil). Most of scientific reports demonstrated that the PTCs irreversibly inhibit the acetylcholinesterase, which is an essential enzyme in the human central nervous system. The inhibitory effects of PTCs on muscular activity and vital organ functions finally result in serious symptoms and may lead to death. Traditional methods for PTC analysis are usually laborious procedures, require highly trained technicians, are time-consuming, and have high risk of errors. Biosensors are novel tools for PTC analysis, which are simple, highly selective, sensitive, cost effective, portable size, and have high response. Therefore, PTC biosensors are considered important tools for recognizing PTCs in food and environmental monitoring.

***Keywords*:** Pesticide, biosensor, enzymatic biosensor, aptasensor, immunosensor, electrochemical, nanoparticle, detection

## Abbreviations

| | |
|---|---|
| FAO | Food and Agriculture Organization of the United Nations |
| FAOSTAT | Food and Agriculture Organization Corporate Statistical Database |
| EC | European Community |
| PTCs | Pesticides |
| PTC-biosen | Pesticide biosensor |
| PTCs-immunosen | Pesticides immunosensor |
| EC | European Community |

*Email*: hoang.tranvinh@hust.edu.vn

Inamuddin and Tariq Altalhi (eds.) Biosensors Nanotechnology, 2nd Edition, (443–468) © 2023 Scrivener Publishing LLC

| | |
|---|---|
| ATCh | Acetylcholine |
| TCh | Thiocholine |
| OP | Organophosphorus compounds |
| 2,4-DB | 2,4-Dichlorophenoxybutyric acid |
| CLB | Clenbuterol |
| AP | Alkaline phosphatase |
| ChP | Chlorpyrifos |
| MLT | Malathion |
| ACP | Acetamiprid |
| FVL | Fenvalerate |
| GA | Glutaraldehyde |
| DTNB | 5,5-Dithiobis(2-nitrobenzoic acid) |
| TML | Tiamulin |
| JUG | 5-Hydroxy-1,4-naphthoquinone |
| JUG-HATZ | [N-(6-(4-Hydroxy-6-isopropylamino-1,3,5-triazin-2-ylamino) hexyl) 5-hydroxy-1,4-naphthoquinone-3-propionamide] |
| Diuron | (3-(3,4-dichlorophenyl)-1,1-dimethylurea) |
| PANi | Polyaniline |
| BSA | Bovine serum albumin |
| ITO | Indium–tin oxide |
| IDEs | Interdigitated electrodes |
| CVD | Chemical vapor deposition |
| SPCE | Screen-printed carbon electrodes |
| SAM | Self-assembled monolayer |
| GCE | Glassy carbon electrode |
| AuE | Gold electrode |
| PtE | Platinum electrode |
| PtNPs | Platinum nanoparticles |
| PdNPs | Palladium nanoparticles |
| CuNPs | Copper nanoparticles |
| AuNPs | Gold nanoparticles |
| Ag/AuNPs | Silver/gold bimetallic nanoparticles |
| AgNPs | Silver nanoparticles |
| GO | Graphene oxide |
| Gr | Graphene |
| LoD | Limit of detection |
| Ag | Antigen |

| | |
|---|---|
| Ab | Antibody |
| AChE | Enzyme acetylcholinesterase |
| HRP | Enzyme horseradish peroxidase |
| OPH | Enzyme organophosphorus hydrolase |
| AP | Enzyme alkaline phosphatase |
| ssDNAs | Single-stranded DNA |
| mAb | Monoclonal antibody |
| pAb | Polyclonal antibody |
| $\alpha$-ATZ | Anti-atrazine antibody |
| EIS | Electrochemical impedance spectroscopy |
| $R_{et}$ | Electron-transfer resistance |
| CV | Cyclic voltammetry |
| SWV | Square wave voltammetry |
| CA | Chronoamperometry |
| SPR | Surface plasmon resonance |
| ppm | Parts per million |
| ppb | Parts per billion |

## 19.1 Introduction

### 19.1.1 Pesticides Analysis

Pesticides (PTCs) can be categorized as follows: insecticides, rodenticides, fungicides, acaricides class, nematicides, herbicides, and rodenticides. They kill, repel or reduce pests, including animal pest, plants or microbes. PTCs are indispensable to modern agriculture and are used to protect crops against attacks from insects, fungi, and rodents; inhibit the growth of weed; and boost agricultural yield, thus increasing crop productivity and reducing post-harvest losses. FAOSTAT (FAO) reported that in the year 2015, a total of 3.42 × 106 tons of PTCs was used worldwide, and 0.36 × 106 tons was used in Europe. Unfortunately, nearly all PTCs, which are released into the environment mainly from agriculture activities, are toxic organic compounds. PTC pollutants are considered major problems because they are present in groundwater and pollute soils, plants, and finally food [1, 2]. PTCs are neurotoxic compounds that irreversibly inhibit the acetylcholinesterase (AChEs) enzyme, thereby promoting the accumulation of acetylcholine (ACh), a neurotransmitter, in nerves, disrupting muscular activity, and vital organs functions and finally producing serious health symptoms. In fact, exposure to high levels of PTCs can lead to acute toxicity, which potentially cause carcinogenicity, long-term health problems, Parkinson's disease, and even death. Some PTCs are selective, impacting only a target organism, whereas some have been designed for a broad range of target organisms. For example, atrazine (ATZ) is widely used to control broadleaf and grassy weeds in agricultural crops. ATZ is considered a PTC with low toxicity, but studies on animals have indicated that exposure to ATZ can cause endocrine disruption or lead to carcinogenicity. ATZ is a persistent environmental contaminant and is usually not absorbed by the soil because of its polarity and percolates through it, leading to groundwater and surface water contamination. Thus, the European Community has set the limits of PTC concentrations in consumer products. In drinking water, regular PTC and total PTC concentrations should be lower than 0.1 and 0.5 ppm, respectively [1–3]. PTCs can be analyzed using conventional techniques, including the GC, HPLC or spectroscopy methods. These conventional techniques require complicated procedures, expert operators, and considerable amount of time and have high risk of errors because of cleanup steps. Hence, they are unsuitable for on-site PTC detection. Recently, biosensor technologies have been used in PTC analysis. These technologies can resolve the problems of conventional analysis techniques and thus have potential applications in PTC detection in environmental, food, or clinical samples.

### 19.1.2 Structures and Principles of Construction of Biosensors for Pesticide Analysis

The IUPAC defined biosensors as analytical devices that combine biological elements as capture probes and physical transducers. Therefore, a pesticide biosensor (PTCs–biosens) can be considered a biosensor that utilizes biomolecules, such as enzymes, antibodies, peptides, proteins, and aptamer sequences as capture probe agents. Output signals are measured before and after binding reactions among biocapture molecules, and PTC target molecules are used for determining the presence or absence of pesticides or their concentrations in samples (Figure 19.1). In fact, PTC–biosens should be designed not only to have simple structures and to be easy to use but also to promote a synergistic effect between a biocapture probe and physical transducer, which decides to the specificity and sensitivity of the PTCs–biosens [4].

As shown in Figure 19.1, PTCs-biosens have three major components: **(1i)** biorecognition elements (also called bioreceptors), such as aptamer sequences, enzymes, and antibodies for recognizing pesticide molecules as analytes, **(2i)** immobilization surfaces, such as functionalized polymers [5–10], nanomaterials [11], sol–gel films [12], and self-assembled monolayers [13] for bioreceptor immobilization and **(3i)** physical transducers for converting capture probe–target recognition as a biochemical reaction into a measurable signal. Transducer are physical components used for transducing biorecognition events between bioreceptors and pesticide molecular targets into measurable physical signals, including electrochemical, electrical, optical, piezoelectric, and thermal signals (Figure 19.2) [14–16]. Currently, electrochemical transducer-based and optical transducer-based PTC–biosens are the most used sensors owing to their simple fabrication and operation and high sensitivity.

Sensors continuously play an important role in every aspect of modern life. PTC–biosens are simple to use and fast and facilitate real-time, continuous, and simultaneous PTC detection in various applications, including bioprocess monitoring, environmental monitoring, clinical diagnostics, and food and agricultural product processing. Therefore, highly selective, sensitive, responsive, and cost-effective sensors for pesticide detection are urgently needed.

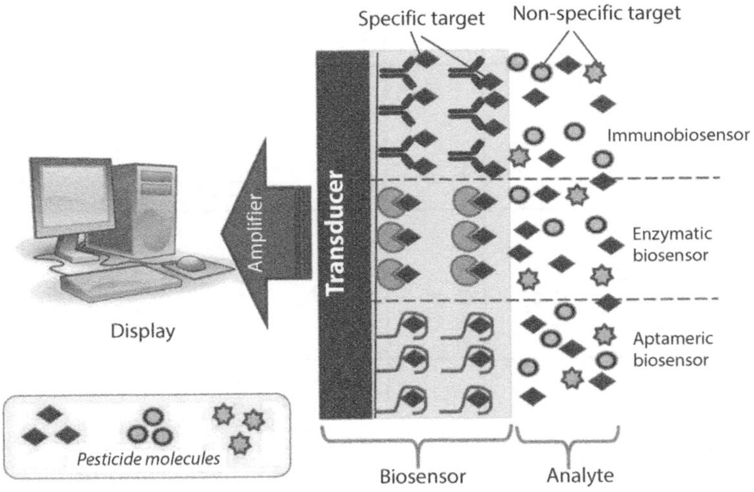


**Figure 19.1** Illustration of pesticide biosensor construction. This figure is redrawn.

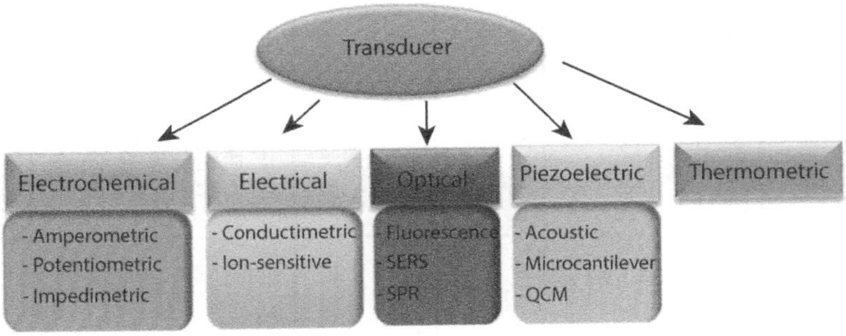


**Figure 19.2** Classification of physical transducers. This figure is redrawn.

## 19.2 Biosensors for Pesticide Detection

Figure 19.1 and Table 19.1 show that enzymes, antibodies or aptamer sequences can be used as bioreceptors for the development of pesticide biosensors that can recognize and capture pesticide analytes on sensor surfaces. Therefore, they are often used as bioreceptors in pesticide biosensors.

### 19.2.1 Enzymatic Pesticides Biosensors

Enzymatic PTC-biosens are constructed to utilize enzymes as bioreceptors, which are immobilized onto physical transducers, such as electrode surfaces. An enzyme bioreceptor degrades a pesticide molecule as a specific substrate into products that generate electrochemical signal. In the detection process, immobilized enzymes degrade pesticide molecules as specific substrates into products that generate electrochemical signals. For example, the PTCs, such as parathion or formaldehyde, can be specifically detected by sulfite parathion

**Table 19.1** Electrochemical pesticide biosensors.

| Pesticide target | Capture probe | Electrode(s) | Detection methods | Immobilization layer | Signaling element | Detection range | LoD | Ref. |
|---|---|---|---|---|---|---|---|---|
| Atrazine (ATZ) | Anti-atrazine monoclonal antibody | Pt microelectrode | SWV | Graphene/ Polyaniline | PANi | $2 \times 10^{-3}$ ppb to 20 ppb | $4.3x10^{-5}$ ppb | [50] |
| Tiamulin (TML) | Anti-tiamulin monoclonal antibody | AuE | DPV | AgNPs/GO/Nf/AuE | AgNPs | 0.05 ppb to 100 ppb | 0.04 ppb | [49] |
| Atrazine (ATZ) | Anti-atrazine monoclonal antibody | GCE | SWV | poly(JUG-co-JUGA) | Quinone groups | 0.1 pM to 10 nM | 1 pM | [10] |
| Atrazine (ATZ) | Anti-atrazine monoclonal antibody | GCE | EIS | poly(JUG-co-JUGA) | Quinone groups | 1 pM to 0.1 μM | 1 pM | [3] |
| Clenbuterol (CLB) | Anti-clenbuterol antibody | SPCE | CA | AuNPs | Alkaline phosphatase enzyme(AP)/ *p*-aminophenyl phosphate (p-APP) | 0.027 to 800 ng $mL^{-1}$. | 0.008 ng mL−1 | [30] |
| Diuron | Anti- diuron antibody | AuE | SWV | Prussian blue-gold nanoparticle (PB-GNP) film | Alkaline phosphatase enzyme (AP)/1-naphthyl phosphate | 1 ppt to 10 ppm | 1 ppt | [35] |
| Malathion (MLT) and chlorpyrifos (CLP) | Acetylcholinesterase (AChE) enzyme | AuE | SWV | ZnS-nanoparticles (ZnSNPs) and poly(indole-5-carboxylic acid) | Acetylthiocholine chloride (ATCl)/ AChE enzyme | MLT: 0.1 to 50 nM; CLP: 1.5 to 40 nM, | 0.1 nM for MLT and 1.5 nM for CLP | [18] |
| Methyl parathion | Acetylcholinesterase (AChE) enzyme | GCE | DPV | MWCNTs/CS/ | 5,5-dithiobis(2-nitrobenzoic acid) (DTNB) | 0.50 μM to 1.0 pM | 0.75 pM | [19] |

*(Continued)*

**Table 19.1** Electrochemical pesticide biosensors. (*Continued*)

| Pesticide target | Capture probe | Electrode(s) | Detection methods | Immobilization layer | Signaling element | Detection range | LoD | Ref. |
|---|---|---|---|---|---|---|---|---|
| Fenvalerate (FVL) | Fenvalerate monoclonal antibody | GCE | EIS | Chitosan | $Fe(CN)_6^{4-/3-}$ | 1 ppb to 10 ppm | 0.8 ppb | [34] |
| Acetamiprid and atrazine (ATZ) | Acetamiprid aptamer and Atrazine aptamer | Interdigitated electrodes (IDEs) | EIS | (3-glycidyloxypropyl) triethoxysilane | Pt NP microwires | ACP: 10 pM to 100 nM; ATZ: 100 pM to 1 μM | ACP: 10 pM, ATZ: 1 pM | [26] |
| Dichlofenthion | Organophosphorus hydrolase enzyme (OPH) | GCE | CA | N/A | Horseradish peroxidase (HRP) | 1 μM to 600 μM | 24 μM (7.6 ppm | [23] |

hydrolase and formaldehyde dehydrogenase, respectively. An electrochemical enzymatic PTC–biosen can be used as an amperometric or potentiometric tool for monitoring enzyme activity through the oxidation or reduction of enzyme substrates, and electrical current or potential is proportionate to PTC concentration. These features facilitate the quantification of a detected molecule [1, 17]. Another approach for enzymatic PTC–biosens development is based on enzyme inhibition reaction, which is named enzyme-selective inhibition. In this case, the specific enzyme is poisoned by the detected substances, thus, a low activity of the enzyme indicates that high concentration of the toxic substance is present. In the literature, the enzyme selective inhibition is the most extensively used for development of an enzymatic PTC–biosens by electrochemically methods. The PTC–biosen-based enzyme-selective inhibition are mainly used the enzyme cholinesterase (ChE) wherein, the organophosphate and carbamate insecticides being the main inhibitors [17]. There are many strategies to organize enzymatic biosensors: mono-enzymatic biosensors (the most common), bi-enzymatic biosensors (theoretically the most sensitive, but difficult to design) or tri-enzymatic biosensors (hardly reproducible) [17]. Despite they have low selectivity, the ChE-based enzymatic PTC–biosens are robust tools in case requiring a very fast toxicity screening. Chauhan [18] has described an amperometry-based enzymatic organophosphorus (OP) biosensor by immobilization the AChE enzyme on a ZnSNPs/Pin5COOH/AuE (Figure 19.3) with its working principle is based on the proportional inhibition of malathion (MLT) and chlorpyrifos (CLP) concentrations onto AChE activity. Under optimum conditions, MLT and CLP concentrations can be found in a ranges from 0.1 to 50 nM and 1.5 to 40 nM, respectively [18]. Similarly, Dong [19] immobilized AChE enzyme onto MWCNTs/CS/GCE for the development of amperometry enzyme-based biosensors for methyl parathion detection. Herein, 5,5-dithiobis(2-nitrobenzoic acid) has been used as an electrochemical transducer and enzyme AChE as a bioreceptor. Through amperometry, AChE/MWCNTs/CS/GCE can detect methyl parathion in a concentration range from 0.50 μM to 1.0 pM with a LoD of 0.75 pM.

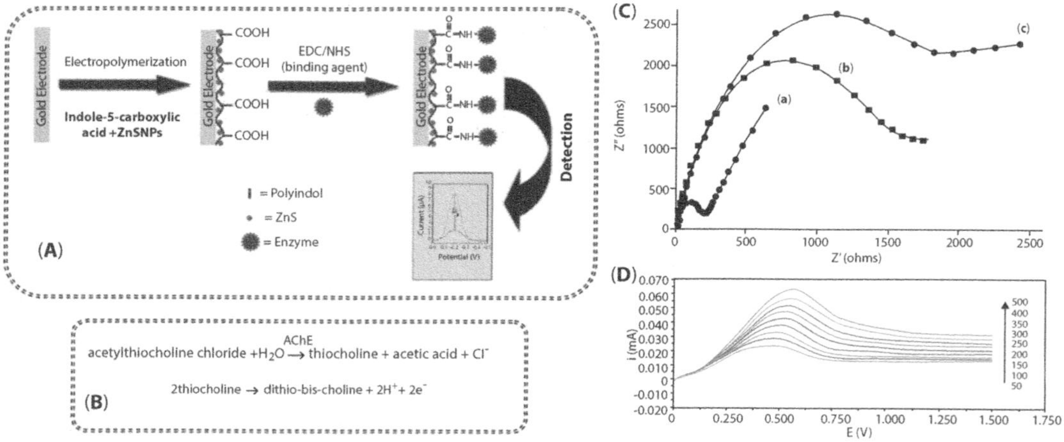


**Figure 19.3** (A) Illustration of an AChE-based selective inhibition enzymatic biosensor for MLT and ChP detection, (B) electrochemical reactions, (C) Nyquist plot of (a) bare AuE, (b) ZnSNPs/Pin5COOH/AuE and (c) AChE/ZnSNPs/Pin5COOH/AuE and (D) SWV responses for the developed enzymatic biosensor at different acetylthiocholine chloride (ATCl) concentrations. This figure is taken from Ref. [18] (copyright 2011 from Elsevier).

Kauffmann [20–22] used a silver disk electrode to develop a enzymatic electrochemical biosensor for small molecule detection. In this approach, an enzyme is used to convert target molecules to other products, which can be detected by direct electrochemical oxidation at the silver disk electrode. Additionally, they designed an electrochemical ATCh sensor by using the silver electrode as a detector for thiocholine (TCh) [22]. AChE is immobilized onto Au substrate, which is then integrated with the silver electrode (Figure 19.4A). Herein, AChE converted acetylthiocholine (ATCh) into TCh, which can be recognized by the silver electrode with the amperometric technique. This developed enzymatic sensor can detect TCh in a concentration range of 10 μM to 10 mM and a LoD of 5.3 μM (Figure 19.4B).

Bi-enzymatic biosensors have been used to detect organophosphorus compounds (OPs) Sahin [23]. This development used tow enzymes, including OPH and HR, to detect a variety of OPs. Figure 19.5 illustrates a working principle of electrochemical signal generation in this developed bi-enzymatic biosensor in dichlofenthion (an insecticide) detection.

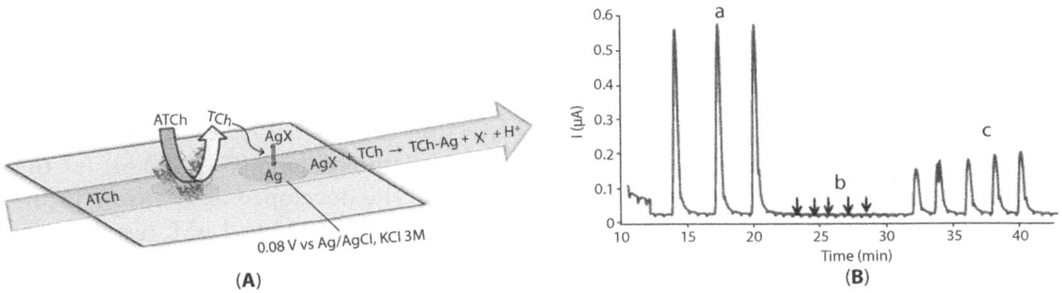


**Figure 19.4** (A) Enzyme-based electrochemical biosensor for TCh detection (ATCh: acetylthiocholine, TCh: thiocholine, AgX: AgCl or Ag(OH)), (B) amperometric response curve of the developed enzymatic biosensor for TCh detection at a silver electrode. This figure is taken from Ref. [22] (copyright 2013 from Elsevier).

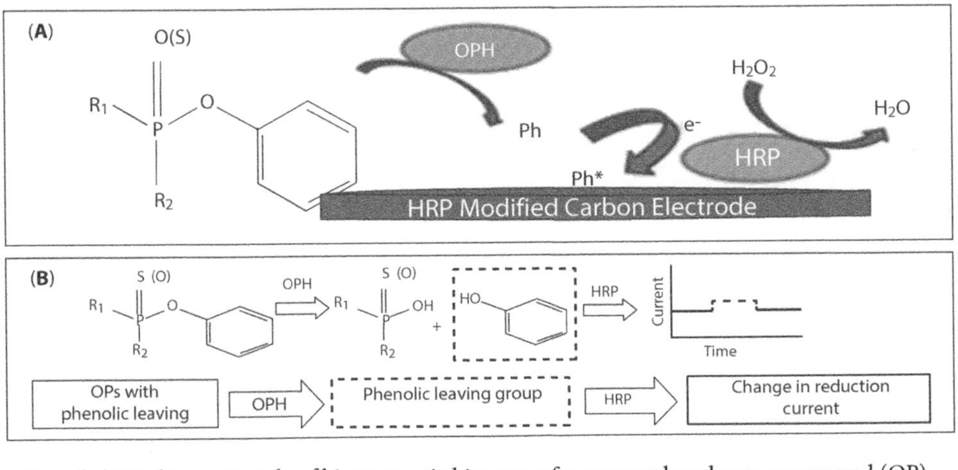


**Figure 19.5** (A) Working principle of bi-enzymatic biosensor for organophosphorus compound (OP) detection by using OPH combined with HRP; (B) hydrolysis scheme of OPs by OPH to produce phenolic compounds, which can be electrochemically detected by HRP in the presence of H2O2. This figure is taken from Ref. [23] (copyright 2011 from Elsevier).

### 19.2.2 Aptameric Biosensors for PTCs Detections

Aptamers are single-strand DNA or RNA, they have approximately 15–100 nucleotides [24] and are highly selective. Aptamer-based pesticide biosensors utilize aptamers as bioreceptor components for pesticide molecule recognition. Thus, they are also called aptasensors. Currently, aptamers are widely used as effective bioreceptors for fabricating various biosensor systems owing to their high affinity and specificity to targets; in addition, aptamers can be easily immobilized on various substrates and integrated into different signal transduction platforms, including colorimetry, fluorometry, electrochemistry, and electrochemiluminescence. Moreover, as synthetic molecules, aptamers can be designed and modified for improving the selective and sensitive to specific pesticide molecules. Currently, PTC aptasensors have various advantages over immunosensors in terms of accuracy and reproducibility. Aptamer chains can bind to specific targets with high affinity as antibodies. To date, a variety of PTCs aptasensors have been established and reported for a wide spectrum of pesticide molecules [1, 17]. Fan [25] used gold nanoparticles (AuNPs) to develop EIS aptasensors for acetamiprid (ACP). In the presence of ACP molecules, a complex of ACP–aptamer can be generated on an electrode surface leading to an increasing of $R_{et}$ (Figure 19.6), which allows ACP sensing in a concentration range of 5 nM to 600 nM.

Madianos [26] have used platinum nanoparticle (PtNP)-decorated electrodes to develop impedimetric aptasensors for ACP and ATZ detection (Figure 19.7). PtNPs were then chemically functionalized with (3-glycidyloxypropyl) triethoxysilane ($C_{12}H_{26}O_5Si$) for the covalent immobilization of specific aptamer sequences. The developed aptasensors facilitate the detection of ACP and ATZ at 10 pM to 100 nM and 100 pM to 1 μM, respectively, LoD values of 1 pM (ACP) 10 pM (ATZ).

PTC aptasensors can be developed using optical transducers. Most colorimetric PTC aptasensors can be developed by using aptamers as specific capture probes, which are combined with noble metal nanoparticles (NPs), such as PtNPs, palladium NPs, copper NPs,

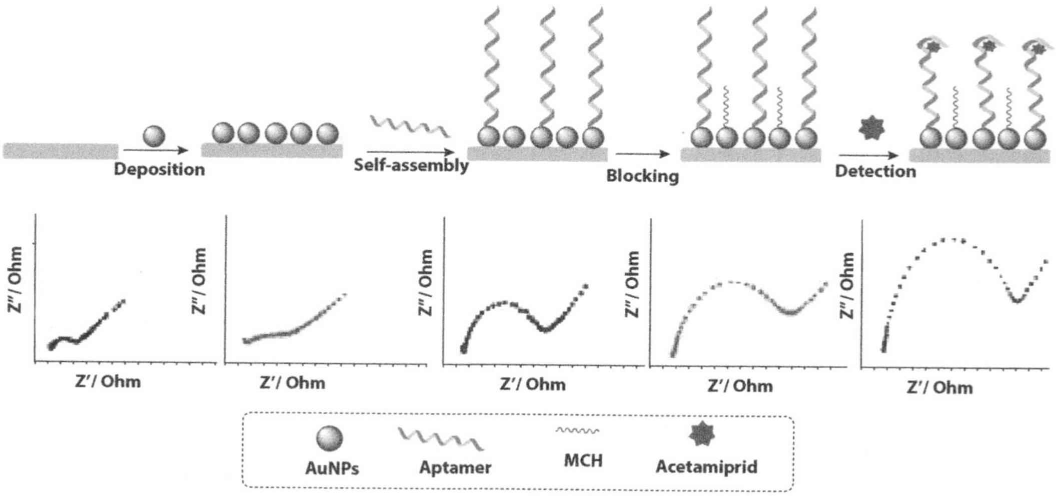


**Figure 19.6** Electrochemical impedance spectroscopy (EIS) aptasensor for acetamiprid detection. This figure is taken from Ref. [25] (copyright 2013 from Elsevier).

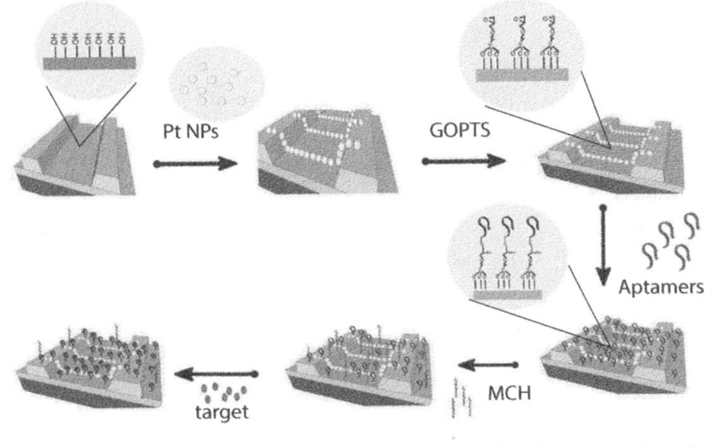


**Figure 19.7** Schematic illustration of EIS aptasensors for acetamiprid and atrazine detection using interdigitated electrodes (IDEs) decorated by PtNPs. This figure is taken from Ref. [26] (copyright 2018 from Elsevier).

silver NPs (AgNPs) and AuNPs, whereas noble metallic NPs with unique SPR properties can be directly utilized as transducers. Noble metallic NP-based colorimetric aptasensors can be developed with various platforms, including lateral flow dipsticks, electrochemical biosensor strips, SERS and paper analytical devices (PADs) for variety applications, including chemistry, bioanalytical technology, food controlling or medical industry.

As shown in Figure 19.8, the detection principle of colorimetric-based PTCs aptasensors are realized through NP agglomeration in the presence of specific pesticide targets. This agglomeration leads to change in color, which can be directly observed by the naked eye or through UV-Vis spectroscopy. These change in turn can be used in measuring readout

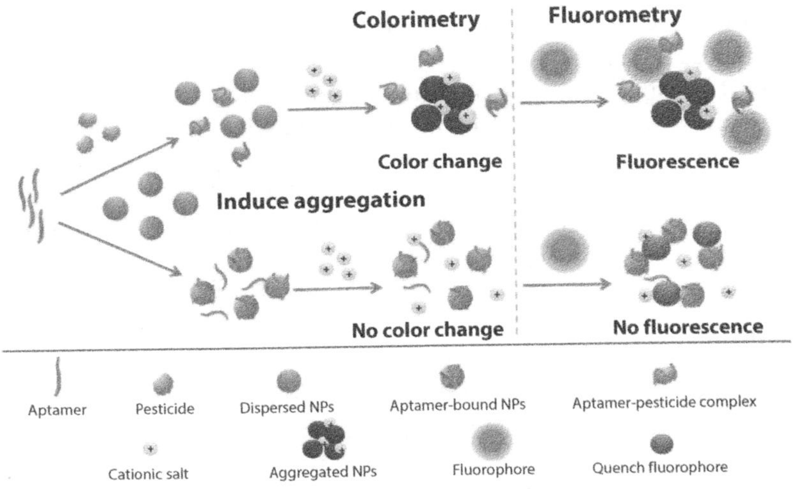


**Figure 19.8** Schematic diagram of AgNP-based colorimetric aptasensors for pesticide detection. The output signal can be measured using colorimetric and fluorometric methods. This figure is taken from Ref. [24] (copyright 2020 from MDPI).

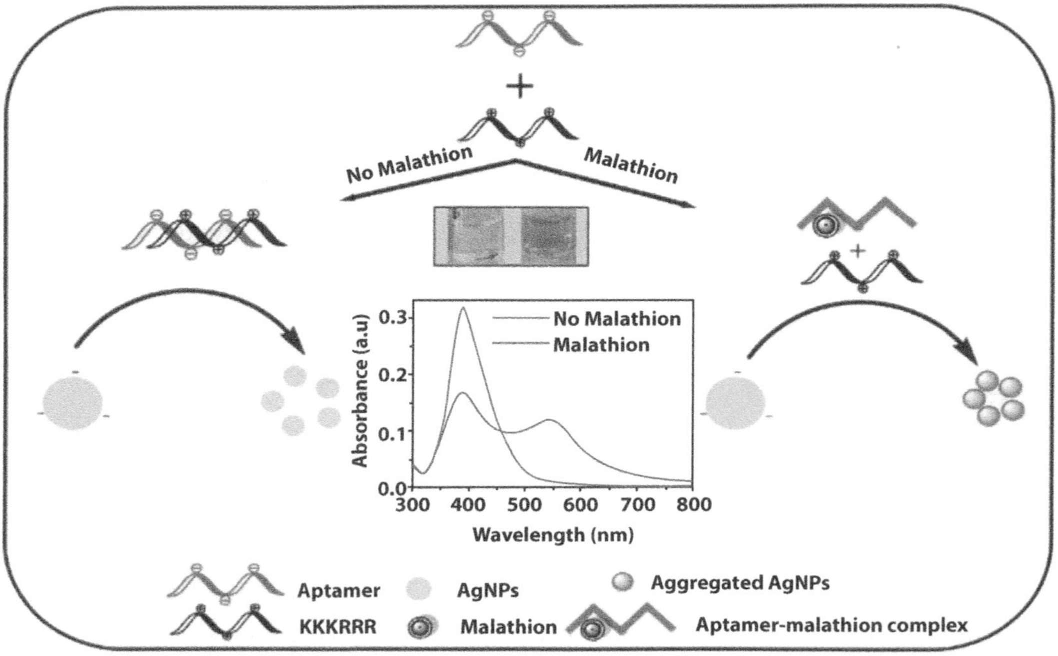


**Figure 19.9** Illustration of working principle of nanoparticle-based colorimetric aptasensors for malathion detection. Herein, the aptamer as capture probe and the color of AgNP solution as an optical transducer were used. This figure is taken from Ref. [27] (copyright 20218 from Elsevier).

signals; a fluorometric method can be used in detecting NP-induced quenching fluorescence signals [24]. Recently, AgNP-based aptasensors widely developed for PTC detection because they facilitate naked-eye detection and robust sensing, increase sensitivity, haves simple fabrication processes and are cost effective, easy to use, and portable.

Bala [27] reported a AgNP-based colorimetric aptasensor for MLT monitoring by utilizing a KKKRRR hexapeptide, MLT-aptamer as bioreceptor and AgNPs as signal probes (Figure 19.9). In this work, the interactions among AgNPs, MLT-aptamer capture probe and KKKRRR peptide result in changes in optical signal responses, which depend on the presence or absence of MLT molecules. This developed AgNP-based colorimetric aptasensor can detect malathion in a concentration range of 0.01–0.75 nM ($R^2$ = 0.9943), with an LoD of 0.5 pM.

### 19.2.3 Antibodies

A biosensor utilizes antibodies (Abs) as biocapture probes and is called an immunosensor. A specific target molecule is called an antigen, and the reaction of an Ab biocapture probe with specific antigen is considered an immune reaction. A typical structure of an IgG antibody is shown in Figure 19.10. Two identical heavy chains (H) and light chains (L) are linked by S–S bridges. Variable regions ($V_H$ and $V_L$) are responsible for antigen binding

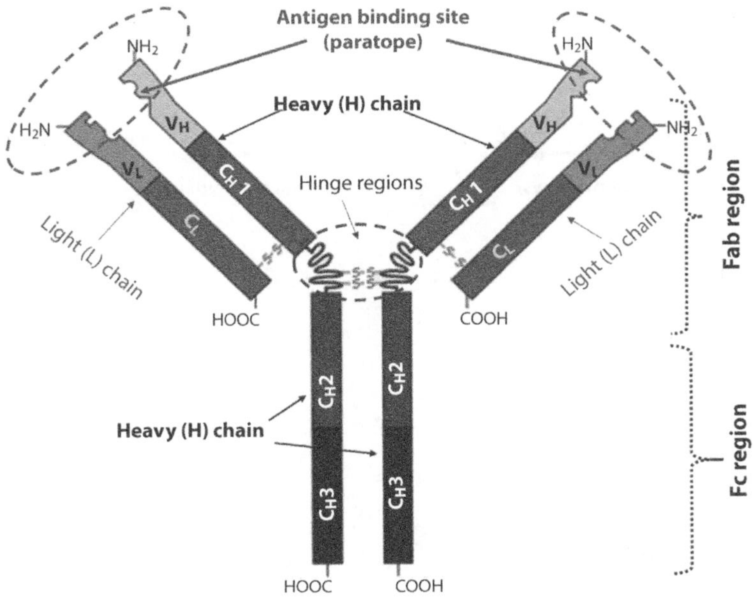


**Figure 19.10** IgG antibody structure. This figure is redrawn.

activity [28]. The working principle of immunosensors are realized through the conversion of Ag–Ab immune reactions into a measurable physical signal.

In fact, Ab–Ag interactions are non-covalent bonds and can be specified by an association constant ($K_a$) ranging from $10^5$ to $10^{13}$ M. pAb directly bind to different binding sites (also called epitopes) on Ag molecules with different $K_a$ values, whereas mAb prefers to only specific epitope on an Ag molecule, implying that mAb-based immunosensor has higher specificity than a pAb-based immunosensor [7]. Specific binding pairs for PTC immunoassays are either pesticide antigens or antigen analogues (also called haptens), which are small molecules each containing a single antigenic site or a specific chemical group that can specifically bind with antibody for its immobilization [1, 17].

Two methods are used in fabricating electrochemical pesticide immunosensors (PTC–immunosens): labeled and label-free detection. Zhang [29] used L-cysteine self-assembled on the AuE for developing the label-free EIS of 2,4-DB immunosensor (Figure 19.11). Anti-2,4-DB antibodies are linked with the amine groups of L-cysteine with GA as a cross-linker. By increasing $R_{et}$ during an immune reaction, the 2,4-DB concentration can be monitored from 0.1 ppb ($1.0 \times 10^{-7}$ g $L^{-1}$) to 1 ppm ($1.0 \times 10^{-3}$ g $L^{-1}$).

Regiart [30] reported a microfluidic electrochemical PTC–immunosen coupled to a AuNP-modified SPCE for clenbuterol (CLB) detection. Herein, an alkaline phosphatase enzyme (AP)- labelled CLB was firstly prepared by the conjugation of AP with CLB molecules (AP-labelled CLB). CLB in a sample competes immunologically with AP-labelled CLB for anti-CLB antibodies (Figure 19.12). In the presence of enzyme AP, *p*-aminophenyl phosphate (p-APP) is converted into a *p*-aminophenol product, which is an electroactive molecule. Therefore, it can be quantified at a potential of +0.1 V.

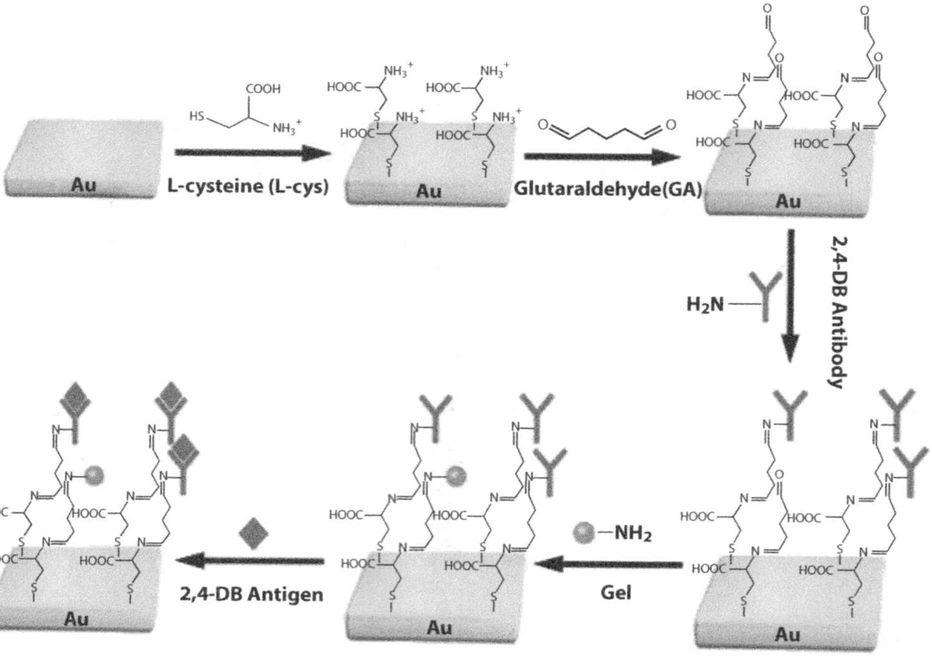


**Figure 19.11** Electrochemical immunosensor for 2,4-DB detection. This figure is taken from Ref. [29] (copyright 2018 from Elsevier).

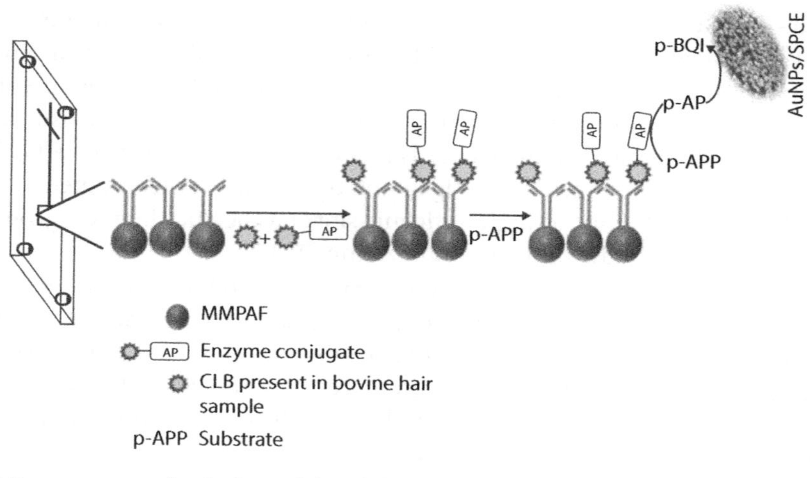


**Figure 19.12** Immunosensor for clenbuterol (CLB) detection. This figure is taken from Ref. [30] (copyright 2013 from Elsevier).

## 19.3 Electrochemical Immunosensors for Pesticide Detection

Electrochemical PTC–immunosens are developed using the most common electrochemical detection method using EIS [3, 31] or square wave voltammetry (SWV) [32] or cyclic voltammetry (CV) techniques. In fact, PTC–immunosens can be categorized according to the mode of electron transfer in transducers: indirect and direct detection.

### 19.3.1 Indirect Detection Mode

Normally, indirect detection mode uses two kinds of antibodies and is called the "sandwich format." Herein, a primary antibody was used to capture target molecules, and secondary antibodies bind to targets, forming a sandwich. The detection of Ab–Ag immune complex is achieved through the labelling of antibodies or antigens with an immunoassay format involving sandwich, competition or capture (Figure 19.13). Indirect electrochemical immunosensors employ electrochemical labels, which are called electrochemical tags or signal tags, which can be prepared with one of the three following approaches: (1) using enzyme-labelled secondary antibody; in this case, substrate and co-substrate must be added for signal measurement, (2) using secondary antibody-conjugated quantum dots or metallic NPs; in this case, after immunoassay, bound quantum dot or metallic NPs are dissolved by acid, and metal ion concentration is measured afterwards, and (3) using electrochemical tags based on the catalytic behavior of nanoparticles, which are conjugated with a secondary antibody, substrate, and co-substrate added for electrochemical measurement. Strategies

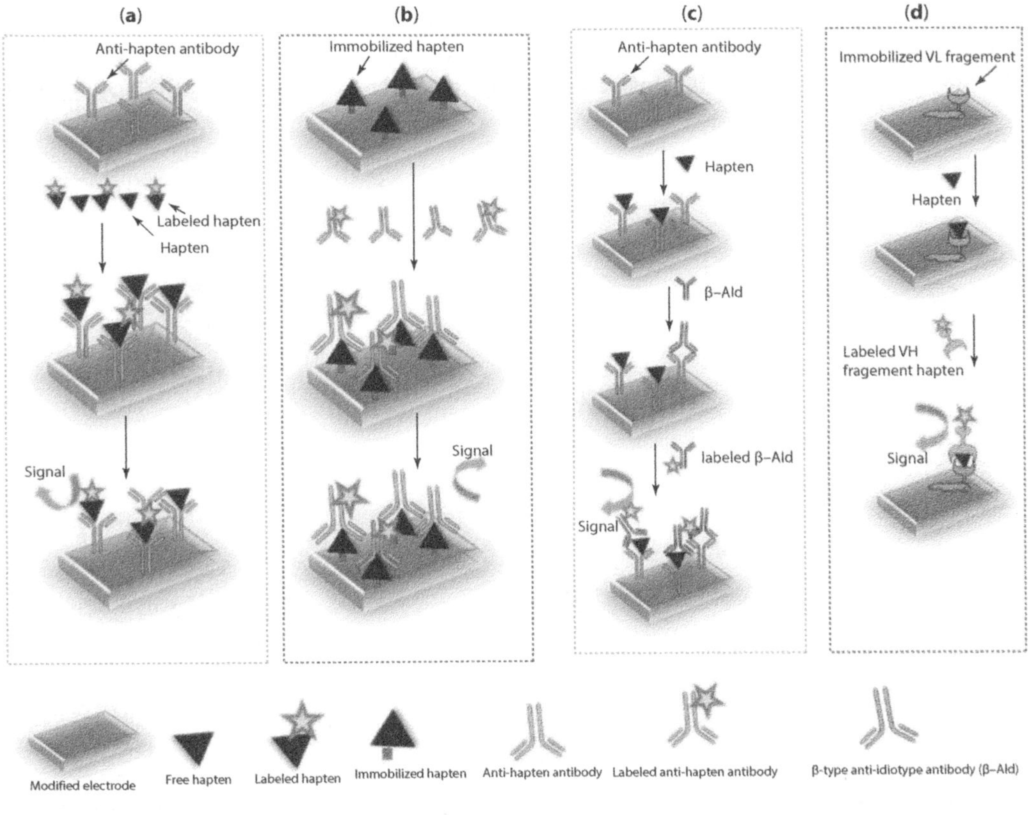


**Figure 19.13** Immunosensors based on indirect detection mode: (a, b) traditional pesticide immunoassay; (c, d) novel immunoassay with pesticide detection potential. This figure is redrawn.

for preparing sandwich electrochemical immunosensors were summarized by Pei [33]. The indirect detection mode is extremely sensitive and selective and can be widely used not only for small molecules but also for proteins and cells. However, the use of labels often indicates increased analysis time and extra steps, thus increasing the complexity of the procedure.

Using indirect detection approach, Wang [34] reported an electrochemical PTC–immunosen for fenvalerate (FVL) in tea product detection was reported by. In this development, the GCE was firstly modified with CS polymer for the anti FVL antibodies immobilization via a crosslinking (Figure 19.14). $Fe(CN)_6^{4-/3-}$ has been used as an electrochemical redox marker for recognition of the Ab–Ag immune reactions, which leads to an increasing on the $R_{et}$ for FVL concentration monitoring in a detection range from 1 ppb to 10 ppm and a LoD of 0.8 ppb.

Sharma [35] was reported an electrochemical immunosensor for diuron determination at concentrations ranging from 1 ppb to 1 ppm with laser-ablated AuE. The AuE was firstly modified with a Prussian blue–AuNP composite and then modified by 3,4-dichloropheny-lurea as a hapten-conjugated bovine serum albumin (hapten–BSA) through drop casting. In the detection step, the diuron target competes immunologically with hapten–BSA for anti-diuron antibodies. AP-labelled rabbit anti-IgG was used as electrochemical reporters, which directly bind anti-diuron antibodies (Figure 19.15). For electrochemical signal generation, 1-naphthyl phosphate was enzymatically transformed into 1-naphthol, which can be monitored using SWV.

An electrochemical for coumaphos detection was constructed with immunoassay using a guanine-rich single-stranded DNA (G-ssDNA) to amplify detection signal and AuNPs [36]. The working principle is illustrated in Figure 19.16. A monoclonal Ab specific to

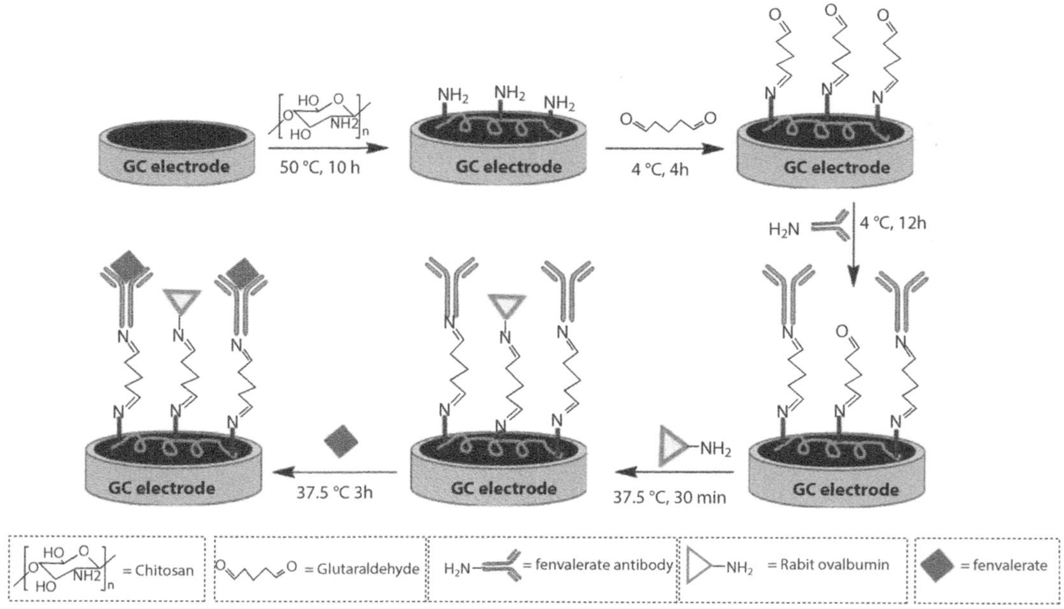


**Figure 19.14** Electrochemical immunosensor based on label-free but indirect detection of fenvalerate. This figure is taken from Ref. [34] (copyright 2013 from Elsevier).

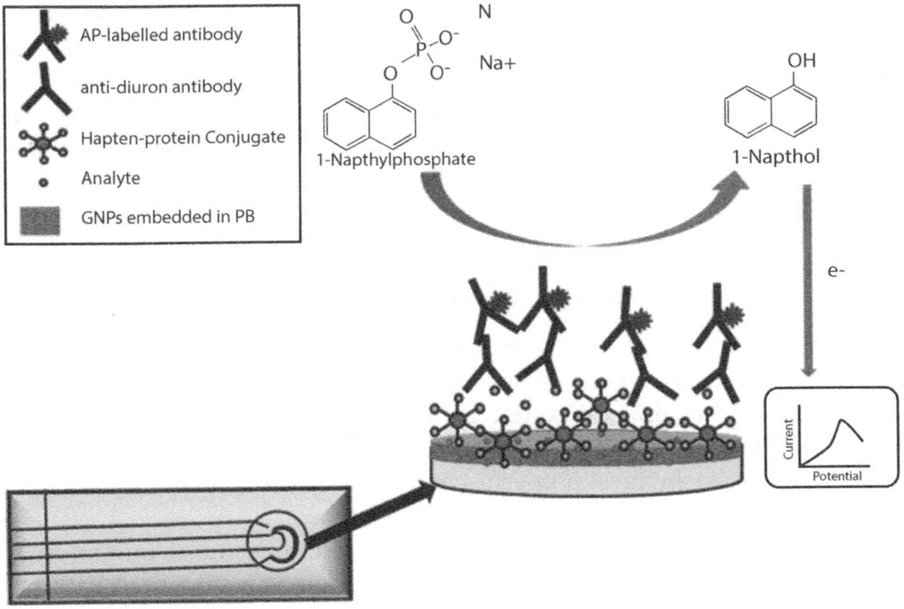


**Figure 19.15** Electrochemical biosensor for diuron detection based on Prussian blue–gold nanoparticle (PB–GNP)-modified laser ablated gold electrodes. This figure is taken from Ref. [35] (copyright 2011 from Elsevier).

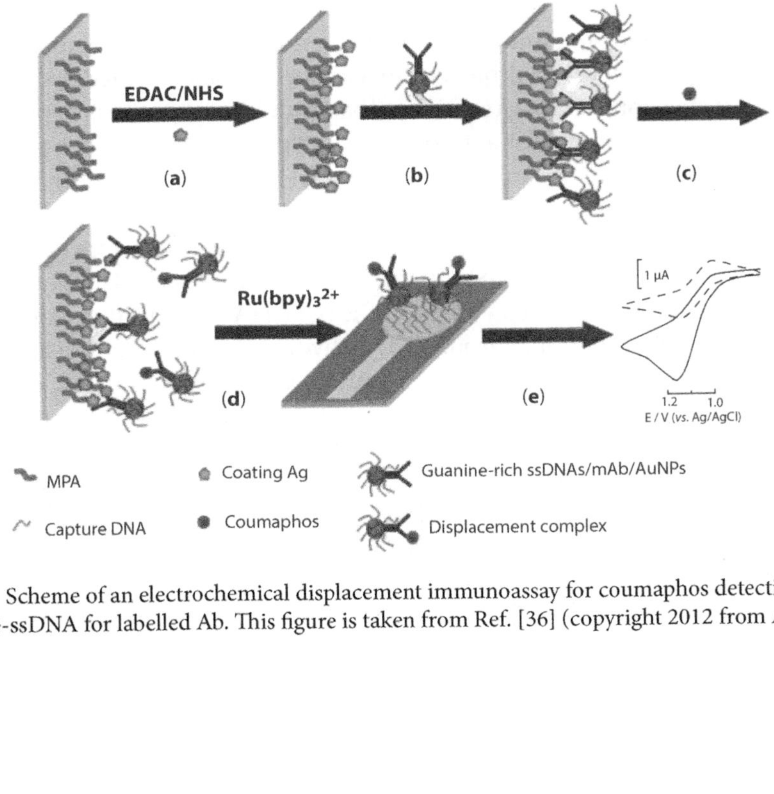


**Figure 19.16** Scheme of an electrochemical displacement immunoassay for coumaphos detection using AuNPs and G-ssDNA for labelled Ab. This figure is taken from Ref. [36] (copyright 2012 from ACS).

organophosphorus compounds was used as a biocapture probe, and a coumaphos–hapten layer was coated on a testing slide to immobilize G-ssDNA-labelled Ab. The detected coumaphos competed immunologically with immobilized Ag coating for the G-ssDNA-labelled Ab. A DNA complementary to the G-ssDNA sensor on an ITO electrode was used in attaching the displaced G-ssDNA-labelled Ab. Finally, $Ru(bpy)_3^{2+}$ was added to bind to DNA and generate an electrochemical signal. This immunoassay can detect coumaphos at a range of 0.5–80 ng $L^{-1}$.

### 19.3.2 Label-Fee and Reagentless Direct Detection Mode

Indirect (and labelled) electrochemical immunosensors are generally sensitive but involve complicated detection architecture, which make them inefficient. Label-free and direct sensors are more efficient. The direct detection approach is called "label-free," in which a specific immunoreaction event between an Ab and a Ag target analyte is monitored according to changes in the physicochemical properties of transducers. Normally, direct detection is called 'reagentless' to distinguish it from labelled detection.

The first used approach for fabricating label-free electrochemical immunosensors is constructed with the following steps. Firstly, electrodes are functionalized with polymer, diazonium, and self-assembled monolayer (SAM; step 1), onto which specific antibodies are immobilized (step 2), which readily capture pesticide molecules (step 3). Pesticide concentration is measured by monitoring electrochemical signal after step (3). By using this approach, Pichetsurnthorn [31] reported a label-free and reagentless EIS immunosensor for ATZ detection using nanoporous alumina membranes (Figure 19.17), which were integrated into a printed circuit board for Ab capture probe immobilization. The sensor was used to trace ATZ in a water sample at a detection range from $10^{-5}$ ppb to 1 ppb.

The second label-free electrochemical format for immunosensor fabrication consists of electrodes functionalized by polymer, diazonium, and SAM (step 1), which have specific groups for further specific Abs immobilized for electrochemical immunosensor fabrication (step 2). Finally, the developed biosensors capture pesticide molecules (step 3). This strategy takes advantage of Ab cross-reactivity. The immobilized probe is hapten (analog of antigen), which has lower affinity for antibodies than Ag. When a target antigen is added, it complexes with the probe and free up the electrode surface. Pesticide concentration is measured by monitoring electrochemical or optical signals. Through this approach, reagentless and label-free electrochemical ATZ immunosensors have been developed [3, 10], which can be used to detect ATZ with the SWV technique. 5-Hydroxy-1,4-naphthoquinone is coupled to hydroxyatrazine for the synthesis of a JUG-HATZ multifunctional monomer comprising three specific functional groups: (1i) a hydroxyl atrazine group as a hapten for anti-atrazine antibody immobilization, (2i) a quinone group with redox electroactivity in an aqueous medium, which serves as an electrochemical transducer, and (3i) a -OH (hydroxyl group) for polymerization to create a poly(JUG-HATZ) polymer film on the electrodes. SWV recording on a poly(JUG-HATZ)/GCE presents a decreasing trend of current when anti-atrazine antibody binds with a hapten and an increasing trend of current in the presence of ATZ (Figure 19.18). This label-free electrochemical immunosensor exhibits excellent sensitivity, with an LoD of 0.2 ng $L^{-1}$ [10].

Based on the poly(JUG-HATZ)/GCE, this a label-free electrochemical format has been extended by using EIS, as descried in Figure 19.19 [3].

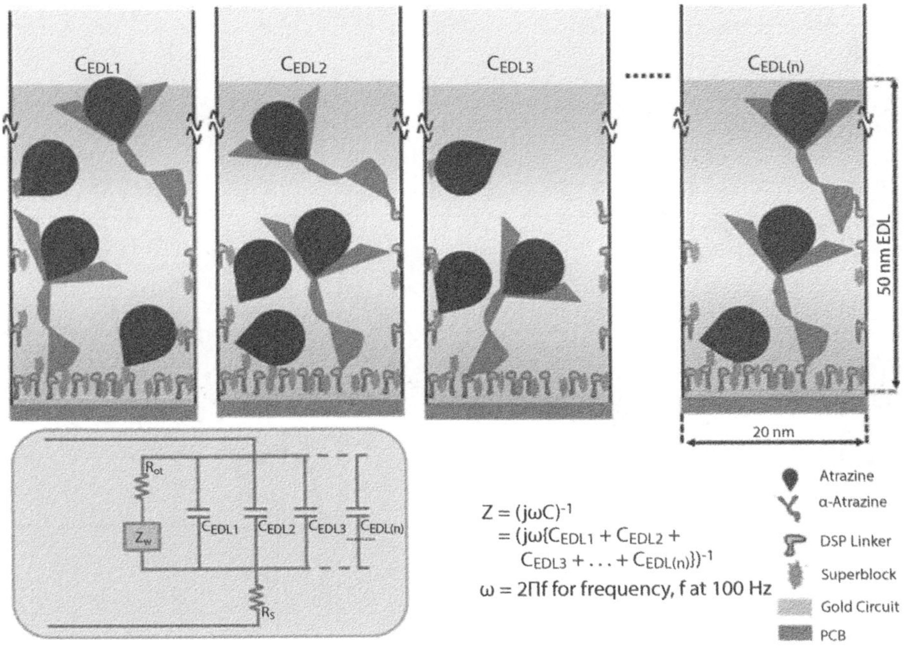


**Figure 19.17** EIS-based immunosenor for ATZ detection using nanoporous alumina. ATZ binding to a specific Ab results in changes in the capacitance of the electrical double layer. This figure is taken from Ref. [31] (copyright 2012 from Elsevier).

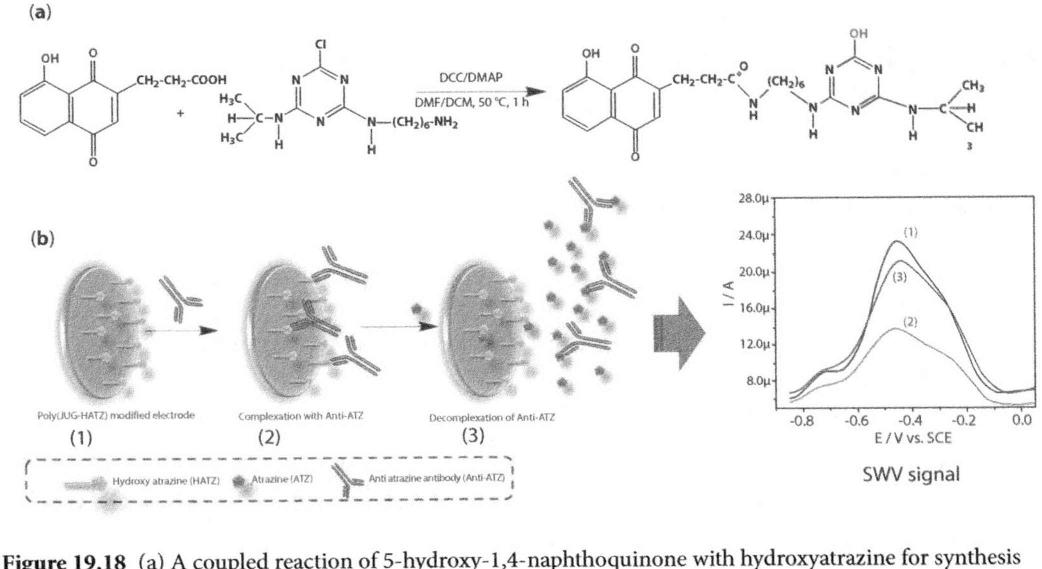


**Figure 19.18** (a) A coupled reaction of 5-hydroxy-1,4-naphthoquinone with hydroxyatrazine for synthesis of JUG-HATZ and (b) working principle of a label-free and reagentless electrochemical ATZ immunosensor based on poly(JUG-HATZ) modified GCE for competitive displacement with an application to ATZ. This figure is redrawn.

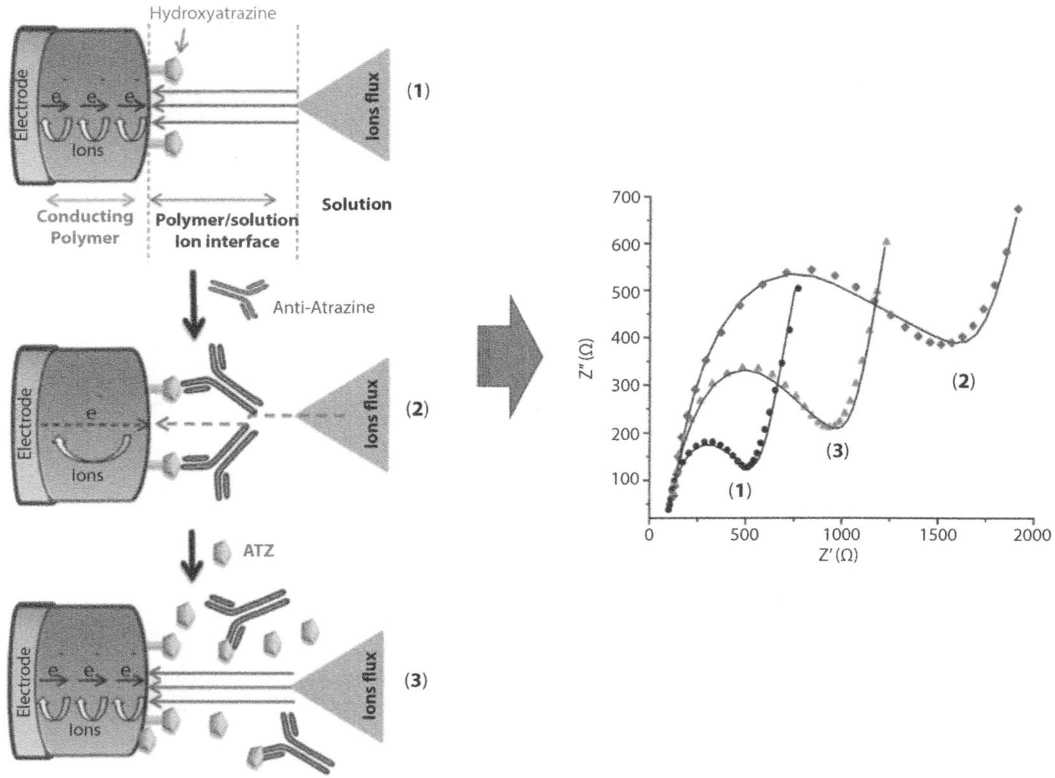


**Figure 19.19** Working principle of a label-free EIS immunosensor for ATZ detection using a competitive displacement strategy. This figure is redrawn.

## 19.4 Applications of Nanomaterials for the Development of Pesticide Immunosensors

Various nanomaterials have been used in the fabrication of PTC–biosens to enhance sensitivity and selectivity. Nanomaterials can be used with two main approaches:

- **(i)** Nano-structuration that enables an electrode to amplify electrochemical signals and improve sensitivity [37–41];
- **(ii)** As an electrochemical signal tag or electrochemical signal reporter for generation response signal [42–44];
- **(iii)** Nanomaterials conjugated with reporters as an electrochemical catalytic tag [42];
- **(iv)** Fabrication of a nanostructured platform that enhances response signal; carbon nanotubes [43, 45], graphene [42, 44, 46], graphene oxide (GO) [47, 48] or metal oxide nanoparticles (such as $Fe_3O_4$ [32]) can be used.

AgNPs with high conductivity and good electrochemical redox property have been utilized in the development of an electrochemical tiamulin (TML) immunosen

(Figure 19.20A) [49]. AgNPs/GO is firstly immobilized onto AuE with a nafion binder (Ag/GO/Nf/AuE). Staphylococcal protein A (SPA) is added to an electrode to form PSA/Ag/GO/Nf/AuE. The anti-TML antibody as a biocapture probe is immobilized by self-assembling with SPA (Figure 19.20B). This probe can be used to sense TML in a detection range of 0.05 to 100 ppb.

Nguyen [50] reported a label-free ATZ electrochemical PTC–immunosens by using polyaniline (PANi)-coated graphene films for improving the sensitivity. Herein, a graphene film (Gr) was fabricated by a thermal CVD and then was transferred to the PANi-predeposited Pt microelectrode (PtME). The anti-ATZ antibody has been immobilized on the PANi/Gr/PtME via using GA cross-linker (Figure 19.21) allowing for ATZ recognition in a concentration range from $2 \times 10^{-3}$ ppb to 20 ppb and a LoD of $4.3 \times 10^{-5}$ ppb [50].

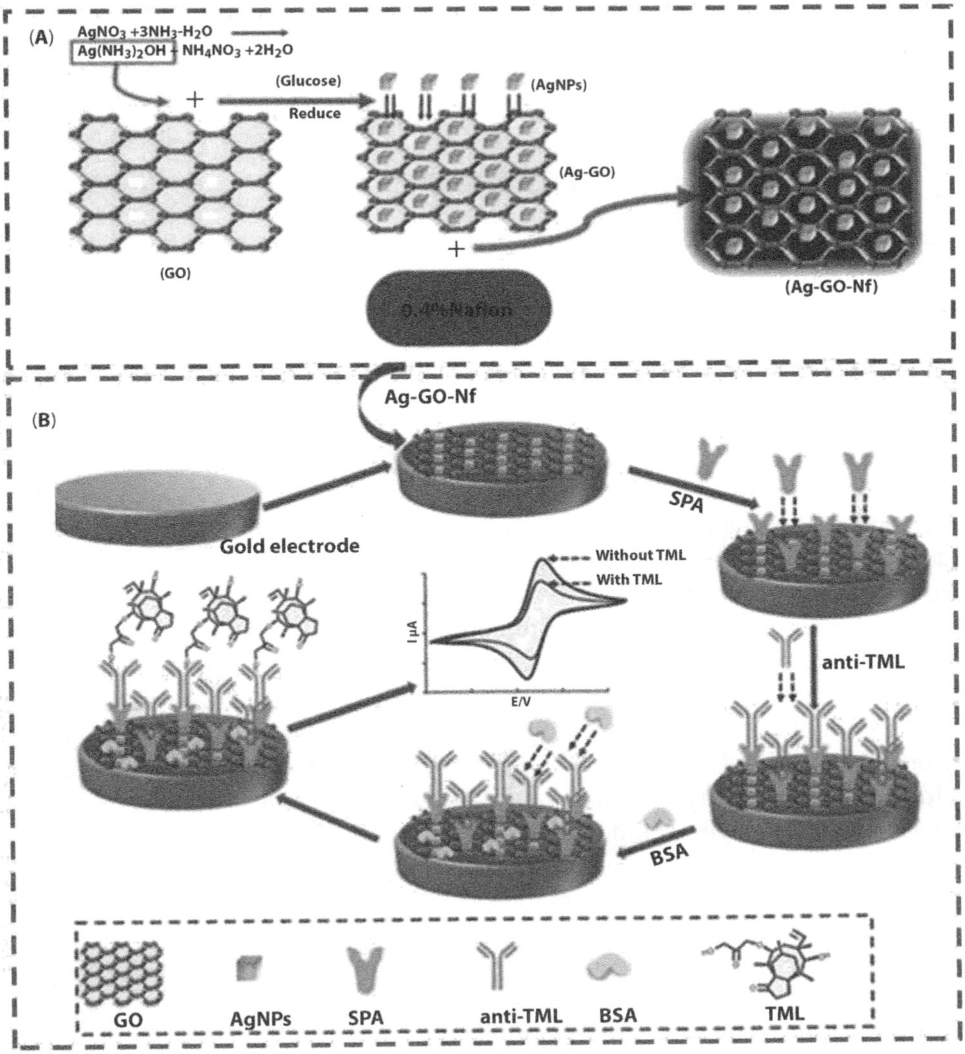


**Figure 19.20** (A) A chemical synthesis of AgNPs/GO and AgNPs/GO/Nf for modification of gold electrode (AuE); (B) process for the fabrication of an electrochemical TML immunosensor based on SPA, AgNPs/GO, and anti-TML antibody. This figure is taken from Ref. [49] (copyright 2020 from Elsevier).

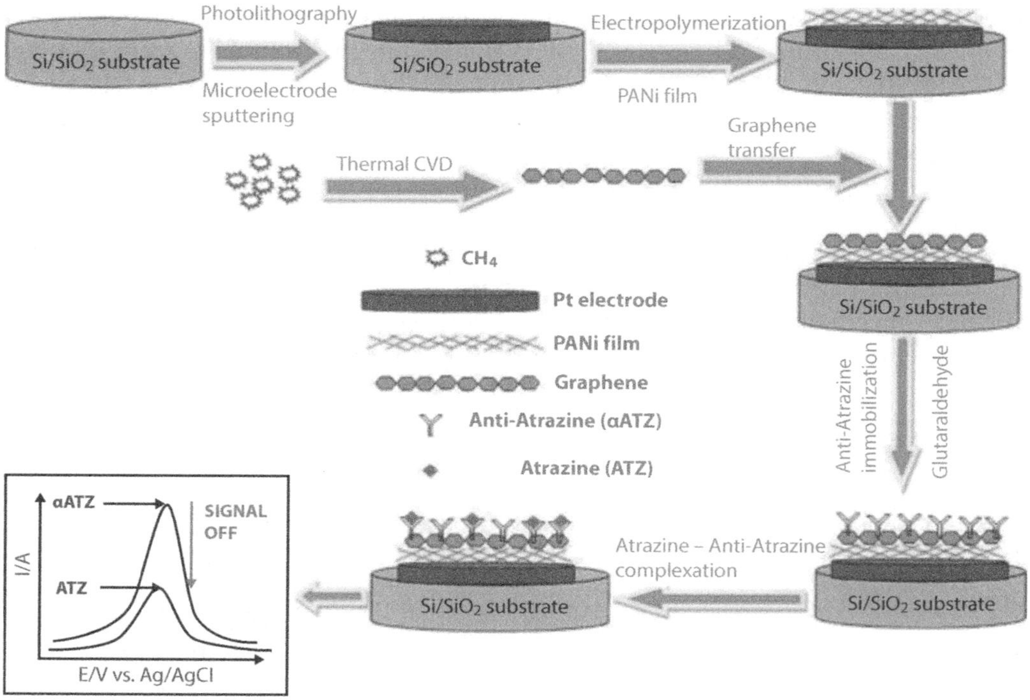


**Figure 19.21** Process for the fabrication of an ATZ electrochemical immunosensor using PANi/Gr modified PtE platform. This figure is taken from Ref. [50] (copyright 2016 from Elsevier).

## 19.5 Conclusion

Electrochemical and colorimetric biosensors can be considered excellent candidates for PTC detection for food, water and environmental applications. Most developed pesticide biosensors used sensitive, selective, and reliable electrochemical or optical transducers, and commonly used biological components are enzymes, antibodies, and aptamers. Immunosensors are constructed according to highly specific Ab–Ag imunoreaction, which enable PTC–immunosens to identify pesticides with extremely high specificity. PTC–immunosens are often use for the specific pesticide identification. Meanwhile, enzymatic biosensors detect a broad range of PTCs, and enzymatic PTC–biosens can be widely used for rapid and general toxicity screening. In the last decade, aptamers constitute a novel class of artificial and short single-stranded DNAs or RNAs, which are used as bio-receptors for PTC molecular recognition, and named aptasensors. Aptasensors are considered excellent tools for developing efficient biosensors for PTC detection because they have high specificity and are designable and synthesizable for various PTC molecule pollutants. Currently, PTC–biosens have selective and robust molecular affinity, and they are widely applied to bioanalysis. However, in the current state, the PTC–biosens are expensive and require a complicated measuring system. To solve problems in the fabrication of PTC–biosens, nanotechnology, nanomaterials and nanoplatforms should be considered. Microtechnologies and nanotechnologies have undergone rapid development, and they are directly applied to the fabrication of efficient PTC–biosens. They have many advantages over conventional biological receptors in terms of cost, portability, wearability, sensitivity, and selectivity. Different types of nanomaterials and specific properties for the fabrication of PTC–biosens have opened new and exciting opportunities and offered encouraging results for improving PTC–biosens performance.

## Acknowledgment

This work was supported by the Vietnam Ministry of Education and Training (under project number CT 2022.04.BKA.03.

## References

1. Badihi-Mossberg, M., Buchner, V., Rishpon, J., Electrochemical biosensors for pollutants in the environment. *Electroanalysis*, 19, 19-20, 2015–2028, 2007.
2. Da Costa Silva, L.M., Melo, A.F., Salgado, A.M., Biosensor for environmental applications, in: *Environmental Biosensors*, V. Somerset (Ed.), pp. 3–16, InTech, Rijeka (Croatia), 2011.
3. Tran, H.V., Reisberg, S., Piro, B., Nguyen, T.D., Pham, M.C., Label-free electrochemical immunoaffinity sensor based on impedimetric method for pesticide detection. *Electroanalysis*, 25, 3, 664–670, 2013.
4. Zhai, C., Sun, X., Zhao, W., Gong, Z., Wang, X., Acetylcholinesterase biosensor based on chitosan/prussian blue/multiwall carbon nanotubes/hollow gold nanospheres nanocomposite film by one-stepelectrodeposition. *Biosens. Bioelectron.*, 42, 124–130, 2013.
5. Gerard, M., Chaubey, A., Malhotra, B.D., Application of conducting polymers to biosensors. *Biosens. Bioelectron.*, 17, 345–359, 2002.
6. Reisberg, S., Piro, B., Noël, V., Pham, M.C., DNA electrochemical sensor based on conducting polymer: Dependence of the "signal-on" detection on the probe sequence localization. *Anal. Chem.*, 77, 10, 3351–3356, 2005.
7. Piro, B., Reisberg, S., Anquetin, G., Duc, H.T., Pham, M.C., Quinone-based polymers for label-free and reagentless electrochemical immunosensors: Application to proteins, antibodies and pesticides detection. *Biosensors*, 3, 1, 58–76, 2013.
8. Piro, B., Zhang, Q.D., Reisberg, S., Noel, V., Dang, L.A., Duc, H.T., Pham, M.C., Direct and rapid electrochemical immunosensing system based on a conducting polymer. *Talanta*, 82, 2, 608–612, 2010.
9. Tran, H.V., Piro, B., Reisberg, S., Tran, L.D., Duc, H.T., Pham, M.C., Label-free and reagentless electrochemical detection of microRNAs using a conducting polymer nanostructured by carbon nanotubes: Application to prostate cancer biomarker miR-141. *Biosens. Bioelectron.*, 49, 0, 164–169, 2013.
10. Tran, H.V., Yougnia, R., Reisberg, S., Piro, B., Serradji, N., Nguyen, T.D., Tran, L.D., Dong, C.Z., Pham, M.C., A label-free electrochemical immunosensor for direct, signal-on and sensitive pesticide detection. *Biosens. Bioelectron.*, 31, 1, 62–68, 2012.
11. Luo, X., Morrin, A., Killard, A.J., Smyth, M.R., Application of nanoparticles in electrochemical sensors and biosensors. *Electroanalysis*, 18, 4, 319–326, 2006.
12. Ansari, A.A., Kaushik, A., Solanki, P.R., Malhotra, B.D., Sol–gel derived nanoporous cerium oxide film for application to cholesterol biosensor. *Electrochem. Commun.*, 10, 9, 1246–1249, 2008.
13. Arya, S.K., Solanki, P.R., Datta, M., Malhotra, B.D., Recent advances in self-assembled monolayers based biomolecular electronic devices. *Biosens. Bioelectron.*, 24, 2810–2817, 2009.

14. Moina, C. and Ybarra, G., Fundamentals and applications of immunosensors, in: *Advances in Immunoassay Technology*, N.H.L.C.a.T.K. Christopoulos (Ed.), pp. 65–80, InTech, Rijeka (Croatia), 2012.
15. De Corcuera, J.I.R., and Cavalieri, R.P., *Encyclopedia of agricultural, food, and biological engineering*, D.R. Heldman (Ed.), pp. 119–123, Marcel Dekker, Inc., New York (U.S.A.), 2003.
16. Mcnaught, A.D. and Wilkinson, A., (The "gold book")-IUPAC- compendium of chemical terminology, in: *Gold Book*, 2nd ed, Blackwell Scientific Publications, Oxford, 1997.
17. Sassolas, A., Simón, B.P., Marty, J.L., Biosensors for pesticide detection: New trends. *A. J. A. C.*, 03, 210–232, 2012.
18. Chauhan, N., Narang, J., Pundir, C.S., Immobilization of rat brain acetylcholinesterase on ZnS and poly(indole-5-carboxylic acid) modified Au electrode for detection of organophosphorus insecticides. *Biosens. Bioelectron.*, 29, 15, 82–88, 2011.
19. Dong, J., Fan, X., Qiao, F., Ai, S., Xin, H., A novel protocol for ultra-trace detection of pesticides: Combined electrochemical reduction of Ellman's reagent with acetylcholinesterase inhibition. *Anal. Chim. Acta*, 761, 25, 78–83, 2013.
20. Blankert, B., Hayen, H., Van Leeuwen, S.M., Karst, U., Bodoki, E., Lotrean, S., Sandulescu, R., Diez, N.M., Dominguez, O., Arcos, J., Kauffmann, J.-M., Electrochemical, chemical and enzymatic oxidations of phenothiazines. *Electroanalysis*, 17, 17, 1501–1510, 2005.
21. Cabanillas, A.G., Diaz, T.G., Salinas, F., Ortiz, J.M., Kauffmann, J.M., Differential pulse voltammetric determination of fenobucarb at the glassy carbon electrode, after its alkaline hydrolysis to a phenolic product. *Electroanalysis*, 9, 12, 952–955, 1997.
22. Parsajoo, C. and Kauffmann, J.M., Development of an acetylcholinesterase immobilized flow through amperometric detector based on thiocholine detection at a silver electrode. *Talanta*, 109, 116–120, 2013.
23. Sahin, A., Dooley, K., Cropek, D.M., West, A.C., Banta, S., A dual enzyme electrochemical assay for the detection of organophosphorus compounds using organophosphorus hydrolase and horseradish peroxidase. *Sens. Actuators, B*, 158, 1, 353–360, 2011.
24. Phopin, K. and Tantimongcolwat, T., Pesticide aptasensors—State of the art and perspectives. *Sensors*, 20, 3, 6809, 2020.
25. Fan, L., Zhao, G., Shi, H., Liu, M., Li, Z., A highly selective electrochemical impedance spectroscopy-based aptasensor for sensitive detection of acetamiprid. *Biosens. Bioelectron.*, 43, 12–18, 2013.
26. Madianos, L., Tsekenis, G., Skotadis, E., Patsiouras, L., Tsoukalas, D., A highly sensitive impedimetric aptasensor for the selective detection of acetamiprid and atrazine based on microwires formed by platinum nanoparticles. *Biosens. Bioelectron.*, 101, 268–274, 2018.
27. Bala, R., Mittal, S., Sharma, R.K., Wangoo, N., A supersensitive silver nanoprobe based aptasensor for low cost detection of malathion residues in water and food samples. *Spectrochim. Acta, Part A*, 196, 268–273, 2018.
28. Omidfar, K., Khorsand, F., Azizi, M.D., New analytical applications of gold nanoparticles as label in antibody based sensors. *Biosens. Bioelectron.*, 43, 336–347, 2013.
29. Zhang, L., Wang, M., Wang, C., Hu, X., Wang, G., Label-free impedimetric immunosensor for sensitive detection of 2,4-dichlorophenoxybutyric acid (2,4-DB) in soybean. *Talanta*, 101, 226–232, 2012.
30. Regiart, M., Fernández-Baldo, M.A., Spotorno, V.G., Bertolino, F.A., Raba, J., Ultra sensitive microfluidic immunosensor for determination of clenbuterol in bovine hair samples using electrodeposited gold nanoparticles and magnetic micro particles as bio-affinity platform. *Biosens. Bioelectron.*, 41, 211–217, 2013.
31. Pichetsurnthorn, P., Vattipalli, K., Prasad, S., Nanoporous impedemetric biosensor for detection of trace atrazine from water samples. *Biosens. Bioelectron.*, 32, 1, 155–162, 2012.

32. Nguyen, B.H., Tran, L.D., Do, Q.P., Nguyen, H.L., Tran, N.H., Nguyen, P.X., Label-free detection of aflatoxin M1 with electrochemical Fe3O4/polyaniline-based aptasensor. *Mater. Sci. Eng. C*, 33, 4, 2229–2234, 2013.
33. Pei, X., Zhang, B., Tang, J., Liu, B., Lai, W., Tang, D., Sandwich-type immunosensors and immunoassays exploiting nanostructure labels: A review. *Anal. Chim. Acta*, 758, 3, 1–18, 2013.
34. Wang, M., Kang, H., Xu, D., Wang, C., Liu, S., Hu, X., Label-free impedimetric immunosensor for sensitive detection of fenvalerate in tea. *Food Chem.*, 141, 1, 84–90, 2013.
35. Sharma, P., Sablok, K., Bhalla, V., Suri, C.R., A novel disposable electrochemical immunosensor for phenyl urea herbicide diuron. *Biosens. Bioelectron.*, 26, 10, 4209–4212, 2011.
36. Dai, Z., Liu, H., Shen, Y., Su, X., Xu, Z., Sun, Y., Zou, X., Attomolar determination of coumaphos by electrochemical displacement immunoassay coupled with oligonucleotide sensing. *Anal. Chem.*, 84, 19, 8157–8163, 2012.
37. Shi, L., Chu, Z., Liu, Y., Jin, W., Chen, X., Facile synthesis of hierarchically aloe-like gold micro/nanostructures for ultrasensitive DNA recognition. *Biosens. Bioelectron.*, 49, 184–91, 2013.
38. Wang, X., Yang, T., Li, X., Jiao, K., Three-step electrodeposition synthesis of self-doped polyaniline nanofiber-supported flower-like Au microspheres for high-performance biosensing of DNA hybridization recognition. *Biosens. Bioelectron.*, 26, 6, 2953–2959, 2011.
39. Du, D., Wang, J., Wang, L., Lu, D., Smith, J.N., Timchalk, C., Lin, Y., Magnetic electrochemical sensing platform for biomonitoring of exposure to organophosphorus pesticides and nerve agents based on simultaneous measurement of total enzyme amount and enzyme activity. *Anal. Chem.*, 83, 10, 3770–3777, 2011.
40. Wang, L., Chen, X., Wang, X., Han, X., Liu, S., Zhao, C., Electrochemical synthesis of gold nanostructure modified electrode and its development in electrochemical DNA biosensor. *Biosens. Bioelectron.*, 30, 1, 151–157, 2011.
41. Malhotra, R., Patel, V., Chikkaveeraiah, B.V., Munge, B.S., Cheong, S.C., Zain, R.B., Abraham, M.T., Dey, D.K., Gutkind, J.S., Rusling, J.F., Ultrasensitive detection of cancer biomarkers in the clinic by use of a nanostructured microfluidic array. *Anal. Chem.*, 84, 14, 6249–55, 2012.
42. Liu, D., Chen, W., Wei, J., Li, X., Wang, Z., Jiang, X., A highly sensitive, dual-readout assay based on gold nanoparticles for organophosphorus and carbamate pesticides. *Anal. Chem.*, 84, 9, 4185–4191, 2012.
43. Liu, H., Xu, S., He, Z., Deng, A., Zhu, J.J., Supersandwich cytosensor for selective and ultrasensitive detection of cancer cells using aptamer-DNA concatamer-quantum dots probes. *Anal. Chem.*, 85, 6, 3385–3392, 2013.
44. Lian, W., Liu, S., Yu, J., Xing, X., Li, J., Cui, M., Huang, J., Electrochemical sensor based on gold nanoparticles fabricated molecularly imprinted polymer film at chitosan-platinum nanoparticles/graphene-gold nanoparticles double nanocomposites modified electrode for detection of erythromycin. *Biosens. Bioelectron.*, 38, 1, 163–9, 2012.
45. Truong, T.N., Tran, D.L., Vu, T.H., Tran, V.H., Duong, T.Q., Dinh, Q.K., Tsukahara, T., Lee, Y.H., Kim, J.S., Multi-wall carbon nanotubes (MWCNTs)-doped polypyrrole DNA biosensor for label-free detection of genetically modified organisms by QCM and EIS. *Talanta*, 80, 3, 1164–9, 2010.
46. Myung, S., Solanki, A., Kim, C., Park, J., Kim, K.S., Lee, K.B., Graphene-encapsulated nanoparticle-based biosensor for the selective detection of cancer biomarkers. *Adv. Mater.*, 23, 2221–2225, 2011.
47. Haque, A.M.J., Park, H., Sung, D., Jon, S., Choi, S.Y., Kim, K., An electrochemically reduced graphene oxide-based electrochemical immunosensing platform for ultrasensitive antigen detection. *Anal. Chem.*, 84, 1871–1878, 2012.
48. Yang, Y., Fang, G., Liu, G., Pan, M., Wang, X., Kong, L., He, X., Wang, S., Electrochemical sensor based on molecularly imprinted polymer film via sol-gel technology and multi-walled carbon

nanotubes-chitosan functional layer for sensitive determination of quinoxaline-2-carboxylic acid. *Biosens. Bioelectron.*, 47, 475–481, 2013.
49. You, X., Zhang, G., Chen, Y., Liu, D., Ma, D., Zhou, J., Liu, Y., Liu, H., Qi, Y., Liang, C., Ding, P., Zhu, X., Zhang, C., Wang, A., A novel electrochemical immunosensor for the sensitive detection of tiamulin based on staphylococcal protein A and silver nanoparticle-graphene oxide nanocomposites. *Bioelectrochemistry*, 141, 107877, 2021.
50. Chuc, N.V., Binh, N.H., Thanh, C.T., Tu, N.V., Huy, N.L., Dzung, N.T., Minh, P.N., Thu, V.T., Lam, T.D., Electrochemical immunosensor for detection of atrazine based on polyaniline/graphene. *J. Mater. Sci. Technol.*, 32, 6, 539–544, 2016.

# 20

# Advances in Biosensor Applications for Agroproducts Safety

**Adeshina Fadeyibi**

*Department of Food and Agricultural Engineering, Faculty of Engineering and Technology, Kwara State University, Ilorin, Kwara State, Nigeria*

***Abstract***
Postharvest change in the properties of Agroproducts during processing and storage is of major concern to the food industry. This can affect consumer acceptance due to the physical and biochemical transformations of the products. Biosensors (BoS) have been used to detect, monitor, and control these changes. In this paper, two types of the BoS were identified for the safety of plant and animal products. Electrochemical and surface plasma BoS were used for the detection and treatment of mycotoxins like ochratoxin in the plant products, such as cereals, fruits, and vegetables during processing. Also, *liposomes* and nanocomposites are employed to help elevate the quality and detect harmful contaminants in the animal products, such as dairy and poultry processing. Novel BoS like nucleic acid, antibodies, and receptors were proposed for monitoring the safety of *fructophilic* lactic acid bacteria against excessive temperature, which is essential for milk processing. This technique can help preserve the product during processing and storage.

***Keywords***: Biosensors, plant products, animal products, processing, storage

## 20.1 Introduction

Agricultural products are raw materials obtain from the processing of plants, animals, and microorganisms [1]. The materials can be in the form of agricultural crops, livestock, such as poultry, dairy, fishery, forestry, and horticultural products. The safety of these products is essential in agricultural value chain so as to ensure proper hygiene and prevent deterioration during handling, preparation, storage and processing [2]. This will ultimately regulate the food hygiene, food pesticides residues and other food safety policy issues between the industry and the market on its way to the consumer [3]. A BoS is an active device which incorporates a transient heat source and a biological element to generate signals or response about a substrate. The biosafety of the substrate, which can be the agricultural product, maybe monitored using the BoS device for effective management of the food products in the postharvest value chain.

*Email*: adeshina.fadeyibi@kwasu.edu.ng
Orcid: https://orcid.org/0000-0002-4538-9246

Inamuddin and Tariq Altalhi (eds.) Biosensors Nanotechnology, 2nd Edition, (469–480) © 2023 Scrivener Publishing LLC

The BoS has found other commercial applications in the food sector, especially in the preservation and processing [4]. The material can promote the shelf-life and retard degradation when added in controlled amounts during food processing [5]. By this action, it can maintain the quality and standard as well as ensure that the food is available in the market during off-season. Normally, the BoS will acts on the food substrate, when it is added as a preservative, to create a low oxygen atmosphere that can prevent the survival of the microorganisms [6–10]. Most of the fungi, bacterial, and other pathogenic organisms are normally destroyed in the process leading to the buildup of excess carbon-dioxide gas concentration and the liberation of a huge quantum of energy [11–13].

Although Zhang *et al.* [14] reviewed some biomaterials for the detection of mycotoxins, pesticides, and antibiotics of the agroproducts, the authors did not consider the application of the BoS for the safety of processed and stored plant and animal products. This review therefore is aimed at providing an overview for the safety of the plant and animal products including cereals, legumes, fruits and vegetables, fodders, dairy, poultry, and other livestock products in the food processing industry. A novel BoS like nucleic acid, antibodies, and receptors were proposed for monitoring the safety of *fructophilic* lactic acid bacteria, which is essential for milk processing, against excessive temperature.

## 20.2 Biosensors for Safety of Plant Products

### 20.2.1 BoS for Cereal Products Safety

The cereal products are known to be the staple food in most countries. They contain more of carbohydrates than proteins. The safety of the cereal products, such as rice, rice bram, maize or corn, wheat, millets, barley, rye, sorghum, and oats, is essential in preservation and processing [15]. During the storage of the cereal crops in silos or bins, the food product can grow mold and get infested by fungi due to poor aeration of the storage facility. This may promote the development of ochratoxins that secretes poisonous carcinogens within the storage system thereby contaminating the stored products. Thus, to control this defect, Yuan *et al.* [16] used a surface plasma-resonance BoS technology to detect and control the incidence of the ochratoxins infested cereals in storage. Also, according to Santovito *et al.* [17], a BoS that is synthesized by incorporating a DNA substrate into paramagnetic micro beads has been proposed for detecting and controlling the ochratoxins with high sensitivity and selectivity properties. There is however the need to further research in the application of the other sensitive materials like the electrochemical and plasma resonance based BoS in cereal products storage.

### 20.2.2 BoS for Legume or Pulse Products Safety

The legumes are grains known to contain more proteins than carbohydrates. The safety of the legume or pulse products, such as peas, beans, vetches, lentils, peanut, and lupins, is essential for an effective storage of the products. The application of the BoS in the quality maintenance of the legumes or pulse products has been proposed but only sparingly reported. A surface plasma resonance biosensor has been used for the detection and treatment of many groups of mycotoxins, such as ochratoxin in pulse and cereals [18].

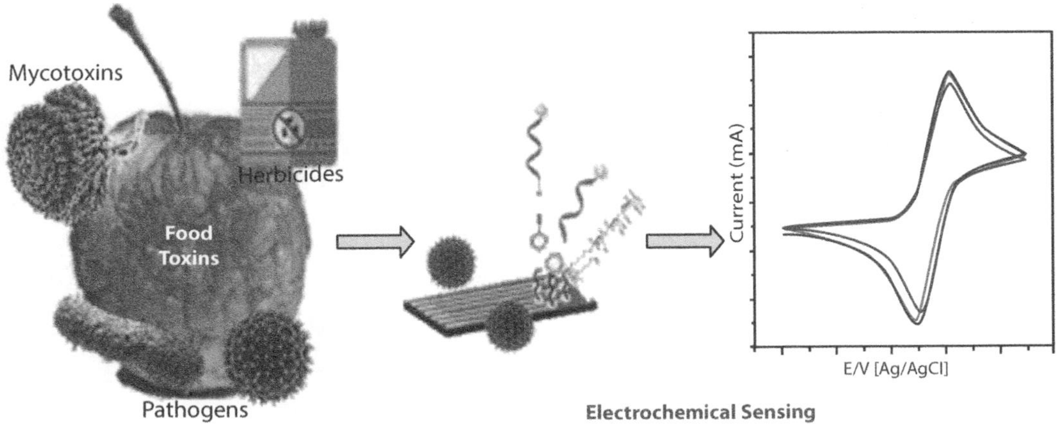


**Figure 20.1** Electrochemical BoS for detection microbial toxins in cereal foods [19].

The technique uses silver nanoparticles to enhance the surface signal of the products to reveal the cellular arrangement with their structure. This provides an avenue for a quantitative and qualitative determination of the mycotoxins in the matrix of the food. In similar studies, Gupta *et al.* [19] and Singh [20] reported the prospect of nanomaterial based electrochemical BoS for application in detecting mycotoxins and protecting the cereal food and pulse products during processing and storage, as was shown in Figure 20.1. Typically, the authors illustrated how the BoS can act on the food to effectively create a shield against the activities of the pathogens and mycotoxins during storage [21]. By this action, the quality of the product is maintained, and deterioration activities are curtailed to ensure their safety in the postharvest value chain.

### 20.2.3 BoS for Fruit and Vegetable Products Safety

Fruits and vegetables are horticultural crops rich in minerals, fibers, and vitamins. The safety of the fruit and vegetable products, such as banana, watermelon, strawberry, cucumber, mush room, spinach, cabbage and so on, is essential during processing and preservation. The application of BoS to monitor and control the pre and postharvest activities of fruits and vegetables is a new research frontier that has gained global attention recently [22–25]. According to Qin *et al.* [26], the BoS can be applied to protect the fruits and vegetables against the health risk associated with use of the fungicides and insecticides before harvest and during shelf storage [19, 27–29]. Although, the mechanism of the action and role of the material on the product is still being studied [30], we can associate this to their ability to create molecular surface to interact with the chemical spray [31], maintain its cellular tissue, and absorb its potency to cause harm to the product and the consumers [32, 33]. The chemical and physiological indices of the fruit juices may in fact be altered, due to an unfavorable condition of processing, packaging, or distribution, but the application of some BoS during the production can help prevent this quality decline [34], as shown in Figure 20.2. Thus, the material presents a simple, portable, and rapid alternative to the tedious analytical methods for the detection of various degradation and spoilage indicators formed in the packaged fruit juices. Other types of the BoS can be applied for detection of nutrients in fruits at the cellular level. For instance, Ribeiro *et al.* [23] described an electrochemical

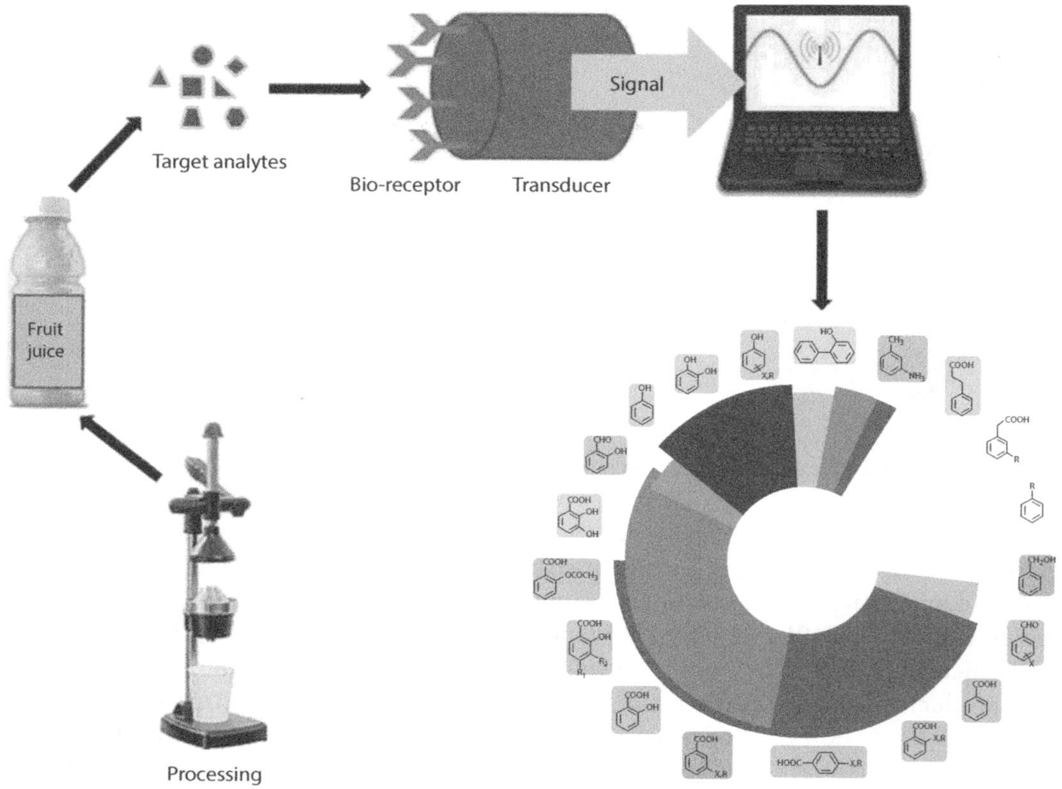


**Figure 20.2** Role of the BoS in preventing quality decline of fruit juices [34].

laccase based BoS for the quantification and detection of formetanate hydrochloride in the fruits. In principle, we can associate this unique behavior on the ability of the BoS to inhibit the laccase catalytic reaction that occurs in the presence of phenolic in the substrate thereby revealing the nutrients in the fruits.

### 20.2.4 BoS for Forestry Products Safety

The safety of the forestry products, such as honey, wild-meat, mushroom, palm wine, palm oil, cola-nuts, timber, and other medicinal products, is very critical during their processing and preservation. The BoS has great potential for application in the treatment of forest plant diseases and preservation of forestry products. However, there are only very few literature reports to demonstrate their abilities. De Beer *et al.* [35] demonstrated the ability of the BoS to monitor and control the quality of honey with an average of 90% compliance to standard. Plasma colorimetric BoS was developed by Zhou *et al.* [36] for rapid determination of chloramphenicol in honey on site with a recoveries rate of 88.0–107.6%. The BoS possesses high sensitivity, good selectivity, low cost, and excellent stability, and could be extended to detect a wide variety of other small molecular analytes, nucleic acids, or proteins [37–40]. To the best of knowledge, the practical and commercial applications of the BoS in the nutrient detection, protection, monitoring and control of the wild-meat, mushroom, palm wine, palm oil, cola-nuts and other forestry products are not known. Hence, this is new frontier of knowledge that needs to be exploited considering the economic values presented by the products.

### 20.2.5 BoS for Fodder Safety

The safety of the fodders used in supplementing animal feeds or livestock feeds, such as hay, legumes, and barley, is essential during their processing and preservation. The formulation of the animal feeds is very critical in animal husbandry especially since composite materials are mixed to get a product with a desired nutritional requirement. The BoS are essentially needed to help monitor and control the product against disease infestation and other foreign materials before they accidentally get into the feed during formulation. The engineering nanomaterial BoS which was reported for detecting anomalies in food products [41] can present potential way-forward if applied in the formulation of feed and preservation of fodder crops. This presents opportunity to realize rapid, sensitive, efficient, and portable detection, thereby overcoming the restrictions and limitations of traditional methods like pretreatment complication, long detection time, and high level of instrumentation.

## 20.3 Biosensors for Safety of Animal Products

### 20.3.1 BoS for Dairy Products Safety

The safety of the dairy products, such as milk, cheese, ice-cream, and butter, is very critical during their processing and preservation. The BoS has great potential to enhance the keeping quality and control the nutritional level of the dairy products during processing [42–44]. The material was successfully applied to monitor the lactose level in milk and cheese using the LC-MS techniques [45]. Ziyaina *et al.* [46] has also proposed a novel technique for detecting microorganisms in dairy products using the BoS technology as show in Figure 20.3. The BoS has been used to detect the presence of nisin in raw milk which

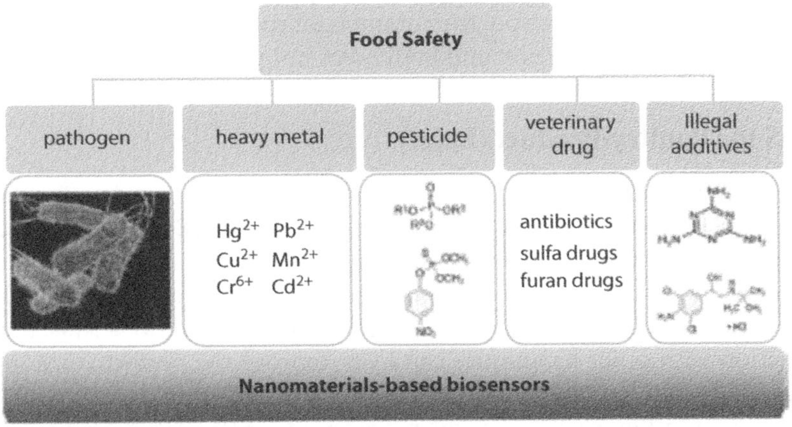


**Figure 20.3** Typical BoS activity for feed safety detection [41].

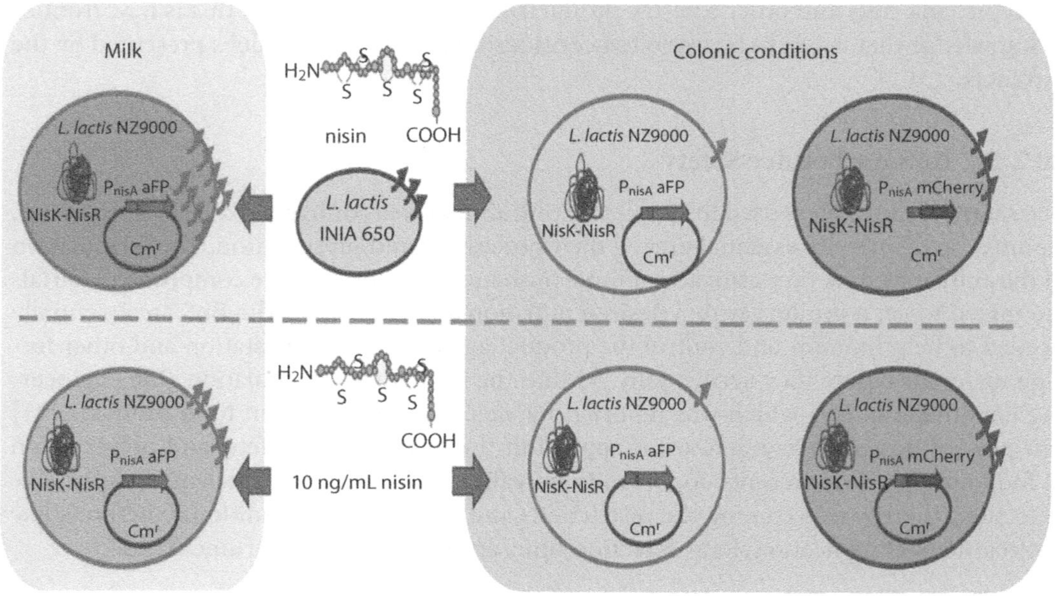


**Figure 20.4** Application of BoS for nisin detection in milk [50].

is a bioactive peptide that can lead to product spoilage if the concentration is higher than the threshold, as shown in Figure 20.4 [47]. The lipolysis reaction which occurs during the ripening of dairy products because of esterase or lipase activity can be controlled using the BoS [48]. To increase the production capacity of the dairy products, a good number of the BoS devices, such as those synthesized from the nanocomposites and liposomes have been deployed to hep control the level of deterioration of the products during processing and preservation [49]. The devices have been tested for quality enhancement, detection of contaminants, and promotion of the product stability against environmental influences that can affect the shelf-life. Also, certain BoS designed by an optical technology was proposed by Indyk *et al.* [12] for the controlling, quantifying, and detecting a bovin serum albumin (BSA) in milk during processing. Although this is applied for milk and cheese quality control, disease monitoring and detection, the commercial application of the material for other dairy products, such as the ice-cream and butter, has not yet been exploited.

### 20.3.2 BoS for Poultry Products Safety

The safety of the poultry products, such as eggs, meat, and manure, is important in their processing and preservation. Although, the application of the BoS in the preservation of poultry products are not well known, other techniques such as the culture, polymerase chain reaction and enzyme-linked immuno-sorbent assay are known for salmonella detection in meat, but are limited in slow reaction, sensitivity, and in field detection. Thus, the BoS alone or in combination with other techniques can help improve the performance in the detection procedure. Typically, a combination of the BoS with an immune-magnetic separation technique, as shown in Figure 20.5, has been proposed for the detection and screening of salmonella in the poultry product [51]. The technique was also successfully applied for the screening of *salmonella* in the poultry, as shown in Table 20.1. This ensures that safe and healthy poultry products are delivered to the end users in the supply chain.

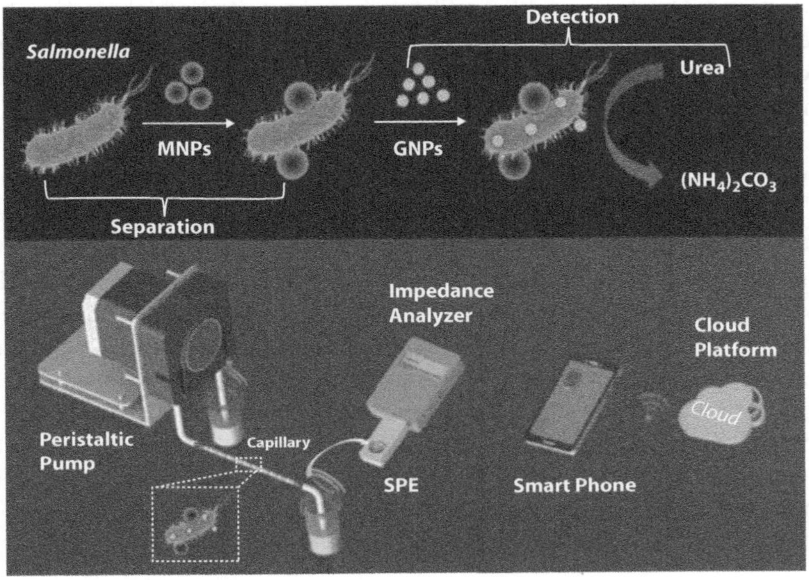


**Figure 20.5** An immune-magnetic separation technique combined with an active BoS for Salmonella screening in poultry products [51].

**Table 20.1** Screening technique for *salmonella* in poultry.

| | | Impedance BoS | | Real-time PCR | | Culture plating | |
|---|---|---|---|---|---|---|---|
| **Sample type** | **s/n** | **−ve** | **+ve** | **−ve** | **+ve** | **−ve** | **+ve** |
| Chicken breast | 6 | 5 | 1 | 6 | 0 | 6 | 0 |
| Chicken carcass | 6 | 6 | 0 | 5 | 1 | 6 | 0 |
| Chicken intestine | 6 | 6 | 0 | 6 | 0 | 5 | 1 |
| Chicken leg | 12 | 12 | 0 | 12 | 0 | 12 | 0 |
| Chicken meat | 8 | 6 | 2 | 7 | 1 | 8 | 0 |
| Chicken wing | 17 | 15 | 2 | 15 | 2 | 17 | 0 |
| Duck breast | 8 | 8 | 0 | 8 | 0 | 8 | 0 |
| Duck leg | 4 | 4 | 0 | 4 | 0 | 4 | 0 |
| Duck liver | 4 | 2 | 2 | 4 | 0 | 4 | 0 |
| Duck wing | 4 | 4 | 0 | 4 | 0 | 4 | 0 |
| Total | 75 | 68 | 7 | 71 | 4 | 74 | 1 |

+ve, −ve mean negative and positive responses, respectively [51].

## 20.4 Biosensors for Safety of Microbes Used in Food Processing and Storage

The safety of the microorganisms, such as *Bacteroides xylanisolvens, Akkermansia muciniphila*, fructophilic lactic acid bacteria, and *Faecalibacterium prausnitzii*, which are beneficial to the food, is essential in processing and storage. The practical production of the BoS for this purpose and its commercial application has not been reported to the best of knowledge. However, to ensure the safety of the microorganisms against excessive temperature of processing and uncontrolled reaction that may cause product spoilage, the application nucleic acid, antibodies, receptors, potable [52] and cell based BoS, which were noted for their fine abilities for food contaminants monitoring and detection [53], can potentially be used for this application. The cell based BoS can inhibits cellular respiration and act as inducer for some specific catalytic protein. Other types of the BoS including the piezoelectric and the thermal sensors, which are noted for their high stability response to sound and heat energy, can be applied for monitoring and controlling the activities of the microbes during thermal processing. Apparently, the application of this technique for the safety of the stored foods is not well known. The few reports focused on the use of the enzymatic BoS based on cobalt phthalocyanine to monitor and control the ageing of beer during storage [54–56]. There is therefore the need to study recent BoS technologies as they apply to monitoring the safety of the microbes assisting food processing and storage.

## 20.5 Prospects and Conclusions

The commercial application of the BoS for agroproducts safety has not been widely studied. A few studies reported the use of the electrochemical, plasma resonance, and novel DNA-based BoS for monitoring and detecting the onset of plant and animal diseases, and in the control of degradation of the food nutrients during the processing. To the best of knowledge, the practical and commercial applications of the BoS in the nutrient detection, protection, monitoring and control of the wild-meat, mushroom, palm wine, palm oil, cola-nuts and other forestry products are not known. Hence, this is new frontier of knowledge that needs to be exploited considering the economic values presented by the products. Although, this is commonly applied for milk and cheese quality control, disease monitoring and detection, there commercial application for other dairy products such as the ice cream and butter has not yet been exploited. There is therefore the need to study recent BoS technologies as they apply to monitoring the safety of the microbes assisting food processing and storage.

## References

1. Fadeyibi, A., Modeling rheological behavior of beef based on time-dependent deformation and packaging. *Gazi Univ. J. Sci.*, 35, 3, 997–1008, 2022.
2. Eyvazi, S., Baradaran, B., Mokhtarzadeh, A., de la Guardia, M., Recent advances on development of portable biosensors for monitoring of biological contaminants in foods. *Trends Food Sci. Technol.*, 114, 712–721, 2021.

3. Sobhan, A., Muthukumarappan, K., Wei, L., Biosensors and biopolymer-based nanocomposites for smart food packaging: Challenges and opportunities. *Food Packag.*, 30, 100745, 2021.
4. Jafarzadeh, S., Mohammadi Nafchi, A., Salehabadi, A., Oladzad-abbasabadi, N., Jafari, S.M., Application of bio-nanocomposite films and edible coatings for extending the shelf life of fresh fruits and vegetables. *Adv. Colloid Interface Sci.*, 291, 102405, 2021.
5. Brodmann, T., Endo, A., Gueimonde, M., Vinderola, G., Kneifel, W., de Vos, W.M., Salminen, S., Gómez-Gallego, C., Safety of novel microbes for human consumption: Practical examples of assessment in the European Union. *Front. Microbiol.*, 8, SEP, 1725, 2017.
6. Fadeyibi, A. and Osunde, Z.D., Effect of thickness and matrix variability on properties of a starch-based nanocomposite supple film. *Food Res.*, 5, 4, 416–422, 2021.
7. Fadeyibi, A., Osunde, Z., Applied, M.Y., Effects of period and temperature on quality and shelf-life of cucumber and garden-eggs packaged using cassava starch-zinc nanocomposite film. *J. Appl. Packag. Res.*, 12, 1, 1–10, 2020.
8. Fadeyibi, A., Osunde, Z.D., Yisa, M.G., Prediction of some physical attributes of cassava starch–zinc nanocomposite film for food-packaging applications. *J. Appl. Packag. Res.*, 3, 1, 35–41, 2019.
9. Fadeyibi, A., Osunde, Z.D., Egwim, E.C., Idah, P.A., Performance evaluation of cassava starch-zinc nanocomposite film for tomatoes packaging. *J. Agric. Eng.*, 48, 3, 137–146, 2017.
10. Fadeyibi, A., Delebo Osunde, Z., GanaYisa, M., Optimization of processing parameters of nanocomposite film for fresh sliced okra packaging. *J. Appl. Packag. Res.*, 11, 2, 1–10, 2019.
11. Lee, D.S. and Wang, H.J., Packaging of milk and dairy products: Packaging materials, technologies and their applications, in: *Encyclo. Dairy Sci.*, pp. 756–765, 2022.
12. Indyk, H.E., Gill, B.D., Woollard, D.C., An optical biosensor-based immunoassay for the determination of bovine serum albumin in milk and milk products. *Int. Dairy J.*, 47, 72–78, 2015.
13. Gimonkar, S., Van Fleet, E.E., Boys, K.A., Dairy product fraud, in: *Food Fraud*, pp. 249–279, 2021.
14. Zhang, C., Jiang, C., Lan, L., Ping, J., Ye, Z., Ying, Y., Nanomaterial-based biosensors for agro-product safety. *TrAC Trends Anal. Chem.*, 143, 116369, 2021.
15. Yisa, M., Fadeyibi, A., Adisa, O., Finite element simulation of temperature variation in grain metal silo. *Res. Agric. Eng.*, 10, 3, 8–17, 2018.
16. Yuan, J., Deng, D., Lauren, D.R., Aguilar, M.I., Wu, Y., Surface plasmon resonance biosensor for the detection of ochratoxin A in cereals and beverages. *Anal. Chim. Acta*, 656, 1–2, 63–71, 2009.
17. Santovito, E., Greco, D., D'Ascanio, V., Sanzani, S.M., Avantaggiato, G., Development of a DNA-based biosensor for the fast and sensitive detection of ochratoxin A in urine. *Anal. Chim. Acta*, 1133, 20–29, 2020.
18. Yuan, J., Deng, D., Lauren, D.R., Aguilar, M.I., Wu, Y., Surface plasmon resonance biosensor for the detection of ochratoxin A in cereals and beverages. *Anal. Chim. Acta*, 656, 1–2, 63–71, 2009.
19. Gupta, R., Raza, N., Bhardwaj, S.K., Vikrant, K., Kim, K.H., Bhardwaj, N., Advances in nanomaterial-based electrochemical biosensors for the detection of microbial toxins, pathogenic bacteria in food matrices. *J. Hazard. Mater.*, 401, 123379, 2021.
20. Singh, P., Electrochemical biosensors: Biomonitoring of food adulterants, allergens, and pathogens, in: *Electro. Biosens.*, pp. 141–192, 2022.
21. Umapathi, R., Ghoreishian, S.M., Sonwal, S., Rani, G.M., Huh, Y.S., Portable electrochemical sensing methodologies for on-site detection of pesticide residues in fruits and vegetables. *Coord. Chem. Rev.*, 453, 214305, 2022.
22. Ahangari, H., Kurbanoglu, S., Ehsani, A., Uslu, B., Latest trends for biogenic amines detection in foods: Enzymatic biosensors and nanozymes applications. *Trends Food Sci. Technol.*, 112, 75–87, 2021.

23. Ribeiro, F.W.P., Barroso, M.F., Morais, S., Viswanathan, S., de Lima-Neto, P., Correia, A.N., Oliveira, M.B.P.P., Delerue-Matos, C., Simple laccase-based biosensor for formetanate hydrochloride quantification in fruits. *Bioelectrochem.*, 95, 7–14, 2014.
24. Zhang, Z., Lou, Y., Guo, C., Jia, Q., Song, Y., Tian, J.Y., Zhang, S., Wang, M., He, L., Du, M., Metal–organic frameworks (MOFs) based chemosensors/biosensors for analysis of food contaminants. *Trends Food Sci. Technol.*, 118, 569–588, 2021.
25. Zhang, C., Huang, L., Pu, H., Sun, D.W., Magnetic surface-enhanced Raman scattering (MagSERS) biosensors for microbial food safety: Fundamentals and applications. *Trends Food Sci. Technol.*, 113, 366–381, 2021.
26. Qin, G., Chen, Y., He, F., Yang, B., Zou, K., Shen, N., Zuo, B., Liu, R., Zhang, W., Li, Y., Risk assessment of fungicide pesticide residues in vegetables and fruits in the mid-western region of China. *J. Food Compost. Anal.*, 95, 103663, 2021.
27. Narenderan, S.T., Meyyanathan, S.N., Babu, B., Review of pesticide residue analysis in fruits and vegetables. Pre-treatment, extraction, and detection techniques. *Food Res. Int.*, 133, 2020.
28. Jain, U., Saxena, K., Hooda, V., Balayan, S., Singh, A.P., Tikadar, M., Chauhan, N., Emerging vistas on pesticides detection based on electrochemical biosensors – an update. *Food Chem.*, 371, 131126, 2022.
29. el Sheikha, A.F., Tracing fruits, and vegetables from farm to fork: Questions of novelty and efficiency, in: *Prod. Manag. Bev.*, pp. 179–209, 2019.
30. Gupta, C. and Prakash, D., Safety of fresh fruits and vegetables, in: *Food Safety and Human Health*, pp. 249–283, 2019.
31. Hua, Z., Yu, T., Liu, D., Xianyu, Y., Recent advances in gold nanoparticles-based biosensors for food safety detection. *Biosens. Bioelectron.*, 179, 113076, 2021.
32. Smart, A., Crew, A., Pemberton, R., Hughes, G., Doran, O., Hart, J.P., Screen-printed carbon-based biosensors and their applications in agri-food safety. *TrAC - Trends Anal. Chem.*, 127, 115898, 2020.
33. Zhang, C., Jiang, C., Lan, L., Ping, J., Ye, Z., Ying, Y., Nanomaterial-based biosensors for agro-product safety. *TrAC - Trends Anal. Chem.*, 143, 116369, 2021.
34. Rai, P., Mehrotra, S., Sharma, S.K., Challenges in assessing the quality of fruit juices: Intervening role of biosensors. *Food Chem.*, 386, 132825, 2022.
35. de Beer, T., Otto, M., Pretoruis, B., Schönfeldt, H.C., Monitoring the quality of honey: South African case study. *Food Chem.*, 343, 128527, 2021.
36. Zhou, C., Sun, C., Zou, H., Li, Y., Plasma colorimetric aptasensor for the detection of chloramphenicol in honey based on cage Au@AuNPs and cascade hybridization chain reaction. *Food Chem.*, 377, 132031, 2022.
37. Se, K.W., Wahab, R.A., Syed Yaacob, S.N., Ghoshal, S.K., Detection techniques for adulterants in honey: Challenges and recent trends. *J. Food Compost. Anal.*, 80, 16–32, 2019.
38. Villalonga, A., Sánchez, A., Mayol, B., Reviejo, J., Villalonga, R., Electrochemical biosensors for food bioprocess monitoring. *Curr. Opin. Food Sci.*, 43, 18–26, 2022.
39. Singh, P., Electrochemical biosensors: Biomonitoring of food adulterants, allergens, and pathogens. *Electrochem. Biosens.*, 1, 141–192, 2022.
40. Lu, L., Zhu, Z., Hu, X., Hybrid nanocomposites modified on sensors and biosensors for the analysis of food functionality and safety. *Trends Food Sci. Technol.*, 90, 100–110, 2019.
41. Lv, M., Liu, Y., Geng, J., Kou, X., Xin, Z., Yang, D., Engineering nanomaterials-based biosensors for food safety detection. *Biosens. Bioelectron.*, 106, 122–128, 2018.
42. Lu, G., Chen, Q., Li, Y., Liu, Y., Zhang, Y., Huang, Y., Zhu, L., Status of antibiotic residues and detection techniques used in Chinese milk: A systematic review based on cross-sectional surveillance data. *Food Res. Int.*, 147, 110450, 2021.

43. Fischer, W.J., Schilter, B., Tritscher, A.M., Stadler, R.H., Contaminants of milk and dairy products: Contamination resulting from farm and dairy practices, in: *Encyclo. Dairy Sci.*, pp. 809–821, 2016.
44. Jangra, S., Development of biosensor-based technology for the detection of pathogenic microorganisms and biomolecules in dairy products, in: *Adv. in Dairy Micro. Prod.*, pp. 377–384, 2022.
45. Yang, J., Rainville, P., Liu, K., Pointer, B., Determination of lactose in low-lactose and lactose-free dairy products using LC-MS. *J. Food Compost. Anal.*, 100, 103824, 2021.
46. Ziyaina, M., Rasco, B., Sablani, S.S., Rapid methods of microbial detection in dairy products. *Food Control*, 110, 107008, 2020.
47. Langa, S., Peirotén, A., Gaya, P., Escudero, C., Rodríguez-Mínguez, E., Landete, J.M., Arqués, J.L., Development of multi-strain probiotic cheese: Nisin production in food and gut. *LWT*, 148, 111706, 2021.
48. García-Cano, I., Rocha-Mendoza, D., Kosmerl, E., Jiménez-Flores, R., Purification and characterization of a phospholipid-hydrolyzing phosphoesterase produced by Pediococcusacidilactici isolated from Gouda cheese. *J. Dairy Sci.*, 103, 5, 3912–3923, 2020.
49. GKP, S., Tiwari, H., Mketo, N., Lakkakula, J., Application of nanomaterials in the dairy industry, in: *Adv. in Dairy Micro. Prod.*, pp. 357–375, 2022.
50. Landete, J.M., Langa, S., Escudero, C., Peirotén, Á., Arqués, J.L., Fluorescent detection of nisin by genetically modified Lactococcus lactis strains in milk and a colonic model: Application of whole-cell nisin biosensors. *J. Biosci. Bioeng.*, 129, 4, 435–440, 2020.
51. Wang, L., Xue, L., Guo, R., Zheng, L., Wang, S., Yao, L., Huo, X., Liu, N., Liao, M., Li, Y., Lin, J., Combining impedance biosensor with immunomagnetic separation for rapid screening of Salmonella in poultry supply chains. *Poult. Sci.*, 99, 3, 1606–1614, 2020.
52. Eyvazi, S., Baradaran, B., Mokhtarzadeh, A., de la Guardia, M., Recent advances on development of portable biosensors for monitoring of biological contaminants in foods. *Trends Food Sci. Technol.*, 114, 712–721, 2021.
53. Huo, B., Hu, Y., Gao, Z., Li, G., Recent advances on functional nucleic acid-based biosensors for detection of food contaminants. *Talanta*, 222, 121565, 2021.
54. Ghasemi-Varnamkhasti, M., Rodríguez-Méndez, M.L., Mohtasebi, S.S., Apetrei, C., Lozano, J., Ahmadi, H., Razavi, S.H., Antonio de Saja, J., Monitoring the aging of beers using a bioelectronic tongue. *Food Control*, 25, 1, 216–224, 2012.
55. Chadha, U., Bhardwaj, P., Agarwal, R., Rawat, P., Agarwal, R., Gupta, I., Panjwani, M., Singh, S., Ahuja, C., Selvaraj, S.K., Banavoth, M., Sonar, P., Badoni, B., Chakravorty, A., Recent progress and growth in biosensors technology: A critical review. *J. Ind. Eng. Chem.*, 109, 21–51, 2022.
56. Dang, Y.T.H., Gangadoo, S., Rajapaksha, P., Truong, V.K., Cozzolino, D., Chapman, J., Biosensors in food traceability and quality, in: *Comp. Foodom.*, pp. 308–321, 2020.

# Index

Printed and bound by CPI Group (UK) Ltd, Croydon, CR0 4YY
05/10/2023
08126488-0002